an outline of organic chemistry

third edition

an outline of organic chemistry:
problems and answers

j. ernest simpson
Professor of Chemistry, California State Polytechnic University, Pomona

corwin hansch
Professor of Chemistry, Pomona College

george k. helmkamp
Professor of Chemistry, University of California, Riverside

McGraw-Hill Book Company

New York   St. Louis   San Francisco   Auckland   Düsseldorf   Johannesburg   Kuala Lumpur   London   Mexico
Montreal   New Delhi   Panama   Paris   São Paulo   Singapore   Sydney   Tokyo   Toronto

AN OUTLINE OF ORGANIC CHEMISTRY:
PROBLEMS AND ANSWERS

  34567890 BABA 798765

This book was set in Press Roman.
The editors were Robert H. Summersgill and Michael LaBarbera;
the designer was Anne Canevari Green;
the production supervisor was Leroy A. Young.
The drawings were done by J & R Services, Inc.
George Banta Company, Inc., was printer and binder.

Library of Congress Cataloging in Publication Data

Simpson, John Ernest, date
     An outline of Organic Chemistry

     A revision of Organic Chemistry: an outline, problems and answers, by C. Hansch and G. Helmkamp.
     Includes index.
          1.   Chemistry, Organic—Outlines, syllabi, etc.   2.   Chemistry, Organic—Problems, exercises, etc.
I.   Hansch, Corwin H., date   Organic chemistry: an outline, problems, and answers.   II.   Hansch, Corwin H., date   joint author.
III.   Helmkamp, George K., joint author.   IV.   Title. [DNLM: 1. Chemistry, Organic—Outlines.   QD256.5 S6130]
QD253.S52   1975   547   74-20988
ISBN 0-07-057436-7

contents

preface

In the third edition of our workbook we have tried to consolidate and reorganize the outline material and problems from the second edition in a way that will be more beneficial and efficient for users of the workbook. At the same time we have added new material which we feel reflects the major changes since the second edition in what is being taught in the core organic chemistry course at various schools. Thus, for example, concepts such as the molecular structure of both aromatic and aliphatic compounds, resonance, and an introduction to stereoisomerism are included in Chapter 2. Previous separate chapters have been combined for study efficiency while adding new concepts in synthesis and reaction mechanisms. Some examples include alkenes and alkynes; aliphatic alcohols and ethers; aliphatic acids, anhydrides, and acyl halides; aliphatic esters and amides; aromatic nitro and sulfur compounds; and aromatic amines and diazonium salts.

Chapter 8 on structure determination by spectroscopic methods (ultraviolet, infrared, and nuclear magnetic resonance) is totally new in this edition and comes after the student has been exposed to several key functional group classes. We feel this is necessary for an easier and more meaningful grasp of spectroscopy. However, this chapter could be covered earlier or later if desired. Additional spectroscopic problems are included in most of the subsequent chapters.

Throughout the workbook the concept of stereochemistry is emphasized. This includes, for example, an introduction in Chapter 2, conformational analysis in Chapter 3, the role of stereochemistry in various reaction mechanisms (Chapters 4 through 7) and in spectroscopy (Chapter 8). Chapter 13 represents a comprehensive study of stereoisomerism in relation to molecular structure, organic synthesis, and reaction mechanisms.

In Chapter 12 on carbanions we have added considerable new material and consolidated the material from various chapters in the previous edition in order to focus attention on the very important role carbanions play in organic chemistry.

The purpose of this book is to provide a large number of drill problems for the beginning student (or for those people who need a concise review of organic chemistry) especially in the areas of nomenclature, reactions, spectroscopy, and syntheses. Introducing the beginning student to organic chemistry is primarily a problem in acquainting the student with a new and strange symbolic language; unless given ample opportunity to use the language the student forgets it. More important, it is necessary to put forth many questions to the student, for often the greatest difficulty the student has is in knowing when he or she "knows" the subject.

An outline of the information necessary for the solution of the problems is presented partly for use in working the problems and also for review. Considerable time may be saved if the student will become familiar with the contents of the outline before the lecture presentation of the same topic.

Organic chemistry as set forth in the problems of this book is a continuous subject; that is, once a given reaction is introduced, it is brought up again and again in the problems of subsequent chapters. The student not only is helped to remember it but becomes increasingly proficient in its use and limitations. As far as possible similar types of problems (nomenclature, single-step reactions, reaction mechanisms, syntheses, structure determination, etc.) are grouped together in each chapter according to the increasing difficulty in each type and likewise within the various parts of a single problem. We have continued the practice of including several problems in each chapter which should challenge the more experienced and better students. In this edition we have provided answers for all problems and have tried not to include any outline material for which there are no problems dealing with that concept.

In Chapters 6, 7, 14, 16, and 17 there are special sections devoted to a description of the approach to organic synthesis and numerous sample solutions are given. These help the uninitiated off to the right start, but developing proficiency in synthesis can come only from trying many, many problems and comparing answers with what is regarded as good practice. Many such answers to the problems in this book are given in the answers section. Heavy emphasis is placed upon synthesis, not only for its own sake, but also because this is the best way to help the student organize reactions into an easily remembered matrix of useful material.

When using this outline for study, the student, after learning the nomenclature, should carefully go over the methods of synthesis and the reactions for a given class of compounds. The best method of learning these is to write out each one until an exact visual picture is retained of the individual atoms and how they fit together. It is also important to be able to place each reaction into its proper mechanistic category. When working a problem to which the solution is not at once apparent, the student should resist turning quickly to the answer section. Instead, the student should look over the reactions in that particular section and even in the preceding chapters for ideas for the solution.

The beginner soon learns that there are often many possible solutions to a given synthetic problem. Space limitations do not permit us to give here more than one or two answers to each problem. Those given have usually been chosen to illustrate the chemistry in the chapter to which a group of problems belongs. Should the student's answer not match the answer given and there is any doubt in the student's mind about the validity of the solution, the instructor should be consulted.

To those who teach organic chemistry courses we would seriously suggest that this *Outline of Organic Chemistry* could well be used as the principal text in their course. The outline format used would allow instructors considerably more flexibility in developing, emphasizing, and embellishing those topics in organic chemistry which they feel are the more important ones and best-suited for their type of student. We feel that the individual instructors can best decide to what extent they want, for example, a biochemical or environmental flavoring in their courses and what topics they want to include. Finally, the smaller size and lower cost of this book should help reduce the initial fright of the beginning organic chemistry student.

We would like to thank all of our students who have class tested the material in the third edition and our colleagues for their helpful suggestions and proofreading. Our special thanks go to Linda Loner and Phyllis Bartosh for typing the final manuscript.

J. Ernest Simpson
Corwin Hansch
George K. Helmkamp

an outline of organic chemistry

1
structure and analysis

1-1 Electronic Formulas

One must have a certain background of information about bond angles and a general feeling about the types of bonds one can expect before a likely electronic structure of an ordinary covalent compound can be formulated. The work presented in this book assumes such knowledge for the common inorganic molecules. This will constitute the foundation from which we shall start building structural organic chemistry. The material in this chapter is primarily review, and in certain instances, consultation of a freshman chemistry text will help in working out electronic formulas.

The following generalizations are of help in drawing electronic structures:

1. Total number of electrons shown in bonds and outer shells must be equal to the sum of valence-shell electrons of all atoms (that is, H = 1, N = 5, O = 6, halogen = 7, etc.).
2. Each covalent bond must contain an even number of electrons.
3. Except in special cases, all electrons are shown in pairs.
4. Each hydrogen atom should show two electrons (one bond) and all other elements in the first row (Li through F) should show eight (octet rule). Occasionally an atom such as boron, illustrated in Sec. 1-5, will have an incomplete shell with no charge.
5. From the number of electrons in bonds and as lone pairs in the outer shell of an atom, the angle of attachment of the neighboring atoms can be estimated. Electrons repel each other and therefore tend to form bonds which will be separated as much as possible from each other. For example, in $CH_3-Mg-CH_3$, a linear structure with the two Mg bonds separated by $180°$ permits the electrons in the magnesium-carbon bonds to be as far from each other as possible. In BCl_3, an angle of $120°$ is optimum, and in CH_4, $109.5°$ gives maximum separation of the four pairs of bonding electrons. For the elements up to Ne, if the bonds are equivalent (i.e., same groups attached), then $180°$, $120°$, and $109.5°$ are the ideal angles expected and actually found. For situations where the bonding pairs of electrons are not equivalent, small deviations from this ideality occur. For example

$\overset{\cdot\cdot}{\underset{H \quad H}{O}}$ has four pairs of electrons and one would thus expect the $\underset{H \qquad H}{\overset{O}{\diagup \diagdown}}$ bond angle to be $109.5°$. Because

of the uneven repulsion of the two sets of lone-pair electrons versus the two sets of electrons in the O–H bonds, the actual angle is found to be $105°$.

Example 1 Ammonia, NH_3. The nitrogen atom has five valence-shell electrons and thus requires three more electrons to complete the shell of eight. Each hydrogen atom has one electron and requires one additional electron. The nitrogen atom then forms three covalent bonds, and each of the three hydrogen atoms forms one covalent bond. These obviously fit in only one fashion. The following formula shows (1) the correct total of eight electrons, (2) all electrons in pairs, (3) two electrons per single covalent bond, and (4) a pyramidal structure with H–N–H bond angles of about $109.5°$. The actual angle in NH_3 is $108°$, slightly different from the ideal.

$$\cdot \overset{\cdot\cdot}{\underset{\cdot}{N}} \cdot + 3\,H\cdot = H\!:\!\overset{\cdot\cdot}{\underset{\underset{\displaystyle H}{\cdot\cdot}}{N}}\!:\!H \equiv \underset{H \qquad\;\; H}{\overset{\displaystyle \overset{\cdot\cdot}{N}}{\diagup | \diagdown}}$$
$$\underset{H}{}$$

Example 2 Nitrous acid, HONO.

H One atom, one bond   Total of 18 electrons showing in structure
N One atom, three bonds
O Two atoms, two bonds each

This combination can be completed by way of a double bond between nitrogen and one oxygen.

or, by simple line formula,

A useful generalization for estimating the shape of molecules with double or triple bonds is that such bonds tend to orient themselves about the central atom in the same fashion as single bonds and lone-pair electrons. (For a more sophisticated analysis, see Chap. 2.)

For nitrous acid we have two nitrogen-oxygen bonds and one lone pair of electrons; therefore we should expect an O—N—O bond angle near $120°$.

Example 3 Nitric acid, $HONO_2$. A new problem often occurs in compounds containing elements such as sulfur, nitrogen, and phosphorus that may exist in high oxidation states. To draw the structure for nitric acid, start with nitrous acid (Example 2) and add an oxygen atom. Obviously, oxygen requires two electrons to complete its shell, but it cannot form bonds by the usual sharing process because none of the atoms in nitrous acid can accept any more electrons. Therefore, if the nitrogen atom furnished *both* electrons to the oxygen atom, an electronic structure could be drawn. Nitrogen, by furnishing both electrons to a shared bond, has become positively charged, and oxygen, by having both electrons furnished, has become negatively charged.

O—N—O bond angle $\cong 120°$

This type of bond is termed *coordinate covalent* (sometimes *dative* or *semipolar*), and in the electronic formulas, formal charge signs are placed on the proper atoms. The charge on any atom may be checked by a simple procedure. Consider the nitrogen atom in nitric acid. The atomic number of nitrogen is seven, and this explicitly indicates a +7 charge on the nucleus. Now, if the nitrogen atom "owned" seven electrons it would be neutral. To total up these nitrogen electrons in HNO_3, add up (1) the unshared electrons in all shells of N (do not forget the shells not shown in the drawing) and (2) half the electrons in covalent bonds (since these are shared and nitrogen has ideally a 50 percent interest in each pair).

Electrons in the K (first) shell	2
Unshared electrons in L shell	0
One-half electrons in covalent bonds	4
	6

This nitrogen atom has a total net charge of +1 because the seven protons in the nucleus are balanced by only six electrons. To be sure you understand this, carry out the process for the three oxygen atoms in the given nitric acid structure. Two of the oxygens have no net charge.

1-2 Electronegativity

The electrons of a covalent bond are shared equally when both atoms are identical. Consequently the two atoms in a molecule such as Cl_2 have no net charge. When the two atoms are different, there can be varying degrees of unequal sharing, resulting in the acquisition of partial charges, δ^+ or δ^-, by the atoms involved. The factors that determine this polarization (resulting in a bond dipole moment) are the nuclear charges of the atoms, the covalent radius of the atom, and the electrical shielding of inner-shell electrons from nuclear effects on outer-shell electrons.

In moving to the right in a periodic series, the nuclear charge increases and outer-shell electrons are attracted more strongly to an atom. Consequently, in a covalent bond such as C—N the nitrogen atom attracts the bond electrons more than carbon, and the bond is polarized, C^{δ^+}—N^{δ^-}. Nitrogen is said to be more electronegative than carbon.

In moving down through a periodic group, the nuclear charge increases, but the atomic radius and shielding effects more than compensate for this. Thus, in group VII, chlorine is more electronegative than iodine, and a compound such as iodine monochloride has a great deal of ionic character:

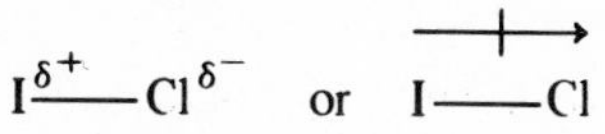

$$I\overset{\delta+}{\rule{1cm}{0.4pt}}Cl^{\delta-} \quad \text{or} \quad I\overset{\rule{1cm}{0.4pt}\!\!\!\!+\rightarrow}{\rule{1cm}{0.4pt}}Cl$$

The following values indicate relative electronegativities of some common atoms:

H	2.1						
C	2.5	N	3.0	O	3.5	F	4.0
Si	1.8	P	2.1	S	2.5	Cl	3.0
						Br	2.8
						I	2.5

1-3 Dipole Moment

The separation of charge between two atoms such as in iodine monochloride above results in what is called a dipole moment. It is defined as follows:

$$\text{Dipole moment} = \mu = e \times d$$

(e is in esu; d is distance between charge centers in angstroms). The dipole moment of a molecule is the vector sum of the individual bond moments.

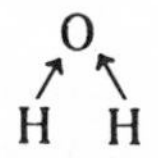

$\mu = 1.84$ debyes 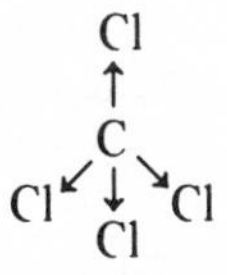 $\mu = 0$ 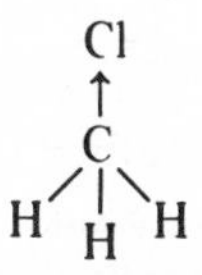 $\mu = 1.86$ debyes

Some individual bond moments are as follows:

Bond	H → N	H → O	C → N	C → O	C → Cl
Moment (in debye units)	1.3	1.5	1.0	1.2	1.9

The head of the arrow represents the negative end of the dipole.

1-4 Resonance

Reflection about the above formula for nitric acid indicates that there are more ways than one in which to arrange the electrons and still not violate the octet rule.

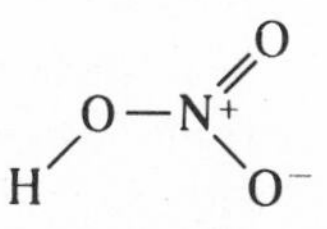

I 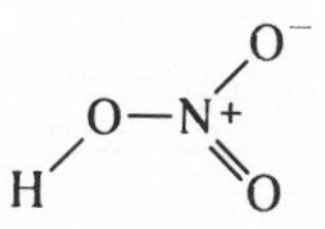   II 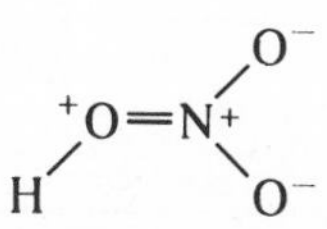   III

Structures I and II are equivalent in every way. Structure III would seem less reasonable since it has an additional charge separation which could be canceled by changing the position of the electrons.

There are good reasons for believing that none of these structures is a satisfactory description of nitric acid, but that it behaves as though it were a hybrid of these with a relatively small amount of character implied by III. In short, we can say that when two or more electronic formulas can be drawn for a given molecule without violating electronic principles, no one formula will be an adequate description of the molecule. Its properties will best be described by a hybrid of all possible "reasonable" formulas.

Generalizations for writing meaningful resonance structures:

1. Elements up to Ne can accommodate a maximum of eight electrons (shared or lone pairs) in their outer shells. (Hydrogen and helium, only two.)
2. Keep the same number of electron pairs in each structure.

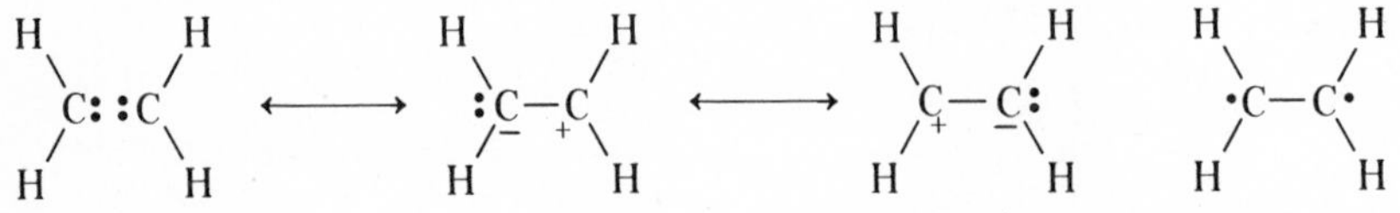

Structures I, II, and III are reasonable. IV, with two unpaired electrons does not contribute.

3. Structures with charge separation (II and III above) are usually less significant than those without. When charge separation between unlike atoms is considered, the negative charge is shown only on the more electronegative atom.

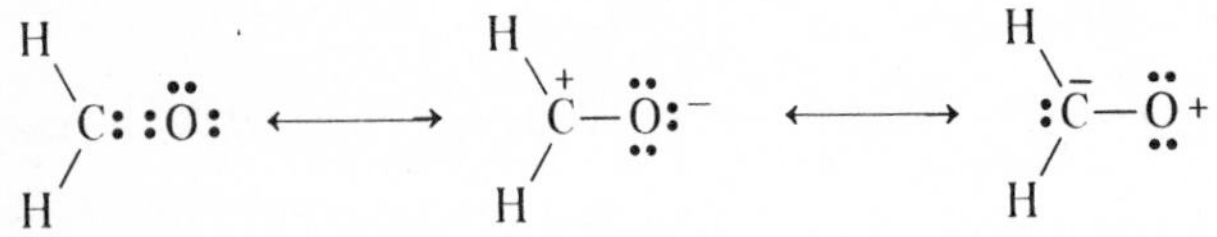

Although II does not violate (1) or (2) above, its contribution to a description of the formaldehyde is small. The chance of electrons leaving electron-attracting oxygen to go to carbon is slight (see Sec. 1-2).

Note that the double-headed arrows used to connect resonance structures in no way imply that the molecule exists first in one form, then in the next, etc. These forms simply mean that a single formula does not properly describe the molecule. Rather the molecule behaves as a hybrid of these extreme possibilities.

1-5 Acids and Bases

The Brönsted-Lowry concept defines an acid as any proton acceptor. Neutralization may be illustrated with acetic acid and ammonia as follows:

$$CH_3COOH + :NH_3 \rightleftharpoons CH_3COO^- + \overset{+}{N}H_4$$

$$Acid_1 \qquad Base_2 \qquad\qquad Base_1 \qquad Acid_2$$

In the above competition for the proton, the base with the greater availability of electrons will be the more successful. $Acid_1$ and $base_1$ are referred to as a conjugate pair, and $base_2$ and $acid_2$ also constitute a conjugate pair. Table 1-1 lists some conjugate acids and conjugate bases.

Table 1-1

Proton + base	Conjugate acid
$H^+ + HO^-$	H_2O
$H^+ + HOH$	H_3O^+
$H^+ + CO_3{}^{2-}$	$HOCO_2{}^-$
$H^+ + (CH_3)_2O$	$(CH_3)_2\overset{+}{O}H$
$H^+ + (CH_3)_3N$	$(CH_3)_3\overset{+}{N}H$

In the *Lewis* theory, a substance is defined as a Lewis acid if it can accept an electron pair and as a Lewis base if it can act as an electron-pair donor. Thus a proton is the most common electron-pair acceptor and HO^- a common electron-pair donor.

Example Draw the electronic structure for the typical Lewis acid, BF_3, and show its reaction with a typical Lewis base, NH_3.

$\cdot \dot{B} \cdot$ One atom, three valence-shell electrons

$: \ddot{F} \cdot$ Three atoms, each needing one electron to complete octet

The structure is drawn with three single covalent bonds and a six-electron shell for boron. Boron compounds of this type can accept a pair of electrons to complete the outer-shell octet.

$$
\begin{array}{c}
F \quad CH_3 \\
| \qquad | \\
F-B + N-CH_3 \\
| \qquad | \\
F \quad CH_3
\end{array}
\longrightarrow
\begin{array}{c}
F \quad CH_3 \\
| \qquad | \\
F-B-N^+ -CH_3 \\
| \qquad | \\
F \quad CH_3
\end{array}
\quad \text{or} \quad
F_3\bar{B}-\overset{+}{N}(CH_3)_3 \equiv F_3B \leftarrow N(CH_3)_3
$$

1-6   Analysis

a.   *Empirical Formulas* A common problem in organic chemistry is the determination of empirical formulas (the relative proportions of each element present in the compound) and then the exact molecular formula. The following example should serve as a review of the techniques used.

Example 1 An unknown organic compound has been found to contain 85.7% carbon and 14.3% hydrogen by weight. Since empirical formulas show the relative number of atoms present in a compound rather than the relative weights, it is necessary to convert the percentage composition values to ratios of atoms. If each of the above values is divided by the respective atomic weights of the elements,

$85.7/12.01 = 7.1$ (carbon) $14.3/1.01 = 14.2$ (hydrogen)

the results indicate that there are 7.1 atoms of carbon for every 14.2 atoms of hydrogen. In order to convert these into integers suitable for use in a formula, divide each value by the lowest value, which gives the empirical formula CH_2.

Element	Atomic ratios	Atomic ratios in integers
Carbon	7.1	$7.1/7.1 = 1.0$ (or simply 1)
Hydrogen	14.2	$14.2/7.1 = 2.0$ (or simply 2)

Example 2 Assume 0.438 g of an organic compound is burned in pure oxygen and the CO_2 and H_2O are measured by absorption in weighed absorbers.

Wt of CO_2 = 1.315 g   Wt of C = $12/44 \times 1.315 = 0.359$ g
Wt of H_2O = 0.640 g   Wt of H = $2/18 \times 0.640 = 0.071$ g

Within experimental error, the total weight of carbon and hydrogen is equal to the total weight of the sample taken. These, then, are the only elements in the compound.

Element	Wt, g	Wt, %	Ratio	Atomic ratio in integers	
Carbon	0.359	83.5	7.0	$7.0/7.0 = 1.00$	$1.00 \times 3 = 3$
Hydrogen	0.071	16.5	16.5	$16.5/7.0 = 2.36$	$2.36 \times 3 = 7$

Note that when each atomic ratio was divided by the smaller value the results were not integers, but they could be converted to integers by multiplication by a small whole number, 3. The empirical formula of this compound is C_3H_7. It may be checked by calculating percentage composition for comparison with the analysis.

Example 3 If organic compounds containing nitrogen are oxidized by cupric oxide, the nitrogen is converted to nitrogen gas. The nitrogen gas may then be measured volumetrically after suitable removal of carbon dioxide and water. Oxygen in organic compounds is usually determined by difference.

When 0.730 g of compound X was oxidized in the presence of air, 1.320 g of CO_2 and 0.630 g of H_2O were obtained. This same weight of sample yielded 112 ml of nitrogen gas (measured under standard conditions, 760 mmHg pressure and 273 K). Qualitative tests on X showed no other elements except possibly oxygen.

Since 1 mol of a gas (28.0 g of nitrogen) occupies 22.4 liters under standard temperature and pressure (STP), 112 ml of nitrogen represents 0.140 g.

Element	Wt, g	Wt/at wt	At wt ratios in integers
Carbon	0.360	0.030	0.030/0.010 = 3
Hydrogen	0.070	0.070	0.070/0.010 = 7
Nitrogen	0.140	0.010	0.010/0.010 = 1
Oxygen	0.160	0.010	0.010/0.010 = 1 (by difference)

Note that the ratio wt/at wt can be used as well as wt %/at wt. The empirical formula is C_3H_7NO.

b. *Formula Evaluation* In Example B, given above, the molecular formula could be C_3H_7 or a multiple of this, C_6H_{14}, C_9H_{21}, etc. Each would give the same combustion analysis. These possibilities may be limited by an elementary rule: only compounds with an odd number of odd-valence atoms have an odd number of hydrogen atoms. The odd-valence atoms include nitrogen (3), halogens (1), phosphorus (3), etc.

The compound $C_5H_9ON_2Cl$ is feasible because there are an odd number of odd-valence atoms and also an odd number of hydrogen atoms. Conversely, $C_6H_{13}NBr$ does not represent a real organic compound because it has an even number of odd-valence atoms but an odd number of hydrogen atoms.

c. *Molecular Weights* The melting point of solvents is depressed in the presence of solutes. The depression brought about by dissolving 1 mol of solute in 1,000 g of solvent is known as the "cryoscopic constant." For example, in the Rast method for determining molecular weight, camphor, mp 179°C, is used as the solvent, and 1 mol of a compound dissolved in 1,000 g of camphor will bring the melting point to 139°C (cryoscopic constant, 40).

Given the following data, show how to determine the molecular weight of the unknown.

Weight of camphor	1.80 g
Weight of unknown	0.054 g
Melting point of solution	165°C

Since the cryoscopic constant is based on a solution containing 1,000 g of camphor, first determine the amount of unknown which would give the same depression (14°) when dissolved in the standard amount. Obviously, any ratio of unknown/solvent which is equal to 0.054/1.80 will have a melting point of 165°C.

0.054/1.80 = X/1,000   X = 30 g unknown

If 30 g gave a depression of 14°, and 1 mol would give a depression of 40°, the molecular weight (MW) is calculated as follows:

30/14 = MW/40   30 x 40/14 = 86 = MW

d. *Equivalent Weights* The equivalent weight (EW) of an acid or base is its molecular weight divided by the number of acid or base groups within the molecule.

	MW	EW
Nitric acid, HNO_3	63	$63/1 = 63$
Sulfuric acid, H_2SO_4	98	$98/2 = 49$
Sodium hydroxide, NaOH	40	$40/1 = 40$
Calcium hydroxide, $Ca(OH)_2$	74	$74/2 = 37$

This equivalent weight of an acid or base is also known as the neutralization equivalent, or NE. Note that the molecular weight is always a simple multiple of the neutralization equivalent.

Example 4 It was found that 0.450 g of compound X neutralized 25.0 ml of 0.100 *N* NaOH. What is its equivalent weight?

Equivalents of NaOH used = $(25.0)(0.100)/1,000 = 0.0025$
$$0.450/0.0025 = 180 = \text{NE of X}$$

The molecular weight of X is 180 or a multiple of 180.

e. *Zerewitinoff Determination* This procedure may be used to find the number of so-called active hydrogens in a given molecule. These more acidic hydrogen atoms are those attached to strong electron-attracting groups or elements such as oxygen, nitrogen, sulfur, and halogen. They react rapidly and quantitatively with methyl-magnesium iodide, CH_3MgI, to yield 1 equiv of methane gas per acidic hydrogen atom.

$$C_2H_5OH + CH_3MgI \longrightarrow CH_4 + Mg(OC_2H_5)I$$

Example 5 When 0.450 g of Z was treated with excess methylmagnesium bromide, 224 ml of methane (STP) was evolved. A Rast determination showed a molecular weight of 90. How many active hydrogen atoms are present per molecule?

$$\frac{224 \text{ ml}}{22,400 \text{ ml/mol}} = 0.01 \text{ mol of methane} \qquad \frac{0.450 \text{ g}}{90 \text{ g/mol}} = 0.005 \text{ mol of Z used}$$

Each mole of Z yielded 2.0 mol of methane: therefore, each molecule of Z had two "acidic" hydrogen atoms.

f. *Molecular Weight of Gases* When 0.345 g of compound Q was vaporized, the gas occupied 232 ml at $100°C$ and 750 mmHg. What was its molecular weight?

$$232 \text{ ml} \frac{750 \text{ mmHg}}{760 \text{ mmHg}} \frac{273°C}{373°C} = 168 \text{ ml gas (standard conditions)}$$

Since 1 mol of any gas occupies 22,400 ml at STP, and 0.345 g of our sample occupies 168 ml, then

$$0.345 \text{ g} \frac{22,400 \text{ ml/mol}}{168 \text{ ml}} = 46 \text{ g/mol}$$

Problems

1-1 Draw electronic formulas, indicating approximate bond angles for each of the following compounds. If in doubt, consult your general chemistry book.

a. Chlorine
b. Carbon dioxide
c. Methane, CH_4
d. CCl_4
e. HCN
f. Carbonic acid
g. CH_3MgI
h. Hypochlorous acid
i. Dinitrogen tetroxide
j. Acetylene, C_2H_2
k. Iodine monochloride, ICl
l. Ethene, C_2H_4
m. Phosgene, $COCl_2$
n. Dinitrogen pentoxide
o. Chloroform, $CHCl_3$

1-2 Taking into account the indicated shape, draw electronic structures for each of following:
  a. SO_2, V-shaped
  b. H_2SO_4, tetrahedral
  c. SF_6, octahedral
  d. Carbon suboxide, C_3O_2, linear
  e. F_2O, $102°$
  f. ClO_4^-, tetrahedral
  g. $\overset{+}{N}O_2$, linear
  h. FNNF, N–N–F = $115°$
  i. $(CH_3)_3NO$, tetrahedral
  j. $\bar{B}H_3\overset{+}{C}O$, B is tetrahedral
  k. $(CH_3)_3\overset{+}{S}\bar{B}r$, S is tetrahedral
  l. C_2N_2, linear

1-3 Point out the most *unlikely* resonance form in each of the following sets. Give a reason for your choice.

a. I   II   III   IV

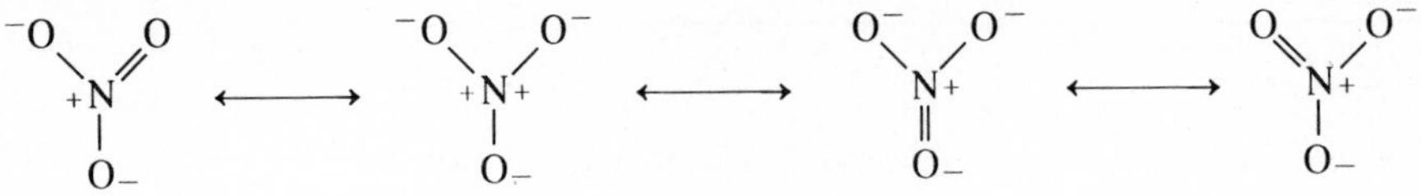

b. I   II   III   IV

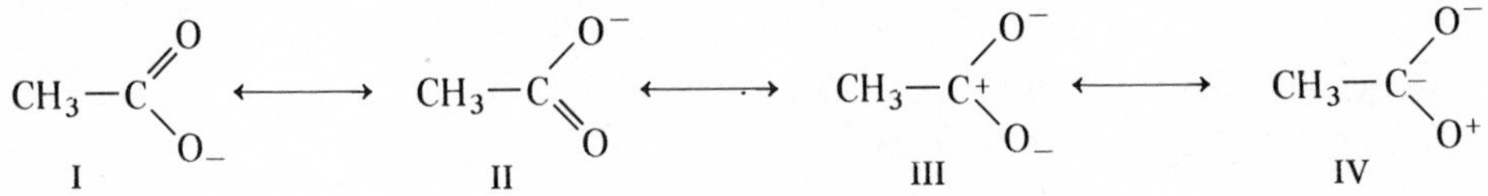

c. H–C≡C–H ⟷ H–$\bar{C}$=$\overset{+}{C}$–H ⟷ H–$\overset{+}{C}$=$\bar{C}$–H ⟷ H–$\dot{C}$=$\dot{C}$–H
       I   II   III   IV

d. H_2C=CH–CH_2^- ⟷ $H_2\overset{+}{C}$–$\bar{C}H$–$\bar{C}H_2$ ⟷ $H_2\bar{C}$–$\overset{+}{C}H$–$\bar{C}H_2$ ⟷ $H_2\bar{C}$–CH=CH_2
        I   II   III   IV

e. H–C≡N: ⟷ H–$\overset{+}{C}$=$\bar{N}$ ⟷ H–$\bar{C}$=$\overset{+}{N}$
       I   II   III

f. I–Cl ⟷ $\overset{+}{I}$=$\bar{Cl}$ ⟷ $\bar{I}$=$\overset{+}{Cl}$
      I   II   III

1-4 Draw the indicated number of equivalent resonance forms for each of the following:
  a. Sulfur dioxide, 2
  b. CH_2=CH–$\overset{+}{C}H_2$, 2
  c. Nitrite ion, 2

  d. I_2, 2
  e. H_2C=CHCH=CH_2, 2
  f. 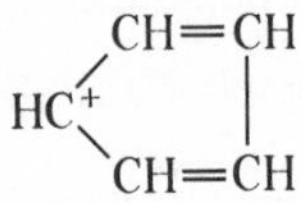 , 5

1-5 Indicate the charge on each atom given, and state whether the structure shown is an ion (give charge) or a compound.

  a. CH_3–$\overset{..}{\underset{..}{Se}}$:, selenium
  b. 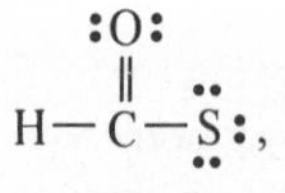 , oxygen and sulfur

  c. CH_3–$\overset{..}{\underset{..}{O}}$–$\overset{..}{N}$=$\overset{..}{\underset{..}{O}}$, nitrogen
  d. 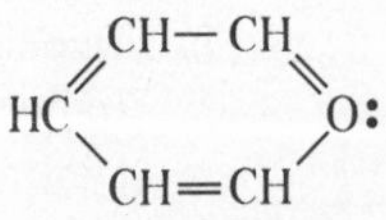 , oxygen

8

e.
$$
CH_3-\overset{\overset{\displaystyle :NH}{|}}{C}=\overset{..}{\underset{..}{O}}, \text{ nitrogen}
$$

f. $:\overset{..}{\underset{..}{Cl}}-\overset{..}{\underset{..}{I}}-\overset{..}{\underset{..}{Cl}}:$, chlorine and iodine

g.
$$
:\overset{..}{\underset{..}{Cl}}-\overset{\overset{\displaystyle :\overset{..}{Cl}:}{|}}{\underset{\underset{\displaystyle :\overset{..}{Cl}:}{|}}{Al}}-\overset{..}{\underset{..}{Cl}}:, \text{ aluminum}
$$

h. $CH_3-N\equiv C:$, nitrogen

i. BF_4, boron

1-6 Calculate the percentage of each element present in each of the following compounds:
   a. Methane, CH_4
   b. Ethanol, C_2H_5OH
   c. Chloroform, $CHCl_3$
   d. Heroin, $C_{21}H_{23}NO_5$
   e. Sulfanilic acid, $C_6H_7NO_3S$
   f. DDT, $C_{14}H_9Cl_5$

1-7 Calculate the empirical formula of each of the following from the combustion data. Assume only C, H, N, and O to be present, and obtain oxygen by difference.

Compound	Sample size, g	CO_2, g	H_2O, g	N_2, ml, STP
A	0.307	0.440	0.275	0
B	0.260	0.880	0.180	0
C	0.6808	1.540	0.363	112
D	0.03213	0.10326	0.03523	0

1-8 For each of the compounds in Prob. 1-7 give the simplest empirical formula that could represent a real organic molecule.

1-9 What would the molecular formula of compound C (Prob. 1-7) be if a solution of 0.034 g in 0.500 g of camphor melted at 159°C?

1-10 When 0.261 g of compound B (Prob. 1-7) was vaporized, the gas occupied 96 5 ml at 80°C and 1.0 atm pressure. What is the correct molecular formula for B?

1-11 It required 32.5 ml of 0.100 N HCl to neutralize exactly 0.442 g of compound C (Prob. 1-7). What is its neutralization equivalent? By what factor does the molecular weight differ (Prob. 1-9) from the neutralization equivalent?

1-12 When 0.124 g of compound A (Prob. 1-7) was treated with an excess of CH_3MgBr, 89.5 ml of methane (STP) was collected. How many "acidic" hydrogen atoms does A have if its simplest molecular formula (Prob. 1-8) is correct?

1-13 The cryoscopic constant for benzene is 4.90°C, and its melting point is 5.51°C. A solution of 0.8160 g of compound M dissolved in 7.500 g of benzene was found to melt at 1.59°C. Find the molecular weight of M. If the empirical formula for M is C_4H_4O, what is the molecular formula?

1-14 Compound N was an acid which could be titrated with aqueous NaOH. It required 24.0 ml of 2.08 N NaOH to neutralize 4.15 g of N. What is the equivalent weight?

1-15 Ethyl alcohol, C_2H_6O, has one hydrogen which will react with CH_3MgBr. If 0.460 g of ethyl alcohol is treated with an excess of the Grignard reagent, how much methane should be expected at 28°C and 740 mmHg?

1-16 Indicate the direction of the dipole moment between each of the following atoms or groups of atoms by means of an arrow, $\longmapsto$, pointing toward the negative end.
   a. C–O
   b. C–Si
   c. CH_3-CF_3
   d. S–O
   e. H–F
   f. C–Cl

9

1-17 Draw the indicated number of the most significant resonance forms for each of the following:

a. $\ddot{:}O:$ over $CH_3\overset{\|}{C}CH_3$, 2

b. $CH_3\ddot{N}=\ddot{O}$, 2

c. CS_2, 3

d. 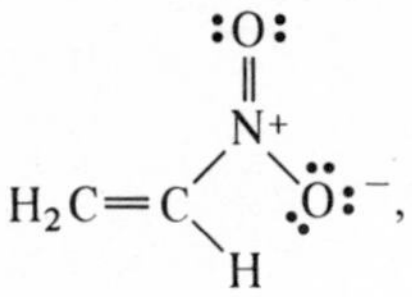 $H_2C=C\overset{\displaystyle \overset{:O:}{\underset{\|}{N^+}}}{\underset{H}{\diagdown}}\ddot{O}:^-$, 3

e. 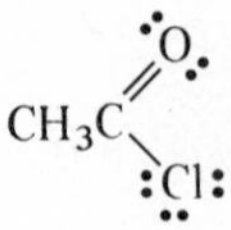 $CH_3C\overset{\displaystyle \cdot\ddot{O}\cdot}{\underset{:\ddot{Cl}:}{\diagup}}$, 3

f. $B(OCH_3)_3$, 4

g. $CH_3C\equiv N\colon$, 2

h. $H_2C=C\overset{\displaystyle \ddot{O}CH_3}{\underset{H}{\diagdown}}$, 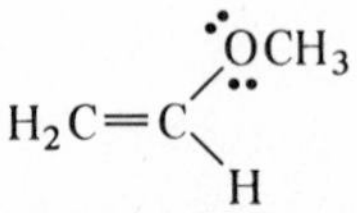 3

i. $HC\equiv CC\equiv CH$, 6

1-18 Complete the reactions between the following acids and bases:
a. $NH_4^+ + NH_2^- \longrightarrow$
b. $BF_3 + CH_3OCH_3 \longrightarrow$
c. $CH_3OH + HCl \longrightarrow$
d. $SO_3 + HCl \longrightarrow$
e. $AlCl_3 + CH_3SCH_3 \longrightarrow$
f. $H_3O^+ + CH_3O^- \longrightarrow$
g. $CH_2=CH_2 + BF_3 \longrightarrow$
h. $Ag^+ + NH_3 \longrightarrow$
i. $HF + CH_3OH \longrightarrow$

1-19 In the following equations indicating a competition between two Brönsted-Lowry- or Lewis-type bases for a proton, indicate which base would be the more successful (i.e., would the equilibrium be more toward the right or the left?). Give a reason for your decision.

a. $CH_3OH + CH_3S^- \rightleftharpoons CH_3SH + CH_3O^-$

b. $^-OH + HO\overset{\displaystyle O}{\overset{\|}{-C}}-O^- \rightleftharpoons HOH + {}^-O\overset{\displaystyle O}{\overset{\|}{-C}}-O^-$

c. $PH_3 + NH_4^+ \rightleftharpoons PH_4^+ + NH_3$

d. $CH_4 + LiNH_2 \rightleftharpoons CH_3Li + NH_3$

e. 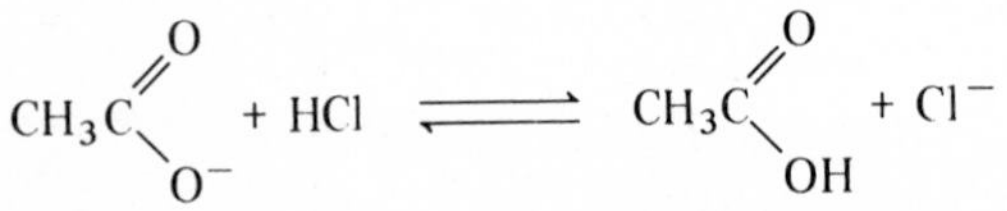 $CH_3C\overset{\displaystyle O}{\underset{O^-}{\diagup}} + HCl \rightleftharpoons CH_3C\overset{\displaystyle O}{\underset{OH}{\diagup}} + Cl^-$

f. $CH_4 + {}^-OH \rightleftharpoons CH_3^- + HOH$

g. $CH_3\overset{+}{\underset{H}{-S}}-CH_3 + NH_3 \rightleftharpoons NH_4^+ + CH_3SCH_3$

h. $Ne + HCl \rightleftharpoons Ne\overset{+}{H} + Cl^-$

10

2
atomic and molecular structure

2-1   Atomic Orbitals

The electrons of an atom may be described in terms of a set of quantum numbers which indicates their principal energy level, the shape of the orbital (i.e., the probable location) they occupy, the orientation of the orbitals, and the electron spin. To each orbital there is ascribed a definite energy, and each orbital may be occupied by a maximum of two electrons having opposite spins. (According to the Pauli exclusion principle, no two electrons in a given atom can have all quantum numbers the same.)

a.   *Main Shell or Energy Level*   The main shell is designated by the *principal quantum number n*, where $n = 1, 2, 3$, etc. (corresponding to the *K, L, M,* etc., shells). The maximum number of electrons in any shell is $2n^2$.

b.   *Subshell*   The subshell is designated by the *azimuthal quantum number l*, where $l = 0, 1, 2 \ldots, n - 1$. Thus when $n = 1$, l can have only the value of 0; when $n = 2$, $l = 0, 1$.

c.   *Orientation of Subshell*   The shell orientations are designated by the *magnetic quantum number m*, where $m = 0, \pm1, \pm2, \ldots, \pm l$. Thus, when $l = 0$, there is only one possible value for m; and when $l = 1$ there are three possible values (orientations) for m: $0, \pm1$.

d.   *Electron Spin*   The *spin quantum number* can have only two values, $s = \pm1/2$. The $l = 0$ and $l = 1$ subshells are of primary interest to organic chemists. The former refers to the spherically symmetric s orbital which has only one orientation (that is, $m = 0$) but which can contain two electrons if their spins are opposite. The latter refers to p orbitals for which there can be three orientations ($m = 0, m = +1, m = -1$), each directed along one axis of a set of three mutually perpendicular axes, and each of which may contain a maximum of two electrons (the assignment of $m = 0, m = +1$, and $m = -1$ to p orbitals directed along the x, y, and z axes, respectively, as shown in Fig. 2-1, is arbitrary).

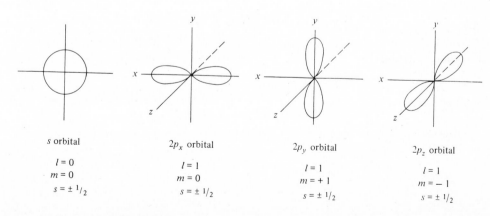

s orbital	$2p_x$ orbital	$2p_y$ orbital	$2p_z$ orbital
$l = 0$	$l = 1$	$l = 1$	$l = 1$
$m = 0$	$m = 0$	$m = +1$	$m = -1$
$s = \pm 1/2$	$s = \pm 1/2$	$s = \pm 1/2$	$s = \pm 1/2$

Figure 2-1.   *s* and *p* atomic orbitals.

11

The atomic orbitals of the elements hydrogen through neon fill by adding one electron at a time to the lowest energy orbital. First the $1s$ shell fills, then the $2s$ and $2p$. At the next level $3s$ and $3p$ fill, after which $4s$ fills before $3d$ begins to fill. The p orbitals are all equivalent in energy but, before a second electron enters a p orbital (for example, p_x), as in oxygen, one electron each is distributed through the $p_x, p_y,$ and p_z orbitals.

Table 2-1 Electronic configurations of elements hydrogen through neon

Symbol	At No.	K shell 1s	L shell 2s	$2p_x$	$2p_y$	$2p_z$	Notation
H	1	↑	○	○	○	○	$1s$
He	2	↑↓	○	○	○	○	$1s^2$
Li	3	↑↓	↑	○	○	○	$1s^2 2s$
Be	4	↑↓	↑↓	○	○	○	$1s^2 2s^2$
B	5	↑↓	↑↓	↑	○	○	$1s^2 2s^2 2p_x$
C	6	↑↓	↑↓	↑	↑	○	$1s^2 2s^2 2p_x p_y$
N	7	↑↓	↑↓	↑	↑	↑	$1s^2 2s^2 2p_x p_y p_z$
O	8	↑↓	↑↓	↑↓	↑	↑	$1s^2 2s^2 2p_x^2 p_y p_z$
F	9	↑↓	↑↓	↑↓	↑↓	↑	$1s^2 2s^2 2p_x^2 p_y^2 p_z$
Ne	10	↑↓	↑↓	↑↓	↑↓	↑↓	$1s^2 2s^2 2p_x^2 p_y^2 p_z^2$

2-2 Molecular Orbitals

Bonds are ordinarily formed by the overlapping of two atomic orbitals. For example, the overlapping and resultant coalescing of two $1s$ orbitals of two hydrogen atoms leads to the molecular orbital in the hydrogen molecule.

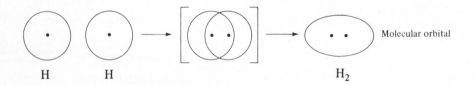

Figure 2-2. H₂ molecular orbital formation.

The overlapping of two s orbitals results in a sigma (σ) bond. The combination of an s and p or two p orbitals also gives a σ bond.

The combination of two atomic orbitals yields two molecular orbitals, one bonding and one antibonding:

12

Electrons in antibonding orbitals lead to repulsions which are about the same strength as the attractive forces in bonding orbitals. Helium does not form a stable molecule as illustrated above for H_2; this can be explained as follows:

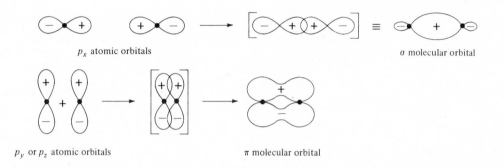

The repulsive forces of the two electrons in the σ^* orbital counterbalance the attractive forces in the σ orbital. In addition to the spherically symmetric atomic s orbitals, the organic chemist is often concerned with p orbitals (Fig. 2-1) which may be combined to give molecular orbitals in two different ways to yield either σ or π bonds (Fig. 2-3).

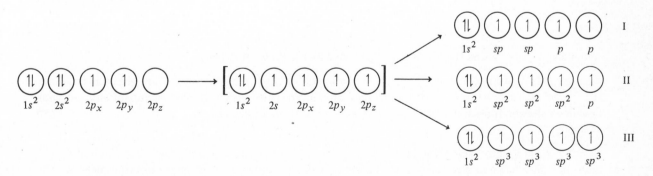

p_x atomic orbitals

σ molecular orbital

p_y or p_z atomic orbitals

π molecular orbital

Figure 2-3. Formation of σ and π molecular orbitals.

The + and − signs do not refer to electric charge, but to different character of the two lobes of the p orbital. Note that in the addition (linear combination) of two p orbitals, lobes of like character combine while lobes of unlike character do not (see Sec. 2-4).

2-3   Hybrid Orbitals

Carbon, the element with which we shall be most concerned, does not use simple s or p orbitals to form bonds; instead it mixes s and p orbitals to form hybrid orbitals which have shapes and properties in between s and p orbitals. Although hybridization does not occur by a stepwise process, this is a convenient way of picturing the overall result. An s electron is visualized as being promoted to the p level and the remaining s electron is then mixed with one, two, or three p electrons as follows:

As mentioned in Sec. 1-1e electrons tend to arrange themselves so as to be as far apart as possible. Hence the two sp orbitals formed in hybridization mode I are 180° apart with the two p orbitals 90° apart. The sp^2 orbitals are 120° apart and the sp^3 109° 28′ apart (Fig. 2-4).

13

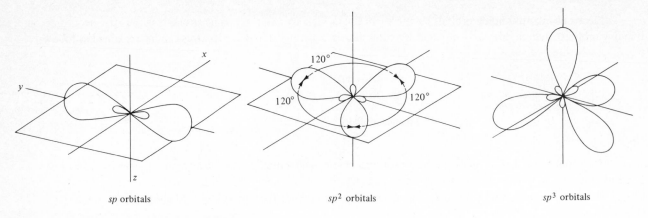

sp orbitals sp^2 orbitals sp^3 orbitals

Figure 2-4. sp, sp^2, and sp^3 hybridized orbitals.

In Fig. 2-5, ethylene (CH_2=CH_2) illustrates how two sp^2 orbitals overlap to form a σ bond and how two p orbitals overlap to form a π bond. The CH bonds have been formed by the overlapping of a carbon sp^2 orbital with a hydrogen s orbital.

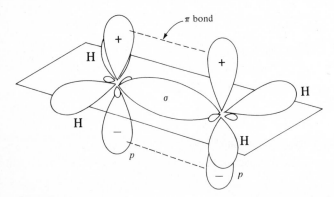

Figure 2-5. Bonding in ethylene.

Other examples are more diagrammatically illustrated in Examples 1 through 4.

Example 1 Acetylene.

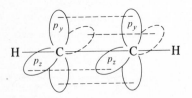

sp bonding between C and H and C and C is indicated by the three centerlines.

Example 2 Methylamine (CH_3NH_2). The carbon atom has four sp^3 σ bonds and the nitrogen atom has three sp^3 σ bonds and a lone pair of electrons in an sp^3 orbital. To clarify the electronic structure of nitrogen, show how one s electron is promoted to the p level and describe the hybridization that results to account for the sp^3 character.

Example 3 Nitric acid. Sketch a molecular orbital picture showing the nitrogen atom with sp^2 hybridization. Show how the remaining p orbital of nitrogen overlaps with a p orbital on each of the two equivalent oxygen atoms to form an extended π orbital. For the remaining oxygen atom (in the hydroxyl group) show how one p orbital overlaps with an s orbital of hydrogen and the other p orbital overlaps with an sp^2 orbital of nitrogen.

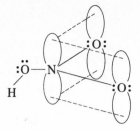

Example 4 Carboxylic acids. In the carboxyl group carbon has sp^2 hybridization.

$$H-C \overset{\ddot{O}-H}{\underset{:\ddot{O}:}{\Big\langle}} \equiv H-C \overset{OH}{\underset{O}{\Big\langle}}$$

2-4 Resonance

Resonance can be discussed in terms of bonds as in Chap. 1 or it can be viewed in terms of molecular orbitals. The molecular-orbital description of resonance in ethylene (see Sec. 1-4) can be depicted as in Fig. 2-6.

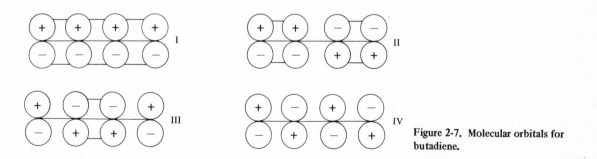

Figure 2-6. **Resonance in ethylene.**

Compounds having an alternating series of single and double bonds are called conjugated. In effect this means that there is a p orbital containing an electron on each atom in the conjugated set. Electrons can move through the series of overlapping p orbitals. Butadiene ($CH_2=CH-CH=CH_2$) is such a compound. Its molecular orbitals resulting from a combination of its atomic orbitals are shown in Fig. 2-7.

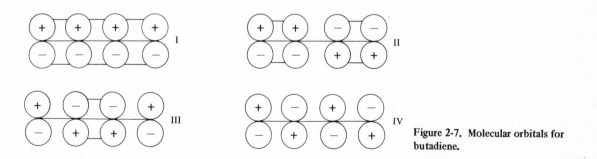

Figure 2-7. **Molecular orbitals for butadiene.**

The most stable of the four molecular orbitals is I, and IV, with all antibonding orbitals, is the least stable.

2-5   Aromatic Systems

Cyclic systems of conjugated bonds are often found to be unusually stable. That is, they form more easily and are less reactive chemically than comparable nonconjugated systems. A precise definition of aromatic character is not possible; however, a feeling for aromatic character comes from the study of compounds having cyclic systems of π electrons. The ideal model of an aromatic system is benzene in which six cyclic carbon atoms are bound together by sp^2 and p orbitals.

In Fig. 2-8 benzene is represented in four different ways. For most conventional discussions, it is usually drawn as in (b) or (d). Form (d) is a shorthand representation of (c). Drawing (c) represents the most stable molecular orbital for benzene.

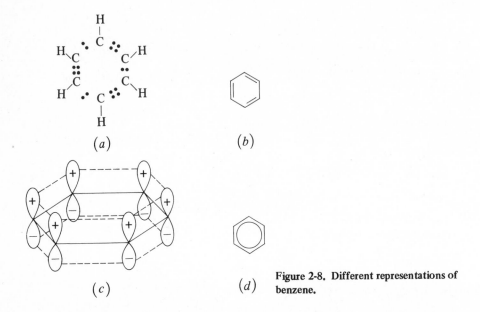

Figure 2-8. Different representations of benzene.

Delocalization of the p electrons in benzene is represented in the valence-bond symbolism used in Chap. 1 as follows (the first two structures are much more important than the last three):

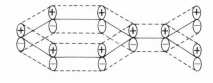

In this symbolism each line represents two paired electrons. Note that electrons are kept paired by moving dashes from position to position. Nitrobenzene serves to illustrate the molecular orbital and valence-bond symbolism used to describe interaction of p electrons in the aromatic ring with those in the attached groups (Fig. 2-9).

Nitrobenzene

Figure 2-9. Molecular orbital drawing and resonance structures for nitrobenzene.

16

2-6   Hückel's Rule

Based on theoretical considerations, Hückel formulated the useful $4n + 2$ rule which states that exceptional resonance stability is to be expected with π-electron systems containing $4n + 2$ π electrons in cyclic structures. Although exceptions to the rule are known, it does predict the relative stability for many such systems. The rule states that a high delocalization energy will be found only for those cyclic systems of π electrons for which n is an integral number, that is, 0, 1, 2, 3, 4, etc. This applies only to a *continuous* series of p orbitals which are capable of effective overlapping.

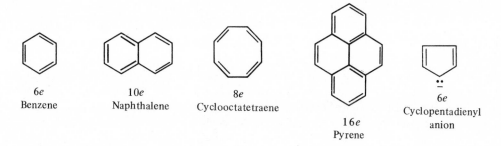

Thus for benzene, $n = 1$, and for naphthalene, $n = 2$, we find especially stable π-electron systems with high resonance energy. For cyclooctatetraene, $n = 1.5$, we find a puckered ring without effective overlap and very little resonance energy over that expected for isolated double bonds. Hückel's rule does not apply to compounds such as pyrene which have π electrons involved other than those in the peripheral cyclic system. Certain ions, such as cyclo-pentadienyl, show unusual stability normally associated with high delocalization of the π electrons and are therefore easily prepared. Others such as cycloheptatrienyl anion, which would have eight π electrons and thus violate the $4n + 2$ rule, have resisted all synthetic attempts.

2-7   Effectiveness of Overlap

The strength of covalent bonds is determined by the degree to which two atomic orbitals may overlap to form a bonding molecular orbital. Thus bonding between two s orbitals is weak, particularly when the two orbitals are at different energy levels (for example, Li + K). A crude way to visualize this is to picture the extent of overlap of the electron clouds (Fig. 2-10).

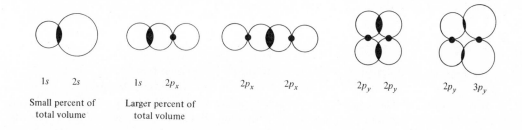

Figure 2-10.   Extent of overlap of atomic orbitals.

The strongest bond between s orbitals would be that between two $1s$ orbitals as in hydrogen. That between two $2s$ orbitals would be much weaker. The more s character a hybrid orbital has, the weaker its bonding strength. A strong σ bond is formed by the end-on overlap of two p orbitals at the same energy level. The bond is much weaker if the orbitals have different principal quantum numbers. The side overlapping of two p orbitals is much less effective and the high activity of the olefin bond indicates the weak interaction of two such orbitals.

If the two p electrons have different energy levels, the strength of the binding approaches zero as the distance between the levels increases. For example, in the halogen derivatives, resonance (as indicated by the structures on p. 18) is almost nil in bromobenzene although it is quite strong in fluorobenzene.

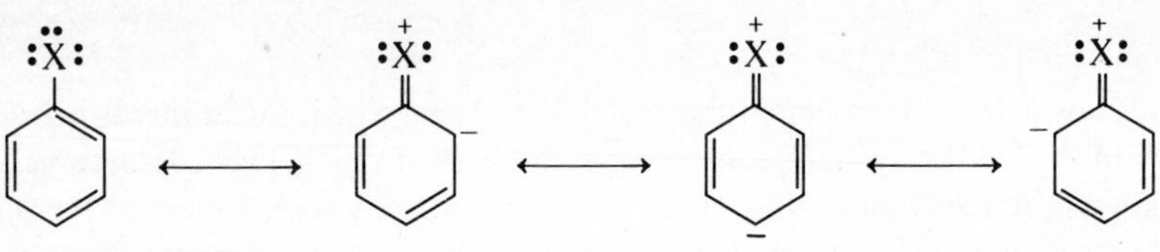

This is because the p electron in carbon is at the two level, whereas that in bromine is at the four level and overlap is not as effective.

2-8 Cis-Trans Isomerism

Because of the overlapping of p orbitals in double bonds as illustrated in Fig. 2-5, the four groups attached to a carbon-carbon double bond are held rigidly in one plane. While more or less free rotation occurs around single bonds such as the σ bonds in ethane, the rigid character of a π bond leads to the following kind of geometric isomers (*different compounds that have the same molecular formula are called isomers*):

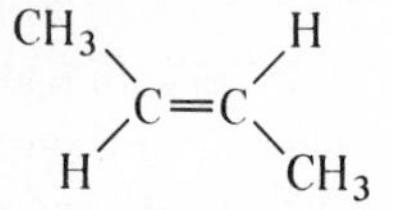

trans-2-Butene

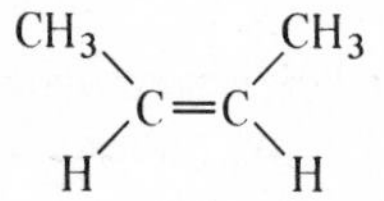

cis-2-Butene

In the above names, "2" refers to the position of the double bond in the four-carbon chain. Since this bond is between carbon atoms 2 and 3 in the chain, its position is specified by the lower number (2). While the terms cis and trans are satisfactory to describe simple compounds such as the 2-butenes, more complicated molecules have names using the descriptors E and Z. E comes from the German *entgegen* meaning opposite and Z comes from *zusammen* meaning together.

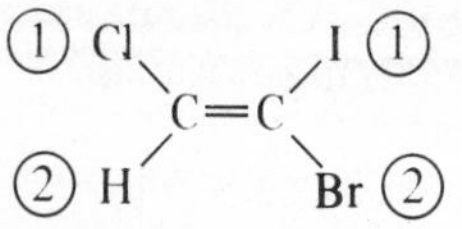

(Z)-1-Bromo-2-chloro-1-iodoethene
I

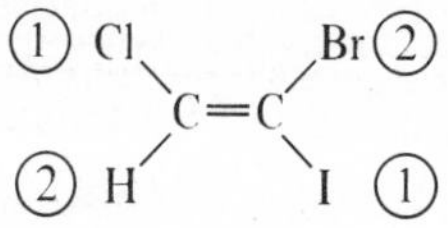

(E)-1-Bromo-2-chloro-1-iodoethene
II

In the above system priorities are assigned, according to Cahn-Ingold-Prelog rules (see Sec. 13-2*b*), to each of the two groups attached to a carbon involved in the C=C unit. The higher the atomic number the higher the priority, that is, I > Br > Cl > F. When the two groups of higher priority are on the same side of the plane bisecting the double bond, as in I (circled numbers indicate priority), Z specifies the configuration. When the two groups of higher priority are on opposite sides as in II, the configuration is designated by E. When groups of more than one atom are involved, comparisons are made between the second atoms, third atoms, etc.

Examples CH_3O has higher priority than HO because C has a higher atomic number than H. $HOCH_2CH_2$— precedes $CH_3CH_2CH_2$— because O has a higher atomic number than C. $(CH_3)_2CH$— precedes $CH_3CH_2CH_2$— because in the former there are two carbons and a hydrogen attached to the first carbon (C—C(H)(C)—) whereas in the latter there is only one carbon and two hydrogens attached to the first carbon (C—C(H)(H)—); here again C has a higher atomic number than H.

2-9 Optical Isomerism

Certain compounds are found to be asymmetric (possess no plane or point of symmetry). These molecules can exist in two different forms differing as the left hand differs from the right hand. The left hand is a mirror image of the right, however it is not possible to superimpose the left hand onto the right hand so that each finger rests on its corresponding member in the other hand. This is a property (called *chirality* or *handedness*) of asymmetric substances and leads to the kind of isomerism shown in bromochloroiodomethane.

Note that form R is the mirror image of form S. The R and S designations come from the Cahn-Ingold-Prelog rules (see Sec. 13-2*b*). Also note that it is not possible to lift R from the plane of the page and to superimpose it on form S. If this is done so that H and I coincide, then Br and Cl do not. If R is turned 180° in the plane of the page and then placed on S, Cl and Br coincide but H and I do not. If R is turned 180° out of the plane of the page none of the groups coincide, for Br and Cl from R will be behind the plane of the page. When placed in a symmetric environment, molecules such as R and S show the same physical properties. When they react with symmetric reagents they also show the same chemical properties. They are distinguished from each other by their effect on plane-polarized light. One isomer always rotates the plane of polarization to the right and the other rotates the plane of polarization an equivalent number of degrees to the left. Molecules which rotate plane-polarized light are called *optically active* or *chiral molecules*. Those which rotate plane-polarized light to the right are termed *dextro* and are symbolized by *d* or +. Their mirror images are called *levo* (*l* or −). In the preparation of organic compounds one usually obtains mixtures containing equal amounts of *d* and *l* forms. Such mixtures are called *racemic* and have no effect on polarized light. The *d* and *l* forms are referred to as optical antipodes, or alternatively, enantiomorphs or enantiomers. Taken together they may be referred to as a *dl* or ± pair.

Note that the geometric isomers in Sec. 2-8 have a plane of symmetry (the plane of the page) and hence are not optically active.

Compounds containing more than one carbon atom to which are attached four *different* atoms or groups (called an *asymmetric carbon*) have larger numbers of stereoisomers (isomers which differ only in the way the atoms are oriented in space but in which the atoms joined are the same). For such compounds the maximum number of stereoisomers possible is 2^n where *n* is the number of asymmetric centers.

Molecule A has two asymmetric centers (marked by *) while molecule B has three such centers. The student should satisfy himself that there are four different stereoisomers of A (2 *dl* pairs) and eight of B (4 *dl* pairs). Molecule C, in addition to having an asymmetric carbon atom, exhibits geometric isomerism. As a result there are four stereoisomers: a cis + and − and a trans + and −.

Problems

2-1 Indicate the maximum number of electrons in a given atom which can occupy each of the following:

a. L shell
b. $n = 3$ shell
c. 2p orbitals
d. 3d orbitals
e. $2p_z$ orbital
f. 2d orbitals

2-2 How many electrons in a given atom can have the following quantum numbers?
  a. $n = 1$
  b. $n = 2, l = 1$
  c. $n = 3, l = 2, m = 0$
  d. $n = 3$
  e. $n = 4, l = 3, m = +2, s = +1/2$
  f. $n = 4$

2-3 Give the electron structure for the following:
  a. Na
  b. Na^+
  c. B
  d. N
  e. F
  f. Cl
  g. I
  h. O^{2-}

2-4 Name the type of orbital used for bonding in the central atom of each of the following and predict the geometric configuration for each.
  a. NH_3
  b. H_3O^+
  c. NH_2^-
  d. CH_4
  e. CH_3^-
  f. CO_3^{2-}
  g. H_2S
  h. BF_4^-

2-5 Give a molecular orbital description of the bonds and the expected geometry of each of the following:
  a. $O=C=C=C=O$
  b. $H_2C=CHC\equiv CH$
  c. H_2NNH_2
  d. HN_3
  e. $H_2C=CHNO_2$
  f. C_2^{2-}
  g. $HC\equiv CC\equiv N$
  h. $CH_3OCH=CH_2$
  i.

2-6 Explain, using the concept of antibonding orbitals, which of the following molecules might be stable enough to be detected and which would not.
  a. H_2^-
  b. H_2^{2-}
  c. He_2^+

2-7 Show how $CH_2=CH_2$ and $HC\equiv CH$ could be formulated using sp^3 hybridized carbon.

2-8 Which of the following systems are aromatic according to Hückel's rule?

  a.
  b.
  c.

  d.
  e.
  f.

2-9 Make molecular orbital drawings showing the electron distribution in the three p orbitals of each of the following:
  a. $CH_2=CH-CH_2^+$
  b. $CH_2=CH-CH_2\cdot$

2-10 Indicate which molecular species in each of the following pairs would have the higher resonance energy and give a reason for your choice. Keep Hückel's rule in mind. Also assume that the more widely a positive or negative charge can be delocalized via a system of overlapping p orbitals the greater the resonance energy.

  a. $CH_3\overset{+}{C}H_2 \quad CH_2=CH\overset{+}{C}H_2$
  b.
  c.
  d.
  e.
  f.

2-11 The blue hydrocarbon azulene

Azulene

is interesting in that it is relatively stable and is polarized so that it displays a dipole moment. Considering Hückel's rule, what resonance structures would you expect to be most important for this substance?

2-12 In the compound, $B(CH_3)_3$, the boron and carbon atoms all lie in the same plane with bond angles of $120°$. Illustrate, using the symbols of Sec. 2-3, how hybridization is achieved in the boron bonds. Use the same technique to rationalize the linear character of CH_3MgCH_3.

2-13 The unstable species $CH_2\colon\colon$ appears to exist in two different electronic forms. In one form the H—C—H angle is expected to be $180°$ with the two electrons *unpaired*. In the other form the angle is about $103°$ with the two electrons paired. What kind of hybridization would you expect for each form?

2-14 Draw structures for all optical and geometric isomers for each of the following. Indicate which isomers would rotate the plane of polarized light.

a. $CH_3CH_2\underset{\underset{D}{|}}{C}HCH_3$

b. $CH_3\underset{\underset{Cl}{|}}{C}H\underset{\underset{Br}{|}}{C}HCH=CH_2$

c. $HC{\equiv}C{-}CH=CH_2$

d. 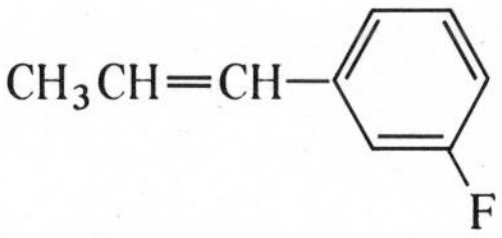 $CH_3CH=CH-$ with F substituent on ring

e. $F{-}\underset{\underset{Cl}{|}}{C}HCH=\underset{\underset{Br}{|}}{C}HCHCH_3$

f. $CH_3\underset{\underset{OH}{|}}{C}H\underset{\underset{OH}{|}}{C}H\underset{\underset{OH}{|}}{C}H\underset{\underset{OH}{|}}{C}HCH_2CH_3$

2-15 Draw the E and Z forms for each of the following:

a. $CH_3CH=CHCH_3$

b. 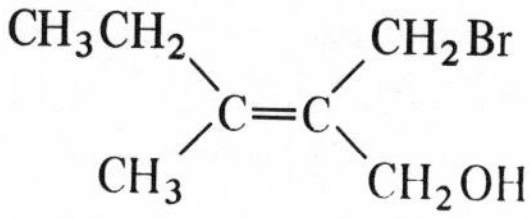 $\underset{CH_3}{\overset{CH_3CH_2}{>}}C=C\underset{CH_2OH}{\overset{CH_2Br}{<}}$

c. $\underset{Cl}{\overset{Br}{>}}C=C\underset{CD_3}{\overset{CH_3}{<}}$

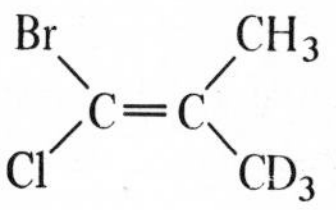

d. $(CH_3)_2CH{-}CH=\underset{\underset{CH_2CH_3}{|}}{C}CH_2CH_2CH_3$

2-16 Would you expect the following 14-membered ring to be aromatic? Can you draw more than one equivalent valence-bond structure?

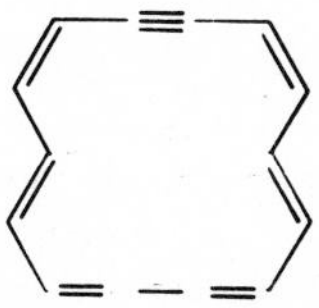

2-17 Show using valence-bond symbolism how the positive charge in the triphenylmethyl ion can be delocalized over nine positions in the benzene rings.

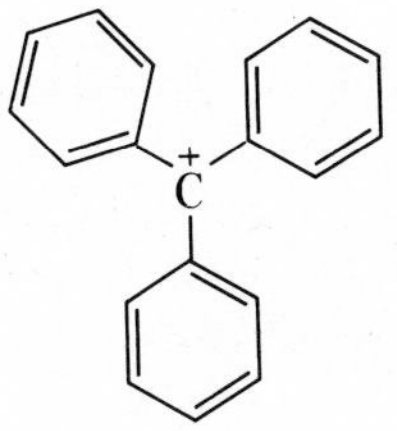

2-18 Compounds such as $\underset{H}{\overset{CH_3}{>}}C=C=C\underset{CH_3}{\overset{H}{<}}$ are optically active. Formulate the orbital structure in the two double bonds and show how this gives rise to asymmetry.

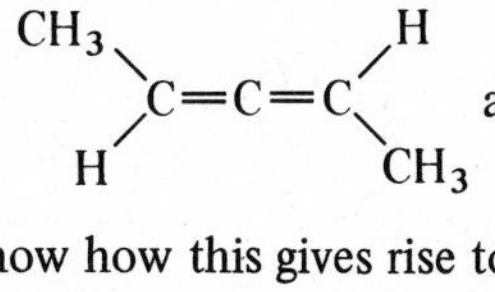

21

3
alkanes (saturated hydrocarbons)

3-1   Homologous Series and Isomerism

a.   *Homologous Series of Alkanes,* C_nH_{2n+2}

Electron formula	Line formula	Condensed molecular formula
Methane 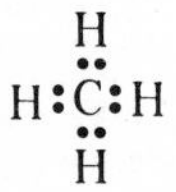 H:C:H (with H above and below)	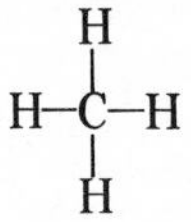 H—C—H (with H above and below)	CH_4
Ethane 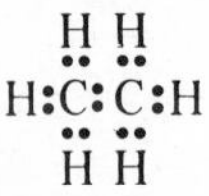 H:C:C:H (with H above and below each C)	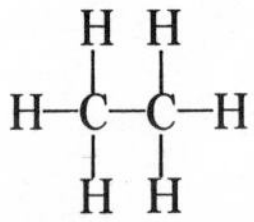 H—C—C—H (with H above and below each C)	CH_3CH_3 (or C_2H_6)
Propane, C_3H_8	Hexane, C_6H_{14}	Nonane, C_9H_{20}
Butane, C_4H_{10}	Heptane, C_7H_{16}	Decane, $C_{10}H_{22}$
Pentane, C_5H_{12}	Octane, C_8H_{18}	Undecane, $C_{11}H_{24}$

b.   *Chain Isomers* Starting with butane, the carbon atoms can be arranged in a continuous chain or as a chain with carbon branches. These arrangements are called *isomers*.

A systematic approach is useful for drawing all the structures of a compound such as hexane, C_6H_{14}. First, start with the isomer having six atoms with no branching. Next, there are two isomers whose longest chain is five carbon atoms. Finally, the shortest possible chain is four atoms, and this leads to two more isomers.

$CH_3CH_2CH_2CH_2CH_2CH_3$ $CH_3CHCH_2CH_2CH_3$ $CH_3CH_2CHCH_2CH_3$
                                                         $|$ $|$
                                                      CH_3 CH_3

A   B 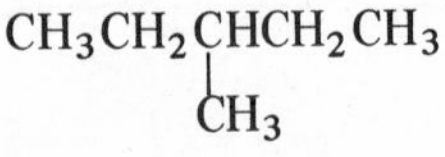   C

$CH_3CH-CHCH_3$ $CH_3CCH_2CH_3$ (with CH_3 above and CH_3 below the central C)
        $|$ $|$
     CH_3 CH_3

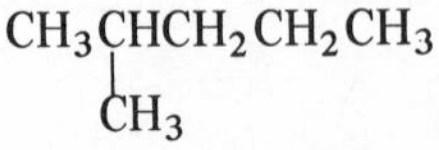

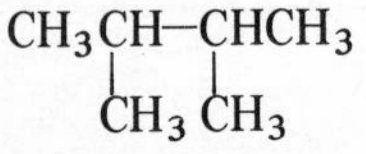

D 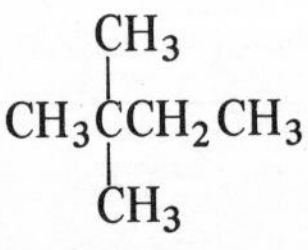   E

3-2 Nomenclature

a. *Common System* The use of *n-* (for normal) before the name of a hydrocarbon indicates an unbranched carbon chain. If a compound has the arrangement $(CH_3)_2CH-$ on one end of a chain with no other branching, the prefix *iso* is added to the name of the compound. Note that the remaining isomer of pentane is named neopentane.

$$CH_3CH_2CH_2CH_2CH_3 \qquad CH_3\overset{\overset{\displaystyle CH_3}{|}}{C}HCH_2CH_3 \qquad CH_3\overset{\overset{\displaystyle CH_3}{|}}{\underset{\underset{\displaystyle CH_3}{|}}{C}}CH_3 \qquad CH_3\overset{\overset{\displaystyle CH_3}{|}}{C}H(CH_2)_2CH_3$$

n-Pentane   Isopentane   Neopentane   Isohexane

Remember that terms such as isobutane, isopentane, etc., refer to compounds containing four, five, etc., carbons in the *total* molecule.

$$\text{Isobutane} = CH_3\overset{\overset{\displaystyle }{|}}{\underset{\underset{\displaystyle CH_3}{|}}{C}}HCH_3 \qquad not \qquad CH_3\overset{\overset{\displaystyle }{|}}{\underset{\underset{\displaystyle CH_3}{|}}{C}}HCH_2CH_3$$

b. *Group Names* Alkanes with one hydrogen removed are called *alkyl groups*. The *ane* ending of the alkane is changed to *yl*.

$$CH_4 \longrightarrow CH_3- \qquad CH_3CH_3 \longrightarrow CH_3CH_2-$$

Methane   Methyl   Ethane   Ethyl

$$CH_3\overset{\overset{\displaystyle }{|}}{\underset{\underset{\displaystyle CH_3}{|}}{C}}HCH_3 \qquad CH_3\overset{\overset{\displaystyle }{|}}{\underset{\underset{\displaystyle CH_3}{|}}{C}}HCH_2- \equiv (CH_3)_2CHCH_2-$$

Isobutane   Isobutyl

In more complex cases, the name depends on whether the hydrogen comes from a primary, secondary, or tertiary carbon atom.

$$CH_3CH_2CH_2CH_2- \qquad CH_3CH_2\overset{\overset{\displaystyle }{|}}{\underset{\underset{\displaystyle CH_3}{|}}{C}}HCH_3 \equiv CH_3CH_2\overset{\overset{\displaystyle }{}}{\underset{\underset{\displaystyle CH_3}{|}}{C}}H- \qquad CH_3\overset{\overset{\displaystyle CH_3}{|}}{\underset{\underset{\displaystyle CH_3}{|}}{C}}- \equiv (CH_3)_3C-$$

n-Butyl   *sec*-Butyl   *tert*-Butyl

A primary carbon ($1°$) is one to which one other carbon atom is attached, a secondary carbon ($2°$) has two other carbon atoms attached, and a tertiary ($3°$) has three. Hydrogen atoms attached to $1°$, $2°$, or $3°$ carbon atoms are $1°$, $2°$, and $3°$ hydrogens, respectively. There are two alkyl groups which may be obtained from propane.

$$CH_3CH_2CH_2- \qquad CH_3\overset{\overset{\displaystyle }{|}}{C}HCH_3 \equiv (CH_3)_2CH-$$

n-Propyl   Isopropyl

Some of the alkyl groups obtained from *n*-pentane, isopentane, and neopentane are:

$$CH_3CH_2CH_2CH_2CH_2- \qquad (CH_3)_2CHCH_2CH_2- \qquad CH_3CH_2\overset{\overset{\displaystyle CH_3}{|}}{C}-CH_3 \qquad CH_3-\overset{\overset{\displaystyle CH_3}{|}}{\underset{\underset{\displaystyle CH_3}{|}}{C}}-CH_2-$$

n-Pentyl   Isopentyl   *tert*-Pentyl   Neopentyl

Amyl is used in place of pentyl in some cases, e.g., *n*-amyl, isoamyl, *tert*-amyl.

In drawing structural formulas the following abbreviations for certain alkyl groups are often used: Me (methyl), Et (ethyl), *n*-Pr (*n*-propyl), and iso-Pr (isopropyl).

c. *Geneva or IUPAC (International Union of Pure and Applied Chemistry) Systematic Nomenclature* To name a hydrocarbon by this system first determine the longest continuous chain of carbons. This establishes the parent name of the hydrocarbon (that is, 4 = butane, 5 = pentane, etc.). The carbon atoms in this chain are numbered consecutively, starting from the end which leads to the smallest possible numbers for the positions of attached groups. The position of each attached group is indicated by a number followed by a hyphen and the name of the substituent.

$$CH_3CHCH_2CH_3 \quad\quad CH_3CH_2\overset{\displaystyle CH_3}{\underset{\displaystyle CH_3}{C}}{-}CH_3 \quad\quad CH_3\overset{\displaystyle CH_3}{\underset{\displaystyle CH_3}{CH}}\overset{\displaystyle CH_3}{\underset{\displaystyle CH_2CH_3}{CH}}CHCH_2CH_3$$

2-Methylbutane

2,2-Dimethylbutane
(not 2-methyl-2-methyl-butane or 2-dimethyl-butane)

2,3,5-Trimethyl-4-ethylheptane

Note the punctuation. Commas separate numbers; numbers are separated from words by hyphens. Each attached group has a number indicating its position. The prefixes *di, tri, tetra, penta,* etc., indicate the number of identical groups attached to the main chain.

Simple alkyl substituents are named as described in Sec. 3-2b above, except that in naming straight-chained alkyl groups (i.e., *n*-alkyl groups) as substituents the *n* prefix is omitted. If a substituent on a chain is complex enough to require its own separate position numbers, these are indicated by calling the carbon atom attached to the main chain "1" and numbering the side chain with its attached groups in the same fashion as the main chain. The name of the side chain is enclosed in parentheses and its position of attachment indicated by a preceding number.

$$CH_3CH_2CH_2CH_2CH_2\overset{\displaystyle CH_2CH(CH_3)_2}{\underset{\displaystyle \underset{\displaystyle CH_3}{CH_3CHCHCH_2CH_3}}{C}}{-}CH_2CH_2CH_2CH_2CH_3$$

6-(1,2-Dimethylbutyl)-6-isobutylundecane

$$CH_3CH_2CH_2CH_2\overset{\displaystyle CH_3CHCH_2CH_3}{\underset{\displaystyle CH_3CH_2CH_2CH_2}{CHCH}}CH_2CH_2CH_2CH_3$$

5-Butyl-6-*sec*-butyldecane

The order of citation of the substituents on the main chain in the full name is either in order of their increasing complexity (roughly this means increasing size or number of carbons in the substituent) or in alphabetical order (the multiplying prefixes *di, tri, tetra,* etc., are omitted for alphabetization purposes except when they occur in names of complex side chains such as the 1,2-dimethylbutyl group in the above example).

d. *Two Types of Naming* The first keeps the respective parts of the names separate (sodium chloride, methyl bromide). This generally holds for the names of organic compounds ending with *ide* or *ate*. The second type places the substituent as a prefix and runs the name together except when modifying numbers occur (chlorobromoiodomethane, 2-methylbutane).

e. *Cycloparaffins (Cycloalkanes),* C_nH_{2n} These compounds are named by placing the prefix *cyclo* before the appropriate paraffin name for the number of carbon atoms in the cyclic chain.

$$\overset{\displaystyle CH_2}{\underset{\displaystyle H_2C-CH_2}{\diagup\diagdown}} \quad \text{or} \quad \triangle$$

Cyclopropane

Cyclohexane

1-Chloro-2,4-dimethylcyclopentane

24

3-3 Conformations of Alkanes

a. *Ethane* Although the bond angles and bond lengths in ethane are clearly defined (H–C–C and H–C–H angles, 109.5°; C–H, 1.10 Å; C–C, 1.54 Å), ethane may exist in an infinite number of *conformations* (called *conformers*) which represent different rearrangements of the atoms in ethane and which can be converted into one another by rotation about the carbon-carbon bond. The conformation of ethane in which the hydrogens on the adjoining carbon atoms are arranged so as to maintain the largest distance from each other is called the *staggered* conformer while the *eclipsed* conformer represents the minimum distance between these hydrogens (see Fig. 3-1). There are more repulsive forces operating in the eclipsed conformer than in the staggered conformer of ethane. Thus, there is a potential energy barrier of about 3 kcal/mol separating these two conformers at room temperature and the rotation about the carbon-carbon bond in ethane is not completely "free." The cause of the energy barrier or of these repulsive forces is not clear for ethane. However, as one or more hydrogens in ethane are replaced by bigger atoms or groups, the major repulsive forces or barrier are due to the unfavorable steric interaction of the larger groups on each carbon called *nonbonded interactions* (see Prob. 3-14 for the conformations of butane).

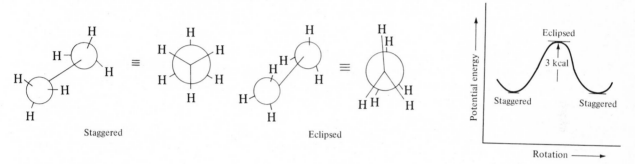

Figure 3-1. Conformations of ethane.

b. *Cyclohexane* Cyclic structures have greater possibilities for repulsion between nonbonded atoms, and restriction to "free" rotation around single carbon-carbon bonds can be very great. For example, with cyclopentane if all carbon atoms are forced into the same plane, then each CH_2 group is forced into an eclipsed position with respect to its neighbor and repulsion is at a maximum. The most probable form for cyclopentane is a puckered one. This is also true of cyclohexane which may exist in a *chair* conformation (see Fig. 3-2) which is the most stable. Note that the hydrogens are all staggered and that there are six hydrogens that lie in the plane of the ring (labeled *e* for *equatorial*) and six hydrogens that lie above or below the plane of the ring (labeled *a* for *axial*). Conversion of one cyclohexane conformer into its mirror image chair conformer (II) proceeds through an infinite number of other possible higher energy conformations, one of which is uniquely designated the boat conformer where the hydrogens are all eclipsed.

3-4 Cis-Trans Isomerism of Substituted Cycloalkanes

Both geometric and optical isomerism are possible in ring compounds with at least two substituents on different carbon atoms. 1,2-Dibromocyclopropane can exist in either cis or trans form. Only the trans isomer is optically active. The cis isomer contains two like asymmetric carbon atoms and as a consequence it has a plane of symmetry which passes through the number 3 carbon and bisects the bond between carbons 1 and 2. Such optically inactive compounds are called *meso*. In compounds such as 1,4-dimethylcyclohexane there are no asymmetric carbon atoms, but one pair of cis-trans isomers is possible.

See Sec. 13-3b for further discussion of the stereochemistry and conformations of cyclic structures.

25

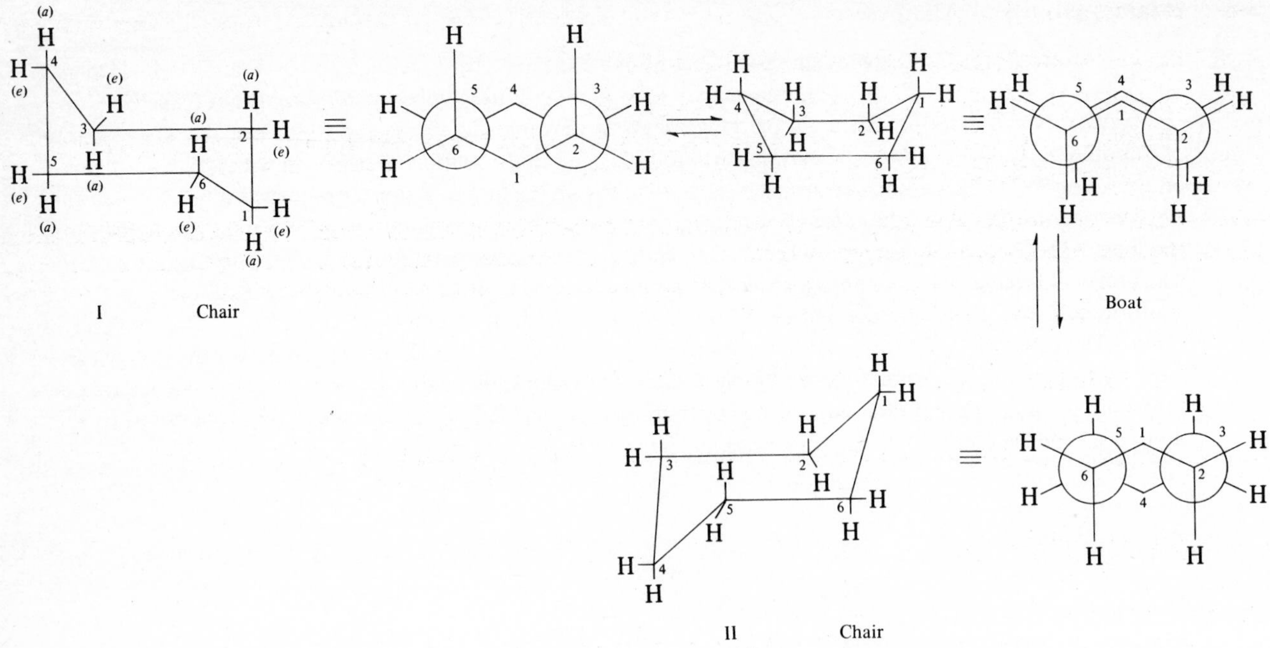

Figure 3-2. Conformations of cyclohexane.

3-5 Methods of Preparation

a. *Reduction of Alkyl Halides* $RX \xrightarrow{[H]} RH + HX$. In this equation X represents any halogen atom, F, Cl, Br, or I. Hydrogen written [H] is taken to mean an "active" form of hydrogen such as that generated from a metal and an acid or simply H_2 in the presence of a special catalyst such as Pt, Pd, or Ni. The reduction may also be accomplished with HI and red phosphorus.

Examples

$$CH_3I + Zn + HCl \longrightarrow CH_4 + ZnClI \qquad CH_3CH_2I + HI \xrightarrow{P, \Delta} CH_3CH_3 + I_2$$
$$(\Delta \text{ is a symbol for heat})$$

$$(CH_3)_2CHCl + H_2 \xrightarrow{Pd} CH_3CH_2CH_3 + HCl$$

b. *Reduction (Hydrogenation of Alkenes* (See Sec. 4-3c.)

$$RCH{=}CHR + H_2 \xrightarrow{Pt, Pd, or Ni} RCH_2CH_2R$$

c. *Grignard Reaction* $RX + Mg \longrightarrow RMgX$

where R = symbol for any alkyl group
     X = Cl, Br, I

The Grignard reagent reacts with "active" hydrogen to give a hydrocarbon (see Sec. 1-6e).

$$RMgX + HOH \longrightarrow RH + MgXOH$$

d. *Wurtz Coupling Reaction* $2RX + 2Na \longrightarrow R{-}R + 2NaX$

26

3-6   Reactions

a.   *Combustion* $2CH_3CH_3 + 7O_2 \longrightarrow 4CO_2 + 6H_2O$

b.   *Nitration* $CH_4 + HNO_3 \xrightarrow{500°C} CH_3NO_2 + H_2O$
In more complex hydrocarbons, the carbon chain may be broken. Thus the nitration of propane yields nitromethane, nitroethane, and the two possible nitropropanes.

c.   *Methylene Insertion (Carbene)* A very reactive methylene can be formed by the photolytic decomposition of diozomethane, CH_2N_2 or ketene, $CH_2=C=O$.

$$\bar{C}H_2 - \overset{+}{N} \equiv N$$
$$CH_2 = O = O$$
$$\xrightarrow[\text{light}]{\text{ultraviolet}} \quad CH_2 \colon \quad \begin{array}{l} + N_2 \\ \\ + CO \end{array}$$

Once generated, this methylene (carbene) can *insert* itself into a carbon to hydrogen bond in alkanes.

$$\left(-\overset{|}{\underset{|}{C}} - H + CH_2 \colon \longrightarrow -\overset{|}{\underset{|}{C}} - CH_2 - H \right)$$

$$CH_3CH_2CH_2CH_3 \xrightarrow[\substack{CH_2N_2 \\ \text{uv light}}]{CH_2=C=O \\ \text{or}} CH_3\underset{\underset{CH_3}{|}}{C}HCH_2CH_3 + CH_3CH_2CH_2CH_2CH_3$$

Insertion of methylene in alkanes occurs in nearly random fashion so that if *n*-pentane is used as the alkane, the insertion products are *n*-hexane, 2-methylpentane, and 3-methylpentane with the respective ratio of yields of 3:2:1.

Methylene can exist in two electronic states, the singlet state and a more stable triplet state. In singlet methylene (see Fig. 3-3) the carbon uses two sp^2 hybridized orbitals to bond to the hydrogens while the third sp^2 orbital contains the unshared pair of electrons leaving a vacant p orbital. The carbon in triplet methylene has two sp hybridized orbitals used in bonding to hydrogen while the unshared pair of electrons reside, one each, in the remaining p_y and p_z orbitals. These two electrons have unpaired spins (the term "triplet state" refers to two unpaired spins vs. a "singlet state" where all electron spins are paired).

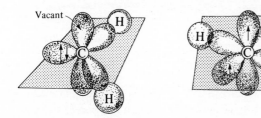

Figure 3-3. Singlet and triplet methylene.

d.   *Chlorination or Bromination* $CH_3CH_3 + Cl_2 \xrightarrow{\lambda} CH_3CH_2Cl + HCl$
$$(Br_2) \qquad\qquad (Br) \quad (Br)$$

where λ = symbol for activating light. In the halogenation of hydrocarbons I_2 is too inert to react, but F_2 reacts explosively and special techniques are needed to control the reaction.

3-7   Reaction Mechanisms. Chlorination of Methane

An integral part of understanding an overall organic reaction is to understand how the reactants proceed to products, i.e., a step-by-step account of the reaction from beginning to end. Such an account is termed the "mechanism of the reaction." While absolute verification of a proposed mechanism for a particular reaction is virtually impossible, such a mechanism does serve as a model with which we can make certain predictions about that reaction and other reactions which are closely related. Our mechanistic model will only be as good as the number of experimental facts about a reaction that it can explain satisfactorily.

In the chlorination of methane the reaction is well established as proceeding through a *free-radical* mechanism consisting of the following steps:

1. $Cl_2 \xrightarrow{\lambda} 2Cl\cdot$
2. $Cl\cdot + CH_4 \longrightarrow HCl + CH_3\cdot$
3. $CH_3\cdot + Cl_2 \longrightarrow CH_3Cl + Cl\cdot$

Steps 2 and 3 continue to repeat until the chain reaction is terminated by the occurrence of a reaction between two radical species, that is,

$$2Cl\cdot \longrightarrow Cl_2 \qquad CH_3\cdot + Cl\cdot \longrightarrow CH_3Cl \qquad \text{or} \qquad 2CH_3\cdot \longrightarrow CH_3CH_3$$

The formation of chlorine free radicals (step 1) requires light energy but once they are formed, steps 2 and 3 proceed quite rapidly.

The course of an organic reaction can be likened to climbing over a mountain: the difficult part is getting to the top. Reacting molecules, like mountain climbers, must ascend a pass before they "roll" down the other side to products. This can best be represented by an energy diagram for which steps 2 and 3 in the chlorination of methane are used for illustration (see Fig. 3-4).

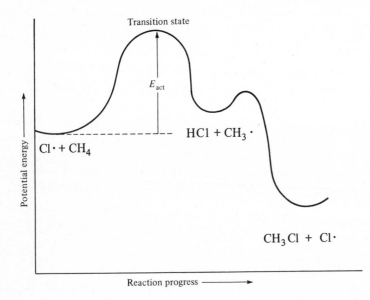

Figure 3-4. Chlorination of methane—progress of reaction diagram.

For step 2 to occur, the collision of a chlorine atom with a methane molecule must be accompanied by a minimum amount of energy, called the *energy of activation* (E_{act}), in order for a successful reaction to occur. Moreover, such a collision must occur with the proper orientation of Cl· and CH_4 with respect to each other. The particular arrangement of Cl· and CH_4 at the pass of the "energy mountain" is called the *transition state*. For the abstraction of a hydrogen in methane by a chlorine-free radical the transition state is depicted as one where the C—H bond is partly broken (dashed line) and the carbon atom has something between sp^3 and sp^2 hybridization. The lone electron in the transition state is shared between the carbon and chloride atoms as indicated by the $\delta\cdot$ symbol.

The concept of a transition state is a useful one although isolation of a particular transition state is not possible by its very fleeting nature. How fast step 2 occurs, i.e., the *rate* of step 2, will depend on the difference in energy between the transition state and the reactants (E_{act}). Because E_{act} for step 2 is greater than that for step 3, we say that step 2 is the slower or *rate-determining* step in the reaction, $Cl\cdot + CH_4 \longrightarrow CH_3Cl\cdot$. For the overall chlorination of methane, however, the rate-determining step is the formation of chlorine-free radicals (step 1).

3-8 Bond Dissociation Energy and Heat of Reaction

When a chemical bond is broken or formed the amount of energy consumed or liberated, respectively, is called the *bond dissociation energy* (D)

$$CH_3-CH_3 \longrightarrow 2CH_3\cdot \qquad D_{CH_3-CH_3} = 84 \text{ kcal/mol}$$

Table 3-1

Bond	Y=H	F	Cl	Br	I	Me	Et	n-Pr	iso-Pr	$(CH_3)_3C$	C_6H_5	$C_6H_5CH_2$
Y—Y	104	37	58	46	36	84						
H—Y	104	135	103	87	71	102	97	97	94	91	102	78
CH_3—Y	102	108	81	67	53	84	82	79	75	74	89	63
Cl—Y	103		58			81	83	77	73	75	86	68
Br—Y	87		46			67	65		59		71	51

From this table we see that the ease of formation of free radicals is: $(CH_3)_3C\cdot > (CH_3)_2CH\cdot > CH_3CH_2\cdot$. The relative ease of formation of free radicals is also their relative order of stability: $3° > 2° > 1° > CH_3\cdot$.

In a chemical reaction where several different bonds are being broken in the reactants and new bonds formed in the products, we can calculate the *change* in the heat content (*enthalpy*) by calculating the total energy required to dissociate or break the necessary bonds in the reactants and subtracting from this the sum of energy released when the necessary new bonds are formed in the products. Such a change in the heat content for a reaction is called the *heat of reaction*, ΔH. When ΔH for a reaction is negative, heat has been liberated or the reaction is exothermic. A positive ΔH for a reaction means that heat is required or the reaction is *endothermic*. For the bromination of ethane:

$$CH_3CH_2-H + Br-Br \longrightarrow CH_3CH_2-Br + H-Br$$
$$D=97 \quad D=46 \qquad\qquad D=65 \quad D=87$$
$$\Delta H = 97 + 46 - (65 + 87) = 143 - 152 = -9 \text{ kcal/mol}$$

Problems

3-1   Give both a common and an IUPAC name for each of the following:

a. $CH_3CH_2CH_3$

b. $CH_3\underset{\underset{\displaystyle CH_3}{|}}{C}HCH_3$

c. $CH_3(CH_2)_7CH_3$

d. $CH_3CH_2CH(CH_3)_2$

e. $CH_3(CH_2)_5CH(CH_3)_2$

f. $CH_3\underset{\underset{\displaystyle CH_3}{|}}{\overset{\overset{\displaystyle CH_3}{|}}{C}}CH_2Cl$

29

3-2   Give the IUPAC name for each of the following:

a. $CH_3CH_2CH_2CH_2Cl$

b. $(CH_3)_2CHCH(CH_3)_2$

c.   a cyclobutane with two CH_3 groups

d. $(CH_3)_3CCH_2CH_3$

e. $(CH_3)_2CHCH_2Br$

f. $CH_3CH_2CH_2CH_2NO_2$

g. $CH_3(CH_2)_2\overset{\displaystyle (CH_3)_3C}{\underset{\displaystyle CH_2CH_2CH_3}{C}}(CH_2)_5CH_3$

h. $(CH_3CH_2)_2CHCH(CH_2CH_3)_2$

i. $(CH_3)_2CHC(CH_3)_2CHC(CH_2CH_3)_2$ with a cyclopropyl and NO_2 group

3-3   Draw structures for each of the following:
a.   4-(1,1-Dimethylethyl)heptane
b.   3-Bromo-4-chloro-5-isopropyl-5-methyloctane
c.   1,7-Dichloro-4-(2-chloroethyl)heptane
d.   1,1,1-Trifluoro-4-(2-methylpropyl)nonane
e.   1-Chloro-2-iodo-3-cyclopropylpentane
f.   1,8-Dibromo-4-(2-iodo-2-methylpropyl)decane

3-4   Using the systematic procedure of Sec. 3-1b, draw all the isomers of C_7H_{16} and give the IUPAC name for each.

3-5   Draw all the isomers for each of the following:
a. C_3H_8 b. C_3H_7Cl c. $C_2H_4Cl_2$ d. C_3H_6ClBr e. C_3H_5ClBrF

3-6   Complete and balance (except f) the following equations:
a.   Hexane + oxygen $\xrightarrow{\Delta}$
b.   2-Bromopropane + sodium $\longrightarrow$
c.   Isopentane + CH_2N_2 $\xrightarrow{\Delta}$
d.   Ethylmagnesium iodide + water $\longrightarrow$
e.   Neopentyl chloride + hydrogen $\xrightarrow{Pt}$
f. n-Butane + nitric acid $\xrightarrow{500°C}$ six products

3-7   When 5.0 g of ethyl iodide is allowed to react with excess sodium in the Wurtz reaction, how many grams of butane would be produced assuming a 60 percent yield?

3-8   When 10.0 liters of methane (at STP) is converted to CCl_4 with excess chlorine, how many grams of carbon tetrachloride would be produced assuming a 100 percent yield?

3-9   One hundred grams of methane is converted to methyl chloride in 40 percent yield. This in turn is treated with sodium to give ethane in 50 percent yield. Bromination of the ethane gave a 60% yield of bromoethane. How many grams of bromoethane were obtained?

3-10   The following projected synthesis for n-pentane is a poor one. Why?
$CH_3CH_2CH_2I + ICH_2CH_3 + 2Na \longrightarrow CH_3CH_2CH_2CH_2CH_3 + 2NaI$

3-11   Indicate by balanced equations the methods for carrying out the following conversions:
a.   Isopropyl iodide to propane via a Grignard reaction
b.   Isobutyl bromide to 2,5-dimethylhexane
c.   Ethane to butane
d.   Methane to chloroethane
e.   Neopentane to 2,2-dimethylbutane
f.   Cyclohexane to cyclohexylcyclohexane

3-12   Why would the chlorination of ethane to ethyl chloride be a more practical synthesis than the chlorination of pentane to 1-chloropentane?

3-13   Calculate the percentages of products that would be expected upon treatment of isobutane with ketene in the presence of uv light if all C—H bonds in isobutane were of equal reactivity. The observed percentages are 86% isopentane and 14% neopentane. What do you conclude about the relative reactivity of a 3° vs. 1° C—H bond in this reaction?

3-14   Draw all the possible staggered and eclipsed conformations of n-butane. Based on the number and type of repulsions present in each conformer, predict the order of increasing potential energy of the conformers.

3-15   Draw all of the possible chair conformations for the following: (a) methylcyclohexane, (b) 1,2-dimethylcyclo-hexane, (c) 1,3-dimethylcyclohexane.

30

3-16   Using the chair and boat conformations of a cyclohexane ring, draw possible structures for the compound [2.2.2]-bicyclooctane, shown below.

$$
\begin{array}{ccc}
CH_2 & \!\!\!\!\!-CH- & \!\!\!\!\!CH_2 \\
| & | & | \\
 & (CH_2)_2 & \\
| & | & | \\
CH_2 & \!\!\!\!\!-CH- & \!\!\!\!\!CH_2
\end{array}
$$

Which conformation of the cyclohexane rings in this compound would appear to be strongly favored?

3-17   Draw all possible geometric and optical isomers for each of the following and indicate which isomers are optically active and which are meso compounds: (a) 1,2- and 1,3-dimethylcyclobutane, (b) 3-bromo-1-methylcyclopentane.

3-18   A hydrocarbon with a molecular weight of 70 exhibits no optical or geometric isomerism and forms only two monochloro substitution products. Suggest a structure for the hydrocarbon.

3-19   Vaporization of 1.0 g of an unknown iodoalkane gave 222 ml of vapor at 740 mmHg and $100°C$. Suggest a formula for the unknown.

3-20   Vaporization of 1.0 g of an unknown chloroalkane gave 418 ml of vapor at 730 mmHg and $110°C$. Suggest two possible formulas for the unknown.

3-21   Calculate the percentage composition of an alkane C_7H_{16} and C_8H_{18}. With an absolute error of 0.5 percent, could a C and H analysis be used to distinguish between the homologs?

3-22   Draw three staggered conformations of 2-methylbutane as viewed from C-2 and C-3 and indicate their relative potential energy.

3-23   The dipole moment of 1,2-dichloroethane increases with temperature. Explain.

3-24   By nuclear magnetic resonance spectroscopy (see Chap. 8) it is possible to show that the methylene hydrogens in *sec*-butyl chloride are different (i.e., nonequivalent) whereas the methylene hydrogens in *tert*-pentyl chloride are not different (i.e., equivalent). Using conformational analysis give an explanation of why this behavior might be expected.

3-25   When 1.37 g of monobromoalkane A was treated successively with magnesium and water, 0.580 g of gaseous hydrocarbon B was obtained. Bromination of A yielded a mixture of three isomeric dibromo compounds. Suggest structures for A and B.

3-26   The compound ferrocene contains 30% iron. What is the minimum molecular weight for this substance?

3-27   Insulin has been found to contain 0.52% zinc. Assuming one atom of zinc per insulin molecule, what would be the molecular weight for insulin?

3-28   When organic compounds are heated in a sealed tube with HNO_3 (Carius method), the halogen present is converted to ionic halide which may be determined by precipitation and weighing as AgX. Sulfur is converted to sulfate, which may be determined by precipitation as $BaSO_4$. When 10 mg of unknown A was oxidized by the Carius method, the chloride ion liberated gave 17.8 mg of AgCl. Combustion analysis indicated A contained 29.81% C and 6.21% H. When treated with excess CH_3MgX, 0.805 g of A yielded 224 ml of methane at STP. Suggest a formula for A consistent with the above information.

3-29   In the chlorination of ethane, suggest three reactions which would result in the loss of free radicals and thus slow the reaction.

3-30   The rate of ease of abstraction of a hydrogen in the chlorination of alkanes follows the order $3° > 2° > 1°$ and the relative rates per hydrogen atom are 5.0:3.8:1.0, respectively, at $25°C$. The expected relative percentage yields of propyl chloride and isopropyl chloride from chlorination of propane can be calculated as follows:

$$\frac{\text{Propyl chloride}}{\text{Isopropyl chloride}} = \frac{\text{No. of } 1° \text{ H}}{\text{No. of } 2° \text{ H}} \frac{\text{reactivity of } 1° \text{ H}}{\text{reactivity of } 2° \text{ H}} = \frac{6}{2} \frac{1}{3.8} = \frac{6}{7.6}$$

or $\frac{44\%}{56\%}$. Calculate the proportions of isomeric products expected from chlorination at $25°C$ of each of the following:

a.   *n*-Butane   *b.*   Isobutane

c.   3,4-Diethylhexane   *d.*   2,3,4-Trimethylpentane

3-31   Calculate the heat of reaction (ΔH) for the following:

    a. $H_2 + F_2 \longrightarrow 2HF$

    b. $CH_4 + I_2 \longrightarrow CH_3I + HI$

    c. $(CH_3)_3CH + Cl_2 \longrightarrow (CH_3)_3CCl + HCl$

    d. $C_6H_5CH_3 + Br_2 \longrightarrow C_6H_5CH_2Br + HBr$

    e. $C_6H_6 + Cl_2 \longrightarrow C_6H_5Cl + HCl$

    f. $CH_3CH_3 + Br_2 \longrightarrow 2CH_3Br$

3-32   For the reaction, $CH_4 + Br\cdot \longrightarrow CH_3\cdot + HBr$, the energy of activation is 18 kcal/mol with 15 kcal/mol of heat absorbed during the reaction. Draw a potential energy diagram labeling the energy of activation, the heat absorbed in the reaction, the transition state, and the potential energy levels of reactants and products for this reaction.

3-33   For the reaction, $R-H + Cl\cdot \longrightarrow R\cdot + HCl$, the following values were obtained for the energy of activation (kcal/mol): $R = CH_3CH_2$, $E_{act} = 1$; $R = (CH_3)_2CH$, $E_{act} = 0.5$; $R = (CH_3)_3C$, $E_{act} = 0.35$. Which will be formed the fastest by this type of reaction: ethyl chloride, isopropyl chloride, or *tert*-butyl chloride? For the same R groups the order of stability of the free radicals $R\cdot$ relative to RH is found to be: $(CH_3)_3C\cdot > (CH_3)_2CH\cdot > CH_3CH_2\cdot$. Draw a potential energy diagram which contrasts the progress of the above reaction when R = ethyl and R = isopropyl.

3-34   Instead of using free chlorine for chlorination of hydrocarbons, sulfuryl chloride, SO_2Cl_2, is sometimes used. It too seems to halogenate by a free-radical mechanism. The products which have been isolated are RCl, RSO_2Cl SO_2, and HCl. Suggest a mechanism to account for these.

3-35   For each of the bonds in the following molecules indicate which atomic orbitals have combined to give the bonding molecular orbitals. Indicate the geometric configuration of each molecule.

    a. Nitromethane   *b.* Methylmagnesium chloride   *c.* Trimethylboron

4
alkenes (olefins) and alkynes (acetylenes)

4-1 Nomenclature

a. *IUPAC System* Hydrocarbons containing double bonds (alkenes) or triple bonds (alkynes) belong to the class of compounds known as unsaturated hydrocarbons. The ending *ane* of the name for alkanes is changed to *ene* for alkenes and *yne* for alkynes. The parent name for the compound is determined by the longest carbon chain containing the unsaturated group. The position of unsaturation is indicated by the number of the lowest-numbered carbon atom in the double or triple bond.

$CH_2=CH_2$ $CH_3CH=CH_2$ $CH_3CH=CHCH_3$

Ethene   Propene   2-Butene

$HC\equiv CH$ $CH_3C\equiv CH$ $CH_3C\equiv CCH_3$

Ethyne   Propyne   2-Butyne

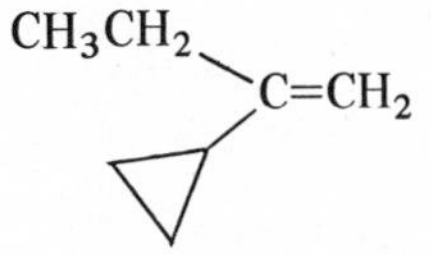

CH_3CH_2
C=CH$_2$ 2-Cyclopropyl-l-butene
                                (not 3-cyclopropyl-3-butene   Cyclohexene
                                or 3-cyclopropyl-4-butene)

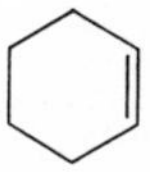

Two multiple bonds are indicated by the endings *adiene*, *adiyne*, or *enyne*; three by *atriene*, *atriyne*, etc.

$CH_2=CHCH_2CH=CH_2$ $HC\equiv CC\equiv CC\equiv CH$ $CH_2=CHCH_2C\equiv CCH_2CH_3$

1,4-Pentadiene   1,3,5-Hexatriyne   1-Hepten-4-yne

b. *Common Names* For olefins the *ane* ending of paraffins is changed to *ylene* for compounds two to five carbon atoms in length. The five-carbon homolog is also called *amylene*. Complex compounds are usually given IUPAC names.

$CH_2=CH_2$ $CH_2=CHCH_3$ $(CH_3)_2C=CH_2$ $CH_3CH_2CH_2CH=CH_2$

Ethylene   Propylene   Isobutylene   Pentylene, or amylene

Certain olefins are often named as derivatives of ethylene.
$CH_3CH=CHCH_3$ *sym*-Dimethylethylene (*sym* = symmetrical)
$(CH_3)_2C=CH_2$ *unsym*-Dimethylethylene
$(CH_3)_2C=CHCH_3$ Trimethylethylene
Acetylenes may be named as derivatives of the simplest member of the series, acetylene.

$HC\equiv CH$ $HC\equiv CNa$ $CH_3C\equiv CH$ $CH_3CH_2C\equiv CCH(CH_3)_2$

Acetylene   Sodium acetylide   Methylacetylene   Ethylisopropylacetylene

c. **Group Names** Certain commonly encountered groups are given trivial names.

$-CH_2-$	CH_2I_2	$-CH_2CH_2-$	$ClCH_2CH_2Cl$
Methylene	Methylene iodide	Ethylene	Ethylene chloride

$CH_2=CH-$	$CH_2=CHBr$	$CH_2=CHCH_2-$	$CH_2=CHCH_2F$
Vinyl	Vinyl bromide	Allyl	Allyl fluoride

$CH_3C\equiv C-$	$HC\equiv CCH_2-$	$HC\equiv CCH_2Br$	$CH_3CH=CH-$	$CH_3C=CH_2$
Propynyl	Propargyl	Propargyl bromide	Propenyl	Isopropenyl

d. *Naming Geometric Isomers* (See Sec. 2-8.)

4-2   Methods of Preparation

a. *Dehydrohalogenation; β elimination* Hydrogen and halogen can be removed from adjacent carbon atoms by bases to yield alkenes or alkynes.

$$CH_3CH_2CH_2Br + CH_3CH_2O^- \xrightarrow{CH_3CH_2OH} CH_3CH=CH_2 + CH_3CH_2OH + Br^-$$

$$BrCH_2CH_2Br \text{ (or } CH_3CHBr_2) + H_2N^- \xrightarrow{\text{liq } NH_3} CH_2=CHBr + NH_3 + Br^-$$

$$CH_2=CHBr + H_2N^- \xrightarrow{\text{liq } NH_3} HC\equiv CH + NH_3 + Br^-$$

The amide ion H_2N^- comes from sodamide formed by the reaction of sodium with ammonia:

$$2Na + 2NH_3 \xrightarrow{Fe^{2+}} 2NaNH_2 + H_2.$$

The reaction mechanism, in which electrons are involved in pairs as illustrated by curved arrows in the following reactions, can be pictured as occurring in three kinds of ways:

1. The proton and halide ion are removed simultaneously in a single, *concerted* step. The reaction rate depends on the concentration of both the base and the organic molecule (termed the *substrate*) because both species appear in the transition state, and hence the reaction is termed $E2$ (elimination, second order). The process occurs most readily when the hydrogen and halogen atoms are trans to each other.

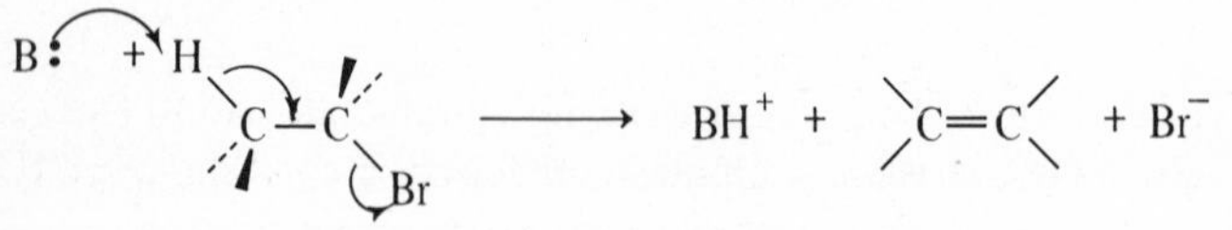

2. The halide ion is removed first (ionization by solvent) in the rate-determining step; subsequently the intermediate (a carbonium ion) loses a proton to a base. The reaction rate depends only on the concentration of the substrate because the base is not involved until after the rate-determining step. The reaction is termed $E1$ (elimination, first order).

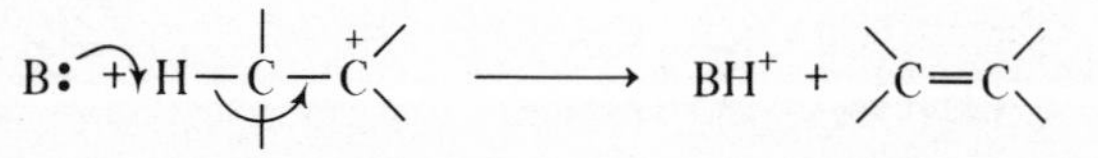

3. The proton is removed first in the rate-determining step; subsequently the intermediate (a carbanion) loses halide ion. The process is found to occur only rarely, if at all.

$$B: + H - \overset{|}{\underset{|}{C}} - \overset{|}{\underset{|}{C}} - Br \;\rightleftharpoons\; \text{slow} \;\; BH^+ + :\overset{|}{C} - \overset{|}{\underset{|}{C}} - Br \;\xrightarrow{\text{fast}}\; BH^+ + _{\diagdown}C = C_{\diagup} + Br^-$$

The mechanism that will be operative in a given situation will depend primarily on the structure of the substrate, the nature of the solvent, the strength of the base, and the nature of the leaving group.

The ionization of a carbon-halogen bond occurs more readily if the carbon atom is highly branched, but the ionization process is not usually observed with methyl or vinyl halides. Ionization, however, can proceed only if an ionizing solvent

$$(CH_3)_3CBr > CH_3CH_2\overset{\displaystyle CH_3}{\overset{|}{C}}HBr > CH_3CH_2CH_2CH_2Br \gg CH_3Br, CH_2=CHBr$$

Tertiary   Secondary   Primary   Methyl, vinyl

(Order of decreasing reactivity in ionization)

is present. Usually such a solvent contains a hydroxyl group (as in water, CH_3OH, or CH_3COOH) which can solvate both the anion and cation that are formed. Dimethyl sulfoxide, $(CH_3)_2SO$, and N,N-dimethyl-formamide, $HCON(CH_3)_2$, are frequently used aprotic solvents which are good solvents in terms of solubility, polarity, and solvating power.

A strong base (HO^-, CH_3O^-, H_2N^-, etc.) will tend to favor the $E2$ process.

The reactivity of halogen as a leaving group is usually in the order $I > Br > Cl \gg F$.

In the elimination of HX from compounds that can form more than one product, that product will predominate which contains the most highly branched sp^2 carbons (Saytzeff rule). The only major exceptions occur when the double bond can be conjugated with another double bond (as with the carbonyl group, C=O, or another carbon-carbon double bond) or when the Saytzeff hydrogen atom cannot be trans to the leaving group.

$$CH_3CH_2\overset{\underset{\displaystyle |}{\underset{\displaystyle Br}{}}}{C}HCH_3 \xrightarrow{CH_3O^-} CH_3CH=CHCH_3 + CH_3CH_2CH=CH_2$$

Major   Minor
(Dialkylethylene)   (Monoalkylethylene)

$$\xrightarrow[\text{to Cl is eliminated}]{\substack{E_2 \\ \text{Only H trans diaxial}}}$$

b. *Dehalogenation* Beta elimination of 1,2-dihalides is brought about by metals such as Mg or Zn.

$$BrCH_2CH_2Br + Mg \longrightarrow CH_2=CH_2 + MgBr_2$$

c. *Dehydration* Alcohols are dehydrated by strong acids (H_2SO_4, H_3PO_4) or hot alumina (Al_2O_3) to yield alkenes. The direction of elimination follows Saytzeff's rule.

$$(CH_3)_2CH\overset{\underset{\displaystyle |}{\underset{\displaystyle OH}{}}}{C}HCH_2CH_3 \xrightarrow[\text{(or } Al_2O_3,\, 250°C)]{H_2SO_4} (CH_3)_2C=CHCH_2CH_3 \text{ (major product)}$$

The acid-induced process proceeds by way of oxonium and carbonium intermediates in a reversible process.

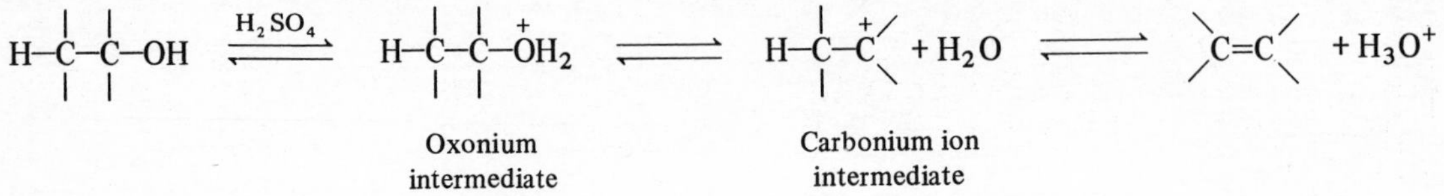

$$H-\overset{|}{\underset{|}{C}}-\overset{|}{\underset{|}{C}}-OH \; \overset{H_2SO_4}{\rightleftharpoons} \; H-\overset{|}{\underset{|}{C}}-\overset{|}{\underset{|}{C}}-\overset{+}{O}H_2 \; \rightleftharpoons \; H-\overset{|}{\underset{|}{C}}-\overset{+}{C} + H_2O \; \rightleftharpoons \; \overset{}{\underset{}{>}}C=C\overset{}{\underset{}{<}} + H_3O^+$$

Oxonium   Carbonium ion
intermediate   intermediate

d. $\quad CaC_2 + 2HOH \longrightarrow HC\equiv CH + Ca(OH)_2$

e. $\quad 6CH_4 + O_2 \xrightarrow{1500°C} 2HC\equiv CH + 2CO + 6H_2$

f. *The Wittig Reaction* (See Chap. 7.)

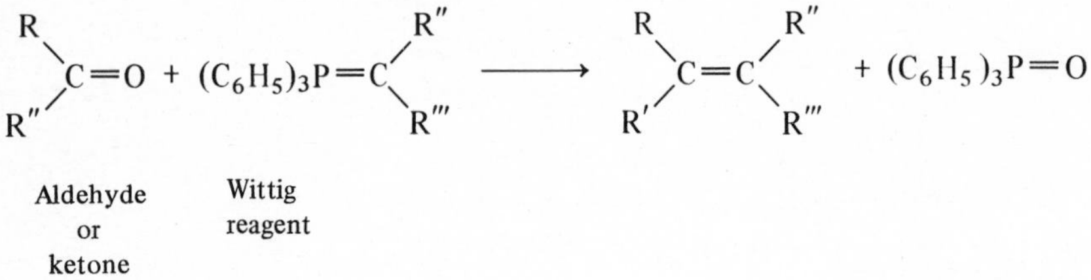

$$\underset{R''}{\overset{R}{>}}C=O + (C_6H_5)_3P=C\underset{R'''}{\overset{R''}{<}} \longrightarrow \underset{R'}{\overset{R}{>}}C=C\underset{R'''}{\overset{R''}{<}} + (C_6H_5)_3P=O$$

Aldehyde   Wittig
or   reagent
ketone

4-3 Reactions

a. *Ionic Additions*

1. Types of additions. The double and triple bonds of hydrocarbons add hydrogen halides (HI, HBr, HCl), halogens (Br_2, Cl_2, and occasionally I_2), and certain other reagents (such as $H-OSO_2OH$, $H-ONO_2$, $I-Cl$, $HO-X$).

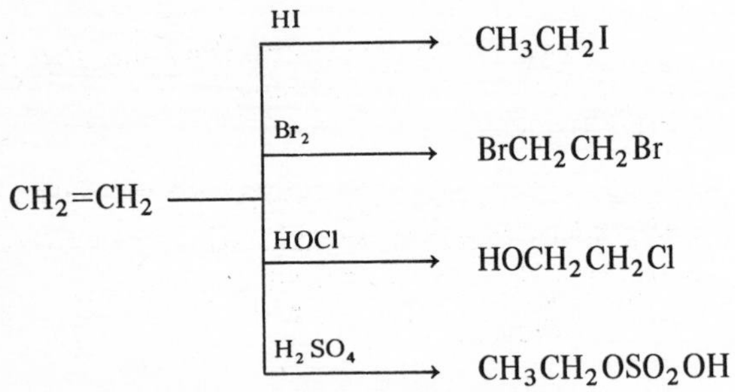

$$CH_2=CH_2 \begin{cases} \xrightarrow{HI} CH_3CH_2I \\ \xrightarrow{Br_2} BrCH_2CH_2Br \\ \xrightarrow{HOCl} HOCH_2CH_2Cl \\ \xrightarrow{H_2SO_4} CH_3CH_2OSO_2OH \end{cases}$$

$$HC\equiv CH \xrightarrow{2Cl_2} CHCl_2CHCl_2$$

2. Mechanism of addition. The multiple bonds of alkenes and alkynes act as electron-donating (nucleophilic) groups that initiate an addition reaction by bonding with an electron-deficient (electrophilic) group to form a *carbonium ion*. The carbonium ion then reacts with any nucleophile present, which will usually be the anion of the acid in a properly designed experiment.

$$CH_2=CH_2 + HX \xrightarrow{slow\ step} CH_3CH_2^+ + X^- \xrightarrow{fast\ step} CH_3CH_2X$$

The direction of addition of "unsymmetric" reagents like HI to unsymmetric unsaturated compounds like $CH_3CH=CH_2$ is governed by the more favorable of two charge distributions in the transition state (that state in which the new C–H bond and the carbonium ion are partially developed). In the reactions under consideration here, it will be assumed that charge stability in the transition state will be governed by the same factors as those involved in stability of the intermediate. Those factors are as follows:

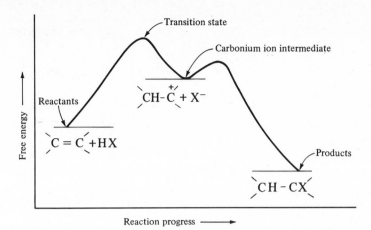

Figure 4-1. Energy diagram for an ionic addition reaction.

Hyperconjugation A positive charge on carbon can be delocalized (stabilized) by adjacent CH bonds on sp^3 carbon. On this basis the reactivity sequence of alkyl halides on p. 35 can be rationalized.

$$\left[\; \begin{matrix} H \\ | \\ -C- \\ | \end{matrix} \overset{+}{C}\diagup_{\diagdown} \longleftrightarrow \overset{H^+}{\begin{matrix} \\ \\ -C= \\ | \end{matrix}} C\diagup_{\diagdown} \;\right]$$

$$CH_3-\underset{\underset{CH_3}{|}}{\overset{\overset{CH_3}{|}}{C}}-Br \longrightarrow \left[CH_3-\underset{\underset{CH_3}{|}}{\overset{\overset{CH_3}{|}}{C}}{}^{+} \longleftrightarrow CH_2=\underset{\underset{CH_3}{|}}{\overset{\overset{H^+ \quad CH_3}{|}}{C}} \longleftrightarrow etc. \right] Br^-$$

Ten resonance structures involved in delocalization of charge

$$CH_3CH_2-\underset{\underset{CH_3}{|}}{CH}-Br \longrightarrow \left[CH_3CH_2-\underset{\underset{CH_3}{|}}{\overset{+}{CH}} \longleftrightarrow CH_3\overset{H^+}{CH}=\underset{\underset{CH_3}{|}}{CH} \longleftrightarrow etc. \right] Br^-$$

Six resonance structures

$$CH_3CH_2CH_2CH_2Br \longrightarrow \left[CH_3CH_2CH_2\overset{+}{CH}_2 \longleftrightarrow CH_3CH_2\overset{H^+}{CH}=CH_2 \longleftrightarrow etc. \right] Br^-$$

Three resonance structures

Conjugation with adjacent C=C π system

$$[CH_2=CH-\overset{+}{CH}_2 \longleftrightarrow \overset{+}{CH}_2-CH=CH_2]$$

Conjugation with adjacent p electrons

$$[-\overset{+}{\underset{|}{C}}-Cl \longleftrightarrow -\underset{|}{C}=\overset{+}{Cl}]$$

Inductive effects

$$\overset{+}{\underset{|}{\text{C}}}\text{--CF}_3 \quad \longrightarrow \quad$$ Unfavorable group dipole moment

$$\overset{+}{\underset{|}{\text{C}}}\text{--C(CH}_3)_3 \quad \longleftarrow \quad$$ More favorable group dipole moment

Based on these factors, the following (termed *Markovnikov* additions) would be predicted:

$$\text{CH}_3\text{CH=CH}_2 + \text{HBr} \longrightarrow [\text{CH}_3\overset{+}{\text{C}}\text{HCH}_3\ \text{Br}^-] \longrightarrow \text{CH}_3\underset{\text{Br}}{\text{CHCH}_3}$$

(Lower energy than
the primary
carbonium ion)

$$\text{CH}_3\text{CH}_2\text{C}{\equiv}\text{CH} + \text{HBr} \longrightarrow \text{CH}_3\text{CH}_2\text{CBr=CH}_2$$

$$\text{CH}_2\text{=CHBr} + \text{HBr} \longrightarrow \text{CH}_3\text{CHBr}_2$$

$$\text{CH}_3\text{CH}_2\text{CH=C(CH}_3)_2 + \overset{\delta+}{\text{I}}{-}\overset{\delta-}{\text{Cl}} \longrightarrow \text{CH}_3\text{CH}_2\underset{\text{I}}{\text{CH}}{-}\underset{\text{Cl}}{\text{C}}\text{(CH}_3)_2$$

Ionic additions are generally trans, suggesting that the intermediate carbonium ion maintains stereochemical integrity.

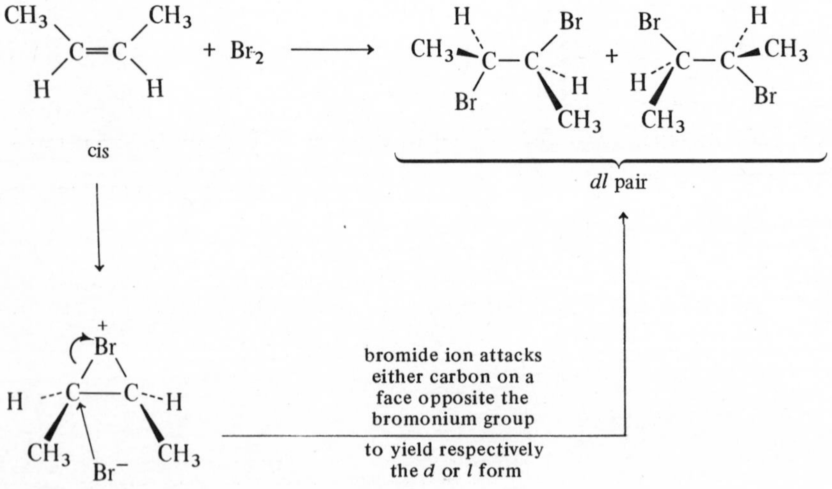

cis

dl pair

bromide ion attacks
either carbon on a
face opposite the
bromonium group

to yield respectively
the *d* or *l* form

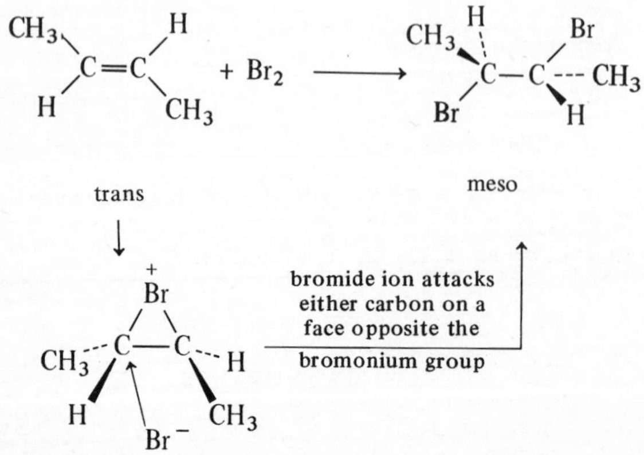

trans   meso

bromide ion attacks
either carbon on a
face opposite the
bromonium group

38

A cyclic bromonium ion intermediate is proposed as the basis for the stereochemical result. If the intermediate had a planar carbonium carbon, the product could not be formed stereospecifically.

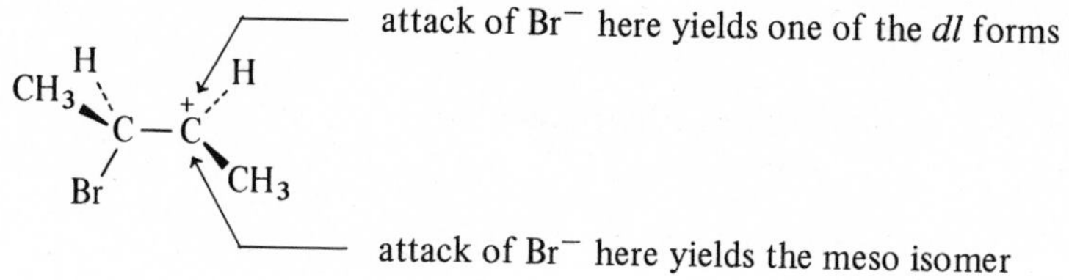

attack of Br^- here yields one of the *dl* forms

$$CH_3 \overset{H}{\underset{Br}{\ddots C}} - \overset{+}{C} \overset{H}{\underset{CH_3}{}}$$

attack of Br^- here yields the meso isomer

Alkynes also undergo trans addition.

$$CH_3C \equiv CCH_3 + Br_2 \longrightarrow \overset{CH_3}{\underset{Br}{}} C = C \overset{Br}{\underset{CH_3}{}}$$

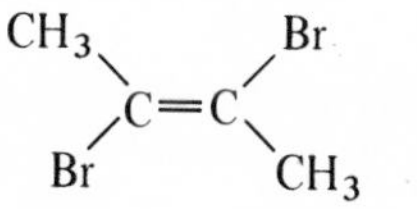

3. Hydroboration. Diborane, B_2H_6, adds to alkenes and alkynes. The addition occurs as if BH_3 added 3 times to three alkenes or alkynes with boron acting as the electrophile. The process is termed *hydroboration*. (See Sec. 6-2e, 1 for uses of this reaction.)

$$6\ CH_3CH=CH_2 + B_2H_6 \longrightarrow 2\ (CH_3CH_2CH_2)_3B$$

4. Addition of water.

$$CH_3CH=CH_2 \xrightarrow{\text{HOH, } H_2SO_4} CH_3\underset{\underset{OH}{|}}{C}HCH_3$$

Alkynes add water in the presence of mercuric ion and a strong acid. The intermediate alcohol, called an *enol*, is unstable and rearranges by the shift of a proton from oxygen to carbon. The product is a ketone (or acetaldehyde, $CH_3CH=O$, in the case of acetylene itself).

$$CH_3C \equiv CH \xrightarrow{\text{HOH, } Hg^{2+}, H_2SO_4} \left[CH_3\underset{\underset{OH}{|}}{C}=CH_2 \right] \longrightarrow CH_3\underset{\underset{O}{\|}}{C}CH_3$$

5. $HC \equiv CH + CH_3COOH \xrightarrow{H_2SO_4} CH_2=CHO\underset{\underset{O}{\|}}{C}CH_3 \qquad$ Vinyl acetate

6. $HC \equiv CH + HCN \xrightarrow{Ba(CN)_2} CH_2=CHCN \qquad$ Acrylonitrile, or vinyl cyanide

7. Reactions of terminal acetylenes. Because of the greater acidity of acetylenic hydrogens compared to olefinic hydrogens, salts are formed by active metals such as Na or K, or strong bases such as $NaNH_2$. The salts react with water to form the original acetylene or with *primary* alkyl halides to give substituted acetylenes. The latter reaction is used for increasing the carbon chain length of a compound.

$$2\ RC \equiv CH + 2\ Na \longrightarrow 2\ RC \equiv C^- Na^+ + H_2$$
$$RC \equiv CH + NaNH_2 \longrightarrow RC \equiv C^- Na^+ + NH_3$$
$$RC \equiv CNa + HOH \longrightarrow RC \equiv CH + NaOH$$
$$RC \equiv CNa + CH_3I \longrightarrow RC \equiv CCH_3 + NaI$$

b. *Free-radical Additions* HBr (not HCl or HI) adds to alkenes in the presence of peroxides. The direction of addition is predictable on the basis of stability of the transition state (again, as reflected in the stability of the intermediate). (See Chap. 3 for a discussion of stability of primary, secondary, and tertiary radicals.)

ROOR (a peroxide) $\xrightarrow{\text{heat}}$ 2 RO·

RO· + HBr $\longrightarrow$ ROH + Br·

— Chain reaction initiation

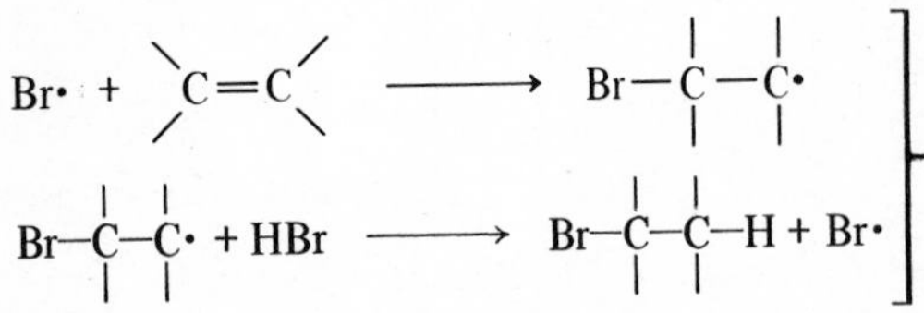

Br· + $\text{C}=\text{C}$ $\longrightarrow$ Br—C—C·

Br—C—C· + HBr $\longrightarrow$ Br—C—C—H + Br·

— Chain reaction propagation

Combination of two radicals of any species in the propagation steps — Chain termination

$\text{CH}_3\text{CH}=\text{CH}_2 \xrightarrow{\text{HBr, peroxide}} \text{CH}_3\text{CH}_2\text{CH}_2\text{Br}$

c. *Addition of Methylene* Cyclopropane and cyclopropene rings can be prepared by the reaction of methylene (see Sec. 3-6c) with an alkene or an alkyne, respectively.

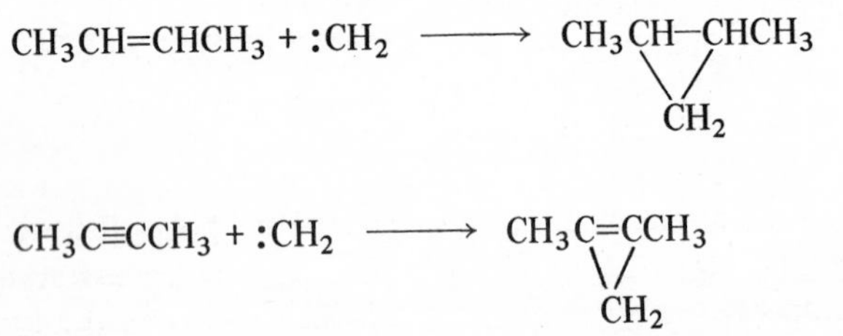

$\text{CH}_3\text{CH}=\text{CHCH}_3 + :\text{CH}_2 \longrightarrow$ CH₃CH—CHCH₃ with CH₂ bridge

$\text{CH}_3\text{C}\equiv\text{CCH}_3 + :\text{CH}_2 \longrightarrow$ CH₃C=CCH₃ with CH₂ bridge

A useful carbenoid source for the transfer of a :CH_2 group is iodomethyl zinc iodide:

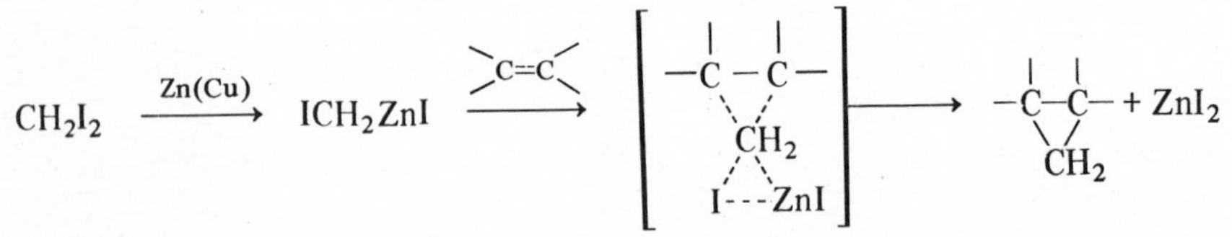

$\text{CH}_2\text{I}_2 \xrightarrow{\text{Zn(Cu)}} \text{ICH}_2\text{ZnI} \xrightarrow{\;>\!\text{C}=\text{C}\!<\;}$ [—C—C— with CH₂ and I---ZnI] $\longrightarrow$ —C—C— with CH₂ + ZnI_2

d. *Hydrogenation*

$\text{CH}_2=\text{CH}_2 + \text{H}_2 \xrightarrow{\text{Pt, Pd or Ni}} \text{CH}_3\text{CH}_3$

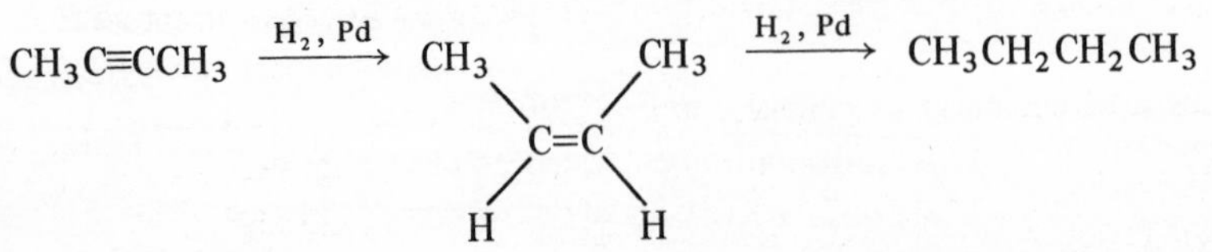

$\text{CH}_3\text{C}\equiv\text{CCH}_3 \xrightarrow{\text{H}_2,\text{ Pd}}$ cis alkene $\xrightarrow{\text{H}_2,\text{ Pd}} \text{CH}_3\text{CH}_2\text{CH}_2\text{CH}_3$

The catalytic addition of H_2 to alkenes or alkynes is usually cleanly cis. Alkynes can be reduced stepwise, so that the synthesis of certain cis alkenes can be carried out. Conversely, the reduction of alkynes with sodium or lithium in liquid ammonia yields the trans alkene.

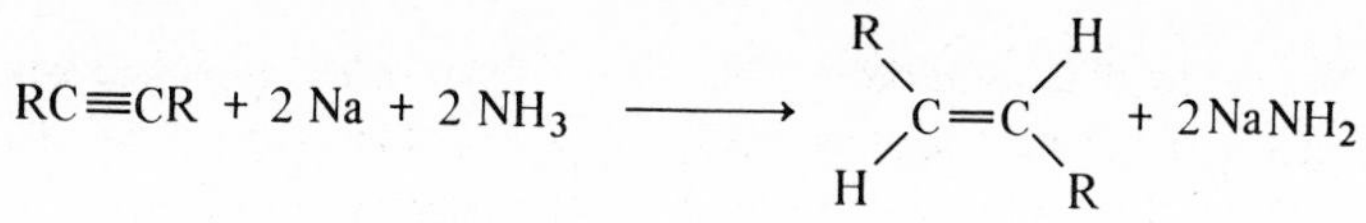

$$RC \equiv CR + 2\,Na + 2\,NH_3 \longrightarrow \underset{H}{\overset{R}{>}}C=C\underset{R}{\overset{H}{<}} + 2\,NaNH_2$$

e. *Oxidations*

1. Ozonolysis. The double or trip bonds are cleaved by ozone, and after hydrolysis the products are aldehydes or ketones (from alkenes) or carboxylic acids (from alkynes).

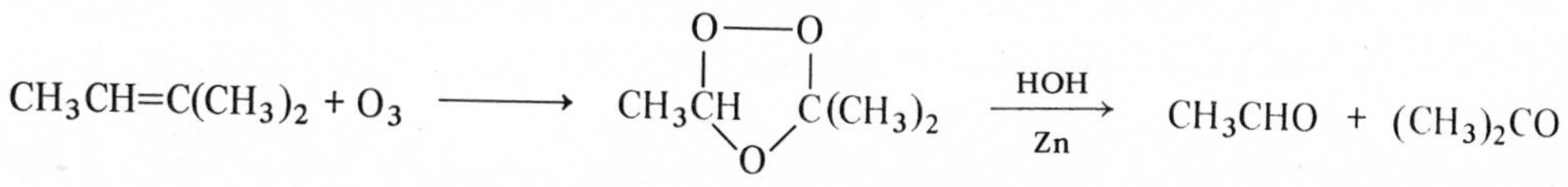

$$CH_3CH=C(CH_3)_2 + O_3 \longrightarrow CH_3CH \underset{O}{\overset{O-O}{<}} C(CH_3)_2 \xrightarrow[Zn]{HOH} CH_3CHO + (CH_3)_2CO$$

Note that each carbon in the multiple bond is, in the products, involved with corresponding bonds to oxygen.

$$CH_2=CHCH_2\overset{CH_3}{\underset{}{C}}=CH_2 \xrightarrow{O_3} \xrightarrow{HOH,\ Zn} 2CH_2=O + O=CHCH_2\overset{CH_3}{\underset{}{C}}=O$$

$$CH_3C \equiv CCH_3 \xrightarrow{O_3} \xrightarrow{HOH,\ Zn} 2CH_3COOH$$

2. Permanganate. Mild oxidation of alkenes with permanganate yields 1,2-diols called *glycols*.

$$CH_3CH=CH_2 \xrightarrow{cold,\ dilute\ KMnO_4} CH_3\underset{OH}{\overset{}{CH}}CH_2OH \qquad \text{Propylene glycol}$$

The glycol is formed by cis addition (see Sec. 13-6d) as opposed to the normal, ionic, trans addition described earlier. OsO_4 can be used for the same purpose.

Vigorous oxidation of alkenes or alkynes with alkaline permanganate or acid dichromate cleaves the double bond. Three types of products are possible: CO_2 from a terminal $=CH_2$ or $\equiv CH$ unit; acids (RCOOH) from $RC\equiv$ or $RCH=$ units; and ketones ($R_2C=O$) from an $R_2C=$ unit. In alkaline solution the CO_2 becomes carbonate, for example, K_2CO_3, and the acid ends up as a salt, for example, RCOOK.

$$R_2C=CHCH_2CH_2CH=CH_2 \xrightarrow{alk\ KMnO_4} R_2C=O + KOOCCH_2CH_2COOK + K_2CO_3$$

$$3RC \equiv CH + 8KMnO_4 + KOH \longrightarrow 3K_2CO_3 + 3RCOOK + 8MnO_2 + 2HOH$$

4-4 Balancing Oxidation-Reduction Equations

Carbon atoms may be assigned oxidation numbers based on the common valences of other atoms in the compound (H, +1; O, −2; halogen, −1; N, −3; etc.)

	Oxidation number of C		Oxidation number of C
Methane, CH_4	−4	Formaldehyde, $H_2C=O$	0
Methyl chloride, CH_3Cl	−2	Formic acid, HCOOH	+2

In figuring the change in oxidation state of C in complex molecules it is usually easiest to consider all of C atoms together rather than as individuals.

$$\underset{\underset{\text{CH}_2=\text{CCH}_2\text{CH}_2\text{CH}=\text{CH}_2}{}}{\overset{\overset{\text{CH}_3}{|}}{}} + \text{KMnO}_4 \longrightarrow 2\text{K}_2\text{CO}_3 + \text{CH}_3\overset{\overset{\text{O}}{||}}{\text{C}}{-}\text{CH}_2\text{CH}_2\text{COOK} + \text{MnO}_2 + \text{H}_2\text{O}$$

First find the oxidation number of carbon in the olefin. Since this contains 12 hydrogen atoms, each of +1, the carbon atoms must be −12 to balance these. Next, consider all of the carbons in the products (note that each terminal CH_2 unit produces an equivalent of carbonate).

2K₂CO₃		Potassium salt of acid	
6O	6(−2) = −12	3O	3(−2) = −6
4K	4(+1) = + 4	7H	7(+1) = +7
Net	= − 8	1K	1(+1) = +1
		Net	+2

The oxidation number for atoms other than carbon is −6 and that for carbon is +6.

The total change for carbon is +18 (from −12 to +6). Since each Mn changes 3 (from +7 to +4), we need 6KMnO_4 to accept the 18 electrons from carbon on the left side of the equation.

$$\underset{\underset{\text{CH}_2=\text{CCH}_2\text{CH}_2\text{CH}=\text{CH}_2}{}}{\overset{\overset{\text{CH}_3}{|}}{}} + 6\text{KMnO}_4 \longrightarrow 2\text{K}_2\text{CO}_3 + \text{CH}_3\overset{\overset{\text{O}}{||}}{\text{C}}\text{CH}_2\text{CH}_2\text{COOK} + 6\text{MnO}_2 + 2\text{H}_2\text{O} + \text{KOH}$$

The equation is now balanced electronically. Next balance the potassium. Any potassium left over from that in carbonate or the salts of organic acids appears as KOH. Next balance oxygen, adding sufficient H_2O to accomplish this. Then check hydrogen. It should be balanced. If not, check your arithmetic.

Dichromates in dilute (4 to 8 M) sulfuric acid also cleave alkenes. Note that under acid conditions CO_2 is produced instead of carbonate and organic acids occur rather than their salts.

$$3\underset{3(-8)}{(\text{CH}_3)_2\text{C}=\text{CH}_2} + 4\text{K}_2\text{Cr}_2\text{O}_7 + 16\text{H}_2\text{SO}_4 \longrightarrow 3\underset{3(-4)}{(\text{CH}_3)_2\text{C}=\text{O}} + 3\underset{3(+4)}{\text{CO}_2} + 4\text{K}_2\text{SO}_4 + 4\text{Cr}_2(\text{SO}_4)_3 + 19\text{H}_2\text{O}$$

The change in oxidation number of C is +8, of Cr, −3. The order of balancing is carbon, chromium, potassium, sulfate, hydrogen, oxygen.

4-5 Qualitative Tests for Carbon-Carbon Unsaturation

a. *Decolorization of a Solution of Bromine in Carbon Tetrachloride* The reddish color of this solution disappears rapidly on addition of an alkene or alkyne.

b. *Discoloration of an Aqueous Solution of KMnO₄* The deep purple color of a permanganate solution is rapidly changed to the brown color of MnO_2 in the presence of an alkene or alkyne.

c. *Salt Formation* 1-Alkynes form precipitates with ammoniacal solutions of cuprous or silver salts. These precipitates are hydrolyzed by acids.

$$\text{RC}{\equiv}\text{CH} \xrightarrow{\text{Ag(NH}_3)_2{}^+} \text{RC}{\equiv}\text{CAg}$$

d. *Alkenes and Alkynes (like Alkanes) Are Highly Insoluble in Water* Most common alkenes and alkynes are liquids. Those with less than five carbon atoms are gases.

4-6 Structure Determination of Ring Compounds

The technique of complete hydrogenation is often used to determine the so-called sites of unsaturation and thus gain information about the number of rings, as well as double and triple bonds, in an unknown compound. For example, an unknown hydrocarbon, $C_{12}H_{10}$, on hydrogenation with platinum gave $C_{12}H_{22}$. Thus, 12 hydrogen atoms were taken up, indicating the presence of six double bonds, three triple bonds, or a combination of these. In comparing $C_{12}H_{22}$ with C_nH_{2n+2} (the general formula for an alkane), it is obvious that the hydrogenated unknown falls short of this by four hydrogens. This implies that two rings are present. (Remember that the general formula for a cycloalkane is C_nH_{2n}.) The original molecule is said to contain eight sites of unsaturation, each site being equivalent to 2H. Thus, cyclopentene would have two sites of unsaturation, one due to the double bond and the other to the ring.

Problems

4-1 Provide a trivial or common name for each of the following:

a. $CH_3CH=CH_2$

b. $CH_3CH_2C\equiv CH$

c. $(CH_3)_2C=CHCH_3$

d. $CH_3CH_2C\equiv CCH_2CH_3$

e. $CH_2=CHCHCH_3$
   |
   CH_3

f. $ClCH_2CH=CH_2$

g. $CH_3C=CH_2$
   |
   Cl

h. $CH_3CH=CHCl$

i. $HC\equiv CCH_2Cl$

j. $(CH_3)_3CC\equiv CH$

k. $BrC\equiv CCH_3$

l. $CH_2=CHC\equiv CH$

4-2 Provide an IUPAC name for each of the following:

a. $(CH_3)_2C=CCH_2CH_3$
   |
   CH_3

b. $CH_3CH_2CH=CHCHCH_3$
   |
   Cl

c. $ClCH_2CH_2C\equiv CCl$

d. $CH_3CH=CHC=CHCH_3$
   |
   CH_3

e. $HC\equiv CC\equiv CC\equiv CCH_3$

f. 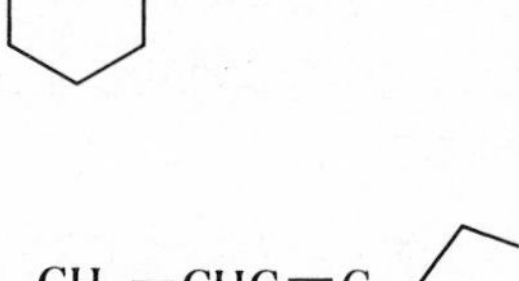 $HC\equiv C-\overset{H}{\underset{\triangle}{C}}-C\equiv CH$

g. (structure: cyclohexene ring with CH_3 and $CH_2CH=CH_2$ substituents)

h. (structure: eight-membered ring with CH_3, C_3H, and triple bond)

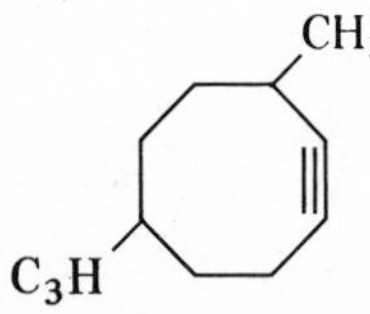

i. 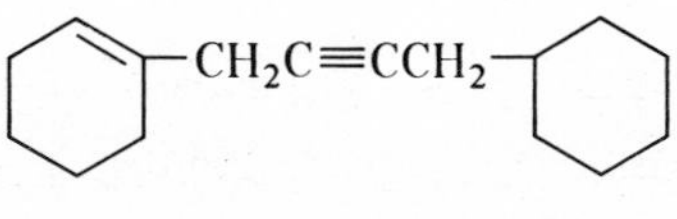 (structure: cyclohexene ring $-CH_2C\equiv CCH_2-$ cyclohexane ring)

j. $CH_2=CHCHCH_2CH_3$
   |
   $CH(CH_3)_2$

k. 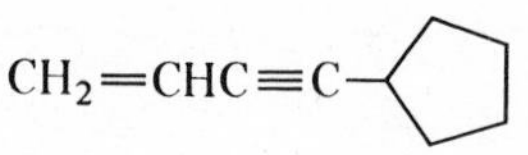 $CH_2=CHC\equiv C-$ (cyclopentane ring)

l. 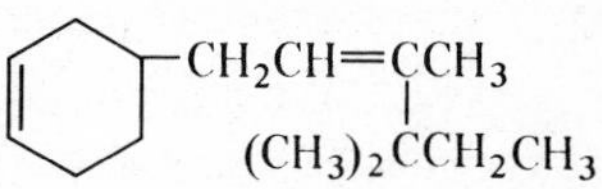 (structure: cyclohexene ring $-CH_2CH=CCH_3$ with $(CH_3)_2CCH_2CH_3$)

4-3 Name each of the following geometric isomers by the E-Z method:

a. 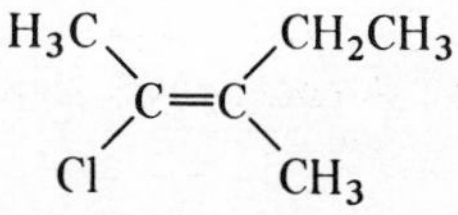 $\underset{Cl}{\overset{H_3C}{>}}C=C\underset{CH_3}{\overset{CH_2CH_3}{<}}$

b. $\underset{F_3C}{\overset{(CH_3)_2CH}{>}}C=C\underset{H}{\overset{CH_3}{<}}$

c. $\underset{F}{\overset{CH_3O}{>}}C=C\underset{NO_2}{\overset{Cl}{<}}$

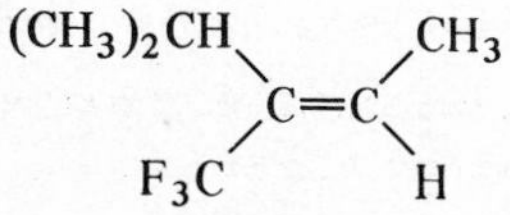

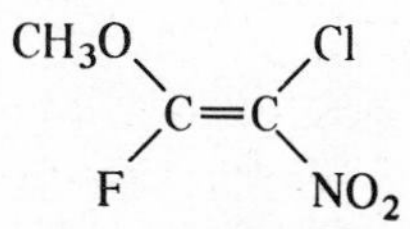

4-4 Give the structure for the products of the following addition reactions (reactants are given in proper stoichiometric proportions). If the product is meso, erythro, threo†, or if it is a cis-trans pair of isomers, provide the proper stereochemical structure.

a. Propene + Br_2
b. Propyne + Br_2
c. Cyclohexene + HOH + H_2SO_4
d. Cyclohexene + Br_2
e. Cyclopentene + O_3
f. 1-Methylcyclohexene + HI
g. 1-Butene + HOCl
h. 3-Methyl-1-butene + B_2H_6
i. 1-Butene + HOH + H_2SO_4
j. 1-Butyne + 2HBr
k. Propyne + $2Br_2$
l. 1-Pentyne + CH_3COOH + H_2SO_4
m. 3-Hexyne + Cl_2
n. 1-Hexene + HBr
o. 1-Hexene + HBr + peroxide
p. 3-Hexyne + $2Cl_2$
q. 1-Hexene + HI
r. 1-Hexene + HI + peroxide
s. cis-2-Pentene + Cl_2
t. trans-2-Pentene + Cl_2
u. 3-Hexyne + HOH + Hg^{2+} + H_2SO_4
v. 3-Heptyne + 2Na + $2NH_3$
w. 3-Heptyne + H_2 + Pd
x. Cyclohexene + D_2 + Pt

4-5 Give the structure for the products of the following elimination reactions. If two or more organic elimination products result, indicate the one that should be formed in highest yield.

a. 2-Bromopropane + hot alcoholic KOH
b. 1-Bromopropane + hot alcoholic KOH
c. 1,1-Dibromopropane + excess KNH_2
d. 1,2-Dibromopropane + excess KNH_2
e. 2,2-Dibromopropane + excess KNH_2
f. An E2 elimination of HCl from cis-1-chloro-2-methylcyclohexane
g. An E2 elimination of HCl from trans-1-chloro-2-methylcyclohexane
h. An E1 elimination of HCl from cis-1-chloro-2-methylcyclohexane
i. An E1 elimination of HCl from trans-1-chloro-2-methylcyclohexane
j. 1-Bromo-2,2-dimethylpropane + hot alcoholic KOH
k. 2-Bromobutane + hot alcoholic KOH
l. 1,2-Dibromohexane + Zn

m. $(CH_3)_2CHC-Cl + R_3N$
$\quad\quad\quad\quad\;\; \overset{\|}{O}$

n. $BrCH_2C-Br + Mg$
$\quad\quad\;\, \overset{\|}{O}$

o. 2-Methyl-2-bromopropane + CH_3COO^-

4-6 Complete and balance each of the following:

a. Ethylene + $KMnO_4$ + heat
b. 2,2,3-Trimethyl-1,4-pentadiene + $KMnO_4$ + heat
c. 1-Methylcyclohexene + $K_2Cr_2O_7$ + H_2SO_4
d. Tetramethylethylene + cold, dilute $KMnO_4$

4-7 Give a convenient method for making the following conversion:

a. Propene to 2-iodopropane
b. 2-Butene to CH_3COOK
c. Propene to 1-bromopropane
d. 2-Chlorobutane to 2,3-dihydroxybutane

† For a structure that has two adjacent asymmetric carbon atoms, each bearing two groups in common, an erythro isomer has the two like groups on the same side and a threo isomer has them on the opposite side as shown below. The nomenclature is derived from the sugar compounds erythrose and threose (see Sec. 13-2a).

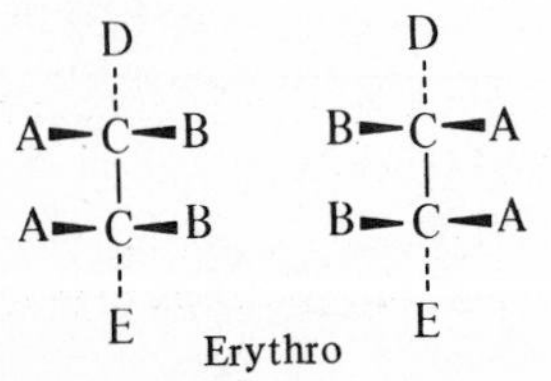

Erythro (dl pair)   Threo (dl pair)

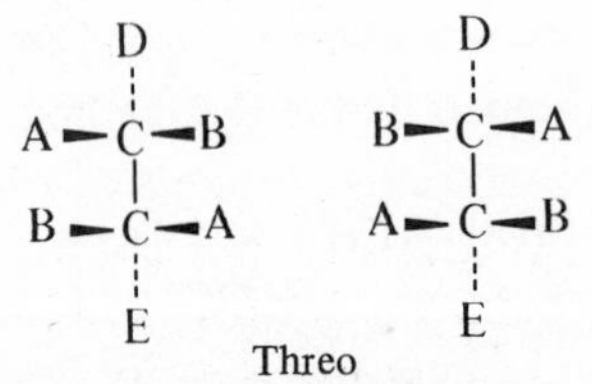

e. 1-Bromobutane to 2-bromobutane
f. 2-Hydroxypropane to 2,3-dimethylbutane
g. 1-Butyne to 1-bromobutane
h. 1-Butyne to octane
i. 1-Bromobutane to 1-butyne
j. 1-Bromopropane to (E)-1-bromopropene
k. Cyclopentene to chlorocyclopentane
l. Bromocyclohexane to 1,2-dibromocyclohexane
m. 3-Hexene to CH_3CH_2CHO
n. Isobutylene to isobutyl bromide
o. *trans*-4-Octene to *trans*-1,2-dipropylcyclopropane
p. 3-Hexyne to 1,2-diethylcyclopropene

4-8 What organic products would be obtained from the ozonolysis of each of the following?

a. 1,3-Butadiene

b. =CH_2

c. 1,2-Dimethylcyclohexene
d. 2-Hexyne
e. Allylacetylene
f. 1,5-Hexadiyne
g. 1,3-Pentadiene
h. $CH_3CH{=}CHCOOH$
i. Divinylacetylene

4-9 Compound A, C_7H_{14}, yielded the same products on ozonolysis as an oxidation with alkaline permanganate. Suggest a structure for A.

4-10 Compound A, C_6H_{12}, gave a positive test with Br_2 in CCl_4. Oxidation of A with alk $KMnO_4$ yielded only B, the potassium salt of an acid. Suggest structures for A and B.

4-11 It required 0.700 g of hydrocarbon A to react completely with 2.00 g of Br_2. Treatment of A with HBr yielded a monobromoalkane, B. The same compound B was formed when A was treated with HBr plus peroxide. Suggest structures for A and B.

4-12 When 10 g of $CHBr_2CHBr_2$ was treated with Zn, 290 ml of acetylene at STP was obtained. What was the percentage yield?

4-13 The silver salt of an unknown acetylene contained 66.8% Ag. Assuming no other functional groups were present except the triple bond, suggest a suitable structure for the silver salt.

4-14 Compound X, C_6H_{10}, gave 2-methylpentane when treated with H_2 and Pd. Treating it with an acid solution of $HgSO_4$ yielded $C_6H_{12}O$. X did not react with ammoniacal CuCl or metallic sodium. Suggest a structure for X.

4-15 When acetylene is treated under pressure with nickel carbonyl catalyst, it is converted into a new hydrocarbon which we shall call X. Vapor-density measurements on X gave a density of 4.643 g/liter at STP. On complete catalytic hydrogenation, 0.2 g of X absorbed 172 ml of hydrogen at STP. Upon ozonolysis, X was converted to a single product OHCCHO. Suggest a structure for X.

4-16 When acetylene is treated with a catalyst consisting of $AlEt_3$ and $TiCl_3$, it is converted to a hydrocarbon we shall call X. Vapor-density measurements on X showed a density of 3.48 g/liter at STP. On complete catalytic hydrogenation, 0.1 g of X absorbed 86 ml of hydrogen at STP. Upon ozonolysis X was converted to a single product, OHCCHO.
a. Suggest a structure for X.
b. What product would you expect from the action of the above catalyst on 5-decyne?
c. What product would you expect from the action of the above catalyst on ethylacetylene?

4-17 Indicate which of the following compounds would not display geometric isomerism. Draw all possible geometric isomers for those that do.
a. 2-Methyl-2-butene
b. 1-(2-Methylcyclopropyl)-1-propene
c. 3-Hexene
d. 1-Chloro-2,3-dimethylcyclohexane
e. 2,4-Hexadiene
f. 1,3-Dimethylcyclobutane
g. 1,3-Butadiene
h. Linoleic acid
[$CH_3(CH_2)_4CH{=}CHCH_2CH{=}CH(CH_2)_7COOH$]
i. 1-Iodo-2-methylcyclopropane
j. 2-Chloro-3-methyl-2-pentene
k. 1-Methylcyclopentene

4-18 When A was vaporized, 0.7013 g occupied 306 ml at 100°C and 1.00 atm. Oxidation of A with alkaline $KMnO_4$ yielded no organic acid salts exclusive of carbonates. Suggest a structure for A.

4-19 The cuprous salt of an unknown acetylene contained 44.6% Cu. Assuming no other functional groups were present except the triple bond, suggest three suitable structures for the salt.

4-20 How many sites of unsaturation would each of the following have?
a. Cyclohexene
b. Cyclohexylacetylene
c. Dicyclopropylacetylene
d. Cycloheptane
e. 1,3,5-Hexatriyne
f. 1,3-Cyclohexadiene

4-21 How many rings do each of the following compounds contain?

a. β-Carotene (a source of vitamin A), $C_{40}H_{56}$ $\xrightarrow{H_2, Pd}$ $C_{40}H_{78}$

b. Benzene, C_6H_6 $\xrightarrow{H_2, Pt}$ C_6H_{12}

c. Naphthalene (mothballs), $C_{10}H_8$ $\xrightarrow{H_2, Pt}$ $C_{10}H_{18}$

d. Bicycloheptadiene, C_7H_8 $\xrightarrow{H_2, Pt}$ C_7H_{12}

4-22 Given acetylene, ethylene, iodomethane, and inorganic reagents, show how you could prepare each of the following:
a. Methylacetylene
b. Dimethylacetylene
c. 1,2-Dichloropropane
d. 1-Butene
e. 2-Iodobutane
f. 1-Bromopropane
g. $CH_3CH_2CH_2CCH_3$
   $\overset{\|}{O}$
h. cis-2-Butene
i. trans-2-Butene

j. CH_3CCH_3
   $\overset{\|}{O}$
k. 1-Iodo-2-chloropropane
l. cis-4-Octene

4-23 Suggest a simple method that will distinguish between each of the following:
a. Hexane and 1-hexene
b. Heptane and water
c. 3-Hexene and NaCl
d. 1-Octene and ethyl alcohol
e. 1-Butyne and 2-butyne
f. 1-Butyne and 1-butene

4-24 In each of the following possible addition reactions draw the resonance forms which would be involved in delocalization of positive charge in the carbonium ion intermediate. Where two modes of addition are possible, consider only the one giving the greater charge delocalization.
a. Ethene + HI
b. Isobutylene + HOCl
c. Cyclobutene + ICl
d. 1-Chloroethene + Br_2
e. $CH_2=CHOCH_3$ + HBr
f. 1,1-Difluoro-1-propene + HCl

4-25 Provide a mechanism which shows how Br_2 might add to 1,3-butadiene to form 1,4-dibromo-2-butene.

4-26 When 1 mol of 1,3-butadiene reacts with 1 mol of HCl, two products are obtained: 3-chloro-1-butene and 1-chloro-2-butene. Rationalize this on the basis of the reaction mechanism. Provide an energy diagram for the competitive processes.

4-27 If ethylene were shaken with a water solution of Br_2 and NaCl, which of the following would be likely to be found among the products: $BrCH_2CH_2Br$, $ClCH_2CH_2Cl$, $BrCH_2CH_2Cl$, $BrCH_2CH_2OH$? Explain.

4-28 When HCl is added to acrolein, $CH_2=CHCH=O$, the addition occurs contrary to Markovnikov's rule to give $ClCH_2CH_2CH=O$. Explain.

4-29 Choose the more reactive compound in each of the following pairs for the given reaction. Give the basis for your choice.
a. 1-Butene or 2-butene, in the addition of Br_2.
b. $CH_2=CH–CH=O$ or $CH_2=CH–CH=CH_2$, in the addition of HBr.
c. Hydroxide ion or amide ion (H_2N^-), in the elimination of HCl from tert-butyl chloride in hot water–ethanol.
d. Hydroxide ion or amide ion, in the elimination of HCl from 1-chlorobutane in hot water–ethanol.

4-30 Draw the indicated number of resonance structures for the following:
a. $CH_3OCH=CH_2$, 2
b. $CH_2=CHC\equiv CH$, 6
c. $HC\equiv CF$, 4
d. $CH_2=CHC\equiv N$, 4
e. $ClCH=CHC\equiv N$, 5
f. $CH_2=CHN(CH_3)_2$, 2

4-31 In each of the following addition reactions draw the intermediate resulting from the attack by the positive part of the reagent and the resonance structures that contribute to its stability.
a. $CH_3C\equiv CH$ + HBr
b. $CH_2=CHF$ + ICl
c. $CH_3C\equiv CF$ + HBr
d. $CH_3CH=CHCF_3$ + HOCl
e. $CH_2=CHCF_3$ + HCl
f. $CH_3OCH=CH_2$ + H_2SO_4

4-32 Two important factors determining the relative acidity of a given hydrogen atom in a molecule are (1) the electron-attracting strength (electronegativity) of the element to which it is attached and (2) the ability of the

46

anion (after the proton is dissociated) to stabilize the negative charge through resonance. Considering these two factors, which compound in each of the following pairs would be the stronger acid? Give a reason for your choice.

a. HI, HCl

b. CH_3OH, CH_3NH_2

c. CH_3CH_3, $HC\equiv CH$

d. $\overset{+}{N}H_4$, NH_3

e. CH_3SH, CH_3OH

f.

g. $CH_2=CHCH_3$

h. Cl_3CH, $ClCH_3$

i. $N\equiv CCH_2C\equiv N$, $CH_2=CHCH_2CH=CH_2$

j. $H-\overset{O}{\overset{\|}{C}}-OH$, CH_3OH

k. $CH_3\overset{O}{\overset{\|}{C}}H$, $CH_3CH=CH_2$

l. OH^-, H_2O

4-33 Cyclooctatetraene is a tub-shaped molecule which shows no aromatic character. However, one of its following ions has been found to be planar and relatively stable. Which one? Why?

$C_8H_8{}^{2-}$ $\qquad$ $C_8H_6{}^{2-}$ $\qquad$ $C_8H_6{}^{2+}$

Draw several resonance forms for the most stable ion.

4-34 Give a molecular orbital description of each of the following ions and molecules. Indicate bond angles, σ and π bonds, and the orbital composition of the σ bonds.

a. $H_2C=NOH$ $\qquad$ b. $FC\equiv N:$ $\qquad$ c. $HC\equiv CCH=CH_2$ $\qquad$ d. $O_2{}^{2-}$

4-35 The diazocyclopentadiene molecule
shows unusual stability for this type of compound.

Presumably this is because the electrons can be arranged in the ring via resonance so that the $4n + 2$ rule is accommodated. Show how this is possible.

4-36 When the molecular weight of an unknown hydrocarbon was determined by the vapor density method, the vapor was found to weigh 3.57 g/liter (STP). When 0.2 g of the unknown was titrated with a solution of 0.2 M bromine in carbon tetrachloride, 25 ml of bromine solution was decolorized. How many double bonds did the unknown contain?

4-37 When 0.092 g of an unknown hydrocarbon was dissolved in 0.5 g of camphor, the melting point of the camphor was lowered 20°. When 0.2 g of the unknown was catalytically hydrogenated, it was found that 169 ml of hydrogen at 30°C and 730 mmHg was absorbed. How many double bonds did the unknown contain? Suggest a possible structure for it.

4-38 The dipole moment for CH_3CH_2Cl is 2.05 debyes while that for $CH_2=CHCl$ is 1.44 debyes. Assuming the negative end of the dipole in each molecule to be toward the chlorine atom, how can the difference be explained?

4-39 a. Draw three pictures that represent conformational energy minima for rotation about the middle carbon atoms of 2-bromobutane.

b. On the basis of a trans elimination of HBr from 2-bromobutane, predict whether cis-2-butene or trans-2-butene will be in excess in the product.

c. On an energy diagram compare the heats of hydrogenation of cis-2-butene (28.6 kcal/mol) and a trans-2-butene (27.6 kcal/mol). Note that the product in each instance is identical.

d. In many trans elimination reactions the relative stability of the olefinic product is a fair measure of its tendency to form. If the heat of hydrogenation of 1-butene is 30.1 kcal/mol, should the butene product contain more 1-butene or trans-2-butene? Explain.

5
halogen derivatives of aliphatic hydrocarbons

5-1 Nomenclature

a. According to the IUPAC system, organic halogen compounds are named as derivatives of the hydrocarbon with the longest continuous chain. The position of the halogen is indicated by a numerical prefix. A position of unsaturation has precedence over a position occupied by halogen in determining the direction of numbering of a chain.

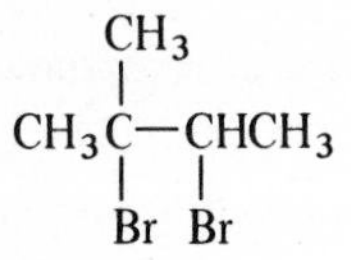

$$
\begin{array}{c}
\text{CH}_3 \\
| \\
\text{CH}_3\text{C}-\text{CHCH}_3 \\
| \quad | \\
\text{Br} \quad \text{Br}
\end{array}
\qquad
\text{HCCl}_3
\qquad
\begin{array}{c}
\text{Br} \quad \text{Cl} \quad \text{F} \\
| \quad\ | \quad | \\
\text{CH}_3\text{CH}_2\text{CH}_2\text{C}-\text{CH}-\text{CHCH}_3 \\
| \\
\text{CH(CH}_3)_2
\end{array}
$$

| 2-Methyl-2,3-dibromobutane | Trichloromethane (Chloroform) | 4-Bromo-3-chloro-2-fluoro-4-isopropylheptane |

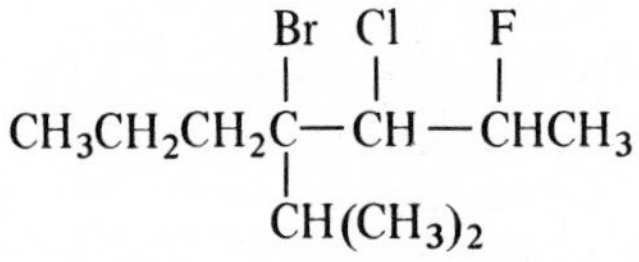

$$\text{FCH}_2\text{CH}_2\text{CH}_2\text{CH}=\text{CH}_2 \qquad \text{5-Fluoro-1-pentene (rather than 1-fluoro-4-pentene)}$$

b. The saturated halogen compounds are also referred to as alkyl halides. (See also the nomenclature of special groups in Chap. 4.)

$$
\text{CH}_3\text{I}
\qquad
\text{CH}_3\text{CHFCH}_3
\qquad
\begin{array}{c}
\text{CH}_3 \\
| \\
\text{CH}_3\text{CH}_2\text{CBr} \\
| \\
\text{CH}_3
\end{array}
$$

| Methyl iodide | Isopropyl fluoride | *tert*-Pentyl bromide or *tert*-Amyl bromide |

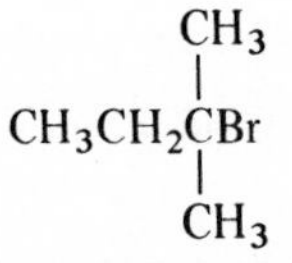

The methyl and ethyl groups are often abbreviated as Me and Et; for example, MeI and EtI.

5-2 Methods of Preparation

(In this section, X stands for Cl, Br, and I.)

a. *From alkanes* (see Sec. 3-5d), *alkenes* (see Sec. 4-3), *and alkynes* (see Sec. 4-3).

b. Although halogenation of hydrocarbons by hydrogen substitution is of little general use because of difficulty of separating the isomers that result when a compound with several carbon atoms is halogenated, there are special instances in which it is useful, as in the following examples which involve substitution in an allylic position which is favored relative to nonallylic radicals because of its stability ($\text{CH}_2=\text{CHCH}_2\cdot \longleftrightarrow \cdot\text{CH}_2\text{CH}=\text{CH}_2$

$$\text{CH}_3\text{CH}=\text{CH}_2 + \text{Cl}_2 \xrightarrow{500°\text{C}} \text{ClCH}_2\text{CH}=\text{CH}_2 + \text{HCl}$$

Allyl chloride

N-Bromosuccinimide (abbreviated NBS) provides chain-reaction bromination in an allylic position of alkenes. The reaction is initiated by light or peroxides.

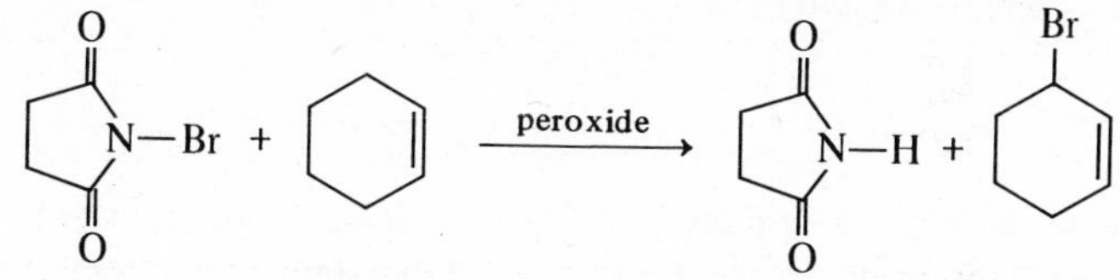

NBS   3-Bromocyclohexene

c. *From alcohols*, $ROH + HX \rightleftharpoons RX + HOH$ (see also the reactions of alcohols with thionyl chloride and phosphorus halides, Sec. 6-4*b*-2 and 6-4*b*-3). The first step in this conversion is the protonation of the alcohol to yield an oxonium intermediate. This may react further by different mechanistic routes (see the later discussion of the mechanism of substitution reactions) to yield substitution or elimination products as follows:

$$CH_3CH_2OH + HX \rightleftharpoons CH_3CH_2\overset{+}{O}H_2X^-$$

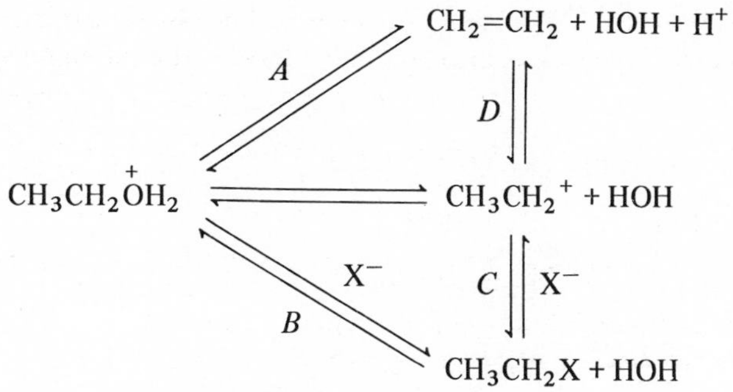

where A = direct elimination route ($E2$)
        B = direct substitution route (S_N2)
        C = carbonium ion substitution route (S_N1)
        D = carbonium ion elimination route ($E1$)

d. Alkyl fluorides are usually made indirectly.

$$3RCl + SbF_3 \longrightarrow 3RF + SbCl_3$$

5-3   Reactions

a. *Nucleophilic Substitution* Electron-rich groups with a pair of electrons available for bonding (in other words, Lewis bases) react with alkyl halides to yield substitution products. The following are typical examples:

1. $OH^- + RX \longrightarrow ROH + X^-$ (forms an alcohol)
2. $SH^- + EtI \longrightarrow EtSH + I^-$ (forms a thiol)
3. $CN^- + CH_3Br \longrightarrow CH_3CN + Br^-$ (forms a nitrile)
4. $RO^- + CH_3CH_2CH_2Cl \longrightarrow CH_3CH_2CH_2OR + Cl^-$ (forms an ether)
5. $RCOO^- + CH_3I \longrightarrow RCOOCH_3 + I^-$ (forms an ester)
6. $NO_2^- + RX \longrightarrow RNO_2 + X^-$ (forms a nitro compound)

These reactions are described as involving attack of a *nucleophile* (Lewis base) on a carbon atom of a *substrate* which has a *leaving group* (in the preceding illustrations, the halide ion). (See also the discussion of mechanism in Sec. 5-5.)

b. *Wurtz Reaction* (See Sec. 3-5*d*.)

c. *Reduction* (See Sec. 3-5*a*.) $4RX + LiAlH_4 \longrightarrow 4RH + LiX + AlX_3$

d. *Elimination Reactions* (See Sec. 4-2*a, b*.)

e.   *Reaction with Magnesium Metal; the Grignard Reaction*

$$RX + Mg \xrightarrow[\text{anhydrous conditions}]{\text{ether solvent}} RMgX \text{ (the Grignard reagent)}$$

$$Me_2CHBr + Mg \xrightarrow{Et_2O} Me_2CHMgBr$$

f.   *α Elimination: Formation of Carbenes* In cases in which base-induced β elimination is not an important competitive process, a compound can lose the elements of hydrogen halide from the same carbon atom to form a carbene.

$$HCCl_3 + HO^- \longrightarrow [Cl_3C^-] + HOH \longrightarrow Cl_2C\text{:} + Cl^- + HOH$$
Dichlorocarbene

5-4   Reactions of the Grignard Reagent

a.   *With "Active" Hydrogen* $RMgX + H^+ \longrightarrow RH + XMg^+$. Any compound that is as acidic as water, an amine (RNH_2), an alcohol, or a terminal alkyne ($RC{\equiv}CH$) is sufficiently reactive to be involved in this reaction. The last reactant is used to prepare an acetylenic Grignard reagent.

$$RC{\equiv}CH + R'MgX \longrightarrow RC{\equiv}CMgX + R'H$$

b.   *With Carbonyl Compounds and with Carbon Dioxide* (See Chap. 6 for reactions with aldehydes, ketones and ethylene oxide.)

$$RMgX + CO_2 \longrightarrow RCOOMgX \xrightarrow{H_3O^+} RCOOH \qquad \text{Carboxylic acid}$$

c.   *With Oxygen* $2RMgX + O_2 \longrightarrow 2ROMgX \xrightarrow{H_3O^+} 2ROH \qquad$ An alcohol

d.   *With Inorganic Halides*

1. $CdCl_2 + 2RMgX \longrightarrow CdR_2 + 2MgXCl \qquad$ Dialkylcadmium formation
2. $SiCl_4 + 4MeMgI \longrightarrow SiMe_4 + 4MgICl \qquad$ Tetramethylsilane formation

5-5   Mechanism of Substitution Reactions

The mechanism of substitution can be described in terms of two extremes, the S_N1 and S_N2 routes. An actual reaction may be proceeding by both routes simultaneously or it may be following a pathway somewhat intermediate between the two. These routes are as follows.

S_N1 (symbol for substitution, nucleophilic, first order):

$$RX \rightleftharpoons R^+ + X^- \xrightarrow{Y^-} RY$$

In this process the first step is rate-determining. Since the nucleophile (Y^-) is not involved, the rate of the reaction depends only on the concentration of the substrate (RX): rate $= k_1 [RX]$. The intermediate carbonium ion (R^+) can react with the leaving group (X^-) to return to starting material, or with added nucleophile (Y^-) to form a product. The transition state for the process is one in which the carbon-halogen bond is partly broken.

S_N2 (symbol for substitution, nucleophilic, second order):

$$RX + Y^- \longrightarrow RY + X^-$$

In this process there is no formation of an intermediate, and the rate of reaction is directly proportional to the concentration of both the nucleophile and the substrate.

$$\text{Rate} = k_2 [RX] [Y^-]$$

The nucleophile initiated the reaction by beginning to bond with the back lobe of the sp^3 orbital that bonds carbon to the leaving group. As the carbon-nuclephile bond forms, the carbon-leaving group bond is disrupted.

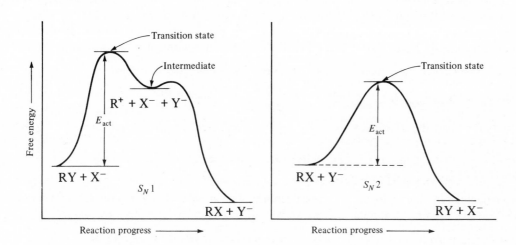

Figure 5-1. Energy diagrams for nucleophilic substitution routes.

a. *Effect of Substrate Structure on Rate* If solvent-induced ionization of the substrate produces a stable carbonium ion, the reaction can be expected to proceed more rapidly than that in which a less stable (higher energy) carbonium ion is formed because the kind of charge delocalization in the transition state is closely related to that in the intermediate. An S_N1 reactivity sequence would be tertiary halide > secondary halide > primary halide > methyl halide. An allylic halide would be very reactive because of good charge delocalization (CH_2=CH—CH_2^+ ⟷ $^+CH_2$—CH=CH_2), but a neopentyl halide (Me_3C—CH_2X) would be very unreactive because no hyperconjugation is possible. Vinyl halides (CH_2=CHX) and acetylenic halides (RC≡CX) are also highly unreactive.

   Bulky groups in the substrate also favor ionization because the 109° bond angle in the starting material approaches 120° in the transition state.

   In the S_N2 process, bulky groups hinder the substitution because of repulsions that increase when five groups are clustered about carbon in the transition state. For this reason a reactivity sequence as follows is observed: methyl halide > primary halide > secondary halide > tertiary halide. As in the S_N1 process, vinyl, acetylenic, and neopentyl halides are unreactive, but an allylic halide reacts readily by either mechanism.

   Electronic factors operate in either situation. In the S_N1 reaction, in which a positive charge is being generated, electron-releasing groups favor the process. Conversely in the S_N2 reaction, in which the transition state is more negative than the starting material, electron-withdrawing groups favor the process.

Z——$C^{\delta+}$ Z——$C^{\delta+}$ Z——$C^{\delta-}$ Z——$C^{\delta-}$

Favorable   Unfavorable   Favorable   Unfavorable

Charge interaction in S_N1 Charge interaction in S_N2
transition state   transition state

b. *Effect of Solvent on Rate* An S_N1 process requires a solvent with good ionizing power. The energy for getting the system to the transition state is the solvation energy between solvent and the charged species that are being generated. Good solvents for S_N1 processes almost always contain hydroxyl groups (as in water, alcohols, or carboxylic acids) because the electronegative oxygen atom interacts with the incipient positive charge, and the electropositive hydrogen atom (through hydrogen bonding) interacts with the incipient negative charge.

51

Favorable solvent orientations

In the S_N2 process, the extent of charge or the localization of charge in the starting materials and transition state can be used to interpret solvent effects on rate. A highly polar solvent will interact more favorably with (and lower the energy of) that state with more charge or with more localized charge. Thus, in all of the reactions shown in Sec. 5-3a, a polar solvent will interact more favorably with starting materials than with the transition state if the process is S_N2, and the reaction will be slower than with nonpolar solvents.

c. *Effect of Nucleophile on Rate* The nucleophilicity of groups can be predicted as follows:

1. For a given atom, an increase in negative charge increases its nucleophilicity: HO^- is more nucleophilic than H_2O.
2. Within a period of the periodic table, nucleophilicity decreases with increasing atomic number: $C^- > N^- > O^- > F^-$, or, $NH_3 > H_2O > HF$.
3. With a group of the periodic table, nucleophilicity increases with increasing atomic number: $I^- > Br^- > Cl^- > F^-$.

In general nucleophilicity is more influenced by polarizability of an atom, ion, or group than by basicity of the same species. Note, for example, that iodide ion is a very strong nucleophile but a very weak base.

d. *Effect of Leaving Group on Rate* Among the halides the order of reactivity is $I > Br > Cl$. Many other leaving groups are common, including sulfate ($SO_4{}^{2-}$), sulfonate ($RSO_3{}^-$), carboxylate ($RCOO^-$), nitrate ($NO_3{}^-$), water, etc. The best leaving groups are those that are least basic. For example, water (from a protonated alcohol, $R\overset{+}{O}H_2$) is a very good leaving group, but direct displacement of hydroxide ion (from an alcohol, ROH) is virtually impossible.

e. *Other Effects* A metal ion such as Ag^+ which has a strong affinity for halide ion can exert a powerful influence in the formation of a carbonium ion intermediate.

$$RX + Ag^+ \longrightarrow [R^+] + AgX\ (ppt) \xrightarrow{\text{nucleophile}} \text{products}$$

5-6 Rearrangement Reactions via Carbonium Ion Intermediates

The migration of hydrogen or an alkyl group from an adjacent carbon atom during formation of a carbonium ion can be expected to occur if the new carbonium ion is more stable. The rearrangement is usually concerted (that is, the migration occurs simultaneously with the formation of the ion).

$$Me_3C\text{-}CH_2Br \xrightarrow{Ag^+} [Me_2\overset{+}{C}\text{-}CH_2Me] \longrightarrow \text{rearranged products}$$

Often it aids the ionization process energetically. The resulting increase in rate over what would otherwise be expected is termed *anchimeric assistance*.

5-7 Competition among Substitution, Elimination, and Rearrangement

S_N1, $E1$ and carbonium ion rearrangement reactions all proceed via a common intermediate. The course of the reaction can often be controlled by specific reaction conditions. In the competition between S_N1 and $E1$, a high concentration of nucleophile favors the former; the presence of a good base favors the latter; and high temperature favors the latter. Rearrangements cannot be controlled if the reaction conditions strongly favor ionization of the substrate, but the fate of the rearranged carbonium ion follows that predicted for any carbonium ion.

Under basic conditions tertiary halides, such as *tert*-butyl chloride, give mainly elimination products. For example, *tert*-butyl chloride gives isobutylene as the major product when warmed with aqueous solutions, regardless of whether a strong base (such as hydroxide ion) or a strong nucleophile (such as cyanide ion) are present. Conversion of the halide to the alcohol can be brought about by the use of Ag_2O in cold water. Because of the very low solubility of silver hydroxide, the solution is very weakly basic. As the silver helps pull off the halogen at low temperature there is a good chance that the resulting carbonium ion will bond with a water molecule before it can lose a proton.

At the other extreme, a primary halide such as 1-bromopropane will tend to undergo substitution rather than elimination unless a very strong base and high temperature are employed.

5-8 Qualitative Tests for Halogens

a. *Beilstein Test* All compounds containing halogens (except fluorine), when placed on a copper wire and thrust into a flame, cause a green coloration.

b. *Alcoholic Silver Nitrate* Except for fluoride, alkyl halides in which only one halogen is attached to a carbon atom give a precipitate when warmed with alcoholic silver nitrate solution. Vinyl halides and halides with more than one halogen atom on the same carbon are inert. The order of reactivity is that expected on the basis of an S_N1 reaction: tertiary $\cong$ allyl > secondary > primary > vinyl, and RI > RBr > RCl.

c. *Sodium Iodide in Acetone* Alkyl bromides or chlorides give a precipitate of sodium bromide or sodium chloride when treated with sodium iodide in acetone. The order of reactivity is that expected for an S_N2 reaction primary $\cong$ allyl > secondary > tertiary > vinyl, and RBr > RCl.

Problems

5-1 *a.* Write structures for the nine dichlorobutanes.
b. Beside each structure that contains one or more asymmetric carbon atoms indicate the total number of stereoisomers that are possible.

5-2 Give an IUPAC and a common name to each of the following:

a. CH_3CHICH_3	*b.* $CH_2=CHCH_2Cl$	*c.* $(CH_3)_2CHF$
d. $CH_2=CHBr$	*e.* $BrCH_2CH_2Br$	*f.* CH_2Cl_2
g. $CH_3CH=CHCl$	*h.* $HC\equiv CCH_2Br$	*i.* $CHCl_3$
j. Me_2CHCH_2Cl	*k.* ⬡—I	*l.* $CH_3CHICH_2CH_3$
m. $CHBr_3$	*n.* CF_4	*o.* $CH_3CHClCH_2Cl$

5-3 Name the following according to the IUPAC system:

a. $(CH_3)_2CHCHClCHFCH_2OH$

b. $CH_2=C(CH_3)CHClCH=CHCH_2Br$

c. $(CH_3)_2CHCH_2CH(CH_2CH_2Cl)_2$

d. $CH_3CH_2CH_2\underset{\underset{CH(CH_3)_2}{|}}{\overset{\overset{F}{|}}{C}}CH(CHClCH_3)_2$

5-4 Give the structure for the products of the following substitution reactions:

a. 1-Bromopropane + HO^-
b. 2-Bromopropane + CH_3O^-
c. Iodomethane + HOH
d. 2MeI + S^{2-}
e. Bromocyclohexane + HS^-
f. 1,5-Dibromopentane + $2CN^-$
g. 2EtBr + $^-SS^-$
h. EtBr + $HC\equiv CNa$
i. $ClCH=CHCH_2Cl + CH_3COO^-$
j. 2MeI + $^-OOC-COO^-$
k. $2CH_3I + NaC\equiv CNa$
l. 1-Iodohexane + CN^-

5-5 In each of the following provide a nucleophilic substitution reaction (attack on carbon) that will lead to the indicated product:

a. CH_3OH
b. CH_3CH_2SH
c. CH_3SCH_3

d. $HC\equiv CCH_2CH_2CH_3$
e. $CH_3CH_2\overset{\overset{O}{\|}}{C}\underset{OCH_2CH_3}{}$
f. $CH_3S-C\equiv N$

g. CH_3NH_2 h. $(CH_3)_3P$ i. $CH_3C\overset{\displaystyle S}{\underset{\displaystyle SCH_3}{\big\|}}$

j. Et_3N k. $CH_3CH_2OCH_2CH_3$ l. Et_4N^+

5-6 a. Provide the electronic structure for the cyanide ion.

b. Cyanide ion can act in two ways as a nucleophile. Give electronic structures for the two different products (nitriles and isonitriles) that could result from the reaction of cyanide ion with methyl iodide.

c. The presence of silver ion promotes reaction of cyanide ion at the more electronegative atom. Formulate the structure of the isonitrile product of the reaction of silver cyanide with methyl iodide.

d. An isonitrile can be formed by the following reactions:

$CHCl_3 \xrightarrow{\text{base (α elimination)}}$ A (a carbene intermediate)

A + $RNH_2 \longrightarrow$ B (an unstable intermediate containing all of the components of A and RNH_2)

B $\xrightarrow{\text{base (loss of 2HCl)}}$ C (an isonitrile)

Provide the structures for A, B and C.

e. Cyanide ion behaves as as an *ambident* nucleophile; that is, it can react in one of two ways. Another group that is an ambident nucleophile is the nitrite ion, $NO_2{}^-$. Give the electronic structure for the nitrite ion and the two products that can result from its reaction with iodomethane.

f. Unlike the cyanide ion situation, the presence of silver ion promotes the reaction at the less-electronegative atom of the nitrite ion. Give the product of the reaction of silver nitrite with 1-bromobutane.

5-7 Iodide ion is both a good nucleophile and a good leaving group.

a. Draw a picture of the S_N2 transition state that results when a molecule of (R)-2-iodobutane reacts with iodide ion.

b. Do the same for (S)-2-iodobutane.

c. Provide an energy diagram for the reaction described in a. What are the relative magnitudes of the activation energy in the forward and reverse directions?

d. When (R)-2-iodobutane is treated with sodium iodide in acetone the optical activity disappears. Explain. What is the product?

5-8 With allylic halides (such as 3-chloropropene) certain substitution reactions are labeled mechanistically as S_N1' or S_N2' when the incoming nucleophile becomes attached to a carbon atom not bonded originally to the leaving group.

a. Give the structure for the S_N2 transition state in the reaction of a nucleophile (Y^-) with 3-chloro-1-butene.

b. Give the structure for the S_N2' transition state in the same reaction.

c. Give the structure for the S_N1 transition state in the same reaction.

d. Give the structure for the S_N1' transition state in the same reaction.

e. Devise a system in which a nucleophile can displace a leaving group from the number 5 position of a carbon chain by attacking the number 1 position of that chain.

5-9 Write balanced equations for each of the following reactions indicating the most likely reaction. Classify each reaction S_N1, S_N2, $E1$, or $E2$.

a. Ethyl iodide + CH_3OK + CH_3OH

b. Methyl chloride + KCN + CH_3OH

c. Isopropyl iodide + H_2O + Δ

d. 1,2-Dibromoethane + Mg

e. Butyl bromide + $LiAlH_4$ + Et_2O

f. *tert*-Butyl iodide + NaOH + H_2O

g. CH_3I + KSCN + EtOH

h. Allyl bromide + sodium acetate + H_2O

i. *tert*-Amyl iodide + Ag_2O + H_2O + Δ

j. 2,3-Dimethyl-2-iodobutane + NaSH + H_2O

k. CH_3I + $NaNO_2$ + dimethylformamide

l. Methyl bromide + sodium acetylide + aq CH_3OH

m. $CH_3CH=CHCl$ + $NaNH_2$ + decane + Δ

n. CH_3MgI + stannic chloride + Et_2O

o. Ethyl iodide + NH_2^- (anhyd)

p. 2,2-Dichloropropane + aq NaOH

q. 2,2-Dichloropropane + alc KOH + Δ

r. 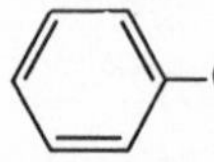 $\langle\!\!\!\bigcirc\!\!\!\rangle$—$CH_2Br$ + aq NaOH

5-10 Give synthetic routes for making the following conversions:

a. 1-Bromobutane to 1-butanol (1-hydroxybutane)

b. 1-Bromobutane to 2-bromobutane

c. 1-Bromopropane to 3-bromopropene   d. Isopropyl bromide to propyl bromide

e. 1-Butene to $CH_3CH_2CH_2CH_2CN$ f. Propene to 1,1,2,2-tetrabromopropane

g. 2-Butyne to $CH_3CH_2\underset{\underset{CH_3}{|}}{C}HSH$ h. Cyclohexane to 3-bromocyclohexene

i. 1-Bromopropane to propyne   j. 1-Bromopropane to 2-hexyne

k. Bromocyclohexane to 1,2-dibromobyclohexane

5-11   Provide a brief explanation for the following experimental observations for nucleophilic substitution reaction conditions which *exclude* an S_N2 process.

a. $CH_2=CHCH_2CH_2Cl$ is less reactive than $CH_3CH=CHCH_2Cl$ toward alc $AgNO_3$.

b. —Cl is very unreactive toward alc $AgNO_3$.

c. $CH_3CH_2-O-\underset{\underset{Cl}{|}}{C}HCH_3$ is more reactive than $CH_3CH_2-O-CH_2CH_2Cl$ with water.

d. $CH_3CH=CHCl$ is very unreactive toward nucleophiles.

e. — CH_2Br is more reactive than —CH_2Br toward aq NaOH.

f. $CH_3CH=CHCH_2Cl$ yields the same products as $CH_3\underset{\underset{Cl}{|}}{C}HCH=CH_2$ with aq NaOH.

5-12   Provide a brief explanation for the following experimental observations for nucleophilic substitution reaction conditions which exclude an S_N1 process.

a. $(CH_3)_3C-CH_2Cl$ is very unreactive toward nucleophiles.

b. $CH_3CH=CHCH_2Cl$ may yield a mixture of isomers with alc KCN.

c. —Cl is very unreactive toward nucleophiles.

d. CH_3CH_2I is more reactive than CH_3CH_2Cl toward KCN.

e. $CH_2=CH\overset{*}{C}H_2Cl$ ($\overset{*}{C}$ represents radioactive carbon) $\xrightarrow{\text{nucleophile}}$ X $\xrightarrow{\text{ozonolysis}}$ formaldehyde which contains C and $\overset{*}{C}$.

5-13   Point out the errors, if any, in the following proposed syntheses.

a. $BrCH_2CH_2Br \xrightarrow[\text{(A)}]{\text{Mg}} BrMgCH_2CH_2MgBr \xrightarrow[\text{(B)}]{(CH_3)_2CHBr} (CH_3)_2CHCH_2CH_2CH(CH_3)_2$

b. *n*-Pentane + $Cl_2 \xrightarrow[\text{(A)}]{\lambda} CH_3CH_2CH_2CHClCH_3 \xrightarrow[\text{(B)}]{CH_3C\equiv CNa} CH_3CH_2CH_2\underset{\underset{CH_3}{|}}{C}HC\equiv CCH_3$

c. Isobutylene + HCl $\xrightarrow[\text{(A)}]{\text{peroxide}} Me_3CCl \xrightarrow[\text{(B)}]{NaCN} Me_3CCN$

d. Propene + HOBr $\xrightarrow{\text{(A)}} CH_3\underset{\underset{Br}{|}}{C}HCH_2OH \xrightarrow[\text{(B)}]{Mg} CH_3\underset{\underset{MgBr}{|}}{C}HCH_2OH$

e. $HC\equiv CH \xrightarrow[\text{(A)}]{HCl} CH_2=CHCl \xrightarrow[\text{(B)}]{NaCN} CH_2=CHCN$

f. $CH_4 + F_2 \xrightarrow{\text{(A)}} FCH_2F \xrightarrow[\text{(B)}]{NaOH} HOCH_2OH$

5-14   When $ClCH_2CH(CH_3)CH_2SH$ is warmed in dilute alkali, a cyclic sulfur-containing compound is formed. Suggest a structure and mechanism for its formation.

5-15   Solvolysis means reaction with solvent. Using isopropyl bromide as a substrate illustrate hydrolysis, ammonolysis, ethanolysis, and acetolysis (reactions, respectively, with water, ammonia, ethanol, and acetic acid). Show both substitution and elimination.

5-16   Write out eight base-catalyzed elimination reactions giving eight different alkenes in which the same combinations of base and leaving group are never used.

5-17   Compound A was found to contain 47.6% chlorine. It decolorized a solution of bromine in carbon tetrachloride and reduced permanganate solution. One gram of A evolved 300.5 ml of methane at STP when treated with excess CH_3MgI. Suggest a structure for A.

5-18   Compound X gave a positive Beilstein test. It was observed that 1.85 g of X, when vaporized, gave 448 ml of vapor at STP. When X was treated with hot alcoholic KOH, Y was formed. Y gave a negative Beilstein test and on oxidation with $KMnO_4$ yielded an acid of NE 74. Suggest structures for X and Y.

5-19   Suggest a simple test to distinguish between:
   a.   Ethyl iodide and ethanol
   b.   Allyl chloride and vinyl chloride
   c.   Chloroform and butyl chloride
   d.   Butyl chloride and butyl fluoride
   e.   1-Bromo-1-butene and 1-bromo-2-butene
   f.   1,1,1-Trichloroethane and ethyl iodide

5-20   When 2-methyl-1,3-butadiene is hydrogenated with 1 mol of hydrogen, the following products are obtained: 3-methyl-1-butene, 2-methyl-1-butene, 2-methyl-2-butene, 2-methylbutane, plus some unreacted 2-methyl-1,3-butadiene. How can you explain this?

5-21   What products would you expect from the addition of 1 mol of bromine to 1 mol of butadiene?

5-22   When 1,3,5,7-cyclooctatetraene is treated with chlorine, one of the products isolated is Assuming attack first by Cl^+ to form a carbonium ion, outline a mechanism for the reaction.

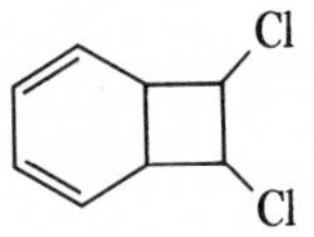

5-23   What kind of electronic structure would you expect for $LiAlH_4$? $NaBH_4$ is a reducing agent like $LiAlH_4$ except that it is not so active. How can this be rationalized?

5-24   What shape would you expect for each of the following?
   a. $(CH_3)_2Be$ b. $HC\equiv CMgI$ c.   ROOR   d. SO_3

5-25   Predict the effect on rate of the change indicated in each of the following and give the basis for your prediction.
   a.   Change to a more polar solvent in an S_N2 reaction between CH_3O^- and EtI.
   b.   Change to a less polar solvent in an S_N2 reaction between Et_3N and EtI.
   c.   Change R from CH_3CH_2- to $CH_3\overset{\text{O}}{\underset{\|}{C}}-$ in an S_N2 reaction between I^- and RCH_2Cl.

   d.   Change R from CH_3- to CF_3- in an S_N1 reaction of R_3CCl.
   e.   Change L (the leaving group) from $-OCH_3$ to $-O\overset{\text{O}}{\underset{\|}{C}}CH_3$ in an S_N2 reaction of $CH_3CH_2CH_2CH_2L$.

   f.   Change from $(CH_3)_2CHOH$ to $(CH_3)_2C=O$ as the solvent in an S_N1 reaction of $(CH_3)_3CCl$.
   g.   Change from CH_3O^- to CH_3COO^- as the base in an $E1$ reaction of 2-chloro-2-methylbutane.
   h.   Change from CH_3O^- to CH_3COO^- as the base in an $E2$ reaction of 1-bromopentane.

5-26   Draw structures for the indicated number of most probable products you would expect in the following reactions.
   a.   Neopentyl alcohol + aq HBr, Δ; 3
   b. $CH_3CH=CHCH_2OH$ + aq HCl; 2
   c.   tert-Butyl bromide + H_2O, 100°C; 2
   d.   3-Bromo-3-methylhexane + alc KOH; 2
   e. $(CH_3)_3CCH(OH)CH_3$ + H^+ + Δ; 2
   f.   2-Bromopropane + CH_3ONa; 2
   g.   Neopentyl iodide + silver acetate; 2
   h.   tert-Butyl bromide + EtOH, 60°C; 2

5-27   The β elimination reactions discussed in this chapter involved removal of the elements of HX by base. However, there are many other β elimination reactions known, some of which are trans, some cis, and some nonstereospecific.
   a.   The reaction of iodide ion with meso-2,3-dibromobutane yields only trans-2-butene. The reaction follows the following kinetic expression:

   Rate = $k[I^-]$[dihalide]

Is this a trans elimination? On what do you base your answer? Provide a mechanism for the process and give a structure for the transition state.

b. Give the structure for a dihalide that, on treatment with iodide ion, will yield *cis*-2-pentene.

c. A trans elimination occurs when an active metal such as Zn or Mg reacts with a 1,2-dihalide. Show by arrows how the electron changes might occur when Mg reacts with 1,2-dibromoethane.

d. When 1,4-dibromo-2-butene is treated with either NaI or Mg it is converted into C_4H_6. Suggest a mechanism for each process.

e. When a β-halo acid, such as $BrCH_2CH_2COOH$, is warmed with base the elimination involves loss of both the carboxyl group (COOH) and the halide ion (yielding ethylene in the example cited). Suggest a mechanism for the process on the basis of a stereospecific trans elimination.

f. Suggest the structure for a β-halo acid that would be converted to *cis*-2-butene when treated with base as in *e*.

g. Certain β elimination reactions are cleanly cis. These usually involve internal attack by a basic atom on a hydrogen atom in the same molecule. Show the transition state and the products that would result from pyrolytic (induced by heating) β elimination in the following compound:

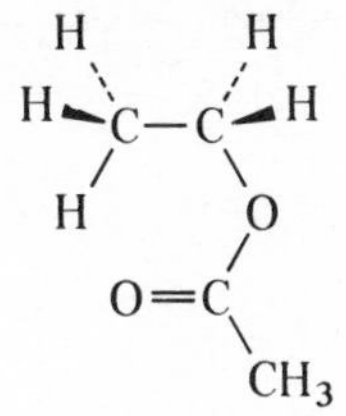

h. Give the product formed via pyrolitic elimination in the following compounds:

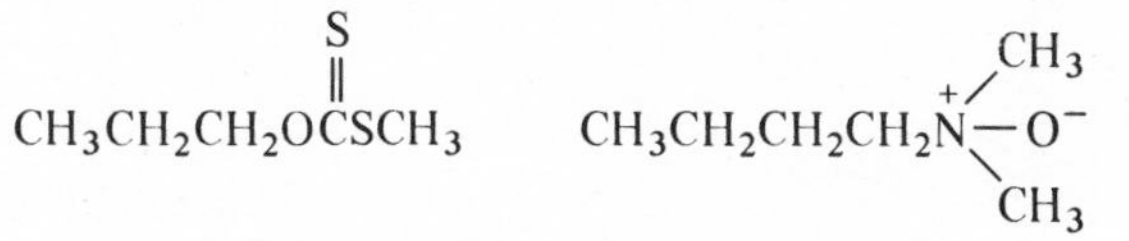

i. In the following example, the compound is subjected to conditions for pyrolitic elimination. Predict the major product and give the basis for your answer.

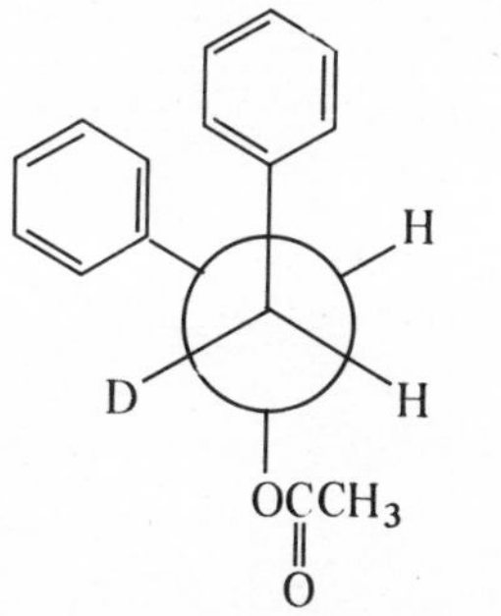

5-28   Based on Prob. 5-27*f* predict the base-catalyzed elimination products from the following and predict which would undergo elimination faster.

a.

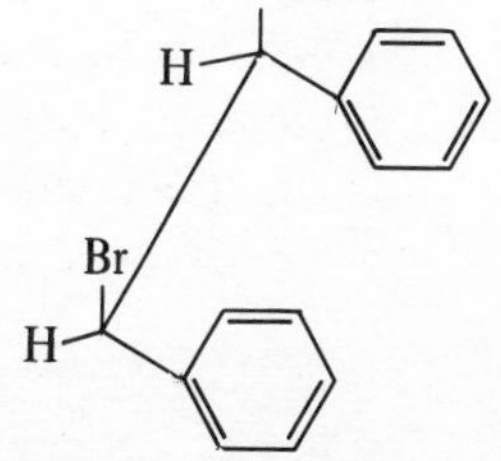

b.

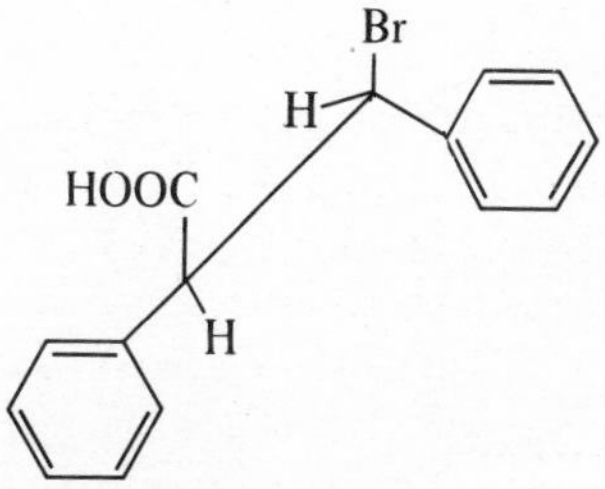

5-29   Predict the major products expected in each of the following elimination reactions and give an explanation.

a.

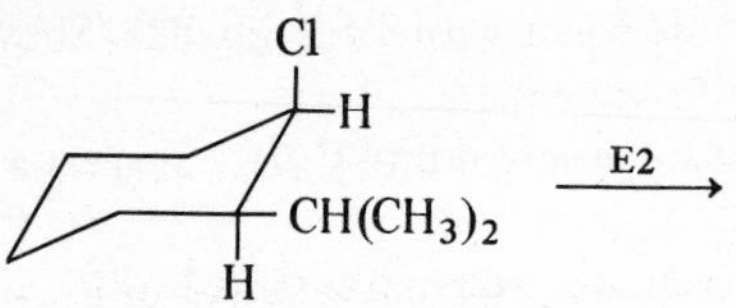

$$\xrightarrow{\text{E2}}$$

b.

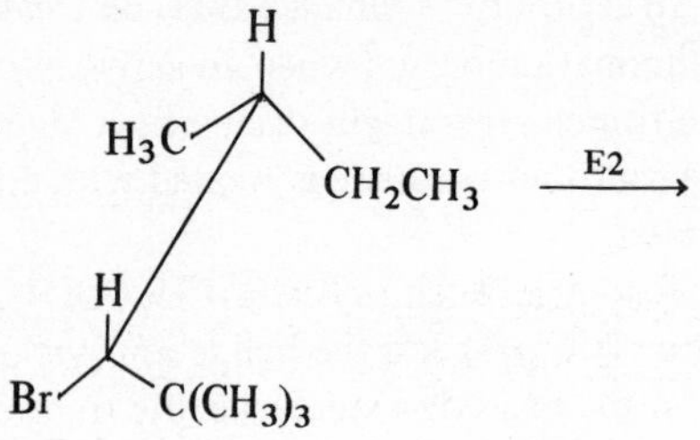

$$\xrightarrow{\text{E2}}$$

c.

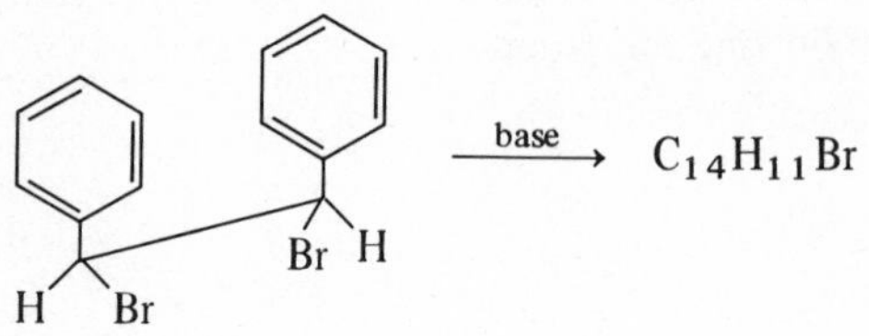

$$\xrightarrow{\text{base}} C_{14}H_{11}Br$$

d.

$$\xrightarrow{\text{base}} C_{14}H_{12}$$

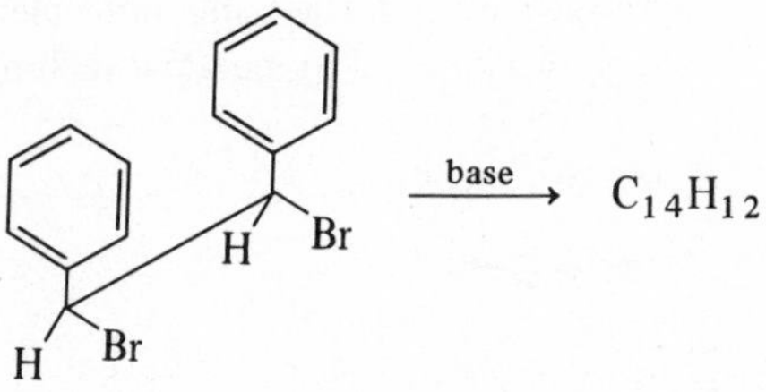

6
aliphatic alcohols and ethers

6-1 Nomenclature

a. *IUPAC System* An alcohol is named as a derivative of the longest continuous chain containing the —OH group. The chain is numbered so as to give the hydroxyl group the lowest number, and the positions of substituents are indicated by number. The *ol* ending is added to designate an —OH group.

CH_3CH_2OH $(CH_3)_2CHCH_2OH$ $CH_2{=}CHCH_2CHCH_3$

Ethanol   2-Methyl-1-propanol   4-Penten-2-ol

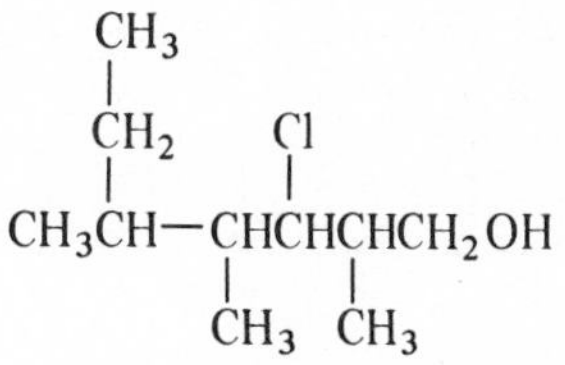

$$CH_3CH-CHCHCHCH_2OH$$

3-Chloro-2,4,5-trimethyl-1-heptanol

Ethers are named as alkoxy (RO) derivatives of hydrocarbons, the larger radical being taken as the parent hydrocarbon.

$CH_3OCH_2CH_2CH_3$ $CH_3C(OCH_3)_3$ $CH_3OCH_2CH-CH_2OCH_2CH_3$

1-Methoxypropane   1,1,1-Trimethoxyethane   1,2-Dimethoxy-3-ethoxypropane

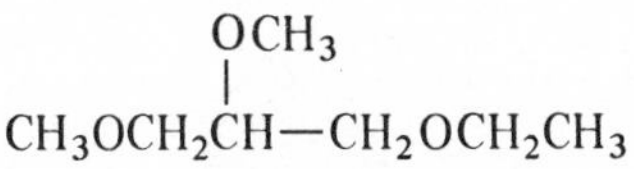

b. *The Alkyl or Common System* Alcohols are named by the addition of the word *alcohol* following the name of the appropriate alkyl group.

CH_3OH CH_3CH_2OH CH_3CHCH_2OH $CH_3CH_2CH_2CH_2OH$

Methyl (wood) alcohol   Ethyl (grain) alcohol   Isobutyl alcohol   Butyl alcohol (a primary alcohol 1°)

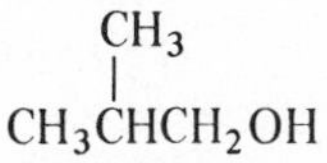

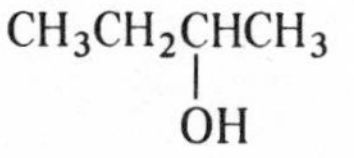

$CH_3CH_2CHCH_3$ / OH CH_3COH

sec-Butyl alcohol (a secondary alcohol 2°) 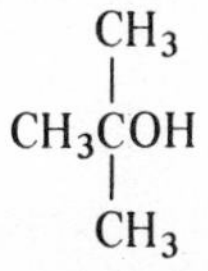   *tert*-Butyl alcohol (a tertiary alcohol, 3°)

For ethers the alkyl groups are named followed by the word *ether*.

CH$_3$OCH$_3$

Methyl ether
(dimethyl ether)

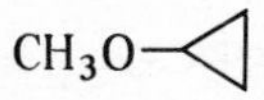

CH$_3$O—◁

Methyl cyclopropyl
ether

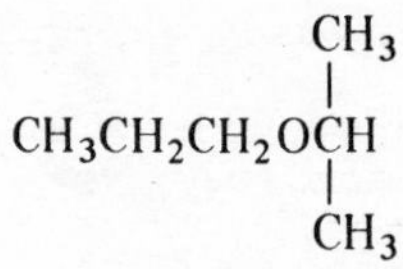

$$CH_3CH_2CH_2O\overset{\overset{\displaystyle CH_3}{|}}{C}H$$
$$\underset{\displaystyle CH_3}{|}$$

Propyl isopropyl
ether

Ethers in which both alkyl groups are the same are called *simple ethers*; when the alkyl groups are different, they are called *mixed ethers*.

c. *The Carbinol System for Alcohols* The alcohol is named as a derivative of CH$_3$OH(methanol or carbinol).

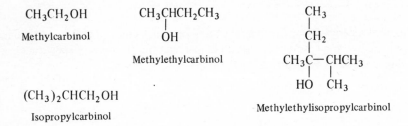

CH$_3$CH$_2$OH

Methylcarbinol

CH$_3$CHCH$_2$CH$_3$
|
OH

Methylethylcarbinol

CH$_3$
|
CH$_2$
|
CH$_3$C—CHCH$_3$
| |
HO   CH$_3$

Methylethylisopropylcarbinol

(CH$_3$)$_2$CHCH$_2$OH

Isopropylcarbinol

d. *Cyclic Alcohols and Ethers* Cyclic alcohols may be named as follows:

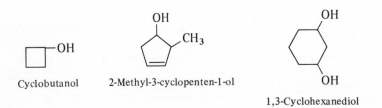

Cyclobutanol

2-Methyl-3-cyclopenten-1-ol

1,3-Cyclohexanediol

Cyclic ethers formed between oxygen and two adjacent carbon atoms are known as oxides or epoxy compounds. As oxides, they are named for the olefin from which they could be derived. They may also be named as epoxy derivatives of hydrocarbons, as indicated in parentheses.

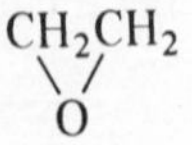

CH$_2$CH$_2$
\ /
O

Ethylene oxide
(epoxyethane)

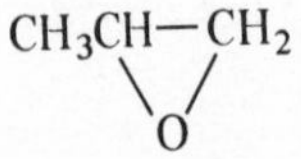

CH$_3$CH—CH$_2$
\ /
O

Propylene oxide
(1,2-epoxypropane)

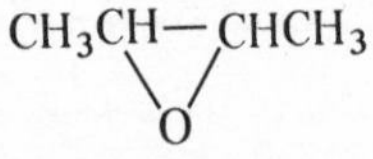

CH$_3$CH—CHCH$_3$
\ /
O

β-Butylene oxide
(2,3-epoxybutane)

e. *Glycols* Compounds containing two hydroxyl groups are known as glycols. They are named either from the olefin from which they could theoretically be made or as polymethylene derivatives (CH$_2$ = methylene group). When they are named as polymethylene derivatives, it is understood that the two hydroxyl groups are on the ends of the chain.

HOCH$_2$CH$_2$OH

Ethylene glycol
Dimethylene glycol

CH$_3$CH(OH)CH$_2$OH

Propylene glycol

HOCH$_2$CH$_2$CH$_2$OH

Trimethylene glycol

CH$_3$CH(OH)CH(OH)CH$_3$

β-Butylene glycol

In the IUPAC system, the positions of the OH groups on the longest continuous chain are designated by numerical prefixes.

$$CH_3CH(OH)CH(CH_3)CH(OH)CH_3 \qquad\qquad HOCH_2CH(OH)CH_2OH$$

3-Methyl-2,4-pentanediol   1,2,3-Propanetriol (Glycerol or glycerin)

6-2   Preparation of Alcohols

a.   Hydration of Olefins

$$CH_3CH{=}CH_2 \xrightarrow{H_2SO_4} \underset{OSO_3H}{CH_3CHCH_3} \xrightarrow{H_2O} \underset{OH}{CH_3CHCH_3} + H_2SO_4$$

b.   Hydrolysis of Alkyl Halides $RX + OH^- \xrightarrow{H_2O} ROH + X^-$. This works well only for allylic, primary, and some secondary alkyl halides; tertiary alkyl halides often give mainly alkenes as products via $E1$ elimination (see Sec. 5-6). This method also has the drawback that many alkyl halides are most easily prepared from the alcohol originally (see Sec. 5-2c and 6-4b).

c.   Reduction of Carbonyl ($\diagdown$C=O) Groups

1. Reduction of aldehydes to *primary* alcohols.

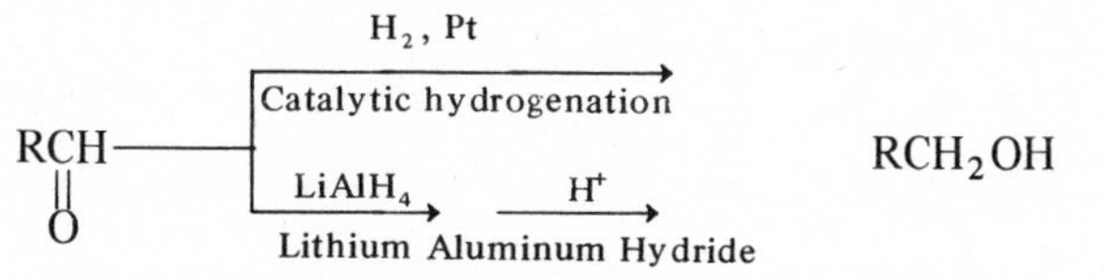

$$\underset{O}{\overset{RCH}{\|}} \begin{array}{l} \xrightarrow[\text{Catalytic hydrogenation}]{H_2,\,Pt} \\ \xrightarrow[\text{Lithium Aluminum Hydride}]{LiAlH_4 \qquad H^+} \end{array} RCH_2OH$$

2. Reduction of ketones to *secondary* alcohols.

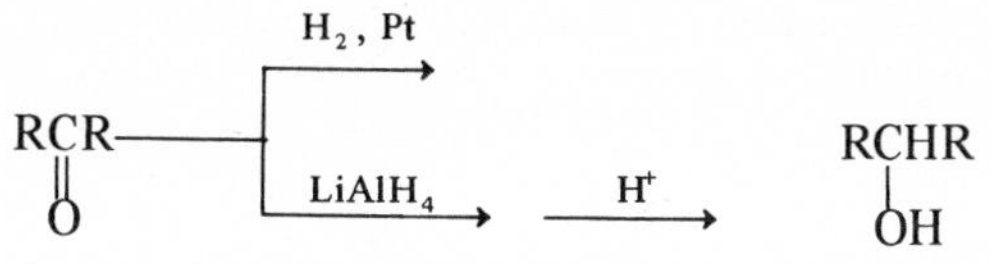

$$\underset{O}{\overset{RCR}{\|}} \begin{array}{l} \xrightarrow{H_2,\,Pt} \\ \xrightarrow{LiAlH_4 \qquad H^+} \end{array} \underset{OH}{RCHR}$$

3. Reduction of esters (see Sec. 11-1 for definition) to *primary* alcohols.

$$\overset{O}{\overset{\|}{R\text{C}OCH_3}} \xrightarrow[H_2,\,250°C]{CuO\text{-}Cr_2O_3} RCH_2OH + CH_3OH$$

$$4\overset{O}{\overset{\|}{R\text{C}OCH_3}} + 2LiAlH_4 \longrightarrow \begin{array}{c} (RCH_2O)_4\,AlLi \\ + \\ (CH_3O)_4\,AlLi \end{array} \xrightarrow{H^+} \begin{array}{c} 4RCH_2OH \\ + \\ 4CH_3OH \end{array}$$

4. Reduction of acids to *primary* alcohols.

$$4RCOOH + 3LiAlH_4 \longrightarrow 2LiAlO_2 + 4H_2 + LiAl(OCH_2R)_4 \xrightarrow{H^+} 4RCH_2OH$$

In the above catalytic reductions, aldehydes and ketones are very easily reduced to alcohols, the reaction going smoothly at room temperature and atmospheric pressure. Esters are much more difficult to reduce and require high temperatures and hydrogen under high pressure. Alternatively, esters may be easily reduced with lithium aluminum hydride, although this is an expensive reagent. Acids are extremely difficult to reduce; the only direct method is that using $LiAlH_4$. Sodium borohydride, $NaBH_4$ may be used to selectively reduce the carbonyl group in aldehydes and ketones which also contain carbon-carbon double bonds, ester groups, or carboxylic acid groups.

d. *Addition of Grignard Reagents to Aldehydes and Ketones* The carbonyl group tends to be polarized as $>C=O \longleftrightarrow >C^+-O^-$ while the Grignard reagent is also polar (R^-Mg^+X). This allows nucleophilic attack of the organic group of the Grignard reagent on the carbonyl carbon of an aldehyde or ketone with the MgX of the Grignard reagent becoming attached to the oxygen of the carbonyl group forming the halomagnesium salt of an alcohol.

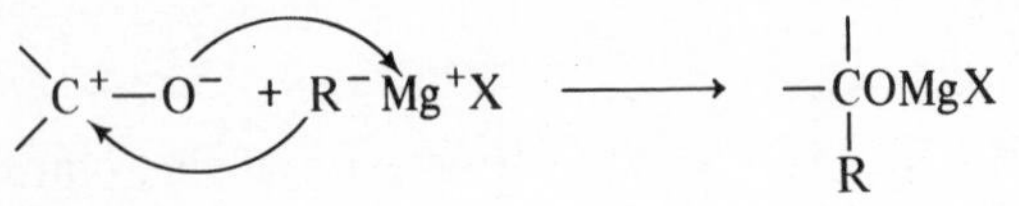

$$>C^+-O^- + R^-Mg^+X \longrightarrow -\overset{|}{\underset{R}{C}}OMgX$$

1. $RMgX + H_2C=O \longrightarrow RCH_2OMgX \xrightarrow{H^+} RCH_2OH$
   (With formaldehyde, *primary* alcohols are obtained.)

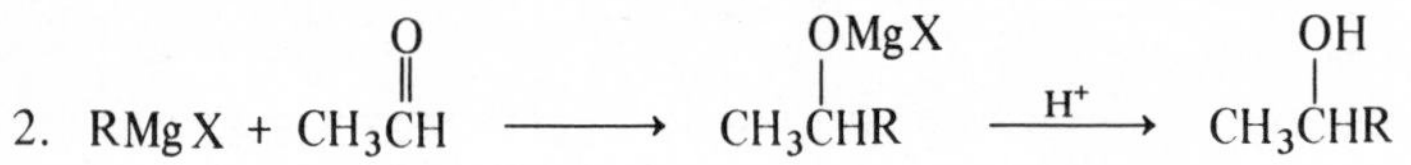

2. $RMgX + CH_3\overset{O}{\overset{\|}{C}}H \longrightarrow CH_3\overset{OMgX}{\overset{|}{C}}HR \xrightarrow{H^+} CH_3\overset{OH}{\overset{|}{C}}HR$
   (With other aldehydes, *secondary* alcohols are obtained.)

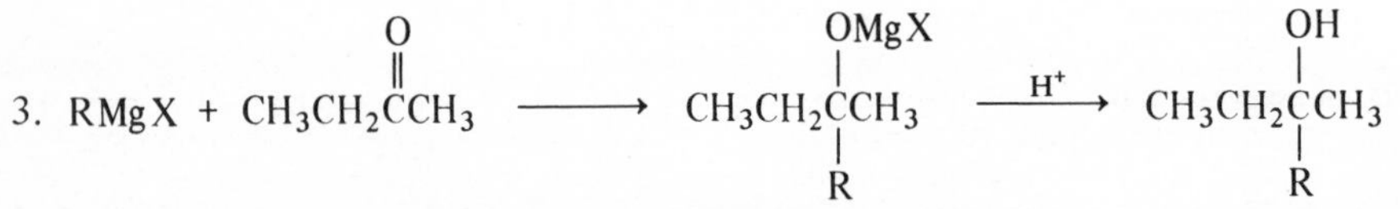

3. $RMgX + CH_3CH_2\overset{O}{\overset{\|}{C}}CH_3 \longrightarrow CH_3CH_2\overset{OMgX}{\underset{R}{\overset{|}{C}}}CH_3 \xrightarrow{H^+} CH_3CH_2\overset{OH}{\underset{R}{\overset{|}{C}}}CH_3$

   (With ketones, *tertiary* alcohols are obtained.)

4. The Grignard reagent reacts with ethylene oxide to give primary alcohols containing *two more carbon atoms*.

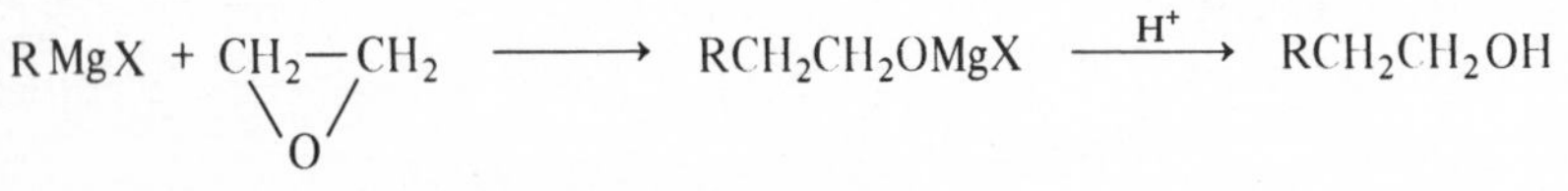

$$RMgX + \underset{\diagdown O \diagup}{CH_2-CH_2} \longrightarrow RCH_2CH_2OMgX \xrightarrow{H^+} RCH_2CH_2OH$$

e. *Hydroboration Methods Alkenes to Alcohols*
   1. Hydroboration-oxidation. (See Sec. 4-3a, 3.)

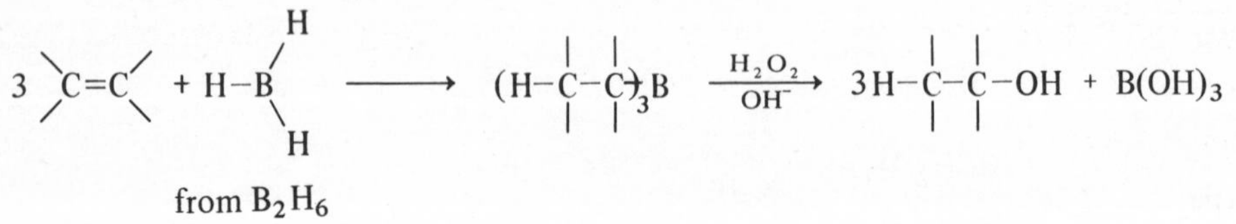

$$3 >C=C< + H-\overset{H}{\underset{H}{B}} \longrightarrow (H-\overset{|}{\underset{|}{C}}-\overset{|}{\underset{|}{C}})_3 B \xrightarrow[OH^-]{H_2O_2} 3H-\overset{|}{\underset{|}{C}}-\overset{|}{\underset{|}{C}}-OH + B(OH)_3$$

from B_2H_6

The cis addition of H—B from diborane across a carbon-carbon double bond takes place via a cyclic four-center reaction:

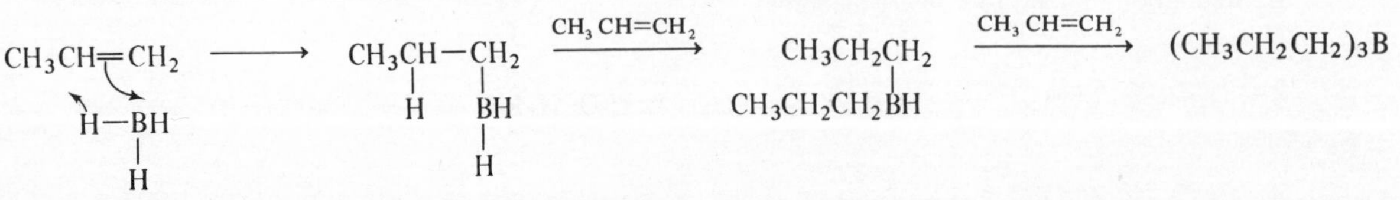

$$CH_3CH=CH_2 \longrightarrow \underset{\underset{H}{\overset{|}{H-BH}}}{CH_3\overset{|}{C}H-\overset{|}{\underset{BH}{C}}H_2} \xrightarrow{CH_3CH=CH_2} \underset{CH_3CH_2CH_2BH}{CH_3CH_2CH_2} \xrightarrow{CH_3CH=CH_2} (CH_3CH_2CH_2)_3B$$

The electron-deficient boron (B is more electropositive than H) attaches itself to the less highly substituted carbon atom of the double bond in a typical Markovnikov addition with the less-positive hydrogen becoming bonded to the other carbon atom of the double bond. Repetition of this kind of addition occurs twice more with two additional molecules of the alkene yielding the trialkylborane, R_3B, as the reaction intermediate. Oxidation of trialkylboranes with alkaline hydrogen peroxide yields the corresponding alcohol, ROH. The mechanism probably involves attack by ^-OOH on boron with migration of the R group from boron to oxygen resulting in a borate ester which is hydrolyzed further.

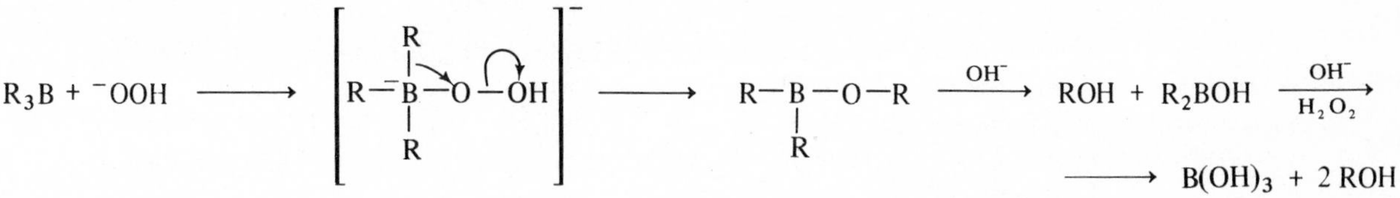

$$R_3B + {}^-OOH \longrightarrow \left[\begin{array}{c} R \\ | \\ R-B-O-OH \\ | \\ R \end{array} \right]^- \longrightarrow \begin{array}{c} R \\ | \\ R-B-O-R \\ | \\ R \end{array} \xrightarrow{OH^-} ROH + R_2BOH \xrightarrow[H_2O_2]{OH^-}$$

$$\longrightarrow B(OH)_3 + 2\,ROH$$

The R group migrates without rearrangement of its carbon skeleton (retention of configuration). The net result of the hydroboration-oxidation sequence is *cis anti-Markovnikov* addition of HOH across the carbon-carbon double bond.

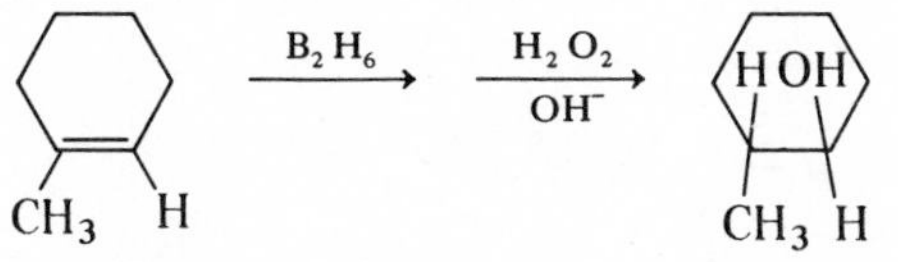

Hydroboration usually proceeds with boron attacking from the sterically less hindered side of the carbon-carbon double bond. Use of a special hydroborating agent allows selective hydroboration as shown below.

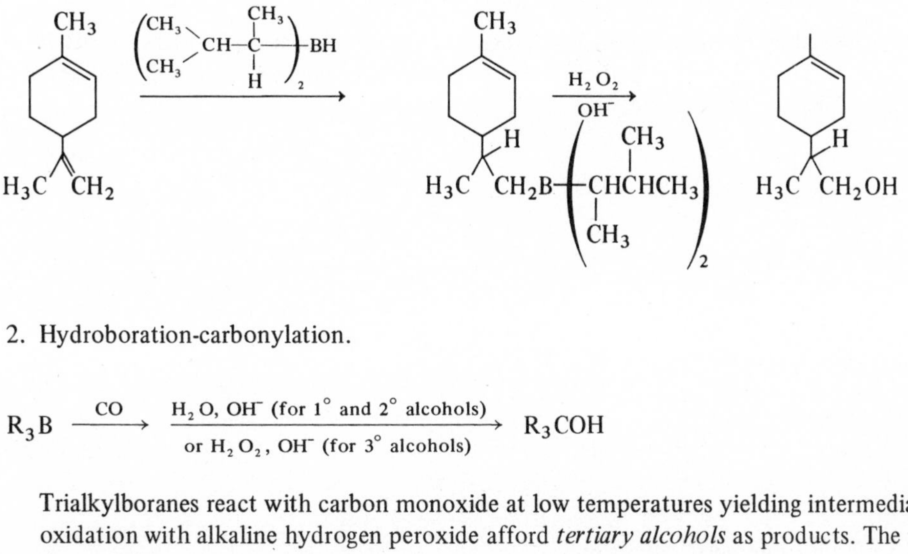

2. Hydroboration-carbonylation.

$$R_3B \xrightarrow{CO} \underset{\text{or } H_2O_2,\ OH^- \text{ (for } 3^\circ \text{ alcohols)}}{\overset{H_2O,\ OH^- \text{ (for } 1^\circ \text{ and } 2^\circ \text{ alcohols)}}{\longrightarrow}} R_3COH$$

Trialkylboranes react with carbon monoxide at low temperatures yielding intermediates which upon oxidation with alkaline hydrogen peroxide afford *tertiary alcohols* as products. The reaction mechanism is shown below:

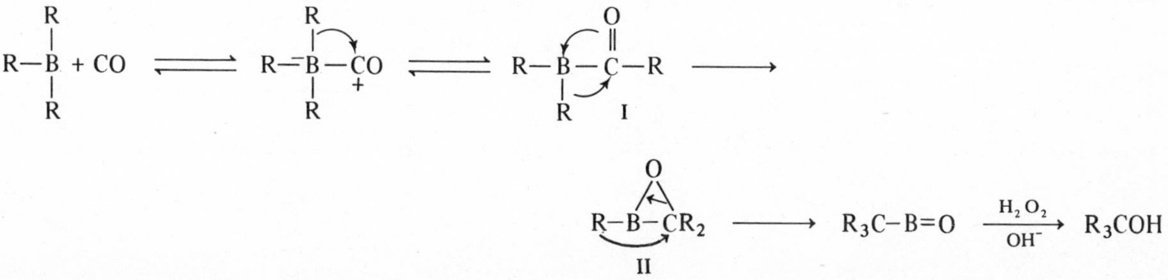

In the presence of lithium borohydride, it is possible to stop the carbonylation reaction at the point where only one alkyl group has transferred from boron to carbon (intermediate I above). Alkaline hydrolysis of I yields the corresponding *primary alcohol*.

$$R_3B + CO \xrightarrow{\text{LiBH}_4} \underset{I}{R_2B\overset{\displaystyle O}{\overset{\|}{C}}R} \xrightarrow[\text{H}_2\text{O}]{\text{OH}^-} RCH_2OH + R_2BOH$$

Use of a special bicyclic hydroborating agent shown below (called 9-BBN for convenience) allows preferential migration of the alkyl group attached to boron rather than the bicyclic group; thus, maximum conversion of an alkene into the corresponding alcohol is permitted (without such special hydroborating agents the best conversion yield would be only 33 percent since only one R group migrates in the reaction with ordinary trialkylboranes).

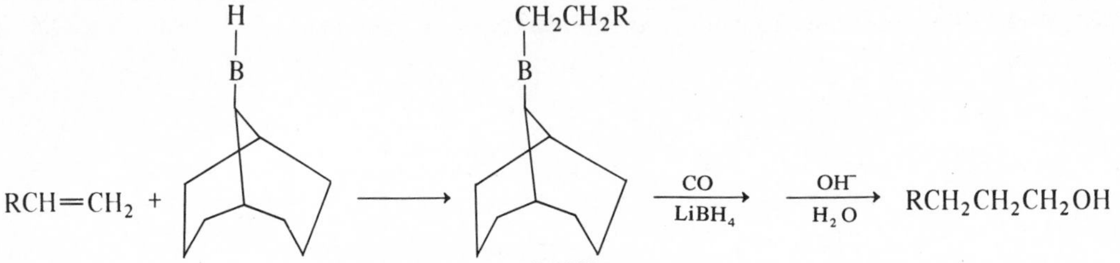

9-Borabicyclo[3.3.1]nonane (also drawn as ⊂ B—H for convenience)

It is also possible to stop the carbonylation reaction at intermediate II by adding an equimolar quantity of water to the reaction mixture. This prevents migration of the third alkyl group and results in the formation of *secondary alcohols* upon alkaline hydrolysis of II.

$$R_3B \xrightarrow[\text{H}_2\text{O}]{\text{CO}} \underset{II}{RB-CR_2 \atop \underset{O}{\diagdown\diagup}} \xrightarrow[\text{H}_2\text{O}]{\text{OH}^-} R_2CHOH + RB(OH)_2$$

Thus, under proper conditions the hydroboration-carbonylation sequence can provide primary, secondary, or tertiary alcohols from alkenes.

f. *Aldol Condensation* (See Sec. 7-3.)

6-3   Preparation of Ethers

a.   Dehydration of alcohols is good for symmetric ethers only

$$ROH \xrightarrow[150°C]{\text{H}_2\text{SO}_4} ROR + H_2O$$

Mechanism:

$$CH_3CH_2OH + H^+ \rightleftharpoons \underset{X}{CH_3CH_2\overset{+}{O}H_2} \rightleftharpoons [CH_3CH_2^+] + H_2O \overset{\text{EtOH}}{\rightleftharpoons}$$

$$\underset{\underset{Y}{H}}{\overset{+}{\text{EtOEt}}} \rightleftharpoons \text{EtOEt} + H^+$$

The proton first reacts with the electron-rich O to give the oxonium compound (X), which on heating gives the carbonium ion ($CH_3CH_2^+$). This ion may react with the electron-rich ethanol to give Y, or it may eliminate a proton to give the olefin. Tertiary carbonium ions eliminate protons most easily, then secondary, and finally primary. Di tertiary ethers cannot be prepared by the above method because of olefin formation.

b. *Williamson Synthesis* $RONa + R'X \longrightarrow ROR' + NaX$. This is a good general method for preparing symmetric and unsymmetric ethers. The R' group in the alkyl halide cannot be tertiary and often secondary alkyl halides react poorly because of the competing $E2$ elimination reaction in the presence of a good base such as RONa. Tertiary alkoxides give slow reactions due to steric hindrance. The Williamson ether synthesis proceeds via an S_N2 mechanism.

c. *Hydroxy Ethers*

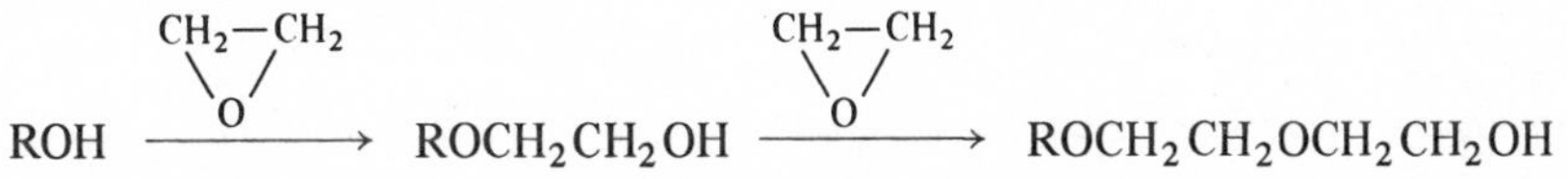

$$ROH \xrightarrow{\underset{O}{\overset{CH_2-CH_2}{\diagdown\diagup}}} ROCH_2CH_2OH \xrightarrow{\underset{O}{\overset{CH_2-CH_2}{\diagdown\diagup}}} ROCH_2CH_2OCH_2CH_2OH$$

d. *Preparation of Epoxides*
   1. Oxidation of olefins with peracetic acid.

$$CH_3CH{=}CHCH_3 + CH_3\underset{\underset{O}{\|}}{C}OOH \longrightarrow CH_3\underset{\underset{O}{\diagdown\diagup}}{CH-CH}CH_3 + CH_3CO_2H$$

   2. Synthesis of ethylene oxide.

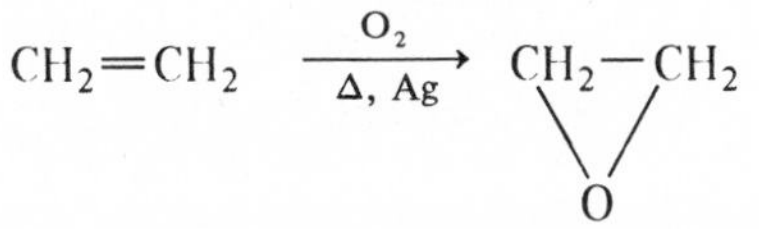

$$CH_2{=}CH_2 \xrightarrow[\Delta,\ Ag]{O_2} \underset{O}{\overset{CH_2-CH_2}{\diagdown\diagup}} \quad \text{(Industrial method)}$$

$$ClCH_2CH_2OH \xrightarrow[\text{NaOH}]{\text{conc.}} \underset{O}{\overset{CH_2-CH_2}{\diagdown\diagup}} + NaCl + H_2O \quad \text{(Laboratory method)}$$

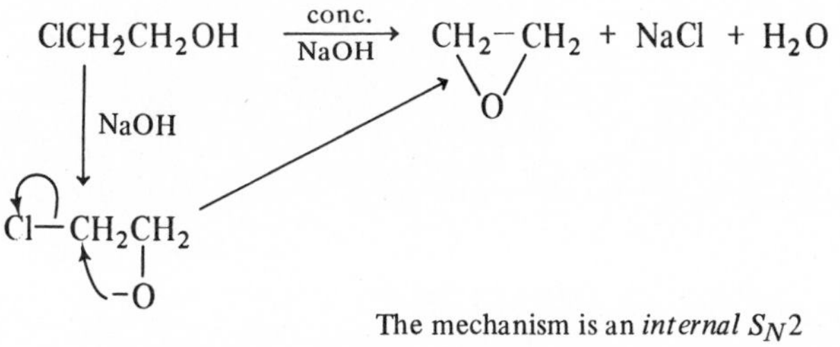

The mechanism is an *internal* S_N2

6-4   Reactions of Alcohols

a. *Salt Formation* Alcohols react directly with active metals such as Na, K, and Ca to form metal salts called alkoxides. These salts are almost completely hydrolyzed by water to the alcohol and the metal hydroxide.

$$2\ CH_3CH_2OH + 2\ K \longrightarrow H_2 + 2\ \underset{\text{Potassium ethoxide}}{CH_3CH_2OK}$$

$$\underset{\text{Sodium isoporpoxide}}{(CH_3)_2CHONa} + H_2O \longrightarrow (CH_3)_2CHOH + NaOH$$

b. *Formation of Alkyl Hlaides*
   1. $ROH + HX \rightleftharpoons RX + HOH$ X = Cl, Br, I. (See Sec. 5-2c for discussion of mechanism.) HI reacts most rapidly and HCl least rapidly. Tertiary alcohols react faster than secondary, and secondary faster than primary.

2. $ROH + SOCl_2$ (thionyl chloride) $\longrightarrow$ $RCl + SO_2 + HCl$. This reaction proceeds by the formation of an alkyl chlorosulfite (A) which dissociates into an ion pair (B) as in an $S_N 1$ mechanism. This "intimate" ion pair pair further decomposes generally by an $S_N i$ mechanism (substitution nucleophilic internal) in which a chloride ion attacks, usually from the same side as the SO_2Cl^- group left originally.

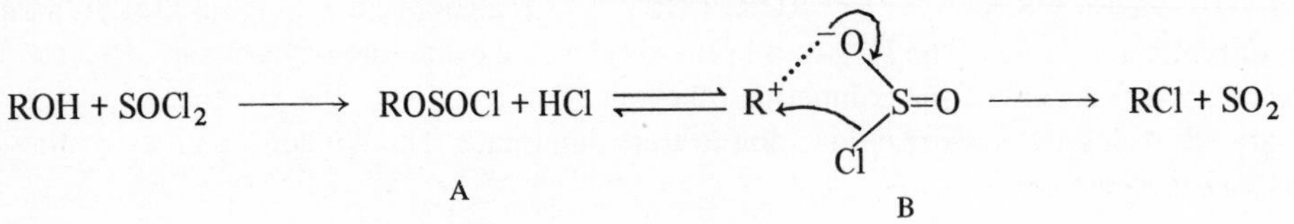

$$ROH + SOCl_2 \longrightarrow ROSOCl + HCl \rightleftharpoons R^+ \overset{O}{\underset{Cl}{\diagdown}} S=O \longrightarrow RCl + SO_2$$

$$\quad\quad\quad\quad\quad\quad\quad\quad\quad\quad\quad\quad\quad A \quad\quad\quad\quad\quad\quad\quad\quad\quad\quad B$$

3. $3ROH + PX_3 \longrightarrow 3RX + P(OH)_3$. X = Cl, Br, I.

c.   *Dehydration* $-\overset{|}{\underset{\underset{H}{|}}{C}}-\overset{|}{\underset{\underset{OH}{|}}{C}}-$ $\xrightarrow[\text{H}_3\text{PO}_4 \text{ or Al}_2\text{O}_3 ,\, \Delta]{\text{H}_2\text{SO}_4}$ $-\overset{|}{C}=\overset{|}{C}-$

Heating alcohols with H_2SO_4 or H_3PO_4 converts them to olefins. However, since the use of strong acids often leads to rearrangement products, a better dehydration method is to pass the alcohol in the vapor phase over Al_2O_3 (see Sec. 4-2c).

d.   *Dehydrogenation*

1. RCH_2OH $\xrightarrow[325°C]{Cu}$ $R\overset{O}{\overset{\|}{C}}H + H_2$

2. $R\overset{OH}{\overset{|}{C}}HR'$ $\xrightarrow[325°C]{Cu}$ $R\overset{O}{\overset{\|}{C}}R' + H_2$

Primary alcohols yield aldehydes; secondary alcohols yield ketones.

e.   *Oxidation*

1. $3RCH_2OH + K_2Cr_2O_7 + 4H_2SO_4 \longrightarrow 3R\overset{O}{\overset{\|}{C}}H + Cr_2(SO_4)_3 + K_2SO_4 + 7H_2O$

2. $3RCH(OH)R' + K_2Cr_2O_7 + 4H_2SO_4 \longrightarrow 3R\overset{O}{\overset{\|}{C}}R' + Cr_2(SO_4)_3 + K_2SO_4 + 7H_2O$

In reaction 1, care must be taken to remove the aldehyde from the reaction mixture as rapidly as it is formed. Otherwise, it will be oxidized to the acid as follows:

$$RCHO \xrightarrow{\text{oxidation}} RCOOH$$

f.   *Ester Formation*

$$ROH + RCOOH \xrightleftharpoons{H^+} RCOOR + H_2O \quad\quad \text{(See Sec. 11-2 for discussion.)}$$

6-5   Reactions of Ethers

The ethers, except for the epoxides, are relatively inert chemically. They do not react with active metals or hot alkali. They do not react with oxidizing agents except under drastic conditions.

a.   *Fission*

$$ROR' + HX \longrightarrow R_2\overset{+}{O}H + X^- \longrightarrow RX + R'OH \xrightarrow{HX} R'X + RX$$

HI and HBr are most effective for fission which proceeds by initial *protonation* of the ether oxygen to form oxonium compounds. This is typical of compounds having unshared electrons (those that contain O, N, or S, or that are unsaturated). As a result ethers are soluble in cold concentrated H_2SO_4. The attack of halide ion on the oxonium intermediate may proceed usually by an S_N1 mechanism if R' is tertiary or by an S_N2 mechanism if R' is primary. Reaction conditions will also be a factor in the mechanism observed.

b.   *Reactions of Epoxides*

$$\underset{\text{O}}{\overset{\text{O}}{\text{RCH—CH}_2}} + \text{Y} \longrightarrow \underset{\overset{|}{\text{O}^-}}{\text{RCHCH}_2\text{Y}} \text{ or } \underset{\overset{|}{\text{Y}}}{\text{RCHCH}_2\text{O}^-} \xrightarrow{\text{H}_2\text{O}} \underset{\overset{|}{\text{OH}}}{\text{RCHCH}_2\text{Y}} \text{ or } \underset{\overset{|}{\text{Y}}}{\text{RCHCH}_2\text{OH}}$$

$$\text{A} \qquad \text{B}$$

$Y = OH^-$, H_2O, HX (I, Br, Cl, F), ROH (see Sec. 6-3c), RO^-, $R^-\overset{+}{Mg}X$ (see Sec. 6-2d), CN^-, NH_3, SH^-, HSO_3^-, R_2CH^- (active methylene compounds – see Chap. 12), and H^- (hydride ion from $LiAlH_4$).

The epoxide ring is highly strained and thus it is readily opened by nucleophilic attack by a variety of reagents at the ring carbon atoms. Depending on the reaction conditions, one of two types of substituted products (A or B) is the major product in the case of unsymmetric epoxides. Under basic or neutral conditions, the reaction is usually an S_N2 reaction and Y attacks at the less highly substituted carbon leading to type A products. Under acidic conditions, the reactive substrate molecule is the protonated epoxide which may react by either an S_N1 or an S_N2 mechanism giving in either case predominantly type B products.

6-6   Organic Synthesis

When dealing with synthetic problems, you will find the OH group to be a flexible joint at which synthetic transformations can be made. The two most general methods for preparation of primary, secondary, and tertiary alcohols are the Grignard method (see Sec. 6-2d) and the hydroboration method (see Sec. 6-2e). Consider the Grignard method first. Recall that OH groups on primary and secondary carbon atoms may be converted (see Sec. 6-4) to carbonyl groups. The Grignard reagent, RMgX, will add across this double bond, the electron-rich R attaching to the electron-poor carbon to form a new C—C bond. Now take a piece of scratch paper and write down the structure for a primary alcohol and then for a secondary alcohol. Remove two H from each of the CHOH units (oxidation or dehydrogenation, see Sec. 6-3). Write down the structures of the first Grignard reagent that comes to mind and add it across each of the carbonyl groups. You have made the halomagnesium salts of two new alcohols. Add H^+, and the ^+MgX units will be replaced to give the alcohols, a secondary and a tertiary (unless the primary alcohol you chose was methanol).

What is the net lesson from the above operation? It is simply that one can attach carbon chains (R groups) to the carbons holding OH groups in primary and secondary alcohols. Run through the following series on scratch paper. Write down methanol and convert it to its corresponding carbonyl derivative (formaldehyde). Now add methylmagnesium bromide to this and then convert the salt so obtained to the free alcohol with dilute acid. What have you done? You have attached a methyl group to the carbon carrying the OH in methanol. Oxidize the alcohol you have just made to the corresponding carbonyl compound and go through the same steps using ethylmagnesium iodide. Now you have a secondary alcohol with methyl and ethyl groups on the CHOH unit. Shall we try for a tertiary? Dehydrogenate the ethylmethylcarbinol to ethyl methyl ketone and add propylmagnesium chloride and then dilute acid. This tertiary alcohol is the end of the line; we cannot convert the COH in a tertiary alcohol to a carbonyl group without getting carbon into the pentavalent state.

Consider next the use of hydroboration in the preparation of alcohols. The first hydroboration method discussed (see 6-2e-1) was the hydroboration-oxidation (abbreviated HB-O hereafter) sequence. Write down the structure of a primary alcohol, of a secondary alcohol, and then of a tertiary alcohol. Recall that the HB-O sequence always starts with an appropriate alkene. Next, for your alcohols write down the most highly substituted alkene that is possible from removal of the OH group and a hydrogen on a carbon atom adjacent to the carbon bearing the OH (this is equivalent to dehydration of your alcohol to yield the Saytzeff alkene, see Sec. 4-2c). Look at the carbon atom of the carbon-carbon double bond which originally had the OH group attached to it. If this carbon atom has more hydrogen atoms attached to it than the other olefinic carbon atom, then the HB-O reaction *will* convert this alkene to your desired alcohol. If it turns out that the other olefinic carbon atom has the same number or more hydrogens attached to it, then your alkene *will not* be converted *solely* to your desired alcohol by a HB-O reaction (except in the case where the alkene is ethylene or a symmetrically substituted alkene, RCH=CHR or $R_2C=CR_2$). Why is this so? Recall that in hydroboration of an alkene, the boron atom becomes attached to the less highly substituted olefinic carbon and that in the subsequent oxidation the OH group, in effect, takes the place of this boron atom in the final alcohol. Thus, for example, if you had chosen 1-pentanol, 3-pentanol, and 2-methyl-2-butanol as your three alcohols, only 1-pentanol could be obtained as the major product from a HB-O reaction (starting with 1-pentene). A mixture of approximately equal parts of 3-pentanol and 2-pentanol would be obtained from a HB-O reaction on 2-pentene. Very little if any 2-methyl-2-butanol would be obtained from HB-O of 2-methyl-2-butene since 3-methyl-2-butanol would be the major product.

Next consider the hydroboration-carbonylation (abbreviated HB-CO hereafter) method (see Sec. 6-2e-2). Using your same three chosen alcohols, look at the alkyl group(s) attached to the carbinol group. In the case of a primary alcohol it can always be prepared by a HB-CO reaction (in the presence of $LiBH_4$) starting with the 1-alkene compound with one less carbon than the desired primary alcohol. For a symmetric secondary alcohol, if both the alkyl groups attached to the carbinol group are either RCH_2CH_2-, R_2CHCH_2- (R groups may be different), or $RCH_2CH(R)-$ (R groups must be the same), then the alkene derived from this alkyl group can be used in a HB-CO reaction (equimolar H_2O required). Unsymmetric secondary alcohols may not be made by a simple HB-CO reaction except when HB-CO can yield the corresponding unsymmetric ketone (see Sec. 7-2g) which can be reduced to the desired secondary alcohol. The HB-CO method will only work for making tertiary alcohols if all three alkyl groups attached to the carbinol group are one of the three types mentioned above for symmetric secondary alcohols.

In summary of the hydroboration methods (HB-O and HB-CO), a variety of primary and secondary alcohols can be made by either one or the other or by both methods. However, tertiary alcohols can be prepared directly from alkenes only by the HB-CO method (except for symmetric alkenes, e.g., $R_2C = CR_2$, which yield 3° alcohols by the HB-O method) and then only if the tertiary alcohol is of the type mentioned above.

The beginner should attempt to design the synthesis of complex molecules by the backing-up process. Write the structure for 1-pentanol, 3-pentanol, and 3-methyl-3-hexanol across the top of a piece of paper. The problem is to build these molecules up from simpler structures via the Grignard or hydroboration methods just discussed. In considering 1-pentanol, note that it is a primary alcohol and that the R group attached to the CH_2OH unit is butyl. This could be made by adding butylmagnesium bromide to a one-carbon carbonyl compound (that is, $H_2C=O$). This process increases the length of the carbon chain by a CH_2 unit. Butylmagnesium bromide is derived from butanol by first treating it with HBr and then with Mg. A similar approach to 1-pentanol could be made from 1-butene via a HB-CO sequence (in the presence of $LiBH_4$). To obtain 1-butene we would dehydrate 1-butanol. If we wish to make 1-butanol from simpler materials, this can be done by adding ethylene oxide to ethylmagnesium iodide (see Sec. 5-2). Now you should be able to draw a series of equations from the bottom of the page up to 1-pentanol, showing its preparation from methanol and ethanol.

Try 3-pentanol. Here we have two ethyl groups attached to the CHOH unit. This secondary alcohol can be made

by adding C_2H_5MgX to a carbonyl group with an ethyl unit attached to it: $(CH_3CH_2\overset{\overset{\displaystyle O}{\|}}{C}H)$. The carbonyl compound is obtained in the usual way from the three-carbon primary alcohol. Show how this alcohol can be built up from methanol and ethylene oxide. Methanol is made commercially by the catalytic hydrogenation of carbon monoxide, which is made from coke by the water-gas reaction, $C + H_2O \rightarrow CO + H_2$. Complete the series of equations from the bottom of the page to the top showing how 3-pentanol can be made from coal and water. Alternatively, we might consider the hydroboration approach to 3-pentanol. Starting from 2-pentene, a HB-O reaction would yield about equal amounts of 2- and 3-pentanol. When possible, you should avoid reactions where such mixtures are obtained. Consider the HB-CO approach. Since 3-pentanol is a symmetric dialkylcarbinol, this method will work quite well. Treatment of ethylene with diborane followed by carbon monoxide plus water and then basic hydrolysis will give 3-pentanol. Overall, this represents a simpler synthesis than the Grignard method. Ethylene can be obtained from the reduction of acetylene which is available via calcium carbide.

The synthesis of 3-methyl-3-hexanol from say coal, air, and water may appear very complicated, but once you learn to make the proper association the problem is much easier. Of course, you do not usually think of 3-methyl-3-hexanol at the mention of water, or even coal. Using the backing-up process you might first make the association that a tertiary alcohol can be prepared from an appropriate alkene via a hydroboration method. However, brief consideration of this method would lead you to the conclusion that no alkene is suitable for a one-step synthesis of the desired alcohol since it is not a symmetric trialkylcarbinol. Another logical association would be that the alcohol could be made from the action of the proper Grignard reagent on a carbonyl compound holding the correct two alkyl groups. In the above example, there are three different sets we could use:

Methyl-$\overset{\overset{\displaystyle O}{\|}}{C}$-Propyl + EtMgX   Ethyl-$\overset{\overset{\displaystyle O}{\|}}{C}$-Propyl + MeMgX   Methyl-$\overset{\overset{\displaystyle O}{\|}}{C}$-Ethyl + PropylMgX

(I)   (II)   (III)

Complete these three equations and note how the same product is obtained in each instance. Now, in our backing-up process, we are faced with the job of getting these reactants. How are ketones made? At this point the only answer is oxidation or dehydrogenation of secondary alcohols. How can secondary alcohols be built up? Except for methylethylcarbinol which is easily available via a HB-O reaction on 2-butene, all the necessary alcohols for the above can be made up by adding an R group via the Grignard reaction to the proper carbonyl compound $R\overset{\text{O}}{\underset{\|}{C}}H$. How is this

type of carbonyl compound made? Oxidation or dehydrogenation of a primary alcohol. Now under set I, write methylpropylcarbinol. Show how this can be made from CH_3CHO and propylmagnesium iodide or from $CH_3CH_2CH_2CHO$ and the methyl Grignard reagent. Now work back on sets I, II, and III. In set I we back up to the point where we need either methanol and 1-butanol to get the secondary alcohol or one molecule of ethanol and one molecule of 1-propanol. In set II we need either ethanol and 1-butanol or 1-propanol, and in set III we need methanol and 1-propanol or ethanol. Now in set I either we can convert 1-butanol to the carbonyl compound and then add the methyl Grignard reagent to get the secondary alcohol, or we can convert ethanol to the carbonyl compound and add the propyl Grignard reagent. Observe that there are two different ways to make each of the three secondary alcohols from which the corresponding ketones I, II, and III are made. We have now backed up to the point where we need the four primary alcohols, methyl, ethyl, propyl, and butyl. At this point the molecules we need are rather simple, and we can start thinking in terms of our starting material, coal. From coke we can get carbon monoxide, which can be hydrogenated to give methanol, or we can get acetylene via calcium carbide. The following associations should be made with acetylene: (1) triple bonds can be converted to double bonds, which can be converted to alcohols, which can be converted to alkyl halides; (2) acetylene can be hydrated to CH_3CHO, which can be reduced to ethanol, which can be converted to ethyl bromide; (3) acetylene can be converted into a Grignard reagent (see Sec. 6-4) which can be used to attach a two-carbon piece to a carbonyl group. We now have methyl and ethyl alcohols. Acetylene can be converted to ethylene, which can be made into ethylene oxide (see Sec. 5-2). The methyl Grignard reagent and ethylene oxide lead to 1-propanol, and the ethyl Grignard reagent and ethylene oxide to 1-butanol. Now fill in the equations from coke at the bottom of your page to the tertiary carbinol at the top. Of course, in practice one would not start with coke. Which of the three methods you might choose would depend on what alcohols and alkyl halides were around the laboratory at the time of synthesis.

Consider the synthesis of tricyclohexylcarbinol. Since this is a symmetric tertiary carbinol the proper association you need to make is that such alcohols can be made from the alkene derived from the alkyl group — in this case cyclohexene — via a HB-CO reaction. Alternatively, the backing-up process for a Grignard approach should lead

you to the decision that you would need dicyclohexyl ketone ⟨◯⟩—$\overset{\text{O}}{\underset{\|}{C}}$—⟨◯⟩ and cyclohexylmagnesium iodide.

Outline all of the steps necessary to prepare these compounds from cyclohexene using the Grignard method. There is no doubt that the HB-CO approach is the better one for the synthesis of tricyclohexylcarbinol.

Try a more difficult problem: the synthesis of 2-methylpentane. Write the molecule across the top of a blank piece of paper and start thinking about how hydrocarbons are made. The Wurtz reaction, hydrogenation of olefins, reduction of alkyl halides, or treating the Grignard reagent with an active hydrogen compound should be associated with the preparation of hydrocarbons. The Wurtz method is out because we are not dealing with a symmetric hydrocarbon, but any of the other methods could be used. Consider the use of an olefin. Remember that olefins are made from alcohols. You should be able to write five different alcohols across the middle of your paper, any one of which could be dehydrated to the olefin, which could in turn be hydrogenated to 2-methylpentane. Thus, the problem of getting the hydrocarbon really boils down to an alcohol synthesis. You can make any of these from coke and water. A hint might be in order for 2-methyl-1-pentanol. This is a primary alcohol and can be made from the action of a secondary Grignard

reagent on $\overset{\text{H}}{\underset{\text{H}}{>}}C{=}O$. Draw out syntheses for all five alcohols. A method you might overlook is the addition of the

allyl Grignard reagent to $CH_3\overset{\text{O}}{\underset{\|}{C}}H_3$. In this instance, you would have a diene to hydrogenate at the end. Of course, any of the above five alcohols could be converted to the corresponding alkyl halide, and this could be converted to 2-methylpentane by reduction or by first forming the Grignard reagent and treating it with an active hydrogen compound — say, water.

How would you make $CH_3\overset{\text{CH}_3\text{O}}{\underset{\big|}{C}}\overset{\text{C}{\equiv}\text{N}}{\underset{\big|}{C}}H_2$? Backing up one step to the alcohol necessary to react with CH_3Br to make the methyl ether linkage would lead you to $CH_3CH(OH)CH_2C{\equiv}N$. Treatment of this alcohol with sodium metal and then CH_3Br would give us the desired compound. Backing up from the alcohol, you should associate $C \equiv N$ with displacement reactions. Thus, if you treat the right bromide (1-bromo-2-propanol) with $NaC \equiv N$, you could get $CH_3CH(OH)CH_2C \equiv N$. How to get the bromopropanol? You might think of starting with the diol and converting this with just the right amount of HBr, but this would give a bad mixture of three products–some dibromide, some 1-bromo-2-propanol, and some 2-bromo-1-propanol. Next, you might think of starting with a dihalide. First add the OH via a displacement reaction with ^-OH and then introduce the $C \equiv N$ in the same way. Here again you would run into the problem of getting mixtures. The correct association to make is as follows: $C \equiv N$ comes from halide; looking at the structure you then see HO Br This should remind you of adding HOBr to an olefin, and the problem is solved.

$$\underset{CH_3CHCH_2}{\overset{HO \;\; Br}{| \;\; |}}$$

   In carrying out synthetic operations, the organic chemist learns to associate certain functional groups with basic groups from which they may be derived or into which they may be converted. Among the many reactions of the OH group there are three that are very important for synthetic purposes: dehydration to the olefin, conversion to the alkyl halide, and dehydrogenation to the carbonyl group.

$$\underset{-CH=CH-}{\qquad} \quad \underset{\underset{X}{|}}{-CHCH_2-} \quad \underset{\overset{\|}{O}}{-CCH_2-}$$

   When an olefin, alkyl halide, or carbonyl compound is flashed before you, one immediate visual association should be that any of these could probably be made from the corresponding alcohol. The OH group is a triple-entry point into these three fields of chemistry.

   What functional groups do you associate with an alkene? Write down five reactions of the olefin group. All five of these compounds can be made from the appropriate alcohol via the olefin. For example, addition of bromine to an olefin gives a dibromide. Treating this with alcoholic KOH yields an acetylene. Hence, acetylenes can be made from alcohols. Building a knowledge of organic chemistry is in large part simply learning to make the proper associations among a great many different types of molecules.

   Note well that in synthetic organic problems of the above type it is impossible to accomplish anything unless you have committed to memory all the basic reactions. It is extremely difficult to make $CH_3CH(OH)CH_2Br$ unless you know about the reaction of olefins with HOBr. There are so many reactions that must be learned, even in an elementary course, that the beginner must be constantly reviewing reactions already learned or he will simply forget them. The easiest way to do this is to work large numbers of problems that involve reactions previously learned as well as new ones.

6-7   Qualitative Tests for Alcohols and Ethers

   a.   *For Alcohols*
   1. Alcohols of low molecular weight (fewer than nine carbon atoms) react rapidly enough with sodium so that one may see bubbles of hydrogen evolve. Other compounds having active hydrogens, such as acetylenes and acids, also give this test.
   2. The hydroxyl group of alcohols reacts rapidly with acetyl chloride, resulting in the evolution of heat and HCl gas.

$$ROH + CH_3\overset{\overset{O}{\|}}{C}Cl \longrightarrow CH_3\overset{\overset{O}{\|}}{C}OR + HCl$$

   3. Lucas test. The Lucas reagent, $HCl + ZnCl_2$, can be used to distinguish between primary, secondary, and tertiary alcohols of *fewer than six* carbon atoms. These small alcohols are soluble in this reagent. The tertiary alcohols react immediately to give the alkyl chloride that is insoluble in the reagent and that separates as a

less-dense oily layer. Secondary alcohols react in 5 to 10 min. Primary alcohols do not react at room temperatures and no oily layer separates. (Exceptions: allyl alcohol and its derivatives behave like tertiary alcohols.)

4. The OH group reacts with CH_3MgX to give methane. Hence the Zerewitinoff determination may be applied to alcohols (see Sec. 1-6e).

5. Primary and secondary alcohols when treated with chromic anhydride, CrO_3, in aqueous sulfuric acid give blue-green opaque solutions within a couple of seconds. Tertiary alcohols do not give this test.

6. Iodoform test. (See Sec. 7-3e.) Alcohols of the type, $RCH(OH)CH_3$, or ethyl alcohol give a yellow precipitate of CHI_3 (iodoform) when treated with iodine and sodium hydroxide.

7. Simple alcohols of fewer than five carbon atoms are soluble in water.

b. *For Ethers* The ether linkage is a very unreactive functional group. Ethers resemble hydrocarbons in many ways but they may be distinguished from alkanes and alkyl halides by adding cold concentrated H_2SO_4. Ethers are soluble in this reagent due to formation of the oxonium salt, while alkanes and alkyl halides are insoluble.

Problems

6-1 Write structural formulas for all alcohols and ethers $C_5H_{12}O$. Name each according to the IUPAC system. In addition, name each of the alcohols according to the carbinol system and each of the ethers as an alkyl ether.

6-2 Name the following according to the IUPAC system and classify each as primary, secondary, or tertiary.

a. $CH_3CH_2CH(OH)CH_2CH_2CH_2CH_3$

b. $(CH_3)_2C(OH)CH(CH_3)_2$

c. $(CH_3)_2CHCH_2CH(CH_3)CH_2OH$

d. $HOCH_2CH=CHCH=CHCH_2OH$

e. $HC\equiv CCH_2OH$

f. $(CH_3)_2CHCH_2CHCH_2OH$
$\quad\quad\quad\quad\quad\quad\quad |$
$\quad\quad\quad\quad\quad\quad CH(CH_3)_2$

g. H_3C- ⬡ $-OH$

h. OH △ OH
$\quad H \quad\quad H$

i. $CH_2=C(CH_3)CHC(CH_3)_2$
$\quad\quad\quad\quad\quad\quad |\quad\quad |$
$\quad\quad\quad\quad\quad Cl \quad OH$

j. $CH_3C\equiv CCH=CHCH_2OH$

6-3 Suggest a suitable name for each of the following:

a. $CH_3OCH_2CH_2OH$

b. $CH_2=CHOCH_2CH=CH_2$

c. CH_3O- △

d. CH_3O ☐ OCH_2CH_3

e. $(CH_3)_2CHOCH_2CH_2CH_2OH$

f. $CH_3OCH_2CH=CHCH_3$

g. $HOCH_2CH_2OH$

h. $CH_3CH(OH)CH_2OH$

i. $(CH_3)_2CHCH_2CH_2OCH_3$

6-4 Write structures for each of the following:

a. 1,4-Cyclohexanediol

b. 2-Cyclopropylcyclopropanol

c. 3,5-Cycloheptadienol

d. Dicyclohexylcarbinol

e. Cyclobutyl isopropyl ether

f. 1,3,5-Trimethoxycyclohexane

6-5 Write balanced equations for each of the following:

a. Ethanol + thionyl chloride

b. Sodium ethoxide + H_2O

c. Calcium + methanol

d. $CH_2=CHCH_2OH$ + conc. HCl

e. $CH_2=CHCH_2OH$ + anhydrous HCl

f. $LiAlH_4$ + acetone

g. 3-Butyn-1-ol + CH_3MgI

h. $LiAlH_4$ + acetic acid

i. Potassium + *tert*-amyl alcohol

j. Isobutylene + B_2H_6

k. $(CH_3)_2CHMgI$ + ethylene oxide

l. Diethyl ether + HI, Δ

m. Propylene + CH_3CO_3H

n. $(CH_3)_2CHO^-Na^+$ + propylbromide

o. Ethanol + concd H_2SO_4, 150°C

p. 1,4-Cyclohexanediol + PCl_3

6-6   How would you expect cyclopentanol to react with each of the following:
       a. K   *b.* Br$_2$ in CCl$_4$ *c.* CH$_3$MgI   *d.* Cu, 350°C

6-7   Arrange CH$_3$CH$_3$, CH$_3$CH$_2$OH, CH$_3$COOH, CH$_3$C≡CH, CH$_3$OCH$_3$ in order of increasing acid strength.

6-8   Show how you would make the following conversions using the compounds indicated plus ethanol and methanol.

a. 2-Propanol to 1-propanol	*b.* Propene to acetone
c. Ethanol to 1-propanol	*d.* Ethanol to 1-butanol
e. 1-Butanol to 2-methyl-2-butanol	*f.* CH$_3$CH$_2$CH$_2$COOH to 2-iodobutane
g. Ethanol to methyl *sec*-butyl ether	*h.* Vinyl chloride to 1,1,2-trimethoxyethane
i. Propene to 1,2-epoxypentane	*j.* Propene to 2,3,4-trimethylpentane
k. (CH$_3$)$_3$COH to 2-methyl-1,2-propanediol	*l.* Ethylene oxide to 2-bromo-1-butene

6-9   What alkene products (major or most stable) would you expect from passing each of the following over hot Al$_2$O$_3$?

a. 2-Propanol	*b.* 1-Butanol
c. 3-Hexanol	*d.* 1,4-Butanediol
e. 2,5-Dimethyl-3-hexanol	*f.* 1,3-Cyclohexanediol

6-10   A substance, C$_5$H$_{12}$O, evolved hydrogen with Na and gave a Lucas test after 5 min standing. Passing the substance over hot alumina converted it to an olefin, which on oxidation gave two products, one a neutral substance and the other an acid of NE 60 ± 1. Suggest a structure for C$_5$H$_{12}$O.

6-11   Compound A gave an immediate Lucas test. When 0.88 g of A was treated with excess methylmagnesium iodide, 224 ml of methane (measured at STP) was evolved. Assuming A to contain only one atom of oxygen per molecule, suggest a structure for A.

6-12   Compound X, C$_5$H$_8$O, gave a negative Lucas test, but it gave a precipitate when treated with ammoniacal silver nitrate. With excess methylmagnesium bromide, 0.42 g of X gave 224 ml of methane at STP. Treatment of X with H$_2$ and Pt, followed by boiling in excess HI, gave *n*-pentane. Suggest a structure for X.

6-13   Compound A, C$_4$H$_8$O, on catalytic hydrogenation gave B, C$_4$H$_{10}$O. Compound A reduced permanganate solution and decolorized a solution of bromine in carbon tetrachloride. Compound B, when refluxed with HBr, yielded C. C was converted to a Grignard reagent, which, when treated with water, gave a gas weighing 1.5 g per 1.12 liters at STP. Suggest a structure for A.

6-14   Compound A, C$_4$H$_{10}$O$_2$, was found to react with sodium to give C$_4$H$_9$O$_2$Na. When this was treated with water, it was converted to A. Boiling A with HBr converted it to a mixture of two bromine-containing compounds. Treating this mixture with KOH yielded two gaseous hydrocarbons. The gas, when treated with ammoniacal cuprous chloride, gave a precipitate, and the remaining gas was found to have a vapor density of 1.25 g/liter. Suggest a structure for A.

6-15   The Zeisel method for the determination of methoxyl groups is conducted by boiling a weighed amount of unknown compound with excess HI and distilling the volatile methyl iodide as it is formed into an alcoholic silver nitrate solution. The resulting precipitate of silver iodide is weighed, and, from the equivalents of AgI formed, the equivalents of CH$_3$O— or the equivalent weight of the unknown may be calculated.

       A sample of 1.34 g of compound A gave 7.04 g of AgI by the Zeisel method. When 0.067 g of A was dissolved in 2 g of camphor, the melting point of the camphor was lowered 10°. How many CH$_3$O— groups are there in A? Suggest a formula for A.

6-16   Point out the errors in the following proposed syntheses:

a. CH$_2$=CHOH $\xrightarrow[\text{(A)}]{\text{HCl}}$ CH$_2$=CHCl $\xrightarrow[\text{(B)}]{\text{Na, }\Delta}$ CH$_2$=CHCH=CH$_2$

b. CH$_2$=CHCH$_2$OH $\xrightarrow[\text{(A)}]{\text{HCl}}$ CH$_2$=CHCH$_2$Cl $\xrightarrow[\text{(B)}]{\text{Pt, H}_2}$ CH$_3$CH$_2$CH$_2$Cl

c. CH$_2$=CHCOOH $\xrightarrow[\text{(A)}]{\text{Pt, H}_2}$ CH$_2$=CHCH$_2$OH $\xrightarrow[\text{(B)}]{\text{HCl}}$ CH$_3$CHClCH$_2$Cl

d. CH$_3$CH$_2$CH$_2$OH $\xrightarrow[\text{(A)}]{\text{Cl}_2}$ ClCH$_2$CH$_2$CH$_2$OH $\xrightarrow[\text{(B)}]{\text{Mg}}$ $\xrightarrow{\text{CH}_2\text{O}}$ $\xrightarrow{\text{H}^+}$ HO(CH$_2$)$_4$OH

e. ICH$_2$CH$_2$I + 2CH$_3$ONa $\longrightarrow$ CH$_3$OCH$_2$CH$_2$OCH$_3$ + 2NaI

f. (CH$_3$)$_3$CBr + CH$_3$ONa $\longrightarrow$ (CH$_3$)$_3$COCH$_3$ + NaBr

g. $(CH_3)_3COH \xrightarrow[150°C]{H_2SO_4} (CH_3)_3COC(CH_3)_3 + H_2O$

h. $CH_3OH + CH_3CH_2OH \xrightarrow[150°C]{H_2SO_4} CH_3OCH_2CH_3 + H_2O$

i. $CH_3C{=}CH_2 \xrightarrow[(A)]{B_2H_6} \left(CH_3\underset{CH_3}{\overset{CH_3}{\underset{|}{\overset{|}{C}}}}{-} \right)_3 B \xrightarrow[OH^-]{H_2O_2} (CH_3)_3COH$

(with CH_3 below the first carbon)

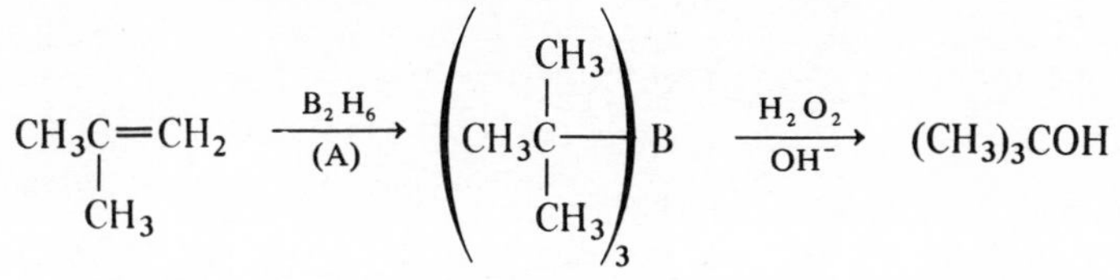

j. $(CH_3)_2C{=}CHCH_3 \xrightarrow[(A)]{B_2H_6} \left(CH_3CH\underset{CH_3}{\overset{CH_3}{\underset{|}{\overset{|}{CH}}}}{-} \right)_3 B \xrightarrow[LiBH_4]{CO} \xrightarrow[(B)]{H_2O,\ OH^-}$

$(CH_3)_2CHCH(CH_3)\underset{\overset{|}{OH}}{CHCH(CH_3)}CH(CH_3)_2$

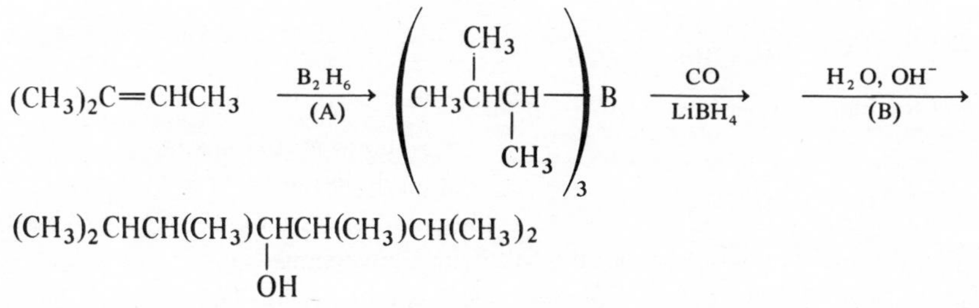

k. $CH_3CH{=}CH_2 \xrightarrow[\underset{(A)}{peroxide}]{HI} CH_3CH_2CH_2I \xrightarrow[(B)]{Pt,\ H_2} CH_3CH_2CH_3 + HI$

l. $(CH_3)_3CBr \xrightarrow[(A)]{HC{\equiv}CNa} (CH_3)_3C{-}C{\equiv}CH \xrightarrow[(B)]{H_2O,\ Hg^{2+}} (CH_3)_3C\overset{\overset{O}{\|}}{C}CH_2CH$

m. $CH_3CH_2OH +$ (cyclohexanol with OH) $\xrightarrow[(A)]{H_2SO_4,\ \Delta}$ (cyclohexane with OEt) $\xrightarrow[(B)]{Cl_2}$ (cyclohexane with OCH$_2$CH$_2$Cl)

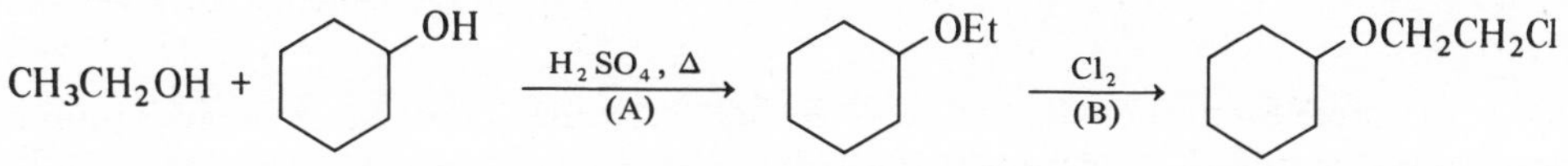

n. (methylcyclohexanol, H$_3$C and OH) $\xrightarrow[(A)]{HBr}$ (methyl bromocyclohexane, H$_3$C and Br) $\xrightarrow[(B)]{NaCN}$ (methyl cyanocyclohexane, H$_3$C and CN)

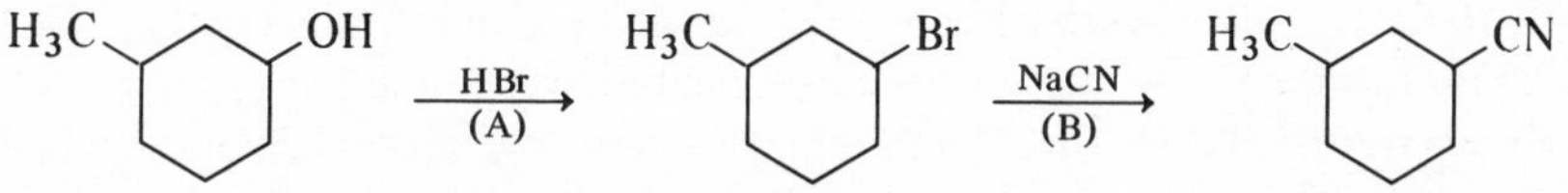

6-17 Suggest a simple test with a readily observable change to distinguish between each of the following:
 a. Hexane and 1-hexanol
 b. 1-Hexyne and 1-butanol
 c. 2-Propanol and 1-propanol
 d. Methanol and 1-hexanol
 e. Ethyl propyl and ethyl allyl ethers
 f. Dibutyl ether and n-octane
 g. 1-Butanol and ethoxyethane
 h. 1-Butyne and vinyl ethyl ether
 i. Butyl iodide and butyl ethyl ether
 j. $(CH_3CH_2)_2 S$ and 1-bromobutane
 k. 2-Propanol and 2-octanol
 l. $FCH_2CH_2OCH_3$ and $ClCH_2CH_2OCH_3$

6-18 Complete the indicated reactions where A = addition reaction, E = elimination reaction, and S = substitution reaction.

 a. 1-Propanol $\xrightarrow{E}$ propene $\xrightarrow{A}$ 1-bromopropane $\xrightarrow{S}$ 1-methoxypropane

 b. 1-Methoxypentane $\xrightarrow{S}$ pentyl bromide $\xrightarrow{E}$ $\xrightarrow{A}$ $\xrightarrow{E}$ 1-pentyne

 c. Ethene $\xrightarrow{A}$ 2-chloroethanol $\xrightarrow[\Delta]{H_2SO_4}$ $\xrightarrow{E}$ divinyl ether

 d. Iodocyclohexane $\xrightarrow{E}$ $\xrightarrow[acid]{peracetic}$ epoxycyclohexane

 e. Ethylene bromide $\xrightarrow{E}$ $\xrightarrow{CH_3MgI}$ $\xrightarrow{A}$ $\xrightarrow{H^+}$ $(CH_3)_2\underset{\overset{|}{OH}}{C}{-}C{\equiv}CH$

 f. Propyne $\xrightarrow{A}$ $\xrightarrow{S}$ $(CH_3)_2C(OCH_3)_2$

73

g. Cyclopentene $\xrightarrow[\text{acid}]{\text{peracetic}}$ $\xrightarrow{S}$ $\xrightarrow{H^+}$ $\xrightarrow{A}$ 2-vinyl-1-cyclopentanol

h. Acetylene $\xrightarrow{S}$ $\xrightarrow{A}$ $\xrightarrow{H^+}$ $(CH_3)_2\underset{\underset{OH}{|}}{C}C\equiv C\underset{\underset{OH}{|}}{C}(CH_3)_2$

i. Cyclohexanol $\xrightarrow{E}$ $\xrightarrow{A}$ $\xrightarrow{S}$ [cyclohexane ring with CN and OH substituents]

j. β-Butylene oxide $\xrightarrow{S}$ $\xrightarrow{H^+}$ $\xrightarrow{E}$ trimethylethylene

k. Propene $\xrightarrow{S}$ $\xrightarrow{A}$ $\xrightarrow{S}$ glycerol

6-19 Given propene, acetylene, and methanol, show how you would make each of the following:
    a. 1,2-Epoxypentane   b. 1,2,5,6-Tetramethoxyhexane
    c. $HC\equiv CCH_2CH_2OH$ d. 2,2-Dimethylbutane
    e. $CH_3CH(OCH_3)CH_2CN$ f. 3,5-Dimethyl-4-sec-butyl-4-methoxyheptane

6-20 Given D_2O, D_2, and $LiAlD_4$, show how you would make the following conversions:
    a. Ethanol to CHD_2CHD_2 b. Acetylene to CH_3CHDOH
    c. CH_4 to CD_4 d. Acetone to $(CH_3)_2CDOD$
    e. Calcium carbide to CD_3CD_3 f. Acetic acid to CH_3CD_2OH

6-21 Write equations showing how an ether might undergo fission by an S_N1 or an S_N2 mechanism. Indicate which would be the more likely mechanism in the cleavage of each of the following:
    a. Dimethyl ether   b. Diallyl ether   c. Diisopropyl ether

    d. Dipentyl ether   e. [benzene ring]$-CH_2OCH_2-$[benzene ring]   f. Di-tert-butyl ether

6-22 Draw the electronic structure for the hypothetical vinyl alcohol. Show the shift of electrons and a proton that must occur for this to rearrange to acetaldehyde ($CH_3CH=O$).

6-23 When allyl alcohol ($CH_2=CHCH_2OH$) is treated with a strong acid, the allyl cation is formed ($CH_2=CHCH_2^+$). Draw two resonance structures for this ion. Draw two for the corresponding anion. Assuming a cation of the above is an intermediate when one treats $CH_3CH=CHCH_2OH$ with HCl, what two alkenyl chlorides would one expect as products? Why is this difficulty not encountered with allyl alcohol?

6-24 Which one of the following three resonance forms for the carbonyl group would be least important? Why?

$>C=O$ $>\overset{+}{C}-O^-$ $>\overset{-}{C}-O^+$

6-25 Recall how ethylene oxide reacts with the Grignard reagent (see Sec. 5-2). What products would you expect from the reaction of ethylene oxide with each of the following:
    a. H_2O b. $HC\equiv CMgX$ c. CH_3OH d. NH_3
    e. CH_3NH_2 f. $HC\equiv N$ g. H_2S h. HBr

6-26 What would be the main product to be expected from isobutylene oxide under the following conditions?
    a. S_N1 reaction with HCl   b. S_N2 reaction with CH_3ONa; then H^+
    c. S_N1 reaction with HCN

6-27 When 0.0120 g of compound X was dissolved in 0.5 g of camphor, the melting point of the camphor was lowered 8°. Combustion analysis of X showed that it contained 70.6% C and 13.8% H. Compound X reacted with acetyl chloride and evolved hydrogen with sodium. When compound X was passed over Al_2O_3 at 350°C, Y was obtained. When Y was treated with alkaline permanganate, two compounds were obtained. One was a neutral compound and the other was an acid with an NE of 74 ± 1. Suggest a structure for X.

6-28 Draw the indicated number of significant resonance forms for each of the following:

    a. $(CH_3)_2BCH=CH_2$, 2   b. $CH_3OCH=CH\overset{\overset{\displaystyle O}{||}}{C}H$, 4

    c. $(CH_3)_2NCH=CHNO_2$, 4   d. $HO-$[benzene ring]$-\underset{\underset{O}{||}}{C}H$, 8

6-29   For the following reactions predict the products expected along with their stereochemistry.

a. 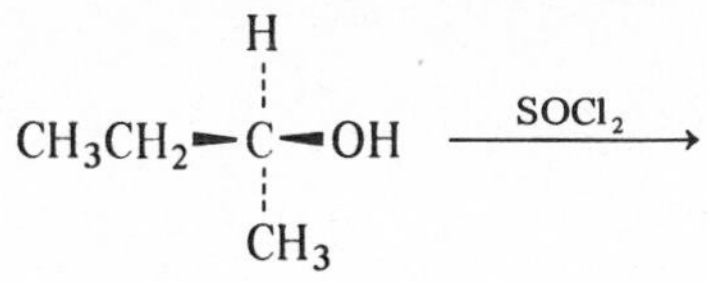
$$\underset{\underset{CH_3}{|}}{\overset{\overset{H}{|}}{CH_3CH_2-C-OH}} \xrightarrow{SOCl_2}$$

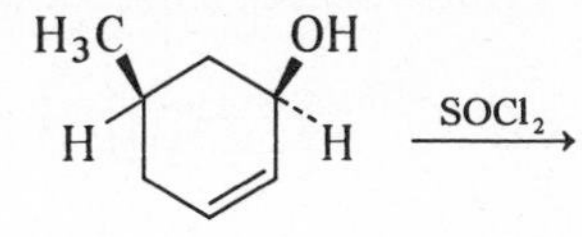

b.
$$\xrightarrow{SOCl_2}$$

c.
$$\xrightarrow{B_2H_6} \xrightarrow[OH^-]{H_2O}$$

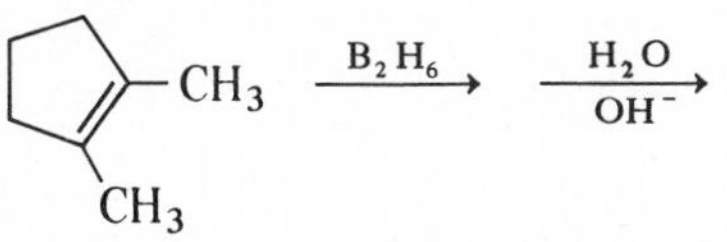

6-30   How would you synthesize the trans isomer of methyl 2-methyl-1-cyclopentyl ether from 1-methylcyclopentene?

6-31   Almost all oxygen-containing compounds form onium salts and, therefore, are soluble when placed in concentrated sulfuric acid.
   a. List the following in order of increasing base strength in such a reaction: CH_3OH, CF_3OH, $CH_2=CHOCH=CH_2$, CH_3OCH_3.
   b. Illustrate how acetic acid (CH_3COOH) could behave as a base.

6-32   Show how you make each of the following using ethylene oxide and any simple organic compounds necessary.
   a. $CH_2=CHC\equiv N$ b. $CH_3OCH_2CH_2OEt$
   c. $N\equiv CCH_2CH_2C\equiv N$ d. 1,2,4-Butanetriol
   e. $HC\equiv CCH_2CH_2CH_2SH$ f. $ICH_2CH_2OCH_2CH_2I$

6-33   In completing the following reaction sequence, note how the relative activity of two differently placed bromine atoms may be put to advantage in substitution reactions.

$$CH_2=CHCH_2Br \xrightarrow{Br_2} (A) \xrightarrow[KOH]{1\ mol} (B) \xrightarrow{(CH_3)_3CMgBr} (C) \xrightarrow{NaNH_2} (CH_3)_3CCH_2C\equiv CH$$

6-34   The oxo process is very important industrially for the production of oxygen-containing compounds. In one variation of this reaction, an olefin is heated (100 to 200°C) under pressure (3,000 lb/in^2) with carbon monoxide and hydrogen with $[Co(CO)_4]_2$ as a catalyst. This produces a carbonyl compound which is then hydrogenated to the corresponding alcohol.

$$CH_3CH=CH_2 + CO + H_2 \xrightarrow[\Delta,\ P]{cat} \text{two aldehydes} \xrightarrow[H_2]{cat} \text{two butyl alcohols}$$

   a. Draw structures for the aldehydes and the alcohols.
   b. What three possible alcohols might one expect using 1-methylcyclohexene?

6-35   Unknown compound A of MW 86 ± 1 was very soluble in water. On catalytic hydrogenation, 0.15 g of A absorbed 78 ml of hydrogen at STP. When 0.15 g of A was treated with excess CH_3MgI, 78 ml of methane at STP was produced. Suggest two possible types of structure for A. Which would be the more reasonable?

6-36   Explain the observed results in the following reactions:

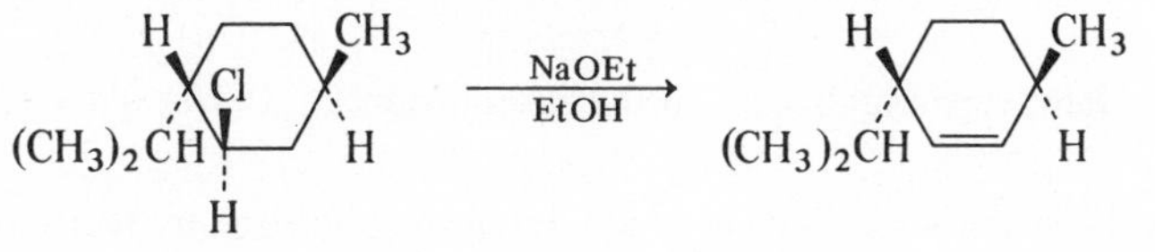

but

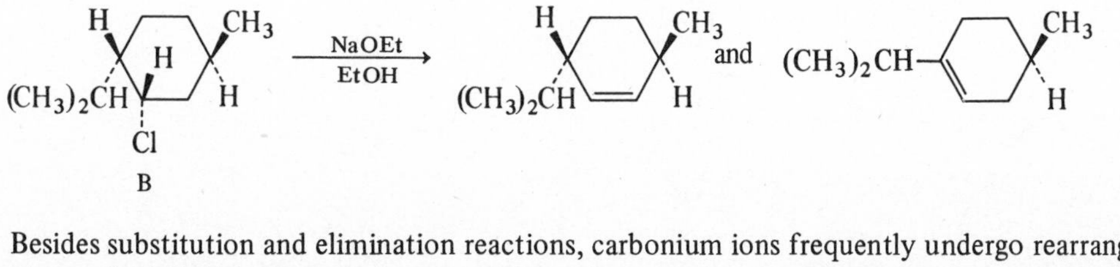

6-37   Besides substitution and elimination reactions, carbonium ions frequently undergo rearrangement reactions by virtue of a 1,2-shift of an alkyl group with its electrons from an adjacent carbon atom to the carbonium ion

75

carbon (see Sec. 5-6). The driving force of such rearrangements is the formation of a more stable carbonium ion $(3° > 2° > 1°)$. Show a mechanism for each of the following $S_N 1, E1$ reactions:

a. $(CH_3)_3CCH_2OH \xrightarrow{HCl} (CH_3)_2C(Cl)CH_2CH_3$

b. 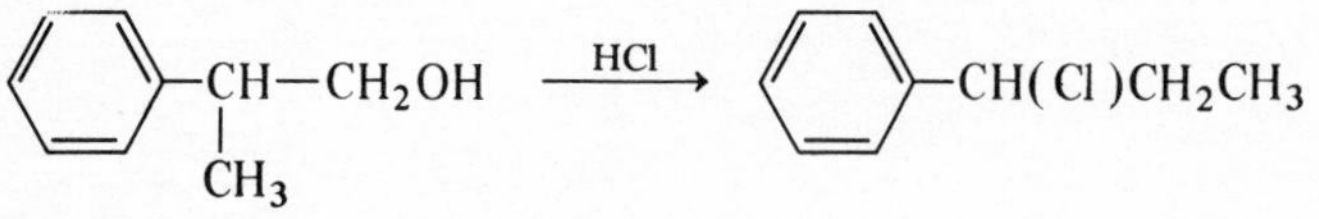 $-CH-CH_2OH \xrightarrow{HCl}$ $-CH(Cl)CH_2CH_3$
  with CH_3

c. 2,2-Dimethyl-1-cyclohexanol $\xrightarrow{H^+}$ 1,2-dimethylcyclohexene + 1-isopropylcyclopentene

d. 2,3-Dimethyl-2,3-butanediol $\xrightarrow{H^+}$ $CH_3CC(CH_3)_3$
  with $\overset{\|}{O}$

e.

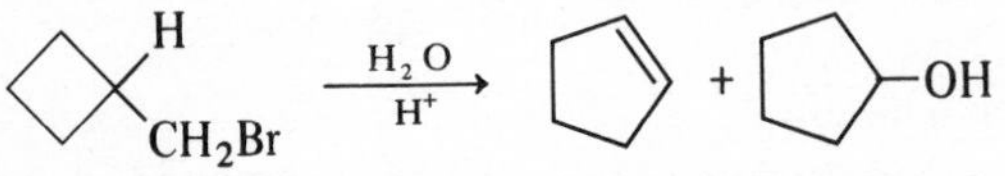

f.

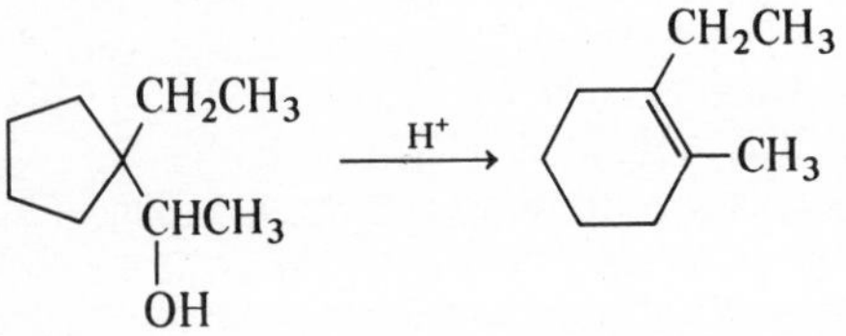

6-38　Draw structures for the geometric isomers of 1-(2-methylcyclopropyl)-2-butene.

6-39　The cyclic diol 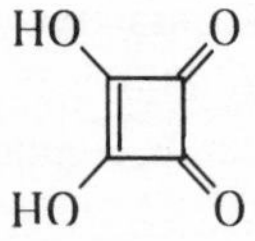 is almost as strong an acid as H_2SO_4. Explain.

6-40　Explain why $-OH$ is a much stronger acid than $-OH$.

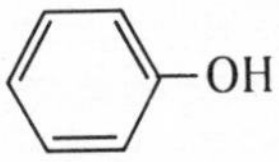

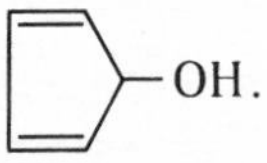

6-41　Fission of an unknown ether with HI gave two different alkyl iodides, A and B. A contained 89.4% iodine and B contained 69.0% iodine. On attempted hydrolysis with concentrated potassium hydroxide, B yielded a gaseous hydrocarbon. Suggest a structure for the ether.

6-42　Explain why 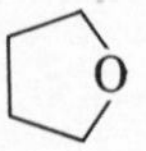 is much more soluble in water than 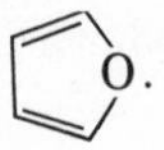 .

6-43　Radicals tend to attack ethers, preferentially removing a hydrogen on the carbon adjacent to oxygen. Using

tetrahydrofuran,  , as an example, show how the resulting radical might be stabilized by resonance with

oxygen.

6-44　Explain why the trans isomer of A reacts much faster than the cis isomer.

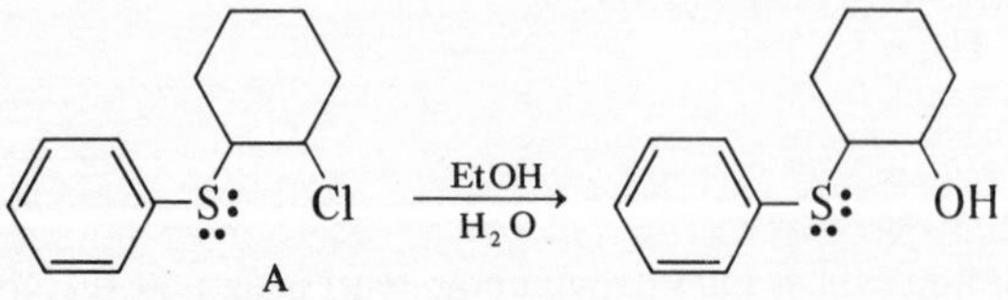

76

6-45 The special hydroborating agent, 9-BBN, mentioned in Sec. 6-2*e*-2, can be prepared from the reaction 1,5-cyclooctadiene with B_2H_6.

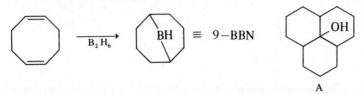

A

a. What polyene compound when treated with B_2H_6 followed by a carbonylation reaction would give compound A?

b. Trialkylboranes corresponding to compound 9-BBN above (H=R) may be prepared by treating an alkene with 9-BBN. The resulting boranes undergo attack by nucleophiles (Y^-) such as OH^- and $BrCHCO_2Et$ (see Chap. 12):

Show how this fact might be used to prepare cyclopropylbenzene from $CH_2=CHCH(Cl)$—⬡ and to

prepare ⬡—CH_2CO_2Et from cyclohexene and $BrCH_2CO_2Et$ ($BrCH_2CO_2Et \xrightarrow[\text{base}]{\text{strong}} Br\bar{C}HCO_2Et$).

7
aliphatic aldehydes and ketones

7-1   Nomenclature

a.   In the IUPAC system, aldehydes are named by dropping the *e* from the name of the hydrocarbon from which they are derived and adding *al*.

$$\underset{\text{Ethanal}}{CH_3\overset{\displaystyle O}{\overset{\|}{C}}H} \qquad \underset{\text{3-Methylbutanal}}{(CH_3)_2CHCH_2\overset{\displaystyle O}{\overset{\|}{C}}H} \qquad \underset{\text{4-Chloro-4-methyl-2-pentenal}}{(CH_3)_2\overset{\displaystyle Cl}{\overset{|}{C}}CH=CH\overset{\displaystyle O}{\overset{\|}{C}}H}$$

b.   Aldehydes are commonly named from the acid they would yield on oxidation (see Sec. 9-1 for acid names). The *ic* ending on the acid is replaced by the word *aldehyde*.

$$\underset{\text{Formaldehyde}}{CH_2O} \qquad \underset{\text{Propionaldehyde}}{CH_3CH_2CHO} \qquad \underset{\text{Isobutyraldehyde}}{(CH_3)_2CHCHO} \qquad \underset{\text{Valeraldehyde}}{CH_3(CH_2)_3CHO}$$

c.   In the IUPAC system, ketones are named by dropping the *e* from the parent hydrocarbon name and adding *one*. When a molecule contains both an aldehyde group (CHO) and a ketone carbonyl group ($\rangle$C=O) the latter is named by the IUPAC system as an *oxo* substituent (common name for such a substituent is *keto*) on the parent aldehyde.

$$\underset{\text{2-Butanone}}{CH_3\overset{\displaystyle O}{\overset{\|}{C}}CH_2CH_3} \qquad \underset{\text{2,5-Hexanedione}}{CH_3\overset{\displaystyle O}{\overset{\|}{C}}CH_2CH_2\overset{\displaystyle O}{\overset{\|}{C}}CH_3} \qquad \underset{\text{3-Hexene-2,5-dione}}{CH_3\overset{\displaystyle O}{\overset{\|}{C}}CH=CH\overset{\displaystyle O}{\overset{\|}{C}}CH_3} \qquad \underset{\text{4-Oxopentanal}}{CH_3\overset{\displaystyle O}{\overset{\|}{C}}CH_2CH_2\overset{\displaystyle O}{\overset{\|}{C}}H}$$

d.   The common name for ketones is formed by naming the two alkyl groups and adding the word *ketone*.

$$\underset{\text{Ethyl methyl ketone}}{CH_3COCH_2CH_3} \qquad \underset{\text{Allyl vinyl ketone}}{CH_2=CHCH_2COCH=CH_2} \qquad \underset{\text{Dimethyl ketone (acetone)}}{CH_3COCH_3}$$

e.   In the IUPAC system, cyclic aldehydes are named by adding the suffix *carbaldehyde* to the name of the ring system. Cyclic ketones are named by dropping the *e* from the ring system's name and adding *one*.

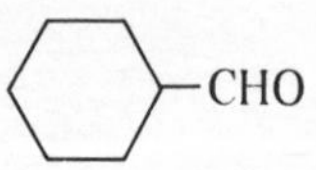

Cyclohexanecarbaldehyde 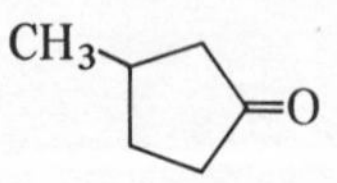   3-Methyl-1-cyclopentanone 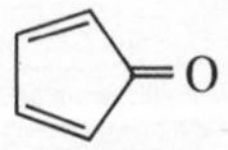   2,4-Cyclopentadien-1-one

7-2 Methods of Preparation

a. *Oxidation and Dehydrogenation of Alcohols* (See Sec. 6.4.)

b. *Pyrolysis of Acids or Acid Salts*

1. $2RCOOH \xrightarrow{\text{400 to 500°C, ThO}_2} RCOR + CO_2 + H_2O$

2. $(RCOO^-)_2 Ca^{2+} \xrightarrow{\Delta} RCOR + CaCO_3$

c. *From Nitriles* ($-C\equiv N$ = nitrile or cyano group.)
1. Nitriles may be reduced to aldehydes with lithium aluminum hydride or stannous chloride. The latter method is known as *Stephen's reduction.*

$$4RC\equiv N + LiAlH_4 \longrightarrow (RCH=N)_4 LiAl \xrightarrow{H_2O} RCHO$$

$$RC\equiv N + SnCl_2 + 4HCl \longrightarrow RCH=NH \cdot H_2 SnCl_6 \xrightarrow{H_2O} RCHO$$

2. Grignard reagents plus nitriles give ketimines, which on hydrolysis give ketones.

$$RC\equiv N + R'MgX \longrightarrow \underset{\underset{R'}{|}}{RC}=NMgX \xrightarrow{H_2O} \underset{\underset{R'}{|}}{R-C}=O$$

d. *Rosenmund Reduction of Acid Chlorides*

$$RCOOH + SOCl_2 \longrightarrow \overset{O}{\overset{\|}{R}CCl} + SO_2 + HCl$$

$$RCOCl + H_2 \xrightarrow{Pd \cdot BaSO_4} RCHO + HCl$$

To reduce the acid chloride a special Pd catalyst is used that does not reduce the aldehyde group to the alcohol.

e. *Ketones from Acid Chlorides* (See Sec. 5-4 for preparation of R_2Cd.)

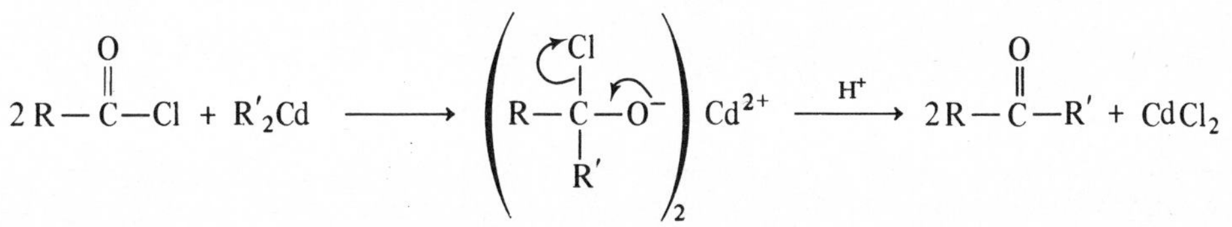

$$2 R-\overset{O}{\overset{\|}{C}}-Cl + R'_2Cd \longrightarrow \left(\underset{\underset{R'}{|}}{R-\overset{Cl}{\underset{|}{C}}-O^-} \right)_2 Cd^{2+} \xrightarrow{H^+} 2 R-\overset{O}{\overset{\|}{C}}-R' + CdCl_2$$

R' must be a primary alkyl group. The reaction represents nucleophilic attack by the alkyl group of the organocadmium compound (R'_2Cd is polarized as $R'_2{}^- Cd^{2+}$) on the slightly positive carbonyl carbon atom

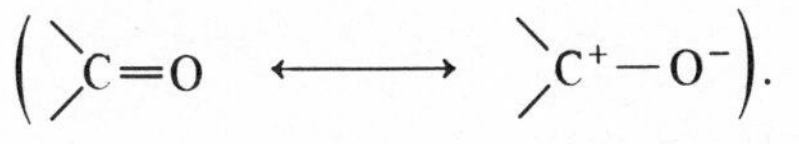

$$\left(\overset{\diagdown}{\diagup}C=O \longleftrightarrow \overset{\diagdown}{\diagup}C^+ - O^- \right).$$

f. *Ozonolysis of olefins* (See Sec. 4-3d.)

g. *Hydroboration Methods: Alkenes or Alkynes to Aldehydes or Ketones*
1. Hydroboration-Oxidation

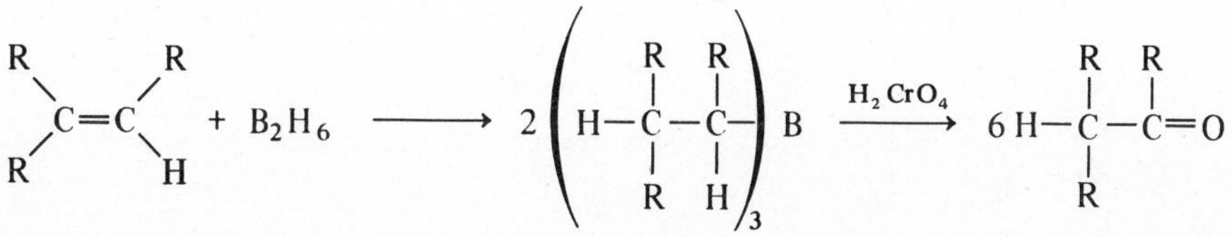

$$\underset{\underset{R}{|}}{\overset{\overset{R}{|}}{C}}=\underset{\underset{H}{|}}{\overset{\overset{R}{|}}{C}} + B_2H_6 \longrightarrow 2\left(\underset{\underset{R\ \ H}{|\ \ |}}{\overset{\overset{R\ \ R}{|\ \ |}}{H-C-C}}-B \right)_3 \xrightarrow{H_2 CrO_4} 6\ \underset{\underset{R}{|}}{\overset{\overset{R\ \ R}{|\ \ |}}{H-C-C}}=O$$

This method differs from the hydroboration-oxidation technique used to prepare alcohols (see Sec. 6-2e-1) only in that a stronger oxidizing agent, chromic acid, is used in place of hydrogen peroxide. Only ketones can be prepared by this method.

Alkynes can be converted to cis alkenes by hydroboration followed by protonation with acetic acid (CH_3COOH). Terminal alkynes can be converted to aldehydes and symmetric alkynes to ketones by hydroboration followed by oxidation (H_2O_2).

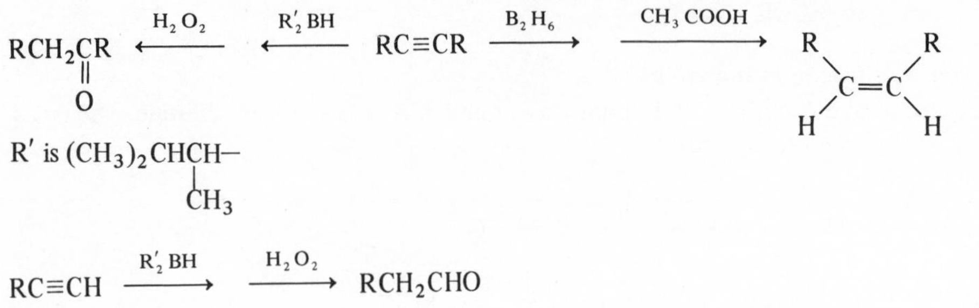

$$RCH_2CR \overset{H_2O_2}{\longleftarrow} \overset{R_2'BH}{\longleftarrow} RC\equiv CR \overset{B_2H_6}{\longrightarrow} \overset{CH_3COOH}{\longrightarrow} \underset{H}{\overset{R}{C}} = \underset{H}{\overset{R}{C}}$$

(with $\underset{O}{\overset{\parallel}{}}$ on the RCH_2CR)

R' is $(CH_3)_2CHCH-$
$\qquad\qquad\quad |$
$\qquad\qquad\; CH_3$

$$RC\equiv CH \overset{R_2'BH}{\longrightarrow} \overset{H_2O_2}{\longrightarrow} RCH_2CHO$$

2. Hydroboration-Carbonylation

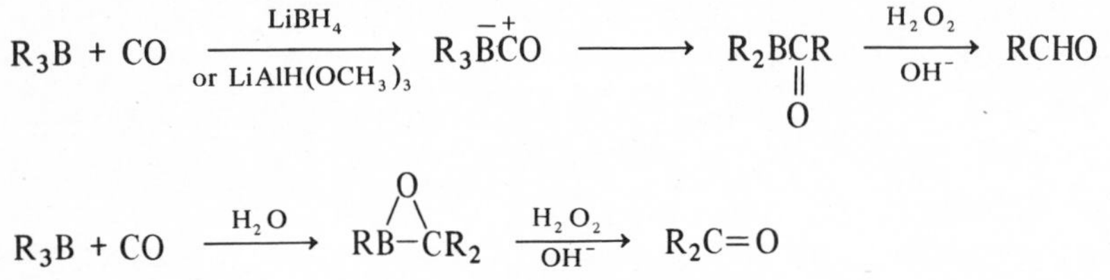

$$R_3B + CO \xrightarrow[\text{or } LiAlH(OCH_3)_3]{LiBH_4} R_3\overset{-}{B}\overset{+}{C}O \longrightarrow \underset{O}{\overset{\parallel}{R_2BCR}} \xrightarrow[OH^-]{H_2O_2} RCHO$$

$$R_3B + CO \xrightarrow{H_2O} RB{-}CR_2 \text{ (epoxide)} \xrightarrow[OH^-]{H_2O_2} R_2C{=}O$$

This technique differs from the method given in Sec. 6-2e-2 for preparation of primary and secondary alcohols in that oxidation with alkaline hydrogen peroxide is used in the final step rather than alkaline hydrolysis. The above reaction is suitable for the preparation of symmetric ketones (R_2CO) but a special hydroborating agent is employed for the synthesis of mixed and cyclic ketones.

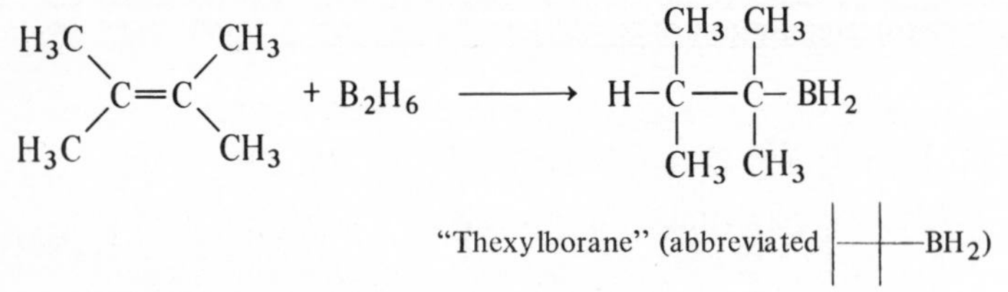

$$\underset{H_3C}{\overset{H_3C}{}}C{=}C\underset{CH_3}{\overset{CH_3}{}} + B_2H_6 \longrightarrow \underset{CH_3\;CH_3}{\overset{CH_3\;CH_3}{H-C-C-BH_2}}$$

"Thexylborane" (abbreviated ⊢⊣—BH_2)

Only the monoalkylborane is formed in this case.

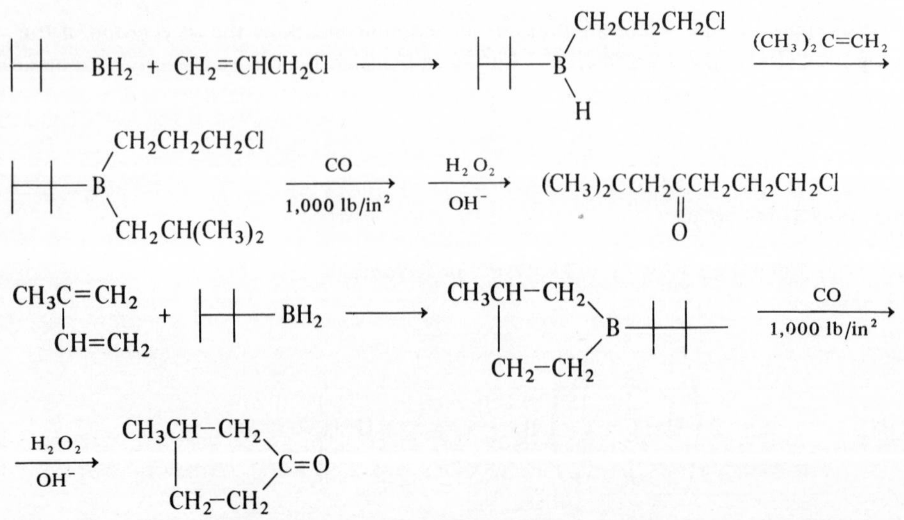

$$\text{⊢⊣}{-}BH_2 + CH_2{=}CHCH_2Cl \longrightarrow \text{⊢⊣}{-}\underset{H}{\overset{CH_2CH_2CH_2Cl}{B}} \xrightarrow{(CH_3)_2C{=}CH_2}$$

$$\text{⊢⊣}{-}\underset{CH_2CH(CH_3)_2}{\overset{CH_2CH_2CH_2Cl}{B}} \xrightarrow[1,000\ lb/in^2]{CO} \xrightarrow[OH^-]{H_2O_2} \underset{O}{\overset{}{(CH_3)_2CCH_2CCH_2CH_2CH_2Cl}}$$

$$\underset{CH=CH_2}{\overset{CH_3C=CH_2}{}} + \text{⊢⊣}{-}BH_2 \longrightarrow \underset{CH_2-CH_2}{\overset{CH_3CH-CH_2}{}}B\text{⊢⊣} \xrightarrow[1,000\ lb/in^2]{CO}$$

$$\xrightarrow[OH^-]{H_2O_2} \underset{CH_2-CH_2}{\overset{CH_3CH-CH_2}{}}C{=}O$$

7-3 Reactions

a. *Addition Reactions* Many of the reactions of aldehydes and ketones may be considered as the addition of a polar reagent across the polar carbonyl group. The mechanism usually involves first a nucleophilic attack by the negative part of the polar reagent at the partially positive carbonyl carbon (the rate-determining step) followed by a rapid second step in which the positive part of the polar reagent attacks the negative oxygen in the reaction intermediate:

$$\left[\begin{array}{c} \overset{O}{\underset{\|}{R-C-H}} \longleftrightarrow \overset{O^-}{\underset{+}{R-\overset{|}{C}-H}}\end{array}\right] + A^-B^+ \xrightarrow[\text{slow}]{} \overset{O^-}{\underset{A}{R-\overset{|}{\underset{|}{C}}-H}} \xrightarrow[\text{fast}]{B^+} \overset{OB}{\underset{A}{R-\overset{|}{\underset{|}{C}}-H}}$$

It should be noted that many common polar reagents which readily add to olefin double bonds do not yield stable products under ordinary conditions with carbonyl double bonds (for example., H_2O, HX, HNO_3, H_2SO_4, X_2).

1. Addition of HCN

$$CH_3CHO + HCN \longrightarrow CH_3\overset{OH}{\underset{|}{CH}}C\equiv N \qquad\qquad CH_3\overset{O}{\underset{\|}{C}}CH_3 + HCN \longrightarrow (CH_3)_2\overset{OH}{\underset{|}{C}}C\equiv N$$

Acetaldehyde
cyanohydrin Acetone
cyanohydrin

2. Addition of Grignard reagents (See Chaps. 5 and 6.)

3. Addition of ROH This is an acid-catalyzed reaction which involves a rapid proton attack on the carbonyl oxygen followed by an attack of ROH at the carbonyl carbon. The first product formed is a hemiacetal, which then reacts to give an acetal.

$$\overset{R-CH}{\underset{\underset{O}{\|}}{}} \underset{}{\overset{H^+}{\rightleftharpoons}} \overset{+}{\underset{\underset{OH}{|}}{R-CH}} \overset{ROH}{\rightleftharpoons} \overset{\overset{+}{ROH}}{\underset{\underset{OH}{|}}{R-\overset{|}{C}H}} \overset{-H^+}{\rightleftharpoons} \overset{OR}{\underset{\underset{OH}{|}}{R-CH}}$$

Hemiacetal

$$\overset{OR}{\underset{\underset{OH}{|}}{R-CH}} \overset{H^+(-H_2O)}{\rightleftharpoons} \overset{OR}{\underset{+}{R-\overset{|}{C}H}} \overset{ROH}{\rightleftharpoons} \overset{OR}{\underset{\overset{+}{ROH}}{R-\overset{|}{C}H}} \overset{-H^+}{\rightleftharpoons} \overset{OR}{\underset{\underset{OR}{|}}{R-CH}}$$

Acetal

Alcohols do not react readily with ketones to give ketals. These compounds are often made indirectly.

$$HCCl_3 + 3CH_3CH_2ONa \longrightarrow HC(OCH_2CH_3)_3 + 2NaCl$$

Triethyl orthoformate

$$CH_3\overset{O}{\underset{\|}{C}}CH_3 + HC(OCH_2CH_3)_3 \longrightarrow (CH_3)_2C(OCH_2CH_3)_2 + H\overset{}{\underset{\underset{O}{\|}}{C}}OCH_2CH_3$$

Neither acetals or ketals are hydrolyzed in basic solution; however, both are converted to the carbonyl compound in acid solution.

4. Addition of $NaHSO_3$ Sodium bisulfite adds to aldehydes, *methyl* ketones, and sterically unhindered cyclic ketones such as cyclohexanone. Ketones with one highly branched group, such as methyl *tert*-butyl ketone, or ketones without a methyl group do not react.

When treated with either acid or base, carbonyl-bisulfite addition compounds decompose to give the aldehyde or ketone from which they were made.

$$
\underset{\substack{\text{O} \\ \|}}{CH_3CH} + :\underset{\substack{\| \\ \text{O}}}{S}\underset{\text{HO}}{O}Na \;\rightleftharpoons\; CH_3\underset{\substack{| \\ \text{H}}}{C}\underset{\text{OH}}{SO_3Na} \xrightarrow{H_3O^+} CH_3CHO + SO_2
$$

Acetaldehyde
sodium
bisulfite

5. Addition of derivatives of ammonia

$$
(CH_3)_2\underset{\substack{\| \\ \text{O}}}{C} + :\underset{\text{H}}{N}HOH \longrightarrow (CH_3)_2\underset{\text{OH}}{C}NHOH \longrightarrow (CH_3)_2C=NOH + H_2O
$$

Hydroxylamine   Acetone oxime

$$
CH_3CH_2CHO + H_2NNH\underset{\substack{\| \\ \text{O}}}{C}NH_2 \longrightarrow CH_3CH_2CH=NNH\underset{\substack{\| \\ \text{O}}}{C}NH_2 + H_2O
$$

Semicarbazide   Propanal semicarbazone

$$
(CH_3)_2CH\underset{\substack{\| \\ \text{O}}}{C}CH_3 + Y-\bigcirc-NHNH_2 \longrightarrow (CH_3)_2CH\underset{\substack{| \\ CH_3}}{C}=NNH-\bigcirc-Y + H_2O
$$

Phenylhydrazine (Y=H) or
2,4-dinitrophenylhydrazine
(Y=NO$_2$)
Methyl isopropyl ketone
phenylhydrazone (Y=H) or
2,4-dinitrophenylhydrazone (Y=NO$_2$)

b. *Oxidation* RCHO $\xrightarrow{\text{oxidation}}$ RCOOH. Ketones are difficult to oxidize and give a mixture of products from the cleavage of the carbon-carbon bond on either side of the carbonyl group. Symmetric cyclic ketones give only one product.

$$
\bigcirc=O \xrightarrow{HNO_3} HOOC(CH_2)_4COOH
$$

c. *Reduction* (See Sec. 6-2c for reduction of carbonyl compounds to alcohols.) Ketones give a bimolecular reduction product when treated with Na · Hg or Mg · Hg. The reaction proceeds via a free-radical intermediate as a result of electron transfer from the active metal.

$$
2R\underset{\substack{\| \\ \text{O}}}{C}R \xrightarrow[\substack{\text{or} \\ Mg \cdot Hg}]{Na \cdot Hg} 2R-\underset{\substack{| \\ O^-}}{\overset{\cdot}{C}}-R \longrightarrow R-\underset{\substack{| \\ {}^-O}}{\overset{\substack{R \\ |}}{C}}-\underset{\substack{| \\ O^-}}{\overset{\substack{R \\ |}}{C}}-R \xrightarrow{H^+} R-\underset{\substack{| \\ HO}}{\overset{\substack{R \\ |}}{C}}-\underset{\substack{| \\ OH}}{\overset{\substack{R \\ |}}{C}}-R
$$

The tetrasubstituted diols are called pinacols.

d. *Pinacol-Pinacolone Rearrangement*

$$
\underset{\substack{| \quad | \\ OH \; OH}}{CH_3\overset{\substack{CH_3 \; CH_3 \\ | \quad |}}{C}-CCH_3} \underset{-H^+}{\overset{H^+}{\rightleftharpoons}} \underset{\substack{| \qquad | \\ +OH_2 \; OH}}{CH_3\overset{\substack{CH_3 \; CH_3 \\ | \quad |}}{C}-CCH_3} \underset{+H_2O}{\overset{-H_2O}{\rightleftharpoons}} \underset{\substack{| \\ OH}}{CH_3\overset{\substack{CH_3 \; CH_3 \\ | \quad |}}{\underset{+}{C}}-CCH_3} \rightleftharpoons
$$

Step A   Step B   Step C

$$
\begin{array}{ccc}
\underset{\substack{|\\CH_3\\|\\CH_3}}{CH_3C}\overset{+}{-}\underset{\substack{|\\OH}}{CCH_3} & \underset{\substack{\xrightarrow{\;-H^+\;}\\[-2pt]\xleftarrow{\;+H^+\;}}}{} & \underset{\substack{|\\CH_3}}{\overset{\substack{CH_3\\|}}{CH_3C}}\underset{\substack{||\\O}}{-CCH_3}
\end{array}
$$

Step D

Since pinacols have two OH groups, either might be lost to yield a carbonium ion as in step B. This determines which alkyl groups will be in a position to migrate to the electron-deficient carbonium ion as in step C. In unsymmetric pinacols, that OH is cleaved which yields the most stable carbonium ion (see Sec. 5-5). In general, of the two R groups which may migrate, the group that does migrate is that with the greatest electron density on the carbon forming the new carbon-carbon bond at the carbonium ion. (For example, ethyl would move before methyl and isopropyl before ethyl.)

e. *Haloform Reaction* This reaction occurs with acetaldehyde or methyl ketones and NaOX. It also occurs with ethanol and secondary alcohols that are oxidized by NaOX to acetaldehyde and methyl ketones, respectively.

$$
\underset{\substack{|\\OH}}{CH_3CH_2CHCH_3} \xrightarrow{\;NaOI\;} \underset{\substack{||\\O}}{CH_3CH_2CCH_3} \xrightarrow{\;NaOI\;} CH_3CH_2\overset{\curvearrowleft}{\underset{\substack{||\\O}}{C}}CI_3 \xrightarrow{\;OH^-\;} \underset{\substack{|\\OH}}{CH_3CH_2\overset{\overset{O^-}{|}}{C}CI_3} \longrightarrow
$$

$$
\underset{\substack{||\\O}}{CH_3CH_2C}-OH \;+\;{}^-CI_3 \longrightarrow \underset{\substack{||\\O}}{CH_3CH_2CONa} + HCI_3
$$

Iodoform

$$
CH_3CH_2OH \xrightarrow{\;NaOBr\;} CH_3CHO \xrightarrow{\;NaOBr\;} Br_3CCHO \xrightarrow{\;NaOH\;} HCBr_3 + HCOONa
$$

Bromoform

Iodoform is a yellow solid with a characteristic odor. Thus, the formation of CHI_3 from an unknown treated with NaOI constitutes a test for methyl ketones and certain alcohols.

f. *Aldol Condensation* Aldehydes and ketones having a hydrogen atom on the α *carbon* (first carbon after the carbonyl group) condense with themselves or with each other when treated with an acid or alkaline catalyst, the latter being used most often. The β-hydroxy compound so formed easily loses water on heating to give an α,β-unsaturated compound.

$$
CH_3CH_2CHO + CH_3CH_2CHO \xrightarrow{\;OH^-\;} \underset{\substack{|\;\;|\\HO\;CH_3}}{CH_3CH_2CHCHCHO} \xrightarrow{\;\Delta\;} \underset{\substack{|\\CH_3}}{CH_3CH_2CH=CCHO} + H_2O
$$

$$
\text{Mechanism:}\quad \underset{\substack{|\\H}}{RCHCHO} \underset{\substack{\xrightarrow{\;OH^-\;}}}{\rightleftharpoons} \left[\underset{\substack{||\\O}}{R\underline{C}HCH} \longleftrightarrow \underset{\substack{|\\O^-}}{RCH=CH}\right] + H_2O
$$

(I)

$$
\left[\underset{\substack{||\\O}}{RCH_2CH} \longleftrightarrow \underset{\substack{|\\O^-}}{\overset{+}{RCH_2CH}}\right] \xrightarrow[\;(I)\;]{\;R\overline{C}HCHO\;} \underset{\substack{|\\R}}{\overset{\overset{O^-}{|}}{RCH_2CHCHCHO}} \underset{H_2O}{\rightleftharpoons} \underset{\substack{|\\R}}{\overset{\overset{OH}{|}}{RCH_2CHCHCHO}} + OH^-
$$

(II)

The function of the basic catalyst is to remove a proton from the α carbon and produce the carbanion I, which adds to the polar carbonyl group of molecule II.

Ketones react much more slowly than aldehydes.

$$(CH_3)_2\overset{\overset{O}{\|}}{C} + \overset{\overset{H}{|}}{CH_2}\overset{\overset{O}{\|}}{C}CH_3 \xrightarrow{OH^-} (CH_3)_2\overset{\overset{OH}{|}}{C}CH_2\overset{\overset{O}{\|}}{C}CH_3 \longrightarrow (CH_3)_2C=CH\overset{\overset{O}{\|}}{C}CH_3$$

Mesityl oxide

Mixed aldol condensations (i.e., between two different carbonyl compounds) are to be avoided if both carbonyl compounds have α hydrogens. Such a reaction would lead to a difficult-to-separate mixture of four different products. Aldehydes such as CH_2O or $HOOCCHO$, which do not have α hydrogens, may be condensed with other aldehydes. A similar reaction occurs between aldehydes or ketones and nitroparaffins.

$$CH_3CH_2\overset{\overset{O}{\|}}{C}H + H\underset{R}{\overset{|}{C}}HNO_2 \xrightarrow{OH^-} CH_3CH_2\underset{R}{\overset{\overset{OH}{|}}{C}H}CHNO_2 \longrightarrow CH_3CH_2CH=\underset{R}{\overset{|}{C}}NO_2 + H_2O$$

g. *Wittig Reaction*

$$R_2C=O + (C_6H_5)_3\overset{+}{P}-\overset{-}{C}R'_2 \longrightarrow R-\underset{\underset{-O}{\overset{|}{C}}}{\overset{\overset{R}{|}}{C}}-\underset{\overset{+}{P}(C_6H_5)_3}{\overset{\overset{R'}{|}}{C}}-R' \longrightarrow R-\overset{\overset{R}{|}}{C}=\overset{\overset{R'}{|}}{C}-R' + (C_6H_5)_3P=O$$

Ylide   Triphenylphosphine
oxide

Betaine

Phenyl group $\equiv C_6H_5- \equiv$ ⬡$-\equiv$ benzene ring as a substituent

Aldehydes or ketones react with phosphorus ylides to give olefins. Phosphorus ylides may be prepared from a reaction of an alkyl halide and a phosphine (e.g., triphenylphosphine) followed by treatment of the intermediate phosphonium salt with a base such as NaOEt or an alkyllithium compound (EtLi).

$$(C_6H_5)_3P + X-\underset{R'}{\overset{|}{C}}HR' \longrightarrow (C_6H_5)_3\overset{+}{P}-\underset{R'}{\overset{|}{C}}HR'X^- \xrightarrow{^-OEt} (C_6H_5)_3\overset{+}{P}-\underset{R'}{\overset{|}{C}}R' \longleftrightarrow$$

Phosphine   Phosphonium salt

$$(C_6H_5)_3P=\underset{R'}{\overset{|}{C}}-R' + EtOH$$

Mechanism:

$$CH_3\overset{\overset{H}{|}}{C}=O \longrightarrow CH_3\underset{|}{\overset{\overset{H}{|}}{C}}-O^- \longrightarrow CH_3\overset{\overset{H}{|}}{C}H-O \longrightarrow$$

$$CH_3CH_2-\underset{(CH_3)_2CH}{\overset{|}{C}}-\overset{+}{P}(C_6H_5)_3 \quad CH_3CH_2\underset{(CH_3)_2CH}{\overset{|}{C}}-\overset{+}{P}(C_6H_5)_3 \quad CH_3CH_2\underset{(CH_3)_2CH}{\overset{|}{C}}-P(C_6H_5)_3$$

$$CH_3CH=\underset{\overset{|}{CH_2CH_3}}{C}CH(CH_3)_2 + (C_6H_5)_3P=O$$

The aldehyde or ketone may be aliphatic, cyclic, or aromatic and it may contain a variety of other functional groups which will not interfere (for example, $C=C$, $C\equiv C$, OH, OR, X, NO_2, NR_2, and CO_2R). The phosphorus ylides can be derived from alkyl halides, which may also contain other noninterfering functional groups, but which have at least one hydrogen on the halogen-bearing carbon. Ylides may also be prepared from trialkylphosphites:

$$(EtO)_3P + CH_3CH_2X \longrightarrow (EtO)_2\overset{\|}{\underset{O}{P}}CH_2CH_3 \xrightarrow{\text{Base}} (EtO)_2\overset{\|}{\underset{O}{P}}-\bar{C}HCH_3$$

Triethylphosphite

Diethyl ethylphosphonate

$$(EtO)_2\overset{\|}{\underset{O}{P}}-\bar{C}HCH_3 + CH_3CH=CHCCH_3 \longrightarrow CH_3CH=CHCCH_3 + (EtO)_2PO_2^-$$

h. *Halogen Replacement of Oxygen*

$$(CH_3)_2CHCHO + PCl_5 \longrightarrow (CH_3)_2CHCHCl_2 + POCl_3$$

$$(CH_3)_2C=O + PBr_5 \longrightarrow CH_3CBr_2CH_3 + POBr_3$$

i. *Hydrogen Replacement of Oxygen* (Wolff-Kishner reduction)

$$\underset{O}{\overset{\|}{CH_3CCH_2CH_3}} + H_2NNH_2 \longrightarrow \underset{\text{}}{\overset{NNH_2}{CH_3\overset{\|}{C}CH_2CH_3}} \xrightarrow[\text{or KOH}]{\text{NaOCH}_3} CH_3CH_2CH_2CH_3 + N_2$$

Hydrazine

j. *Cannizzaro Reaction* Aliphatic (and aromatic) aldehydes containing no α hydrogens (thus no aldol reaction is possible with these aldehydes), when treated with sodium hydroxide or other strong bases, undergo a reaction where one molecule of aldehyde oxidizes another to the acid and is itself reduced to the primary alcohol. If two different aldehydes are used the reaction is called a crossed-Cannizzaro. One useful variation of this employs formaldehyde and another aldehyde. Formaldehyde reduces the other aldehyde to the alcohol while formaldehyde is oxidized to formic acid.

$$2(CH_3)_3CCHO \xrightarrow{\text{NaOH}} (CH_3)_3CCH_2OH + (CH_3)_3CCOONa$$

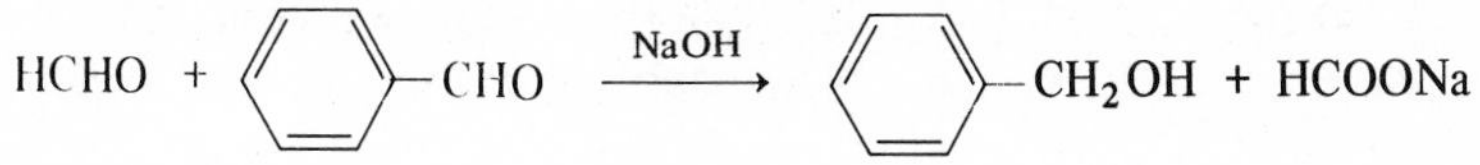

$$HCHO + \bigcirc\!\!-CHO \xrightarrow{\text{NaOH}} \bigcirc\!\!-CH_2OH + HCOONa$$

The mechanism of the Cannizzaro reaction involves a shift of a hydrogen atom with its electrons (a hydride shift):

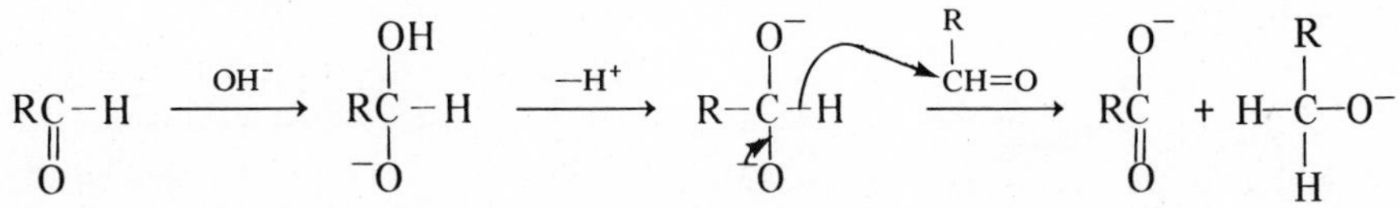

$$\underset{O}{\overset{\|}{RC}}-H \xrightarrow{OH^-} \overset{OH}{\underset{\bar{O}}{R\overset{|}{\underset{|}{C}}-H}} \xrightarrow{-H^+} \underset{\bar{O}}{R-\overset{}{\underset{}{C}}\overset{}{\underset{}{\ _{H}}} \overset{R}{\underset{}{CH=O}} \longrightarrow \underset{O}{\overset{O^-}{R\overset{}{\underset{}{C}}}} + H-\overset{R}{\underset{H}{\overset{|}{C}-O^-}}$$

$$RCH_2O^- \xrightarrow{H_2O} RCH_2OH$$

7-4 Qualitative Tests for Aldehydes and Ketones

a. Aldehydes and unhindered methyl ketones give precipitates with saturated sodium bisulfite solution (see Sec. 7-3*a*-4).

b. Alcoholic solutions of aldehydes or ketones give precipitates when warmed with phenylhydrazine or semicarbazide.

c. Aldehydes (but not ketones) react with Fehling's solution, Tollens' reagent, and Schiff's reagent.

1. Fehling's solution is a basic solution of copper sulfate stabilized by tartrate. Aldehydes reduce the copper to a red precipitate of Cu_2O.
2. Tollens' reagent is ammoniacal silver nitrate, which oxidizes aldehydes to acids; the silver is reduced to metallic silver, which deposits as a mirror on the walls of the test tube.
3. Schiff's reagent is a colorless solution of the dye pararosaniline. It gives a pink to violet color with aldehydes.

   d.   *Iodoform Test*   (See Sec. 7-3*e*.)

7-5   Organic Syntheses

The carbonyl group of aldehydes and ketones is one of the most active of all functional groups and therefore one of the most interesting and exciting from the point of view of the synthetic chemist. It gives a large number of addition reactions that can be used to introduce new functions into a molecule, and we have already seen how new R groups can be hooked on via the Grignard reagent at the C=O group. It can be reduced in good yield to the alcohol group; hence, all of the reactions of the alcohol group can be utilized after reduction of a carbonyl group. Of course, the alcohol can be converted to the olefin and this in turn to an acetylene, which means that utilization of all the reactions of these functional groups is possible via a carbonyl group! In addition, the carbonyl group can be converted into a C=C linkage with an increase in the number of carbon atoms, and do not forget hydroboration methods which allow conversion of alkenes into alcohols, aldehydes, or ketones. Write a sample carbonyl compound in the center of a piece of scratch paper. Now, to the right of this write all the products that can be made by addition reactions to the carbonyl group. To the left show the conversions carbonyl ⟶ alcohol ⟶ alkyl halide ⟶ olefin ⟶ acetylene. Above the alkyl halide show six different displacement reactions leading to six new functional groups. Show as many additional conversions of the olefin and acetylene as you can. Now, below the carbonyl compound show how it can undergo an aldol condensation and how this product can be dehydrated to the α,β-unsaturated carbonyl compound. One of the displacement products you should have shown from the alkyl halide is the nitrile (see Sec. 8-2). Write the equation for the conversion of this to a new carbonyl group. Now you have a new carbonyl compound of one more carbon atom, and you could write all the reactions you have just completed below for this. This chart gives you a rough idea why organic chemists hold the carbonyl group in such high esteem.

Now let's dissect a few synthetic problems.

Synthesize $CH_3CH_2CH(CH_3)CH_2CH(OH)C\equiv N$ from carbon. The first problem is to recognize what kind of molecule you are dealing with. In this case it is a cyanohydrin (OH and CN attached to the same carbon). Now start asking questions and backing up. How are cyanohydrins made? From aldehydes. Write 3-methylpentanal and show its conversion to the above cyanohydrin. How are aldehydes made? From primary alcohols or nitriles with the same number of carbon atoms or via hydroboration-carbonylation-oxidation (see Sec. 7-2*g*-2; abbreviated HB-CO-O hereafter) from a 1-alkene with one less carbon. Assume the necessary primary alcohol and show its conversion to the aldehyde. How can we build up primary alcohols from smaller carbon compounds? One way would be by the action of the Grignard reagent on ethylene oxide. Assume the necessary reactants, *sec*-butylmagnesium bromide and ethylene oxide. Use the approach outlined in Sec. 6-6 to make *sec*-butyl alcohol from carbon. Another approach to 3-methyl-pentanal from *sec*-butyl alcohol would be oxidation to 2-butanone, treatment with $(C_6H_5)_3\overset{+}{P}-\overset{-}{C}H_2$ in a Wittig reaction, and followed by a HB-CO-O sequence on the resulting 2-methyl-1-butene. What if we had decided to make 3-methylpentanal from the nitrile rather than the alcohol? In this case the starting material would be 1-bromo-2-methylbutane.

Synthesize 2,4-dimethyl-1-nitropentane. Remember that the general method for introducing a nitro group is to use a displacement reaction on an alkyl halide. Assume 2,4-dimethyl-1-bromopentane and draw its conversion to the nitro compound. Of course, you can get the bromide from the alcohol, but note that this alcohol contains two points of branching. The $-CH_2-$ between the functional group and the first point of branching can be introduced by the action of formaldehyde on the Grignard reagent from 2-bromo-4-methylpentane. To get this bromide we need a secondary alcohol. This can be made by the action of isobutylmagnesium bromide on the two-carbon aldehyde. The isobutyl unit can be built up by the same kind of process.

One approach to the above compound that may have occurred to you is a Wittig reaction between the ylide prepared from isobutyl bromide and nitroacetone. This would yield 2,4-dimethyl-1-nitro-2-pentene. The problem then becomes one of reducing the carbon-carbon double bond without affecting the NO_2 group. Most reducing agents are not suitable, e.g., catalytic reduction or $LiAlH_4$, because they will reduce the NO_2 group to NH_2 in addition to reducing the olefinic linkage. Diborane, B_2H_6, is a suitable reducing agent here because alkenes may be hydroborated with B_2H_6 followed by a hydrolysis of the trialkylborane with a carboxylic acid such as acetic acid yielding the corresponding alkene. The NO_2 group would be unaltered if the proper conditions were used.

The aldol condensation is an extremely useful reaction when in the hands of a person who understands its possibilities as well as its limitations. This is a handy way to get *two* functional groups into a molecule. One of these will be an —OH group one carbon removed from the carbonyl or nitro group. Since these molecules are very easily dehydrated to the olefin, the aldol condensation affords an easy route to unsaturated compounds. Whenever you face the synthesis of a molecule with two functional groups, especially if one is unsaturated, it is worth considering the aldol condensation for its preparation. Consider the following:

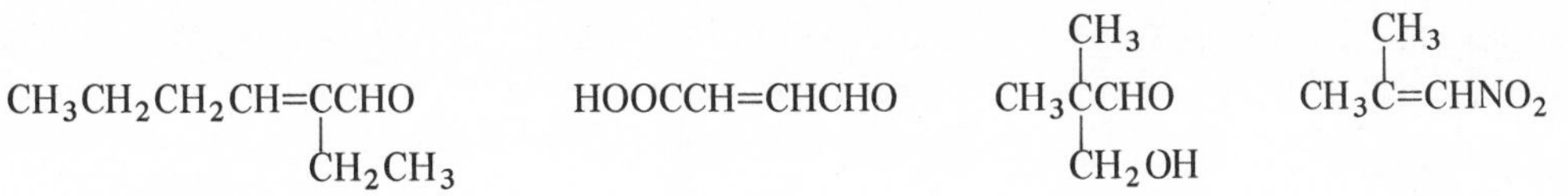

$$CH_3CH_2CH_2CH=CCHO \qquad HOOCCH=CHCHO \qquad CH_3\overset{\displaystyle CH_3}{\underset{\displaystyle CH_2OH}{C}}CHO \qquad CH_3\overset{\displaystyle CH_3}{C}=CHNO_2$$

$$\underset{\displaystyle CH_2CH_3}{}$$

$$\qquad\qquad I \qquad\qquad\qquad\qquad II \qquad\qquad\qquad\qquad III \qquad\qquad\qquad\qquad IV$$

All of the above molecules can be made by variations of the aldol condensation. Before you read further it would be worthwhile to play a bit on a piece of scratch paper and see if you can show how they are derived.

In considering the preparation of I, II, and IV, a technique which is very improper from the point of view of the mechanism of the reaction but which is useful in seeing quickly how unsaturated compounds are made by the aldol method, is to employ what is sometimes called *lasso chemistry*.

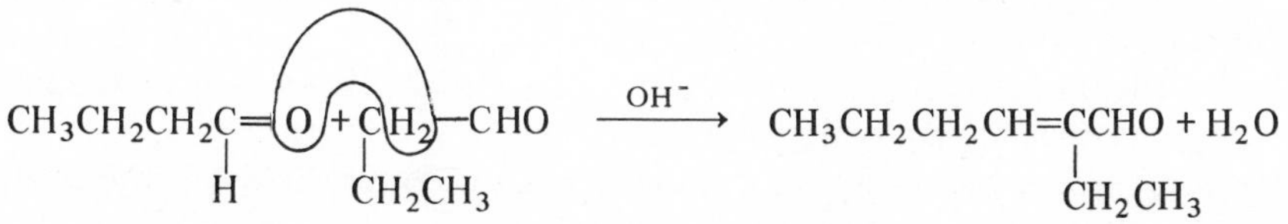

$$CH_3CH_2CH_2C=\!\!\left(O\right)+C\!\left(H_2\right)\!-CHO \xrightarrow{\ OH^-\ } CH_3CH_2CH_2CH=CCHO + H_2O$$

$$\underset{\displaystyle H}{}\underset{\displaystyle CH_2CH_3}{} \qquad\qquad\qquad\qquad \underset{\displaystyle CH_2CH_3}{}$$

Thus, when you see a problem that involves a double-bond α,β to a carbonyl group or nitro group, this molecule can be put together (visually) by "lassoing" the elements of water from the two compounds. Using this type of thinking, you see that for II above you need HOOCCHO and CH_3CHO and for IV you need $(CH_3)_2C=O$ and CH_3NO_2. As you know, the first step involved in preparations I, II, and IV is the formation of the β-hydroxy compound, which then, on heating, easily loses water to give the unsaturated compound. Compound III is a simple aldol produced from the reaction of formaldehyde and isobutyraldehyde. Products II and III are properly made by mixed aldol condensations in which one of the carbonyl compounds has no hydrogen on the α carbon.

Now let's play with some syntheses in which the aldol condensation may be used but in which the final product is greatly changed from the aldol step. Consider $CH_3CH=CH-C(CH_3)=CH_2$, dealing with a molecule having two functions which are 1,3 to each other or, in effect, separated by one carbon. By now you should be thinking of dehydration of a diol. Aldol condensation of a three-carbon carbonyl gives you the proper carbon skeleton. Hydrogenation of the aldehyde function gives you the diol, which on dehydration yields the diene. The aldol condensation of a molecule with itself always gives a product with an even number of carbon atoms. Except for acetaldehyde one always gets a branching in the carbon chain at the α carbon. Now show the complete synthesis of the diene using propionaldehyde for the aldol condensation. What about a synthesis of the diene which does not involve an aldol condensation? Remember that alkenes can also be made by a Wittig reaction. Outline the synthesis of the diene starting from acetaldehyde and isobutylene.

Jot down the structure for 3-methylheptane. Note that the carbon skeleton could be formed from 2 four-carbon units. Draw the aldol condensation for butyraldehyde. Now all you have to do is remove the oxygens and you have the desired product. There are a variety of paths to take at this point. You could convert to the diene as shown above and then hydrogenate, or you could form the unsaturated aldehyde, then hydrogenate to the saturated alcohol, and then convert this to the hydrocarbon by a variety of ways. You might elect to use the Wolff-Kishner reduction to reduce the aldehyde function to the methyl group, and then hydrogenate the double bond catalytically. If you did not want to use an aldol condensation, you might think of a number of monoalkenes which upon catalytic hydrogenation will give the desired alkane. The necessary alkene could be made by a Wittig reaction. The problem then becomes one of which ylide and which carbonyl compound to use. The best method would involve 2 four-carbon units. Outline the Wittig synthesis of 3-methylheptane from butyl alcohol.

Consider the problem of making $CH_3C(NO_2)=CHCH=C(NO_2)CH_3$. There are two double bonds that might be introduced by the aldol condensation, and we have also two nitro groups. A few minutes of reflection should bring the revelation that this molecule can be put together from two molecules of nitroethane and one molecule of the two-carbon dialdehyde called glyoxal. A Wittig reaction between glyoxal and the ylide prepared from 1-bromo-1-nitroethane would also be satisfactory.

A problem that involves an added subtlety is the synthesis of crotonic acid ($CH_3CH=CHCOOH$). Recognizing that the COOH group can be made by the oxidation of CHO, your first impulse may be to say "that's easy." Simply condense acetaldehyde, dehydrate to get crotonaldehyde ($CH_3CH=CHCHO$), and oxidize this to the acid. And, of course, that is exactly right, except that one must be very careful about the oxidation. A strong reagent such as potassium permanganate or dichromate would also attack the double bond, and one would get very little of the desired acid. A very selective reagent is necessary: Tollens' reagent is good for such a job. The ammoniacal silver is reduced to metallic silver and the aldehyde function is oxidized to the acid without disturbing the double bond. A Wittig reaction

$$\overset{O}{\overset{\|}{}}$$

between acetaldehyde and the ylide prepared from $BrCH_2COCH_2CH_3$ followed by acid hydrolysis of the α,β-unsaturated ester product will yield crotonic acid also.

The relative sensitivity of various functions to oxidizing and reducing agents is a serious problem the organic chemist constantly encounters. Of the groups studied so far, aldehydes are the most easily oxidized, followed very closely by olefins, acetylenes, and then alcohols. Ethers and alkyl halides are quite resistant. In hydrogenation (reduction), the double bonds of both a carbonyl group and of an olefin are both easy to reduce, but by the proper choice of conditions either one can be reduced without touching the other. Lithium aluminum hydride will usually reduce a carbonyl without attacking a double or triple bond in the same molecule. Alkyl halides are very easily reduced, and it is almost impossible to reduce an olefin or carbonyl double bond without breaking a carbon-halogen bond in the same molecule.

To see more clearly the above difficulties, try the synthesis of 2-bromo-2-methylpentane. The carbon skeleton for this can easily be put together by condensing two molecules of propionaldehyde and dehydrating to $CH_3CH_2CH=C(CH_3)CHO$. You must first reduce the aldehyde group before adding HBr or else the Br will be removed in the subsequent reduction. In this instance you could use the Wolff-Kishner reduction and then add HBr to the olefin. You could have reduced the aldehyde directly to the alcohol using lithium aluminum hydride, but the CH_2OH would be difficult to convert to a methyl group. You might think of making the bromide with HBr, then converting this to a Grignard reagent, and treating it with water to get the olefinic hydrocarbon. All of these difficulties can be avoided in the present case by employing a Wittig reaction. Starting with acetone and propyl bromide outline the synthesis of 2-bromo-2-methylpentane.

Finally, consider the synthesis of cyclopentyl isobutyl ketone from any organic compounds containing four carbons or less. Backing up using the Grignard approach you might try cyclopentylmagnesium bromide and 3-methylbutanal or cyclopentanal and isobutylmagnesium bromide. Using a hydroboration approach cyclopentene and 1-butene could be hydroborated with thexylborane followed by carbonylation and oxidation. The problem then becomes one of obtaining a cyclopentane ring. What methods do we know for forming cyclic compounds? So far for five-membered or larger ring compounds the only general method would involve hydroboration of some alkadiene with thexylborane followed by carbonylation and oxidation to give a cyclic ketone. Now complete the synthesis.

7-6   Structure Determination of Ring Compounds Containing N, O, and S

In Sec. 4-6 the problem of ring structure elucidation by complete hydrogenation was introduced for rings containing only carbon and hydrogen. Rings on which there are substituents with elements other than carbon must be compared with parent compounds having the same elements. Oxygen and sulfur do not change the carbon/hydrogen ratio. For example, consider the hydrogenation of 2,4-cyclopentadien-1-one, C_5H_4O. Complete hydrogenation gives cyclopentanol, $C_5H_{10}O$. Six hydrogens were taken up showing three sites of unsaturation, two carbon-carbon double bonds, and one carbon-oxygen double bond. The general formula for open-chain saturated oxygen-containing compounds is $C_nH_{2n+2}O_x$. Since $C_5H_{10}O$ falls short of this by two hydrogens the presence of one ring is indicated. For open-chain saturated nitrogen compounds with one nitrogen atom (for example, $CH_3(CH_2)_xNH_2$) the general formula is readily seen to be $C_nH_{2n+3}N$.

Example Unknown X, $C_{11}H_7N$, on complete hydrogenation gives $C_{11}H_{21}N$. Since 14 hydrogens were taken up, 7 sites of unsaturation are present. Since $C_{11}H_{21}$ falls four hydrogens short of that predicted by the general formula, the unknown also contains two rings.

7-1   Give an IUPAC name and a common name (wherever possible) for each of the following:

a. $(CH_3)_2CHCHO$

b. $(CH_3)_2CHCH_2COCH_3$

c. $CH_2=CHCHO$

d.   OHCCHO

e. $CH_3COCOCH_3$

f. $CH_3CH(OH)CH_2CHO$

g. 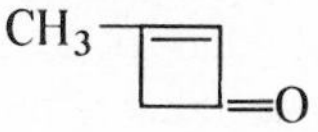   CH$_3$ ▢=O

h. 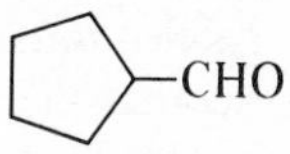   ⬠-CHO

i. $CH_2=CHCH=CHCHO$

j. $(CH_3)_2C=CHCOCH_3$

7-2   Draw structures for the following compounds:

a.   3-Hepten-5-yn-2-one

b.   *sec*-Butyl isopropyl ketone

c.   3-Hexene-2,5-dione

d.   *trans*-2-Chloro-4-methoxy-1-cyclopentanone

e.   6-Hydroxy-4-nitro-3-hexenal

f.   1-(4-Bromocyclohexyl)-3-penten-2-one

g.   6-Oxo-3-heptenal

h.   2,2-Dimethyl-1-cyclobutanecarbaldehyde

7-3   Write balanced equations for the following reactions:

a.   Acetone + semicarbazide

b. PCl_5 + valeraldehyde

c.   2,2-Dimethylpropanal + OH^- + Δ

d.   EtOH + NaOH + Br_2

e.   Tri-*sec*-butylborane + H_2CrO_4

f.   Cyclobutanone + hydroxylamine

g.   Diethyl ketone + Mg·Hg

h.   Propyl cyanide + $SnCl_2$ + HCl

i.   Butanal + $(C_6H_5)_3P=C(CH_3)_2$

j.   Calcium acetate + Δ

k.   1-Propyne + $HgSO_4$ + H_3O^+

l.   Thexylborane + $2CH_2=CHCH_2CO_2Et$

m.   Acetone + triethyl orthoformate

n.   2-Butanone + NaOCl

o.   Triphenylphosphine + 2-bromobutane

p.   2-Methyl-1,2-propanediol + H^+ + Δ

q. $LiAlH_4$ + methyl allyl ketone

r.   Di-*tert*-butyl ketone + $NaHSO_3$

s.   Cyclohexanone + Mg·Hg

t.   2-Propanol + KOH + I_2

u.   Thexylborane + 2-methyl-1,3-butadiene

v. $CH_2=CHCOCl$ + H_2 + Pd·$BaSO_4$

w.   Acrolein + Tollens' reagent

x. $(CH_3)_2CHBr$ + triethylphosphite

y.   Propynal + $(C_6H_5)_3P=CHCH=CH_2$

z.   Formaldehyde + 2,2-dimethylpropanal + NaOH

7-4   Show how you would make the following conversions:

a. $CH_3CH_2CO_2H$ to pentane

b. $CH_3CH_2CO_2H$ to propanal

c. CH_3CHO to triethyl orthoformate

d.   Propanal to methylacetylene

e.   Ethanal to 2-butenal

f.   Butyraldehyde to 2-ethyl-1,3-hexadiene

g.   Cyclohexene to 1,2-epoxycyclohexane

h.   Acetone to $CH_2=C(CH_3)C≡N$

i.   Propanal to 2-methyl-1,3-pentanediol

j.   Cyclohexanone to ⬡=CH$_2$

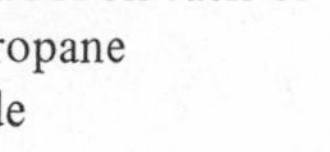

k. CH_3CH_2COOH to 2-pentene

l.   Butanal to 2-ethyl-1-hexanol

m.   Acetone to 4-methyl-1,3-pentadiene

n.   Acetone to 2,3-dimethyl-1,3-butadiene

o. $CH_3COCH(CH_3)CHO$ to 1,2-dimethylcyclohexane

p.   Isobutylene to 2,6-dimethyl-4-isobutylheptane

7-5   Possibly the most important general class of reactions is that which involves the making of new carbon-carbon bonds. So far we have covered about a dozen of these. How many examples can you give?

7-6   Write the equation (where a reaction occurs) for the action of hot concentrated NaOH on each of the following:

a.   1,1-Diethoxyethane

b.   2-Methyl-2-iodopropane

c.   1-Butyne

d.   Neopentyl chloride

e.   *tert*-Butyl alcohol

f.   Vinyl chloride

g.   Di-*tert*-butyl ketone

h.   Isobutyraldehyde

7-7   Give a simple chemical test to distinguish between:

a.   Acetic acid and propanal

b.   Acetaldehyde and propanal

c.   2- and 3-Pentanone

d.   Butanal and 1-butanol

e.   Ethyl ether and pentane

f.   Ethanol and methanol

g.   1-Fluoropentane and ethyl ether

h.   Hexachloroethane and 1-chloro-1-pentene

i.   Acetaldehyde and acetone

j.   4-Pentyn-2-one and 3-pentyn-2-one

7-8 Four bottles with lost labels contain isopropyl iodide, acetone, propionaldehyde, and heptane. What tests could you use to aid in relabeling the bottles?

7-9 A low-boiling compound gave an iodoform test and a Schiff test, but did not evolve methane when treated with methylmagnesium iodide. Suggest a structure.

7-10 A compound of MW 74 did not react with acetyl chloride, $KMnO_4$, phenylhydrazine, or Schiff's reagent. It dissolved in concentrated H_2SO_4. Suggest three possible structures.

7-11 Show the delocalization of the electrons in acrolein; i.e., draw the resonance forms. HCl adds to acrolein to give $ClCH_2CH_2CHO$. Remembering that alcohols of the vinyl type are unstable and rearrange, illustrate the mechanism for the addition of HCl to acrolein. How might acrolein be expected to react with each of the following: CH_3MgX, CH_3SH, $HC\equiv N$, NH_3?

7-12 Dehydration of 3,3-dimethyl-2-butanol by heating with a strong acid gives chiefly 2,3-dimethyl-2-butene. Draw a mechanism for this reaction.

7-13 Draw resonance structures for acrylonitrile ($CH_2=CHC\equiv N$). In what two ways might this substance react with RMgX?

7-14 Point out the errors, if any, in the following proposed syntheses:

a. $CH_2=CHCH_2OH \xrightarrow[\text{(A)}]{K_2Cr_2O_7} CH_2=CHCHO \xrightarrow[\text{(B)}]{HBr} BrCH_2CH_2CHO \xrightarrow[\text{(C)}]{H_2,\ Pt} BrCH_2CH_2CH_2OH$

b. $CH_3COCH_2CH_2Cl \xrightarrow[\text{(A)}]{Mg} CH_3COCH_2CH_2MgCl \xrightarrow[\text{(B)}]{CH_2O} \xrightarrow[H_2O]{H^+} CH_3COCH_2CH_2CH_2OH$

c. $(CH_3)_3CCl \xrightarrow[\text{(A)}]{NaCN} (CH_3)_3CCN \xrightarrow[\text{(B)}]{EtMgBr} \xrightarrow[H_2O]{H^+} (CH_3)_3CCOCH_2CH_3$

$\xrightarrow[\text{(C)}]{NaOI} \xrightarrow{H^+} (CH_3)_3CCOOH$

d. $(CH_3)_2CHBr + BrCH=CH_2 \xrightarrow[\text{(A)}]{Na} (CH_3)_2CHCH=CH_2 \xrightarrow[\text{(B)}]{I_2} (CH_3)_2CHCHICH_2I$

e. $2HOCH_2CH_2COOH \xrightarrow[\text{(A)}]{ThO_2,\ 450°C} (HOCH_2CH_2)_2CO \xrightarrow[\text{(B)}]{NaHSO_3} (HOCH_2CH_2)_2C(OH)SO_3Na$

f. $BrCH_2CH_2OH \xrightarrow[\text{(A)}]{CH_3ONa} CH_3OCH_2CH_2OH \xrightarrow[\text{(B)}]{K_2Cr_2O_7} CH_3OCH_2COOH$

g. $HC\equiv CMgX \xrightarrow[\text{(A)}]{H_2,\ Pt} H_2C=CHMgX \xrightarrow[\text{(B)}]{Br_2} BrCH_2CHBrMgX \xrightarrow[\text{(C)}]{CH_2O} \xrightarrow{H^+} BrCH_2CHBrCH_2OH$

h. $\underset{\underset{OH}{|}}{(CH_3)_2C}-\underset{\underset{OH}{|}}{C(CH_3)_2} \xrightarrow[\text{(A)}]{HBr} \underset{\underset{Br}{|}}{(CH_3)_2C}-\underset{\underset{Br}{|}}{C(CH_3)_2} \xrightarrow[\text{(B)}]{Mg} \xrightarrow{H^+} (CH_3)_2CHCH(CH_3)_2$

i. $(C_6H_5)_3P=C(CH_3)CH_2CHO \xrightarrow[\text{(A)}]{CH_3COCH_2CH_3} CH_3CH_2C(CH_3)=C(CH_3)CH_2CHO$

$\xrightarrow[\text{(B)}]{NaBH_4} \xrightarrow{H^+} CH_3CH_2C(CH_3)=C(CH_3)CH_2CH_2OH$

j.

k. $(CH_3)_2CHCOOH \xrightarrow[\text{(A)}]{SOCl_2} (CH_3)_2CHCOCl \xrightarrow[\text{(B)}]{Pd,\ H_2} (CH_3)_2CHCHO$

$\xrightarrow[\text{(C)}]{H_2CO,\ OH^-} (CH_3)_2CHCH_2OH + HCOO^-$

90

7-15 Given the fact that the carbonyl group in aldehydes is usually more reactive than the carbonyl group in ketones, what products would you expect from an aldol condensation of acetone and acetaldehyde? Which would be the main product?

7-16 Given acetylene as the only organic compound, show how you would prepare the following *without* using the Grignard reagent.
a. $CH_3CH_2CH_2COOH$ b. Acrylonitrile
c. 2,3-Butanedione   d. $NCCH_2CH_2CH(OH)CN$
e. 3,5-Dimethylheptane   f. 4-Propyl-4-heptanol

7-17 Given acetylene as the only organic compound, show how you would prepare the following using the Grignard reagent.
a. 3-Ethyl-4-heptyn-3-ol   b. 3-Ethyl-2,4-heptadiene
c. 3,4-Diethyl-2,4-hexadiene   d. $(CH_3)_3CCOOH$
e. 2,6-Octadiyne-4,5-diol   f. *tert*-Butyl methyl ketone

7-18 Given methanol, acetylene, 1-propanol, cyclopentanol, triphenylphosphine, and 2,3-dimethyl-2-butene as the only organic compounds, show how you would make each of the following:
a. 1-Cyanocyclopentene   b. 3-Methyl-3-ethylpentane
c. 2-Cyclopentyl-1-cyclohexanone   d. 4-Ethyl-3-octanol
e. $(HOCH_2)_2C(CH_3)CHO$ f. 2-Methyl-1-nitro-2-propanol
g. Vinyl propyl ketone   h. $(CH_3)_2CHCH_2\underset{\underset{\displaystyle CH_3}{|}}{C}(OH)CN$

i. 2-Cyano-1-cyclopentanol   j. 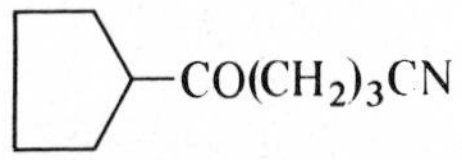 $\boxed{}\!\!>\!\!-CO(CH_2)_3CN$

k. $(CD_3)_3CCOOH$ l. $CD_3CDOHCH_3$
m. (Z)-4-Deuterio-4-octene   n. 4-Octanone

7-19 Write the equation, where a reaction occurs, for the action of hot 50% sulfuric acid on each of the following:
a. 2,2-Diethoxypropane   b. 2-Isopropoxypropane
c. 2-Methyl-2,3-butanediol   d. 2,3-Epoxy-2,3-dimethylbutane
e. Ethylene oxide   f. 3,3-Dimethyl-2-butanol

7-20 The methyl group in crotonaldehyde will undergo many of the same reactions as that in acetaldehyde. Draw resonance structures for the carbanion obtained by treating it with a strong base. Give structures for the products obtained from an aldol condensation between crotonaldehyde and (*a*) itself (*b*) formaldehyde.

7-21 *One* of the important factors affecting the rate at which the carbonyl group reacts with nucleophilic reagents is the degree to which the carbon of the carbonyl is electron-deficient (positively charged). Groups that tend to increase this electron deficiency increase the relative rate of reaction, and groups that delocalize the positive charge through resonance slow the rate of reaction. Arrange the following carbonyl compounds in order of increasing reactivity toward an electron-rich reagent:

a. Acetaldehyde   b. Propionaldehyde   c. CH_3SCH_2CHO
Acetone   Chloroacetaldehyde CH_3OCH_2CHO
Trifluoroacetaldehyde   Bromoacetaldehyde $NCCH_2CHO$
Crotonaldehyde   Acrolein CH_3SeCH_2CHO
Methyl vinyl ketone   2,2-Difluoropropanal $HSCH_2CH_2CHO$

7-22 Taking advantage of the fact that acetals are hydrolyzed in acidic solution but not in basic solution, show how, by protecting the aldehyde group as an acetal, $CH_2=CHCHO$ can be converted to $HOCH_2CHOHCHO$.

7-23 Explain why HCN will not add to the double bond in $RCH=CHR$, but will add to the double bond in $CH_2=CHCOOH$.

7-24 Explain why only the α hydrogen atoms in aldehydes and ketones are involved in the aldol condensation.

7-25 An unknown compound C_7H_5N on catalytic hydrogenation gave $C_7H_{15}N$. Hydrolysis of the unknown converted it into an acid which on fusion with NaOH yielded C_6H_6. On catalytic hydrogenation this hydrocarbon gave C_6H_{12}. Suggest a suitable structure for the unknown.

7-26 Suggest a mechanism for each of the following reactions:

a. $CH_3COCH=CH_2 + HC(NO_2)_3 \xrightarrow{\text{OH}^-} CH_3COCH_2CH_2C(NO_2)_3$

b. $(CH_3)_2C(OH)CH_2I \xrightarrow{\text{HgO}} CH_3COCH_2CH_3$

c. 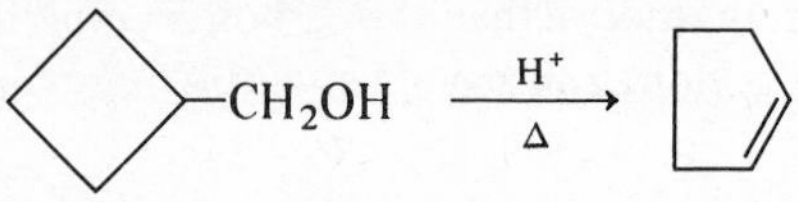 $-CH_2OH \xrightarrow[\Delta]{H^+}$

d. $CH_3COCH_3 + CH_2=CHC\equiv N \xrightarrow{OH^-} CH_3COCH_2CH_2CH_2CN$

e. $CH_3CH=CHCH=CHCO_2CH_3 + CH_2(CN)_2 \xrightarrow{OH^-} CH_3\underset{\underset{CH(CN)_2}{|}}{C}HCH_2CH=CHCO_2CH_3$

f. $(C_6H_5)_3\overset{+}{P}-CH_2(CH_2)_3COC_6H_5\ Cl^- \xrightarrow{CH_3Li}$ $-C_6H_5 + (C_6H_5)_3PO$

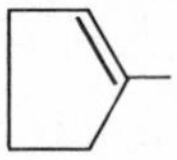

g. $OHCCHO \xrightarrow{OH^-} HOCH_2CO_2^-$

h. 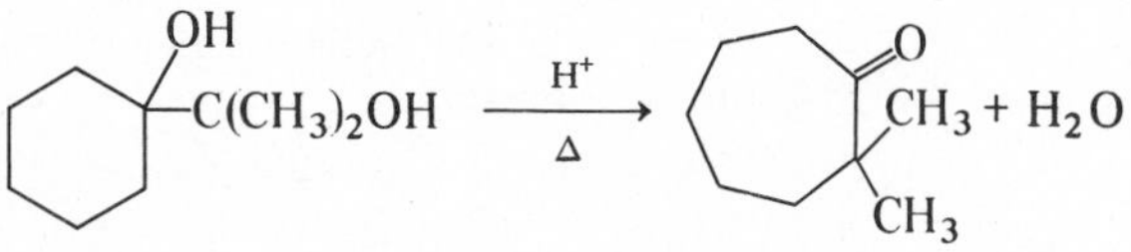 $\underset{-C(CH_3)_2OH}{\overset{OH}{|}} \xrightarrow[\Delta]{H^+}$ $CH_3 + H_2O$

i. $(C_6H_5)_3\overset{+}{P}-CH_2(CH_2)_2CH_2Br\ Br^- \xrightarrow{CH_3Li} (C_6H_5)_3\overset{+}{P}-\underset{\underset{CH_2-CH_2}{|}}{C}H-CH_2\ Br^-$

7-27 Ketene, $CH_2=C=O$, can be prepared by passing acetone vapor over a coiled resistance wire at $700°C$. Ketene will react readily with a variety of ROH and RNH_2 compounds.
  a. Write a mechanism for the reaction of ketene and CH_3OH to yield the ester, CH_3COOCH_3.
  b. Predict the products of the reaction of ketene with CH_3NH_2 and with CH_3CO_2H.
  c. Propose a free-radical mechanism for the pyrolysis of acetone, $CH_3COCH_3 \xrightarrow{\Delta} CH_2=C=O + CH_4$.

7-28 Aldehydes and ketones (less readily) react with ethylene glycol under acid catalysis to form cyclic acetals and ketals, respectively.

$$R_2C=O + HOCH_2CH_2OH \longrightarrow R_2C\underset{O-CH_2}{\overset{O-CH_2}{<}}\ |$$

Using this as a protecting device, outline the synthesis of the following.
  a. $OHC(CH_3)C=CHCH=C(CH_3)_2$ from $(CH_3)_2C=CHCH_2Br$ and CH_3COCHO
  b. $FCH_2C(CH_3)=CHCH_2COCH_3$ from FCH_2COCH_3 and $BrCH_2CH_2COCH_3$

7-29 From the mechanism of the Wittig reaction what would be the expected stereochemistry of the resulting trialkylphosphine oxide if an optically active phosphonium salt was treated as shown below?

$$CH_3-\underset{\underset{CH_2C_6H_5}{|}}{\overset{\overset{C_6H_5}{|}}{P^+}}-CH_2CH_3\ I^- \xrightarrow{C_6H_5Li} \xrightarrow{C_6H_5CHO} CH_3CH_2\underset{\underset{O}{||}}{P}(CH_3)C_6H_5 + C_6H_5CH=CHC_6H_5$$

7-30 Using the Wittig reaction and hydroboration methods, outline the synthesis of the following compounds starting from any organic compound containing four carbons or less, triphenylphosphine, and any inorganic reagents needed.

  a. 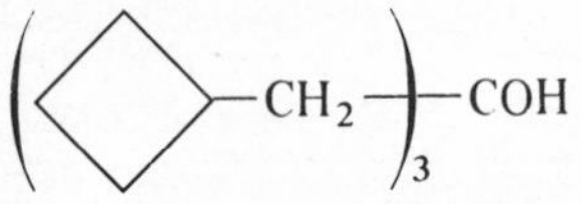 $\left(\text{}-CH_2\right)_3-COH$

  b. $(CH_3)_2CHCH_2\underset{\underset{O}{\overset{\overset{N-OH}{||}}{}}}{C}CH(CH_3)CH_2CH_3$

c. 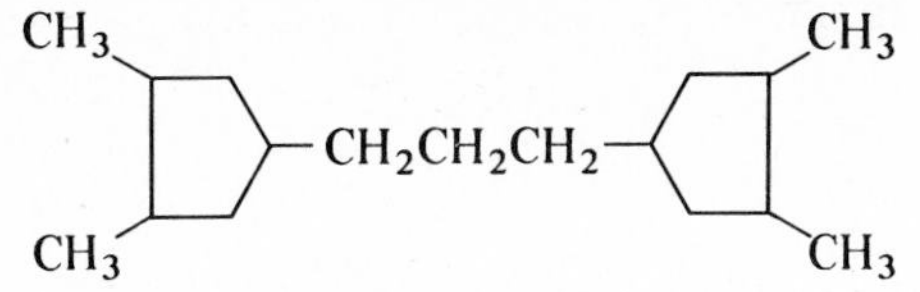
$$CH_3\text{-ring-}CH_2CH_2CH_2\text{-ring-}CH_3$$

(with CH3 substituents as shown)

d. 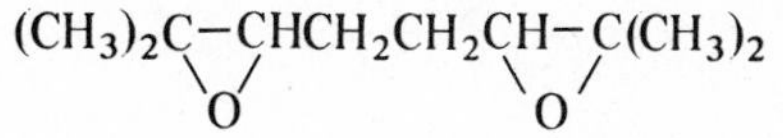 $(CH_3)_2C\text{—}CHCH_2CH_2CH\text{—}C(CH_3)_2$ with two epoxide (O) bridges

e. 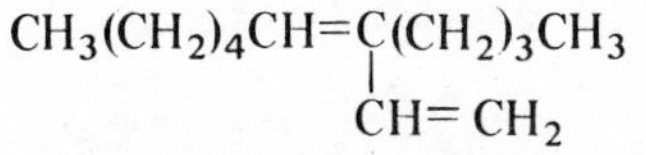 $CH_3(CH_2)_4CH{=}C(CH_2)_3CH_3$
  $\qquad\qquad\qquad CH{=}CH_2$

f. 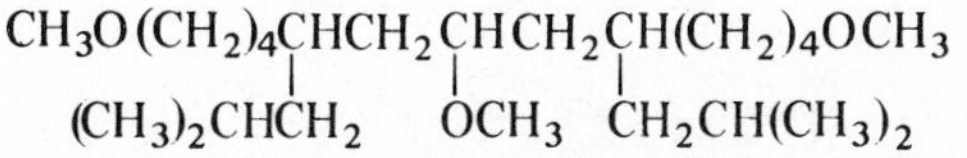 $CH_3O(CH_2)_4CHCH_2CHCH_2CH(CH_2)_4OCH_3$
  $\qquad (CH_3)_2CHCH_2 \quad OCH_3 \quad CH_2CH(CH_3)_2$

7-31 Show how each of the following groups of atoms could be arranged into an aromatic heterocyclic system (a cyclic structure having atoms other than carbon in the ring). Keep in mind the $4n + 2$ rule and the fact that a cyclic system of overlapping p orbitals is necessary.
 a. C_4H_4O b. C_5H_7N c. $C_3H_3N_3$ d. $C_3H_{12}B_3N_3$

7-32 An interesting cyclic system showing some aromatic character occurs in the sydnones. The name derives from the University of Sydney in Australia, where they were first studied.

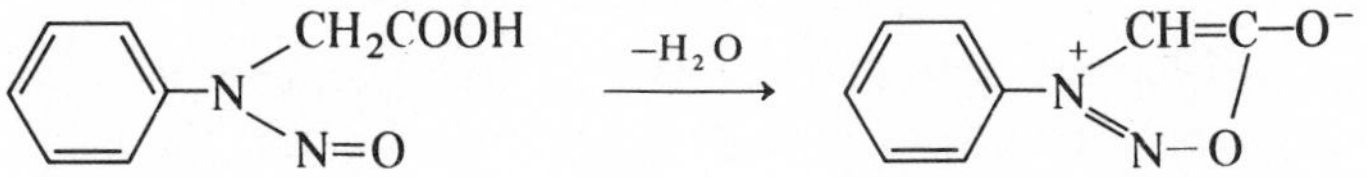

$$\text{phenyl-N(CH}_2\text{COOH)(N{=}O)} \xrightarrow{-H_2O} \text{phenyl-N}^{+}\text{(CH{=}C{-}O}^{-}\text{)(N{-}O)}$$

N-Phenylsydnone

What are the possibilities for resonance in the heterocyclic ring? Does the molecule fit the $4n + 2$ rule? What atomic orbitals would be used in forming the molecular orbitals?

7-33 The ketone tropone, (tropone structure), is exceptionally basic in character, forming much more stable salts with HCl than simple ketones. How can this be explained?

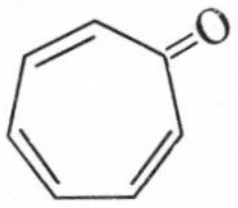

7-34 Pentaerythritol, $C(CH_2OH)_4$, can be obtained by condensing acetaldehyde and formaldehyde together in the presence of sodium hydroxide. Write a mechanism for this reaction.

7-35 For the reactions of alkynes mentioned in Sec. 7-2g-1, draw structures for the intermediates formed after hydroboration and prior to treatment with H_2O_2 or CH_3COOH. What product is expected for hydroboration $-$ H_2O_2 treatment of 4,4-dimethyl-2-pentyne (see Sec. 12-1b)?

8
structure determination by spectroscopic methods

8-1   The Electromagnetic Spectrum

There are many types of electromagnetic radiation ranging from cosmic rays at the high-energy end of the spectrum to radio waves at the low-energy end of the spectrum with wavelengths of about 500 m (see Fig. 8-1). The energy E in ergs associated with electromagnetic radiation of a particular frequency ν is given by the expression, $E = h\nu$, where h is Planck's constant (6.62×10^{-27} erg-s), or by $E = hc/\lambda$ where c is the speed of light (3×10^{10} cm/s) and λ is wavelength (in cm).

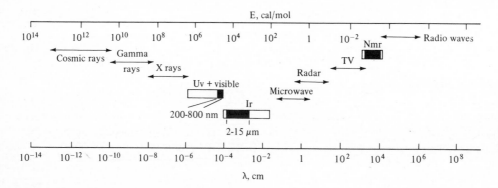

Figure 8-1. The electromagnetic spectrum.

Molecules exhibit many ways in which they can absorb electromagnetic radiation or energy but, according to quantum theory, a molecule will only absorb radiation whose frequency ν corresponds to the energy difference ΔE between two discrete energy levels in the molecule so that $\Delta E = h\nu$. The energy absorbed might result in promoting an electron in the molecule from a lower to a higher electronic energy level or it may cause a change from a lower to a higher vibrational or rotational energy state. The electronic, vibrational, and rotational levels which are allowed in a molecule are determined by the structure of the molecule.

For organic chemistry, there are three particularly useful regions of the electromagnetic spectrum. They are: (1) the ultraviolet and visible region, where the energy of the radiation or light is such that absorption by an organic molecule will promote an electron to a higher electronic energy level; (2) the near infrared and infrared (ir) region, where absorption of light in this energy range will cause vibrational excitation of organic molecules; and (3) nuclear magnetic resonance (nmr), where the energy absorbed causes transitions between different spin orientations of nuclei (protons in particular) when an organic molecule is placed in a magnetic field.

8-2   Ultraviolet and Visible Spectroscopy

a.   *Recording of UV Spectra* For this region (200 to 800 nm is the most useful range) spectra of organic compounds are recorded as absorbance A versus wavelength λ in nanometers (nm, where 10^{-7} cm = 1 nm). The absorbance (also called optical density) is dependent on the nature and the concentration of the absorbing species and the path length which the light travels in passing through the sample. This relationship is expressed mathematically by Beer's law, $A = \epsilon Cl$, where A is the absorbance, ϵ is called the molar extinction coefficient and is a constant which is characteristic of the absorbing species, C is the *molar* concentration of the absorbing species in solution, and l is the path length in centimeters (sample cell path lengths are usually 1.0 cm). The wavelength which corresponds to the maximum absorbance is called λ_{max} and the molar extinction coefficient which is associated with λ_{max} is called ϵ_{max}. The latter is a measure of how strong or how "allowed" the particular transition is.

b.   *Modes of Electronic Excitation*   The types of electronic excitation in the ultraviolet and visible region can best be understood in terms of molecular orbital energy diagrams. Molecules contain electrons which are involved directly in its bonding (called *bonding electrons*) and which are in sigmas (σ) or pi (π) molecular orbitals. *Antibonding orbitals,* σ^* and π^*, are unstable orbitals where the electron density between the nuclei is very low. In addition, many molecules, especially those containing O, S, N, Br, Cl, F, and I atoms, have electrons which are not directly involved in bonding (called *nonbonding electrons*) and which are in unhybridized nonbonding orbital(s) n. Figure 8-2 shows the various possible transitions of these bonding and nonbonding electrons from their respective energy levels. Of these possible excitation modes (designated $\sigma \rightarrow \sigma^*$, $\sigma \rightarrow \pi^*$, $\pi \rightarrow \pi^*$, $n \rightarrow \sigma^*$, and $n \rightarrow \pi^*$), the $n \rightarrow \pi^*$ and $\pi \rightarrow \pi^*$ transitions are the most observed and useful transitions in organic molecules. It might seem that these discrete transitions would lead to very sharp, simple line spectra. However, Fig. 8-3 illustrates that this is not the case because each electronic state (E) has a series of vibrational levels (V)

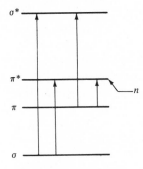

Figure 8-2. Molecular orbital energy diagram.

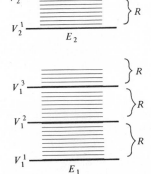

Figure 8-3. Electronic states with associated vibrational and rotational levels.

associated with it which are separated by relatively small energies and which in turn each have a series of rotational states (R) associated with them. Since a large number of transitions can occur between rotational levels of a lower vibrational state and rotational levels of a higher vibrational state, a broad absorption curve corresponding to the superposition and overlap of all such transitions is observed in the uv spectra for organic compounds in solution (see Fig. 8-4).

A molecular or functional group which can give rise to $\pi \rightarrow \pi^*$ and/or $n \rightarrow \pi^*$ transitions is called a chromophore (color bearer). Typical chromophores are C=C, C≡C, C=O, N=O, C=S, and aromatic rings. All of these contain π electrons and/or nonbonded electrons. Other groups such as –OH, –NH$_2$, –SH, and the halogens which contain unshared electrons do not usually absorb light above 200 nm, but when placed in conjugation with or attached to a chromophoric group, they cause the chromophore's absorption to move to longer wavelengths (called a *bathochromic shift* vs. a shift to shorter wavelength, which is called a *hypsochromic shift*). Such groups are given the term *auxochrome* (that is, they increase color). Conjugation of two or more chromophores, for instance, C=C–C=O, will also cause a bathochromic shift.

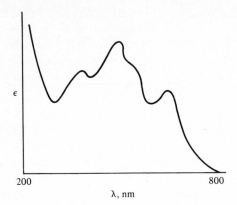

Figure 8-4. Typical uv spectrum.

The carbonyl group of aldehydes and ketones is a good example of a chromophore which can undergo $\pi \to \pi^*$ and $n \to \pi^*$ transitions:

$$\overset{\diagdown}{\diagup}C=\overset{..}{\underset{..}{O}}. \longrightarrow \overset{\diagdown}{\diagup}C-\overset{..}{\underset{..}{O}}: \qquad \overset{\diagdown}{\diagup}C=\overset{..}{\underset{.}{O}}: \longrightarrow \overset{\diagdown}{\diagup}C=\overset{..}{\underset{.}{O}}.$$

$$\pi \to \pi^* \qquad\qquad\qquad n \to \pi^*$$

The $\pi \to \pi^*$ transition involves an electron going from a bonding π orbital to a less stable antibonding orbital π^*. The $n \to \pi^*$ transition occurs when a nonbonded electron of oxygen is promoted to a less stable antibonding orbital π^*. For simple aldehydes and ketones, the more intense $\pi \to \pi^*$ transition occurs around 190 nm ($\epsilon \cong 10^2$ to 10^3) and the weaker $n \to \pi^*$ transition around 275 to 295 nm ($\epsilon \cong 10$ to 100).

A simple alkene group will undergo a $\pi \to \pi^*$ transition; however, this usually occurs below 200 nm and is not very useful. A conjugated diene or polyene system shifts this transition to longer wavelength, for example, 1,3-butadiene, $\lambda_{max} = 217$ nm ($\epsilon_{max} = 21{,}000$) and 1,3,5-hexatriene, $\lambda_{max} = 268$ nm ($\epsilon_{max} = 56{,}000$).

c. *Correlation of Structure with UV Spectra* The main value of ultraviolet and visible spectra lies in the detection of conjugation of various chromophores and in determining the nature of this conjugation. The following are some classes of compounds where uv spectra are useful in determining structures.

1. Conjugated dienes and trienes in cyclic systems. For *conjugated* polyenes in cyclic systems certain guidelines have been proposed by which one can predict the λ_{max} of similar compounds. The predictions start by taking a base value for the parent system of 214 or 253 nm, depending on whether the compound is a heteroannular diene (both double bonds not in the same ring) or a homoannular diene (both double bonds in the same ring), respectively. To this base value are added various increments (in nm), depending on the nature of other structural features present in the compound. Some examples are: +5 nm for each alkyl substituent or ring residue attached to the diene moiety, +30 nm per extension of the conjugated diene system by another C=C, +5 nm for each C=C which is exocyclic to a ring (one carbon of the C=C is in the ring with the double bond extending outside of the ring).

Predicted λ_{max} values using these guidelines agree well enough with the observed λ_{max} values for these kinds of compounds that they are useful in distinguishing between structures for compounds. For example, structures I and II have been proposed for a compound A whose observed λ_{max} is 282 nm. Using the above values, which would you propose as the most likely structure for A?

Homoannular diene	253
Substituents (marked S),	
4 x (5)	20
Exocyclic C=C, 2 x (5)	10
Calculated λ_{max}	283 nm

I

II

Heteroannular diene	214
Substituents (marked S),	
3 × (5)	15
Exocyclic C=C	5
Calculated λ_{max}	234 nm

Thus, structure I is the more likely structure for A.

2. α,β-Unsaturated ketones and aldehydes. Similar rules can be used to predict λ_{max} values for α,β-unsaturated ketones and aldehydes using a base value of 215 nm for the parent unsaturated ketone and 210 nm for the parent unsaturated aldehyde. To these base values are added the following increments: +30 nm per extension of the conjugation by a C=C; +10 mμ, +12 mμ, and +18 mμ for an α, β, and γ (or higher) alkyl group or ring residue, respectively; +39 nm for the presence of a homocyclic group (two C=C in the same ring); and +5 for exocyclic double bonds. Some examples:

Base value	215
Extended conjugation	30
Exocyclic C=C	5
Substituents (marked S), γ or higher, 2 × (18)	36
Calculated λ_{max}	286 nm
Observed λ_{max}	290 nm

Base value	215
Extended conjugation	60
Homodienic component	39
Substituents (marked S), β, 1 × 12	12
γ or higher, 3 × (18)	54
Exocyclic C=C	5
Calculated λ_{max}	385 nm
Observed λ_{max}	388 nm

3. Aromatic or benzenoid systems. Benzene uv spectrum shows three bands attributed to various $\pi \rightarrow \pi^*$ transitions: 184 nm (ϵ = 50,000), 204 nm (ϵ = 7,400), and 255 nm (ϵ = 200). If the benzene ring has an additional unsaturated group, (for instance, C=O, C=C) which is attached to the ring and conjugated with the benzene ring's π-electron system, then an additional overlapping band will appear in the 200 to 250 nm region. Attachment of groups such as OH, NH_2, and other auxochromes which have unshared electrons will cause an intensification (larger ϵ) and a bathochromic shift of the benzene band at 255 nm. However, it is generally found that accurate correlation of structure and uv spectra for benzene derivatives and other aromatic compounds is extremely difficult and works well only in very closely related series of compounds.

4. Derivatives of aldehydes and ketones. 2,4-Dinitrophenylhydrazone derivatives of aldehydes and ketones are usually nice yellow or red crystalline compounds and are found to have λ_{max} at 360 nm, ϵ = 2 × 10^4 (for these derivatives of α,β-unsaturated ketones or aldehydes λ_{max} = 375–385 nm, ϵ_{max} = 3 × 10^4). Oximes usually absorb at about 40 nm longer wavelength than do the parent aldehyde or ketone.

8-3 Infrared Spectroscopy

a. *Recording of IR Spectra* Infrared spectra are most commonly determined for the region of 2.0 to 16.0 micrometers (μm). The ir spectrometer records the percentage transmittance (%T) of incident light through the sample (transmittance is related to absorbance A by the equation, $A = \log_{10} 1/T$) vs. the wavelength of light expressed in micrometers (1 μm = 10^{-4} cm), or vs. the frequency of incident light expressed in wave numbers. The wave number is the reciprocal of the wavelength expressed in centimeters, and it is directly proportional to the energy of the light absorbed.

b. *Modes of Vibrational Excitation* The absorption of infrared radiation results in the excitation of organic molecules through conversion of this energy into increased vibrational motion of the bonded atoms. Molecular vibrations are of two types. First, there are stretching vibrations which involve movement of bonded atoms along the bond axis thus lengthening or compressing the interatomic distance. Second, there are bending vibrations which consist of movement of atoms, but away from the bonding axis. The various types of stretching and bending vibrations are illustrated for a methylene group:

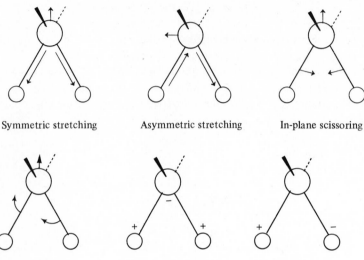

Symmetric stretching   Asymmetric stretching   In-plane scissoring

Bending, rocking   Out-of-plane wagging   Bending, twisting

For a molecule containing n atoms there are $3n - 6$ normal stretching and bending modes ($3n - 5$ for linear molecules). Not all of these vibrations will be observed in the ir spectrum since for such a vibration to result in an ir absorption it must be accompanied by a change in the dipole moment of the molecule. Thus, vibrational modes such as C=C stretching in a symmetric molecule like ethylene will not result in an ir band. Other reasons for the failure to see all of the theoretical number of bands in the ir spectrum of a molecule are: (1) vibrational frequencies not in the normal 2 to 15 nm range, (2) extremely weak modes, (3) coalescence of two or more close-lying vibrational frequencies into broad bands. In addition to the fundamental vibrational modes mentioned above, there may be less intense bands that appear at twice or three times the frequency of a particular fundamental mode. These are called *overtones*. Also, bands due to the combination of two different bands or bands corresponding to the difference in frequencies between two vibrational modes may appear.

c. *Correlation of Structure with IR Spectra* A typical ir spectrum will contain many sharp and broad bands. Even a person very experienced in the interpretation of ir spectra will not be able to assign every band to specific vibrational modes. The portion of an ir spectrum between 1300 and 909 cm^{-1} (7.7 to 11.0 μm) is frequently quite complex. This is appropriately called the *fingerprint* region, and it can be very helpful in matching a spectrum of an unknown compound with one for which the compound's structure is known. If two such spectra match peak for peak in the fingerprint region, one can be very sure that the two compounds are the same.

      Fortunately for the organic chemist, many functional groups have characteristic absorptions whose frequencies are relatively independent of the structure of the rest of the molecule. The identification of the presence or absence of a functional group in a molecule is one of the primary uses of ir spectra since this is possible by rather simple inspection. Complete determination of the entire structure of a compound by ir spectroscopy is not done very often since there are other methods which can yield the same additional structural information usually much faster and easier than depending solely on the ir spectrum. One of these methods is nuclear magnetic resonance, which is discussed in Sec. 8-4.

      Table 8-1 lists some of the more characteristic ir bands for various functional groups, some of which have not been discussed in preceding chapters. Keep in mind that the frequency for a particular group's ir band(s) may fall outside the value or range given in the table in certain situations. As new functional groups are discussed in subsequent chapters, the student should refer back to Table 8-1 for reference. Ir spectra for some are shown in Fig. 8-8 along with an explanation of key absorption bands.

Table 8-1   Characteristic ir absorptions of functional groups

Group and vibrational type	Most useful bands, μm, cm^{-1} (intensity)†	Others, μm, cm^{-1} (intensity)†
I. Alcohols (ROH), phenols (ArOH)		
A. O–H stretching		
Free OH	2.74–2.79, 3650–3584 (*s, sh*)	
H-bonded, intermolecular (dimer)	2.82–2.90, 3550–3450 (*s, sh*)	
H-bonded, intermolecular (polymer)	2.94–3.13, 3400–3200 (*s, b*)	
H-bonded, intramolecular	2.80–2.90, 3570–3450 (*m, sh*)	
B. O–H bending, 1°, 2°, 3°, and aromatic		7.1–7.9, 1410–1260 (*s*)
C. C–O stretching		
1° alcohols	9.22–9.52, 1085–1050 (*s*)	
2° alcohols	8.90–9.20, 1124–1087 (*s*)	
3° alcohols	8.30–8.90, 1205–1124 (*s*)	
Phenols	7.93–8.48, 1260–1180 (*s*)	
II. Aldehydes (RCHO)		
A. C=O stretching		
Aliphatic, saturated	5.75–5.81, 1740–1720 (*s*)	
Aliphatic, α,β-unsaturated	5.87–5.95, 1705–1680 (*s*)	
Aromatic	5.83–5.90, 1715–1695 (*s*)	
B. C–H stretching (2 bands)	3.53–3.71, 2830–2695 (*w-m*)	
III. Alkanes (RH)		
A. C–H stretching	3.38–3.51, 2962–2853 (*m-s*)	
B. C–H bending		
–CH–		$\simeq$ 7.46, $\simeq$ 1340 (*w*)
–CH$_2$–	6.74–6.92, 1485–1445 (*m*)	
–CH$_3$	7.25–7.30, 1380–1370 (*s*)	6.80–7.00, 1470–1430 (*m*)
gem-Dimethyl, –C(CH$_3$)$_2$ (2 bands)	7.22, 7.30, 1385, 1370 (*s*)	
tert-Butyl (2 bands)	7.20, 7.30, 1389, 1370 (*m, s*)	
IV. Alkenes 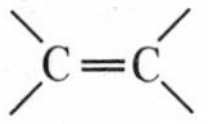 $\diagdown C = C \diagup$		
A. C=H stretching	3.24–3.32, 3086–3012 (*m*)	
B. C–H bending		
Monosubstituted, RCH=CH$_2$ (3 bands)	10.10, 11.00, 990, 909 (*s*)	7.04–7.09, 1420–1410 (*s*)
Cis disubstituted 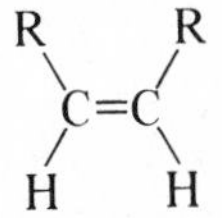	$\simeq$ 14.5, $\simeq$ 690 (*s*)	
Trans disubstituted 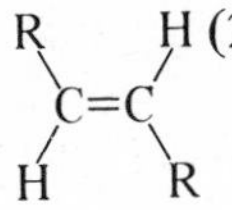 (2 bands)	10.31–10.42, 970–960 (*s*)	7.64–7.72, 1310–1295 (*m*)
Geminal disubstituted 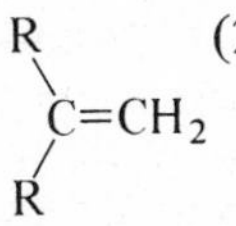 (2 bands)	11.17–11.30, 895–885 (*s*)	7.04–7.09, 1420–1410 (*s*)
Trisubstituted, R$_2$C=CHR	11.90–12.66, 840–790 (*m-s*)	
C. C=C stretching		6.00–6.10, 1660–1640 (*w-m*)
Allenes, C=C=C (2 bands)	5.1, 9.4, 1950, 1060 (*m*)	
V. Alkynes (–C≡C–)		
A. C–H stretching	$\simeq$ 3.03, $\simeq$ 3300 (*m-s*)	
B. C–H bending		$\simeq$ 15.9, $\simeq$ 630 (*s*)

Table 8-1 (*Continued*)   Characteristic ir absorptions of functional groups

Group and vibrational type	Most useful bands, μm, cm⁻¹ (intensity)†	Others, μm, cm–₁ (intensity)†
C. C≡C stretching		
Monsubstituted	4.67–4.76, 2140–2100 (*w-m*)	
Disubstituted		4.42–4.57, 2260–2190 (*w*)
VI. Amides (RCONH₂)		
A. N–H stretching		
1°, free (2 bands)	≃ 2.86, 2.94, ≃ 3500, 3400 (*m*)	
1°, bonded (2 bands)	≃ 2.99, 3.15, ≃ 3350, 3180 (*m*)	
2°, free	≃ 2.90, ≃ 3448 (*m*)	
2°, bonded	≃ 3.15, ≃ 3175 (*m*)	
B. N–H bending		
1°, dilute solution	6.17–6.29, 1620–1590 (*s*)	
2°, dilute solution	6.45–6.62, 1550–1510 (*s*)	
C. C=O stretching		
1°, concentrated solution or solid	≃ 6.06, ≃ 1650 (*s*)	
1°, dilute solution	≃ 5.92, ≃ 1690 (*s*)	
2°, concentrated solution or solid	5.95–6.14, 1680–1630 (*s*)	
2°, dilute solution	5.88–5.99, 1700–1670 (*s*)	
3°, dilute or concentrated solution	5.99–6.14, 1670–1630 (*s*)	
VII. Amines (RNH₂)		
A. N–H stretching		
1°, free (2 bands)	≃ 2.86, 2.94, ≃ 3500, 3400 (*m*)	
2°, free	2.86–3.02, 3500–3310 (*m*)	
Amine salts	3.0–3.3, 3300–3030 (*s, b*)	
B. N–H bending		
1° and 2°	6.06–6.45, 1650–1550 (*v*)	11.1–14.3, 900–700 (1° only, broad)
Amine salts (2 bands)	≃ 6.30, 6.67, 1587, 1500 (*s*)	
C. C–N stretching		
1°, 2°, 3° aromatic	7.3–8.0, 1370–1250 (*s*)	
Aliphatic		8.2–9.8, 1220–1020 (*m-w*)
VIII. Anhydrides (RCO₂)₂O		
A. C=O stretching (2 bands)	≃ 5.40, 5.76, ≃ 1852, 1761 (*s*)	
B. C–O stretching		
Acyclic	≃ 9.55, ≃ 1047 (*s*)	
Cyclic (2 bands)	≃ 10.75, 8.0, ≃ 930, 1250 (*s*)	
IX. Substituted aromatic rings		
A. C–H stretching		
B. C–H bending		
Monosubstituted (2 bands)	≃ 13.3, 14.3, ≃ 750, 700 (*s*)	Overtones and combination bands appear at 5–6 μm, 2000–1670 cm⁻¹ Weak — normally not helpful
Ortho disubstituted	≃ 13.3, ≃ 750 (*s*)	
Meta disubstituted (2 bands)	≃ 11.3, 12.8, ≃ 780, 880	
Para disubstituted	≃ 12.0, ≃ 830 (*s*)	
C. C=C stretching (several bands)		6.25–6.67, 1600–1500 (*v*)
X. Carboxylic acid (RCO₂H)		
A. C=O stretching		
Aryl	5.88–5.95, 1700–1680 (*s*)	
α,β-Unsaturated and saturated aliphatic	5.80–5.92, 1725–1690 (*s*)	
B. O–H stretching (several bands)	3.4–4.0, 2940–2500 (*w-m*)	
XI. Esters (RCO₂R′)		
A. C–O stretching		7.7–9.5, 1300–1050 (*s*)

Table 8-1 (*Continued*) Characteristic ir absorptions of functional groups

Group and vibrational type	Most useful bands, μm, cm^{-1} (intensity)†	Others, μm, cm^{-1} (intensity)†
B. C=O stretching		
Saturated acyclic	5.71–5.76, 1750–1735 (*s*)	
δ-(or larger ring) Lactones	5.71–5.76, 1750–1735 (*s*)	
α-Lactones	5.62–5.68, 1780–1760 (*s*)	
β-Lactones	≃ 5.5, ≃ 1820 (*s*)	
α,β-Unsaturated and aromatic	5.78–5.82, 1730–1717 (*s*)	
α-Keto esters	5.70–5.75, 1755–1740 (*s*)	
β-Keto esters (enolic)	≃ 6.06, ≃ 1650 (*s*)	
XII. Ethers (R_2O)		
A. C–O–C stretching	8.7–9.35, 1150–1070 (*s*)	
XIII. Halides, acyl (RCOX)		
A. C=O stretching		
Chlorides	≃ 5.57, ≃ 1795 (*s*)	
Bromides	≃ 5.53, ≃ 1810 (*s*)	
B. C–X stretching (alkyl or acyl)		
C–F	7.1–10.0, 1400–1000 ⎱	C–Br and C–I absorb
C–Cl	12.5–16.6, 800–600 ⎰	below 16 μm (500 cm^{-1})
XIV. Ketones ($R_2C=O$)		
A. C=O stretching		
Saturated acyclic and six-membered cyclic	5.80–5.87, 1725–1705 (*s*)	
Five-membered cyclic	5.71–5.75, 1750–1740 (*s*)	
Four-membered cyclic	≃ 5.63, ≃ 1775 (*s*)	
α,β-Unsaturated, acyclic and six-membered cyclic	5.94–6.01, 1685–1665 (*s*)	
α,β-Unsaturated, five-membered cyclic	5.80–5.85, 1725–1708 (*s*)	
Aryl and diaryl	5.88–6.02, 1700–1660 (*s*)	
α-Diketones	5.73–5.85, 1730–1710 (*s*)	
β-Diketones (enolic)	6.10–6.50, 1640–1540 (*s*)	
1,4-Quinones	5.92–6.02, 1690–1660 (*s*)	
XV. Nitriles (RC≡N)		
A. C≡N stretching	4.42–4.50, 2260–2220 (*m*)	
Isocyanates R–N=C=O	4.40–4.46, 2275–2240 (*m*)	
XVI. Nitro compounds (RNO_2)		
A. NO_2 stretching		
RNO_2 and $ArNO_2$ (2 bands)	6.37–6.67, 7.30–7.70, 1570–1500, 1370–1300 (*s*)	
XVII. Phenols (see alcohols)		
XVIII. Sulfur compounds		
A. S–H stretching (mercaptans)	3.85–3.92, 2600–2550 (*s*)	
B. S–O stretching	11.1–14.2, 900–700 (*s*)	
C. S=O and SO_2 stretching		
Sulfoxides ($>$S=O)	9.43–9.62, 1090–1020 (*s*)	
Sulfones 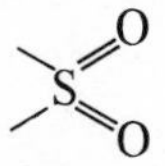 (S with two =O), 2 bands	7.42–7.53, 8.62–8.93, 1350–1310, 1160–1120 (*s*)	
Sulfonic acids (–SO_2OH, 2 bands)	7.93–8.70, 9.26–9.90, 1260–1150, 1080–1010 (*s*)	

Table 8-1 (*Continued*)   Characteristic ir absorptions of functional groups

Group and vibrational type		Most useful bands, μm, cm^{-1} (intensity)†	Others, μm, cm^{-1} (intensity)†
	Sulfonates ($-SO_2OR$, 2 bands)	7.04–7.52, 8.33–8.73, 1420–1330, 1200–1145 (s)	
	Sulfonamides ($-SO_2NR_2$, 2 bands)	7.30–7.52, 8.47–8.62, 1370–1330, 1180–1160 (s)	
	Sulfonyl chlorides ($-SO_2Cl$, 2 bands)	8.44–8.59, 7.30–7.46, 1185–1165, 1370–1340 (s)	
D.	C=S stretching	8.33–9.52, 1200–1050 (s)	
XIX.	Miscellaneous nitrogen compounds		
A.	Azo compounds, N=N stretching	6.14–6.35, 1630–1575 (v)	
B.	Imides		
	Cyclic, six-membered ring (2 bands)	$\simeq$ 5.82, 5.93, $\simeq$ 1718, 1686 (s)	
	Cyclic, five-membered ring (2 bands)	$\simeq$ 5.62, 5.86, $\simeq$ 1779, 1706 (s)	
C.	Imines, oximes, C=N stretching	5.96–6.12, 1678–1634 (v)	
D.	Ureas, C=O stretching	$\simeq$ 6.02, $\simeq$ 1660 (s)	
E.	Urethanes, C=O stretching	5.75–5.92, 1740–1690 (s)	

† s = strong, m = medium, w = weak, v = variable, b = broad, sh = sharp.

8-4   Nuclear Magnetic Resonance Spectroscopy

a.   *Basic Theory*   Nuclear magnetic resonance (nmr) deals with the transitions of nuclei from one spin state to another spin state by absorption of energy. Nuclei, like electrons, have a mechanical spin which, along with the charge of the nucleus, generates a magnetic field. In effect, such nuclei are similar to tiny bar magnets. Usually when a spinning nucleus is placed in a magnetic field H_0, the magnetic moment of the nucleus can assume $2I + 1$ orientations with respect to the applied field where I is the nuclear spin quantum number. The value of I depends on the nucleus involved. Nuclei such as ^{12}O and ^{16}O have $I = O$ and are nonmagnetic. The hydrogen nucleus or the proton has $I = \frac{1}{2}$; thus, it can assume two possible orientations — with or against the external magnetic field. Alignment with the field is the more stable, lower-energy spin state. Energy is required to cause the hydrogen nucleus in this spin state to "flip" over to the higher-energy spin state where the alignment of the nuclear magnetic moment is against the field. The energy required is dependent on the strength of the external magnetic field. Stronger field strengths require greater energies and higher frequencies of radiation to flip the proton over. This relationship is expressed by the equation $\nu = 2\mu H_0/h$, where ν is the frequency of exciting radiation, μ is the magnetic moment of the nucleus, H_0 is the external magnetic field strength, and h is Planck's constant. When H_0 is about 14,000 g, the frequency of radiation required to cause a nuclear transition is about 60 MHz (60 x 10^6 c/s).

A set of hydrogen nuclei will undergo transitions from the lower to the higher spin state by absorption of energy until such time as the populations of the two spin states are equal. When this point is reached, no more absorption will occur (saturation). To begin with, the population distribution of spin states for hydrogen at room temperature is such that a very slight excess (about 10^{-3} percent) of the nuclei are in the lower spin state. If it were not for this slight excess, we would observe no absorption at all, since the probability of a transition from a lower to a higher spin state (absorption) is equal to the probability of a transition from a higher to a lower spin state (emission). Additional pathways are available, however, by which a proton in the higher spin state can return to the lower spin state. These pathways are called relaxation processes, and they are of two general types. The first is spin-spin relaxation which consists of the exchange of spin states between neighboring nuclei. The second is spin-lattice relaxation which occurs when the spin energy of a nucleus in a higher spin state is lost to the lattice (the aggegate of other atoms or molecules in the system) by increasing the transitional, vibrational, and rotational energy of the lattice. The rates at which these two processes occur (called *relaxation times*) affect the width of the absorption line observed — the longer a nucleus remains in an excited spin state the broader its absorption line will be.

102

Nuclear transitions in hydrogen nuclei can be made to occur by keeping the applied magnetic field strength constant and varying the frequency of a rotating magnetic field whose magnetic field vector is perpendicular to the applied magnetic field. When the frequency of the rotating magnetic field becomes equal to the frequency needed to flip the proton over, then *resonance* occurs. Another method of obtaining proton resonance which lends itself more easily to instrument design is to keep the frequency of the rotating magnetic field constant (often at 60 Mc) and to vary the strength of the applied magnetic field.

b. *The Chemical Shift* All protons in a molecule do not have the same resonance frequencies. This is due to the dependence of a particular proton's resonance frequency on the molecular environment in which it resides. The electrons, which are around a proton's environment and the induced circulation of these electrons in the presence of a magnetic field, provide a magnetic field of their own which then changes the "effective" magnetic field that the proton actually feels in its particular environment. This effect is called *shielding.* Protons in different environments will differ in the strength of the applied magnetic field which is necessary before the magnetic field actually felt by the different protons is large enough to induce a nuclear transition.

In practice, the relative position of a proton's absorption to that of a standard proton such as tetramethylsilane (TMS), $(CH_3)_4Si$ is determined. It gives a single, sharp absorption line at a higher frequency than most organic compounds, and it is chemically inert. The distance expressed in hertz (Hz) [cycles per second (cps)] between the absorption of TMS and the absorption of a particular proton is called the *chemical shift.*

The symbol delta (δ) is also used to designate resonance positions and can be calculated using the equation

$$\delta = \frac{(\text{chemical shift in Hz}) \times 10^6}{\text{spectrometer frequency}}$$

Note that δ is a dimensionless number. Unless designated otherwise, all nmr spectra in this book are for a spectrometer frequency of 60×10^6 Hz. The third symbol tau (τ) is used sometimes to denote absorption positions. It is related to δ by the equation, $\tau = 10 - \delta$. The TMS peak is arbitrarily adjusted to $\delta = 0$ ($\tau = 10$) on the chart paper when recording nmr spectra. It is possible to use different sweep widths in scanning the range of applied magnetic field strengths. The most common sweep width is 500 Hz, which corresponds on the chart paper to δ values of 0 to 8.3.

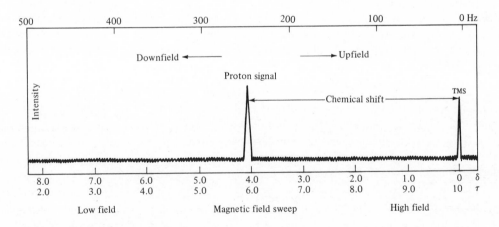

Figure 8-5. Sample nmr spectrum.

c. *Spin-Spin Interactions* If one considers a molecule such as ethanol, CH_3CH_2OH, it is readily seen that there are three types of protons — the methyl protons, the methylene protons, and the hydroxyl proton — all of which are in a different environment and which should give rise to three absorption peaks. At low resolution, the nmr spectrum of ethanol does show three broad peaks: the hydroxyl proton absorbs at $\delta = 5.2$, the methylene protons absorb at $\delta = 3.5$, and the methyl protons absorb at $\delta = 1.1$. The areas under these peaks are found to be in the ratio of 1:2:3, respectively. This is the same ratio as the number of protons in each respective group. This is a particularly useful feature of nmr spectra: by measurement of the areas under different peaks (the nmr

103

spectrometer uses an electronic integrator), one is able to determine the relative ratio of protons which are in different chemical environments. Furthermore, if one is able to make a definite assignment of a particular kind of proton to a given observed peak (e.g., the hydroxyl proton in an alcohol containing one OH group), then determination of the absolute ratio of the different protons from the relative ratio becomes possible.

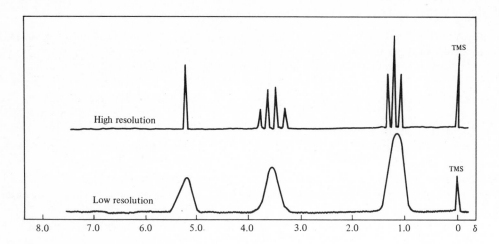

Figure 8-6. Nmr spectrum of ethanol.

The high-resolution spectrum of acidified ethanol shows that the rather broad peaks due to the methylene and methyl protons in the low-resolution spectrum have respectively split into a set of four peaks (a quartet) centered at $\delta = 3.6$ and a set of three peaks (a triplet) centered at $\delta = 1.2$. The total areas under the quartet (due to CH_2) and the triplet (due to CH_3) are still in the ratio of 2:3. This observed splitting is due to what is called *spin-spin interaction*. The methylene protons in ethanol are affected by the spin arrangements of the methyl protons, and the methyl protons are influenced by the spin arrangements of the methylene protons. Consider the diagram below in which a proton H_A is adjacent to a methylene group, both of whose hydrogens have the same chemical shift which is greatly different from the chemical shift of H_A.

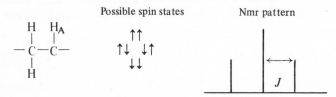

The possible spin-state combinations of the methylene protons that H_A might "see" and be influenced by are: $+\frac{1}{2}$ (↑↑); $+\frac{1}{2}$ and $-\frac{1}{2}$ (↑↓) or $-\frac{1}{2}$ and $+\frac{1}{2}$ (↓↑); and both $-\frac{1}{2}$ (↓↓). Any of these combinations is equally probable and thus H_A gives rise to the three-line pattern shown where the relative intensities of the lines (i.e., relative peak areas) are 1:2:1 (explanation of J given below).

Consider the situation where H_A is adjacent to a methyl group as shown below.

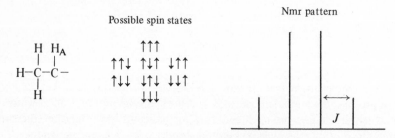

The different combinations of spins of the methyl hydrogens are shown along with the four-line pattern or quartet expected for a proton H_A in such an environment. The line intensities are 1:3:3:1. It should be mentioned that the above splitting pattern for a proton such as H_A will not be as symmetric in the actual nmr

104

spectra as indicated in the previous discussion. One observes rather that the line in H_A's quartet that is nearest to the peaks of the coupled hydrogens (CH_3 in this case) will be larger than the line farthest removed from the CH_3 absorption. For example, an ethyl group, CH_3CH_2-, will show a pattern something like

(CH_2) (CH_3),

rather than a totally symmetric one. This can be helpful in deciding which peaks in an nmr spectrum represent coupled sets of hydrogens.

Extension of the above discussion would show that the number of peaks or *multiplicity* observed for a given proton (or equivalent protons) is equal to $n + 1$, where n is the number of protons on adjacent atoms. Some examples illustrating the concept of multiplicity are shown below:

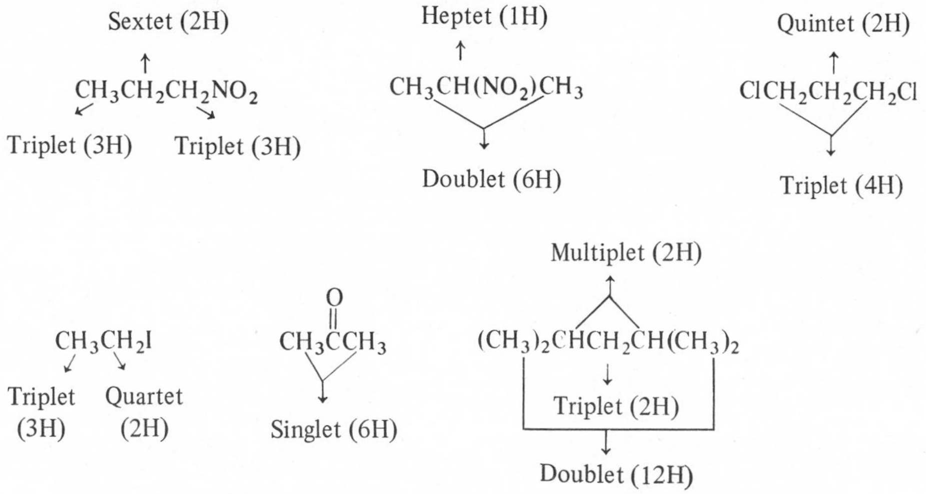

Note: Predicted multiplicities will be observed only if the difference in chemical shifts of the two types of coupled protons is much larger than their spin coupling constant J (see below).

d. *Spin Coupling Constants* The separation between lines (measured in Hz) in a multiplet (doublet, triplet, etc.) is called the spin coupling constant J. The magnitude of J depends on how efficiently the proton(s) in question are spin-coupled with adjacent proton(s), i.e., influenced by the spin states of adjacent protons and thus it can be helpful in determining the nature of the coupled hydrogens.

Some typical values for J are listed in Table 8-2. It should be noted that J is independent of the applied magnetic field strength and is a constant for coupled hydrogens in a given molecule. The ranges given for J

values for various types are the result of often being dependent on the dihedral angle,  H, and/or the types of substituents present on the carbon(s) bearing the coupled hydrogens.

e. *Shielding Mechanisms* Protons in molecules experience various types of magnetic shielding due to induced circulations of electrons in the molecule by the applied magnetic field.

1. Local diamagnetic shielding. This is due to the circulation of the hydrogen's own electrons and hence will be highly dependent on the electron density at the hydrogen atom. Any electronegative group (see Table 8-3) will tend to lower the electron density at the hydrogen atom and will cause a shift in the position of the hydrogen's absorption to lower field (higher δ values). Other things being equal, the more electronegative groups there are attached to a carbon holding a hydrogen, the greater will be the shift downfield: CH_3I ($\delta = 2.16$), CH_2I_2 ($\delta = 4.09$). This effect is not always strictly additive. The chemical shifts of methyl, methylene, and methine protons generally increase in that order:

$CH_3CH_3(\delta_{CH_3} = 0.9)$, $CH_3CH_2(\delta_{CH_2} = 1.25)$, $CH_3CH(\delta_{CH} = 1.5)$

Table 8-2

Group	J Hz	Group	J Hz
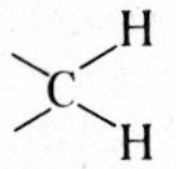 $\diagup C \diagdown^H_H$ geminal	12–15	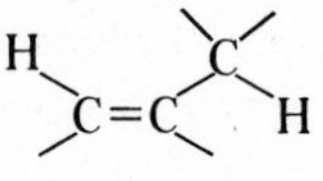 $H\diagdown C=C\diagup_H$	0.5–3
$\diagdown CH-CH\diagup$ vicinal	2–9	$\diagdown C=CH-CH=C\diagup$	9–13
CH_3CH_2X	6–8	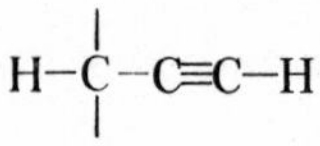 $H-\overset{\vert}{\underset{\vert}{C}}-C\equiv C-H$	2–4
$(CH_3)_2CHX$	5.5–7	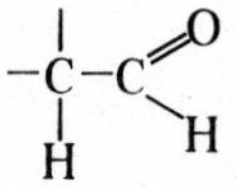 $-\overset{\vert}{\underset{H}{C}}-C\overset{O}{\diagdown}_H$	1–3
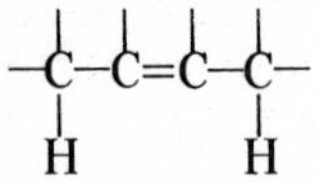 $-\overset{\vert}{C}-\overset{\vert}{\underset{H}{C}}=\overset{\vert}{C}-\overset{\vert}{\underset{H}{C}}-$	0–1.6	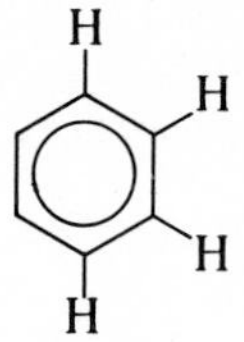 benzene ring (ortho, meta, para)	ortho 7–10 / meta 2–3 / para 0–1
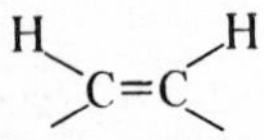 $\underset{}{^H\diagdown}C=C\diagup^H$ cis	6–14	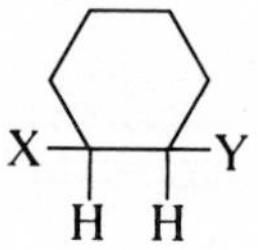 cyclohexane X—Y	axial, axial 5–10 / axial, equatorial 2–4 / equatorial, equatorial 2–4
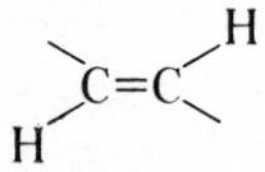 $^H\diagdown C=C\diagup^H$ trans	12–18	cyclopropane	cis 7–11 / trans 4–8
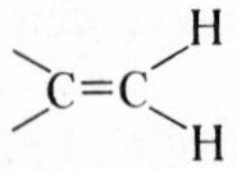 $\diagdown C=C\diagdown^H_H$ geminal	1–3	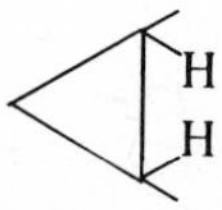	
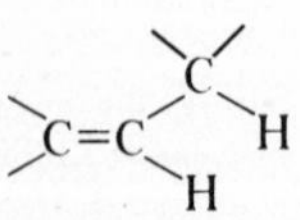 $\diagdown C=C\diagup^H_H$	4–10		

Table 8-3

	CH_4	NH_3	HOH	HI	HBr	HCl	HF
δ	0.9	2.2	3.5	2.16	2.68	3.05	4.26
Electronegativity	2.5	3.0	3.5	2.5	2.8	3.0	4.0

2. Neighboring diamagnetic and paramagnetic shielding. Induced circulations of the electrons about two neighboring atoms in such groups as C=C, C=O, C≡C can give rise to diamagnetic shielding (shift to lower δ or upfield) or paramagnetic deshielding (shift to higher δ or downfield) depending on where the hydrogen in question is spatially with respect to these groups. Figure 8-7 illustrates various spatial relationships of the proton and the direction of the shift observed. As a result of these effects, aldehydic protons (II) absorb far downfield ($\delta = 10$), olefinic protons (I) are farther upfield ($\delta = 5$), and acetylenic protons (III) are the farthest upfield ($\delta = 2.5$).

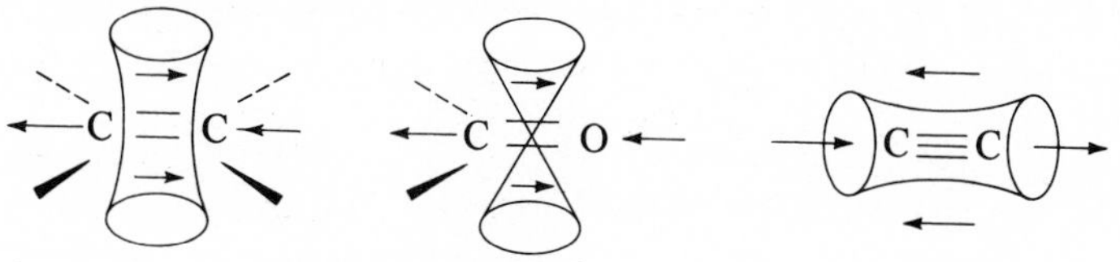

Figure 8-7. Diamagnetic and paramagnetic shielding.

3. Interatomic diamagnetic and paramagnetic shielding. This occurs in molecules where electrons may circulate very readily over two or more atoms. The π-electron system of a benzene ring is a good example of where strong electron currents are induced by a magnetic field. Consequently, the ring protons of benzene which are in the plane of the ring will be shielded and will absorb downfield. Aromatic ring protons absorb at $\delta = 6$ to 8, while the protons of a methyl group attached to benzene rings absorb at $\delta = 2.5$, versus an aliphatic CH_3C- which absorbs at $\delta = 1.0$. Any proton in a relatively rigid molecule which is held above or below the plane of the benzene ring will feel a shielding effect due to the ring current effect and will absorb farther upfield as a result.

$\delta_{CH_3} = 1.77$

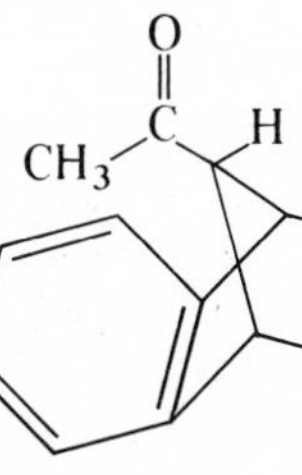

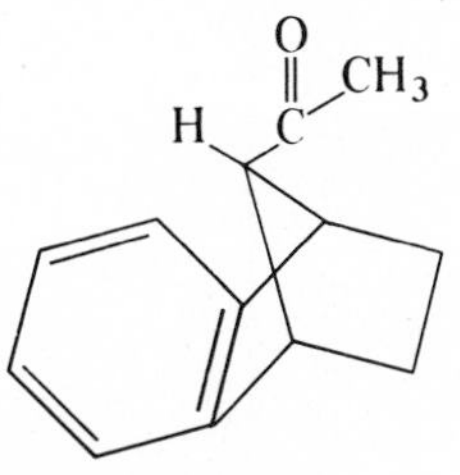

$\delta_{CH_3} = 2.31$

f. *Exchange of the Chemical Environment of Protons* Consider the situation where two protons are in different environments which may be exchanged by rotation about a carbon-to-carbon bond as shown below.

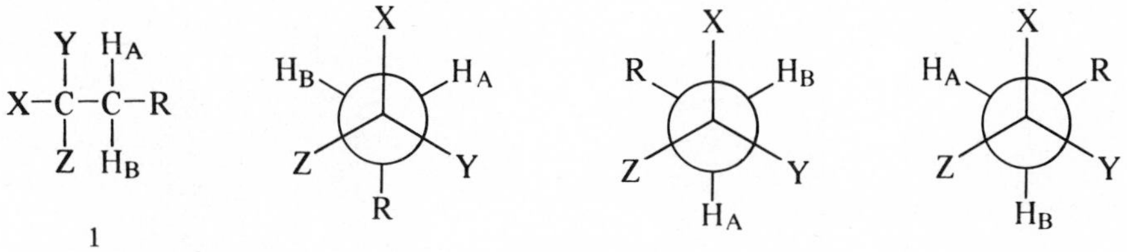

If the rotation about the carbon-to-carbon single bond in 1 were slow enough to allow the nmr spectrometer to "see" the proton H_A (or H_B) in each of its three distinct environments (a through c), then H_A would give rise to three peaks. If, however, the rotation about the carbon-to-carbon bond is very rapid, then the spectrometer will see H_A in a single time-averaged position and only one sharp resonance peak will be observed. If the frequency of rotation is intermediate, then a broad resonance peak will be observed for H_A. In summary, the kind of peak(s) that one will observe for a proton such as H_A will depend on how fast its frequency of interconversion of different chemical environments is compared to the difference in the chemical shifts (measured in Hz) for H_A in its three distinct environments.

Table 8-4 Characteristic absorption positions for protons in different structural environments.

Structural type	Normal range (δ)
$CH_3\!-\!\overset{\displaystyle |}{\underset{\displaystyle |}{C}}\!-$, saturated	0.95–0.8
R_2NH, RNH_2 (inert solvent) R_2NH 1.6–0.4, RNH_2	1.5–1.1
$CH_3\!-\!\overset{\displaystyle |}{\underset{\displaystyle |}{C}}\!-\!\overset{\displaystyle |}{\underset{\displaystyle |}{C}}\!-\!X$ (X = N, O, OR, OH, C=O, halogen)	1.1–0.90
$-CH_2\!-\!\overset{\displaystyle |}{\underset{\displaystyle |}{C}}\!-$, saturated	1.4–1.2
RSH, ArSH	1.4–1.1, 4.0–3.0
$-\overset{\displaystyle |}{\underset{\displaystyle |}{C}}\!-\!H$, saturated	1.6–1.4
$CH_3\!-\!\overset{\displaystyle |}{\underset{\displaystyle |}{C}}\!-\!X$ (X = halogen, N, OH, OR, OAr)	1.9–1.2
$CH_3\!\overset{\displaystyle |}{C}\!=\!\overset{\displaystyle |}{C}\!-$	1.9–1.6
$CH_3\overset{\displaystyle |}{C}\!=\!O$	2.6–2.1
CH_3Ar	2.5–2.2
CH_3X (X = S, N)	2.9–2.1

δ values

14.0 13.0 12.0 11.0 10.0 9.0 8.0 7.0 6.0 5.0 4.0 3.0 2.0 1.0 0

RSH   ArSH

108

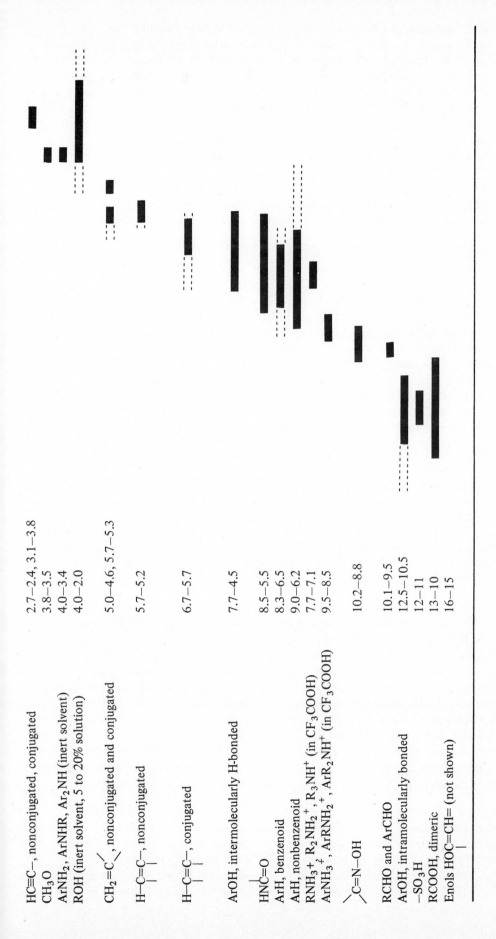

HC≡C–, nonconjugated, conjugated	2.7–2.4, 3.1–3.8
CH_3O	3.8–3.5
$ArNH_2$, $ArNHR$, Ar_2NH (inert solvent)	4.0–3.4
ROH (inert solvent, 5 to 20% solution)	4.0–2.0
$CH_2=C\diagdown$, nonconjugated and conjugated	5.0–4.6, 5.7–5.3
$H-C=C-$, nonconjugated	5.7–5.2
$H-C=C-$, conjugated	6.7–5.7
ArOH, intermolecularly H-bonded	7.7–4.5
$HN\overset{|}{C}=O$	8.5–5.5
ArH, benzenoid	8.3–6.5
ArH, nonbenzenoid	9.0–6.2
RNH_3^+, $R_2NH_2^+$, R_3NH^+ (in CF_3COOH)	7.7–7.1
$ArNH_3^+$, $ArRNH_2^+$, ArR_2NH^+ (in CF_3COOH)	9.5–8.5
$\diagup$C=N–OH	10.2–8.8
RCHO and ArCHO	10.1–9.5
ArOH, intramolecularly bonded	12.5–10.5
$-SO_3H$	12–11
RCOOH, dimeric	13–10
Enols $HOC=CH=$ (not shown)$\overset{|}{|}$	16–15

Another type of chemical exchange is intramolecular or intermolecular exchange of protons. For example, in acidified ethanol, the intermolecular exchange of hydroxyl protons, catalyzed by trace amounts of acid (or base), is very rapid. In fact, this exchange is so rapid at room temperature that the nmr spectrometer views a particular hydroxyl proton in a single time-averaged position over many ethanol molecules rather than on a single molecule. The methylene protons also view the hydroxyl proton as undergoing rapid exchange, and the hydroxyl proton itself does not see the methylene protons of a single ethanol molecule but of many such molecules. The result of this rapid exchange is *spin decoupling*. The expected multiplicities for the methylene proton coupled to the hydroxyl proton and vice versa are not observed (see Sec. 8-4*b*).

There are several techniques available by which rotation about bonds or intra- and intermolecular exchange of protons can be slowed down sufficiently for the nmr spectrometer to "take a picture" of a proton in an unique environment. The use of extremely pure ethanol with no trace impurities of acid or base will slow the hydroxyl exchange enough so that one can observe the expected triplet for the hydroxyl proton coupled with the methylene protons, and the methylene protons will appear as a multiplet (not a simple quintet as predicted by the $n + 1$ rule because the coupling constants, J_{CH_2OH} and $J_{CH_2CH_3}$, are not equal and in theory should give rise to eight lines). Intermolecular exchange of protons, as in alcohols or compounds containing acidic hydrogens, can also be slowed down by dilution with a solvent; thus, determination of the nmr spectrum at different concentrations in the same solvent will often indicate whether such chemical exchange is occurring. By far one of the most useful techniques in slowing down chemical exchange is to lower the temperature of the sample. For example, the spectrum of methanol at $-40°C$ shows a quartet due to the hydroxyl proton and a doublet due to the methyl group. As the temperature is raised from -40 to $-4°C$, the quartet and the doublet both gradually become broader and less distinct as a quartet and doublet until at $-4°C$ only two broad singlets are observed. Such a temperature at which the individual lines of a multiplet broaden into a singlet is called the *coalescence temperature T_c*. It is possible to measure how fast the proton is exchanging per second in a situation like this by using the formula, $\pi J/\sqrt{2}$, where J is the coupling constant (in Hz) where essentially no exchange is occurring (e.g., at $-40°C$ for methanol). At $-4°C$ the hydroxyl proton of methanol is changing to another methanol molecule about 12 times per second on the average.

g. *Magnetic Properties of Other Nuclei* Fluorine (^{19}F) is similar to the proton in its magnetic properties and can give rise to nmr spectra. ^{19}F absorption, however, is far upfield from normal proton absorption and will not appear in a proton nmr spectrum. Splitting of hydrogens by fluorine will follow the same multiplicity rules for proton coupling. Thus, a molecule such as CH_3CF_3 will show a normal 1:2:2:1 quartet for the methyl protons. J_{HF} values are usually much greater than J_{HH} in comparable situations. ^{13}C nuclei (natural abundance about 1 percent), unlike ^{12}C, can couple with protons giving rise to absorption which appears as *very* weak coupled satellite bands (unless the compound has been enriched in ^{13}C by labeling) on either side of the main proton absorption peak. Some other nuclei for which useful nmr studies can be obtained, given the right kind of magnetic spectrometer, are ^{31}P, ^{11}B, and ^{14}N.

h. *Deuterium Labeling and Double Irradiation* Several techniques are available for simplifying very complicated proton nmr spectra. One is *deuterium labeling*. Deuterium absorbs at a much higher field than hydrogen; therefore, replacement of a hydrogen by deuterium will remove any peaks or coupling due to that hydrogen. For example, CH_3CD_3 would show only a singlet in its proton nmr spectrum due to the methyl hydrogens. Another technique is double resonance or *double irradiation*. Here two radiofrequency beams are used to irradiate the sample. One beam is that which corresponds to the resonance frequency of one absorbing proton (for example, H_A) and the other much stronger beam corresponds to the absorbing frequency of a set of protons to which H_A is coupled (for example, H_B). This causes the H_B protons to change their spin states so rapidly that H_A "sees" a single time average of the possible spin states of the H_B protons. Thus, H_A will show a singlet absorption peak rather than a normal triplet. The effect then is one of spin decoupling H_A from H_B. Note, however, that H_B in this case will show the normal splitting by H_A. Of course, it is possible to reverse the situation and spin-decouple H_B from H_A by using the stronger radiofrequency beam to irradiate H_A at its resonance frequency.

i. *Correlation of Structure and NMR Spectra* Table 8-4 gives some characteristic absorptions of different kinds of protons. The ranges given are for a certain kind of proton or functional group in situations that it is usually found. Occasionally, one may find cases where the observed absorption is outside of the given range.

Unless it is stated otherwise, the spectra throughout this book are run at 500-Hz sweep width on a 60-Mc spectrometer using TMS ($\delta = 0$) for the reference peak. Some typical nmr spectra are shown in Fig. 8-9.

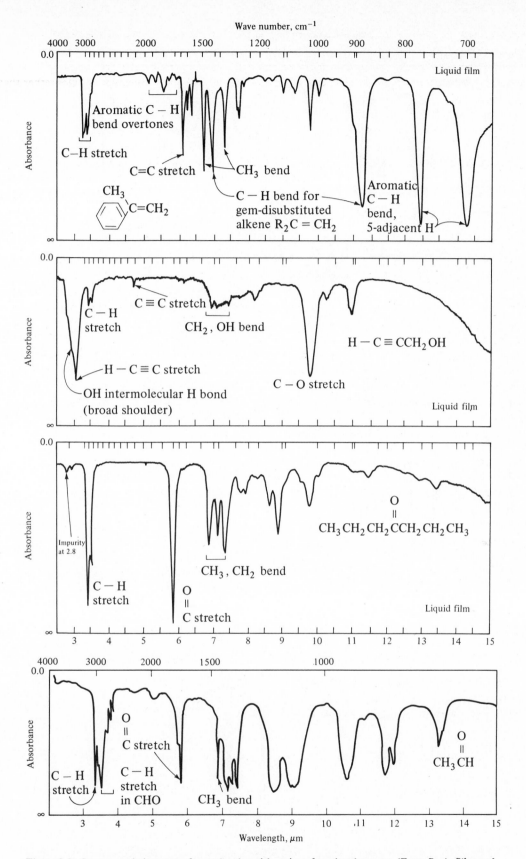

Figure 8-8. Some sample ir spectra for molecules with various functional groups. *(From R. A. Silva and R. Spenger, "The Quantitative Organic Analysis Spectra Kit," McGraw-Hill Book Company, New York, 1969. By permission of the publishers.)*

111

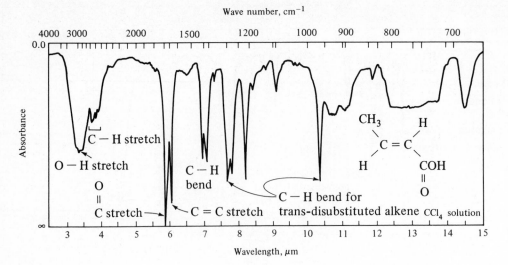

Wave number, cm⁻¹

Figure 8-8. *(Continued.)*

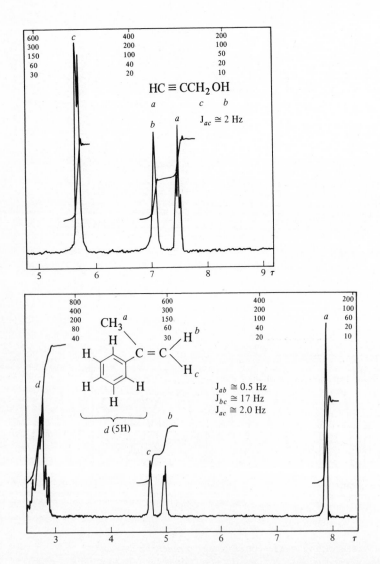

Figure 8-9. **Some sample nmr spectra for molecules with various functional groups.** *(From R. A. Silva and R. Spenger, "The Quantitative Organic Analysis Spectra Kit," McGraw-Hill Book Company, New York, 1969. By permission of the publishers.)*

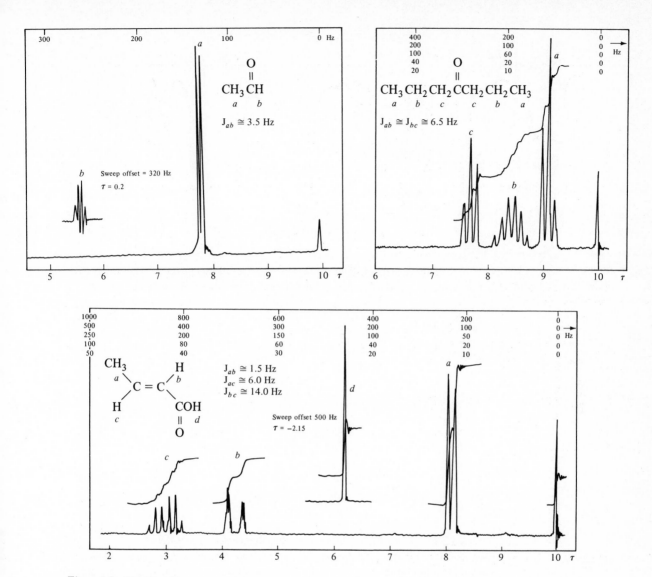

Figure 8-9. (*Continued.*)

Problems

8-1   What would be the expected λ_{max} values for each of the following compounds?

    a.   3-Methyl-3-buten-2-one   *b.*   2-Cyclopenten-1-one

    c.   3-Methyl-3-penten-2-one   *d.*   2-Butenal

    e.   The 2,4-dinitrophenylhydrazone of acetone   *f.* $CH_3(CH=CH)_2CHO$

    g.

    h.

    i.

    j.

113

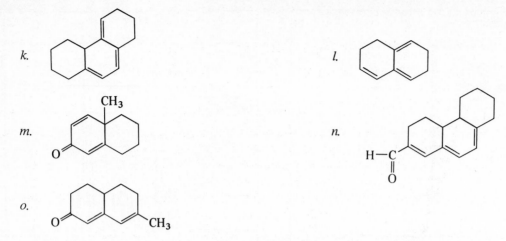

k.

l.

m.

n.

o.

8-2   The λ_{max} values for five compounds (A through E) were measured. The following values were obtained: 258, 278, 284, 293, and 317. For each λ_{max}, identify the compound most likely to give that value.

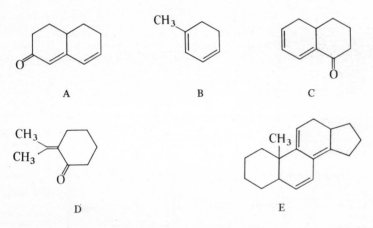

A

B

C

D

E

8-3   A 1.75×10^{-5} M solution of 1,3-butadiene gives a maximum absorbance reading of 0.735 at 217 nm when placed in a cell whose path length is 2.0 cm. Calculate ϵ_{max}.

8-4   a. How many different stretching and bending vibrational modes would be expected for water and carbon dioxide?

b. Which of the following molecules or specific vibrational modes would you not expect to give rise to an observed ir band: H_2, CO, the symmetric C–C stretching mode in acetylene, the symmetric C–N stretching mode in CH_3CN, and the asymmetric C–O–C stretching mode in diethyl ether?

c. If the C=O stretching frequency for an aldehyde occurs at 5.8 μm (A = 0.8), what is the absorption frequency expressed in wave numbers and in cycles per second? What is the percent transmittance?

8-5   Propose a structure for each of the following compounds which is consistent with the ir and other data given. All ir bands for a particular compound are not necessarily given.

a. Compound A, $C_8H_{12}O$, has λ_{max} = 238 nm and an ir band at 1686 cm^{-1} (5.93 μm). It decolorizes Br_2 in CCl_4 and gives a positive iodoform test. Ozonolysis of A gives one product only.

b. Compound B, MW = 60, is transparent in the uv but has strong ir bands at 3400 cm^{-1} (2.94 μm), 2995 cm^{-1} (3.34 μm), 1385 cm^{-1} (7.22 μm), 1380 cm^{-1} (7.25 μm), 1090 cm^{-1} (9.18 μm), and 950 cm^{-1} (10.53 μm).

c. Compound C, C_6H_{10}, has ir bands at 2990 cm^{-1} (3.34 μm), 2250 cm^{-1} – very weak band (4.44 μm), 1430 cm^{-1} (7.0 μm), and 1370 cm^{-1} (7.30 μm). It decolorizes Br_2 in CCl_4, gives no precipitate with $Ag(NH_3)_2^+$, and yields only one carboxylic acid upon ozonolysis.

d. Compound D is a liquid (MW = 82), insoluble in water, decolorizes Br_2 in CCl_4 and reacts with sodium metal. Ir spectrum shows bands at 3325 cm^{-1} (3.01 μm), 2900 cm^{-1} (3.45 μm), 2119 cm^{-1} – medium intensity (4.72 μm), 1471 cm^{-1} (6.8 μm), 1383 cm^{-1} (7.23 μm), 1250 cm^{-1} (8.0 μm), and 1101 cm^{-1} (9.08 μm).

e. Compound E (MW = 98) can be converted to compound F (MW = 100) by treatment with Pt and H_2. Compound F contains an active hydrogen. Vigorous oxidation of E or F leads to one product only. Ir spectrum for E has bands at 2990 cm^{-1} (3.34 μm), 1710 cm^{-1} (5.85 μm), 1450 cm^{-1} (6.9 μm), 1430 cm^{-1} (7.0 μm), 1350 cm^{-1} (7.41 μm), 1314 cm^{-1} (7.61 μm), and 1220 cm^{-1} (8.2 μm). In addition, E has λ_{max} = 285 nm (ϵ = 14). Compound F exhibits ir absorption at 3333 cm^{-1} — broad band (3.0 μm), 2941 cm^{-1} (3.4 μm), 1449 cm^{-1} (6.9 μm), 1355 cm^{-1} (7.38 μm), 1080 cm^{-1} (9.26 μm), and 996 cm^{-1} (10.35 μm).

f. Compound G, C_4H_8O, shows ir absorption at 2793 cm^{-1} (3.58 μm), 2710 cm^{-1} (3.69 μm), 1718 cm^{-1} (5.82 μm), 1466 cm^{-1} (6.82 μm), 1377 cm^{-1} (7.26 μm), 1115 cm^{-1} (8.97 μm), and 769 cm^{-1} (13.0 μm).

g. Compound H, $C_6H_{10}O$, gives a positive Tollens' test and discolors aqueous $KMnO_4$. Ir bands at 3077 cm^{-1} (3.25 μm), 2703 cm^{-1} (3.70 μm), 1736 cm^{-1} (5.76 μm), 1650 cm^{-1} (6.06 μm), 1420 cm^{-1} (7.04 μm), 1299 cm^{-1} (7.70 μm), 995 cm^{-1} (10.05 μm), and 914 cm^{-1} (10.94 μm). It also shows very weak uv absorption at 290 nm.

h. Compound I, $C_6H_{12}O$, gives a positive iodoform test. When compound I is treated with Pt and H_2, compound J is formed; upon heating with aqueous sulfuric acid solution compound J yields compound K as the major product. The latter does not show geometric isomerism and yields only one organic product upon ozonolysis. Compound I exhibits ir absorption at 3000 cm^{-1} (3.33 μm), 1710 cm^{-1} (5.85 μm), 1470 cm^{-1} (6.80 μm), 1400 cm^{-1} (7.15 μm), and 1368 cm^{-1} (7.31 μm).

i. Compound L, $C_8H_{14}O$, gives a negative iodoform test, decolorizes Br_2 and CCl_4, has very weak uv absorption at 280 nm, and has ir absorption bands at 2703 cm^{-1} (3.70 μm), 1736 cm^{-1} (5.76 μm), 1661 cm^{-1} (6.02 μm), and 838 cm^{-1} (11.93 μm). Complete catalytic hydrogenation of compound L requires 2 mol of H_2 and ozonolysis of L yields as one of two organic products, C_3H_6O, which gives a positive iodoform test.

j. Compound M (MW = 82) is a water-insoluble liquid which discolors aqueous $KMnO_4$ and upon ozonolysis yields only one organic product (MW = 114). Compound M absorbs at 2941 cm^{-1} (3.4 μm), 1429 cm^{-1} (7.0 μm), 1136 cm^{-1} (8.8 μm), 909 cm^{-1} (11.0 μm), 877 cm^{-1} (11.4 μm), and 714 cm^{-1} (14.0 μm).

k. Compound N, $C_6H_{10}O$, has a calculated λ_{max} = 232 nm. Ozonolysis of compound N yields as one of the organic products a neutral compound (MW = 58) which gives a negative iodoform test. Compound N absorbs in the ir at 2970 cm^{-1} (3.37 μm), 2870 cm^{-1} (3.48 μm), 2690 cm^{-1} (3.72 μm), 1685 cm^{-1} (5.94 μm), 1625 cm^{-1} (6.15 μm), 1460 cm^{-1} (6.85 μm), 1405 cm^{-1} (7.11 μm), 1380 cm^{-1} (7.25 μm), 1360 cm^{-1} (7.36 μm), 1305 cm^{-1} (7.66 μm), 1220 cm^{-1} (8.20 μm), 1050 cm^{-1} (9.52 μm), 1000 cm^{-1} (10.0 μm), and last but not least, 840 cm^{-1} (11.9 μm).

l. Compound O contains 70.53% C, 13.81% H, and 15.66% O. It is a neutral liquid compound which is soluble in cold concentrated H_2SO_4, fails to give a positive Lucas test, and does not react with either phenylhydrazine or hot aqueous $KMnO_4$. Heating compound O with HI forms only one compound which contains 21.19% C, 4.15% H, and 74.66% I. Compound O has ir bands at 2857 cm^{-1} (3.50 μm), 1410 cm^{-1} (7.09 μm), 1380 cm^{-1} (7.25 μm), and 1120 cm^{-1} — a broad, strong band (9.0 μm).

8-6   *a.* The chemical shift for a particular proton is 366 Hz. Express its absorption position in terms of δ and τ.
b. Between what limits of δ and τ would a nmr spectrum correspond if a 100-, 250-, or 1,000-Hz sweep width is used? Does the chemical shift of a particular proton depend on what sweep width is used?

8-7   Considering electronegativities, arrange each of the following compounds in order of the expected decreasing τ values for the protons underlined.
a. $CH_3C\underline{H}_2CH_3$, $CH_3C\underline{H}_3$, $(CH_3)_3C\underline{H}$
b. $CH_3C\underline{H}_2OH$, $CH_3C\underline{H}(OH)_2$, $CH_3C\underline{H}_3$, $CH_3C\underline{H}I_2$, $CH_3C\underline{H}_2I$
c. $C\underline{H}_3Cl$, $(C\underline{H}_3)_3P$, $(C\underline{H}_3)_4Si$, $(C\underline{H}_3)_2S$, $C\underline{H}_3F$

8-8   Sketch the expected nmr line pattern for $CH_3OCH_2CH_2CH_2OCH_3$ and indicate the following:
a. The relative number of protons to which each line or set of lines corresponds.
b. The relative intensities of the lines in a given set.
c. Draw all the possible spin-state combinations with adjacent protons to which each set of methylene protons is coupled.
d. Give the approximate magnitude of any *J* values and the distance in the spectrum from which *J* could be determined.

8-9 Indicate the coupled protons and the approximate J value(s) expected in the following compounds:

a.

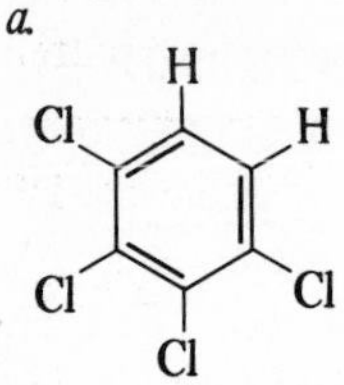

b. $(CH_3)_2CHC{\equiv}CH$

c. $CH_3CH_2OCH_2\overset{\displaystyle O}{\underset{\displaystyle \|}{C}}H$

d.

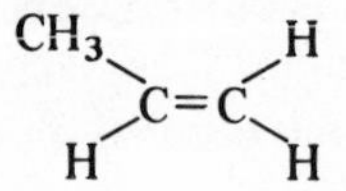

8-10 For each of the following structures, indicate the number of different nmr signals that should be obtained by labeling the magnetically different sets of hydrogen atoms as a, b, c, d, etc. Underline the hydrogen atom(s) whose nmr signal should be the farthest downfield. For each different set of hydrogen atoms, indicate the multiplicity that would be expected in its signal.

a. $CH_3CH_2CH_2Br$

b. Cl_2CHCH_2Cl

c. $CH_3\overset{\displaystyle O}{\underset{\displaystyle \|}{C}}H$

d. $Cl(CH_2)_3Cl$

e.

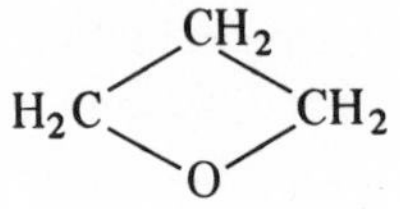

f. $(CH_3)_2CHNO_2$

g. $CH_2{=}C(Cl)CH_2Cl$

h. $CH_3CH_2CH_2OH$

i. $HC{\equiv}CCH_2OH$

j.

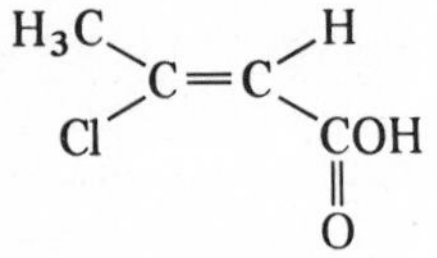

k. $CH_3CH_2CH(Br)CO_2H$

l. $(CH_3)_2C{=}CHCl$

m. $(CH_3)_2C(OH)C{\equiv}N$

n. $CH_3(CH_2)_2CHO$

o. $CH_3CO_2CH_2CH_3$

p. $(CH_3)_3CNH_2$

q. $CH_2{=}CHCH_2OCH_2CH{=}CH_2$

r.

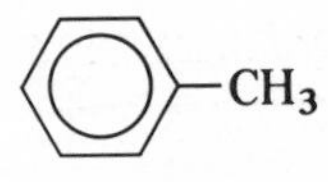

s.

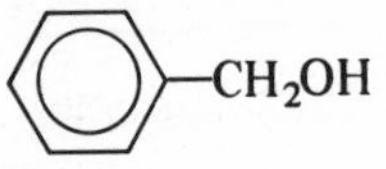

t. $(C_6H_5)_3SiH$

u.

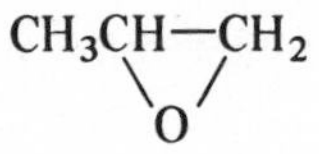

8-11 In the following questions, the results (positive or negative) of some qualitative tests on each compound are given along with the nmr pattern. The kind of nmr signal is shown as singlet (*s*), doublet (*d*), triplet (*t*), quartet (*q*), quintet (*p*), sextet (*sx*), septet (*sp*), and multiplet (*m*) followed by the respective ratio of hydrogens producing each signal. Draw structures for compounds A through Z which are consistent with the data given.

Compound	Lucas test	Iodoform test	$K_2Cr_2O_7$ oxidation	Tollens' test	2,4-DNP test	Nmr pattern
A. $C_5H_{12}O$	+ (5 min)	+	+ Yields compound B, $C_5H_{10}O$			s,d,sp; 3:6:1 for compound
C. $C_5H_{12}O$		−				s,s,t,q; 1:6:3:2
D. $C_5H_{12}O$	−	−	−			s,s; 3:1
E. $C_5H_{10}O$	−			−	+	t,q; 3:2

116

Compound	Lucas test	Iodoform test	$K_2Cr_2O_7$ oxidation	Tollens' test	2,4-DNP test	Nmr pattern
F. $C_5H_{10}O$				+	+	s,s; 1:9
G. $C_4H_{10}O$	−	−	−		−	s,d,sp; 3:6:1
H. MW = 74			−		−	s,s; 1:9
I. $C_6H_{12}O$		+	−			s,d,d,m; 3:6:2:1
J. $C_6H_{14}O$	−	−	−		−	t,t,sx; 1.5:1:1
K. C_7H_8O	+ (< 1 min)	−	+		−	s,s,s; 5:2:1
L. MW = 44			+			d,q; 3:1
M.	+ (5 min)		+	−		s,d,sp; 1:6:1
N. $C_4H_{10}O_2$						s,s; 3:2
O. MW = 74			−	no reaction with Na		t,q; 3:2
P. MW = 90	stable in base; P $\xrightarrow[\Delta]{H^+}$ an aldehyde					s,d,q; 6:3:1

Compound	Reaction with Na	Br_2 in CCl_4	Reaction with $Ag(NH_3)_2^+$	Products from ozonolysis (O_3) or $KMnO_4$ oxidation (after H^+)	Nmr pattern
Q. $C_{12}H_{20}$		+	−	$KMnO_4 \rightarrow$ Q' = $C_6H_{10}O_3$, a keto acid	For Q' d,sp,s,s; 6:1:2:1
R. MW = 84	+ (+ Lucas test)	+	+		$s,s,s,$; 6:1:1
S. $C_4H_9NO_3$	+ (− Lucas test)		−	$\xrightarrow{KMnO_4}$ $C_4H_7NO_4$	s,s,s; 6:2:1
T. MW = 84		+		$\xrightarrow{O_3}$ a ketone	s
U. Contains C and H only		+		$\xrightarrow{KMnO_4}$ a ketone	s,t,q; 3:3:2
V. $C_{14}H_{10}$		+		$\xrightarrow{O_3}$ an acid, MW = 122	m (10H)
W. MW = 216. Contains Br					s,s; 3:1
X. MW = 147 ± 1. Contains Cl					s,s; 3:2
Y. MW = 114 ± 1. Contains F and Cl					t,t; 3:2
Z. MW = 135 ± 1. Contains F and Cl					t (2H)

8-12 Compound A (MW = 100) gives upon catalytic hydrogenation compound B (MW = 102) which when passed over hot alumina gives compound C (MW = 84) as the major product. Ozonolysis of C gives two products only one of which gives a positive iodoform test. The nmr and some ir data on A are given below. Nmr (τ): 9.00, t; 8.87, d; 7.87, q; 6.48, m. The respective areas under these nmr signals as obtained from the integration curve are 7.1 units, 13.9 units, 4.5 units, and 2.3 units. Ir cm^{-1}, μm: 1712, 5.84; 1383, 7.23; 1368, 7.31. Suggest structures for A, B, and C.

8-13 A (MW = 72) + B (MW = 62) $\xrightarrow{H^+}$ C (MW = 116). Compound A is water-soluble and reacts with phenylhydrazine. Compound B is water-soluble and evolves H_2 when treated with Na. Compound C is stable in aqueous sodium hydroxide, but heating in aqueous acid yields A and B. The nmr spectrum of C shows: (δ) 0.88, d, 6H; 1.65, m, 1H; 3.81, m, 4H; 4.48, d, 1H. Suggest structures for A, B, and C.

8-14 Make a list of some different types of hydrogens whose nmr absorption position will change rather dramatically as a function of sample concentration. As more dilute solutions are used, in which direction will the absorption of these types of hydrogens shift?

8-15 Suggest methods of synthesis from readily available materials for the following solvents which are very useful in nmr work.

a. DCCl$_3$ *b.* CD$_3$COCD$_3$ *c.* C$_6$D$_6$ (See Prob. 4-16).

8-16 *a.* Starting from calcium carbide and acetone outline the synthesis of a properly deuterium-labeled 3-methyl-2-bromobutane which would allow you to look at only the spin-spin coupling of proton B with protons A without interference from other spin-spin couplings.

$$\underset{\underset{A}{}\,\underset{B}{}}{CH_3CH(Br)CH(CH_3)_2}$$

b. Suggest another method not involving labeling which would suitably decouple proton B from all protons except the A protons.

8-17 Indicate how you would make each of the following chemical conversions and whether there would be a bathochromic (B) or a hypsochromic (H) shift in λ_{max} for the uv absorption of the organic compound accompanying each step.

a. Acetaldehyde(I) $\longrightarrow$ 2-butenal(II)

b. Propanal(I) $\longrightarrow$ 1,3-hexadiene(II)

c. 3-Bromo-2-methylpentanal(I) $\longrightarrow$ 2-methyl-2-pentenal(II) $\longrightarrow$ 2-methylpentanal(III) $\longrightarrow$ 2-methylpentane(IV)

d. 1,5-Dihydroxy-3-pentanone(I) $\longrightarrow$ 1,4-pentadien-3-one(II) $\longrightarrow$ 3-methylpentane(III)

e. C$_6$H$_5$Br(I) $\longrightarrow$ C$_6$H$_5$CH$_2$CH$_2$OH(II) $\longrightarrow$ C$_6$H$_5$CH=CH$_2$(III) $\longrightarrow$

C$_6$H$_5$CH$_2$CH$_2$COCH$_2$CH$_2$CH$_3$(IV)

8-18 In the following problems information about a compound's nmr spectrum is shown according to the format: δ or τ value, number of peaks or multiplicity in each signal (see Prob. 8-11 for abbreviations used), and the number of hydrogens responsible for the signal or the area under the peak(s) as obtained from the integration curve. The ir data on a compound (all significant bands are not necessarily given) are presented with the absorption position in cm^{-1} followed by the corresponding μm value. Propose a structure in each case which is consistent with the data given.

a. C$_9$H$_9$OBr nmr (δ): 3.78, s, 7.9 units; 3.82, s, 8.1 units; 7.21, s, 19.9 units. Ir band at 1730 cm^{-1}, 5.78 μm.

b. C$_7$H$_{14}$O can exist as two stereoisomers; it decolorizes Br$_2$ in CCl$_4$ but gives no reaction with sodium metal. Nmr (τ): 8.83, t (J = 7.0 Hz), 3H: 8.75, d (J = 7.0 Hz), 3H; 8.32, m, 3H; 6.62 m, 1H; 6.2, q (J = 7.0 Hz), 2H; 5.1, m, 2H.

c. C$_7$H$_{12}$O has no asymmetric carbon atoms and decolorizes Br$_2$ in CCl$_4$. Nmr (δ): 1.10, s, 20.2 units: 1.82, s, 10.1 units; 2.63, s, 3.3 units; 3.23, s, 6.3 units. Ir cm^{-1}, μm: 3590, 2.79; 2180, 4.59 (weak); 1390, 7.2; 1370, 7.3; 1050, 9.52.

d. C$_7$H$_{12}$O is a closely related isomer of the compound in (*c*). Nmr (δ): 1.19, s, 6H; 1.75, t (J = 2.5 Hz), 3H; 2.28, q (J = 2.5 Hz), 2H; 2.43, s, 1H. Ir cm^{-1}, μm: 3570, 2.8; 3450, 2.9; 2930, 3.41; 1460, 6.85; 1380, 7.25; 1355, 7.38; 1155, 8.66.

e. MW = 113, positive Beilstein test. Nmr (δ): 2.20, p, 5.3 units; 3.70, t, 10.8 units.

f. C$_{12}$H$_{15}$Cl can exist as one dl pair. Nmr (τ): 8.5, broad multiplet, poorly resolved multiplet, 14.4 units; 7.5, m, 1.7 units; 5.45, d, 1.8 units; 2.94, s, 9.2 units.

g. C_4H_7NO Nmr (δ): 1.63, *s*, 6H; 3.25, broad singlet, 1H.

h. $C_7H_{16}O$ is a neutral compound, soluble in cold concd H_2SO_4. Nmr (δ): 1.12, *d* (J = 7.0 Hz), 6.9 units; 1.18, *s*, 10.3 units; 3.78, *m* (J = 7.0 Hz), 1.1 units.

i. MW = 119, positive Beilstein test. Nmr (τ): 6.15, *d* (J = 2.5 Hz), 9.3 units; 7.45, *t* (J = 2.5 Hz), 4.7 units. Ir cm^{-1}, μm: 3333, 3.00; 3000, 3.33; 2128, 4.7; 1429, 7.0; 1200, 8.33; 952, 10.5.

j. $C_{18}H_{18}O$ can exist as one *dl* pair. Nmr (δ): 9.12, *d* (J = 7.0 Hz), 1H; 7.4 to 7.1, *m*, 10H; 2.28, *d* (J = 7.0 Hz), 1H; 1.51, *s*, 3H; 1.09, *s*, 3H. Ir cm^{-1}, μm: 3077, 3.25; 2924, 3.42; 2665, 3.49; 1686, 5.93; 1597, 6.26; 1513, 6.61; 1493, 6.70; 1130, 8.85; 1111, 9.00; 694, 14.4.

8-19   When compound A, $C_{31}H_{32}O_2$, is heated with alumina at 150°C, compound B, $C_{31}H_{28}$, is obtained. Treatment of B with hot $KMnO_4$ solution yields two compounds: C, $C_5H_8O_4$; and D, $C_{13}H_{10}O$. The following are the nmr (note τ values are used) and some ir data on A through D.

Compound A   Nmr (τ): 2.4 to 3.1, *m*; 5.98, *s*; 7.32, *s*; 9.55, *s*; respectively area ratios, 10:1:2:3. Ir cm^{-1}, μm: 3425, 2.92; broad absorption, 1250 to 1111, 8.0 to 9.0; 772, 12.95; 743, 13.46; 669, 14.30.

Compound B   Nmr (τ): 2.81 to 3.0, *m*; 4.05, *s*; 8.98, *s*; respective area ratios, 10:1:3. Ir cm^{-1}, μm: 3077, 3.25; 3021, 3.31; 2865, 3.49; 1592, 6.28; 1445, 6.92; 1381, 7.24; 1361, 7.35; 885, 11.3; 746, 13.41; 694, 14.4.

Compound C   Nmr (τ): −1.5, *s*; 8.7, *s*; respective area ratio, 1:3. Ir cm^{-1}, μm: broad absorption 2680 to 2525, 3.73 to 3.96; 1725, 5.80.

Compound D   Nmr (τ): 2.1 to 2.7, *m*. Ir cm^{-1}, μm: 1653, 6.04; 760, 13.2; 695, 14.4. Uv: λ_{max} = 252 μm, ϵ = 23,700.

Suggest structures for A through D.

8-20   The nmr spectrum for bromocyclohexane at room temperature shows a distinct singlet (1H) at δ = 4.16. Upon cooling the sample to −75° this singlet separates into two peaks at δ = 4.64 and δ = 3.97 which have a respective area ratio of 1.0:4.6 and are equivalent to a total of one proton. Explain the change in the spectrum upon cooling. How fast is the proton on the bromine-substituted carbon exchanging its environment at room temperature?

8-21   For a proton which is rapidly exchanging its environment as in cyclohexane, the observed chemical shift for that proton is a weighted average of the various chemical shifts for the proton when it is only in the axial (δ_a) and when it is only in the equatorial (δ_e) position:

$$\delta_{observed} = \delta_a N_a + \delta_e N_e$$

where N_a and N_e are the respective mole fractions of the two conformers. Assume that a *tert*-butyl group on a cyclohexane ring must be almost exclusively in the equatorial position because of its size (i.e., $N_a \cong 0$). The room temperature nmr spectrum of *trans*-4-*tert*-butyl-1-bromocyclohexane (both *tert*-butyl and Br prefer equatorial in this case) shows a peak (1H) at δ = 3.83; the corresponding cis isomer (where the *tert*-butyl prefers to be equatorial, thus Br must be axial) shows a similar peak (1H) at δ = 4.63. At room temperature *cis*-4-methyl-1-bromocyclohexane shows a similar downfield peak (1H) at δ = 4.49. Assuming that the 4-*tert*-butyl and 4-methyl groups exert the same *direct* effect on the chemical shift of the proton on C-1, what percentage of *cis*-4-methyl-1-bromocyclohexane at room temperature has the bromine group in the axial position?

8-22 .   For the sets of methylene protons A and B shown below, which would be expected to absorb farther upfield? Explain.

8-23   Compound A, C_8H_{10}, has been shown to be capable of existing as two structurally degenerate isomers, I and II, shown below. The rapid rearrangement (called a Cope rearrangement) of I $\longrightarrow$ II is an example of "fluxionalism" or valence tautomerism. The nmr spectral data of A at various temperatures are given below and they are evidence that fluxionalism does occur. For the nmr spectrum at each temperature assign each observed τ value to specific hydrogens and explain the changes in the spectrum as the temperature is raised from $-50°C$ to $180°C$.

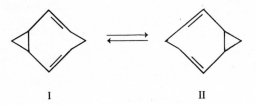

| I | II |

$T = -50°C$	$T = 25°C$	$T = 180°C$
τ	τ	τ
4.3 4H	4.2 4H	4.3 2H
7.2 2H	7–10 6H	6.4 4H
8–9 2H	(broad peak)	8.2 2H
9.7 2H		8.9 2H

Identify the compounds in Probs. 8-24 through 8-32. Either the ir or nmr (or both) spectra are given.

8-24   Compound contains C, H, O and halogen. MW = 109 ± 1. Ir and nmr spectra are shown.

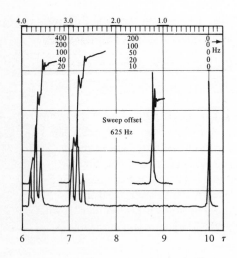

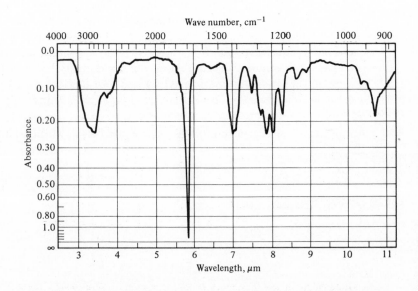

Figure P8-24. Ir and nmr spectra for Prob. 9-24. *(From R. A. Silva and R. Spenger, "The Quantitative Organic Analysis Spectra Kit," McGraw-Hill Book Company, New York, 1969. By permission of the publishers.)*

8-25   Compound has MW = 88 ± 2. Ir and nmr spectra are shown.

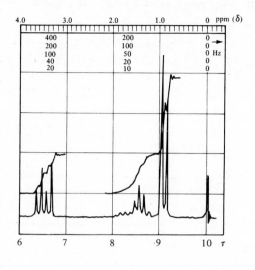

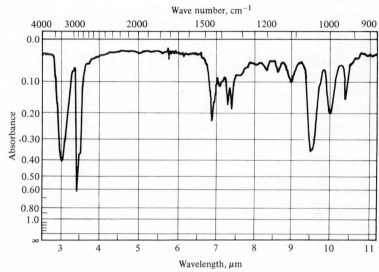

Figure P8-25. Ir and nmr spectra for Prob. 8-25. *(From R. A. Silva and R. Spenger, "The Quantitative Organic Analysis Spectra Kit,"*
McGraw-Hill Book Company, New York, 1969. By permission of the publishers.)

8-26   Compound contains C, H, O and its nmr spectrum shows singlets at 400 Hz (9.6 integration units) and 228 Hz
          (13.9 integration units). Ir spectrum is shown.

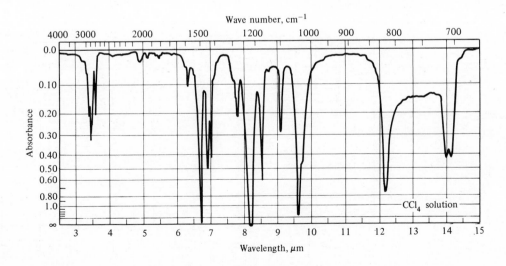

Figure P8-26. Ir spectrum for Prob. 8-26. *(From R. A. Silva and R. Spenger, "The Quantitative*
Organic Analysis Spectra Kit," McGraw-Hill Book Company, New York, 1969. By permission of the
publishers.)

121

8-27 Halogen containing compound with MW = 150 ± 1. Nmr spectrum is shown.

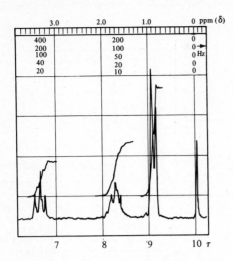

Figure P8-27. Nmr spectrum for Prob. 8-27.
*(From R. A. Silva and R. Spenger, "The
Quantitative Organic Analysis Spectra Kit,"
McGraw-Hill Book Company, New York, 1969.
By permission of the publishers.)*

8-28 Compound contains C, H, N with MW = 108 ± 2. Ir and nmr spectra are shown.

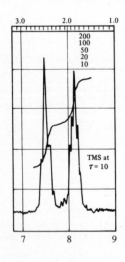

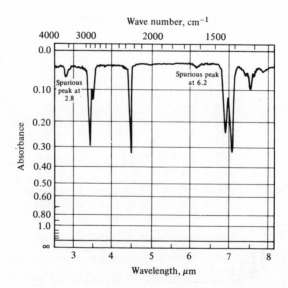

Figure P8-28. Ir and nmr spectra for Prob. 8-28. *(From R. A. Silva and R. Spenger, "The
Quantitative Organic Analysis Spectra Kit," McGraw-Hill Book Company, New York,
1969. By permission of the publishers.)*

122

8-29   Compound's ir spectrum shows bands at 3.02 μm (3311 cm^{-1}, strong), 4.7 μm (2128 cm^{-1}, weak), and 8.47 μm (1180 cm^{-1}) among others. MW = 125 $\pm$ 1. Nmr spectrum is shown.

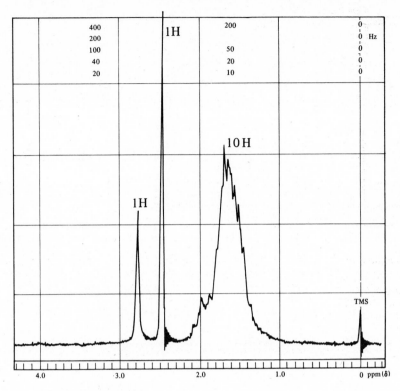

Figure P8-29. Nmr spectrum for Prob. 8-29. *(From "High Resolution NMR Spectra Catalog," vol. 1, Varian Associates, 1962. By permission of the publishers.)*

8-30   Compound (MW = 71 $\pm$ 1) shows ir bands at 5.90 μm (1695 cm^{-1}), 7.70 μm (1299 cm^{-1}), and 10.36 μm (965 cm^{-1}) among others. Nmr spectrum is shown.

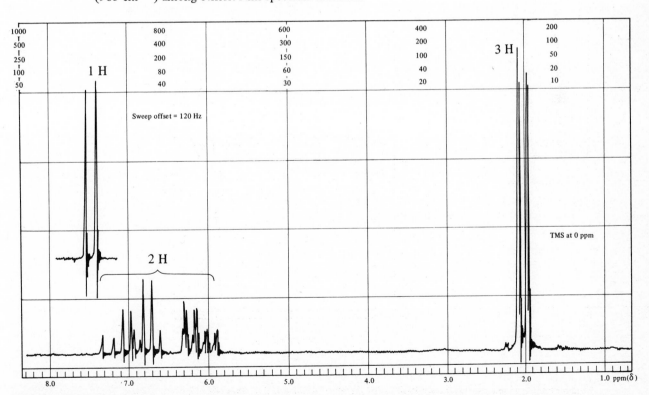

Figure P8-30. Nmr spectrum for Prob. 8-30. *(From "High Resolution NMR Spectra Catalog," vol. 1, Varian Associates, 1962. By permission of the publishers.)*

8-31 Compound has MW = 180 ± 1. Ir and nmr spectra are shown.

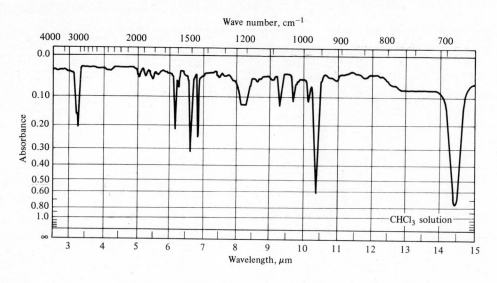

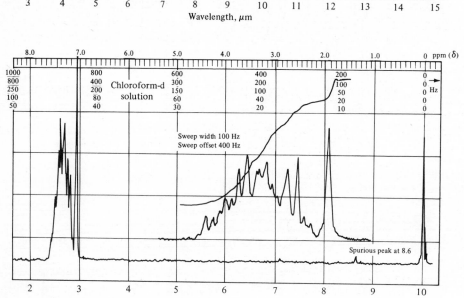

Figure P8-31. Ir and nmr spectra for Prob. 8-31. *(From R. A. Silva and R. Spenger, "The Quantitative Organic Analysis Spectra Kit," McGraw-Hill Book Company, New York, 1969. By permission of the publishers.)*

8-32   Compound has MW = 133 ± 1. Contains C, H, O only and reacts with phenylhydrazine. Ir and nmr spectra are shown.

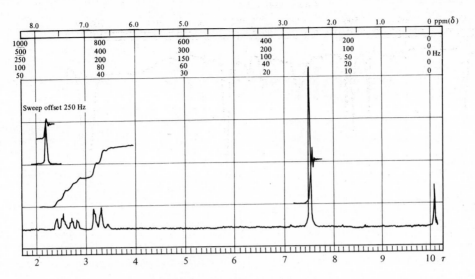

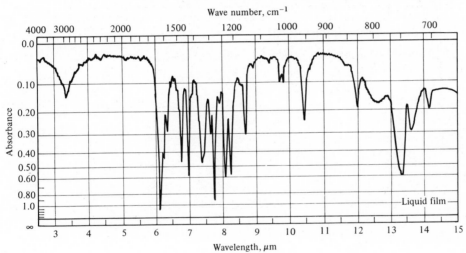

Figure P8-32. Ir and nmr spectra for Prob. 8-32. *(From R. A. Silva and R. Spenger, "The Quantitative Organic Analysis Spectra Kit," McGraw-Hill Book Company, New York, 1969. By permission of the publishers.)*

9
aliphatic acids,
anhydrides, and acyl halides

9-1 Nomenclature

a. Acids

1. In the IUPAC system the ending *oic* is used. The acid is named as a derivative of the longest chain containing the carboxyl group. Numbering is started with the carbon atom in the carboxyl group. Acids with the COOH group attached to a cycloalkane ring are called cycloalkanecarboxylic acids.

CH_3COOH $CH_3CH=CHCH(NO_2)CHFCOOH$ $(CH_3)_2CHCHBrCH_2COOH$

Ethanoic acid   2-Fluoro-3-nitro-4-hexenoic acid   3-Bromo-4-methylpentanoic acid

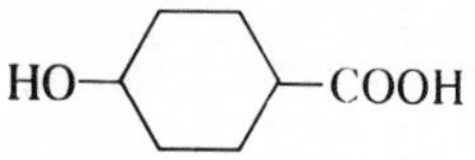

4-Hydroxy-1-cyclohexanecarboxylic acid

2. The common names of some simple carboxylic acids are given in the following list:

HCOOH	Formic	$CH_3(CH_2)_6COOH$	Caprylic
CH_3COOH	Acetic	$CH_3(CH_2)_8COOH$	Capric
CH_3CH_2COOH	Propionic	$CH_3(CH_2)_{10}COOH$	Lauric
$CH_3(CH_2)_2COOH$	Butyric	$CH_3(CH_2)_{12}COOH$	Myristic
$CH_3(CH_2)_3COOH$	Valeric	$CH_3(CH_2)_{14}COOH$	Palmitic
$CH_3(CH_2)_4COOH$	Caproic	$CH_3(CH_2)_{16}COOH$	Stearic

Greek letters are used in the common system to indicate the position of substituents. The first carbon *after* the functional group is called alpha (α), the next beta (β), then gamma (γ), delta (δ), and epsilon (ϵ), in that order.

$CH_3CH_2CH_2CHFCH(CH_3)COOH$ $ICH_2CHBrCCl_2COOH$

α-Methyl-β-fluorocaproic acid α,α-Dichloro-β-bromo-γ-iodobutyric acid

3. Acids with branched chains are sometimes referred to by the *iso* terminology. Acids may also be named as derivatives of acetic acid in the same fashion that alcohols are named as derivatives of carbinol (see Sec. 6-1c).

$(CH_3)_2CHCOOH$ Isobutyric acid (dimethylacetic acid)

$(CH_3)_2CHCH_2COOH$ Isovaleric acid (isopropylacetic acid)

4. Polyfunctional acids: dicarboxylic, hydroxy, keto, and unsaturated. In the IUPAC system dicarboxylic acid

names are formed by adding the ending *dioic acid* to the name of the longest chain containing both carboxyl groups.

$$HOOCCH(CH_3)CH(CH_3)COOH \qquad HOOC(CH_2)_5COOH$$

2,3-Dimethylbutanedioic acid   Heptanedioic acid

Common names for the simple dicarboxylic acids, $HOOC(CH_2)_nCOOH$, having from 2 to 10 carbon atoms are: oxalic (2), malonic (3), succinic (4), glutaric (5), adipic (6), pimelic (7), suberic (8), azelaic (9), sebacic (10).

Hydroxy acids are usually named after the acids from which they are derived or they have other common names: $HOCH_2COOH$, hydroxyacetic (glycolic) acid; $CH_3CH(OH)COOH$, α-hydroxypropionic (lactic) acid; $HOOCCH_2CH(OH)COOH$, hydroxysuccinic (malic) acid; $HOOCCH(OH)CH(OH)COOH$, α,β-dihydroxysuccinic (tartaric) acid.

The common keto acids are referred to by their trivial names, for instance, $CH_3COCOOH$, pyruvic (α-ketopropionic) acid; CH_3COCH_2COOH, acetoacetic (β-ketobutyric) acid; $CH_3COCH_2CH_2COOH$, levulinic (γ-ketovaleric) acid. See Chap. 12 for other examples.

Simple unsaturated carboxylic acids are usually referred to by common names, e.g., acrylic acid, $CH_2=CHCOOH$; crotonic acid, *trans*-$CH_3CH=CHCOOH$; isocrotonic acid, *cis*-$CH_3CH=CHCOOH$; propiolic acid, $HC\equiv CCOOH$; maleic acid and fumaric acid, *cis*- and *trans*-$HOOCCH=CHCOOH$.

b. *Acyl Halides* Acyl halides are derived from acids by replacing the hydroxyl group by a halogen. The *oic* ending of the IUPAC name of the parent acid is changed to *oyl* (for cycloalkanecarboxylic acids *oxylic* is replaced by *onyl*) and the appropriate halide name replaces the word *acid*. Similarly, the *ic* ending of the common name of the acid is changed to *yl*.

	IUPAC name	Common name
CH_3COBr	Ethanoyl bromide	Acetyl bromide
$(CH_3)_2CHCOCl$	2-Methylpropanoyl chloride	Isobutyryl chloride
$CH_3CH=CHCOCl$	2-Butenoyl chloride	Crotonyl chloride
⬡—COCl	Cyclohexanecarbonyl chloride	

c. *Anhydrides* The name of the parent acid precedes the word anhydride.

	IUPAC name	Common name
$CH_3CH_2\overset{\text{O}}{\overset{\|}{C}}O\overset{\text{O}}{\overset{\|}{C}}CH_2CH_3$	Propanoic anhydride	Propionic anhydride
$(ClCH_2CH_2CH_2CO)_2O$	4-Chlorobutanoic anhydride	γ-Chlorobutyric anhydride

9-2 Construction of Organic Names

By now you are aware of the complexity of organic names. Although the layman is impressed by the ease with which chemists spell these frightening mixtures of numbers and letters, the insider who knows a few rules finds their formation easy. In addition to the rules learned to date, you need a few more on when and how to divide names.

In the IUPAC system names are run together without breaks, e.g., bromochlorofluoroiodomethane. When numbers or letters are necessary for the location of substituents, the words and numbers are joined by hyphens, except where two or more numbers occur consecutively; then these are separated by commas, for instance, 2-chloro-2,3,4-trimethyl-1-heptanol.

Words ending in *ide, ite,* or *ate* are written separately, e.g., ethylmagnesium bromide, sodium acetate, ethyl nitrite. Words describing a functional group such as alcohol, ether, oxide, acid, anhydride, cyanide, hydroxide, or ketone are written separately, e.g., ethyl alcohol, chloroacetic anhydride, ethylene oxide. An exception to this is the amine, e.g., butylamine, diethylamine. When two different groups are attached to the same or different functions, the common names are written as individual words, e.g., ethyl methyl ether (not ethylmethylether) or methyl isobutyl ketone. Again the amines are an exception to the rule, e.g., ethylmethylamine.

Remember that in naming a compound it is not good practice to mix systems, i.e., IUPAC and common terms: 3-chloropropionic acid or α-bromopentanoic acid.

9-3 Preparation of Acids

a. *Oxidation of Primary Alcohols, Aldehydes, Olefins, and Acetylenes* (See Sec. 6-4 for primary alcohols, Sec. 7-3 for aldehydes, and Sec. 4-3 for olefins and acetylenes.)

b. *Hydrolysis*
1. Cyanides (nitriles). $RC{\equiv}N + 2H_2O \xrightarrow{H^+ \text{ or } OH^-} RCOO^- + \overset{+}{N}H_4$.

2. Esters (Chap. 11). $RCOOR' + H_2O \xrightarrow{H^+ \text{ or } OH^-} RCOOH + R'OH$.

c. *Carbonation of Grignard Reagent* (See Sec. 5-4.)

d. Formic acid is prepared by a special commercial method.

$$NaOH + CO \xrightarrow{pressure} HCOONa \xrightarrow{H^+} HCOOH$$

e. *Haloform Reaction* (See Sec. 7-3.)

9-4 Preparation of Acyl Halides and Anhydrides

a. *Acyl Halides* These are prepared by treating the corresponding acid with phosphorus halides (PCl_3, PBr_3, PCl_5) or thionyl chloride ($SOCl_2$).

$$3CH_3COOH + PBr_3 \longrightarrow 3CH_3COBr + H_3PO_3$$
$$CH_3CH_2COOH + SOCl_2 \longrightarrow CH_3CH_2COCl + SO_2 + HCl$$
$$CH_2{=}CHCOOH + PCl_5 \longrightarrow CH_2{=}CHCOCl + POCl_3 + HCl$$

Acetyl fluoride may be prepared by treating acetyl chloride with sodium fluoride and hydrofluoric acid. Formic acid yields only carbon monoxide and hydrogen chloride when treated with $SOCl_2$ or PCl_5. *Formyl chloride* does not exist, nor does *formic anhydride*.

b. *Anhydrides*
1. Acyl halides react with salts of organic acids to yield anhydrides.

$$CH_3COCl + CH_3COONa \longrightarrow (CH_3CO)_2O + NaCl$$

$$(CH_3)_2CHCOCl + (CH_3)_3COOTl \longrightarrow (CH_3)_2CHC\overset{O}{\overset{\|}{C}}O\overset{O}{\overset{\|}{C}}(CH_3)_3 + TlCl$$

2. Ketene ($CH_2{=}C{=}O$) reacts with acids to yield anhydrides. Ketene can be prepared by passing acetone vapor over a hot wire at $700°C$.

$$CH_3COCH_3 \xrightarrow{700°C} CH_2{=}C{=}O + CH_4$$

$$CH_2{=}C{=}O + CH_3CH_2COOH \longrightarrow CH_3CH_2\overset{O}{\overset{\|}{C}}O\overset{O}{\overset{\|}{C}}CH_3$$

128

3. Thallium(I) salts of carboxylic acids react with thionyl chloride to give symmetric anhydrides.

$$2RCOOTl + SOCl_2 \longrightarrow [(RCOO)_2 SO] + 2TlCl \longrightarrow (RCO)_2O + SO_2$$

4. When dicarboxylic acids such as succinic, glutaric, or adipic acid are heated, cyclic anhydrides result. The longer-chain dicarboxylic acids give linear polymeric anhydrides

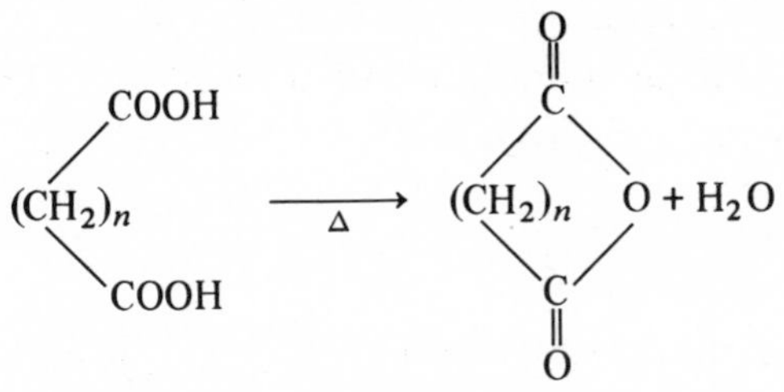

$$n = 2,3,4$$

9-5 Reaction of acids

a. *Ionization* Organic acids, when placed in polar solvents such as water, dissociate. Usually the degree of dissociation is not great, and for most carboxylic acids the equilibrium constant K is between 1×10^{-3} and 1×10^{-5}.

$$K = \frac{[RCOO^-][H_3O^+]}{[RCOOH][H_2O]}$$

The relative degree of dissociation of various proton acids in water can usually be predicted on the basis of the extent of delocalization of the negative charge in the anion. Important considerations are resonance and inductive effects.

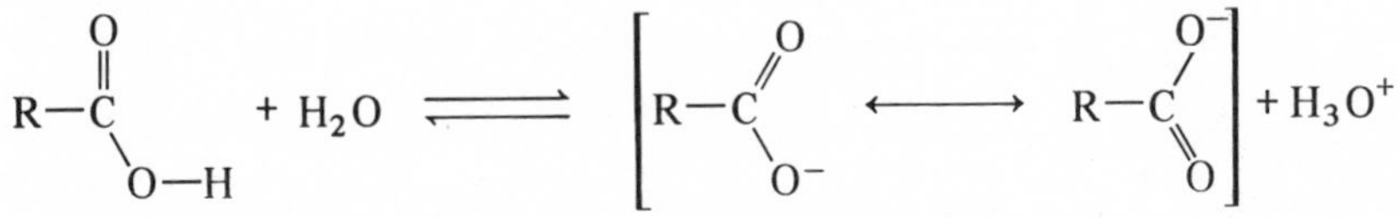

Delocalized charge, $K \sim 10^{-5}$

$$RCH_2OH + H_2O \rightleftharpoons RCH_2O^- + H_2O^+ \qquad \text{Localized charge, } K \sim 10^{-15}$$

The introduction of an electron-attracting group or atom into an acid will result in a partial neutralization of the anionic negative charge. Consequently such acids will be stronger than the corresponding unsubstituted compounds. Compare chloroacetic and acetic acids.

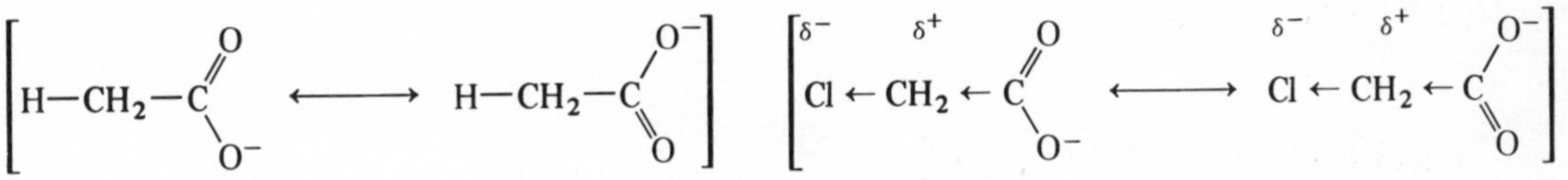

Dipole moment of Cl–C bond

One can estimate the comparative strengths of acids by considering the kinds of atoms or groups of atoms near the carboxyl group. It is not surprising that chloroacetic acid is a stronger acid than iodoacetic acid. Alkyl groups are the only common groups that release electrons by the inductive effect. This is illustrated by the fact that propionic acid is a weaker acid than acetic (that is, CH_3 is electron-releasing with respect to H). The inductive effect falls off rapidly with the distance; for example, β-chloropropionic acid is much weaker than α-chloropropionic acid. The electron-attracting strength of the common groups in the inductive effect is as follows:

$$-NO_2 > -CN > -F > -Cl > -Br > -I > -H > -CH_3 > -CH_2CH_3 > -CH(CH_3)_2 > -C(CH_3)_3$$

b. *Salt Formation* Carboxylic acids form salts with bicarbonates, carbonates, hydroxides, and oxides. Thallium(I) salts of carboxylic acids may be formed by treating the free acid with thallium(I) ethoxide in ethanol.

c. *Esterification* (See also Chap. 11.)

$$RCOOH + R'OH \xrightarrow{H^+} RCOOR' + H_2O$$

d. *Conversion of Acids to Amides* (See also Chap. 11.)

$$RCOOH \xrightarrow{NH_3} RCOO^- \overset{+}{N}H_4 \xrightarrow{\Delta} \underset{\underset{O}{\|}}{R}CNH_2$$

e. *Decarboxylation* $RCOOH + 2NaOH \xrightarrow{fuse} RH + Na_2CO_3 + H_2O.$

f. *Hell-Volhard-Zelinsky (HVZ) Reaction* Substitution occurs on the α *carbon* when acids having an α hydrogen are treated with chlorine or bromine in the presence of phosphorus.

$$CH_3CH_2COOH \xrightarrow{Br_2, P} \underset{\underset{Br}{|}}{CH_3CH}COOH \xrightarrow{Br_2, P} \underset{\underset{Br}{|}}{CH_3}\overset{\overset{Br}{|}}{C}COOH$$

The carbonyl group of aldehydes and ketones also promotes the above reaction.

g. *Acyl Halides from Acids* (See Sec. 9-4a.) α-Halo acyl halides can be prepared in one step with free halogen + PX_3. $3CH_3COOH + 3Br_2 + PBr_3 \longrightarrow 3BrCH_2COBr + P(OH)_3 + 3HBr$

h. *Conversion of Acids to Ketones* (See Sec. 7-2.)

i. *Reduction with LiAlH₄* (See Sec. 6-2.)

j. *Oxidation* The simple organic acids with the exception of *formic* (which easily decolorizes $KMnO_4$ solution) are stable to oxidizing agents except under drastic conditions.

k. *Kolbe Synthesis*

$$2RCOONa + 2H_2O \xrightarrow{electrolysis} RR + 2CO_2 + 2NaOH + H_2$$

Works well only for acids with no α branching.

l. *Hunsdiecker Reaction*

$$RCH_2COOAg + Br_2 \longrightarrow RCH_2Br + CO_2 + AgBr$$

A modified Hunsdiecker reaction occurs with thallium(I) carboxylic acid salts.

$$2RCH_2COOTl + 3Br_2 \longrightarrow 2RCH_2Br + 2CO_2 + Tl_2Br_4$$

9-6 Reaction of Acyl Halides and Anhydrides

The reactions of the acyl halides and anhydrides are quite similar. Both classes of compounds react readily with nucleophilic reagents resulting in nucleophilic substitution at the acyl carbon.

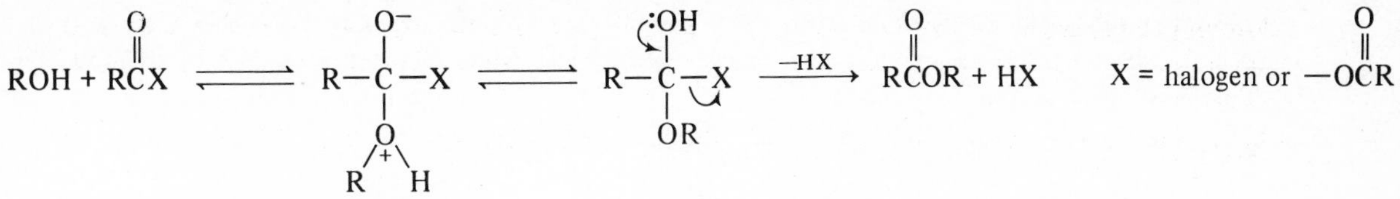

$$\text{ROH} + \text{RCX} \rightleftharpoons \underset{\underset{\overset{+}{R}\quad H}{\underset{|}{O}}}{\overset{\overset{O^-}{|}}{R-C-X}} \rightleftharpoons \underset{\underset{OR}{|}}{\overset{\overset{:OH}{|}}{R-C-X}} \xrightarrow{-HX} \text{RCOR} + \text{HX} \qquad X = \text{halogen or } -\text{OCR}$$

a. *Hydrolysis*

$$\text{RCOCl} + \text{H}_2\text{O} \longrightarrow \text{RCOOH} + \text{HCl} \qquad (\text{RCO})_2\text{O} + \text{H}_2\text{O} \longrightarrow 2\text{RCOOH}$$

b. *Alcoholysis*

$$\text{CH}_3\text{COBr} + \text{CH}_3\text{CH}_2\text{OH} \longrightarrow \text{CH}_3\text{COOCH}_2\text{CH}_3 + \text{HBr}$$
$$(\text{CH}_3\text{CO})_2\text{O} + \text{CH}_3\text{OH} \longrightarrow \text{CH}_3\text{COOCH}_3 + \text{CH}_3\text{COOH}$$

Tertiary alcohols may react with acid halides to give alkyl halides along with some ester. Usually a mixture of products results.

$$(\text{CH}_3)_3\text{COH} + \text{CH}_3\text{COCl} \longrightarrow (\text{CH}_3)_3\text{CCl} + \text{CH}_3\text{COOH}$$

Anhydrides often dehydrate tertiary alcohols to give olefins.

$$(\text{CH}_3)_3\text{COH} + (\text{CH}_3\text{CO})_2\text{O} \longrightarrow (\text{CH}_3)_2\text{C}=\text{CH}_2 + 2\text{CH}_3\text{COOH}$$

c. *Ammonolysis*

$$\text{CH}_3\text{CH}_2\text{COCl} + 2\text{NH}_3 \longrightarrow \text{CH}_3\text{CH}_2\text{CONH}_2 + \text{NH}_4\text{Cl}$$
$$(\text{CH}_3\text{CH}_2\text{CO})_2\text{O} + 2\text{NH}_3 \longrightarrow \text{CH}_3\text{CH}_2\text{CONH2} + \text{CH}_3\text{CH}_2\text{COONH}_4$$

d. *Reduction to Alcohols*

$$2\text{CH}_3\text{COCl} + \text{LiAlH}_4 \longrightarrow \text{LiAlCl}_2(\text{OCH}_2\text{CH}_3)_2 \xrightarrow{\text{H}^+} 2\text{CH}_3\text{CH}_2\text{OH}$$

$$(\text{CH}_3\text{CO})_2\text{O} + \text{LiAlH}_4 \longrightarrow \text{LiAlO}(\text{OCH}_2\text{CH}_3)_2 \xrightarrow{\text{H}^+} 2\text{CH}_3\text{CH}_2\text{OH}$$

e. *Reactions of Cyclic Anhydrides* Alcoholysis or ammonolysis of *cyclic anhydrides* leads to formation of the half ester–half acid and the half amide–half acid, respectively. Reduction of half ester–half acids with sodium and alcohol yields hydroxy acids.

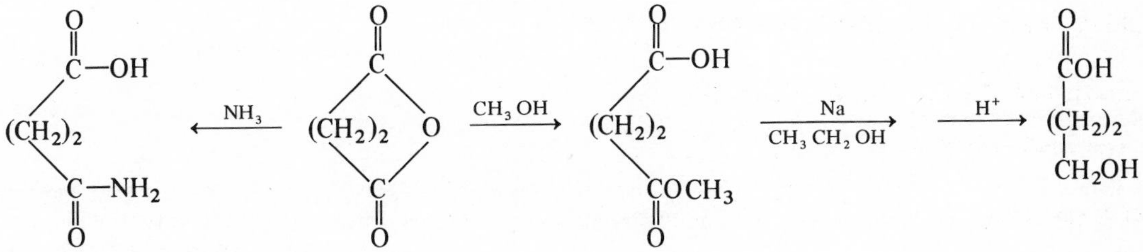

f. *Rosenmund Reduction of Acyl Halides* (See Sec. 7-2.)

g. *Reaction of Acyl Halides with Organocadmium Compounds* (See Sec. 7-2.)

9-7 Qualitative Tests

a. In aqueous solution, organic acids will turn blue litmus red.

b. Organic acids are stronger acids than carbonic acid; thus, when placed in sodium bicarbonate solution, they will form sodium salts, releasing CO_2 and dissolving. The fizzing of the CO_2 is easily observed except with the high-molecular-weight acids such as stearic. These react very slowly. Also, their sodium salts are only slightly soluble in water.

c. Low-molecular-weight aliphatic acyl halides react vigorously with water or alcohols, liberating heat. Anhydrides react more slowly.

d. Acyl halides yield an immediate precipitate with alcoholic silver nitrate.

9-8   Kuhn-Roth Oxidation or C-Methyl Determination

When aliphatic compounds are oxidized by boiling with a sulfuric acid solution of CrO_3, oxidation proceeds to carbon dioxide and water, except when a CH_3C- grouping is present in the compound; then acetic acid will be produced as an intermediate. The acetic acid so formed is fairly stable toward oxidation and the amount formed can be determined by steam distillation from the mixture and titration. For example, each mole of stearic acid, $CH_3(CH_2)_{16}COOH$, yields 1 mol of CH_3COOH in the Kuhn-Roth C-methyl determination and each mole of $CH_3CH_2CH(CH_3)CH_2CH(CH_3)CH_2OH$ would yield 3 mol of acetic acid. In theory, then, each CH_3C- grouping present in a compound will yield one equivalent of acetic acid. In practice, the yield of acetic acid is often less than predicted. Groups such as $(CH_3)_2CH-$ and $(CH_3)_3C-$ give only one equivalent of acetic acid per group.

9-9   Separation of Organic Compounds

The practicing chemist is constantly bumping into the problem of separating mixtures of organic compounds. Often it can be solved by physical methods such as distillation or crystallization. However, many times it is advantageous to separate mixtures by the differences in solubility of the various components. This is done by taking advantage of differences in polarity of the molecules in question. A useful rough rule of thumb is that one functional group (olefins, acetylenes, halogens, and nitro excepted) will confer water solubility on a molecule of up to five carbon atoms. For the above rule, we define solubility to mean about 3 g/100 g of solvent. There are many exceptions to the above rule, but it is nevertheless valuable to the beginner in thinking about separations. Thus, for example, one could predict that ethyl alcohol could be separated from 1-octanol by extraction with water.

Often, great differences in polar character of molecules can be induced where this difference does not exist. For example, decanoic acid contains 10 carbon atoms and therefore is not soluble in water. Its separation from dibutyl ether could be accomplished by extraction of the mixture with dilute sodium hydroxide. This would convert the acid to a sodium salt. This highly polar salt would then dissolve in the polar water phase, leaving the nonpolar butyl ether. After the butyl ether is removed, the water phase may be acidified with a mineral acid to regenerate the decanoic acid from its sodium salt.

A similar technique could be used to separate valeraldehyde from dibutyl ketone. If this mixture is treated with sodium bisulfite solution, the aldehyde will form a sodium bisulfite addition complex (see Sec. 7-3), which is a highly polar salt. If a saturated solution of bisulfite is used, the salt is insoluble, and in this case the nonpolar nine-carbon ketone could be washed away from the salt with a solvent such as ether. The aldehyde could be recovered by decomposing the bisulfite complex with a dilute acid or base.

Apply the above thinking to a more complex situation. How would you separate a mixture of:

Methanol   Octanoic acid   2-Heptanone   Hexane

In general, it is best to separate the molecules in order of decreasing polarity. Note that only one is present that has less than five carbons per functional group and is therefore water-soluble. Therefore, first extract the methanol with water. Now the octanoic acid can be made highly polar by conversion to the sodium salt, so extract with dilute sodium hydroxide. Next, 2-heptanone can be made very polar with sodium bisulfite and the hydrocarbon separated from the salt with ether. Evaporation of the ether gives hexane.

9-10   Classification of Organic Reactions

From your study of organic chemistry up to this point it should be clear that the way in which organic molecules react is greatly influenced by charges present in the molecules or those induced by the use of polar solvents or catalysts. The negative, electron-rich centers are usually referred to as nucleophilic sites and the positive, electron-deficient points as electrophilic sites. The three broad categories into which organic reactions fall are addition, substitution, and elimination. Addition and substitution reactions may further be subdivided as nucleophilic substitution(S_N), nucleophilic addition (A_N), electrophilic substitution (S_E), and electrophilic addition (A_E). In nucleophilic additions an electron-rich reagent adds to a π-electron system:

$$\underset{\overset{\|}{O}}{RCH} + {}^-CN \longrightarrow \underset{\overset{|}{O^-}}{RCHCN} \xrightarrow{H^+} \underset{\overset{|}{OH}}{RCHCN}$$

$$\left[\underset{\overset{\|}{O}}{CH_2\!\!=\!\!CHCH} \longleftrightarrow \underset{\overset{|}{O^-}}{\overset{+}{C}H_2CH\!\!=\!\!CH} \right] \xrightarrow{CH_3S^-} \left[\underset{\overset{|}{O^-}}{CH_3SCH_2CH\!\!=\!\!CH} \longleftrightarrow \underset{\overset{\|}{O}}{CH_3SCH_2\bar{C}HCH} \right] \xrightarrow{H^+}$$

$$CH_3SCH_2CH_2CHO$$

In these reactions the CN^- and CH_3S^- are conventionally regarded as the reagents; hence the reactions are considered to be nucleophilic rather than electrophilic, which would be the case when viewed from electron-deficient carbon on the carbonyl group. We have considered many examples of electrophilic addition in which an electron-deficient reagent initiates reaction on an electron-rich site, e.g., addition of HX to olefins. The following example in which electron-deficient hydrogen attacks electron-rich carbon illustrates electrophilic substitution.

$$RMgX + HOH \longrightarrow RH + MgXOH$$

Problems

9-1 Draw structures for and name (IUPAC and common) all of the acid chlorides and anhydrides that may be derived from $C_4H_8O_2$ acids.

9-2 Give an IUPAC and a common name for each of the following:
a. CH_3CBr_2COOH
b. CH_3COCl
c. F_3CCOOH
d. $(CH_3CO)_2O$
e. $CH_3\underset{\overset{\|}{O}}{C}O\underset{\overset{\|}{O}}{C}CH(CH_3)_2$
f. $CH_3CH\!\!=\!\!CHCOCl$

g. △—$\overset{\overset{\displaystyle CH_3}{|}}{C}HCOOH$
h. $[(CH_3)_3CCO]_2O$

9-3 Give IUPAC names for each of the following:
a. $CH_3C\!\!\equiv\!\!CCOBr$
b. $HOOCCH(CH_2CH_3)CH_2COOH$
c. $ClOC(CH_2)_3COCl$
d. $CH_3C\!\!\equiv\!\!CCH_2CH\!\!=\!\!CHCOOH$
e. $[(CH_3CH_2)_3CCO]_2O$
f. $CH_2\!\!=\!\!CHCH\!\!=\!\!CHCOOH$

g. ⬡—COCl
h. HOOC—⬡—COOH

i. $CH_3COCH\!\!=\!\!CHCOOH$

9-4 Draw structures for each of the following:
a. Methylvinylacetic acid
b. α,β,γ-Tribromocapric acid
c. γ,δ-Difluorovaleric anhydride
d. Fumaric acid
e. Cyclohexanecarboxylic acid
f. Myristyl chloride
g. Triisobutylacetyl bromide
h. Acetoacetic acid
i. (E)-3-Bromo-3-chloropropenoic acid
j. Glutaric anhydride

9-5 Point out the errors, if any, in each of the following names:
a. Methylvinyl ketone
b. Methylmagnesiumiodide
c. Ethylamine
d. sec-Butanol
e. 2,Chloro-4-bromoheptane
f. α-Methyloctanoic acid

9-6 Complete each of the following and classify the reaction from the point of view of how the reagent or the underlined part of the reagent is *acting*: nucleophilic substitution (S_N), nucleophilic addition (A_N), electrophilic substitution (S_E), electrophilic addition (A_E), or elimination (E).

a. $CH_3CH_2CH_2COONa + \underline{H}Cl \longrightarrow$ b. $EtI + CH_3COOAg \longrightarrow$

c. $\underline{CH_3(CH_2)_4CH_2}MgI + CO_2$ d. $CH_3\overline{CH_2CH_2}COOH + CH_3CH_2\underline{MgBr} \longrightarrow$

e. $\underline{CH_2}=CHCH_2CH_3 + HO\underline{Br} \longrightarrow$ f. $CH_2=CHCHO + \underline{H}Cl \longrightarrow$

g. $CH_3CHBrCHBrCH_3 + NaNH_2 \longrightarrow$ h. $CH_3COCH_3 + H\underline{CN} \longrightarrow$

i. $CH_3CH_2COOH + PBr_3 \longrightarrow$ j. $(CH_3CH_2CH_2CO)_2O + LiAl\underline{H}_4 \longrightarrow$

k. $(CH_3)_3CBr + \underline{NaOH} \longrightarrow$ l. $CH_3COCl + 2\underline{NH}_3 \longrightarrow$

9-7 Complete and balance the following:

a. Acrylic acid + HBr   b. β-Bromobutyric acid + NaOH

c. Crotonic acid + HCOOOH   d. 2-Oxopropanal + Tollens' reagent

e. Glyoxal + Tollens' reagent   f. Sodium acetylide + acetone

g. Ketene + chloroacetic acid   h. Nitroacetic acid + $SOCl_2$

i. Oxalic acid + PCl_5 j. Glycolic acid + PCl_5

k. Methylmagnesium iodide + CS_2 l. CH_3CN + dry HCl

m. Acetyl chloride + AgCN   n. Succinyl chloride + $(CH_3)_2Cd$

o. Succinic anhydride + EtOH   p. Ketene + ethylene glycol

q. Crotonic acid + HBr   r. Sodium acetylide + *tert*-butyl chloride

s. Tl(I) salt of palmitic acid + Br_2 t. Succinyl chloride + CuCN

u. Propiolyl chloride + NH_3 v. Rosenmund reduction of CH_3CH_2COCl

w. Acetic butyric anhydride + H_2O x. Formic acid + $COCl_2$

y. Tl(I) salt of trimethylacetic acid + $SOCl_2$ z. 4-Hydroxyhexanoic acid + PBr_3

9-8 Show how you would make the following conversions:

a. Hexanoic acid to pentane   b. Hexanoic acid to decane

c. 1-Propanol to propionyl chloride   d. 1-Propanol to isobutyric acid

e. Acetone to acetic anhydride   f. Acetic acid to $BrCH_2COCl$

g. Acetic acid to acetaldehyde oxime   h. Acetic acid to F_3CCOBr

i. Ethylene oxide to acrylonitrile   j. Butyric acid to 2-butenoic acid

k. Acetic acid to $(CH_3)_2C(OH)COOH$ l. CH_3CHO to β-hydroxybutyric acid

m. Propionic acid to propylene oxide   n. Butyric anhydride to 1-butene

o. 1-Butanol to butanoyl fluoride   p. Acetyl chloride to acetyl iodide

q. Lactic acid to propiolic acid   r. Pyruvic acid to lactic acid

s. Levulinic acid to valeric acid   t. Acetylene to glycolic acid

u. 2-Butanone to propionyl chloride   v. Butyryl bromide to 1-butyne

w. Acetic acid to 1,1-diethoxyethane   x. Propionic acid to 2-pentyne

y. Malic acid to tartaric acid   z. Butyl iodide to valeric acid

a'. CH_3COCl to 3,4-dimethyl-3,4-hexanediol   b'. 2-Butene to 2-methylbutanoic acid

c'. Butanoic acid to 2-ethyl-2-hexenal   d'. Cyclohexanol to $HOOC(CH=CH)_2COOH$

e'. Acetoacetic acid to 3,4-dimethyl-3-hexenedioic acid   f'. 3-Cyano-1-propene to azelaic acid

9-9 An acid, A, was found to have NE 104. It reacted with acetyl chloride but not with phenylhydrazine or NaOI. Oxidation of A gave a new acid, B (NE 59), whose nmr spectrum showed only two different signals. Suggest structures for A and B.

9-10 Compound A contains C,H,O, and Cl (MW 108 ± 1) and reacts vigorously with water to yield a chlorine-free compound (MW 90 ± 1). The nmr spectrum of A shows two singlets with an area ratio of 3:2. Suggest a structure for A.

9-11 An unknown compound, A, was found to react with semicarbazide and with NaOI to yield CHI_3. A 0.29-g sample of A required 25 ml of 0.1 N KOH for neutralization. The nmr spectrum of A showed a quartet, a doublet, and two singlets in a 1:3:3:1 ratio, respectively. Suggest a structure for A.

9-12 Compound A, $C_5H_6O_3$, reacted with ethanol to yield two compounds, B and C. Either B or C, when treated successively with $SOCl_2$ and ethanol, yielded D. Suggest structures for A through D.

9-13 Suggest a simple chemical test to distinguish between:

a. Formic and acetic acids   b. Crotonic and butyric acids

c. Hexanoic acid and 1-hexyne   d. 1-Propanol and 1-octanol

e. Levulinic and valeric acids   f. Lactic and malic acids

g. Acetic acid and acetyl chloride   h. Acetic acid and caprylic acid

i. Caproic anhydride and capryl alcohol   j. Ketene and ethene (both are gases)

9-14 Suggest a procedure for separating each of the following mixtures without using crystallization or distillation.

a. Hexane
   Methanol
   Hexanoic acid
   Heptanal

b. 2-Hexanone
   3-Hexanone
   Formic acid
   Triethylacetic acid

c. Succinic acid
   Suberic acid
   Isopropyl ether
   Valeraldehyde

d. Hexane
   1-Hexyne
   Hexanal
   Hexanoic acid

9-15 Match the proper pK value with each acid. [pK = log $(1/K)$, K = ionization or dissociation constant.]

a. pK values: 1.7, 2.7, 2.8, 4.1, 4.8, 4.9. Acids: nitroacetic, propionic, fluoroacetic, β-chloropropionic, acetic, α-chloropropionic.

b. pK values: 0.23, 3.1, 3.7, 3.8, 4.8, 4.9. Acids: acetic, isobutyric, trifluoroacetic, CH_3SCH_2COOH, iodoacetic, formic.

c. pK values: 0.7, 1.3, 2.8, 2.9, 3.1, 4.8. Acids: trichloroacetic, bromoacetic, chloroacetic, dichloroacetic, iodoacetic, acetic.

d. pK values: 1.2, 3.4, 3.9, 4.2 (pK values for dicarboxylic acids). Acids: lactic, malic, oxalic, succinic.

e. pK values: 2.5, 3.3, 3.8, 4.4. Acids: formic, acetoacetic, pyruvic, levulinic.

9-16 Write equations for the dissociation of acetic acid in methanol and in acetonitrile as a solvent. Write the expression for K for each of the dissociations.

9-17 List the following in order of decreasing activity toward alcoholic silver nitrate:

a. 5-Chloro-2-pentanone, isovaleryl chloride, 1-chloro-1-penten-3-ol, 5-chloro-3-penten-1-ol.

b. Carbon tetrachloride, butanoyl chloride, 2-methyl-2-chloropropane, 2-chlorobutane, 2-iodopentane, allyl chloride.

9-18 If the rate of addition of bromine to a double bond is dependent on the ease of transfer of a bromonium ion from Br_2 to C=C and thus is dependent on the electron density and availability of the C=C, arrange the following in order of decreasing ease of bromination.

a. Ethylene, vinyl iodide, propene, isobutylene.

b. Acrylic acid, crotonic acid, allylacetic acid, fumaric acid.

9-19 Indicate which compound in each of the following pairs would hydrolyze more rapidly.

a. Ethyl chloride
   Acetyl chloride

b. Ethyl chloride
   Ethyl iodide

c. Acrylyl chloride
   Propionyl chloride

d. Vinyl chloride
   Allyl chloride

e. Acetic anhydride
   Acetyl chloride

f. Acetyl fluoride
   Acetyl bromide

g. Acetyl chloride
   CH_3COOCH_3

h. Crotonyl chloride
   Butanoyl chloride

9-20 The methods of preparing dicarboxylic acids are generally similar to the methods used to prepare monocarboxylic acids. However, oxalic acid is prepared from the fusion of sodium formate (HCOONa) with sodium hydroxide at 350°C, and succinic acid can be prepared from maleic anhydride which is readily obtained from benzene:

$$C_6H_6 + O_2 \xrightarrow{V_2O_5,\ 450°C} \text{maleic anhydride}$$

Given methanol and ethanol, show how you would make the following conversions:

a. CO to oxalic acid
b. Benzene to succinic acid
c. Acetic to malonic acid
d. Succinic to adipic acid
e. Malonic to glutaric acid
f. Cyclohexanone to succinic acid

9-21 Given methanol, ethanol, and 1-propanol as the only available organic compounds, and any necessary inorganic reagents, show how you would prepare the following (for those marked with an asterisk do not use a Grignard reagent in your synthesis):

a. 4-Methylpentanoic acid
b. 2,4-Dimethylpentanoic acid
c. 2,7-Dimethyloctane
d. Isobutyric anhydride
e. Acetic isobutyric anhydride
f. *2-Methyl-2-pentenoyl chloride
g. 2,6-Dimethyl-3-heptanone
h. 3-Ethylheptanoic acid

Triethylacetyl bromide via pinacol reaction
j. Nitroacetyl fluoride
k. *2-Amino-3-methylbutanoic acid
l. *2,2-Dimethylpropanoyl chloride

9-22 Starting from any organic compounds containing four carbons or less, thexylborane, and triphenylphosphine, outline the synthesis of the following compounds using hydroboration and the Wittig reaction wherever possible.

a.

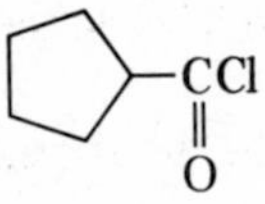

b. 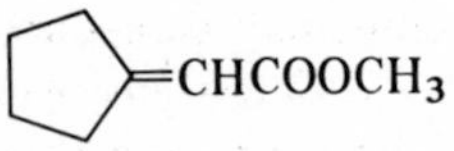 $\bigcirc$=CHCOOCH$_3$

c. CH$_3$OOC(CH$_2$)$_3$CO(CH$_2$)$_3$COOCH$_3$
d. 1,4-Dicyclohexylbutane
e. 2,4,4,6-Tetramethyl-2,5-heptadiene
f. Cyclohexanecarboxylic anhydride

9-23 Write equations for the following reactions:

$$CH_2\text{=}CHCH_2CN \xrightarrow{CH_3\,MgX} (A) \xrightarrow{H_3O^+} (B) \xrightarrow[CH_3\,ONa]{H_2\,NNH_2} (C) \xrightarrow{HOCl} (D) \xrightarrow[NaOH]{conc} (E) \xrightarrow{CH_3\,MgX} (F)$$

9-24 Monoprotonation of a compound having more than one possible reaction site yields the most stable conjugate acid. In each of the following instances give the monoprotonated product and justify your choice.

a. HCOOH
b. HOCl
c. H$_2$N$\overset{\overset{\displaystyle NH}{\|}}{C}NH_2$
d. HONH$_2$
e. $^-$OCH$_2$C≡C$^-$
f. CH$_3$CN
g. HOCH$_2$CH$_2$OCH$_3$
h $^-$OCH$_2$CH$_2$S$^-$

9-25 In the following equations indicating a competition between two bases for an acid, indicate which base would be the more successful. Give a reason for your choice.

a. CH$_3$CH$_2$Na + HC≡CH $\rightleftharpoons$ CH$_3$CH$_3$ + NaC≡CH
b. (NC)$_2$CH$^-$ + CH$_3$CN $\rightleftharpoons$ (NC)$_2$CH$_2$ + $^-$CH$_2$CN

c. 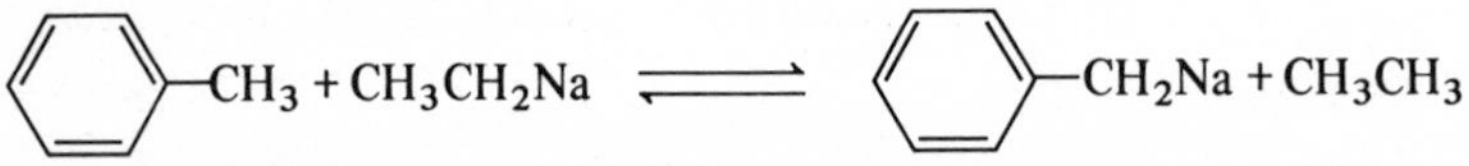 $\bigcirc$—CH$_3$ + CH$_3$CH$_2$Na $\rightleftharpoons$ $\bigcirc$—CH$_2$Na + CH$_3$CH$_3$

d. NH$_3$ + CH$_3\overset{+}{O}$H$_2$ $\rightleftharpoons$ NH$_4$$^+$ + CH$_3$OH
e. C$_4$H$_9$Li + NH$_3$ $\rightleftharpoons$ C$_4$H$_{10}$ + LiNH$_2$
f. CH$_3$ONa + HCOOH $\rightleftharpoons$ CH$_3$OH + HCOONa
g. CH$_3$CHO + CH$_3$CH$_2$$^-$ $\rightleftharpoons$ $^-$CH$_2$CHO + CH$_3$CH$_3$

h. 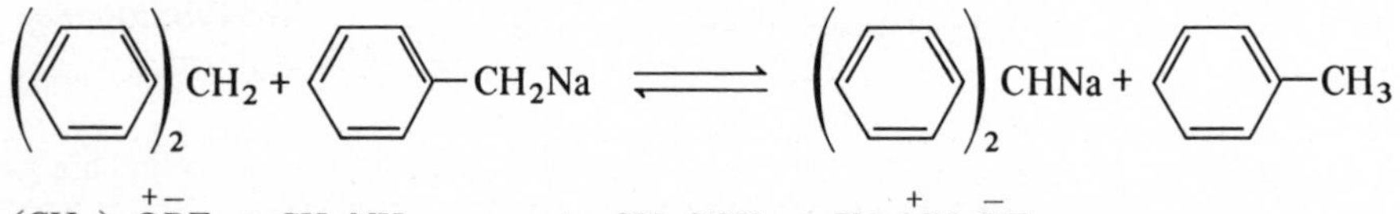 $\left(\bigcirc\right)_2$CH$_2$ + $\bigcirc$—CH$_2$Na $\rightleftharpoons$ $\left(\bigcirc\right)_2$CHNa + $\bigcirc$—CH$_3$

i. (CH$_3$)$_2\overset{+\,-}{O}$BF$_3$ + CH$_3$NH$_2$ $\rightleftharpoons$ CH$_3$OCH$_3$ + CH$_3\overset{+}{N}$H$_2\overset{-}{B}$F$_3$

9-26 Draw the indicated number of resonance forms for each of the following:
a. Acetyl chloride, 3
b. Acetic anhydride, 5
c. Acrylic acid, 5
d. 3-Methoxypropenoic acid, 7

9-27 Give a molecule orbital description of each of the following:

a. CH$_2$=CHCOOH
b. 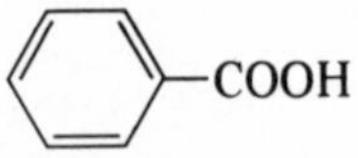 $\bigcirc$—COOH

9-28 In each of the following examples write the equation for the solvolysis (reaction with solvent) of the underlined molecule with the solvent beneath at the indicated temperature.

a. $\dfrac{CH_3COONa}{H_2O}$
b. $\dfrac{(CH_3)_3CI}{aq\ CH_3OH}$, 80°C

c. $\dfrac{HCOOH}{CH_3OH}$, 100°C
d. $\dfrac{CH_3\overset{\overset{\displaystyle OH}{|}}{C}HCH_2CHO}{HCOOH}$, 100°C

$$e. \quad \frac{CH_3COOCH_3}{H_3O^+}, 100°C \qquad\qquad f. \quad \frac{\overset{\displaystyle Cl}{\overset{|}{CH_3CHOCH_3}}}{H_2O}, 100°C$$

9-29 Most alkyl halides are insoluble in concentrated sulfuric acid making this a useful reagent for distinguishing them from oxygen-containing compounds. Some secondary and tertiary bromides and iodides do dissolve slowly in this reagent. How can this be explained?

9-30 In each of the following reactions generating a charged reaction site, indicate which of the listed reagents could cause the change.

   a. $(CH_3)_3COH \longrightarrow [(CH_3)_3C +]$
   NaOH, BF_3, $HClO_4$, Ag^+, Al_2O_3, CH_3COOH

   b. $CH_3CHO \longrightarrow [^-CH_2CHO]$
   OH^-, EtO^-, BF_3, CH_3NH_2, $HCOO^-$, $(CH_3)_3C^+$

9-31 Illustrate the mechanism of the reaction of butyric anhydride and ammonia.

9-32 When an anhydride is heated with a carboxylic acid in the presence of a mineral acid catalyst, an exchange reaction occurs. Describe how one could take advantage of this to convert valeric acid to valeric anhydride, using acetic anhydride. Illustrate a possible mechanism for the exchange reaction.

9-33 How many equivalents of acetic acid would one expect from a Kuhn-Roth oxidation of each of the following:

   a. Lauric acid
   b. $(CH_3)_3CCOOH$
   c. Maleic acid
   d. $CH_3(CH_2)_{17}\overset{\displaystyle CH_3}{\overset{|}{CH}}CH_2\overset{\displaystyle CH_3}{\overset{|}{C}}\!=\!=\!\overset{\displaystyle CH_3}{\overset{|}{C}}COOH$

   e. 3-Methyl-1-cyclopentanecarboxylic acid
   f. 2,2-Dimethylbutanoic acid

9-34 Acid X, $C_8H_{12}O_2$, gave $C_8H_{14}O_2$ upon catalytic hydrogenation. Oxidation of X with potassium permanganate led to only one compound Y, an acidic compound with MW = 172. The nmr spectrum of Y showed two singlets, a doublet, and a quintet with a respective area ratio of 1:6:4:1. Both compounds X and Y gave approximately 2 mol of acetic acid on Kuhn-Roth oxidation. Suggest structures for X and Y.

9-35 When 0.482g of unknown A was treated with $AgNO_3$, an immediate precipitate of 0.9 g of AgCl resulted. On warming with water, A was converted to acid B. B decolorized $KMnO_4$ solution and on catalytic hydrogenation at 30°C and 740 mmHg, 1.0 g of B absorbed 221 ml of hydrogen. Heating B caused the evolution of water and the formation of a new compound, C. Suggest structures for A, B, and C.

9-36 Compound A (MW = 102) gives a negative Lucas test. Its ir spectrum shows strong absorption bands at 3350 and 1710 cm^{-1}, and its nmr spectrum shows four different signals, one a singlet at $\tau = 0.5$. Careful oxidation of A yields compound B. A 0.198-g sample of B requires 20 ml of 0.15 N NaOH for neutralization. The nmr spectrum of B shows two singlets only. Suggest structures for A and B.

9-37 Neutral compound A, MW = 70, reacted with water to give compound B, MW = 88. Compound A shows only one signal in its nmr. Compound B shows a broad ir band at 3.4 μm (2940 cm^{-1}) and a sharp band at 5.95 μm (1680 cm^{-1}). Its nmr spectrum shows a doublet at $\delta = 1.2$, a multiplet at $\delta = 2.6$, and a singlet at $\delta = 11.8$. The integration of the spectrum shows 12.3, 2.1, and 2.0 units, respectively. Suggest structures for A and B.

9-38 Compound Z (MW = 100) contains only C, H, and O. Its nmr spectrum showed only one singlet, at $\tau = 7.0$. Its ir spectrum shows bands at (μm, cm^{-1}): 3.35, 2985; 5.35, 1869; 5.62, 1779; 8.2, 1220; 9.5, 1053; 11.0, 909. Suggest a structure for Z.

9-39 A mixture of two compounds gave a positive 2,4-dinitrophenylhydrazine test. When the mixture was extracted with ether and the ether layer treated with a concentrated solution of $NaHSO_3$, compound M (MW = 100) was obtained upon evaporation of the ether layer while compound N (MW = 112) was obtained upon acidification of the $NaHSO_3$ layer. Compound M does decolorize alkaline $KMnO_4$ at 25°C while N does not. However, both M and N underwent vigorous oxidation with hot alkaline $KMnO_4$ to yield, after acidification, single organic compounds O (NE = 72 ± 1) and P (NE = 80 ± 1), respectively. Of compounds M, N, O, and P, only P contains an asymmetric carbon atom. The following are some spectral data on compounds M and N.

Compound M Ir bands at (μm, cm^{-1}): 3.0, 3333; 3.42, 2924; 6.89, 1451; and 9.2, 1087. Nmr peaks: $\tau = 5.8$, singlet, 1H; broad unresolved peaks at $\tau = 8.2$, 1H; 8.7, 4H; and 9.0, 6H.

Compound N Ir bands at (μm, cm^{-1}): 3.42, 2924; and 5.87, 1704. Nmr peaks: $\tau = 9$, doublet ($J = 7$ Hz), 3H; broad unresolved peaks at $\tau = 7.7$, 4H; and $\tau = 8.0$ to 8.8, 5H. Suggest structures for M, N, O, and P.

9-40   Compound A (MW = 136) shows singlets at $\tau = -1.8$, 2.85, and 6.5 in the respective ratio 1:5:2. Its ir spectrum is shown. Suggest a structure for A.

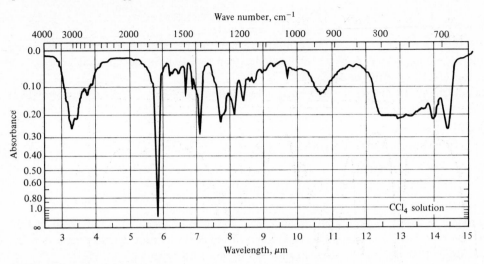

Figure P9-40. Ir spectrum for Prob. 9-40. *(From R. A. Silva and R. Spenger, "The Quantitative Organic Analysis Spectra Kit," McGraw-Hill Book Company, New York, 1969. By permission of the publishers.)*

9-41   Compound B has MW = 88. Its ir and nmr spectra are shown. Suggest a structure for B.

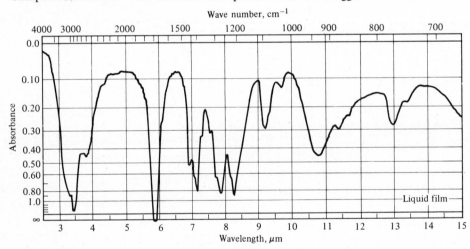

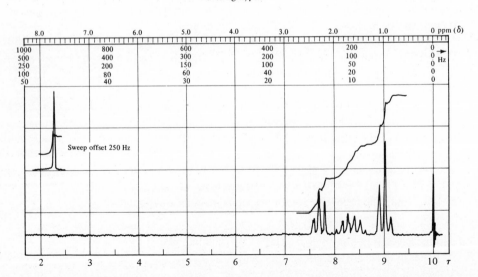

Figure P9-41. Ir and nmr spectra for Prob. 9-41. *(From R. A. Silva and R. Spenger, "The Quantitative Organic Analysis Spectra Kit," McGraw-Hill Book Company, New York, 1969. By permission of the publishers.)*

9-42   Compound C (MW = 128) yields compound D (MW = 154) when treated as shown below. Ir and nmr spectra for C and D are shown. Suggest structures for C and D.

$$(C) \xrightarrow{\text{P, Br}_2} \xrightarrow[\text{KOH}]{\text{Alc}} \xrightarrow{\text{HBr}} \xrightarrow{\text{NaCN}} \xrightarrow{\text{H}^+, \text{H}_2\text{O}} \xrightarrow{\Delta} (D)$$

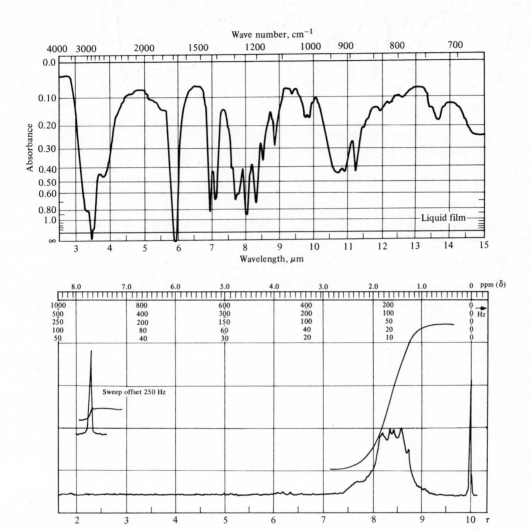

Figure P9-42. Ir and nmr spectra for Prob. 9-42, compound C. *(From R. A. Silva and R. Spenger, "The Quantitative Organic Analysis Spectra Kit," McGraw-Hill Book Company, New York, 1969. By permission of the publishers.)*

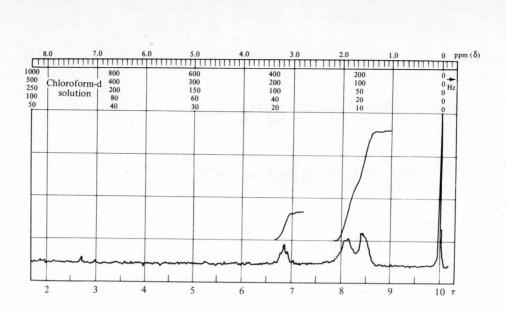

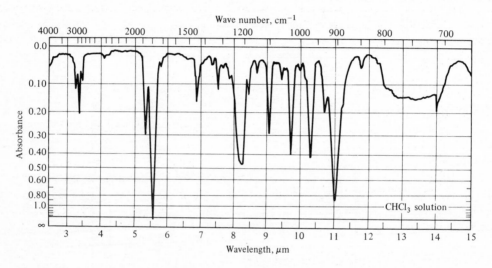

Figure P9-42. Ir and nmr Spectra for Prob. 9-42, compound D. *(From R. A. Silva and R. Spenger, "The Quantitative Organic Analysis Spectra Kit," McGraw-Hill Book Company, New York, 1969. By permission of the publishers.)*

10
aliphatic amines

10-1 Nomenclature

a. The amines are divided into three classes. If one alkyl group is attached to the N atom, the amines are called primary; if two are attached, secondary; and if three are attached, tertiary.

b. The common system of naming this class of compounds is to name the alkyl groups attached to the N atom and add the ending *amine.*

$CH_3CH_2NH_2$
Ethylamine

$CH_3CH_2CHCH_3$
|
NH_2
sec-Butylamine

$CH_3NHCH_2CH_3$
Methylethylamine

$CH_3CH_2NCH(CH_3)_2$
|
$C(CH_3)_3$
Ethylisopropyl-*tert*-butylamine

c. In the IUPAC system primary amines are named by dropping the *e* from the parent hydrocarbon's name and adding *amine.* Secondary and tertiary amines are named as *N*-substituted derivatives of a primary amine choosing the most complex alkyl group attached to the nitrogen as the parent primary amine. Although not recommended by the IUPAC, many amines are named as aminoalkanes (1°), *N*-alkylaminoalkanes (2°), and *N,N*-dialkylaminoalkanes (3°).

$CH_3(CH_2)_4CH_2NH_2$
1-Hexanamine
(1-Aminohexane)

$CH_3CH_2CH_2CH_2CH_2NHCH_3$
N-Methylpentylamine
[1-(*N*-Methylamino)pentane]

$CH_3CH_2CH_2CH_2N(CH_3)CH_2CH_3$
N-Ethyl-*N*-methylbutylamine
[1-(*N*-Ethyl-*N*-methylamino)butane]

d. Diamines may be named as polymethylene derivatives.

$H_2NCH_2CH_2CH_2CH_2NH_2$
Tetramethylenediamine
1,4-Butanediamine (IUPAC)

$H_2N(CH_2)_6NH_2$
Hexamethylenediamine
1,6-Hexanediamine (IUPAC)

$(HOOCCH_2)_2NCH_2CH_2N(CH_2COOH)_2$
Ethylenediaminetetraacetic acid
(EDTA)

10-2 Quaternary Ammonium Salts

These highly polar, water-soluble compounds are named as derivatives of the ammonium ion.

CH_3
|
$CH_3-N^+-CH_3 \ Cl^-$
|
CH_3

Tetramethylammonium chloride

CH_3
|
$CH_3CH_2-N^+-CH_2CH_3 \ OH^-$
|
CH_3

Dimethyldiethylammonium hydroxide

10-3 Preparation of Primary Amines

a. Alkylation

$$RX + NH_3 \longrightarrow R\overset{+}{N}H_3X^- \xrightarrow{NH_3} NH_4X + RNH_2 \xrightarrow{RX} R_2\overset{+}{N}H_2X^- \xrightarrow{NH_3}$$

$$NH_4X + R_2NH \xrightarrow{RX} R_3\overset{+}{N}HX^- \xrightarrow{NH_3} NH_4X + R_3N \xrightarrow{RX} R_4\overset{+}{N}X^-$$

While this reaction theoretically may be used to make almost any amine, it is of little practical value since such a large number of compounds result that it is exceedingly difficult to purify any one of them.

b. Reduction

1. Nitro compounds $RNO_2 \xrightarrow{Pt, H_2} RNH_2 + 2H_2O$.

2. Nitriles (cyanides) $RC\equiv N \xrightarrow{Pt, H_2} RCH_2NH_2$.

3. Oximes $R_2C=NOH \xrightarrow{Pt, H_2} R_2CHNH_2 + H_2O$.

c. Reductive Alkylation
Aldehydes and ketones have a tendency to react with ammonia and amines. Although this equilibrium reaction is often far on the side of the reactants rather than that of the products, if the reaction is carried out in the presence of hydrogen and a catalyst, the imines formed are hydrogenated and good yields of primary, secondary, or tertiary amines may be had (see Sec. 10-4b and 10-5b).

$$R_2C=O + NH_3 \rightleftharpoons R_2\overset{\overset{\displaystyle OH}{|}}{C}NH_2 \rightleftharpoons R_2C=NH + H_2O \xrightarrow[\text{cat (Pt or Ni)}]{H_2} R_2CHNH_2$$

d. Hofmann Hypobromite Reaction

$$\underset{\underset{O}{\|}}{RCNH_2} + NaOBr + 2NaOH \longrightarrow RNH_2 + Na_2CO_3 + NaBr + H_2O$$

The reaction is believed to proceed via a rearrangement of the R group to an electron-deficient nitrogen in an intermediate (B) resulting in the formation of an isocyanate (C), which under the reaction conditions is hydrolyzed to the amine. Both proposed intermediates, A and C, have been isolated in the reaction.

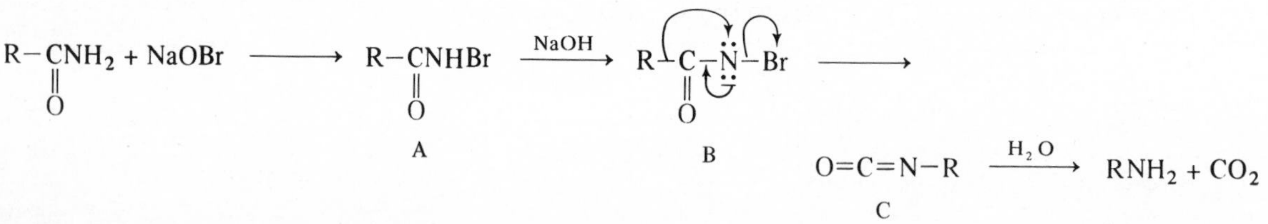

$$\underset{\underset{O}{\|}}{R-CNH_2} + NaOBr \longrightarrow \underset{\underset{O}{\|}}{R-CNHBr} \xrightarrow{NaOH} \underset{\underset{O}{\|}}{R-C-\overset{..}{N}-Br} \longrightarrow$$

$$\qquad\qquad\qquad\qquad\qquad\qquad A \qquad\qquad\qquad\qquad B$$

$$O=C=N-R \xrightarrow{H_2O} RNH_2 + CO_2$$

$$C$$

e. Gabriel's Method for Primary Amines

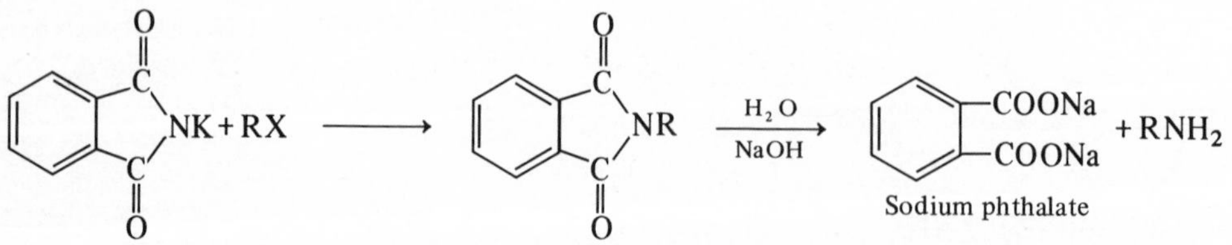

Potassium phthalimide   N-Alkylphthalimide   Sodium phthalate

f. Amination of Trialkylboranes

$$R_3B + 2H_2NOSO_3H \xrightarrow{4NaOH} 2RNH_2 + 2Na_2SO_4 + RB(OH)_2 + 2H_2O$$

Hydroxylamine-O-sulfonic acid

The yields of primary amines from this reaction are only about 60 percent, apparently due to the failure of $RB(OH)_2$ to react with more H_2NOSO_3H.

10-4 Preparation of Secondary Amines

a. *From Sodium Cyanamide*

$$2RX + Na_2NC{\equiv}N \longrightarrow R_2NC{\equiv}N \xrightarrow[H^+ \text{ or } OH^-]{H_2O} NH_3 + (R_2NCOOH) \longrightarrow R_2NH + CO_2$$

This works well only for the synthesis of simple symmetric secondary amines.

b. *Reductive Alkylation*

$$RCHO + H_2NCH_3 \rightleftharpoons \left[\begin{array}{c} \overset{\displaystyle OH}{\underset{\displaystyle |}{RCHNHCH_3}} \end{array} \right] \rightleftharpoons RCH{=}NCH_3 + H_2O \xrightarrow[\text{Ni or Pt}]{H_2} RCH_2NHCH_3$$

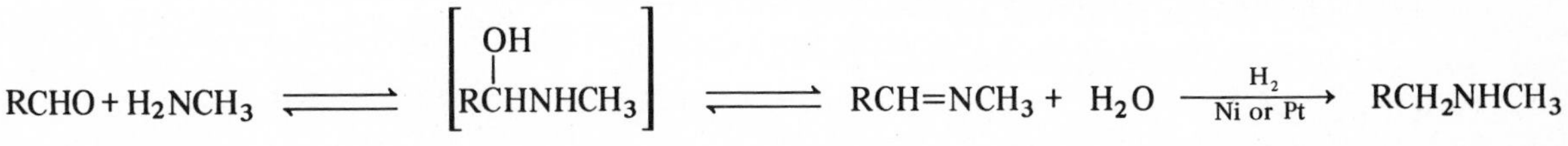

10-5 Preparation of Tertiary Amines

a. *Alkylation of Secondary Amines*

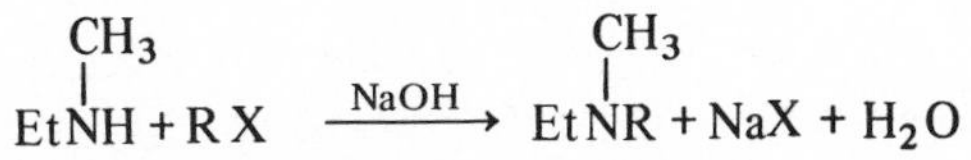

$$\overset{\displaystyle CH_3}{\underset{\displaystyle |}{Et\,NH}} + RX \xrightarrow{\text{NaOH}} \overset{\displaystyle CH_3}{\underset{\displaystyle |}{Et\,NR}} + NaX + H_2O$$

b. *Reductive Alkylation*

$$R_2NH + CH_2O \xrightarrow[\text{Ni}]{H_2} R_2NCH_3 + H_2O$$

10-6 Preparation of Quaternary Ammonium Salts

a. *Halides*

$$CH_3CH_2NH_2 + 3CH_3I \xrightarrow{2\text{NaOH}} CH_3CH_2\overset{+}{N}(CH_3)_3I^- + 2NaI + 2H_2O$$

b. *Hydroxides*

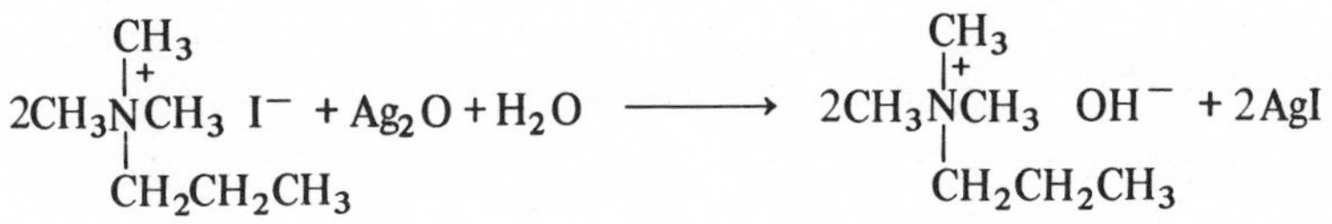

$$2CH_3\overset{\overset{\displaystyle CH_3}{\displaystyle |+}}{\underset{\underset{\displaystyle CH_2CH_2CH_3}{\displaystyle |}}{N}}CH_3 \ I^- + Ag_2O + H_2O \longrightarrow 2CH_3\overset{\overset{\displaystyle CH_3}{\displaystyle |+}}{\underset{\underset{\displaystyle CH_2CH_2CH_3}{\displaystyle |}}{N}}CH_3 \ OH^- + 2AgI$$

10-7 Reactions of Amines

a. *Salt Formation* The N atom of primary, secondary, and tertiary amines possesses an unshared pair of electrons. This pair of electrons will form a bond with a proton or a Lewis acid.

$$RNH_2 + HY \longrightarrow R\overset{+}{N}H_3\overset{-}{Y} \ (Y = \text{halogen, } OSO_2OH, ONO_2, OOCR)$$

b. *Hydroxide Formation* All amines form hydroxides by hydrogen bonding with water.

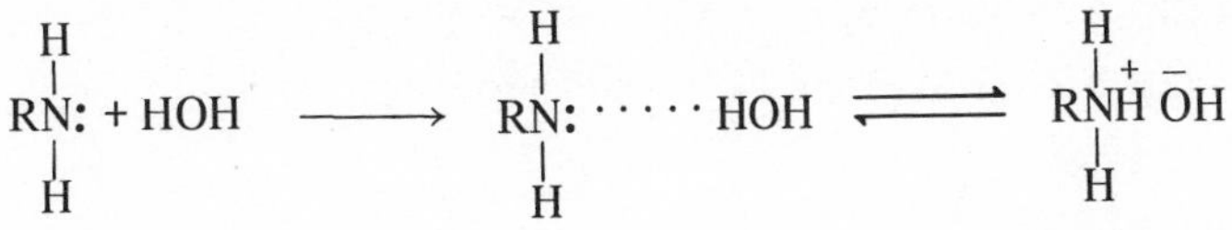

$$\overset{\overset{\displaystyle H}{\displaystyle |}}{\underset{\underset{\displaystyle H}{\displaystyle |}}{RN}}{:} + HOH \longrightarrow \overset{\overset{\displaystyle H}{\displaystyle |}}{\underset{\underset{\displaystyle H}{\displaystyle |}}{RN}}{:} \cdots HOH \rightleftharpoons \overset{\overset{\displaystyle H}{\displaystyle |}}{\underset{\underset{\displaystyle H}{\displaystyle |}}{R\overset{+}{N}H}} \ \overset{-}{OH}$$

c. *Anhydrides and Acyl Halides* These react with ammonia (see Sec. 9-6c) and with primary or secondary amines to give amides. Tertiary amines (i.e., those without an H on the N atom) do not yield amides.

$$(RCO)_2O + CH_3NH_2 \longrightarrow RCONHCH_3 + RCOOH$$
$$RCOCl + (CH_3)_2NH \longrightarrow RCON(CH_3)_2 + HCl\dagger$$

† The HCl formed reacts with the amine to give the amine hydrochloride unless the HCl is removed as fast as it is formed by a stronger base such as sodium hydroxide.

d. *Oxidation* Primary and secondary amines do not give simple oxidation reactions. Tertiary amines, however, may be smoothly converted to oxides.

$$R_3N \xrightarrow{\text{H}_2\text{O}_2} R_3N^+{-}O^-$$

e. *Reaction with Nitrous Acid*
   1. Primary amines. Nitrous acid with primary amines gives first the carbonium ion and nitrogen. This ion may then give any of its typical reactions (see Sec. 5-6) with the nucleophiles present. In dilute solution, water is present in very large excess and often an alcohol is the principal product.

$$RCH_2NH_2 + HONO \longrightarrow RCH_2^+ + N_2 \xrightarrow{\text{HOH}} RCH_2\overset{+}{O}H_2 \rightleftharpoons RCH_2OH + H^+$$

   2. Secondary amines. $R_2NH + HONO \longrightarrow R_2NN{=}O$ Dialkylnitrosoamine (yellow compounds).
   3. Tertiary amines do not react (except for salt formation).

10-8 Qualitative Tests for Amines

a. A simple test for amines that are not soluble in water is to treat the compound in question with dilute acid. Most amines form water-soluble salts with acids. The low-molecular-weight, water-soluble amines can easily be recognized by their ammonialike odor. Water solutions of these substances turn red litmus blue.

b. *Hinsberg Reaction* This reaction is carried out with an amine, benzenesulfonyl chloride, and an excess of sodium hydroxide. The reaction provides a means for differentiating primary, secondary, and tertiary amines. The amide formed from the reagent and primary amine is soluble in sodium hydroxide solution, the amide from the secondary amine is not, and the tertiary amine does not react.

$$RNH_2 + C_6H_5SO_2Cl \xrightarrow{\text{NaOH}} C_6H_5SO_2\overset{\text{H}}{\underset{|}{N}}R \xrightarrow{\text{NaOH}} C_6H_5SO_2\overset{-}{N}R\overset{+}{N}a \xrightarrow{\text{HCl}} C_6H_5SO_2NHR + NaCl$$

Water-soluble   Water-insoluble

$$R_2NH + C_6H_5SO_2Cl \xrightarrow{\text{NaOH}} C_6H_5SO_2NR_2$$

Water-insoluble

Thus, when one shakes a *primary amine* with the Hinsberg reagent and sodium hydroxide, one gets a clear solution. If one acidifies the solution with a strong acid, the sodium salt is converted to the free amide, which is usually a white solid. With *secondary amines* a solid insoluble in sodium hydroxide is formed. With *tertiary amines* there is no reaction except the slow conversion of the benzenesulfonyl chloride to the sodium salt of benzenesulfonic acid, which is very soluble in water. The tertiary amine is left unchanged. This can be confirmed by adding a strong acid, which forms the salt of the amine, rendering it soluble in water. Benzenesulfonic acid is very soluble in H_2O.

10-9 Decomposition of Quaternary Ammonium Hydroxides

Heating quaternary ammonium hydroxides that have at least one alkyl group larger than methyl causes them to split, giving a tertiary amine and an olefin.

$$CH_3CH_2\overset{\overset{\displaystyle CH_3}{|+}}{\underset{\underset{\displaystyle CH_3}{|}}{N}}CH_3 \ \underset{OH^-}{} \xrightarrow{\Delta} CH_2{=}CH_2 + (CH_3)_3N + H_2O$$

The mechanism of the reaction seems to involve the ^-OH pulling off a β hydrogen with the simultaneous breaking of the C—N bond. Of the common alkyl groups, ethyl is the one most easily split off as the olefin.

The splitting of amines by this technique is known as the *Hofmann* degradation and is very useful in breaking down complex amines into simpler products for the purpose of elucidating the structure of an unknown amine. If the unknown amine

$$CH_3CHCH_2$$
$$|$$
$$\qquad\qquad NCH_2CH_3$$
$$|$$
$$CH_3CHCH_2$$

is treated with excess methyl iodide and then silver oxide, the quaternary ammonium hydroxide is formed. On heating

this breaks down to give ethylene and

$$CH_3CHCH_2$$
$$|$$
$$\qquad NCH_3$$
$$|$$
$$CH_3CHCH_2$$

. If the above process is repeated, one obtains

$$CH_2=C-CHCH_2NCH_3$$ with substituents H_3C, CH_3, CH_3. Repeating the procedure again with this compound, one obtains $CH_2=C-C=CH_2$ with substituents H_3C, CH_3 and

$(CH_3)_3N$. These are relatively simple molecules, and their structures can be established by comparison with known molecules. Or, one might further degrade the diolefin by ozonolysis and establish the structure of these products (that is, CH_2O and CH_3C-CCH_3 with two $=O$ groups), and from this information deduce the structure of the diolefin and in turn that of the

original amine. Quite often in degrading cyclic olefins after the introduction of the first double bond by elimination, the molecule is catalytically hydrogenated so that after the next elimination a monoolefin is obtained rather than a diolefin. Dienes polymerize much more easily and therefore are harder to characterize.

Problems

10-1 Give IUPAC and common names for each of the following. Indicate whether the amine is primary, secondary, or tertiary.

a. $(CH_3)_2CHNHCH_3$
b. $CH_2=CHCH_2NH_2$
c. $CH_3CH_2CH(CH_3)NH_2$
d. $H_2NCH_2CH_2CH_2NH_2$
e. $(CH_3)_2CHN(CH_3)_2$
f. $HOCH_2CH_2NH_2$
g. H_2NCH_2COOH
h. $(CH_3)_3CNH_2$
i. $CH_2=CHN(CH_3)_2$
j. $(Et)_2NCH_2CH_2CH_2CH$ with $=O$
l. $(H_2NCH_2CH_2)_2C=O$
k. H_2N-⬡$-CH_2COOH$

10-2 Write structures for each of the following:

a. 3-Dimethylaminopropanal
b. 1-Dimethylamino-3-diethylaminopropane
c. 4-Methylethylamino-2-pentenoic acid
d. 3-Diallylamino-2-hexanone
e. Dimethylaminoacetic acid
f. 1-Amino-2-pentyne
g. Tetraisobutylammonium nitrate
h. *meso*-1,2-Cyclohexanediamine
i. 4-(*N*-Butyl-*N*-vinylamino)-1-pentanol
j. (*Z*)-1-Amino-2-butene

10-3 Complete and balance the following:

a. Acetic acid + ethylamine
b. BF_3 + *tert*-butylamine
c. Nitrous acid + methylamine
d. Acetic anhydride + 2-aminoethanol
e. Tetraethylammonium iodide + Ag_2SO_4
f. 3-Pentanone oxime + H_2 + Pt
g. Sodium cyanamide + ethyl iodide
h. Butanedial + CH_3NH_2 + H_2 + Pt
i. Sodium phthalimide + hexyl iodide
j. $CH_3CH_2CONH_2$ + NaOBr + NaOH
k. Nitrous acid + trimethylamine
l. $N\equiv CCH_2CH_2C\equiv N$ + H_2 + Pt
m. Ketene + methylamine
n. CH_3NH_2 + CH_3MgI

o. Trihexylborane + H_2NOSO_3H $\xrightarrow{\qquad}$ $\xrightarrow{\text{NaOH}}$
p. Acetic anhydride + $HSCH_2CH_2NHCH_3$
q. Diethyl cyanamide + hot aqueous NaOH
r. Acrylonitrile + H_2 + Pt
s. $(CH_3CH_2CH_2)_4\overset{+}{N}OH^-$ + heat
t. Diallylamine + $C_6H_5SO_2Cl$ + NaOH
u. $(CH_3)_3B$ + ammonia
v. Hexamethylenediamine + acetone + H_2 + Pt

10-4   Indicate how you would make the following conversions:
   a. Acetic acid to methylamine   *b.* Acetone to isopropylamine (two ways)
   c. Methyl chloride to ethylamine   *d.* Propylene to isobutylamine
   e. Acetone to 1-amino-2-methyl-2-propanol   *f.* Ethane to ethylamine
   g. Ethanol to diethylnitrosoamine   *h.* Ethene to 1,4-butanediamine
   i. $(Et)_2NH$ to tetraethylammonium hydroxide   *j.* 1-Butanamine to 1-butene (two ways)
   k. $(CH_3)_2NH$ to trimethylamine oxide   *l.* Diethylacetic acid to 3-pentanamine
   m. CH_3CHO to 4-amino-2-butanol   *n.* Acetone to 3,3-dimethyl-2-butanamine
                                                      p. Ethane to 2-methyl-2-amino-1,3-propanediol

   o. 1-Methylcyclopentene to

   q. Isobutylene to 2,6-dimethyl-4-(*N*-methylamino)heptane
   r. Isobutyraldehyde to 2,2,4-trimethyl-1-(*N*-isobutylamino)pentane
   s. Butanal to 2-ethyl-1-(*N*-butylamino)hexane   *t.* Divinylacetic acid to HO—〈 〉—CH_2OH

10-5   Suggest a convenient chemical test to distinguish between each of the following:
   a. Ethylamine and dimethylamine   *b.* Triethylamine and butylamine
   c. NH_4Cl and dimethylammonium chloride   *d.* Dibutylamine and pentyl cyanide
   e. Diethylamine and diamylamine   *f.* 1-Hexanol and hexylamine
   g. Dibutylamine and tributylamine   *h.* $CH_3CH_2NH_3{}^+Cl^-$ and butyl chloride

10-6   Suggest procedures for separating the following mixtures chemically:
   a. Heptane, 2-octanone, tributylamine, and propylene glycol
   b. 1-Hexanol, 2-hexanone, trimethylamine, and heptylamine
   c. 1-Octanol, 2-octanamine, and octanoic acid

10-7   Amines may be titrated to the equivalence point with strong acids. Calculate the equivalent weight for each of the following:
   a. Triethylamine   *b.* 2-Aminopropanoic acid
   c. Ethylenediamine   *d.* 1-Dimethylamino-2-aminoethane
   e. Dimethyldiethylammonium hydroxide

10-8   What would be the principal products to be expected from the action of heat on each of the following:
   a. Dimethyldiethylammonium hydroxide   *b.* Methylethylpropylsulfonium hydroxide
   c. Dimethylethylisopentylammonium hydroxide   *d.* Trimethylcyclopentylammonium hydroxide
   e. Diethylpropylsulfonium hydroxide   *f.* Dimethylpropylisobutylammonium hydroxide

10-9   Recalling the nature of the *E*2 reaction (see Sec. 4-2*a*-1), explain why the following order of increasing ease of elimination of groups in the decomposition of quaternary ammonium hydroxide is observed.

$$(CH_3)_2CHCH_2- < (CH_3)_2CHCH_2CH_2- < CH_3CH_2- < ClCH_2CH_2-$$

10-10   Recalling that the base strength of amines depends on their ability to yield a pair of electrons to a proton or other electron acceptor, arrange the following in order of increasing base strength. Review Sec. 9-5 for electron-attracting power of various groups.
   a. Ammonia, methylamine, dimethylamine, chloramine ($ClNH_2$)
   b. Dimethylvinylamine, 3-dimethylaminopropenal, trimethylamine
   c. Hydroxylamine, hydrazine, ammonia, ethylamine

   d. 〈 〉—NH_2,   〈 〉—$NHCH_3$,   〈 〉—NH_2,   〈 〉—NH—〈 〉

10-11   Trimethylamine is a weaker base than dimethylamine despite the fact that it has three methyl groups that have a tendency to release electrons toward nitrogen. This statement is based on a study of the reaction:

$$(CH_3)_2NH\colon + HOH \qquad (CH_3)_2\overset{+}{N}H_2OH^-$$

   Instead of measuring the base strength against a proton, if we use a Lewis acid, such as BF_3, we find that trimethylamine appears even weaker. Offer an explanation.

10-12 A compound, $C_5H_{13}N$, gave $C_5H_{13}NO$ on oxidation with hydrogen peroxide. Suggest three possible structures.

10-13 What product(s) would you expect from each of the following when subjected to exhaustive methylation (i.e., Hofmann degradation)?

a. *tert*-Butylamine   b. 3-Dimethylaminopentane   c. Butylpropylamine

d. 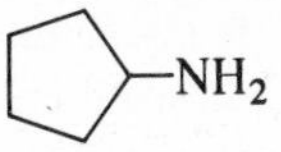 —NH_2 e. 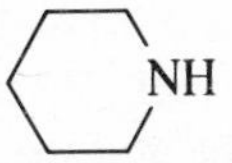 NH   f. 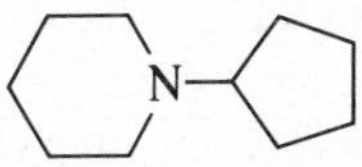 N—

10-14 In each of the following instances, a *primary* amine was subjected to exhaustive methylation and the resulting olefin subjected to ozonolysis. From the products listed suggest a structure for the original amine.

a. Acetone and 2,2-dimethylpropanal   b. Hexanedial
c. Formaldehyde and propanal   d. $2CH_2O$ and $CH_3CH_2CH_2CH(CHO)_2$
e. Ethanal and propanal   f. $CH_3COCH_2CH_2CH(CH_3)CHO$

10-15 A tertiary amine was subjected to the Hofmann degradation and the resulting olefin to ozonolysis. The products were 3 mol of CH_2O and 1 mol of $CH(CHO)_3$. Suggest a structure for the amine.

10-16 The Cope reaction is related to the Hofmann degradation. It involves the pyrolysis (100 to 150°C) of tertiary amine oxides and goes smoothly enough to be a useful preparatory method for olefins. The mechanism is apparently an internal elimination via a cyclic transition state. What two organic products would you expect from the pyrolytic elimination of dimethylisohexylamine oxide? Show a likely mechanism for this reaction.

10-17 Explain why, when H_2NNH_2 is treated with 2 mol of CH_3I, the product is $H_2NN(CH_3)_2$ rather than $CH_3NHNHCH_3$. *Hint:* this occurs by a displacement reaction.

10-18 Point out the errors, if any, in the following proposed syntheses:

a. CH_2CH_2 (epoxide, O) $\xrightarrow[\text{(A)}]{\text{HCl}}$ $ClCH_2CH_2OH$ $\xrightarrow[\text{(B)}]{\text{Cu, 350°C}}$ $ClCH_2CHO$ $\xrightarrow[\text{(C)}]{CH_3NH_2,\ Pt,\ H_2}$ $ClCH_2CH_2NHCH_3$

b. $(CH_3)_2C=CH_2$ $\xrightarrow[\text{(A)}]{\text{HCl}}$ $(CH_3)_3CCl$ $\xrightarrow[\text{(B)}]{NH_3}$ $(CH_3)_3CNH_2$ $\xrightarrow[\text{(C)}]{H_2O_2}$ $(CH_3)_3CN^{+}-O^{-}$ (with H above and H below N)

c. $CH_3CH=CH_2$ $\xrightarrow[\text{(A)}]{\text{HCN}}$ $(CH_3)_2CHCN$ $\xrightarrow[\text{(B)}]{H_2,\ Pt}$ $(CH_3)_2CHCH_2NH_2$ $\xrightarrow[\text{(C)}]{\text{HONO}}$ $(CH_3)_2CHCH_2OH$

d. $H_2NCH_2CH_2OH$ $\xrightarrow[\text{(A)}]{K_2Cr_2O_7}$ H_2NCH_2COOH $\xrightarrow[\text{(B)}]{\text{EtI, NaOH}}$ Et_2NCH_2COOH $\xrightarrow[\text{(C)}]{H_3O^+}$ $Et_2NH + HOCH_2COOH$

10-19 Compound A, C_3H_9ON, when titrated with HCl was found to have a NE of 75 ± 1. A reacted with benzenesulfonyl chloride to give a product that was soluble in dilute NaOH. With excess acetyl chloride, A was converted to B, $C_5H_{11}NO_2$. When 0.75 g of A was treated with excess methylmagnesium iodide, 448 ml of methane was evolved. Suggest structures for A and B.

10-20 Compound A, C_3H_9ON, was basic in aqueous solution and reacted with benzenesulfonyl chloride to give an alkali-insoluble derivative. When 0.75 g of A was treated with excess methylmagnesium iodide, 448 ml of methane was evolved. Suggest a structure for A.

10-21 Compound A, $C_5H_{10}N_2$, dissolved in water, gave a solution basic to litmus paper. When 0.98 g of A was titrated with 0.5 N HCl, 20 ml was required for neutralization. Hydrogenation of A gave B, $C_5H_{14}N_2$, which was found to have a NE of 51 ± 1 when titrated with HCl. Compound A did not react with benzenesulfonyl chloride; however, on boiling with strong HCl, A was converted to compound C, $C_5H_{12}O_2NCl$, a compound that was very soluble in water. The nmr spectrum for compound A indicates the presence of three different sets of hydrogens in a ratio of 3:1:1. Suggest structures for A, B, and C.

10-22 Compound A, $C_{12}H_{28}N_2$, was soluble in dilute acid, but did not react with benzenesulfonyl chloride. Compound A was treated with excess methyl iodide, then silver oxide, and the resulting product heated. Thus B, C_8H_{14}, was formed. B was treated with ozone and the ozonide decomposed with dilute acid in the presence of zinc dust. Two products were isolated. One was a volatile water-soluble compound that gave a positive Fehling's test. The other substance, C, was a neutral, slightly water-soluble substance. When C was treated with NaOI, iodoform, and compound D were formed. Compound D (NE 59 ± 1) showed only three peaks in its nmr spectrum — a quartet, a doublet, and a singlet in the ratio 1:3:2, respectively. Suggest structures for A, B, C, and D.

10-23 A useful test for primary amines is to treat them with $CHCl_3$ and NaOH resulting in the formation of a carbylamine, a class of compounds which has an extraordinarily repulsive odor. Compound A, C_5H_9N, the carbylamine obtained from butylamine, when catalytically hydrogenated or treated with $LiAlH_4$ yields compound B, $C_5H_{13}N$, which is soluble in dilute HCl and yields a NaOH-insoluble derivative with benzenesulfonyl chloride. When compound A is treated with HgO, compound C, C_5H_9NO, is obtained. Compound C has a characteristic ir band at 2260 cm^{-1} (4.42 μm). It may also be isolated as an intermediate in the treatment of the amide of valeric acid with NaOBr. When compound A is heated in the presence of aqueous acid, compound D is obtained (MW 88). The nmr spectrum of D shows a singlet at 713 Hz, a triplet centered at 140 Hz, a sextet centered at 100 Hz, and a triplet centered at 59 Hz. The areas under the peaks are in the respective ratio of 1:2:2:3. Suggest structures for A, B, C, and D.

10-24 An unknown amine X was found to have a NE of 99 ± 1. X did not react with the Hinsberg reagent. Complete Hofmann degradation yielded a hydrocarbon Y whose vapor density was 3.03 g/liter. On catalytic hydrogenation 0.2 g of Y absorbed 131.8 ml of hydrogen at STP and was converted to n-pentane. Suggest structures for X and Y.

10-25 Unknown compound A had a MW of 104 ± 2. Microcombustion analysis: % C, 69.2; % H, 3.9; % N, 26.9. On catalytic hydrogenation, A was converted into a new compound whose analysis was % C, 63.1; % H, 12.3; % N, 24.6. Compound A was soluble in dilute acid. On boiling with NaOH solution, A was converted to the sodium salt of an acid which on fusion with NaOH was decarboxylated to B, C_5H_5N. B gave reactions typical of a tertiary amine and on catalytic hydrogenation gave $C_5H_{11}N$. Give structures for A and B.

10-26 Give any organic compound containing three carbons or less, triphenylphosphine, thexylborane, and phthalimide as starting materials, outline the synthesis of each of the following. Where indicated use the specified starting material.

a. Acetone to 2-isopropylaminoethanol
b. CH_3CHO to $(CH_3)_2CHCH_2CH(CH_3)NCH_3$
c. CH_3CHO to 1-amino-2-propanol (two ways)
d. CH_3CHO to 1-amino-2,4-pentanediol
e. 2,7-Dimethyl-3,6-octanediamine
f. 3-Amino-1,4-pentadiene
g. 2-Butene to $CH_3CH(OH)CH(CH_3)N(CH_3)_2$
h. Butanal to 2-amino-1,3-hexanediol
i. Acetone to $(CH_3)_3CCONH_2$
j. tert-Butyl alcohol to neopentylamine

k. tert-Butyl alcohol to 2,6-dimethyl-4-aminoheptane
l. Cyclohexanol to

m. CH_3CHO to $(HOCH_2)_3CCH_2NH_2$
n. $(CH_3)_2CHCHO$ to $(CH_3)_2CHCH_2NHCH_2CH_2OH$
o. $CH_2=CHCH_2Br$ to $(H_2NCH_2CH_2CH_2)_2C=CHCH=CH_2$

p. 1,3-Butadiene to

10-27 Keeping in mind the rearrangement of carbonium ions, what rearranged products would you expect when each of the following is treated with nitrous acid?

a. Isobutylamine
b. $(CH_3)_2C-C(CH_3)_2$ with H_2N OH
c.

d.
e. $(CH_3)_2C-CHCH_3$ with H_2N OH
f. Neopentylamine

10-28 Draw electronic resonance forms for methyl azide (CH_3N_3). The nitrogen atoms have a linear structure.

10-29 Show how you would make ethylenimine, . Predict its reaction with each of the following:

a. H_2S *b.* Acrylonitrile   *c.* Ketene

d. When ethylenimine is treated with an acid chloride, an onium intermediate is first formed. If alkali is present, the reaction continues to give $RC\overset{\displaystyle O}{\overset{\|}{-}}N\langle$. If no alkali is present, the product is $RCONHCH_2CH_2Cl$. Rationalize.

10-30   Review Prob. 6-26, then predict the main product expected in each of the following:
       a. Ethylene oxide + NH_3
       b. 2,3-Epoxybutane + $(CH_3)_2NH$
       c. 1,2-Epoxybutane + CH_3SH, S_N2
       d. 1,2-Epoxybutane + CH_3OH, S_N1
       e. 2-Methyl-1,2-epoxybutane + ethylene glycol, S_N1
       f. Isobutyleneimine + H_2O, S_N1
       g. Isobutylene sulfide + CH_3SH, S_N2
       h. Ethylene oxide + ethylene glycol + H^+, Δ

10-31   Describe the bonding of N in methylethylpropylisopropylammonium chloride. How would you account for the fact that this compound is capable of showing optical activity?

10-32   Ethylenediaminetetraacetic acid (EDTA), $(HOOCCH_2)_2NCH_2CH_2N(CH_2COOH)_2$, is frequently used as a complexing reagent for various metal ions. Show how EDTA could bind to an Al^{3+} ion so that the aluminum is bound to the EDTA molecule six times in an octahedral arrangement with the Al in the center of the octahedron.

10-33   A 1.43-g sample of neutral compound A, $C_7H_{13}NO_2$, was heated with 40.0 ml of 0.5 N NaOH. Ammonia gas was evolved. After cooling the reaction mixture, the excess NaOH was titrated with 0.1 N HCl requiring 100.0 ml to reach the phenolphthalein end point. Compound B, $C_7H_{14}O_4$, was isolated from this procedure. Compound A shows a characteristic ir band at 4.44 μm (2250 cm^{-1}). It does not react with phenylhydrazine or acetyl chloride. On heating with aqueous acid, compound A is converted to compound C, $C_5H_8O_3$, whose nmr spectrum shows a singlet (3H), two triplets (4H), and a singlet far downfield (1H) at $\tau = -1.3$, and its ir spectrum shows a finely split doublet in the 1725 to 1700 cm^{-1} region. Suggest structures for A, B, and C.

10-34   Liquid X has elemental composition 40.0% C, 13.3% H, and 46.6% N. It reacts readily with aldehydes and ketones. Its nmr spectrum shows singlets at $\delta = 2.3$ and 3.1 in a 3:1 area ratio, respectively. Suggest a structure for X.

10-35   Compound A (MW = 88 ± 1) shows two characteristic peaks in its ir spectrum at 2.85 and 2.93 μm (3509 and 3413 cm^{-1}). Its nmr spectrum shows singlets at $\delta = 0.8$, 1.1, and 2.35 with the integration curve showing 18.4, 4.2, and 4.0 units, respectively, for the peak areas. Suggest a structure for A.

10-36   A basic liquid Y (MW = 88 ± 1) reacts with acetyl chloride to give a basic liquid Z (MW = 131 ± 1). Both X and Y contain C, H, N, and O only. The ir spectrum of Y shows a broad absorption band around 3 μm (3333 cm^{-1}). Its nmr spectrum shows singlets at $\delta = 4.6$ and 2.2 and triplets centered at $\delta = 3.6$ and 2.4 in the area ratio 1:6:2:2, respectively. Suggest structures for Y and Z.

10-37   Compounds A through E are related by the following transformations:

              1. $LiAlH_4$ 1. $LiAlH_4$
            ⟋   → (B)   ⎡   → (D)
              2. H_2O, H^+ ⎢ 2. H_2O, H^+
(82.0% O, 6.0% H,   (A)   ⎢
and 12.0% N)   ⟍ H^+, H_2O, Δ ⎣ 1. $Ca(OH)_2$ 2. Δ
                           → (C)   → (E) neutral, MW = 210

The following are some spectral data on compounds A through E:

Compound A   Ir: 4.42 μm (2260 cm^{-1}). Nmr: singlets at $\tau = 2.8$ and 6.5 (area ratio 5:2, respectively).
Compound B   Ir: 2.84 μm (3521 cm^{-1}) and 2.96 μm (3378 cm^{-1}). Nmr: singlets at $\tau = 2.9$ and 9.1 and two overlapping triplets (same J values) at approximately $\tau = 7.3$ (area ratio 5:2:2:2, respectively).
Compound C   Ir: broad absorption 3.45 to 4.0 μm (2900 to 2500 cm^{-1}) and 5.85 μm (1709 cm^{-1}). Nmr: singlets at 727, 430, and 210 Hz in a 1:5:2 are ratio, respectively.
Compound D   Ir: strong absorption at 2.94 μm (3400 cm^{-1}). Nmr: singlets at $\tau = 2.9$ and 5.9 and triplets at $\tau = 6.4$ and 7.3 in an area ratio 5:1:2:2, respectively.
Compound E   Ir: 5.8 μm (1724 cm^{-1}). Nmr: singlets at $\tau = 2.9$ and 6.5 in a 5:2 area ratio, respectively. Uv: $\lambda_{max} = 258$ nm with a weaker absorption at 288 nm.

10-38   Liquid A (bp 106 to 107°C, MW = 86 ± 1) gives a yellow liquid B (bp 217 to 218°C) when treated with nitrous acid. The nmr spectrum of A is given. Suggest structures for A and B.

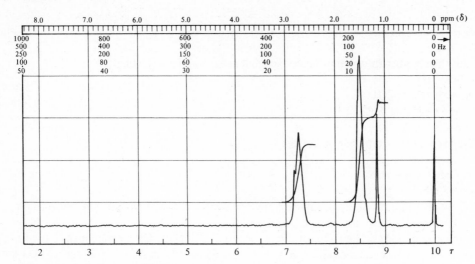

Figure P10-38. Nmr spectrum for Prob. 10-38, compound A. *(From R. A. Silva and R. Spenger, "The Quantitative Organic Analysis Spectra Kit," McGraw-Hill Book Company, New York, 1969. By permission of the publishers.)*

10-39   Liquid X (MW = 86 ± 1) yields finally, upon repeated Hofmann degradation, trimethylamine and a low-boiling liquid Y which upon catalytic hydrogenation yields liquid Z (MW = 75 ± 1). The nmr spectrum of X is given. Suggest structures for X, Y, and Z.

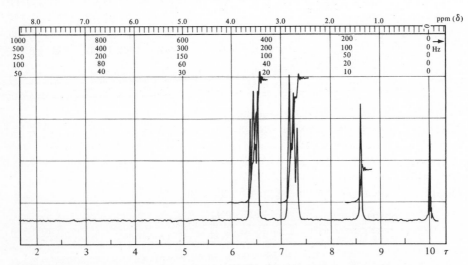

Figure P10-39. Nmr spectrum for Prob. 10-39, compound X. *(From R. A. Silva and R. Spenger, "The Quantitative Organic Analysis Spectra Kit," McGraw-Hill Book Company, New York, 1969. By permission of the publishers.)*

10-40 The ir and nmr spectra for compound C (MW = 136 ± 1) are given. Suggest a structure for compound C.

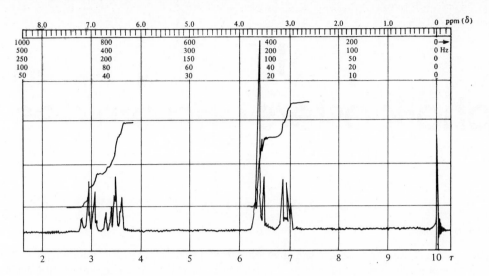

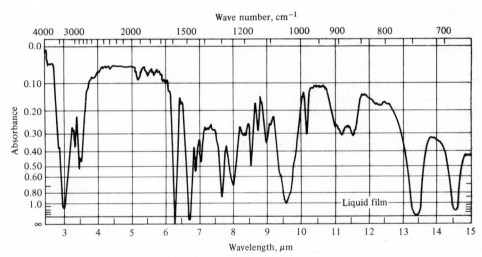

Figure P10-40. Ir and nmr spectra for Prob. 10-40. *(From R. A. Silva and R. Spenger, "The Quantitative Organic Analysis Spectra Kit," McGraw-Hill Book Company, New York, 1969. By permission of the publishers.)*

11

aliphatic esters and amides

11-1 Nomenclature of Esters

a. An ester is a compound derived from an acid and an alcohol. In the IUPAC and in the common system the alkyl group from the alcohol is named first, followed by the name of the acid (IUPAC or common) with the *ic* ending changed to *ate*. Esters of cycloalkanecarboxylic acids have the ending *carboxylate*.

	IUPAC name	Common name
CH_3COOCH_3	Methyl ethanoate	Methyl acetate
$(CH_3)_2CHCH_2COOCH_2CH_3$	Ethyl 3-methylbutanoate	Ethyl isovalerate
$CH_3OOCCH_2CH_2COOCH_3$	Dimethyl butanedioate	Methyl succinate
$CH_2{=}CHCOOCH_2CH(CH_3)_2$	Isobutyl propenoate	Isobutyl acrylate
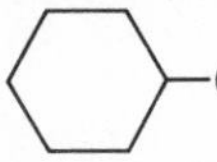 ⬡—$COOCH_3$	Methyl cyclohexanecarboxylate	Methyl cyclohexylformate

b. In complex systems esters may be named as derivatives of hydrocarbons.

$-COOCH_3$ Carbomethoxy group
$-COOCH_2CH_3$ Carbethoxy group

CH_2COOCH_3
|
$CHCOOCH_3$ 1,2,3-Tricarbomethoxypropane
|
CH_2COOCH_3

$RCOO-$ Acyloxy group
CH_3COO- Acetoxy group

CH_3COOCH_2
|
CH_2 1,3-Diacetoxypropane or trimethylene glycol diacetate
|
CH_3COOCH_2

c. *Esters of inorganic acids* are named like salts.

$CH_3OSO_2OCH_3$ or $(CH_3)_2SO_4$ $(CH_3)_2CHCH_2CH_2ONO_2$ $(CH_3CH_2O)_3P{=}O$

Dimethyl sulfate[†]   Isopentyl nitrate   Triethyl phosphate[†]

d. *Orthoesters* are named after a hypothetical orthoacid, such as orthoformic acid $HC(OH)_3$ or orthocarbonic acid.

$HC(OC_2H_5)_3$ $C(OC_2H_5)_4$

Triethyl orthoformate[†]   Tetraethyl orthocarbonate[†]

e. *Fats* and *oils* are esters composed of glycerol (1,2,3-trihydroxypropane) and fatty acids. For *fats* the acid component(s) are mostly *saturated* fatty acids, such as stearic acid, and for *oils* the acid component(s) are mostly *unsaturated* fatty acids such as oleic acid, $CH_3(CH_2)_7CH{=}CH(CH_2)_7COOH$.

† The prefixes *di, tri*, etc., are frequently omitted in cases like these.

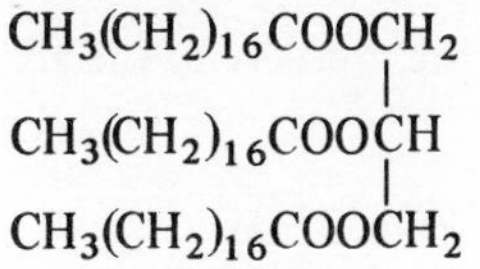

$$CH_3(CH_2)_{16}COOCH_2$$
$$CH_3(CH_2)_{16}COOCH$$
$$CH_3(CH_2)_{16}COOCH_2$$

Glyceryl tristearate
(tristearin)

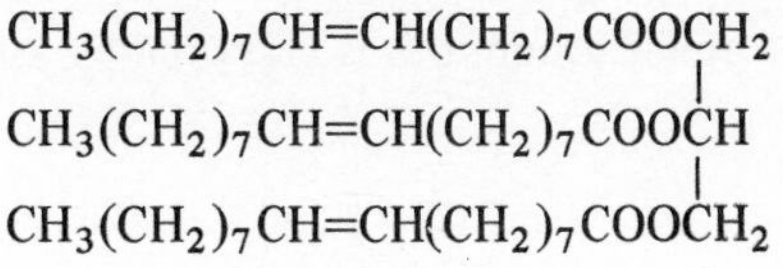

$$CH_3(CH_2)_7CH=CH(CH_2)_7COOCH_2$$
$$CH_3(CH_2)_7CH=CH(CH_2)_7COOCH$$
$$CH_3(CH_2)_7CH=CH(CH_2)_7COOCH_2$$

Glyceryl trioleate
(triolein)

f. *Waxes* are esters composed of fatty acids and long-chain or complex alcohols such as cetyl alcohol, $CH_3(CH_2)_{14}CH_2OH$.

$$CH_3(CH_2)_{14}COOCH_2(CH_2)_{14}CH_3$$

Cetyl palmitate (spermaceti wax)

g. *Cyclic esters* are either *lactones* (obtained from β, λ, or δ-hydroxy acids) or *lactides* (obtained from α-hydroxy acids such as lactic acid).

The IUPAC system names lactones by adding *olide* to the name of the parent (nonhydroxylated) hydrocarbon. The common names for certain lactones are derived from the common names for the parent (nonhydroxylated) acid. Lactides are named by changing *ic* to *ide* in the parent acid name and adding the prefix *di* to denote the number of molecules involved.

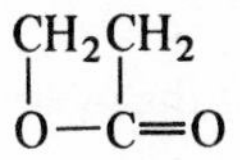

3-Propanolide
β-Propiolactone

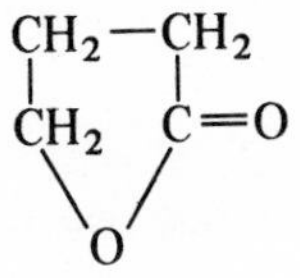

4-Butanolide
γ-Butyrolactone

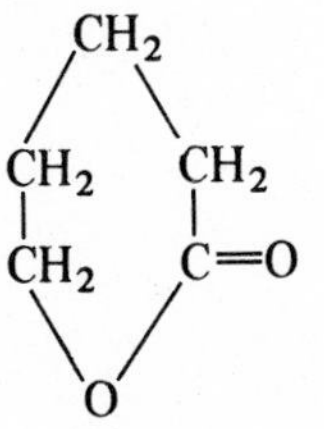

5-Pentanolide
δ-Valerolactone

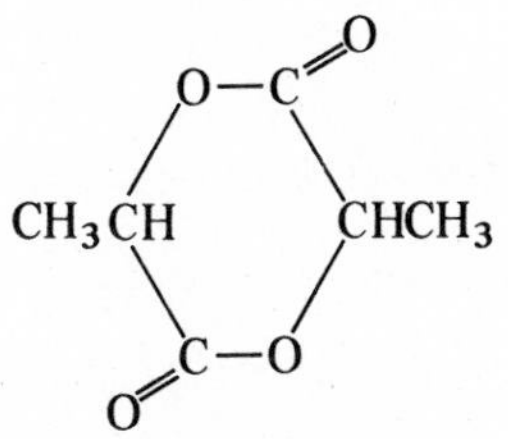

Dilactide

11-2 Nomenclature of Amides, Ureas, and Related Compounds

a. *Primary Amides*

1. In the IUPAC system, the suffix *amide* replaces the *e* in the name of the parent hydrocarbon (or the *oic* in the IUPAC name for the parent acid). In the common system, the suffix *amide* replaces the *ic* ending of the common name of the parent acid. Amides of cycloalkanecarboxylic acids have the ending *carboxamide*.

	IUPAC name	Common name
$HCONH_2$	Methanamide	Formamide
$(CH_3)_2CHCONH_2$	2-Methylpropanamide	Isobutyramide
$CH_2=CHCONH_2$	Propenamide	Acrylamide
$H_2NOCCH_2CH_2CONH_2$	1,4-Butanediamide	Succinamide
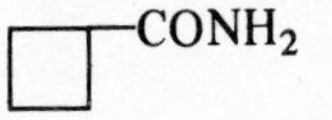 $\square$—$CONH_2$	Cyclobutanecarboxamide	Cyclobutylformamide

2. When alkyl groups replace H on the amide group, this is indicated by attaching the prefix *N* to the alkyl group.

$$HCON(CH_3)_2$$

N,N-Dimethylformamide (or *N,N*-dimethylmethanamide)

$$(CH_3)_2CCONHCH(CH_3)_2$$
$$|$$
$$Br$$

N-Isopropyl-*a*-bromoisobutyramide
(or *N*-Isopropyl-*2*-bromo-*2*-methylpropanamide)

3. Cyclic primary amides are called lactams and are commonly named after the *β*-, *γ*-, or *δ*-amino acids from which which they may be obtained. The IUPAC system names lactams by adding *lactam* to the name of the parent hydrocarbon without the amino group.

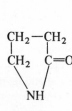

γ-Butyrolactam
4-Butanelactam

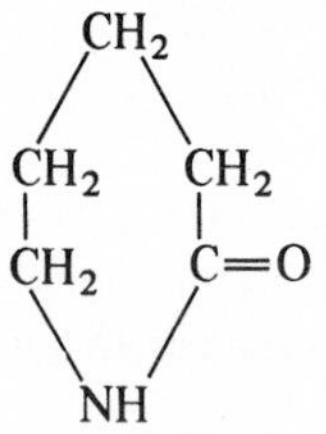

δ-Valerolactam
5-Pentanelactam

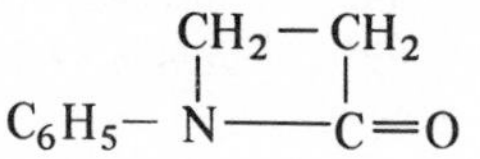

$$CH_2-CH_2$$
$$C_6H_5-N \qquad C=O$$

N-Phenyl-*β*-propiolactam
N-Phenyl-3-propanelactam

b. *Secondary and Tertiary Amide*

1. The simple amides result from the replacement of one H of NH_3. When two or three of the hydrogens are replaced by acyl groups, secondary and tertiary amides result.

$$(CH_3CO)_2NH \qquad (CH_3CO)_3N$$

Diacetamide   Triacetamide

2. Secondary amides, particularly cyclic ones, are often called imides.

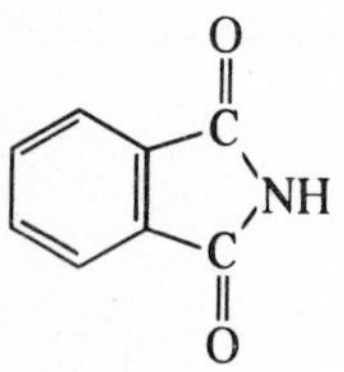

Phthalimide 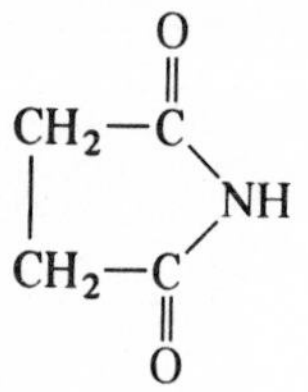   Succinimide

c. *Urea* H_2NCONH_2. This is a diamide of carbonic acid. The alkyl-substituted ureas are named as derivatives of urea. Often when identical alkyl groups are involved, the terms symmetric and unsymmetric are used to indicate the mode of attachment. The position of the alkyl group may also be specified by using the notation *N* and *N'* to refer to the two nitrogen atoms.

$$\overset{O}{\overset{\|}{CH_3NHCNH_2}} \qquad \overset{O}{\overset{\|}{CH_3NHCNHCH_3}} \qquad \overset{O}{\overset{\|}{(CH_3)_2NCNH_2}} \qquad \overset{O}{\overset{\|}{CH_3CH_2N(CH_3)CNHCH_3}}$$

Methylurea   *sym*-Dimethylurea   *unsym*-Dimethylurea   *N,N'*-Dimethyl-*N*-ethylurea
N-Methylurea   *N,N'*-Dimethylurea   *N,N*-Dimethylurea

d. *Acyl Derivatives of Urea* These compounds are known as ureides.

$$\overset{O \quad O}{\overset{\| \quad \|}{H_2NCNHCCH_3}} \qquad \overset{O \quad O}{\overset{\| \quad \|}{H_2NCNHCCH_2CH_3}}$$

Acetureide or *N*-acetylurea   Propionureide or *N*-propionylurea

154

Cyclic ureides may be formed from diesters and urea. The most important is the ureide of malonic acid called barbituric acid.

Keto form   Enol form

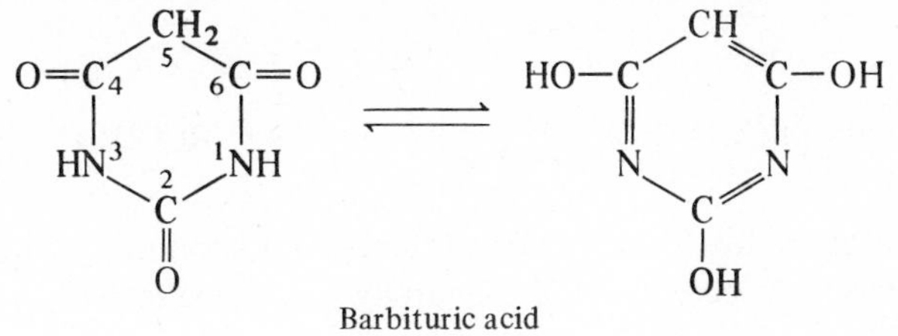

Barbituric acid

e. *Carbamates (Urethanes)* The urethanes are the amide esters of carbonic acid, H_2NCOR. They are named as derivatives of carbamic acid, and although this acid does not exist, its derivatives do.

$$CH_3NHCOCH_2CH_3 \qquad (CH_3CH_2)_2NCOCH(CH_3)_2$$

Ethyl *N*-methylcarbamate   Isopropyl *N,N*-diethylcarbamate

f. Two other classes of compounds which may also be considered as derivatives of carbonic acid are H_2NCNH_2, thiourea, and H_2NCNH_2, guanidine. Their derivatives are named in the same fashion as those of urea.

11-3 Preparation of Esters (Esterification)

a. *From Alcohols and Acyl Halides or Anhydrides* (See Sec. 9-6.)

b. *From Salts and Alkyl Halides* (Compare Williamson synthesis in Sec. 6-3*b*.)

$$CH_3COOAg + C_2H_5Br \longrightarrow CH_3COOC_2H_5 + AgBr$$
$$HCCl_3 + 3C_2H_5ONa \longrightarrow HC(OC_2H_5)_3 + 3NaCl$$

c. *From Acids and Alcohols (Fischer Method)* The reaction of alcohols with carboxylic acids is a completely reversible acid-catalyzed reaction.

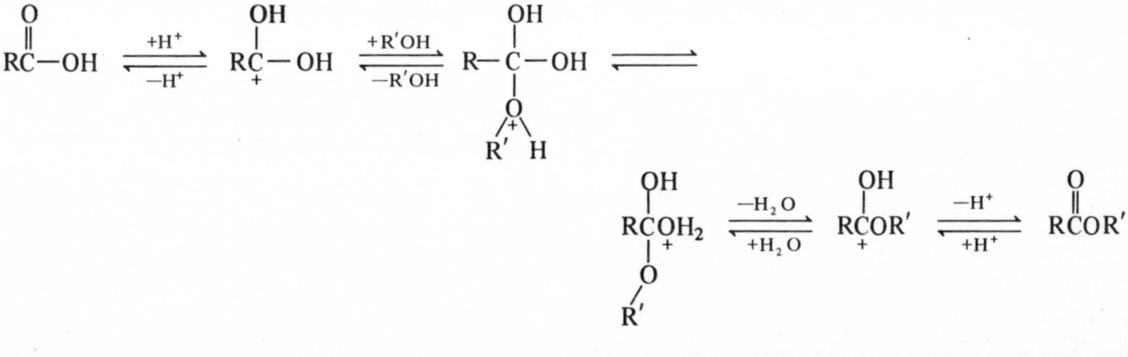

The rate of attainment of equilibrium is decreased markedly by highly branched α carbon atoms of acids and carbinol carbon atoms of alcohols. In instances of such hindrance, alternate synthetic routes must be used, for example, those in Sec. 11-2*a* or *b*. In many cases, low-boiling esters may be converted to higher-boiling esters by *transesterification:*

The lower-boiling alcohol $R'OH$ is distilled off as it forms, thus driving the equilibrium to the right.

d. *Acetates from Alcohols and Ketene* $CH_2=C=O + ROH \longrightarrow CH_3COOR$.

e. *Methyl Esters from Diazomethane* $RCOOH + CH_2N_2 \longrightarrow RCOOCH_3 + N_2$. The mechanism of the reaction is:

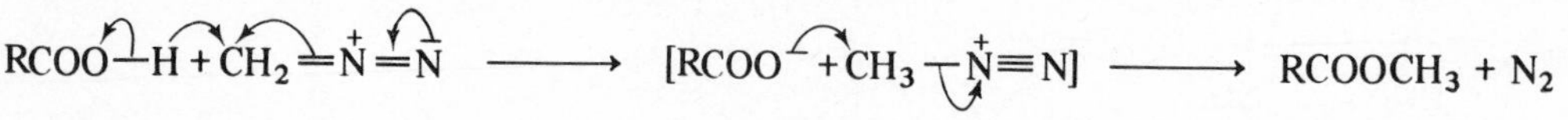

$$RCOO\!-\!H + CH_2\!=\!\overset{+}{N}\!=\!N \longrightarrow [RCOO^- + CH_3\!-\!\overset{+}{N}\!\equiv\!N] \longrightarrow RCOOCH_3 + N_2$$

f. *Preparation of Lactones and Lactides* β-Hydroxy acids undergo elimination upon heating to give α,β-unsaturated acids; γ-hydroxy acids yield *lactones* while α-hydroxy acids yield *lactides*.

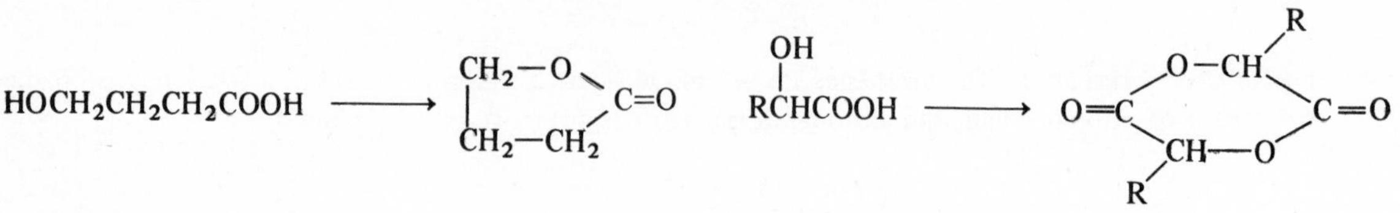

$$HOCH_2CH_2CH_2COOH \longrightarrow \begin{array}{c} CH_2\!-\!O \\ | \qquad\quad \rangle C\!=\!O \\ CH_2\!-\!CH_2 \end{array} \qquad \overset{OH}{\underset{|}{RCHCOOH}} \longrightarrow \begin{array}{c} O\!-\!CH\!-\!R \\ O\!=\!C \qquad\quad C\!=\!O \\ R\!-\!CH\!-\!O \end{array}$$

11-4 Preparation of Amides

a. *Heating Ammonium Salts* $RCOONH_4 \xrightarrow{\Delta} RCONH_2 + H_2O$

b. *From Nitriles*

1. $RC\!\equiv\!N \xrightarrow[NaOH]{H_2O_2} RCONH_2 + \frac{1}{2}O_2$

2. $RC\!\equiv\!N \xrightarrow[H^+ \text{ or } ^-OH]{H_2O} RCONH_2 \longrightarrow RCOOH$

The hydrolysis can be controlled so as to yield the amide or it can be carried to the acid.

c. *From Acyl Halides or Anhydrides*

1. $R'COCl + 2HNR_2 \longrightarrow R'CONR_2 + R_2\overset{+}{N}H_2Cl^-$
2. $(RCO)_2O + 2NH_3 \longrightarrow RCONH_2 + RCOONH_4$

d. *Ammonolysis of Esters* $RCOOR' + NH_3 \longrightarrow RCONH_2 + R'OH$.

e. *Acetamides from Amines and Ketene* $RNH_2 + CH_2=C=O \longrightarrow CH_3CONHR$.

f. *Beckman Rearrangement* Heating oximes of ketones with a Lewis acid, such as PCl_5 or H_2SO_4 causes them to rearrange to amides. The reaction along with the proposed mechanism is shown below:

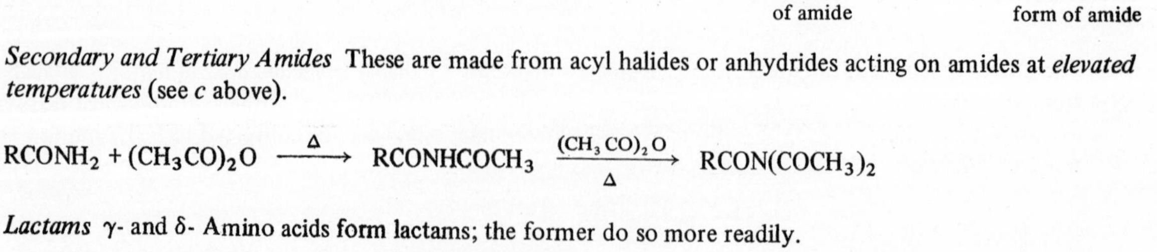

$$\underset{\substack{\text{R group trans} \\ \text{to OH migrates}}}{\overset{\displaystyle :N\!-\!OH}{\underset{\displaystyle R\!-\!CR'}{\|}}} \xrightarrow{H^+} \overset{\displaystyle :N\!-\!\overset{+}{O}H_2}{\underset{\displaystyle R\!-\!C\!-\!R'}{\|}} \longrightarrow R\!-\!N\!=\!\overset{+}{C}\!-\!R' \xrightarrow{H_2O}$$

$$\underset{+OH_2}{\overset{\displaystyle R\!-\!N\!=\!C\!-\!R'}{|}} \xrightarrow{-H^+} \underset{\substack{H\!-\!O \\ \text{Enol form} \\ \text{of amide}}}{R\!-\!N\!=\!C\!-\!R'} = \underset{\substack{H \quad O \\ \text{More stable keto} \\ \text{form of amide}}}{R\!-\!N\!-\!C\!-\!R'}$$

g. *Secondary and Tertiary Amides* These are made from acyl halides or anhydrides acting on amides at *elevated temperatures* (see c above).

$$RCONH_2 + (CH_3CO)_2O \xrightarrow{\Delta} RCONHCOCH_3 \xrightarrow[\Delta]{(CH_3CO)_2O} RCON(COCH_3)_2$$

h. *Lactams* γ- and δ- Amino acids form lactams; the former do so more readily.

$$H_2NCH_2CH_2CH_2COOH \xrightarrow{\Delta} \begin{array}{c} CH_2CH_2CH_2C\!=\!O + HOH \\ |\rule{2cm}{0pt}| \\ \rule{0.3cm}{0pt}NH\rule{0.3cm}{0pt} \end{array}$$

γ-Butyrolactam

156

11-5 Preparation of Ureas, Ureides, and Carbamates

a. *Ureas*

1. $CO_2 + NH_3 \xrightarrow{\text{pressure}} HO\overset{\displaystyle O}{\overset{\|}{C}}NH_2 \xrightarrow{NH_3} H_2N\overset{\displaystyle O}{\overset{\|}{C}}NH_2 + H_2O$

2. Urea, or symmetric alkyl ureas, may be made from ammonia or amines and phosgene.

$4NH_3 + Cl\overset{\displaystyle O}{\overset{\|}{C}}Cl \longrightarrow H_2N\overset{\displaystyle O}{\overset{\|}{C}}NH_2 + 2NH_4Cl$

$4(CH_3)_2NH + Cl_2CO \longrightarrow (CH_3)_2NCON(CH_3)_2 + 2(CH_3)_2\overset{+}{N}H_2Cl^-$

3. Isocyanates, which are made by heating the alkylcarbamyl chlorides formed from phosgene and *primary* amines, react with amines to yield substituted ureas (symmetric or unsymmetric ureas may be prepared in this way).

$2RNH_2 + Cl_2CO \longrightarrow R\overset{+}{N}H_3Cl^- + RNHCOCl \xrightarrow{\Delta} RN=C=O + HCl$

An isocyanate

$CH_3N=C=O + HNHCH(CH_3)_2 \longrightarrow CH_3NHCONHCH(CH_3)_2$

Methyl isocyanate   *N*-Methyl-*N'*-isopropylurea

b. *Ureides*

1. Urea reacts with esters, acyl halides, or anhydrides to give ureides.

$H_2NCONH_2 + CH_3COCl \longrightarrow H_2NCONHCOCH_3 + HCl$

Acetylurea or acetureide

$(CH_3CO)_2O + H_2NCONH_2 \longrightarrow CH_3CONHCONHCOCH_3 + 2CH_3COOH$

sym-Diacetylurea or *sym*-diacetureide

2. Cyclic ureides are usually made by the action of esters of dicarboxylic acids on urea or its derivatives. The most interesting of these is barbituric acid and its derivatives, many of which are important hypnotics.

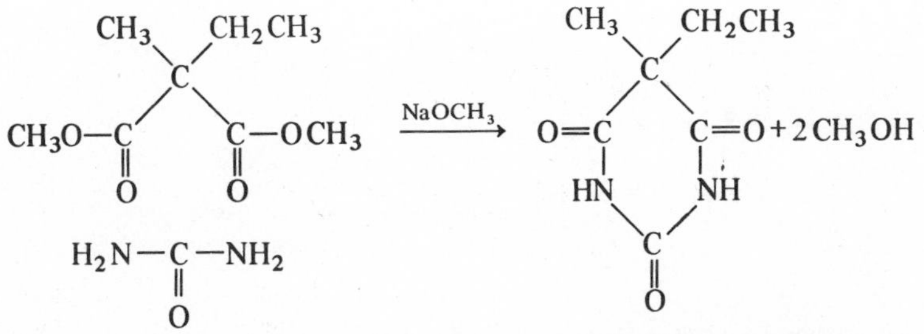

5-Methyl-5-ethylbarbituric acid

c. *Alkyl Carbamates (Urethanes)* These compounds, which contain both an amide and an ester group of carbonic acid, may be prepared by the reaction of alcohols with isocyanates or by the action of ammonia or amines on alkyl chloroformates.

$RN=C=O + (CH_3)_2CHOH \longrightarrow RNHCOOCH(CH_3)_2$

Isopropyl *N*-alkylcarbamate

$Cl_2CO \xrightarrow{CH_3OH} CH_3OCOCl \xrightarrow{2CH_3NH_2} CH_3OCONHCH_3 + CH_3\overset{+}{N}H_3Cl^-$

Methyl *N*-methylcarbamate

11-6 Reactions of Esters

a. *Hydrolysis* As shown in the Fischer esterification reaction, acids catalyze the reversible hydrolytic reaction. The use of alkaline solutions (saponification) prevents reversal by removal of the acid through salt formation.

$$CH_3CH_2COOCH_3 + NaOH \longrightarrow CH_3CH_2COONa + CH_3OH$$

b. *Reaction with Grignard Reagents*

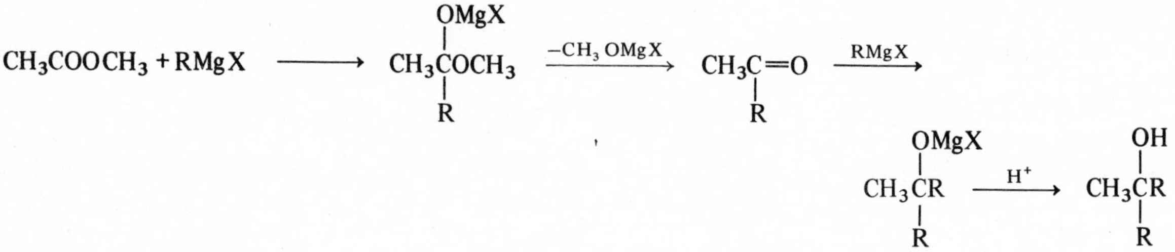

$$CH_3COOCH_3 + RMgX \longrightarrow \underset{\underset{R}{|}}{CH_3\overset{\overset{OMgX}{|}}{C}OCH_3} \xrightarrow{-CH_3\,OMgX} \underset{R}{\overset{|}{CH_3C}}{=}O \xrightarrow{RMgX}$$

$$\underset{\underset{R}{|}}{CH_3\overset{\overset{OMgX}{|}}{C}R} \xrightarrow{H^+} \underset{\underset{R}{|}}{CH_3\overset{\overset{OH}{|}}{C}R}$$

The intermediate ketone is more reactive toward Grignard reagents than the original ester, and hence the reaction is not good for ketone synthesis. Tertiary alcohols are the normal product.

c. *Reduction*

1. Bouveault-Blanc method. $RCOOR' \xrightarrow{Na\,+\,alcohol} RCH_2OH + R'OH$
2. Lithium aluminum hydride.

$$2RCOOR' + LiAlH_4 \longrightarrow LiAl(OCH_2R)_2(OR')_2 \xrightarrow{H^+} 2RCH_2OH + 2R'OH$$

3. Catalytic hydrogenation at elevated pressure and temperature.

$$RCOOR' + 2H_2 \xrightarrow{CuCr_2O_4\ \Delta,\ pressure} RCH_2OH + R'OH$$

d. *Ammonolysis* $RCOOR' + NH_3 \longrightarrow RCONH_2 + R'OH$

e. *Reactions of Orthoesters* (See Sec. 7-3a-3 for preparation of orthoesters.)
   1. Hydrolysis.

$$HC(OR)_3 + 2H_2O \xrightarrow{H^+} HCOOH + 3ROH$$

Orthoesters, like acetals, are not hydrolyzed in alkaline solution.
   2. Preparation of ketals. (See Sec. 7-3a-3.)

f. *Claisen Ester Condensation* (See Sec. 12-2b-3.)

11-7 Reactions of Amides

a. *Hydrolysis*

1. $RCONHCH_3 + H_2O \xrightarrow[or\ OH^-]{H^+} RCOOH + CH_3NH_2$

2. $H_2NCON(CH_3)_2 + 2H_2O \xrightarrow[or\ OH^-]{H^+} NH_3 + (CH_3)_2NH + [HOCO_2H] \longrightarrow CO_2 + H_2O$

3. 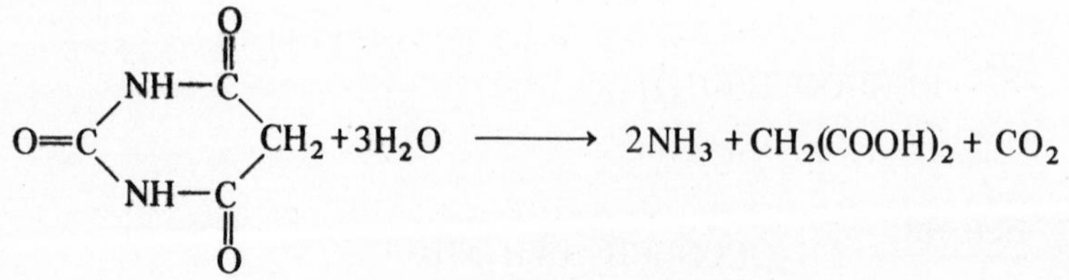 $O{=}C\overset{\displaystyle NH-\overset{\overset{O}{\|}}{C}}{\underset{\displaystyle NH-\underset{\underset{O}{\|}}{C}}{\diagup\!\!\diagdown}}CH_2 + 3H_2O \longrightarrow 2NH_3 + CH_2(COOH)_2 + CO_2$

b. *Salt Formation* Primary amides are very weakly basic compounds, insoluble in dilute acids. They will form salts under anhydrous conditions, but the salts are hydrolyzed by the stronger base water. The hydrogen of secondary amides is more acidic and stable salts are formed with aqueous sodium hydroxide. Recall Gabriel's synthesis, Sec. 10-3*e*.

c. *Nitrous Acid*

1. $RCONH_2 + HONO \longrightarrow RCOOH + N_2 + H_2O$
2. $H_2NCONH_2 + 2HONO \longrightarrow 2N_2 + CO_2 + 3H_2O$

3. $H_2NCONHCH_3 + HONO \longrightarrow H_2NCON\overset{\overset{\displaystyle CH_3}{|}}{}-N{=}O$

d. *Dehydration* $RCONH_2 + P_2O_5 \longrightarrow RC{\equiv}N + 2HPO_3$

e. *Reduction* $RCONHCH_3 + LiAlH_4 \longrightarrow \xrightarrow{H_2O} RCH_2NHCH_3$. Note that primary, secondary, or tertiary amines may be made by reduction of the proper type of amide. Reduction may also be accomplished catalytically but drastic conditions are necessary and yields are low.

f. *Hofmann Hypobromite Reaction* (See Sec. 10-3*d*.)

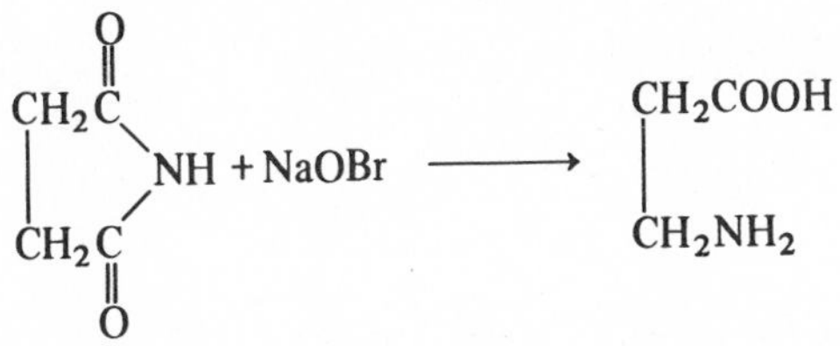

g. *N-Bromosuccinimide* At $0°C$, NaOBr reacts with succinimide to give the *N*-bromo derivative:

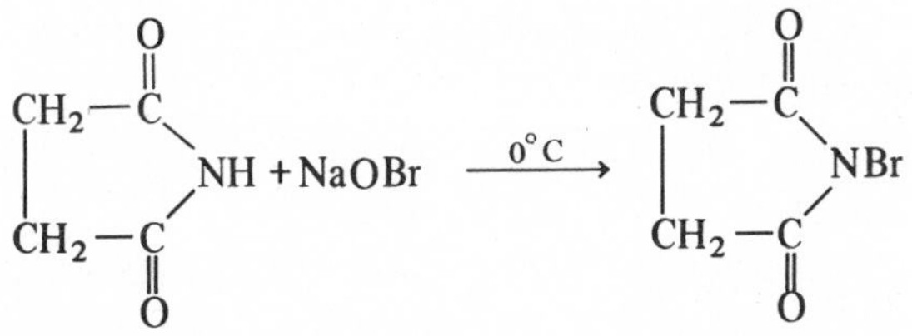

N-Bromosuccinimide (often abbreviated as NBS) is a very useful and selective reagent for the bromination of methyl or methylene groups adjacent to double or triple bonds (see Sec. 5-2*b*).

$$NBS + CH_3CH{=}CHCH_2R \longrightarrow CH_3CH{=}CHCHR + \text{succinimide}$$
$$\overset{\quad\quad\quad\quad\quad\quad\quad\quad\quad\quad\quad|}{\quad\quad\quad\quad\quad\quad\quad\quad\quad\quad Br}$$

Note that methylene groups react more easily than methyl groups. Active hydrogen on OH or COOH reacts more rapidly than H on an allylic carbon.

11-8 Qualitative Tests for Amides and Esters

a. Esters, acid chlorides, and anhydrides all react with hydroxylamine, H_2NOH, to yield hydroxamic acids which form intensely colored (blue, red, or violet) complexes with ferric salts. Under more drastic conditions nitriles and amides will also give a positive test.

$$RCOOR' + H_2NOH \longrightarrow \underset{\text{Hydroxamic acid}}{RCONHOH} \xrightarrow{Fe^{+3}} \underset{\text{Ferric hydroxamate}}{(RCONHO)_3Fe}$$

b. Primary amides upon treatment with nitrous acid evolve nitrogen that is readily observed.

c. On boiling with sodium hydroxide, primary amides evolve ammonia which can be easily detected by its odor. Nitriles also give off ammonia under these conditions. Ammonium salts give off ammonia with *cold* sodium hydroxide.

11-9 Calculations Using Equivalent Weights

Although most organic reactions do not go to completion and give yields of 100 percent, there are a number that do. These so-called quantitative reactions are very useful in determining the equivalent weights of unknown compounds from which one can make a guess at the molecular weight.

For example, the neutralization equivalent (NE), as discussed in Sec. 1-6, is the amount of an unknown acid that will react with 40 g (1 mol) of sodium hydroxide.

$$CH_3COOH + NaOH \longrightarrow CH_3COONa + H_2O$$

| 60 g | 40 g | 82 g | 18 g |

$$HOOCCH_2COOH + NaOH \longrightarrow NaOOCCH_2COONa + H_2O$$

| 52 g | 40 g | 74 g | 18 g |

In the case of acetic acid, 1 mol of it is equivalent to 1 mol of sodium hydroxide. With malonic acid, since there are two acidic hydrogens, $\frac{1}{2}$ mol of it will neutralize 1 mol of sodium hydroxide, and, hence, its "equivalent" weight will be $\frac{1}{2}$ (MW) or 52. Although the equivalent weight is simply the amount of acid that will react with 1 equiv of any base, it is useful to remember that its molecular weight will be some multiple of it. The number of acidic hydrogens times the equivalent weight will be the molecular weight.

Example 1 If 0.060 g of an acid requires 10.0 ml of 0.10 N sodium hydroxide for neutralization, the equivalent weight of the acid is given by the expression:

$$EW = \frac{\text{wt acid}}{\text{equiv of acid}} = \frac{(1,000)(\text{wt acid})}{\text{ml base} \times N\text{base}} = \frac{(1,000)(0.060)}{(10)(0.1)} = 60$$

The molecular weight of the acid could be 60 or some whole number times 60. A carboxyl unit (COOH) represents 45 weight units, and this subtracted from 60 leaves 15 units to be accounted for. This could be contributed by a methyl group, and the acid could be CH_3COOH. If the acid contained two carboxyl groups, its molecular weight would be twice 60, or 120. We would then reason as follows:

MW (120) minus 2COOH (90) = residue (30)

The 30 units could represent the group, −CHOH, and the compound might have been hydroxymalonic acid, $HOOCCH(OH)COOH$.

Saponification Equivalent $CH_3COOCH_2CH_3 + NaOH \longrightarrow CH_3COONa + CH_3CH_2OH$. The saponification equivalent (SE), like the neutralization equivalent, is the number of grams of ester that will react with 1 mol of sodium hydroxide. In the above example with ethyl acetate, the SE will be the same as the molecular weight.

Esters do not react immediately with sodium hydroxide as acids do; hence, one cannot directly titrate esters. The saponification equivalent is determined indirectly by boiling the ester in excess standard sodium hydroxide and then titrating the excess sodium hydroxide with standard acid.

Example 2 An unknown ester (1.78 g) was boiled for 1 h with 30 ml of 1.0 N NaOH. The mixture then required 40 ml of 0.25 N HCl for titration to a phenolphthalein end point: 30 meq (milliequivalents) (30 x 1) of NaOH (originally) minus 10 meq (40 x 0.25) of HCl used (which equals the meq of unused NaOH) is equal to 20 meq of NaOH which reacted with the ester. Therefore, the saponification equivalent is equal to 20 meq of NaOH which reacted with the ester. Therefore, the saponification equivalent is equal to (1,000)(1.78)/20, or 89.

Ordinarily, the accuracy of this method is not better than ±1 percent, so the value may be given as 89 ± 1. The simplest possible ester, $HCOOCH_3$, represents 60 weight units. If only one ester group is present, then 89 less 60 gives 29 ± 1 units to be accounted for. For example, this could be accounted for by two methylene units (28), a −CHOH unit (30), or a −OCH_2 unit (30). Some structural possibilities: CH_3COOEt, $CH_3CH_2COOCH_3$, $HCOOCH_2CH_2CH_3$, $HCOOCH(CH_3)_2$, $HOCH_2COOCH_3$, $HCOOCH_2CH_2OH$, and $HCOOCH_2OCH_3$.

The saponification equivalent for primary amides may be determined by boiling the amide in excess standard sodium hydroxide until no more ammonia is evolved, and then titrating the excess sodium hydroxide with standard acid.

Example 3 When 0.200 g of neutral compound X, $C_4H_6ON_2$, was boiled for several hours with 50 ml of 0.20 *N* KOH, ammonia was evolved. This aqueous solution was then distilled until the smell of NH_3 disappeared (i.e., all NH_3 was evaporated). The remaining solution was titrated to the neutral point with 0.200 *N* sulfuric acid requiring 30.0 ml. On treatment with nitrous acid, compound X yielded an acid of NE 99 ± 1. Hydrolysis of this acid, Y, gave a new acid, Z, of NE 59. Heating Z gave a new compound, which on warming with water was converted to Z. Solution: 10 meq (50 x 0.2) of KOH (originally) minus 6.0 meq of (30 x 0.2) of H_2SO_4 used to titrate the unused or excess KOH equals 4.0 meq of KOH used to react with X. Since 200 mg of X is equal to 4 meq of X, then the equivalent weight of X is 50 and there are 2 equiv X/mol X. Compound X has three sites of unsaturation (see Sec. 7.6) and cannot be a diamide (only one oxygen in X). Thus, X must contain some other group which upon heating with NaOH is also converted to a carboxylic acid group. One such group is C≡N. Primary amides react with HONO to give carboxylic acids. Assuming that the EW (equivalent weight) of Y = MW of Y, then 99 − 45 (COOH) = 54 unit residue which could be a C_3H_4N grouping; if a C≡N group is present then the residue would be $-CH_2CH_2CN$. Compound Z must be a diacid (MW = 2 x 59 or 118) resulting from the hydrolysis of the C≡N in Y. Thus, X is $N≡CCH_2CH_2CONH_2$, Y is $N≡CCH_2CH_2COOH$, and Z is succinic acid.

Active Hydrogen Determination by Zerewitinoff Method (see Sec. 1-6) or LiAlH$_4$

$$CH_3CH_2OH + CH_3MgI \longrightarrow CH_3CH_2OMgI + CH_4$$
$$CH_3CH_2OH + LiAlH_4 \longrightarrow CH_3CH_2OAlLiH_3 + H_2$$

Note that in the above equations an active hydrogen gives 1 mol of methane with methylmagnesium iodide and 1 mol of hydrogen with lithium aluminum hydride. Thus, one can calculate the equivalent weight of an unknown molecule in terms of methane or hydrogen evolved. For example, if 4.0 g of an unknown compound, when treated with excess methylmagnesium iodide, yielded 2.0 g of methane (that is $\frac{1}{8}$ mol CH_4), it would take 8 x 4 to yield 1 mol of methane (16 g), and the equivalent weight of the compound would be 32. Thus, the molecular weight of the unknown would be 32 if it contained one active hydrogen, 64 if it contained two, 96 if three, etc.

Determination of Unsaturation by Means of Hydrogen or Bromine

$$RCHBrCH_2Br \xleftarrow{Br_2} RCH=CH_2 \xrightarrow{H_2, Pt} RCH_2CH_3$$

Note that, as indicated above, 1 mol (160 g) of bromine reacts with one double bond and that 1 mol of hydrogen (2 g or 22.4 liters) reacts with one double bond. One can calculate the equivalent weight of an unknown in terms of double or triple bonds, or, if one knows the molecular weight, one can calculate the number of unsaturated bonds. The *iodine number* of an unsaturated compound (usually a fat or an oil) is defined as the number of grams of iodine that will react with 100 g of the unknown compound. Since iodine does not give a stable addition product with olefins, ICl is used in practice, and the value obtained is multiplied by I_2/ICl to convert it into terms of pure iodine.

Example 4 An acid, A, was found by the Rast method to have a molecular weight of 156 ± 1. Titration of 0.300 g of A with 0.100 *N* NaOH required 38.4 ml for neutralization. When 0.1 g of A was treated with excess LiAlH$_4$, 43.1 ml of hydrogen measured at STP was evolved. When 0.156 g of A was catalytically hydrogenated, it absorbed 44.8 ml of hydrogen measured at STP. Compound A did not show reactions of an OH or carbonyl group. On heating, A evolved carbon dioxide to give a new acid, B, of NE 112 ± 1. Compound A did not contain elements other than carbon, hydrogen, and oxygen. Suggest a structure for A.

First, calculate the NE for A.

$$NE = \frac{(1,000)(0.3)}{(38.4)(0.1)} = 78$$

Since the molecular weight is given as 156, the compound has two acidic hydrogens. Now calculate the equivalent weight in terms of H_2 evolved. If 0.1 g of A gives 43.1 ml of hydrogen, then X g will give 1 mol (22,400 ml).

$$\frac{0.100}{43.1} = \frac{X}{22,400} \qquad X = 52$$

161

Hence, there are three active hydrogens if the molecular weight is 156. The equivalent weight in terms of hydrogen absorbed would be:

$$\frac{0.156}{44.8} = \frac{X}{22,400} \qquad X = 78$$

Hence, the compound contains either two double bonds or one triple bond.

Now put together a simple molecule that will satisfy the above data. Since the unknown contained two acid hydrogens (titratable), we assume two carboxyls: 2COOH = 90 weight units. We know that the compound has three active hydrogens with respect to $LiAlH_4$. Two of these would be carboxyl hydrogens, but we must form a working hypothesis about the third. It cannot be OH since A does not react with acetyl chloride, so the only other possibility, among functional groups we have covered, is an acetylenic hydrogen. We also have the information that the compound absorbs hydrogen, which reinforces the idea that we have a HC≡C− unit accounting for 25 weight units. Since the compound is easily decarboxylated, we can assume that both carboxyl groups are attached to the same carbon atom (see Sec. 14-3). Now we can write the partial structure HC≡C−C(COOH)$_2$, which would have a molecular weight of 127. A has a molecular weight of 156 ± 1, so we still have 29 ± 1 units to account for. An ethyl group would satisfy this, and we can write HC≡C−C(COOH)$_2$.
$$\qquad\qquad\qquad\qquad\qquad\qquad\qquad\qquad\qquad\qquad\qquad\qquad\qquad\qquad\qquad\quad | $$
$$\qquad\qquad\qquad\qquad\qquad\qquad\qquad\qquad\qquad\qquad\qquad\qquad\qquad\qquad\qquad CH_2CH_3$$

Problems

11-1   Give both IUPAC and common names for each of the following:
    a. $CH_3(CH_2)_3CONH_2$ b. $CH_3CH_2CH_2COOCH_3$ c. $CH_3CH_2COOCH_2CH_3$
    d. $CH_3CH=CHCONH_2$ e. $CH(OCH_3)_3$ f. $CH_2=CHCOOC(CH_3)_3$
    g. $CH_3CH_2CONHBr$ h. $CH_3OOC(CH_2)_2COOCH_3$ i. $CH_3NHOCCH_2CONHCH_3$
    j. $HC≡CCONH_2$ k. $HO(CH_2)_3COOCH=CH_2$ l. $CH_3COCON(CH_3)CH_2CH_3$

    m. 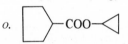   n. ⌐O⌐C=O   o. ⬠−COO−◁

11-2   Draw structures for each of the following:
    a. Butyronitrile   b. Allyl isocyanate
    c. Glyceryl triacetate   d. Isopropyl crotonate
    e. sym-Diethylurea   f. 1,2,3-Tricarbomethoxypropane
    g. Propyl nitrite   h. Methyl orthocarbonate
    i. N-Methyl-δ-valerolactam   j. 5-Pentanolide
    k. unsym-Diethylthiourea   l. Cetyl oleate
    m. 1-Ethylbarbituric acid   n. sym-Diacetureide
    o. Dilactide   p. Divinylcarbonate
    q. Triethylnitrosourea   r. 3-Carbethoxyglutaraldehyde

11-3   Complete and balance the following:
    a. Butyryl chloride + isobutyl alcohol   b. Acetic anhydride + 2-propanol
    c. Propionyl chloride + sec-butylamine   d. Butyric anhydride + allylamine
    e. Succinamide + P_2O_5 f. Isovaleramide + Br_2 + NaOH
    g. Diazomethane + malonic acid   h. Ketene + 1,3-propanediol
    i. 2-Butanone + triethyl orthoformate + H^+ j. Methyl valerate + $2CH_3MgI$
    k. Thiourea + nitrous acid   l. Methyl formate + CH_3NH_2
    m. Glutaramide + Δ   n. unsym-Dimethylurea + NaOH solution, Δ
    o. Glyceryl tributyrate + H_2 + $CuCr_2O_4$, Δ   p. Sodium methoxide + Cl_3CCH_3
    q. Silver propionate + ethyl iodide   r. Methyl caproate + Na + EtOH
    s. Ammonium laurate + Δ   t. Butanamide + K
    u. Allyl butyrate + H_2NOH + $FeCl_3$ v. Ketene + allylamine
    w. Methyl cyclohexanecarboxylate + butyl alcohol + Δ + H^+
    x. 1,5-Dimethylbarbituric acid + hot NaOH solution
    y. Diacetamide + cold NaOBr   z. Lactamide + CH_3MgI
    a'. Diethyl sulfite + NaOH + Δ   b'. Acetyl chloride + tert-butyl alcohol

162

c'. Succinimide + LiAlH$_4$ d'. Butyronitrile + H$_2$O$_2$ + NaOH
e'. NBS + cyclohexene f'. Guanidine + NaOH solution, Δ
g'. Cyclopentanone oxime + PCl$_5$ h'. *sym*-Dimethylthiourea + NaOH solution, Δ
i'. 4-Hydroxybutanoic acid + Δ j'. Cyclopentyl formate + 2EtMgI

11-4   Give a convenient chemical test for distinguishing between the following:
a. Caproic acid and ethyl butyrate   *b.* Butyl bromide and ethyl acetate
c. Urea and *sym*-diethylurea   *d.* Hexanamide and valeronitrile
e. 2-Propanol and dimethyl succinate   *f.* Hexanal and isobutyl acetate
g. *sym*-Dimethylurea and *N*-methylhexanamide   *h.* Urea and semicarbazide
i. Urea and tetraethylammonium hydroxide   *j.* Hexanamide and hexylamine
k. Butyl acetate and dibutyl ether   *l.* 2-Hexanone and ethyl trimethylacetate
m. Vinyl butyrate and ethyl butyrate   *n.* Hexanamide and ammonium hexanoate
o. Nonanamide and acetamide   *p.* Octanamide and dibutyramide
q. Ethyl carbonate and ethyl carbamate   *r.* 5-Butylbarbituric acid and valeramide

11-5   Up to this point we have described at least seven ways to shorten a carbon chain by breaking carbon-carbon bonds. See how many ways you can recall.

11-6   Show how you would make the following conversions:
a. Propionitrile to ethylamine   *b.* Cyclohexene to 1,3-cyclohexadiene
c. Ethyl propionate to propionitrile   *d.* Propanal to 2-hydroxybutanamide
e. Acetone to *N*-methylacetamide   *f.* Cyclohexene to H$_2$N(CH$_2$)$_5$COOH
g. Ethene to H$_2$NCH$_2$CH$_2$COOH   *h.* Ethyl acetate to 3-methyl-3-pentanol
i. Butyronitrile to butyl butyrate   *j.* Cyclobutene to γ-butyrolactam
k. Acetone to isopropyl acetate   *l.* Methyl isobutyrate to isobutylamine

m. Cyclohexene to   *n.* Trimethylacetic acid to *tert*-butyl isocyanate

o. Methyl 3-butenoate to 4-butanolide   *p.* Propionamide to *sym*-diethylurea
q. 1-Butene to butyl carbamate   *r.* Ethanal to 1,1-diethoxy-2-butene

11-7   Calculate the saponification equivalents of each of the following expressing the values in whole numbers. Calculate the number of grams of these compounds that would react with 50 ml of 0.50 N sodium hydroxide.
a. Ethyl acetate   *b.* Ethyl isovalerate
c. Diethyl succinate   *d.* Butyramide
e. Malonamide   *f.* 1,2,3-Tricarbethoxypropane
g. Butyric anhydride   *h.* Tributyl phosphate
i. CH$_3$OOCCH$_2$CH$_2$CONH$_2$

11-8   What would be the equivalent weight of a compound of which:
a. 0.200 g evolved 22.4 ml (STP) of hydrogen on treatment with LiAlH$_4$?
b. 0.92 g evolved 672 ml (STP) of CH$_4$ when treated with CH$_3$MgI?
c. 0.54 g reacted with 3.2 g of bromine?
d. 0.0930 g absorbed 7.74 ml of hydrogen at STP?

11-9   Suggest a structure for each of the following:
a. Compound has two ester groups, a SE of exactly 100, and no C=C bonds.
b. Compound contains N, has a SE of 52 ± 1, and on hydrolysis gives an acid of NE 45.
c. Compound has a SE of 116 and yields one product when reduced with LiAlH$_4$.
d. Compound has one ester group, a SE of exactly 72, and does not decolorize Br$_2$ in CCl$_4$.

11-10 $\bullet$ Give the reaction of methyl acetate with each of the following reagents. Indicate whether the reaction of the reagent in this instance is electrophilic (E) or nucleophilic (N).
a. LiI   *b.* BF$_3$ *c.* CH$_3$NH$_2$ *d.* AlCl$_3$ *e.* CH$_3$SH

11-11   The $^-$OSO$_2$OCH$_3$ unit is a very good leaving group. Bearing this in mind, what products would you expect from the reaction of dimethyl sulfate with each of the following?
a. NaCN   *b.* CH$_3$CH$_2$MgI   *c.* LiAlH$_4$ *d.* CH$_3$C$\equiv$CNa   *e.* LiI

11-12   After refluxing 2.90 g of a colorless liquid with 35.0 ml of 1.00 N sodium hydroxide for 3 h, 20.0 ml of 0.500 N sulfuric acid was required to neutralize the excess base. After calculating the saponification equivalent, decide which of the following compounds fit the data to ±1 percent: isopropyl propionate, dimethyl succinate, dimethyl decanedioate, and diethyl 1,4-cyclohexanedicarboxylate. Suggest a simple chemical procedure for distinguishing between the compounds that fit the SE data.

11-13   It required 33.5 ml of 1.00 N sodium hydroxide to saponify 3.42 g of compound A. Reduction of A with $LiAlH_4$ yielded two compounds, both of which gave a positive iodoform test. Suggest a structure for A.

11-14   Compound X, C_4H_7ON, was neutral to litmus in aqueous solution. On boiling with sodium hydroxide, X evolved ammonia, and on treatment with nitrous acid, X gave off nitrogen. X reduced alkaline permanganate solution and possessed a SE of 86 ± 2. Suggest three structures for X.

11-15   How many milliliters of nitrogen, collected under standard conditions, would you expect from the action of excess nitrous acid on (*a*) 0.4 g acetamide, (*b*) 0.3 g of urea?

11-16   Compound X, C_4H_7ON, was neutral to litmus. On boiling with sodium hydroxide solution, X evolved ammonia. X was found to have a SE of 86 ± 2. X did not react with permanganate or nitrous acid. Refluxing X with HBr gave $C_3H_5O_2Br$. Suggest a structure for X.

11-17   Compound A, $C_6H_{12}O_2N_2$, on heating was converted to $C_6H_9O_2N$(B). On treatment with NaOBr, B gave $C_5H_{11}O_2N$(C). C reacted with HONO to give an acid, D, $C_5H_{10}O_3$. D, on treatment with NaOBr, gave the sodium salt of the acid $C_4H_6O_4$(E). E possessed a NE of 58 ± 2 and on heating evolved CO_2. Suggest structures for A through E.

11-18   Compound A, $C_4H_{11}ON$, reacts with acetyl chloride to give compound B that contains 8.09% nitrogen. Compound A reacts with nitrous acid without the evolution of nitrogen and, on treatment with NaOI, yields iodoform. Suggest structures for A and B.

11-19   Compound A contains 59.34% C, 10.96% H, 15.82% O, and 13.88% N. When A is heated with PCl_5 it gives compound B (MW 101 ± 1, strong ir absorption at 1685 cm^{-1}) which upon boiling with hydroxide solution is converted into C (MW 46 ± 1) and D (MW 74 ± 1). Nmr spectral data for A, C, and D are given below. Suggest structures for A, B, C, and D.

Compound A   Singlet at δ = 9.9, quartet at δ = 2.3, triplet at δ = 1.1; area ratio, 1:4:6.
Compound C   Quartet at δ = 2.6, singlet at δ = 1.5, triplet at δ = 1.0; area ratio, 2:2:3.
Compound D   Singlet at δ = 11.8, quartet at δ = 2.25, triplet at δ = 1.0; area ratio, 1:2:3.

11-20   Compound X gives a positive hydroxamic test. The nmr spectrum of X shows a quartet at τ = 5.9, a singlet at τ = 6.8, and a triplet at τ = 8.8 in a 2:1:3 area ratio, respectively. After boiling compound X with hydroxide solution, compounds Y (MW 103 ± 1) and Z (MW 45 ± 1) are isolated. The nmr spectrum for Y shows two singlets with equal areas, one of which is at approximately τ = −2. Compound Z shows a triplet, a quartet, and a singlet in a 3:2:1 respective ratio in its nmr spectrum. Suggest structures for X, Y, and Z.

11-21   Compound M contains 48.5% C, 5.1% H, and 14.2% N. Its nmr spectrum shows singlets at δ = 3.5 (9.1 integration units) and δ = 3.8 (13.4 integration units). When boiled with aqueous acid, compound M yields compound N with NE = 60 ± 1. Suggest structures for M and N.

11-22   Point out the errors, if any, in the following proposed syntheses:

a. $ClCH_2CH_2COOCH_3 \xrightarrow[\text{(A)}]{\text{NaOH soln}} ClCH_2CH_2COONa \xrightarrow[\text{(B)}]{(CH_3)_3CCl} ClCH_2CH_2COOC(CH_3)_3$

b. $CH_3COOH \xrightarrow[\text{(A)}]{\text{EtOH}} CH_3COOEt \xrightarrow[\text{(B)}]{Br_2, P} BrCH_2COOEt \xrightarrow[\text{(C)}]{LiAlH_4} BrCH_2CH_2OH$

c. $HCOOH \xrightarrow[\text{(A)}]{SOCl_2} HCOCl \xrightarrow[\text{(B)}]{CH_3OH} HCOOCH_3 \xrightarrow[\text{(C)}]{CH_3MgI} (CH_3)_2CHOH + MgIOCH_3$

d. $RCOOH \xrightarrow[\text{(A)}]{SOCl_2} RCOCl \xrightarrow[\text{(B)}]{NH_3} RCONH_2 \xrightarrow[\substack{NaOH \\ \text{(C)}}]{CH_3I} RCONHCH_3$

e. $\underset{\underset{NH_2}{|}}{RCHCOOH} \xrightarrow[\text{(A)}]{SOCl_2} \underset{\underset{NH_2}{|}}{RCHCOCl} \xrightarrow[\text{(B)}]{CH_3OH} \underset{\underset{NH_2}{|}}{RCHCOOCH_3}$

f. $(CH_3)_2NCH_2CH_2OH \xrightarrow[\text{(A)}]{\text{NaOH}} (CH_3)_2NH + HOCH_2CH_2OH \xrightarrow[\text{(B)}]{SOCl_2} ClCH_2CH_2Cl$

11-23   Given methanol, ethanol, 1-propanol, *N*-bromosuccinimide (NBS), triphenylphosphine, and any necessary inorganic reagents, show how you would make each of the following:
a.   Ethyl formate
b.   Methyl *cis*-2-butenoate

c. Isopropyl butyrate
e. Ethyl 2-methylbutanoate
g. Crotanamide
i. $CH_3NHCOOCH_2CH_2NHCONHCH_3$
k. $CH_2=C(CH_2CN)_2$
m. Triethyl phosphate
o. Ethyl orthoformate
q. cis-$CH_3CH=CHCH(OCH_3)_2$
s. $CH_2=C(CH_3)CONHCH_3$
u. 5-Methyl-4-nitro-1-hexen-3-ol

d. 1,2-Diacetoxyethane
f. sec-Butyl trimethylacetate
h. $O=C=N(CH_2)_3N=C=O$
j. $(BrCH_2)_2C=C(CH_2Br)_2$
l. $CH_3CON(CH_3)CH_2CH_2OH$
n. $CH_3COOCH_2CH_2Br$
p. 1,6-Heptadien-4-ol
r. N-Ethylsuccinimide
t. $(CH_3)_2C(OH)COCH_3$

v. 1,3,5,7-Cyclooctatetraene starting from

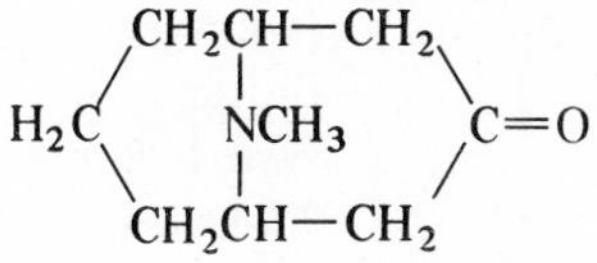

w.

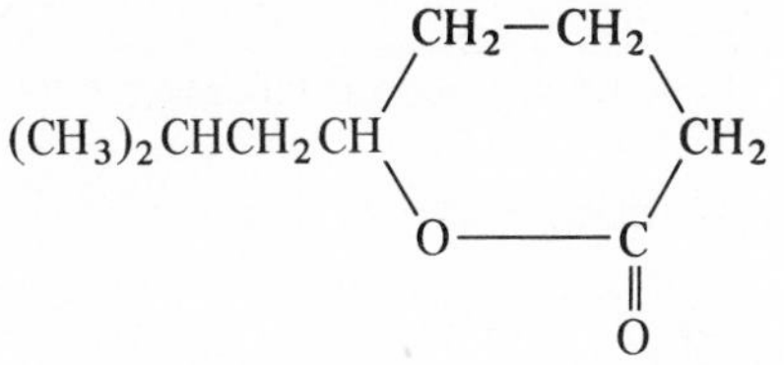

11-24   Propose a mechanism for each of the following reactions:

a. $CH_3CH_2\overset{\|}{\underset{O}{C}}{}^{18}OCH_2CH_3 \xrightarrow{H^+, H_2O} CH_3CH_2\overset{\|}{\underset{O}{C}}OH + CH_3CH_2{}^{18}OH$

b. $CH_3COOC(CH_3)_3 + H_2{}^{18}O \xrightarrow{H^+} (CH_3)_3C^{18}OH + CH_3COOH$

c. $CH_3COOCH_3 + (CH_3)_2CHCH_2CH_2OH \xrightarrow{H^+} CH_3COOCH_2CH_2CH(CH_3)_2 + CH_3OH$

d. $(CH_3)_2C=C=O + CH_3OH \longrightarrow (CH_3)_2CHCOOCH_3$

e.

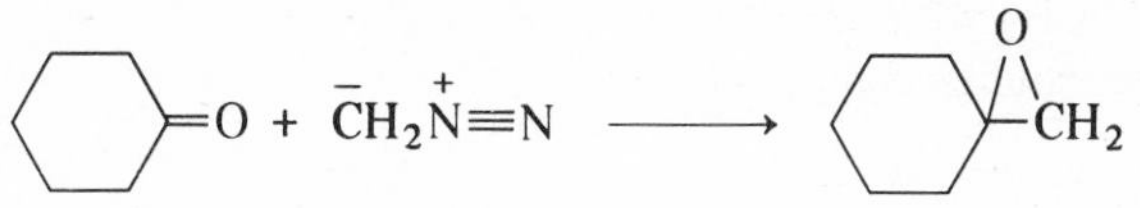

f.

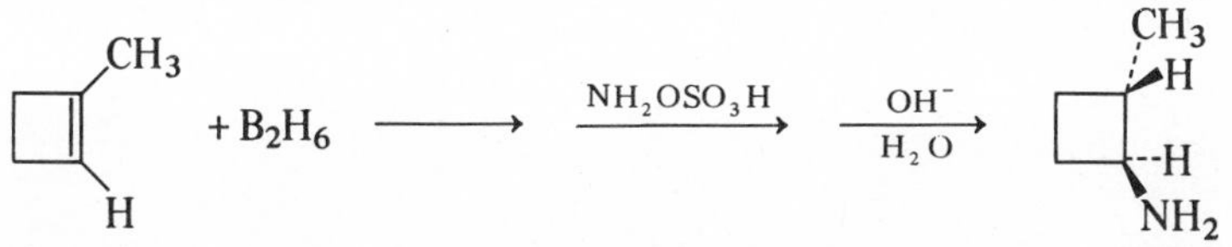

g.

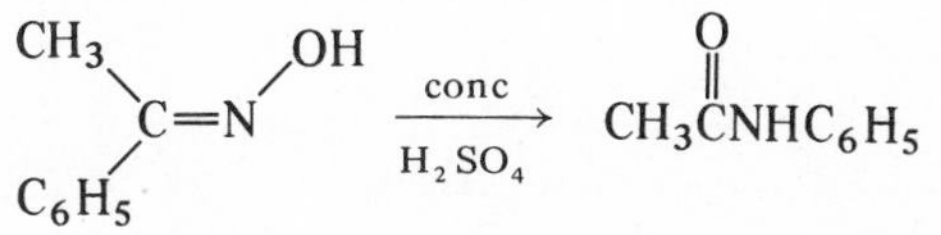

h.

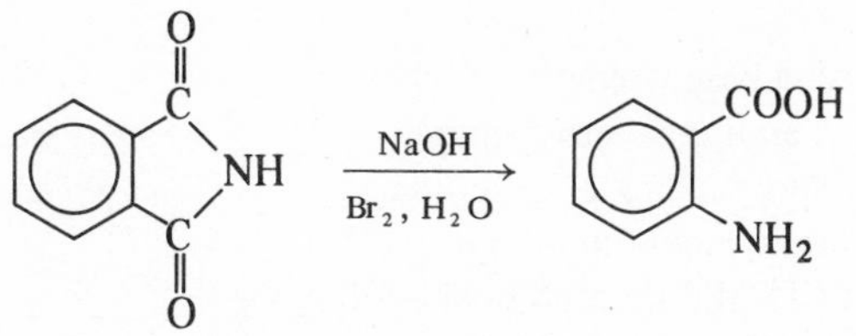

i. $C_6H_5\overset{\|}{\underset{O}{C}}CH_3 + CF_3\overset{\|}{\underset{O}{C}}OOH \longrightarrow C_6H_5O\overset{\|}{\underset{O}{C}}CH_3 + CF_3COOH$

165

11-25 Propose a mechanism which would explain why *cis*-2-hydroxy-1-cyclohexanecarboxylic acid forms a lactone while the trans isomer does not.

11-26 The ester group has much less ability to activate an α hydrogen atom for an aldol type condensation than an aldehyde group has. Suggest a reason in terms of resonance.

11-27 The relative rates of esterification of the following acids with methanol were determined by Sudborough and coworkers. Match the proper rate constant in the second column with the correct acid in the first column. The larger the constant the faster esterification occurs. Steric hindrance and relative electron deficiency of the carbon of the carbonyl group will be the two most important factors affecting the rates of reaction.

Formic	112	Isobutyric	104
Acetic	33.6	Crotonic	91.9
Propionic	1.2	2-Methylpropenoic	41.9
Butyric	0.8	2,3-Dimethylbutanoic	1.3

11-28 Explain why $CH_3CH=CHCOOC(CH_3)_3$ might not give only the normal type of ester reaction with a Grignard reagent.

11-29 Explain the difference between a fat and an oil from a structural viewpoint. How would this account for the lower melting points of oils vs. fats?

11-30 Write resonance forms for each of the following:
a. Dimethyl carbonate
b. Vinyl acetate
c. Guanidine
d. BrC≡N
e. $CH_3N=C=O$
f. Phosgene
g. CH_3OOCNH_2
h. Diazomethane
i. Nitroethane
j. Succinimide
k. Phenyl cyanide
l. Methyl phenyl ether

11-31 Show how atomic orbitals overlap to form molecular orbitals in each of the following:
a. $CH_3N=C=O$
b. Acrylamide
c. $N_2CHCOOCH_3$
d. Urea
e. $(CH_3)_3N \rightarrow O$

11-32 Offer an explanation for each of the following observations:
a. Methyl carbonate is hydrolyzed in alkaline solution. Methyl orthocarbonate is not.
b. Diethyl malonate will not react with urea unless a very strong base (for example, CH_3ONa) is present.
c. Guanidine is a much stronger base than urea.
d. The reaction of $(CH_3)_2NCH_2CH_2CH_2CH=CH_2$ with Br_2 does not give the expected normal addition product, but instead gives a highly polar, water-soluble compound.

e. The molecule   is unusually basic and forms stable salts with strong acids.

f. The base strength of the following compounds decreased in the order $R-NH_2 > R_2C=NH > RC≡N$ despite the fact that in each compound a lone pair of electrons is available.

11-33 Which of the following would you expect to be Lewis acids, bases, or both?
a. Acetamide
b. Nitromethane
c. $AlCl_3$
d. Acetic acid
e. Methylamine
f. Urea
g. BF_3
h. $(NC)_2C=C(CN)_2$

11-34 Arrange the following compounds in order of increasing base strength:
a. Acetamide, *N,N*-dimethylacetamide, methylamine, ammonia, ethylamine.
b. Acetamide, succinimide, acetic acid, *N,N*-dimethylpropanamide, ammonia.
c. Ammonia, trimethylamine, dimethyl ether, tetramethylammonium hydroxide.

11-35 What products would you expect from the action of CH_3MgI on each of the following?
a. $HCON(CH_3)_2$
b. $CH_3N=C=O$
c. Phosgene
d. Dimethyl carbonate

11-36 What products would you expect from the reaction of methyl isothiocyanate, $CH_3N=C=S$, with each of the following?
a. H_2O
b. CH_3NH_2
c. CH_3SH
d. $CH_2=CHCH_2OH$

11-37   Amidines and imido esters are made according to the following reaction schemes. What structures would you expect for these compounds?

a. $RC \equiv N + 1$ mol dry HCl $\longrightarrow$ $\xrightarrow{NH_3}$ amidine

b. $RC \equiv N + 1$ mol dry HCl $\longrightarrow$ $\xrightarrow{ROH}$ imido ester

11-38   From the structure of the segments of nylon and polyurethane polymers below, show how each could be made using cyclohexene as raw material.

a. $-(CH_2)_6 NHCO(CH_2)_4 CONH(CH_2)_6 NHCO(CH_2)_4-$ Nylon

b. $-(CH_2)_6 OCONH(CH_2)_6 NHCOO(CH_2)_6 OCONH-$ Polyurethane

11-39   Show how you would make the biologically important acetylcholine molecule $(CH_3)_3 \overset{+}{N}CH_2 CH_2 OCOCH_3 Cl^-$ from alcohol and methanol via ethylene oxide.

11-40   a. Outline the synthesis of 5-ethyl-5-phenylbarbituric acid (phenobarbitol) from starting materials containing 10 carbons or less.

b. Heating aminoacetic acid (glycine) results in the formation of a new compound B (MW 114). Saponification of B gives back only glycine. Propose a structure for B.

11-41   Compound A, $C_7H_{11}O_3Cl$, yielded no methane when it reacted with methylmagnesium bromide. When 3.56 g of A was refluxed with an excess of 2.0 N sodium hydroxide, 29.9 ml of the base was neutralized. The reaction mixture yielded two organic compounds, an alcohol that failed to give an iodoform test, and an acid. The acid had a NE of 59; it did not decarboxylate on heating, and it did not reduce permanganate. Suggest a structure for A.

11-42   Compound A, $C_{10}H_{14}O_4$, decolorized bromine in carbon tetrachloride. It had a SE of 97 ± 2. Hydrolysis of A yielded only two organic compounds, B and C. B was a neutral compound that gave a positive reaction with acetyl chloride but did not decolorize bromine in carbon tetrachloride. C was an acid with a NE of 72. Suggest structures for A, B, and C.

11-43   Nicotinamide, $C_6H_6N_2O$, is an important vitamin. On hydrolysis it is converted into an acid with the evolution of ammonia. The acid on decarboxylation gives an amine X, C_5H_5N, which on catalytic hydrogenation gives Y, $C_5H_{11}N$. Treatment of Y with excess CH_3I, followed by Ag_2O and heat and then catalytic hydrogenation, gave Z, $C_7H_{17}N$. Further exhaustive methylation of Z converted it into C_5H_{10} which on ozonolysis gave formaldehyde and butyric acid. Suggest three possible structures for nicotinamide.

11-44   A colorless crystalline compound, A, was found to contain C, H, O, and N. It was soluble in water but insoluble in nonpolar solvents. An aqueous solution of A was acidic and gave a NE of 124 ± 1. On treatment with nitrous acid, a gas was evolved, part of which was soluble in alkali and part of which was not. Evaporation of the solution from the nitrous acid treatment to dryness yielded only inorganic salts: NaCl, $NaNO_2$, and $NaNO_3$. On boiling A with sodium hydroxide solution, ammonia was evolved, and evaporation of this solution to dryness gave only inorganic salts. Suggest a structure for A.

11-45   Compound X (SE 99 ± 1) shows only a septet and a doublet in its nmr spectrum. When the alkaline solution from the hydrolysis was distilled, the distillate contained a water-soluble compound Y whose nmr spectrum showed a doublet at $\delta = 1.2$, a singlet at $\delta = 1.6$, and a septet at $\delta = 4.0$. Compound Z was isolated from the alkaline solution and its nmr spectrum showed only one singlet far downfield. When 0.1 g of X was catalytically hydrogenated (H_2, Pt) it absorbed 22.6 ml of hydrogen at STP. X did not form a phenylhydrazone or react with acetyl chloride, and it did not evolve methane when treated with CH_3MgI. Suggest structures for X, Y, and Z.

11-46   When 0.500 g of neutral compound X was treated with 0.020 mol of methylmagnesium iodide, it evolved 199.8 ml of methane at STP. Treating the solution with excess water caused it to evolve 147.8 ml of methane. When 0.100 g of X was catalytically hydrogenated, it absorbed 60.0 ml of hydrogen at STP. The ir spectrum of X shows characteristic bands at 3450 to 3550 cm^{-1} (broad), 2120 cm^{-1}, and 1720 cm^{-1}. Its nmr spectrum shows a series of three overlapping triplets and a doublet in the region $\delta = 2.5$ to 3.5 with a hydrogen ratio of 1:2:2:2 and a singlet at $\delta = 5.0$ (1H). Suggest a structure for X.

11-47   Compound A yields compound B when treated with lithium aluminum hydride and compound C when treated with aqueous acid. Compounds A, B, and C are all water-soluble. The following are some spectral data on A, B, and C.

Compound A   Ir (cm^{-1}, μm): 3500, 2.86; 3400, 2.94; 3000, 3.33; 1685, 5.93. Nmr (δ, multiplicity, integration units): 6.5, s, 9.7; 2.2, q, 9.6; 1.2, t, 144.
Compound B   Ir: 3480, 2.87; 3420, 2.92. Nmr: 2.6, t, 8.4; 1.5, s, 8.2; 1.3, m, 8.3; 0.9, t, 12.4.

Compound C Ir: several bands 2900 to 2600, 3.45 to 3.85; 1708, 5.85. Nmr: 11.8, *s*, 2.8; 2.25, *q*, 5.7; 1.0, *t*, 8.3.

Suggest structures for A, B, and C.

11-48 Compound M (MW 87) shows three singlets in its nmr spectrum at τ = 7.0, 7.2, and 8.1 with equal areas. When heated with aqueous acid, an acid (MW 60) and a gas (MW 45) with an ammonialike odor are obtained. Suggest a structure for M and an explanation for its observed nmr spectrum.

11-49 Compound X (MW 100 ± 1) when treated with a cold solution of sodium hydroxide and bromine yields a bromine containing compound Y (MW 178 ± 1). If X is heated with a sodium hydroxide–bromine solution, compound Z (MW 90 ± 1) is obtained. Compound Y shows only a singlet at δ = 2.97 in its nmr spectrum while compound Z shows two triplets with equal areas if it is first treated with D_2O. Suggest structures for X, Y, Z and an explanation for Z's observed nmr spectrum.

11-50 Compound A (SE 210 ± 2) when treated with $LiAlH_4$ yields only one product, B (MW 108). The nmr spectrum of B shows singlets at δ = 7.3, 4.6, and 2.4. Suggest structures for A and B.

11-51 Compound C, $C_6H_{12}O_2$, shows strong ir bands at 2950 cm^{-1} (3.39 μm), 1724 cm^{-1} (5.80 μm), and 1176 cm^{-1} (8.5 μm). Its nmr spectrum is shown. Suggest a structure for C.

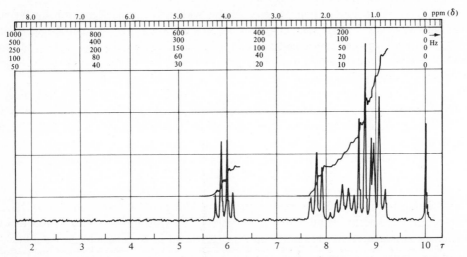

Figure P11-51. Nmr spectrum for Prob. 11-51. *(From R. A. Silva and R. Spenger, "The Quantitative Organic Analysis Spectra Kit," McGraw-Hill Book Company, New York, 1969. By permission of the publishers.)*

11-52 Compound D contains C, H, and O only. It has a molecular weight of 118 and can exist as a *dl* pair. Its ir and nmr spectra are shown. Suggest a structure for D.

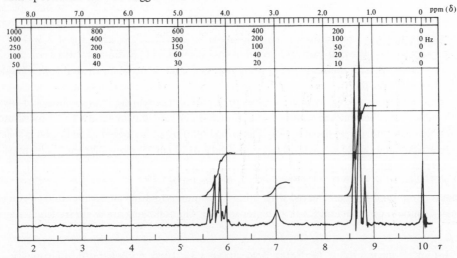

Figure P11-52. Ir and nmr spectra for Prob. 11-52. *(From R. A. Silva and R. Spenger, "The Quantitative Organic Analysis Spectra Kit," McGraw-Hill Book Company, New York, 1969. By permission of the publishers.)*

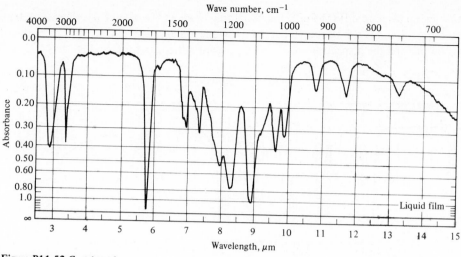

Figure P11-52 *Continued*.

11-53 Compound E (MW 86) contains C, H, and O only, does not decolorize Br_2 in CCl_4, and does not absorb hydrogen when treated with H_2 and Pt. However, upon treatment with $LiAlH_4$, E does react to give compound F (MW 90) which shows only three signals in its nmr spectrum. The ir and nmr spectra of compound E are shown. Suggest structures for E and F.

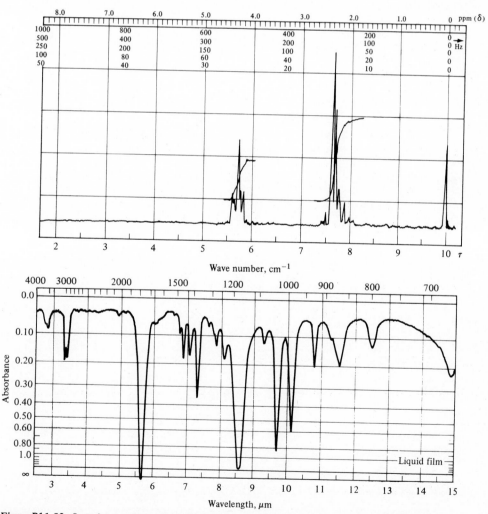

Figure P11-53. Ir and nmr spectra for Prob. 11-53, compound E. *(From R. A. Silva and R. Spenger, "The Quantitative Organic Analysis Spectra Kit," McGraw-Hill Book Company, New York, 1969. By permission of the publishers.)*

11-54 Two isomeric compounds, G and H (MW 172), contain only C, H, and O. They both show a triplet at $\tau = 8.7$ and a quartet at $\tau = 5.8$, but G shows a singlet at $\tau = 3.2$ while H shows a singlet at $\tau = 3.75$ in their nmr spectra. Based on this and the ir spectrum of G shown, suggest structures for G and H.

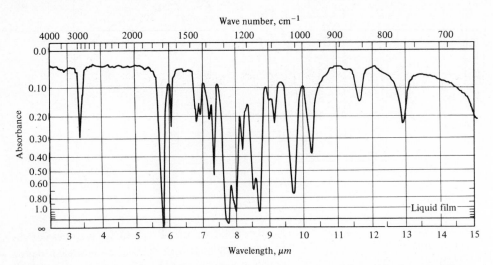

Figure P11-54. Ir spectrum for Prob. 11-54, compound G. *(From R. A. Silva and R. Spenger, "The Quantitative Organic Analysis Spectra Kit," McGraw-Hill Book Company, New York, 1969. By permission of the publishers.)*

12

carbanion chemistry

12-1 Enolate Anions

a. *Formation from Active Methylene Compounds* Hydrogens attached to a saturated carbon atom which has one or more strong electron-attracting groups bonded to it such as $-COOR$, $-COR$, $-NO_2$, or $-C{\equiv}N$ are acidic enough to react with strong bases to form carbanions or enolate anions.

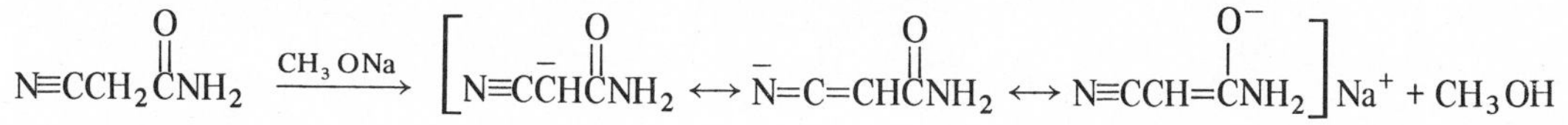

$$N{\equiv}CCH_2\overset{O}{\overset{\|}{C}}NH_2 \quad \xrightarrow{CH_3ONa} \quad \left[N{\equiv}C\overset{\bar{}}{C}H\overset{O}{\overset{\|}{C}}NH_2 \longleftrightarrow \overset{\bar{}}{N}{=}C{=}CH\overset{O}{\overset{\|}{C}}NH_2 \longleftrightarrow N{\equiv}CCH{=}\overset{O^-}{\overset{|}{C}}NH_2 \right] Na^+ + CH_3OH$$

Cyanoacetamide

Compounds containing these types of acidic $-CH_2-$ groups are called active methylene compounds. The following table lists some examples of this type of compound and the approximate acidities of the methylene (or methyl) hydrogens.

Name	Structure	pK_a (underlined H)
Ethyl cyanoacetate	CH$\underline{_2}$(CN)COOEt	9
Acetylacetone	CH$\underline{_2}$(COCH$_3$)$_2$	9
Nitromethane	CH$\underline{_3}$NO$_2$	10
Ethyl acetoacetate	CH$_3$COCH$\underline{_2}$COOEt	11
Malononitrile	CH$\underline{_2}$(CN)$_2$	11
Ethyl malonate	CH$\underline{_2}$(COOEt)$_2$	13
Acetone	CH$\underline{_3}$COCH$_3$	20
Ethyl acetate	CH$\underline{_3}$COOEt	24

b. *Keto-Enol Tautomerism*

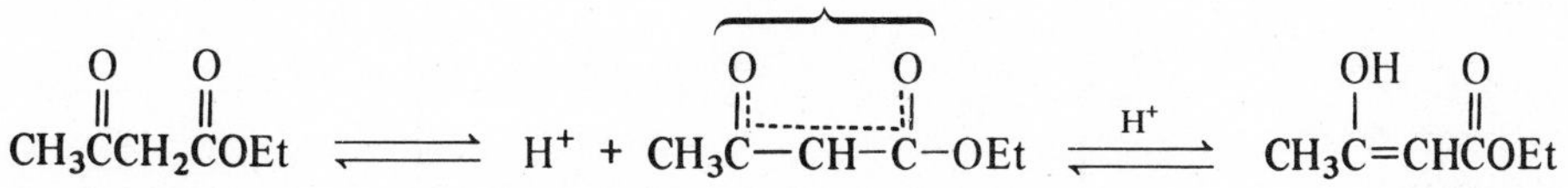

$$\underset{\text{92\% Keto}}{CH_3\overset{O}{\overset{\|}{C}}CH_2\overset{O}{\overset{\|}{C}}OEt} \rightleftharpoons H^+ + CH_3\overset{O}{\overset{\|}{C}}{-}\overset{\bar{}}{C}H{-}\overset{O}{\overset{\|}{C}}{-}OEt \overset{H^+}{\rightleftharpoons} \underset{\text{8\% Enol}}{CH_3\overset{OH}{\overset{|}{C}}{=}CH\overset{O}{\overset{\|}{C}}OEt}$$

Ketones undergo isomerization as depicted above. In ordinary ketones such as acetone only about 2.5×10^{-4} percent of the enol form is present. However, when the enol form has two or more conjugated double bonds, stabilization by resonance in the enol form occurs. Hydrogen bonding may also stabilize the enol form. Easily interconverted isomers which differ only in the position of one atom or group (most often H) are called tautomers.

The following example with acetone indicates how keto-enol isomerization may be catalyzed by base or acid.

$$CH_3\overset{\overset{O}{\|}}{C}CH_3 + :B \rightleftharpoons \left[CH_3\overset{\overset{O}{\|}}{C}CH_2^- \longleftrightarrow CH_3\overset{\overset{O^-}{|}}{C}=CH_2 \right] \overset{HB}{\rightleftharpoons} CH_3\overset{\overset{OH}{|}}{C}=CH_2 + :B$$

$$CH_3\overset{\overset{O}{\|}}{C}CH_3 + HB \rightleftharpoons CH_3\overset{\overset{OH}{|}}{\underset{+}{C}}CH_3 + B^- \rightleftharpoons CH_3\overset{\overset{OH}{|}}{C}=CH_2 + HB$$

β-Keto esters which are an equilibrium mixture of an alcoholic and a ketonic form give reactions of both ketones and alcohols. For example,

$$CH_3\overset{\overset{O}{\|}}{C}CH_2\overset{\overset{O}{\|}}{C}OCH_3 + CH_3COCl \longrightarrow CH_3\overset{\overset{OOCCH_3}{|}}{C}=CHCOOCH_3$$

$$\downarrow NH_2OH$$

$$CH_3\overset{}{\underset{\underset{OH}{N}}{C}}CH_2\overset{}{\overset{\curvearrowleft OCH_3}{C}}=O \longrightarrow CH_3\overset{}{\underset{\underset{O}{N}}{C}}-CH_2 \underset{C=O}{} + CH_3OH$$

12-2 Reactions Involving Enolate Anions or Carbanions

a. Reactions with Alkylating Reagents

1. Acetoacetic ester synthesis. Enolate ions formed from the compounds listed in the table above may act as nucleophiles in typical S_N reactions. For example, ethyl acetoacetate may undergo mono- or dialkylation with alkyl halides (it may also undergo acylation with acyl halides — see Sec. 12-2b-7) in the acetoacetic ester synthesis. The products of this synthesis are β-keto esters which upon acid hydrolysis yield unstable acids, such as A or B, which when heated will decarboxylate to yield ketones.

Carboxyl groups attached to a carbon carrying a strong electron-attracting group such as –COR, –CHO, –CN, –NO$_2$, and –COOH are easily decarboxylated on heating.

$$CH_3COCH_2COOEt \xrightarrow[\text{2. RX}]{\text{1. EtONa}} CH_3CO\overset{\overset{R}{|}}{C}HCOOEt \xrightarrow[\text{2. R'X}]{\text{1. EtONa}} CH_3CO\overset{\overset{R}{|}}{\underset{\underset{R'}{|}}{C}}COOEt \xrightarrow{H^+}$$

$$CH_3COCH_2R + CO_2 \overset{\Delta}{\longleftarrow} CH_3CO\overset{\overset{R}{|}}{C}HCOOH \qquad CH_3CO\overset{}{\underset{\underset{R'}{|}}{C}}HR + CO_2 \overset{\Delta}{\longleftarrow} CH_3CO\overset{\overset{R}{|}}{\underset{\underset{R'}{|}}{C}}COOH$$

$$\qquad\qquad\qquad\text{A} \qquad\qquad\qquad\qquad\qquad\qquad\qquad\qquad\qquad\qquad \text{B}$$

On heating with concentrated alkali, β-keto esters are hydrolyzed and a nucleophilic cleavage also occurs on the carbonyl group.

$$2OH^- + CH_3COCR_2COOEt \longrightarrow \left[CH_3\overset{\overset{O}{\|}}{C}OH + {}^-CR_2COOH + EtO^- \right] \longrightarrow$$

$$CH_3COO^- + R_2CHCOO^- + EtOH$$

An acetoacetic ester synthesis followed by the above nucleophilic cleavage (called the *acid split*) is useful for making substituted acetic acids such as RCH$_2$COOH and R$_2$CHCOOH.

172

Diketones and dicarboxylic acids, such as $CH_3COCHR(CH_2)_xCHRCOCH_3$ and $HOOCCHR(CH_2)_xCHRCOOH$, may be formed by linking two acetoacetic ester units as shown below:

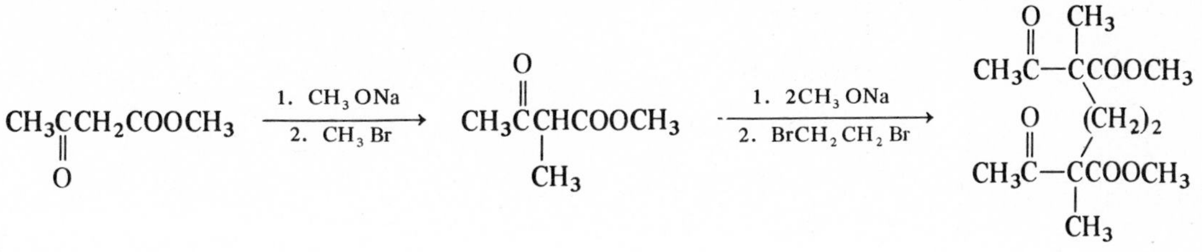

$$CH_3CCH_2COOCH_3 \xrightarrow[\text{2. } CH_3Br]{\text{1. } CH_3ONa} CH_3CCHCOOCH_3 \xrightarrow[\text{2. } BrCH_2CH_2Br]{\text{1. } 2CH_3ONa}$$

(with carbonyl O on first structure; second structure has CH$_3$ substituent)

$$\begin{array}{c} O \quad CH_3 \\ \| \quad | \\ CH_3C-CCOOCH_3 \\ O \quad (CH_2)_2 \\ \| \quad / \\ CH_3C-CCOOCH_3 \\ | \\ CH_3 \end{array}$$

I

$$NaOOCCH(CH_3)CH_2CH_2CH(CH_3)COONa \xleftarrow[\text{NaOH}]{\text{concd}} (I) \xrightarrow{H^+, H_2O} CH_3COCH(CH_3)CH_2CH_2CH(CH_3)COCH_3$$

2. Malonic ester synthesis. The above alkylation reactions of acetoacetic ester may be applied to malonic esters.

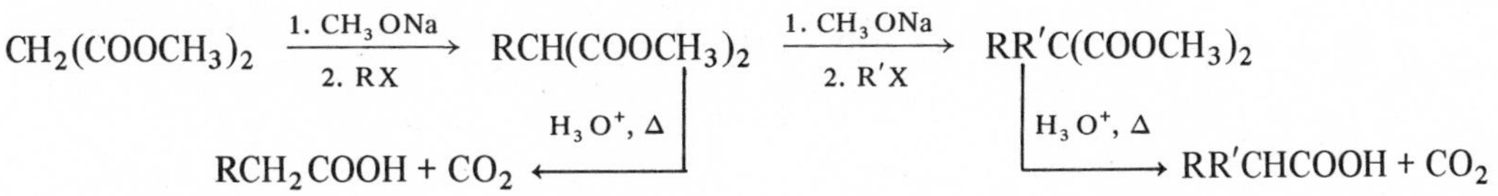

$$CH_2(COOCH_3)_2 \xrightarrow[\text{2. RX}]{\text{1. } CH_3ONa} RCH(COOCH_3)_2 \xrightarrow[\text{2. R'X}]{\text{1. } CH_3ONa} RR'C(COOCH_3)_2$$

$$RCH_2COOH + CO_2 \xleftarrow{H_3O^+, \Delta} \qquad \qquad \downarrow H_3O^+, \Delta$$
$$\qquad \qquad \qquad \qquad \qquad \qquad \qquad \rightarrow RR'CHCOOH + CO_2$$

Malonic ester gives better yields in the synthesis of acids than acetoacetic ester.

b. *Addition of Enolate Anions to Carbonyl Groups*
   1. Aldol condensation. (See Sec. 7-3f.)
   2. Wittig reaction. (See Sec. 7-3g.) In this reaction the ylide $[(C_6H_5)_3 \overset{+}{P}-\overset{-}{C}R_2 \longleftrightarrow (C_6H_5)_3P=CR_2]$ is the carbanion that attacks the carbonyl group in an aldehyde or ketone.
   3. The Claisen ester condensation. This condensation is similar to the aldol condensation (see Sec. 7-3f), but a stronger base is necessary to promote the reaction; sodium ethoxide is commonly used. The α hydrogen of the ester is involved.

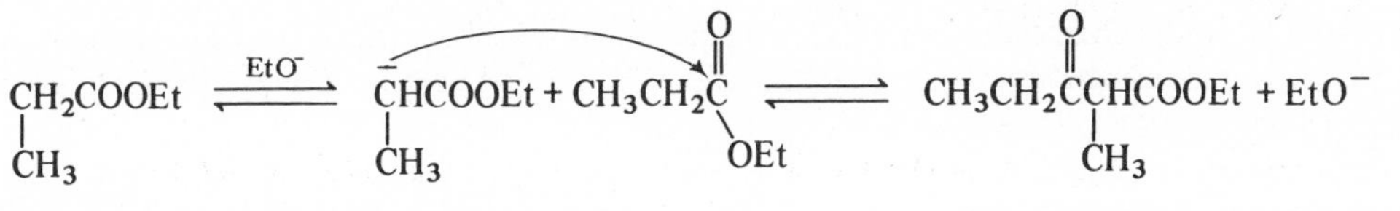

$$\begin{array}{c} CH_2COOEt \\ | \\ CH_3 \end{array} \underset{}{\overset{EtO^-}{\rightleftharpoons}} \begin{array}{c} CHCOOEt + CH_3CH_2C \\ | \qquad \qquad \qquad \| \\ CH_3 \qquad \qquad \quad OEt \end{array} \rightleftharpoons \begin{array}{c} CH_3CH_2CCHCOOEt + EtO^- \\ \| \quad | \\ O \quad CH_3 \end{array}$$

Ethyl 2-methyl-3-ketopentanoate

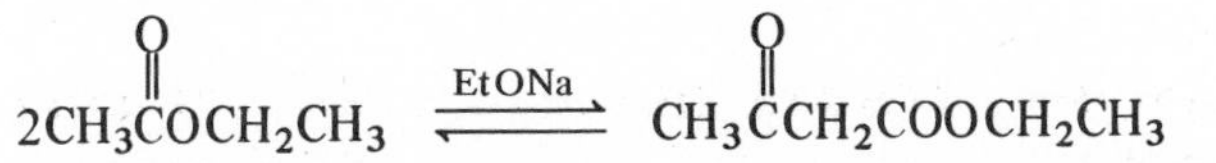

$$2CH_3COCH_2CH_3 \xrightarrow{EtONa} CH_3CCH_2COOCH_2CH_3$$

Acetoacetic ester or ethyl acetoacetate

For esters having an alkyl group in place of one α hydrogen atom, a base stronger than RONa is necessary. Triphenylmethyl sodium or potassium is often used.

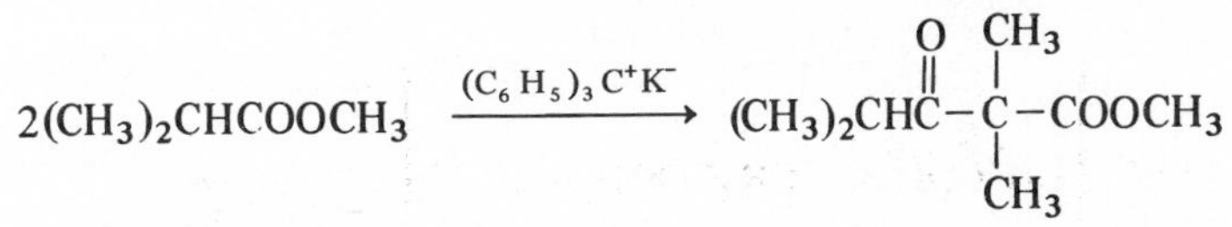

$$2(CH_3)_2CHCOOCH_3 \xrightarrow{(C_6H_5)_3C^+K^-} (CH_3)_2CHC-C-COOCH_3$$
$$\qquad \qquad \qquad \qquad \qquad \begin{array}{c} O \quad CH_3 \\ \| \quad | \\ \qquad \quad | \\ \qquad \quad CH_3 \end{array}$$

Normally the Claisen condensation is carried out between two molecules of the same ester. If two different esters, both having α hydrogen atoms, are used, then four different products are possible. This means that the yield of any one isomer is low and the job of separating the four products is very difficult.

   Mixed Claisen ester condensations can be *used* profitably when *only one* of the esters has an active hydrogen.

173

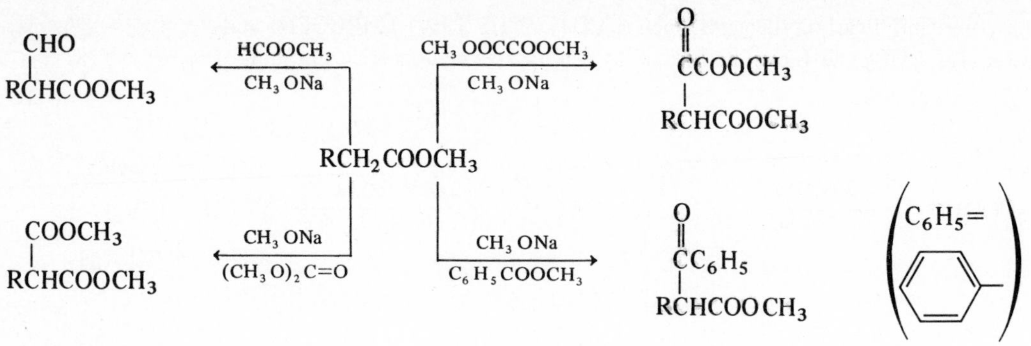

An intramolecular Claisen condensation (called the *Dieckmann reaction*) is useful for making cyclic molecules and works best when five- or six-membered rings are formed.

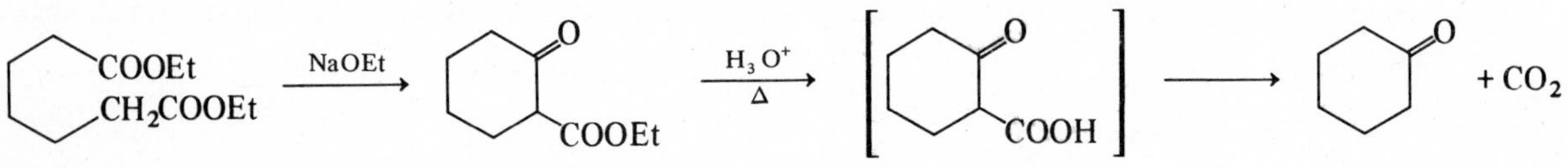

4. Reformatsky reaction.

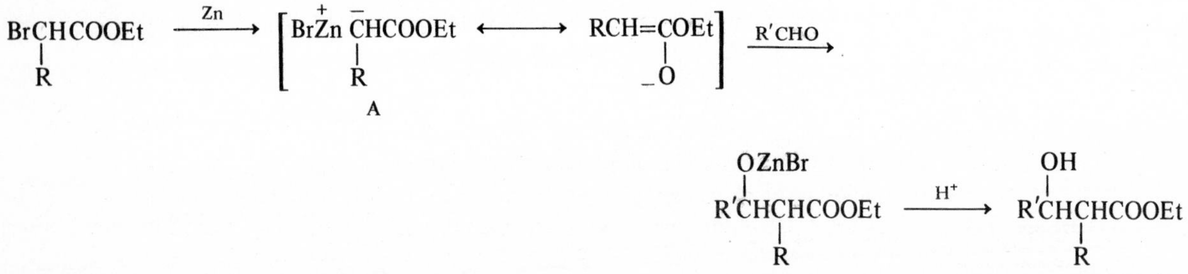

This reaction is similar to the Grignard reaction. The organozinc intermediate (A) is the enolate anion. It is less active than a Grignard reagent and hence does not react with the ester group. The carbonyl compound may be either an aldehyde or a ketone. The halide must be an α-halo ester.

5. Stobbe condensation. In this reaction an enolate anion of a succinic ester adds to a ketone or aldehyde. The reaction proceeds through an intermediate lactone (A) which opens up via base-catalyzed elimination.

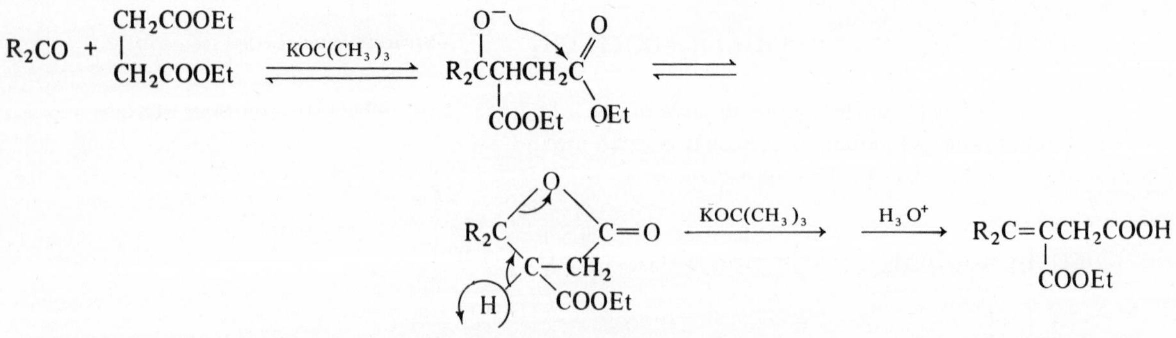

6. Darzens glycidic ester condensation. An enolate anion from an α-halo ester attacks a ketone or aldehyde. This is followed by an intramolecular S_N reaction to form the ester epoxide I (called a *glycidic ester*) which upon saponification, acidification, and heating decarboxylates yielding an aldehyde.

174

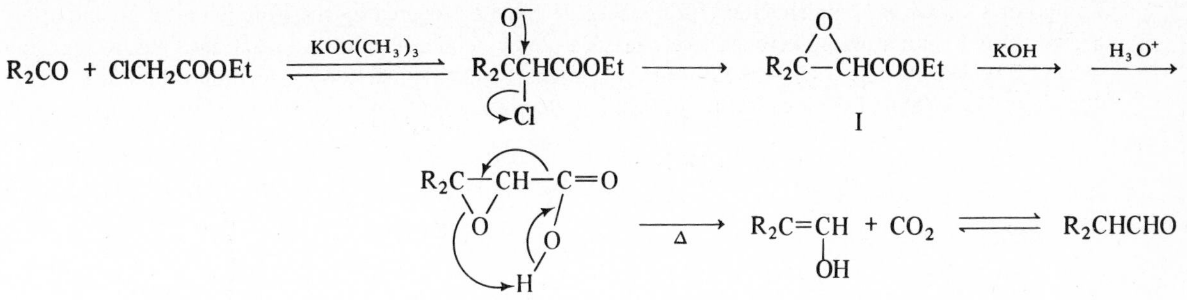

$$R_2CO + ClCH_2COOEt \xrightleftharpoons{KOC(CH_3)_3} \underset{\underset{Cl}{\overset{\downarrow}{|}}}{R_2C\overset{\overset{O^-}{\downarrow}}{\underset{}{C}}HCOOEt} \longrightarrow \underset{I}{R_2C\overset{O}{\overset{\triangle}{\diagdown}}CHCOOEt} \xrightarrow{KOH} \xrightarrow{H_3O^+}$$

$$\underset{H}{R_2C\overset{}{\diagup}CH-C=O} \xrightarrow{\Delta} \underset{OH}{R_2C=CH} + CO_2 \rightleftharpoons R_2CHCHO$$

7. Reactions with acyl chlorides.

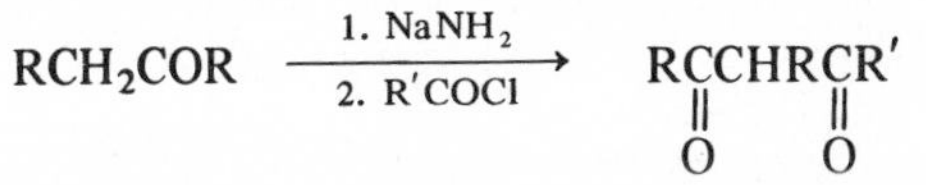

$$RCH_2COR \xrightarrow[\text{2. R'COCl}]{\text{1. NaNH}_2} \underset{O\quad\;\;O}{R\overset{\|}{C}CHR\overset{\|}{C}R'}$$

8. Reactions with aromatic aldehydes. A number of reactions of enolate anions with aromatic aldehydes (see Chap. 19) are listed below. Remember that aromatic aldehydes have no α hydrogens.

Claisen-Schmidt Reaction

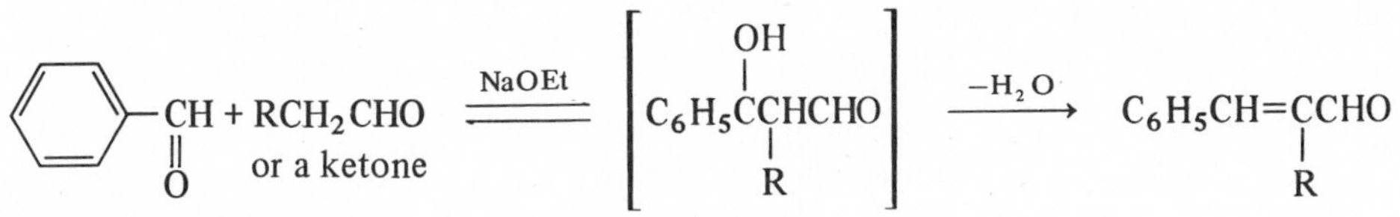

$$\underset{O}{\text{⬡}-\overset{\|}{C}H} + RCH_2CHO \xrightleftharpoons{NaOEt} \left[\underset{R}{C_6H_5\overset{\overset{OH}{|}}{C}CHCHO}\right] \xrightarrow{-H_2O} \underset{R}{C_6H_5CH=\overset{}{C}CHO}$$
or a ketone

Benzaldehyde (C_6H_5CHO)

Knoevenagel Reaction

$$C_6H_5CHO + CH_2(COOEt)_2 \xrightarrow{R_2NH} \xrightarrow{-H_2O} C_6H_5CH=C(COOEt)_2$$
or acetoacetic ester

Perkin Condensation

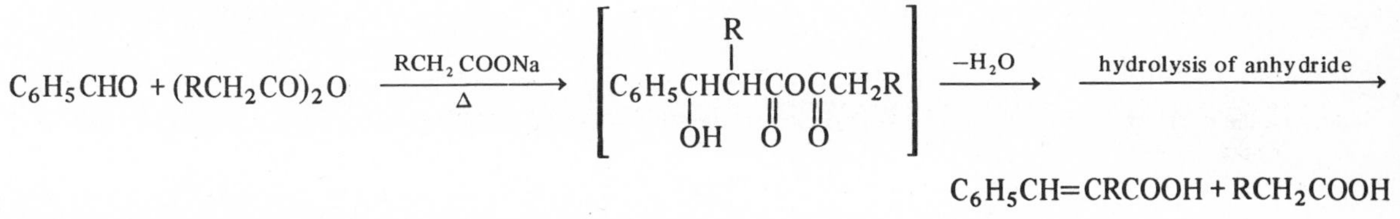

$$C_6H_5CHO + (RCH_2CO)_2O \xrightarrow[\Delta]{RCH_2COONa} \left[\underset{OH\;\;O\;O}{C_6H_5\overset{\overset{R}{|}}{C}HCHCOCCH_2R}\right] \xrightarrow{-H_2O} \xrightarrow{\text{hydrolysis of anhydride}}$$

$$C_6H_5CH=CRCOOH + RCH_2COOH$$

c. *Addition of Enolate Anions to α,β-Unsaturated Carbonyl Compounds; the Michael Addition Reaction* Carbanions of almost any active methylene compound may undergo nucleophilic addition to activated double bonds (see Sec. 9-10).

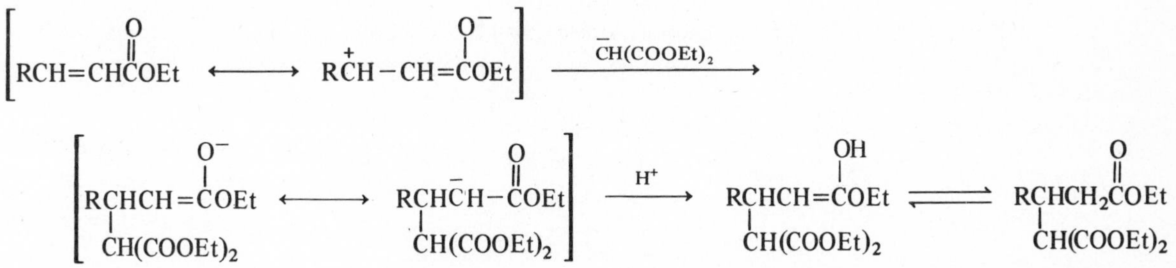

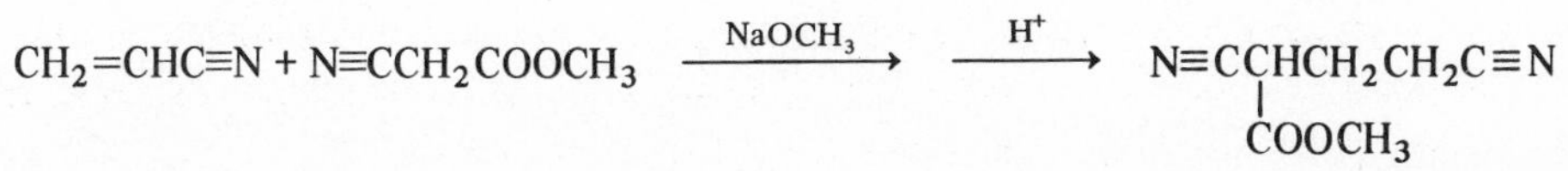

$$\left[\underset{O}{RCH=CH\overset{\|}{C}OEt} \longleftrightarrow R\overset{+}{C}H-CH=\overset{\overset{O^-}{|}}{C}OEt\right] \xrightarrow{\overset{-}{C}H(COOEt)_2}$$

$$\left[\underset{CH(COOEt)_2}{RCHCH=\overset{\overset{O^-}{|}}{C}OEt} \longleftrightarrow \underset{CH(COOEt)_2}{RCH\overset{-}{C}H-\overset{\|}{C}OEt}\right] \xrightarrow{H^+} \underset{CH(COOEt)_2}{RCHCH=\overset{\overset{OH}{|}}{C}OEt} \rightleftharpoons \underset{CH(COOEt)_2}{RCHCH_2\overset{\|}{C}OEt}$$

$$CH_2=CHC\equiv N + N\equiv CCH_2COOCH_3 \xrightarrow{NaOCH_3} \xrightarrow{H^+} \underset{COOCH_3}{N\equiv CCHCH_2CH_2C\equiv N}$$

175

Compounds known as Mannich bases (for instance, I) may be prepared by the Mannich reaction and upon heating yield α,β-unsaturated ketones which may be used in a Michael addition reaction. Thus, for example, methyl vinyl ketone may be generated in situ in a Michael addition reaction by decomposition of I. This is advantageous since methyl vinyl ketone and other α,β-unsaturated ketones tend to polymerize upon standing.

$$(CH_3)_2C=O + CH_2O + R_2NH \xrightarrow{\;HCl\;} CH_3COCH_2CH_2\overset{+}{N}HR_2Cl^- \xrightarrow{\;\Delta\;} CH_3COCH=CH_2 + R_2\overset{+}{N}H_2Cl^-$$

or other ketones   I

One very useful application of Mannich bases is illustrated below where a new ring is formed (called Robinson annulation) via a Michael addition reaction followed by an intramolecular aldol condensation.

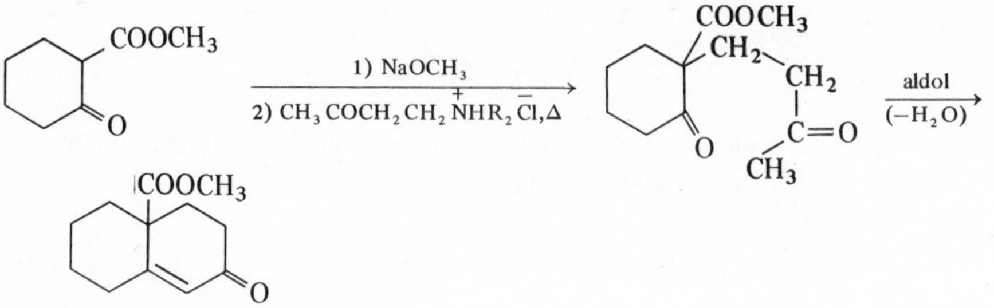

d.   *Enamines as a Source of Carbanions*
     1. Preparation. Reaction of aldehydes or ketones with a primary or a secondary amine (e.g., pyrrolidine) yields enamines

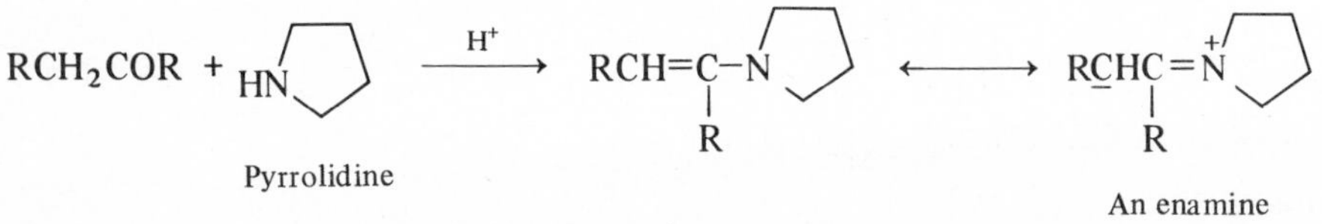

Pyrrolidine   An enamine

     2. Reactions of enamines. From the resonance structures drawn above for an enamine, it can be seen that enamines can serve as a source of carbanions in many of the reactions discussed in Secs. 12-2a–c.

$$(CH_3)_2CHCHO \xrightarrow[\;]{RNH_2,H^+} (CH_3)_2C=CHNHR \xrightarrow[2.\ C_6H_5CH_2Cl]{1.\ EtMgBr}$$

$$(CH_3)_2C=CH\overset{+}{N}HR + (CH_3)_2\underset{CH_2C_6H_5}{C}CH=NR \xrightarrow{\;H_3O^+\;} (CH_3)_2\underset{CH_2C_6H_5}{C}CHO + RNH_2$$

                   $\underset{CH_2C_6H_5}{}$

            Minor (A)   Major (B)

EtMgBr is used to promote C alkylation (B) rather than N alkylation (A) which can predominate in alkylation reactions of aldehydes or ketones. Some other uses of enamines are shown below:

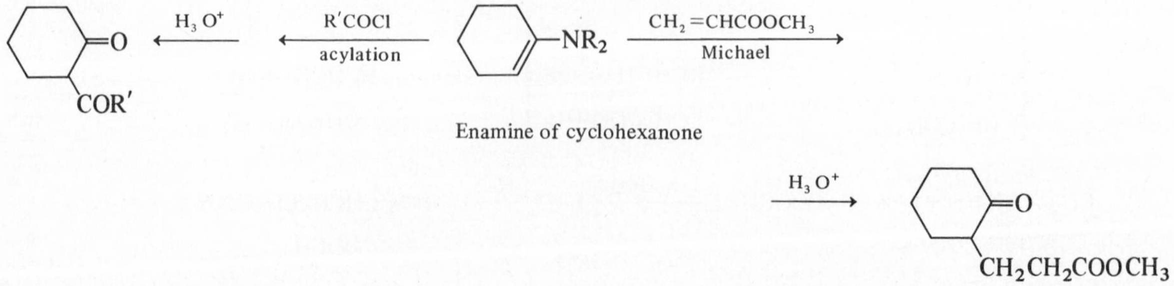

Enamine of cyclohexanone

e. *Organoboranes as Alkylating Agents* Alkylation of α-halo-substituted esters and ketones with organoboranes (see Sec. 6-1*e*) may be accomplished using potassium *tert*-butoxide as a base.

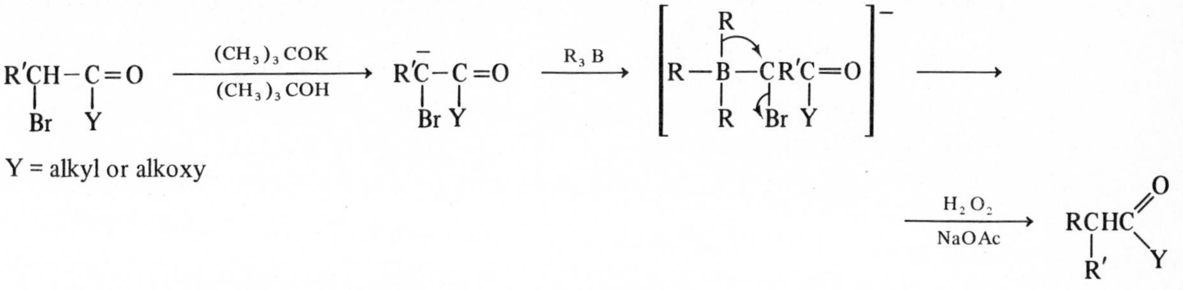

Y = alkyl or alkoxy

$$(Isobutyl)_3B + BrCH_2COOEt \xrightarrow{(CH_3)_3COK} (CH_3)_2CHCH_2CH_2COOEt$$

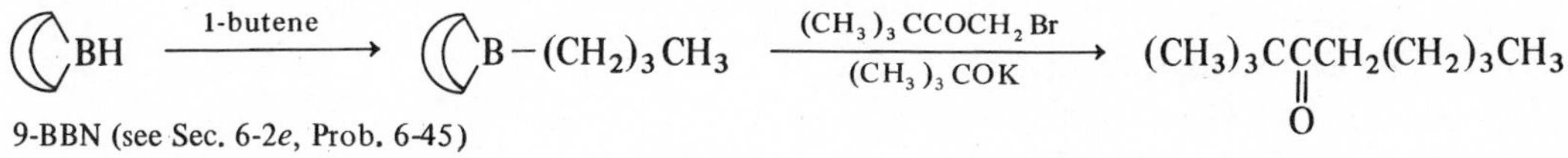

9-BBN (see Sec. 6-2*e*, Prob. 6-45)

12-3 Nomenclature of Derivatives of Malonic and Acetoacetic Esters

The alkyl derivatives of these compounds are often referred to without indicating the position of the alkyl group. In these instances the alkyl groups are understood to be attached to the active methylene group.

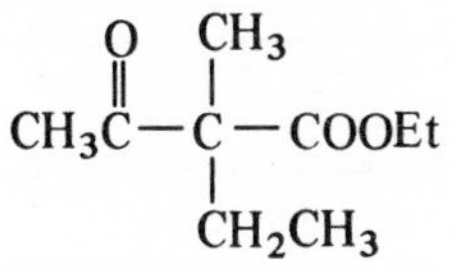

Ethyl ethylmethylacetoacetate 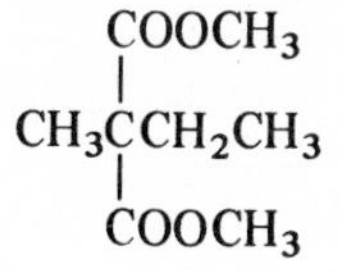   Methyl ethylmethylmalonate

12-4 Qualitative Tests for β-Keto Esters

β-Keto esters give reddish color with ferric chloride solution. This is due to the *enolic form* and is characteristic of the group $-\underset{\underset{OH}{|}}{C}=C-$. Hence, this result is not given by a β-keto ester in which there is no H that can tautomerize to give the enolic form (e.g., methyl dimethylacetoacetate). Phenols (see Chap. 18) also give positive tests with ferric chloride.

Problems

12-1   Draw structures for each of the following:
   a. *sec*-Butyl methylethylacetoacetate
   b. β-Propionylpropionic acid
   c. Cyanoacetic acid
   d. Methyl dimethylmalonate
   e. 1,2-Dicarbomethoxycyclopentane
   f. β-Ketoglutaric acid
   g. Methyl dimethylaminomalonate
   h. Cyclopropyl 2-cyclopentyl-3-ketobutanoate

12-2   Give the formula for the S_N reaction product of each of the following with sodium methyl acetoacetate. Also, give the products you would expect from the S_N reaction products (1) upon acid hydrolysis and (2) upon concentrated alkaline hydrolysis.
   a. Allyl bromide
   b. Methyl bromoacetate
   c. Bromoacetone
   d. Propionyl chloride
   e. 1,2-Dibromoethane
   f. Methyl α-bromosuccinate

12-3   Complete the following:

   a. $EtCH(COOCH_3)_2 \xrightarrow{CH_3ONa} \xrightarrow{CH_3Br}$

   b. $\bar{C}H(COOCH_3)_2 + CH_3COCl$

c. $(CH_3)_2CHCHO + OH^-$

d. $(C_6H_5)_3P=CHCH_3 + CH_3COCH_2COOEt$

e. $CH_3CH_2COOEt + NaOEt$

f. Ethyl acetoacetate + $LiAlH_4 \xrightarrow{\quad} \xrightarrow{H^+}$

g. $CH_3COCH_2COOEt + CH_3NH_2 + H_2 + Pt$

h. Methyl malonate + $NH_3 + \Delta$

i. $(CH_3)_2CO + CH_3CHBrCOOEt + Zn$

j. C_6H_5CHO + propanal + NaOEt

k. $C_6H_5CHO + (CH_3CO)_2O + CH_3COONa$

l. Ethyl fumarate + $CH_3NO_2 + EtO^-$

m. Cyclohexanone + $CH_2O + R_2NH \xrightarrow{H^+} \xrightarrow{\Delta}$

n. Cyclohexanone + ethyl succinate + $(CH_3)_3COK$

o. $Et_3B + BrCH_2COOEt + (CH_3)_3COK$

p. $C_6H_5CHO + CH_3COCH_2CO_2Et + R_2NH$

q. Cyclohexanone + $ClCH_2COOEt + EtO^-$

r. Cyclohexanone + pyrrolidine + H^+

s. Product from (r) $\xrightarrow{\text{acrolein}} \xrightarrow{H^+}$

t. Isobutylene $\xrightarrow[\text{(see Sec. 12-2}e)]{\text{9-BBN}} \xrightarrow[(CH_3)_3COK]{C_6H_5COCH_2Br}$

u. Ethyl crotonate + malononitrile

v. Acrolein + $(CH_3COO)_2CH^-$

w. $CH_3COCH_2COOEt + Br_2$

x. CH_3COCH_2COOEt + hydrazine

y. Ethyl cyclobutanecarboxylate + EtO^-

z. $\overset{-}{C}H(COOEt)_2 + \underset{\diagdown_S\diagup}{CH_2-CH_2}$

a'. $CH_3CO\overset{-}{C}HCOOEt$ + ethylene oxide

b'. Hexanedial + OH^-

c'. $CH_2(COOEt)_2 + 2EtO^- + Br(CH_2)_5Br$

d'. Acetone + $2CH_2=CHCN + EtO^-$

e'. $CH_2=CHCN + CHCl_3 + (CH_3)_3COK$

f'. [cyclopentanone with =CH₂ at C-2] + $CH_3COCH_2COOEt + EtO^-$

12-4 Complete the following:

Reactant(s)	Name of reaction	Product
a. _____	Hofmann	$C_6H_5CH_2NH_2$
b. _____	Beckmann	$CH_3CH_2CONHCH_2C_6H_5$
c. _____	Wittig	$CH_3OOC(CH_2)_4CH=CHCOOCH_3$
d. $C_6H_5COCH_2CH_3 + CH_2O + CH_3NH_3Cl$	Mannich	_____
e. $CH_3CH_2NO_2 + C_6H_5CH=CHCOCH_3$	Michael	_____
f. $(EtOOC)_2CH(CH_2)_2CH(COOEt)_2$	Dieckmann	_____
g. $EtOOCCH_2COCOOEt + BrCH_2COOEt + Zn$	_____	_____
h. $(CH_3)_2C=O + Ba(OH)_2$	_____	_____
i. $C_6H_5COCH_3$ + ethyl succinate	Stobbe	_____
j. _____	Knoevenagel	$C_6H_5CH=C(CN)COOEt$
k. _____	Cannizzaro	$C_6H_5CH_2OH + C_6H_5COONa$
l. _____	Darzens	$\underset{\diagdown_O\diagup}{EtCH-CHCOOEt}$
m. _____	Mixed Claisen	Ethyl ethylmalonate

12-5 Suggest a simple chemical test to distinguish between each of the following:
  a. Ethyl acetoacetate and ethyl malonate
  b. 2,4-Hexanedione and 2,5-hexanedione
  c. Acetoacetic ester and ethyl diethylacetoacetate
  d. 2,6-Octanedione and 3,6-octanedione
  e. Ethyl valerate and ethyl diethylacetoacetate
  f. Methyl malonate and 2-pentanol
  g. Butyl malonate and diamyl ether
  h. Methyl malonate and malononitrile

12-6 Suggest a chemical procedure for the separation of each of the following mixtures:
  a. Butyl acetoacetate, hexanamide, tributylamine, octanoic acid
  b. 1-Octanol, 1-nitrobutane, 2-octanone, butyl hydrogen malonate

12-7 Write structures for the indicated number of products from the Claisen condensation in each of the following:
  a. Ethyl acetate and ethyl butyrate, 4
  b. Methyl formate and methyl acetate, 2
  c. Ethyl oxalate and ethyl propionate, 2
  d. Ethyl carbonate and butyronitrile, 1

e. Methyl malonate and CH_3COOCH_3, 4   *f.* $EtOOC(CH_2)_4COOEt$, 1

g. Ethyl isobutyrate and acetonitrile, 1   *h.* Methyl acetate and $CH_3OOCC{\equiv}CCOOCH_3$, 2

12-8 Primary and secondary alkyl halides give good yields in the alkylation of sodiomalonic and sodioacetoacetic esters. Why wouldn't you expect this to be true of tertiary and vinyl halides?

12-9 The labels to bottles containing the following chemicals were lost: 2-heptanol, 2-aminoheptane, 1-methylaminohexane, methyl cyclohexylmalonate, hexyl acetoacetate, hexanamide, ammonium hexanoate, valeronitrile. Describe how you would identify each compound.

12-10 Compound A did not evolve methane when treated with methylmagnesium iodide, nor did it react with acetyl chloride or phenylhydrazine. It was found to have a SE of 80. An acid was isolated from its saponification which had a NE of 66. On heating, this acid evolved carbon dioxide to give a new acid of NE 88. Suggest a structure for A.

12-11 Compound X, $C_5H_8O_2$, formed a dioxime, gave a ferric chloride test, and, when treated with hydrazine and potassium hydroxide, was converted into pentane. Suggest a structure for X.

12-12 Compound A, $C_6H_8O_5$, was soluble in water, acidic to litmus paper, and reacted with phenylhydrazine but not with Schiff's reagent. On heating, A was converted to a new acid, B, of NE 115 ± 2. When B was treated with NaOBr and the reaction made acidic, C was isolated. C possessed a NE of 60 ± 1. Suggest structures for A, B, and C.

12-13 Draw resonance forms of each of the following:
a. Enol form of 2,4-pentanedione   *b.* Enolate anion of methyl cyanoacetate

12-14 When nitromethane is alkylated according to the equation, $CH_3NO_2 \xrightarrow{\text{NaOH}} \xrightarrow{\text{RX}}$, two isomeric products may result. How can you account for this?

12-15 Arrange the following in order of increasing acid strength (tendency to ionize to give H^+):
a. Ethanol, hexane, acetic acid, acetoacetic ester, 3-acetyl-2,4-pentanedione.
b. Cyclopentadiene, cyclopentane, 1-butanol, nitromethane, formic acid, CF_3COOH.

12-16 Draw the indicated number of tautomers (consider only H to migrate) for each of the following:
a. Acetic acid, 3   *b.* Acetamide, 3
c. CH_3NO_2, 2   *d.* CH_3COCH_2NO, 5

12-17 The following barbiturates are employed as hypnotics and pain relievers. Given any necessary one-, two-, and three-carbon alcohols, phosgene, and malonic ester show a synthetic route for each (see Sec. 11-5*b*).
a. Veronal, 5,5-diethylbarbituric acid   *b.* Nembutal, 5-ethyl-5-(2-pentyl)barbituric acid

12-18 Ketones may be condensed with esters in a Claisen condensation although yields are not high (30 to 50 percent) since the ketones undergo the aldol condensation concurrently. What products would you expect in each of the following?
a. Acetone + ethyl acetate   *b.* Ethyl propionate + methyl isobutyl ketone

12-19 Compound A, $C_5H_8O_4$, was found to have a NE of 66. On heating, A evolved CO_2 and gave a new acid, B, of NE 88. Suggest two possible structures for A.

12-20 An acid of NE 72 did not react with bromine in the presence of phosphorus. On heating, it was converted to a new acid of NE 100. Give structures for the two acids.

12-21 A hydrocarbon, C_5H_8, decolorized a solution of Br_2 in CCl_4. On oxidation with permanganate, an acid, A, of NE 66 was isolated. A, on heating, gave a new compound, $C_5H_6O_3$, which on warming with water was changed to A. Suggest two possible structures for the hydrocarbon.

12-22 Given any necessary compounds containing three carbons or less, succinic acid, and triphenylphosphine as starting materials, show how you would make the following conversions using specific methods as part of your synthesis when indicated.
a. Ethyl butyrate to 4-heptanone   *b.* Ethyl adipate to 2-methyl-1-cyclopentanol
c. Ethyl succinate to $HOOCCH_2CH_2COCOOH$ *d.* Malononitrile to $(N{\equiv}C)_2CHCH(OH)C{\equiv}N$
e. n-C_3H_7COOEt to n-$C_3H_7CH(NHCH_3)CH(Et)COOH$ *f.* Propene to $CH_3COCH_2COCOOEt$
g. CH_3CHO to β-hydroxybutyric acid   *h.* Pimelic acid to *cis*-1,2-cyclohexanediol
i. $CH_2{=}CHCHO$ to tetramethylenediamine   *j.* $CH_2{=}CHCN$ to γ-nitrobutyronitrile
k. $CH_2{=}CHCN$ to $(CH_3CO)_2CHCH_2CH_2CN$ *l.* CH_3COCH_3 to $CH_3COCH{=}CH_2$

m. Propene to
$$(CH_3)_2CHC{-}C(CH_3)_2$$

(structure with fused barbiturate-like ring: $N{=}$, $C{=}O$, N, H)

n. Ethyl adipate to (structure: two cyclopentanone rings joined by CH_2CH_2, each with O)

179

o. Ethyl acetoacetate to $(CH_3CO)_3CCH_3$

p. Ethyl butyrate to $CH_3CH_2CH_2COCOOH$

q. Acetone to 3-methyl-1,3-butanediol (Reformatsky)

r. 2-Pentanone to 3-methyl-2-hexanone (Darzens

s. CH_3CH_2COOEt to 4-methyl-6-hepten-3-one

t. Butanal to $OHCCH(Et)CH_2CH_2COOH$
(Michael reaction using an enamine)

u. $EtOOCCH_2CCH_2COOEt$ to
(with O below the central C)

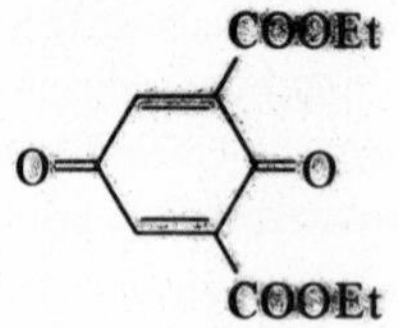

v. 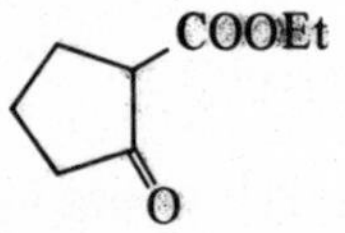 (cyclopentanone with COOEt) to 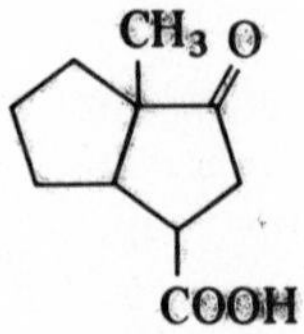 (Stobbe)

w. 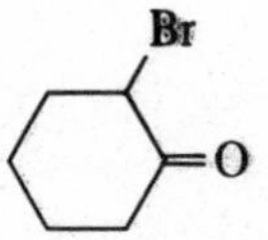 (2-bromocyclohexanone) to (methylene cyclohexane with $CH_2CH_2CH_3$)
(Organoboranes as alkylating agents)

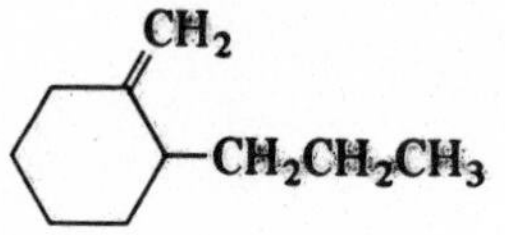

x. (2-bromocyclohexanone) to (Et decalone)
(Robinson annulation using a Mannich base)

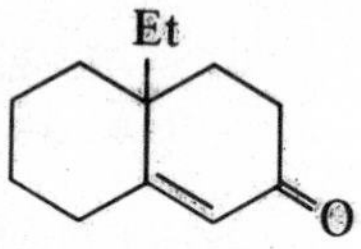

12-23 Show how to prepare cinnamic acid, $C_6H_5CH=CHCOOH$, by each of the following reactions starting from benzaldehyde and any other necessary compounds.
a. Claisen-Schmidt   *b.* Knoevenagel   *c.* Reformatsky
d. Perkin   *e.* Wittig   *f.* Claisen ester

12-24 When the sodium salt of an active methylene compound is treated with I_2, a so-called coupling reaction occurs to join two molecules together. This probably involves first reaction with I_2, followed by an S_N reaction. Illustrate the mechanism of this reaction and show how it can be used to prepare (a) 2,5-hexanedione and (b) a,a'-diethylsuccinic acid.

12-25 Suggest a mechanism to account for the following transformations:

a. 2-Carboethoxy-1-cyclohexanone $\xrightarrow{NaOH, \Delta}$ sodium ethyl pimelate

b. $CH_2(CN)_2 + CH_2O \xrightarrow{OH^-} (NC)_2CHCH_2CH(CN)_2$

c. $CH_3COCH=CHCH_3 + CH_3-$(cyclohexanone) $\xrightarrow{EtNOa}$ (octalone with H_3C, CH_3)

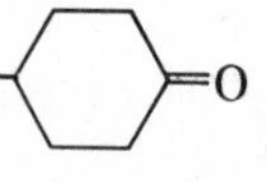

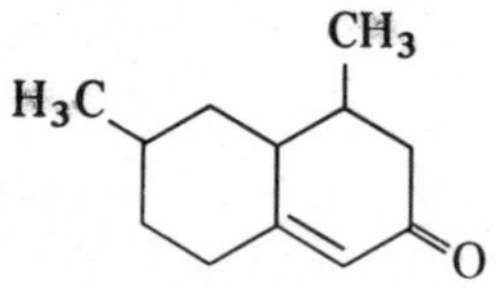

d. $CH_2=CHCH_2COOEt \xrightarrow{EtONa} CH_3CH=CHCOOEt$

e. $CH_3(CH=CH)_2COOEt + CH_2(COOEt)_2 \xrightarrow{EtONa} CH_3CHCH_2CH=CHCOOEt$
with $CH(COOEt)_2$ below

f. $CH_3COCH_3 + Br_2 \xrightarrow{OH^-} CH_3COCH_2Br$

The rate at which bromination occurs is directly proportional to the concentration of acetone, but independent of the concentration of bromine. This also applies to part *g* below.

g. $CH_3COCH_3 + Br_2 \xrightarrow{H^+} CH_3COCH_2Br$

h. $ClCH_2CH_2CH_2CN \xrightarrow{OH^-, \Delta} \xrightarrow{H^+} C_4H_6O_2$ (an acid)

i. 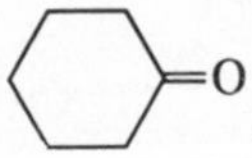 (cyclohexanone) $=O + (EtO)_2CO \xrightarrow{EtONa}$ (cyclohexyl)$-COOEt$ + (cyclohexanone)$-COOEt$
Minor   Major

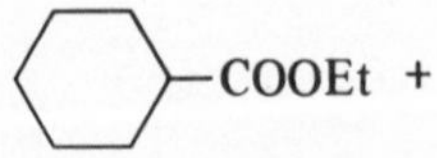

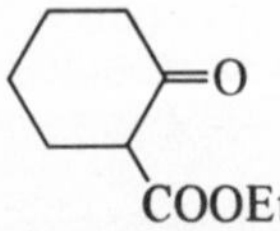

180

j. $(CH_3)_2CO + BrCH_2CH=CHCOOEt \xrightarrow{Zn} \xrightarrow{H_2O} (CH_3)_2C(OH)CH_2CH=CHCOOEt$

k. $HOCH_2C(CH_3)_2COOH \xrightarrow{H^+} CH_3CH=C(CH_3)COOH$

l. Tartaric acid $\xrightarrow{H^+}$ pyruvic acid

m. $EtOOCCH_2CH_2COOEt + CH_3COCH_3 \xrightarrow{EtONa}$

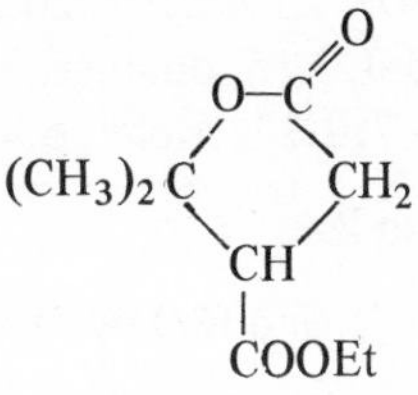

n. $CH_2=CBrCOOEt + CH_2(COOEt)_2 \xrightarrow{EtONa} \xrightarrow{H^+}$

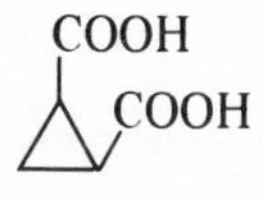

o. $CH_3CH=CHCHO \xrightarrow{OH^-} CH_3CHO$

p. $EtOOCCH_2COCH=CH_2 + $ 2-methyl-1-cyclohexanone $\xrightarrow{EtONa}$

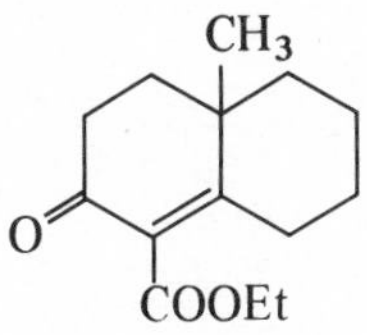

q. 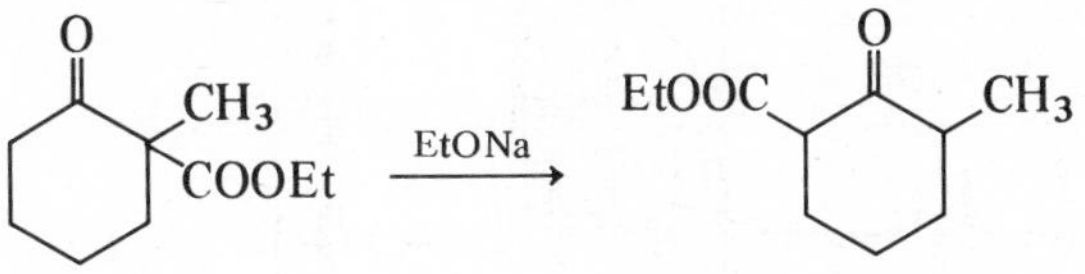 $\xrightarrow{EtONa}$

r. $C_6H_5CH=CHCOCH=CHC_6H_5 + N\equiv CCH_2COOMe \xrightarrow{CH_3ONa}$

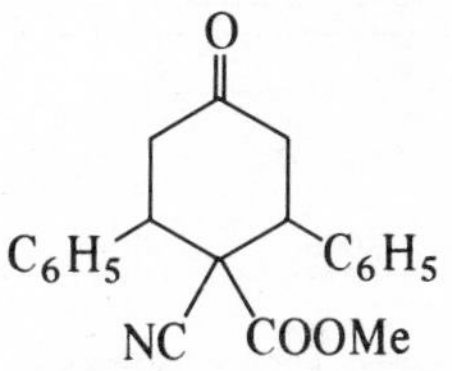

s. 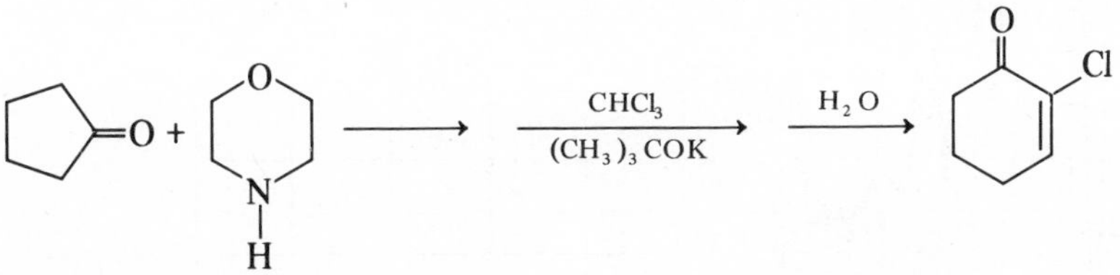 $\xrightarrow{CHCl_3}{(CH_3)_3COK} \xrightarrow{H_2O}$

t. 2-Methyl-1-cyclopentanone $+ HC\equiv CCOCH_3 \xrightarrow{Na} C_{11}H_{14}O \qquad uv_{max} = 224$ nm

12-26 Compound X (MW 130) can exist in two tautomeric forms. The nmr spectrum of the predominant tautomer shows the following peaks (δ, multiplicity, peak integration units): 1.2, triplet, 18.0; 2.2, singlet, 17.9; 3.4, singlet, 6.1; 4.2, quartet, 5.9. Suggest a structure for X.

12-27 Compound A contains 53.1% C, 6.2% H, 12.4% N, and 28.3% O. Its nmr spectrum contains a triplet ($\delta = 1.3$), a singlet ($\delta = 3.6$), and a quarter ($\delta = 4.25$) with a peak area ratio of 3:2:2, respectively, and ir bands (selected) at 4.45 μm (2247 cm^{-1}) and 5.75 μm (1739 cm^{-1}). Suggest a structure for A.

12-28 Compound B (MW 88) shows only a singlet in its nmr spectrum at $\delta = 4.4$ and ir bands (selected) at 5.5 μm (1818 cm^{-1}) and 5.65 μm (1770 cm^{-1}). Upon warming with aqueous acid B yields CO_2 and compound C (MW 62). Suggest structures for B and C.

12-29 Compound D (MW 97) can exist in tautomeric forms. The nmr spectrum of liquid D shows the following singlets (δ, peak integration units); 1.8, 34.0; 2.0, 6.0; 3.6, 2.0; 5.6, 5.7; 15.2 (broad), 5.7. Suggest a structure for D. Determine the percentages of each tautomer present and assign the nmr peaks to specific hydrogens.

12-30 Compound M (MW 160) when treated with EtONa and bromine-containing compound N (MW 171) gives compound O ($C_{14}H_{18}O_4$). Compound O upon heating with aqueous acid gives compound P (MW 150). The following are nmr spectral data for compounds M through P (δ, multiplicity, peak integration units):

Compound M   1.2, triplet, 24.0; 3.3, singlet, 16.1; 4.15, quartet, 15.9.
Compound N   4.25, singlet, 8.0; 7.2, singlet, 20.0.
Compound O   1.1, triplet, 15.0; 3.15, doublet, 5.0; 3.55, triplet, 2.5; 7.2, singlet, 12.5
Compound P   2.55, triplet, 6.7; 2.85, triplet, 6.6; 7.2, singlet, 16.0; 12.0, singlet, 3.2

Suggest structures M, N, O, and P.

12-31 Compound Y contains only C, H, and O (MW 188). Its nmr and ir spectra are shown below. Suggest a structure for Y.

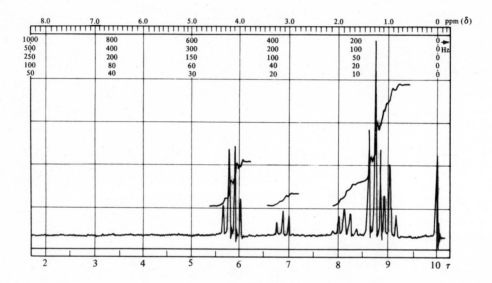

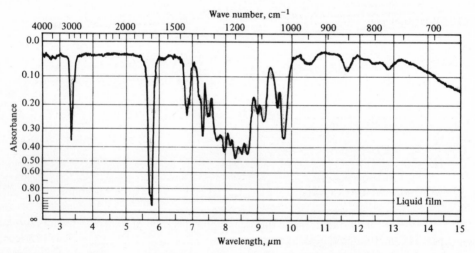

Figure P12-31. **Ir and nmr spectra for Prob. 12-31.** *(From R. A. Silva and R. Spenger, "The Quantitative Organic Analysis Spectra Kit," McGraw-Hill Book Company, New York, 1969. By permission of the publishers.)*

13
stereoisomerism

Up to this point several types of isomerism (different compounds having the same molecular formula) have been encountered:

Chain isomers		Position isomers	

CH₃CHCH₂CH₂CH₃
    $\|$
    CH₃
	CH₃CH₂CHCH₂CH₃
	       $\|$
	       CH₃
		CH₃CH₂CH₂OH	CH₃CHCH₃
			    $\|$
			    OH

2-Methylpentane	3-Methylpentane	1-Propanol	2-Propanol

Functional group (or constitutional) isomers		Tautomers (dynamic isomerism)	
$CH_3CH_2CH_2OCH_3$	$CH_3CH_2CH_2CH_2OH$	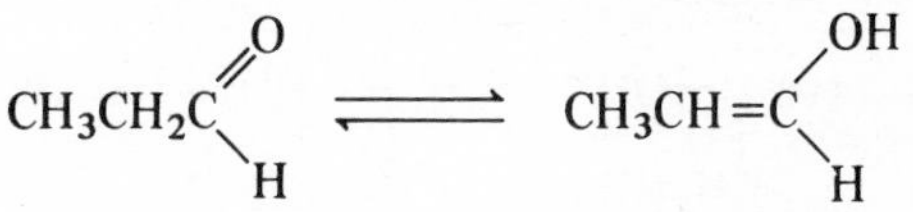 $CH_3CH_2C{\overset{O}{\underset{H}{}}}$ ⇌ $CH_3CH=C{\overset{OH}{\underset{H}{}}}$	
1-Methoxypropane	1-Butanol	Keto form	Enol form

A fifth type of isomerism has also been mentioned before, namely, "stereoisomerism." Stereoisomers differ only in the arrangement in space of atoms or groups of atoms. Stereoisomers can be further divided into geometric isomers and optical isomers (review Secs. 2-8, 2-9, and 3-4).

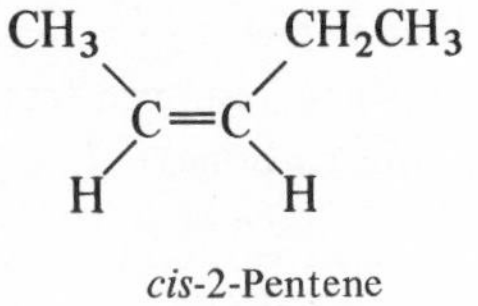

cis-2-Pentene
(Z)-2-Pentene

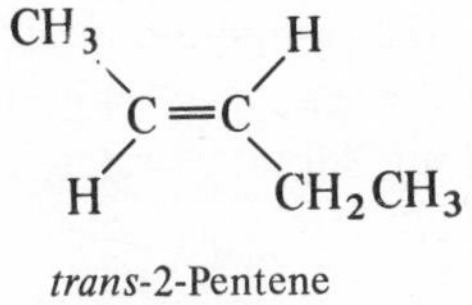

trans-2-Pentene
(E)-2-Pentene

Geometric isomers

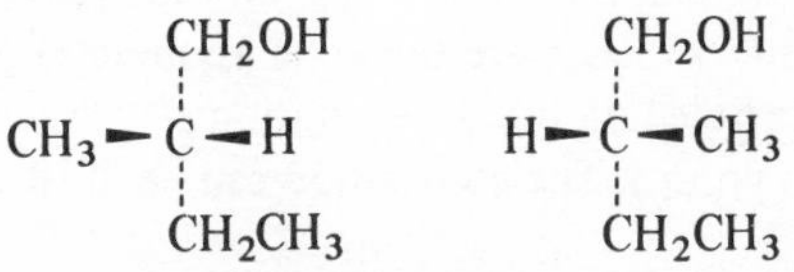

(S)-(−) and (R)-(+) 2-Methyl-1-butanol
Optical isomers

An interesting example of geometric isomerism can be observed in the oximes. The oxime of an aldehyde or an unsymmetric ketone can exist in two modifications termed *syn* and *anti*, analogous to cis and trans. For aldoximes these terms refer to the arrangement of OH and H, and for ketoximes to the OH and the smaller R group in the parent unsymmetric ketone.

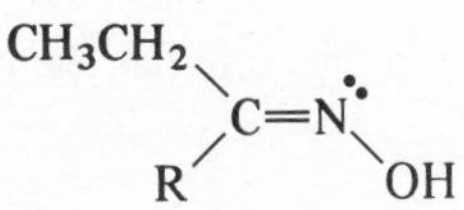

Syn

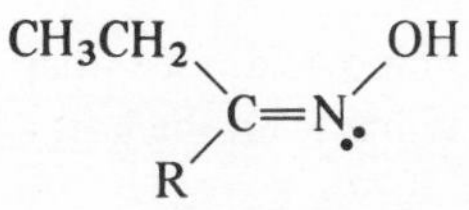

Anti

R=H, *syn*- and *anti*-propionaldoxime

R=CH₃, *syn*- and *anti*-methylethylketoxime

183

Using the IUPAC E and Z descriptions (see Sec. 2-8) the above syn and anti isomers would be (E) and (Z) isomers, respectively.†

13-1 Measurement of Optical Activity

Enantiomers are distinguished by the direction they rotate plane-polarized light (d or positive isomers rotate such light to the right while l or negative isomers rotate it to the left). The amount of optical rotation is measured by a polarimeter in which the enantiomer is placed and plane-polarized light (obtained by passing ordinary light through a Nicol prism) is allowed to pass through the sample. Optical activity is reported as *specific rotation* $[\alpha]$, which is defined by the equation, $[\alpha]_{\lambda}^{T} = \alpha/(l \times C)$, where T is the temperature of the sample (°C), λ is the wavelength of the plane-polarized light used (usually the D line of sodium which has λ = 5870 Å), α is the observed rotation in degrees, l is the length of the sample tube in decimeters, and C is the concentration of the sample in grams per milliliter. Values of $[\alpha]$ are reported as being measured "neat" (pure liquid enantiomer) or in a given solvent at a specified concentration, for example, $[\alpha]_{D}^{20}$ = +41.0° (5% CHCl$_3$).

13-2 Specification of Configuration in Stereoisomers

Each stereoisomer has its own unique arrangement of atoms which is called its configuration. There are two methods of specifying the configuration about an asymmetric carbon atom: the Fischer convention and the Cahn-Ingold-Prelog system. The latter system has been extended to specify configuration in geometric isomers (E, Z descriptors, see Sec. 2-8).

a. *Fischer Convention* In this system used primarily for sugars (see Chap. 21) optical isomers are compared structurally to the simplest sugar glyceraldehyde in its two enantiomeric forms. Using Fischer projection formulas, the D and L isomers of glyceraldehyde are shown below. Note that the D or L does not refer to the sign of rotation of light which is indicated by a (+) or (−). As we shall see, D or L isomers may be either positive or negative depending on the particular compound involved.

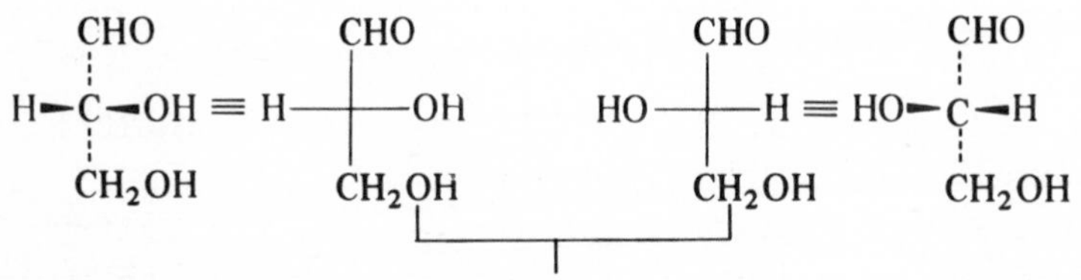

D(+)-Glyceraldehyde   Fischer projection formulas   L(−)-Glyceraldehyde

A sugar isomer is said to belong to the D series if, when it is written with the aldehyde-containing group (or a related group such as COOH) attached to the asymmetric carbon at the top of the molecule in the Fischer projection formula and when the other carbon-containing group attached to the asymmetric carbon is written at the bottom, the OH group is on the right side. Its enantiomer must then belong to the L series. When a compound has more than one asymmetric carbon atom, the compound is said to belong to the D or L family or series based on the configuration at the highest numbered asymmetric carbon atom (farthest removed from the CHO group). The convention can be illustrated by the four aldotetroses (sugars with four carbons).

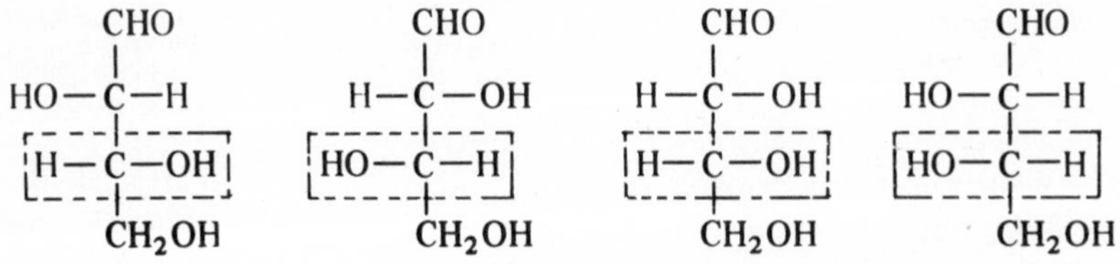

I	II	III	IV
D(−)-Threose	L(+)-Threose	D(−)-Erythrose	L(+)-Erythrose

Using Fischer projection formulas as shown below for amino acids (see Chap. 20) a D-amino acid has the NH$_2$ group at the α *carbon* on the right while an L-amino acid has it on the left.

† For a discussion of IUPAC stereochemistry nomenclature see *J. Organic Chemistry*, vol. 35, p. 2849, 1970

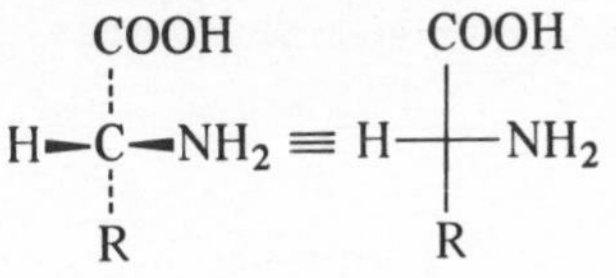
COOH
H
NH2
R
COOH
H
NH2
R

D-Amino acid

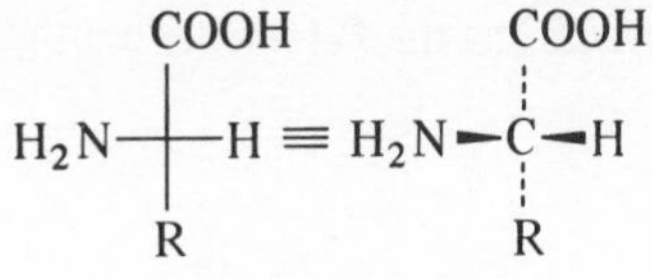
COOH
H2N
H
R
COOH
H2N
H
R

L-Amino acid

b. *Cahn-Ingold-Prelog System* In this system the groups attached to an *asymmetric carbon atom* (C^*) are assigned an order of priority (let *a, b, c, d* represent the priority of groups with $a > b > c > d$) according to the following sequence rules.

1. Atoms attached to C^* with higher atomic number precede those with lower atomic number. If two or more atoms attached directly to C^* have the same atomic number, then proceed out to the next atoms, always working out toward atoms of higher atomic number until a decision can be reached, for example, $-CH_2CH_2OCH_3$ precedes $-CH_2CH_2OH$. In this procedure if double or triple bonds are encountered, they will be treated as shown below for a $>C{=}O$ group (1), a $-C{\equiv}CH$ group (2), a $-C{\equiv}N$ group (3), and a phenyl (a benzene ring with one hydrogen replaced, C_6H_5) group (4).

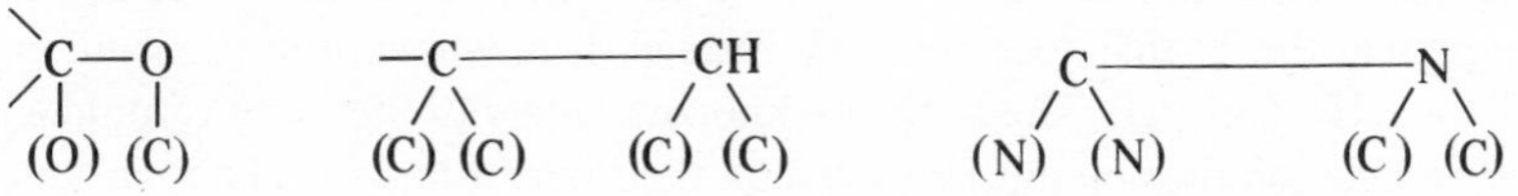
C—O
(O) (C)
—C
CH
(C) (C)
(C) (C)
C
N
(N) (N)
(C) (C)

(1) (2) (3)

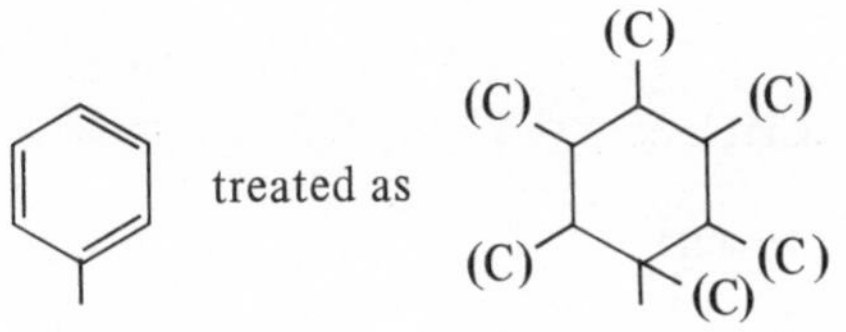
treated as
(C)
(C)
(C)
(C)
(C)
(C)

(4)

Examples $-CHO$ has priority over $-CH_2OH$ but not over COOH.

$-CH{=}CHCH_3 > -CH(CH_3)CH_2CH_3 > -CH{=}CH_2 > -CH_2CH{=}CH_2$

$-C{\equiv}CH >$ phenyl $(C_6H_5) > CH_3\underset{|}{C}{=}CH_2 >$ cyclohexyl

2. When two groups attached to a C^* differ only in their geometric configuration, cis has priority over trans, e.g.,

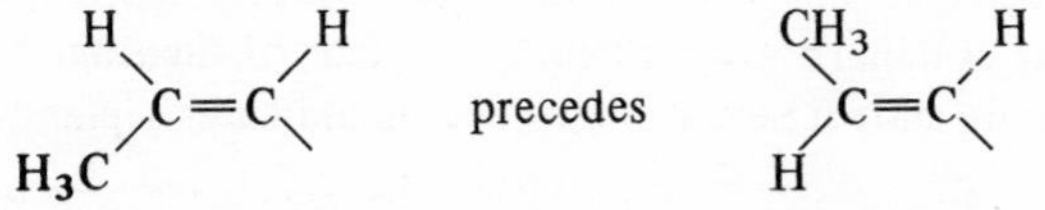
H
H
C=C
H3C
precedes
CH3
H
C=C
H

3. When two groups differ only in that they are enantiomeric, R has priority over S, e.g.,

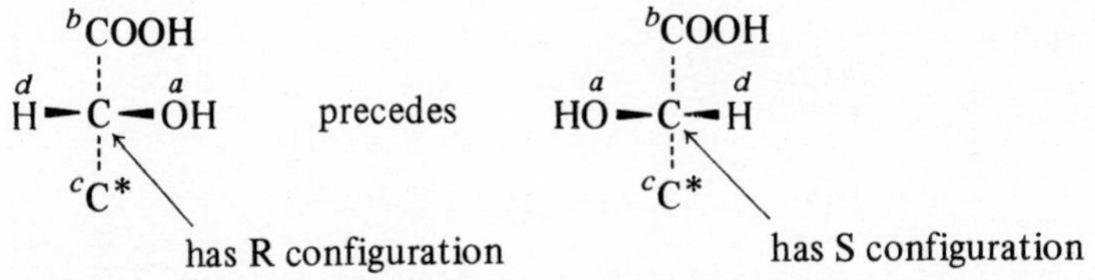
bCOOH
d
a
H
C
OH
cC*
precedes
bCOOH
a
d
HO
C
H
cC*
has R configuration
has S configuration

See below for the assignment of R and S.

4. When two groups differ only isotopically, higher mass number precedes lower mass number. Thus, $T > D > H$ and $^{18}O > ^{17}O > ^{16}O$.

Once the order of priority of groups is decided by the above rules, then one views the molecule *down* the bond from the C^* to the lowest priority group (*d*). From this vantage point one traces an arc from groups *a* to *b* to *c*. If this arc is clockwise, then C* has the R (rectus) configuration, if counterclockwise, the configuration is S (sinister).

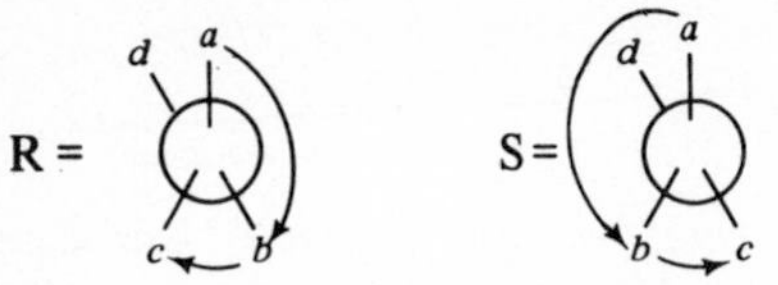

$$R = \qquad S =$$

13-3 Compounds with Two or More Asymmetric Carbon Atoms

a. *Open-Chain Compounds* If an open-chain compound contains more than one *different* asymmetric carbon atom, the maximum number of stereoisomers is given by 2^n, where *n* is the number of these atoms. The isomers that are not enantiomers are known as *diastereomers*. Enantiomers have identical physical properties except with respect to their effect on plane-polarized light, but diastereomers have different properties. The four aldotetroses shown in Sec. 13-2*a* are examples of enantiomeric vs. diastereomeric relationships: enantiomers – I,II; III,IV; diastereomers – I,III; I,IV; II,III; II,IV. When two or more similar or like asymmetric carbon atoms are present in a compound, the 2^n rule does not hold. Instead there will exist *dl* pairs and one or more compounds that are optically inactive (achiral). The latter are known as *meso isomers* and contain a plane of symmetry. Note that meso compounds can thus be superimposed on their mirror images. The four stereoisomers of 2,3,4-pentanetriol illustrate these points (the 2^n rule would predict eight stereoisomers for this compound):

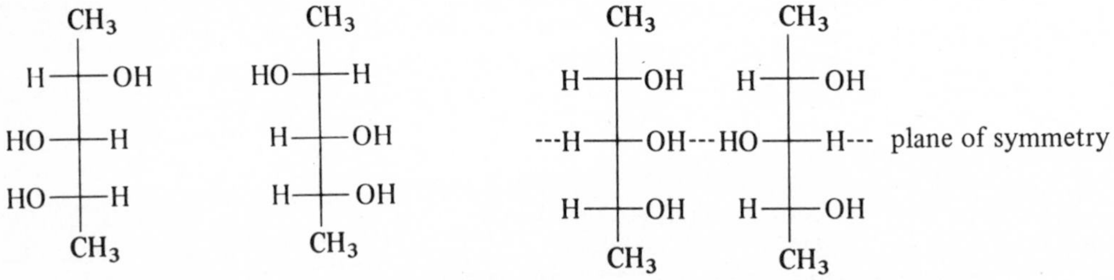

CH₃	CH₃	CH₃	CH₃
H—OH	HO—H	H—OH	H—OH
HO—H	H—OH	---H—OH---	HO—H--- plane of symmetry
HO—H	H—OH	H—OH	H—OH
CH₃	CH₃	CH₃	CH₃
dl pair		Meso	Meso

C-3 in the *dl* pair is asymmetric because it has four different groups attached to it; but because this carbon lies in a plane of symmetry for both meso structures, C-3 is referred to as a pseudoasymmetric carbon atom.

b. *Cyclic Structures*
   1. Determining possible stereoisomers. For the purpose of drawing all the possible stereoisomers, optical and geometric, for substituted cyclic compounds (see Sec. 3-4) it is best to consider the ring as a flat structure (although it rarely is — see below) and the substituents as being above or below the plane of the ring. The four possible stereoisomers for 1,2,4-trichloropentane are shown below (a dashed line indicates a plane of symmetry in the meso isomers):

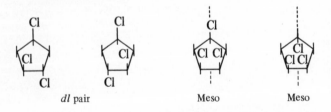

   dl pair   Meso   Meso

   2. Conformational analysis in substituted cyclohexanes. The chair form for cyclohexane (see Sec. 3-36) and its derivatives is normally the more stable. Substituents on the cyclohexane ring prefer to be in the equatorial position. Consider the chair conformers of methylcyclohexane (see Fig. 13-1). When CH₃ is equatorial there is much less 1,3-nonbonded repulsion between the CH₃ and the axial H atoms on C-3 and C-5 than when CH₃ is axial as shown by the Newman projection formula for I. Considering the interconversion of conformers, such as I and II, as an equilibrium process, a quantitative measure of the preference of a substituent for the equatorial position can be obtained from the difference in free energy (ΔG) between the

186

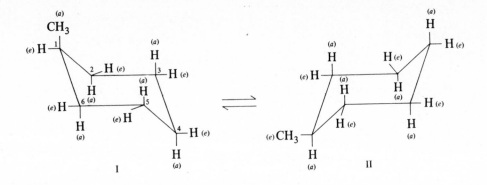

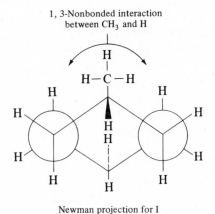

1, 3-Nonbonded interaction
between CH₃ and H

Newman projection for I

Figure 13-1. Chair conformers of methylcyclohexane.

axial and equatorial conformers ($\Delta G = -2.3\, RT \log K$, where K is the equilibrium constant, R is the gas constant $= 2$ cal-mol^{-1} deg^{-1} and T is temperature in kelvins. The larger ΔG is (i.e., a greater negative value), a greater percentage of the more stable equatorially substituted conformer will be present at equilibrium. The ΔG values for several substituted cyclohexanes are given below along with the percent of equatorial isomer present at equilibrium *at 25°C.*

Group	Cl, Br, I	OH(OR)	Me	Et
$-\Delta G$, kcal/mol	0.5	0.7	1.7	1.8
~% equatorial isomer	70	77	95	95–96

Group	iso-Pr	*t*-Bu	C₆H₅—(phenyl)
$-\Delta G$, kcal/mol	2.1	5.5	2.7
~% equatorial isomer	96–97	99.99	99

cis-1,2-Dimethylcyclohexane (see Fig. 13-2) has one methyl axial and the other equatorial while *trans*-1,2-dimethylcyclohexane has both methyl groups equatorial in the most stable conformation (see Prob. 13-21 for *cis*- and *trans*-1,3- and 1,4-dimethylcyclohexanes).

In compounds having two or more cyclohexane rings fused together, such as *cis*- and *trans*-decalin and steroids (see Fig. 13-3), the more stable conformation for the cyclohexane rings is the chair. Certain bicyclic compounds, such as [2.2.1] bicycloheptane, and *cis*-1,4-di-*t*-butylcyclohexane, are more stable with the cyclohexane ring in the boat form.

cis-1, 2

trans-1, 2

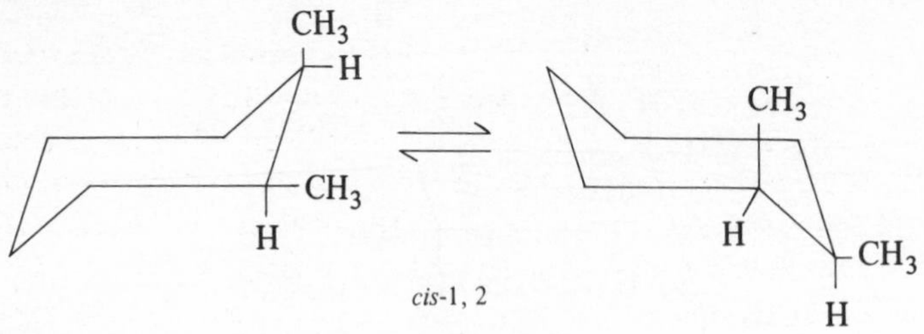

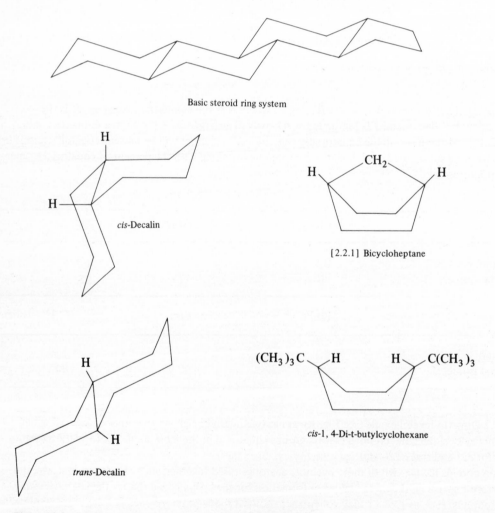

Figure 13-2. Chair conformations of *cis* and *trans*-1,2-dimethylcyclohexane.

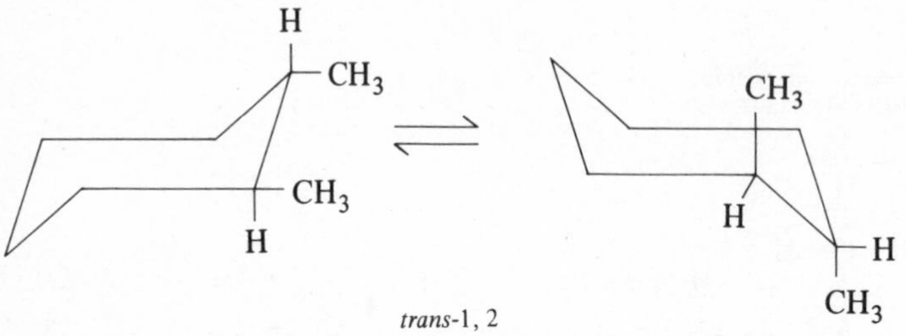

Basic steroid ring system

cis-Decalin

[2.2.1] Bicycloheptane

trans-Decalin

cis-1, 4-Di-t-butylcyclohexane

Figure 13-3. Some examples of compounds with cyclohexane rings.

13-4 Molecular Asymmetry

Any time a molecule can pass "freely" through one or more conformations (i.e., where the potential energy barrier to rotation is low as in ethane or butane) which have a plane (or point) of symmetry, then that compound will not be optically active. This is why it is better to ignore conformations when attempting to draw all the possible stereoisomers (see Sec. 13-3b-1). However, if the potential energy barrier for passing through a conformation which has a plane or point of symmetry is high enough (usually 16 kcal/mol or more at room temperature), then certain rotational conformers may indeed rotate plane-polarized light. Such asymmetry or optical activity which is due to restricted rotation about some bond in the molecule is called *molecular asymmetry*. Some different types of compounds that can exhibit molecular asymmetry are shown below.

1. Allenes. Allene, $CH_2=C=CH_2$, has two CH_2 groups which lie in planes at right angles to one another (compare relative positions of H atoms in allene to those of tetrahedral methane). Substituted allenes of the type $ABC=C=CAB$ (where $A \neq B$) can exist in enantiomeric forms.

2. Spiranes. An arrangement similar to that of the allenes is possible with bicyclic compounds called *spiranes* which can exist in optically active forms without having any asymmetric carbon atoms.

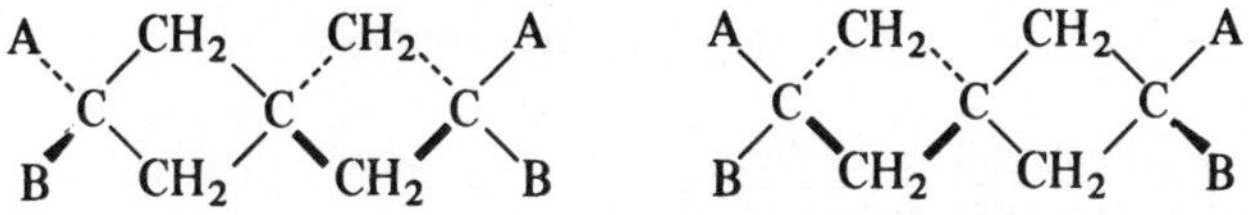

3. Substituted biphenyls. Biphenyl compounds which have appropriate substituents in the ortho positions in each ring (positions 2,6 and 2',6') may exhibit optical activity if the substituents are large enough to prevent free rotation about the 1,1' bond (thus allowing the rings to become coplanar) and if the ortho substituents in any one ring are not identical, that is, $A \neq B$. Common groups in order of their increasing ability to restrict rotation in such compounds include OH, COOH, NH_2, CH_3, Cl, NO_2, Br, and I.

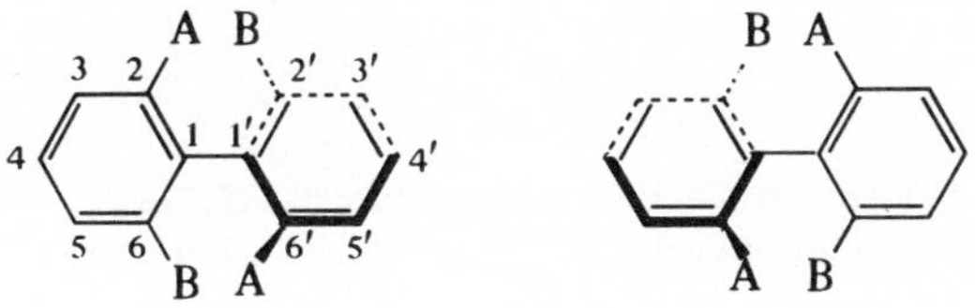

dl pair

Certain types of bridged aromatic compounds may also exhibit molecular asymmetry.

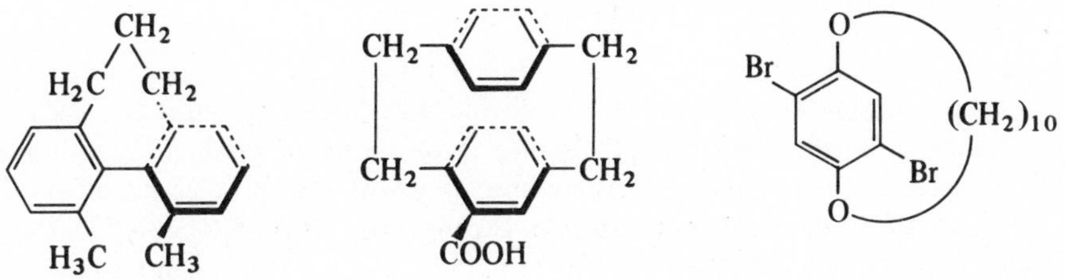

13-5 Production and Resolution of Racemic Mixtures

a. *Racemization* The process of converting a pure enantiomer into equal amounts of the *d* and *l* isomers, i.e., a racemic mixture, is called *racemization*. In some cases racemization of asymmetric molecules may be induced easily while for others it may be caused only with difficulty, but the process usually involves one or more asymmetric carbon atoms. Normally this can be brought about by converting the asymmetric carbon atom into a trigonal sp^2 hybridized state such as in the formation of a flat carbonium ion which may form a bond from either side to give rise to a racemic mixture. Racemic mixtures may result if instead the asymmetric carbon is converted to a flat carbanion or a free radical intermediate. Even if these intermediates are tetrahedral (see Sec. 13-6c-2) they can be expected to behave like ammonia and its derivatives in which N atoms are constantly inverting $\geq$N: $\rightleftharpoons$:N$\leq$. Thus reactions of tetrahedral carbanions or free radicals may give racemic mixtures.

b. *Resolution* Separation of a racemic mixture into its *d* and *l* isomers is called resolution. Three common methods are available for the resolution of enantiomers.

1. Spontaneous crystallization. This was the method that enabled Pasteur to make the first separation of a *dl* pair into its component parts in 1848. The method has mainly historical significance because it is rarely encountered.

2. Reaction with optically active reagents. In all reactions of a *dl* pair with an optically *inactive* compound, each enantiomer will react at the same rate and the product of the reaction will be optically inactive. However, if this same *dl* pair were to react with an optically active compound, the spatial relationships between the reactant and each enantiomer would be different, and different reaction rates could be expected. The most common illustration of this type occurs among enzymatic reactions. Enzymes, either isolated or in organisms, are optically active and may cause preferential synthesis or destruction of one or two enantiomers.

$$\text{DL-Tartaric acid} \xrightarrow{\text{Penicillium glaucum}} \text{D-tartaric acid}$$

In this case, the L isomer is destroyed.

3. Diastereomer formation. Synthetic conversion of enantiomers into diastereomers will produce compounds having different physical properties, particularly solubility. An example is salt formation.

$$\begin{array}{l} \text{CH}_3\text{CH(OH)COOH} \\ \text{DL-Lactic acid} \end{array} + d \text{ base} \longrightarrow \begin{array}{l} \text{D-CH}_3\text{CH(OH)COO}^- \ d \text{ base}^+ \\ \overline{\text{L-CH}_3\text{CH(OH)COO}^- \ d \text{ base}^+} \end{array}$$

Note that the two salts formed are not enantiomers. They are stereoisomers that are not mirror images and hence are called *diastereomers*. They may be separated by fractional crystallization, and the pure D and L-lactic acid may be obtained by displacing the weak *d* base with a strong base such as NaOH. The *d* base can then be separated from the mixture (e.g., extract with ether). The lactic acid isomers can then be generated from their sodium salts with strong acid and purified by extraction or crystallization.

13-6 Stereochemical Considerations in Organic Reactions

Looking at the stereochemical properties of the products in a given reaction, we say a *stereoselective* synthesis is one in which only one of two stereoisomers is formed predominantly or exclusively (or where two or more stereoisomers greatly predominate over other stereoisomers). If in a reaction one stereoisomer gives one product while a different product is obtained if another stereoisomer is used, the reaction is said to be *stereospecific*. The following discussion of the stereochemistry of various reactions is divided according to the basic type of reaction involved (e.g., addition, elimination, and substitution), the type of mechanism involved (for example, S_N1, S_N2, $E1$, $E2$), and whether or not a migration of a group has occurred (e.g., rearrangement reactions).

a. *Elimination Reactions*

1. The $E2$ reaction, e.g., dehydrohalogenation of alkyl halides (see Sec. 4-2a-1), proceeds through trans elimination. This term does not imply anything about the geometry of the alkene product but only that the atoms or groups being eliminated are in a preferred trans conformation with respect to each other in the transition state. The stereospecific $E2$ dehydrohalogenation of the meso isomer (R,S) of 3,4-dichlorohexane gives (E)-3-chloro-3-hexene exclusively while either the R,R or S,S isomers give (Z)-3-chloro-3-hexene.

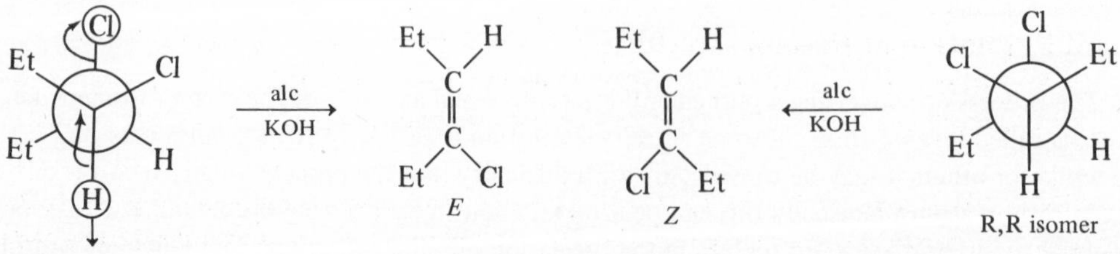

Meso or R,S isomer

R,R isomer

See Prob. 13-25*e* and *f* for $E2$ reactions of cyclohexane derivatives.

190

2. $E1$ reactions (see Sec. 4-2a-2) are not stereospecific because a planar carbonium ion is formed in the rate-determining step.
3. Cis elimination reactions usually involve pyrolysis reactions, e.g., the pyrolysis of acetates, which are believed to proceed through a six-membered cyclic transition state.

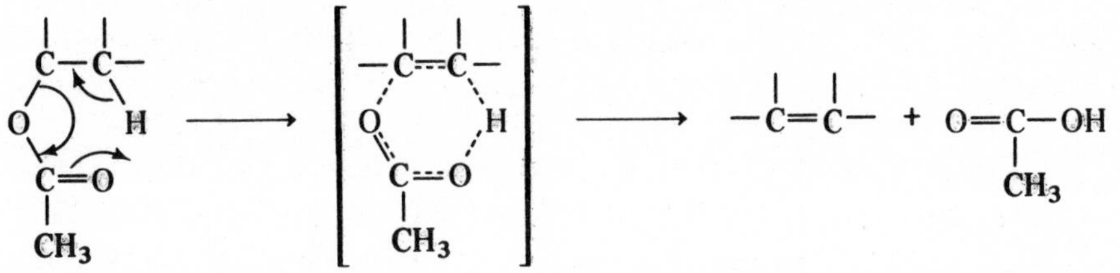

b. *Nucleophilic Substitution*   Reactions at an asymmetric carbon atom in which the configuration of that atom is changed are said to occur with *inversion* of configuration (sometimes called *Walden inversion*). This can be indicated by a circle on the reaction arrow (—⊖→). Reactions in which the configuration at an asymmetric carbon atom does not change are said to occur with *retention* of configuration.

1. S_N2 reactions (see Sec. 5-5) occur stereospecifically with inversion of configuration. In going from I $\longrightarrow$ II, the configuration at the asymmetric carbon (hereafter designated C*) has been inverted.

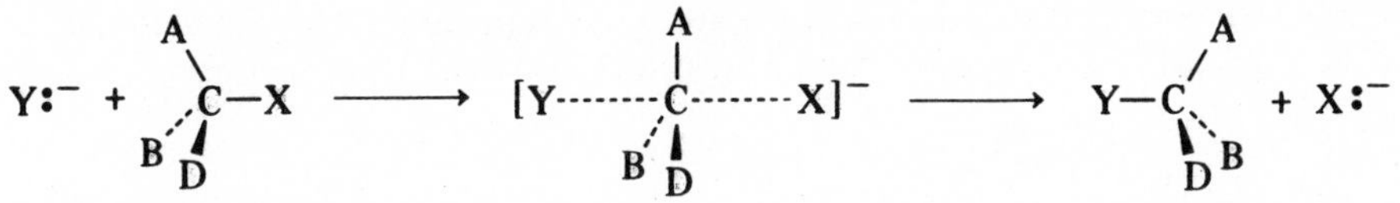

Remember that just because the sign of optical rotation (*d* or *l*) changes during a given transformation, it does not necessarily mean that an inversion of configuration has occurred since the sign of rotation of a compound does not alone determine its absolute configuration.

2. In S_N1 reactions at C* (see Sec. 5-5) a carbonium ion is produced which may be attacked equally well from both sides leading to racemization at C*. This occurs provided the carbonium ion is truly a free one, i.e., the leaving group is far enough removed so as not to block the incoming nucleophile on that side thus leading to some inversion of configuration.

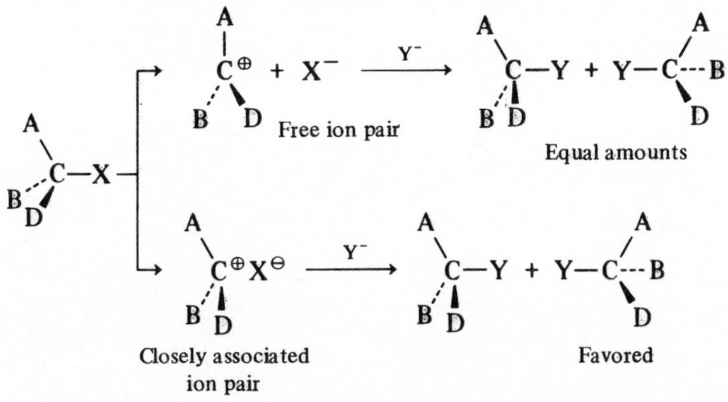

3. S_Ni reactions (see Sec. 6-4b-2) at C* proceed with retention of configuration. Note that PX_3 and PX_5 on R*OH usually give inversion of configuration.
4. Neighboring group participation (NGP) is involved in certain substitution reactions which occur with retention of configuration at C* and which proceed at a rate faster than expected. Such reactions involve the presence of a group, usually with an unshared pair of electrons (for instance, $-OCR$, $-COOR$, $-COO^-$,

$-OH, -O^-, -SH^-, -S^-, -OR, -NR_2, Br, I, \underset{\displaystyle}{>}C:^-$), situated so that it may participate in pushing the departing group out by formation of a cyclic intermediate (usually a three-, five-, or six-membered ring) which is then opened up by nucleophilic attack. The retention of configuration results because both the formation and opening of the cyclic intermediate proceed with inversion at C* as shown below.

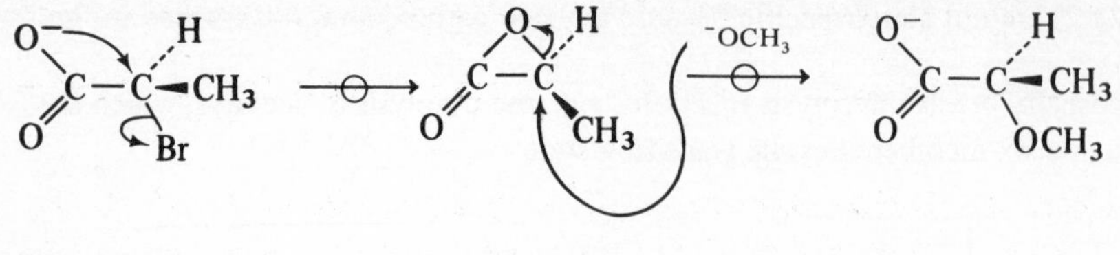

c. Electrophilic Substitution

1. S_E2

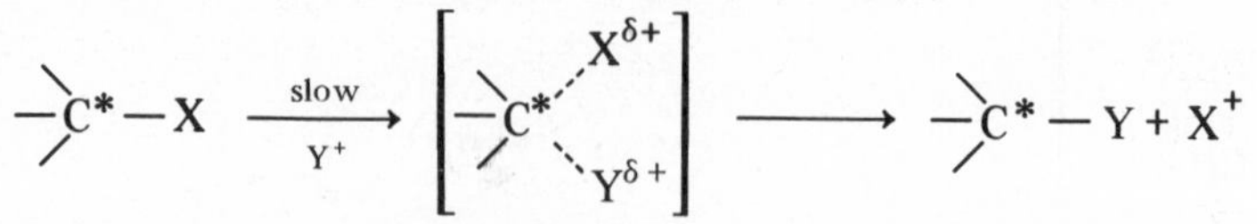

Front-side attack by Y^+

(See Prob. 13-35 for the stereochemistry of this reaction.)

2. S_E1

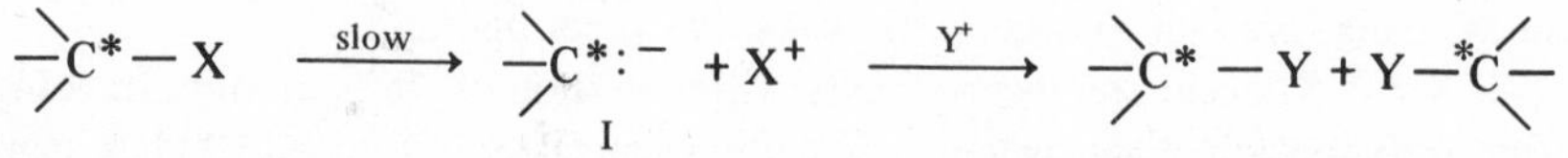

If the carbanion I is planar then racemization should occur at C^*. If I is pyramidal then retention of configuration would be expected unless I undergoes rapid inversion in which case racemization would be expected $\left(-\overset{*}{\underset{}{C}}:^- \underset{\longleftarrow}{\overset{fast}{\longrightarrow}} \ ^-: \overset{*}{C}-\right)$.

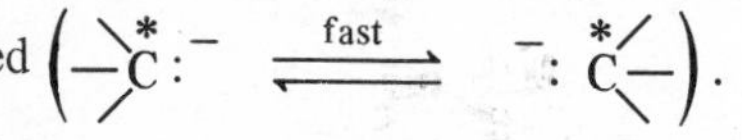

      Best evidence indicates that carbanions stabilized by resonance are planar (though not always) and give rise to racemization. Simple alkyl carbanions give rise to racemization but whether this means they are planar or undergo rapid inversion is not certain.

3. Organometallic compounds. With rare exceptions the reactions of halides with magnesium lead to racemic RMgX. However, in the reaction of *vinyl* halides with Li, the stereochemistry of the starting material is retained.

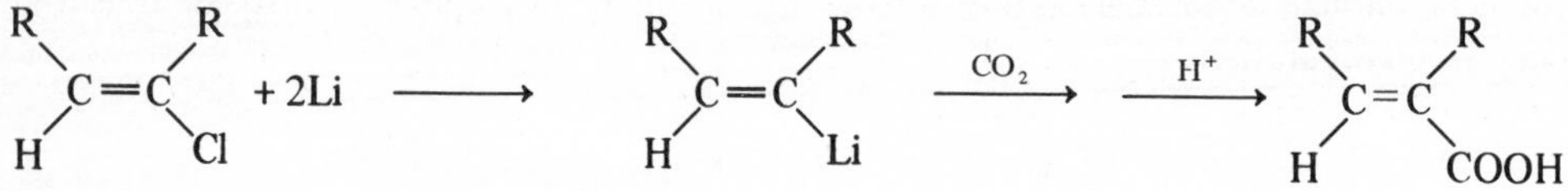

d. Addition to Alkenes and Alkynes

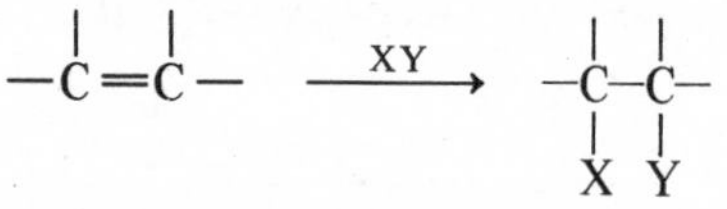

1. Electrophilic addition.

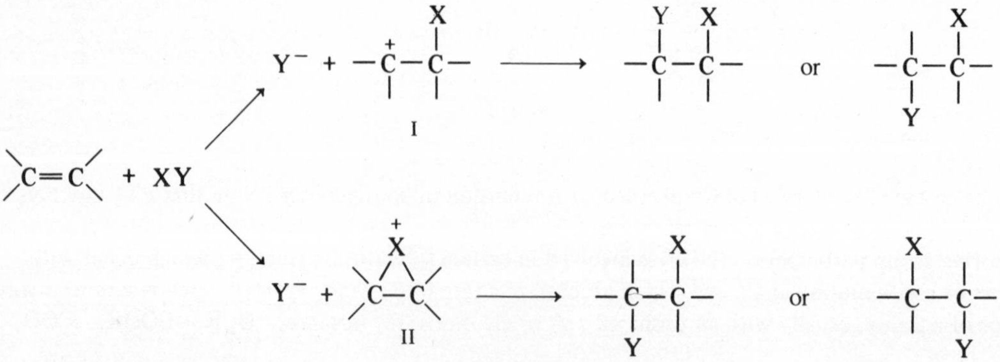

Studying the stereochemistry of electrophilic addition reactions can be very helpful in deducing whether the addition is cis or trans and whether a cyclic intermediate such as II or an open carbonium ion such as I is involved. Reactions in which X = Br (see Sec. 4-3a-2) or I (for example, ICl) or RS apparently involve the

192

formation of intermediates such as II. When X = Cl, the evidence for the existence of II is not as certain and when X = H then only intermediates of type I or a π complex of the proton and the alkene,

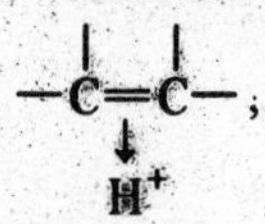

$$-\overset{|}{\underset{\underset{H^+}{\downarrow}}{C}}=\overset{|}{C}-\,,\;\text{are}$$

apparently involved. Other *ionic* electrophilic addition reactions of alkenes usually give trans addition. Some reactions involving cis addition to alkenes and alkynes which have been mentioned previously are catalytic hydrogenation (see Sec. 4-3d), epoxidation (see Sec. 6-3d), addition of carbenes (see Sec. 4-3c), hydroboration (see Sec. 6-2e), and mild $KMnO_4$ oxidation to form glycols (see Sec. 4-3e). The latter presumably goes through a cyclic intermediate such as A. Similar results are obtained with OsO_4.

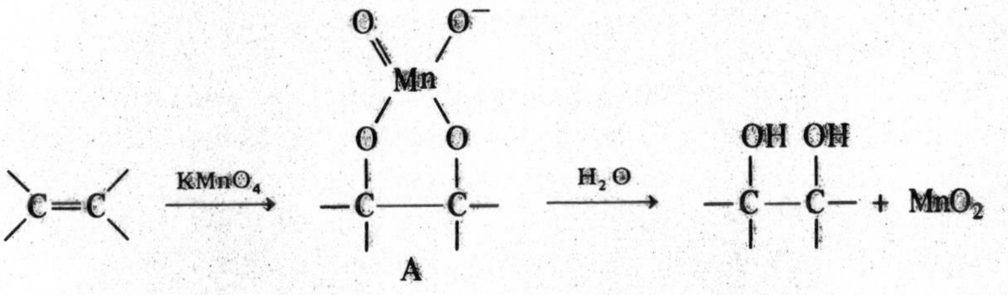

Cis addition to cycloalkenes usually proceeds from the less-hindered side of the double bond.

2. **Nucleophilic addition.**

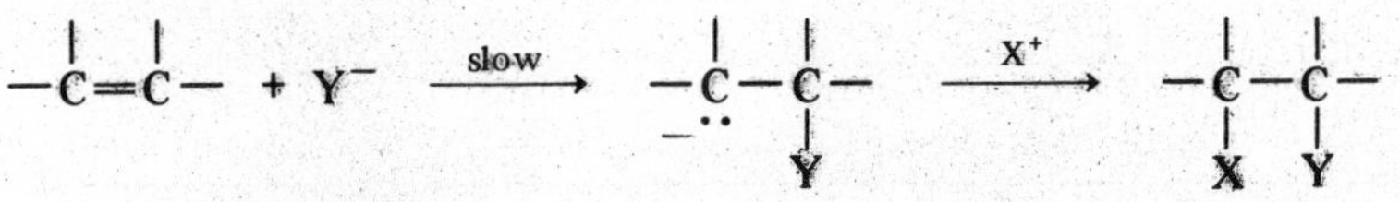

The stereochemistry of this type of reaction has not been studied extensively. If the carbanion mechanism is operating, then the reaction should be nonstereospecific; but it may be stereoselective. Polyhalo- or polycyanoolefins and most alkynes undergo nucleophilic rather than electrophilic addition. Michael-type reactions (see Sec. 12-2c) represent another large category of nucleophilic addition reactions, some of which are stereoselective.

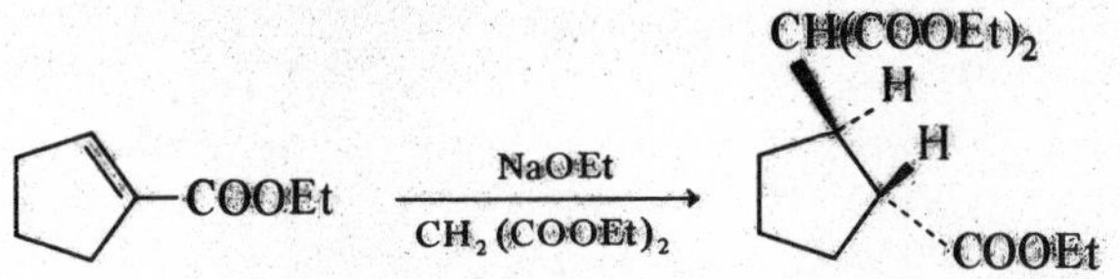

(trans only)

e. *Addition to Carbonyl Group* (See Chap. 7.)

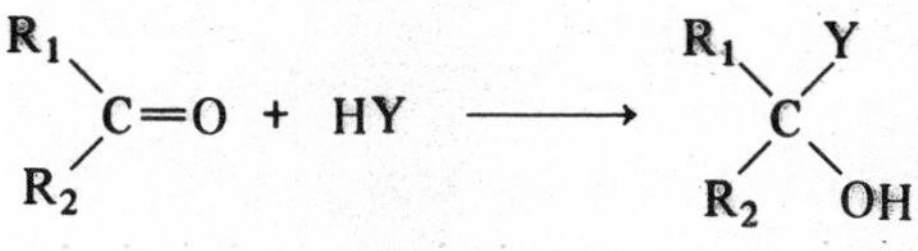

If there is no asymmetry in R_1, R_2, or in the reagent HY, these reactions will yield racemic mixtures. Asymmetric α carbons in R_1 or R_2 can lead to some stereoselectivity in the products (see Prob. 13-29).

f. *1,2 Rearrangement Reactions with R Group Migration*

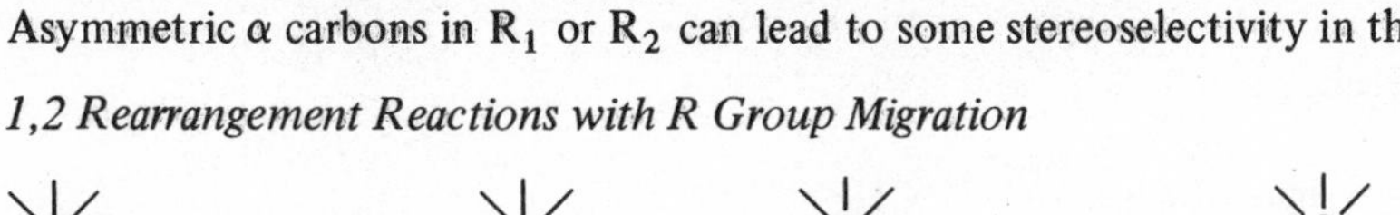

Several reactions have been discussed already which involve intramolecular migration of an R group from a carbon or boron atom to an electron-deficient carbon, oxygen, or nitrogen atom. In all of these reactions, if the migrating R group is attached via an asymmetric carbon atom, the latter's configuration is retained in the migration.

1. Migration to electron-deficient C: Dehydration of alcohols (see Sec. 4-2c), pinacol rearrangement (see Sec. 7-3d), and hydroboration-carbonylation of alkenes (see Sec. 6-2e).

2. Migration to electron-deficient N: Hofmann reaction (see Sec. 10-3*d*), Beckmann reaction (see Sec. 11-4*f*), and hydroboration-amination of alkenes (see Sec. 10-3*f*).

3. Migration to electron-deficient O: Hydroboration-oxidation of alkenes (see Sec. 6-2*e*) and Baeyer-Villiger rearrangement of ketones to esters (see Prob. 11-24*i*).

See Prob. 13-34 for rearrangements involving migration of a group to an electron-rich site.

13-7 Symmetry Elements and Point Groups

Molecules (specifically molecular conformations) are often classified by point groups. A molecule's point group or type of symmetry is determined by various combinations or absence of three symmetry elements: (1) a *plane of symmetry* (called σ) which divides the molecule into two equal mirror image halves; (2) a *simple axis of rotation* (called C_n, where $n = 1, 2, 3$, etc., and is the order of the rotational axis) which may be defined as an axis passing through the molecule so that a rotation of the molecule by $360°/n$ about this axis leads to the molecule's original conformation; (3) a *rotation-reflection axis* (called S_n where $n = 1, 2, 3$, etc.) which may be defined as an axis through the molecule whereby reflection of the molecule in a plane perpendicular to the S_n axis followed by a rotation of $360°/n$ about that axis will yield the original structure. Molecular conformations lacking an S_n axis have no reflection symmetry and are called *dissymmetric* or *chiral*. Those which do have an S_n axis or are superimposable on their mirror image have reflection symmetry and are called *nondissymmetric* or *achiral*. A listing of some symmetry or point group classifications (along with the group's symbol) based on the above symmetry elements is given below along with an example of each type.

Point group	Symmetry elements

a. Dissymmetric
  1. C_n
  2. D_n
	(No σ present)
	C_n axis only ($n > 1$)
	C_n axis + nC_2 axes

b. Nondissymmetric
  1. C_s
  2. S_2 or C_i
	σ but no C_n ($n > 1$)
	S_2 but no σ
	
	! *has point of symmetry:*

3. S_4
  4. C_{nv}
 	
  5. C_{nh}
	S_4 but no σ
	C_n + nσ intersecting
	at C_n (called $σ_v$)
	C_n + one σ perpendicular
	to C_n (called $σ_h$)

6. D_{nd} 7. D_{nh}	C_n + nC_2 + $nσ_v$ (which bisect angles between C_2 axes) Same as above except one $σ_h$ present

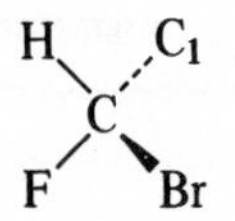

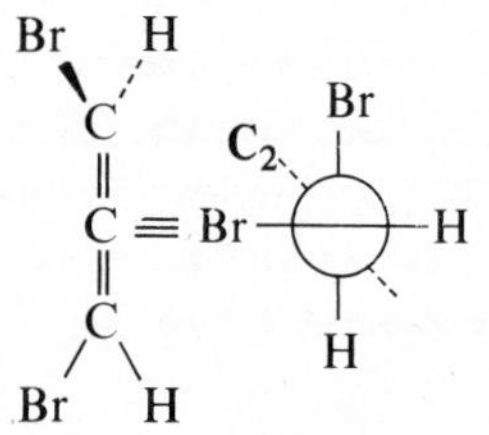

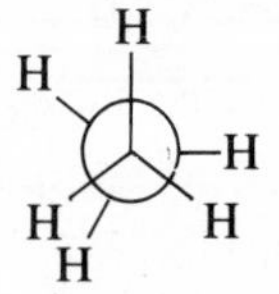

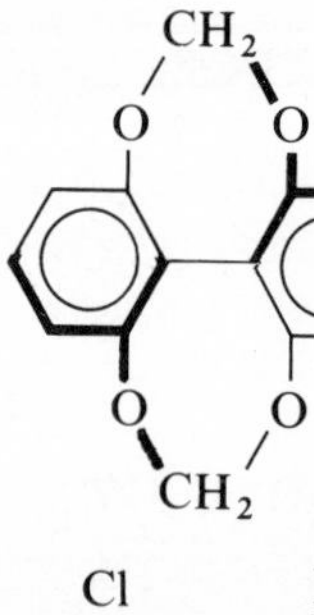

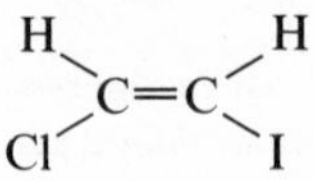

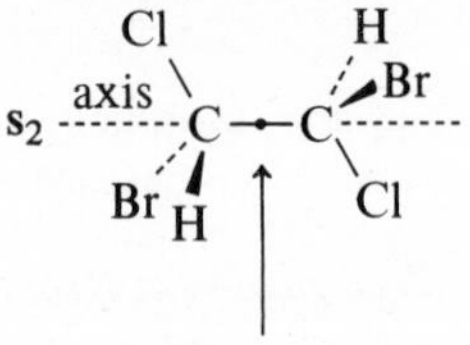

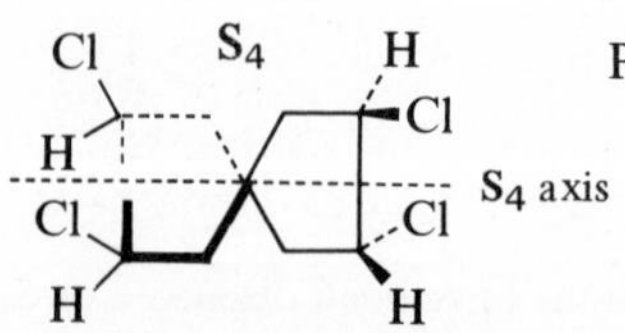

Other examples (see Prob. 13-37): cis-1,2-dibromoethene (C_{2v}), methyl bromide (C_{3v}), *trans*-1,2-dibromoethene (C_{2h}), allene (D_{2d}), staggered conformation of ethane (D_{3d}), cyclopropane (D_{3h}, $σ_h$ is plane of ring), and benzene (D_{6h}).

Problems

13-1   Draw structures for all 32 isomers of C_4H_8O including optical and geometric, but excluding tautomers and conformers. Give a suitable name for each. For optical isomers designate the configuration about any asymmetric carbon atoms (R or S) and for geometric isomers designate the configuration about any C=C bonds (E or Z).

13-2   Draw structures for all optical and geometric isomers for each of the following. Indicate which isomers would be expected to rotate plane-polarized light.

    a. Lactic acid

    c. 2-Butene

    e. $CH_3CHClCHBrCH_3$

    g. Acetaldoxime

    i. 4-Chloro-2-pentene

    k. 2,4,6-Nonatriene

    m. $C_6H_5CHCHCHC_6H_5$
                 H_3C Cl CH_3

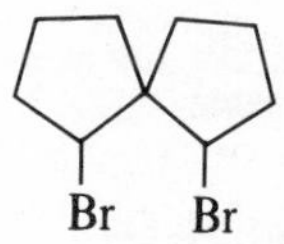

    o. $CH_3CH_2CH=C=CHCH_3$

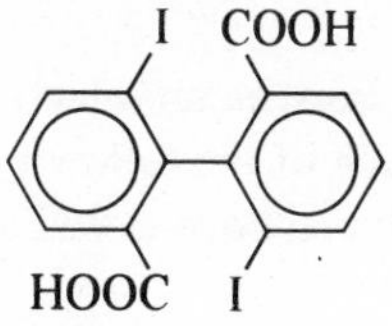

    q. 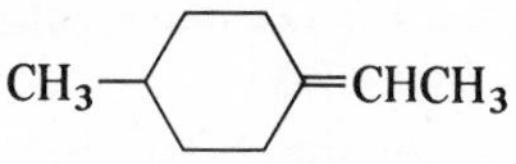 CH_3-⬡$=CHCH_3$

    s. $CH_3CH=C=CHCHCH_2Br$
                         CH_3

    b. 1-Butene

    d. Malic acid

    f. Acetone oxime

    h. 2,4-Hexadiene

    j. Tartaric acid

    l. *sec*-Butyl α-methylbutyrate

    n.

    p.

    r. $C_6H_5N=NC_6H_5$

    t.

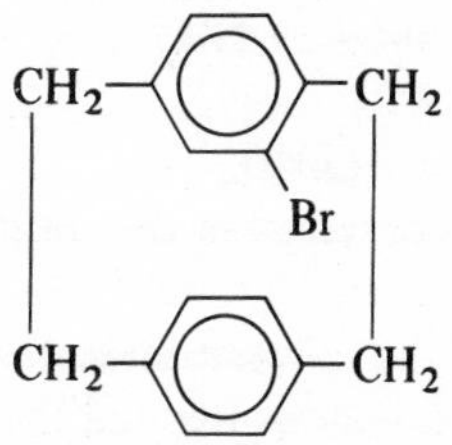

    u.

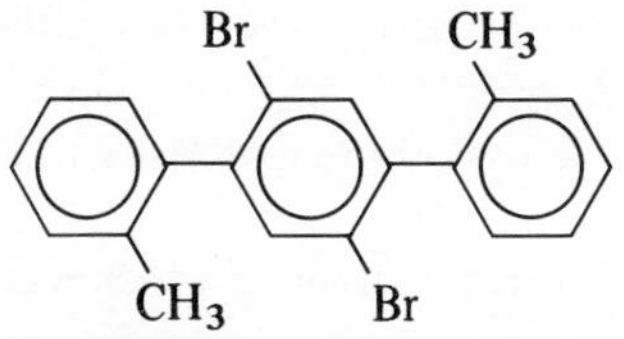

    v. $HC{\equiv}CC{\equiv}CCH=C=CHCH=CHCH=CHCH_2COOH$

               The antibiotic mycomycin

13-3   Draw simple formulas indicating all the stereoisomers of each of the following. For each stereoisomer indicate R or S configuration for each asymmetric carbon (and r and s for each pseudoasymmetric carbon; see answer to Prob. 13-2*m*). Where appropriate indicate the family by D or L and show each pair of enantiomers. Give an example of a pair of diastereomers in each case.

    a. $HOCH_2(CHOH)_3CHO$

        An aldopentose

    c. $HOCH_2(CHOH)_4CHO$

        An aldohexose

    b. $CH_3CH_2CH{-}CHCOOH$
                     CH_3 NH_2

        Isoleucine, an amino acid

    d. $HOOC(CHCl)_3COOH$

13-4   Using the method in Sec. 13-3*b*, draw all the possible structures for the following. Indicate which isomers are optically active.

    a. 1,2-Cyclobutanedicarboxylic acid

    c. 1,2,3,4-Tetrachlorocyclobutane

    e. 2,4-Diphenyl-1,3-cyclobutanedicarboxylic acid

    b. 1,3-Cyclobutanedicarboxylic acid

    d. 1,2,3-Trichlorocyclobutane

    f. 1,2,3,4,5,6-Hexachlorocyclohexane

13-5 Show by means of three-dimensional formulas the structures for the principal product(s) expected in each of the following. Where possible indicate whether the product has the erythro, threo, or meso configuration. [The term erythro implies configuration like that in erythrose (see Sec. 13-2a) with both like groups on the same side. Threo implies configuration like threose.] Where appropriate indicate whether product is the E or Z isomer.

a. *cis*-2-Butene + HOBr
b. *trans*-2-Butene + ICl
c. (Z)-3-Hexene + CH_3COOOH
d. *meso*-2,3-Dibromobutane + Mg
e. (3R:4R)-3,4-Dibromohexane + alc KOH
f. Cyclohexene + Br_2
g. 1,2-Dimethylcyclohexene + $KMnO_4$ + H_2O
h. $CH_3C{\equiv}CCOOH$ + HBr
i. (2S:3R)-2,3-Dibromohexane + alc KOH
j. 1,2-Dimethylcyclopentene + B_2H_6
k. 2-Pentyne + H_2 + Pt
l. 2-Pentyne + Li + NH_3
m. (2S:3S)-2-Acetoxy-3-methylpentane + Δ
n. (2S:3R)-2-Acetoxy-3-methylpentane + Δ

13-6 Each of the following reactions can be used to synthesize a *dl* mixture from a compound without an asymmetric atom. Illustrate each, indicating the asymmetric atoms with an asterisk.

a. Hell-Volhard-Zelinsky reaction
b. Grignard preparation of 3° alcohol from ketone
c. Addition reaction with propene
d. Oxidation of a cyclic ketone
e. Aldol condensation
f. Hydroboration-oxidation of an alkene
g. Oxidation of a 3° amine with H_2O_2
h. Reformatsky reaction
i. Michael reaction
j. Reductive alkylation of a ketone

13-7 Are all stereospecific reactions stereoselective? Are all stereoselective reactions stereospecific?

13-8 In which compound in each of the following reactions would an inversion of configuration (Walden inversion) probably occur? In which reactions would a racemic mixture be the main result? Keep in mind the nature of the S_N1 and S_N2 reactions.

a. (R)-2-Butanol + Na
b. Sodium 2-butoxide + (S)-*sec*-butyl chloride
c. Acetaldehyde + sodium acetylide
d. (S)-*sec*-Amyl alcohol + HBr
e. (S)-*sec*-Amyl bromide + $(CH_3)_2NH$
f. 2-Methylbutanoyl chloride + (R)-2-butanol
g. (2S,3S)-2,3-Epoxybutane + NH_3
h. (R)-2-Iodobutane + NaI
i. Propylene oxide + KI
j. Propylene oxide + H_3O^+
k. (R)-2-Bromobutane + $LiAlD_4$
l. Acetic anhydride + (S)-2-butanol

13-9 Compare the products obtained in the formation of the monoethyl ester of *meso*-tartaric acid, D-tartaric acid, and DL-tartaric acid.

13-10 Give the structure for the simplest saturated hydrocarbon that might exist as a *dl* mixture, for the simplest olefin; for the simplest monodeuterated hydrocarbon.

13-11 Compound A, $C_5H_6O_2$, liberated CO_2 from sodium bicarbonate and was optically active. On hydrogenation, it yielded $C_5H_8O_2$, which was optically active. Suggest a structure for A.

13-12 Compound A, $C_5H_6O_2$, liberated CO_2 from $NaHCO_3$ and was optically active. On hydrogenation, it yielded $C_5H_{10}O_2$, which was optically inactive. Suggest a structure for A.

13-13 Compound A, $C_5H_8O_2$, liberated CO_2 from $NaHCO_3$. It existed in two forms, neither of which was optically active. On hydrogenation, it yielded $C_5H_{10}O_2$, which could be resolved into enantiomorphs. Suggest a structure for A.

13-14 When 0.90 g of compound A, $C_4H_{10}O_2$, was treated with methylmagnesium iodide, it yielded 448 ml of methane (STP). Compound A could be separated into two fractions by crystallization. Only one of these fractions could be resolved into optically active enantiomorphs. Suggest a structure for A.

13-15 Compound A, $C_6H_8O_2$, liberated CO_2 from $NaHCO_3$ and was optically active. On hydrogenation, it gave $C_6H_{12}O_2$, which was optically inactive. Suggest two structures for A.

13-16 Isomeric compounds A and B (MW 88 ± 1) contain chlorine. Only A is chiral. The following spectra data were obtained:

Compound A   Nmr (δ): 2.52, $d(J = 2.5$ Hz), 5.1 units; 1.73, $d(J = 6.9$ Hz), 15.6 units; 4.58, quartet $(J = 6.9$ Hz) and doublet $(J = 2.5$ Hz), 5.3 units $(q + d)$.

Compound B   Nmr (δ): 1.88, $t(J = 2$ Hz), 15.7 units; 4.22, $q(J = 2$ Hz), 10.4 units.

Characteristic ir bands   A, 3300 cm^{-1} (3.03 μm) and 2109 cm^{-1} (4.74 μm); B, 2225 cm^{-1} (4.49 μm). Suggest structures for A and B.

13-17 Optically active compound X contains 73.5% C, 10.2% H, and 16.3% O. Its nmr (δ) spectrum: 3.35, s, 7.4 units; 2.35, s, 7.6 units; 1.62, q, 14.6 units; 1.38, s, 22.4 units; 1.0, t, 22.4 units. Its ir spectrum has characteristic bands at (cm^{-1}, μm): 3571, 2.80; 3311, 3.0; 2104, 4.75; 1466, 6.82; 1370, 7.3; and 1190, 8.4. Suggest a structure for X.

13-18 Draw the cyclic intermediate involved in each of the following reactions and predict the stereochemistry of the products (meso, R, S, *dl* mixture, diastereomers, or no asymmetric centers).

a. Cyclopentene + Br$_2$ b. (*E*)-2-Butenoic acid + Br$_2$
c. Maleic acid + Br$_2$ d. Fumaric acid + Br$_2$ + Cl$^-$

13-19 Suggest a chemical and/or physical method to separate each of the following mixtures:

a. (R)-2-Octylamine   b. Maleic acid   c. (R)-2-Octanol
   (S)-2-Octylamine   Fumaric acid   (S)-2-Octanol

13-20 Give an explanation for each of the following:

a. Addition of HBr to acetylenedicarboxylic acid gives

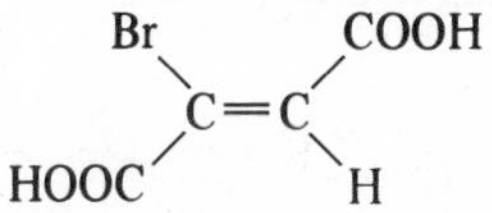

$$\underset{HOOC}{\overset{Br}{\diagdown}}C=C\underset{H}{\overset{COOH}{\diagup}}$$

b. When 3,6-dimethyl-4-octanone is shaken with alkaline solution, one asymmetric center is racemized, but the other is not.

c. Neopentyl chloride does not react with hot NaOH, but reacts with Ag$_2$O + H$_2$O.

d. Treatment of neopentyl alcohol with HCl causes much rearrangement with only a low yield of neopentyl chloride. Using SOCl$_2$ instead gives a good yield of neopentyl chloride.

$$e.\ (S)\text{-}C_6H_5\underset{CH_3}{\overset{O}{\overset{\|}{C}}}HOCCl \xrightarrow{\Delta} (S)\text{-}C_6H_5\underset{CH_3}{C}HCl + CO_2$$

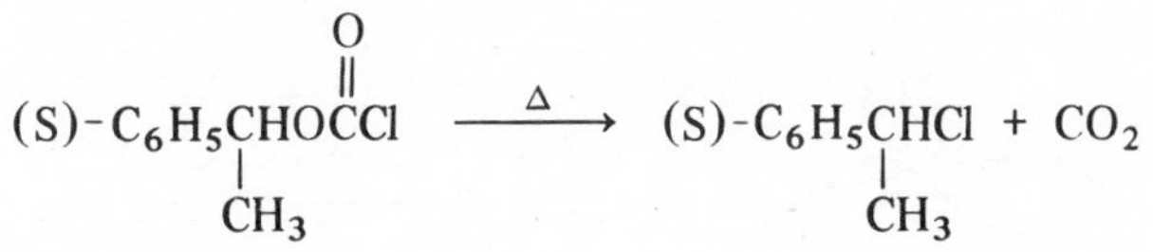

f. 3-Methylhexane is difficult to racemize, whereas 2-methylbutanoic acid is easily racemized with acidic or basic catalysis.

g. D-Chlorosuccinic acid $\xrightarrow[\text{H}_2\text{O}]{\text{Ag}_2\text{O}}$ D-hydroxysuccinic acid

13-21 a. Using Newman projections and remembering that substituents in the equatorial position cause the least nonbonded interaction, decide whether the substituents in each of the dimethylcyclohexanes (*cis*- and *trans*-1,2; 1,3; and 1,4) prefer *aa, ee,* or *ae.*

b. Assuming *each* butane-gauche-type nonbonded interaction as shown below raises the energy of that conformer 0.9 kcal and that each 1,3 diaxial interaction of methyl groups raises the energy by 1.8 kcal, calculate the *energy difference* between the most favorable and the least favorable chair conformers for each compound in part *a* and between the cis and trans isomers in each case.

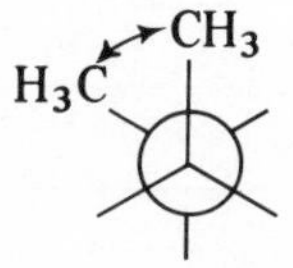

Butane-gauche interaction 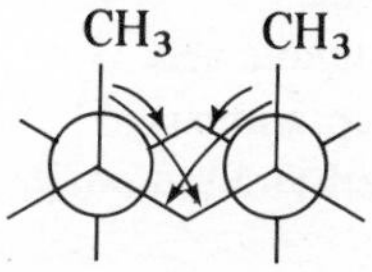   One 1,3 diaxial methyl interaction
                                   Four gauche interactions

c. Assuming planar cyclohexane rings, decide whether the isomers in part *a* would be expected to exist as meso compounds or as a *dl* pair.

d. Using Newman projections for the possible *chair* conformers in part *a*, decide whether each conformer is dissymmetric or nondissymmetric (see Sec. 13-7). Explain why none of the dimethylcyclohexanes are optically active even though some can exist in dissymmetric chair conformations.

13-22 Calculate the equilibrium constant at 25°C for the interconversion of the chair conformations in each of the following (see Sec. 13-3*b*-2).

a. Ethylcyclohexane   b. *cis*-4-Methyl-1-cyclohexanol

13-23 What conformation, chair or boat, would you expect to be the more stable conformation for *trans*-1,3-di-*tert*-butylcyclohexane?

13-24 The two reactions shown below were carried out starting with the Z isomer of $C_6H_5CH=CHD$ with the almost exclusive product (96 percent) being the Z isomer of $C_6H_5C(N_3)=CHD$. Based on these results, which of the proposed intermediates A or B was more likely involved? Illustrate the mechanism of the two reactions.

$$C_6H_5CH=CHD \xrightarrow{IN_3} C_6H_5\underset{\underset{N_3}{|}}{C}HCHDI \xrightarrow[\text{ether}]{(CH_3)_3COK} C_6H_5\underset{\underset{N_3}{|}}{C}=CHD$$

$$C_6H_5\overset{+}{\underset{\diagdown\diagup}{CH-CHD}} \qquad C_6H_5\overset{+}{CH}\underset{\underset{I}{|}}{CHD}$$
$$\qquad\quad I$$
$$\qquad A \qquad\qquad\qquad B$$

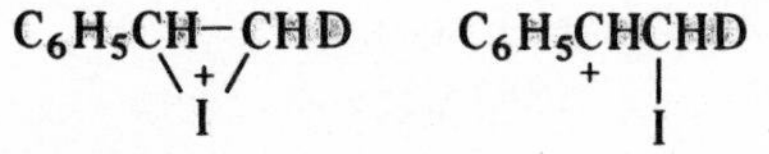

13-25 Using three-dimensional formulas, trace the steric course of each of the following sequences (the tosyl group, Ts, is $CH_3-C_6H_4-SO_2-$):

a. $HC\equiv CH \xrightarrow{2NaNH_2} \xrightarrow{2EtI} \xrightarrow[H_2]{Pd} \xrightarrow{Cl_2} C_6H_{12}Cl_2 \xrightarrow[KOH]{alc} C_6H_{11}Cl \xrightarrow{Li} \xrightarrow{CO_2}$

$\xrightarrow{H^+} C_7H_{12}O_2$

b. (R)-2-Butanol $\xrightarrow[\text{(TsCl)}]{CH_3C_6H_4SO_2Cl}$ ester $\xrightarrow{LiI}$ halide $\xrightarrow{KOH}$ olefin + alcohol

c. (S)-$C_6H_5CH(OH)CH_3 \xrightarrow{PCl_5} \xrightarrow{KOAc}$ ester $\xrightarrow{NaOH}$ alcohol $\xrightarrow{PBr_3}$

d. 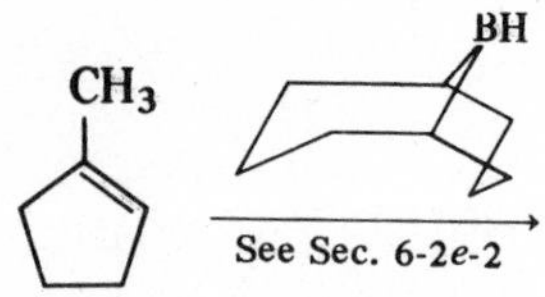 $\xrightarrow[\text{See Sec. 6-2e-2}]{} \xrightarrow{CO} \xrightarrow[H_2O]{NaOH} C_7H_{14}O \xrightarrow{Na} \xrightarrow[(R)\text{-}CH_3\overset{\overset{Br}{|}}{C}HCH_2CH_3]{} C_{11}H_{22}O$

e. cis-1-Bromo-4-t-butylcyclohexane $\xrightarrow[KOH]{alc}$

f. trans-1-Bromo-4-t-butylcyclohexane $\xrightarrow[KOH]{alc}$

13-26 Given methanol, ethanol, d-mandelic acid $[C_6H_5CH(OH)COOH]$, d-α-phenylethylamine $[C_6H_5CH(CH_3)NH_2]$, and phthalic anhydride, show how you would make the following conversions. In certain instances it will be necessary to resolve racemic mixtures. Explain in detail just how you would do this.

a. (R)-CH_3CHDCH_2Br to (R)-CH_3CHDCH_2COOH
b. (S)-2-Methylbutanoic acid to (S)-2-aminobutane
c. Cyclohexane to cis-1,2-dichlorocyclohexane
d. (R)-2-Butanol to (R)-2-aminobutane
e. $HC\equiv CH$ to meso-$CH_3CDClCDClCH_3$
f. (R)-2-Octanol to (R)-2-methyl-1-aminooctane
g. (R)-2-Butanol to (2R:3R)-2,3-dimethylpentanoic acid
h. 1-Butene to (2R:5S)-2-hydroxy-5-methylheptanoic acid
i. 1-Butene to (R)-2-deutero-2-butyl (E)-2-methyl-2-butenoate

j. 2-Butyne to (R)

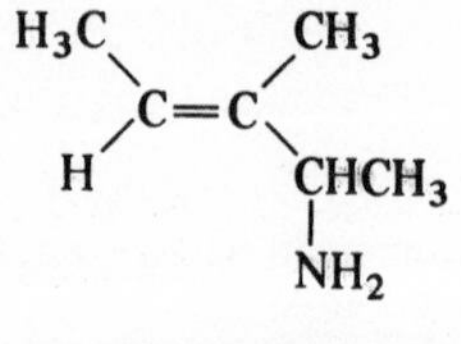

13-27 In each of the following, participation by a neighboring group plays a part in determining the steric course of the reaction. Explain, using three-dimensional drawings.

a. trans-2-Chloro-1-cyclohexanol $\xrightarrow{NaOH}$
trans-1,2-Cyclohexanediol

b.

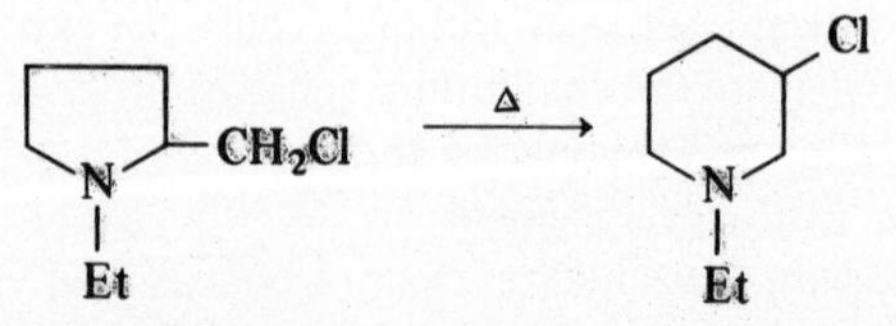

c. (D)-RCHCOOH $\xrightarrow{\text{HONO}}$ (D)-RCHCOOH   *d.*
$\quad\quad\quad$ | $\quad\quad\quad\quad\quad\quad\quad\quad$ |
$\quad\quad$ NH$_2$ $\quad\quad\quad\quad\quad\quad\quad\quad$ OH

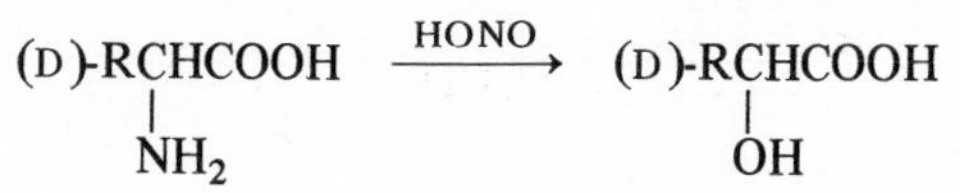

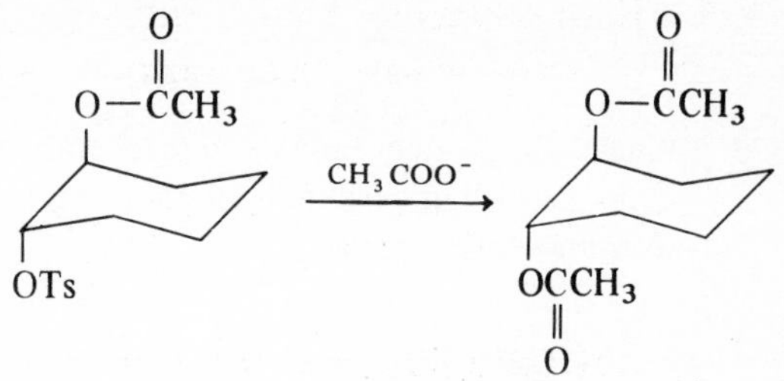

13-28   Alkyl fluorides are known to undergo ionization in the presence of antimony pentafluoride (SbF$_5$) and SO$_2$:

$$RF \xrightarrow[\text{SO}_2]{\text{SbF}_5} R^+ \, SbF_6^-$$

a. When ICH$_2$CH$_2$F is treated with SbF$_5$/SO$_2$, the nmr spectrum of the reaction mixture (at $-60°$C) shows only a singlet at $\delta = 5.8$; if this mixture is added to methanol and K$_2$CO$_3$, 1-methoxy-2-iodoethane is isolated. Suggest a structure for the intermediate ion.

b. Based on the structure in *a*, show a mechanism for the expected reaction of erythro-*dl*-2-iodo-3-fluorobutane when treated with SbF$_5$/SO$_2$ followed by CH$_3$OH/K$_2$CO$_3$.

13-29   Cram's rule of asymmetric induction states that in the addition of a reagent A$^-$B$^+$ to a carbonyl group holding three different groups on the α carbon (hence it is asymmetric), classified as large, medium and small, a preponderance of that isomer is formed in which A$^-$ approaches that flat carbonyl group from the least hindered flank (S side rather than M side). It is assumed that the most stable conformation for the carbonyl oxygen is in the staggered form with the O between the two smallest groups.

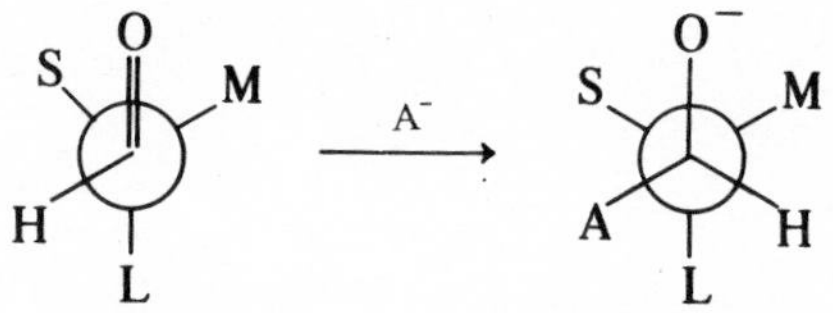

Predict whether the main isomer in each of the following would be threo or erythro:

a. (S)-2-Phenylpropanal + CH$_3$MgX $\longrightarrow$

b. (S)-3-Phenyl-2-butanone + LiAlH$_4$ $\longrightarrow$

13-30   For a pair of diastereomers corresponding hydrogens are in diastereomeric environments and will have different chemical shifts. If this difference is large enough to allow separation of the nmr signals for such diastereomeric hydrogens, then integration of the peaks will be in the same ratio as the two diastereomers present.

a. Draw the structures for the expected products when *dl*-1-phenylethylamine is reacted with pure (R)-0-methylmandelyl chloride, C$_6$H$_5$CH(OCH$_3$)COCl.

b. If one runs the above reaction starting with the amine being 80 percent optically pure (the rest being racemic mixture), the nmr spectrum, in part, shows three sets of peaks: two singlets at $\delta = 4.28$ and 4.21, two singlets at $\delta = 2.67$ and 2.62, and two doublets centered at $\delta = 0.99$ and 0.86. Assign each set of peaks to a set of hydrogens and give the expected ratio for the peak areas in a given set (for example, $\delta = 4.28$ and 4.21).

c. If the peak areas at $\delta = 4.28$ and 4.21 were found to be in a 4:1 ratio, the optical purity of the starting amine had to be what? How could a 1:1 ratio be obtained?

13-31   Pure (S)-2-methyl-1-butanol and (S)-1-chloro-2-methylbutane have specific rotation values of $-5.76°$ and $+1.64°$, respectively. Predict the specific rotation of the purified product obtained in each of the following:

a. (R)-2-Methyl-1-butanol (optically pure) $\xrightarrow{\text{HCl}}$

b. (R)-2-Methyl-1-butanol ($[\alpha]_D^{20} = +3.25$) $\xrightarrow{\text{HCl}}$

c. (S)-2-Methyl-1-butanol (optically pure) $\xrightarrow{\text{HBr}}$ 3.0 g of purified product is dissolved in 10 ml of solvent and the observed rotation of the solution in a 2.0-dm polarimeter cell is $+1.20°$.

13-32   Predict the stereochemistry of the products formed in the following hydroboration reactions of 1-methylcyclopentene (I).

a. (I) $\xrightarrow[\text{OH}^-]{\text{B}_2\text{H}_6 \quad \text{H}_2\text{O}_2}$ *b.* (I) $\xrightarrow{\text{B}_2\text{H}_6 \quad \text{H}_2\text{NOSO}_3\text{H}}$

c. (I) $\xrightarrow[\text{(Sec. 6-2e-2)}]{\text{9-BBN}}$ $\xrightarrow[\text{LiBH}_4]{\text{CO}}$ $\xrightarrow[\text{H}_2\text{O}_2]{\text{HO}^-}$ *d.* (I) $\xrightarrow{\text{9-BBN}}$ $\xrightarrow[\text{(CH}_3)_3\text{COK}]{\text{BrCH}_2\text{CO}_2\text{Et}}$

13-33 Predict the product (and its stereochemistry where appropriate) expected in the following synthetic sequences each of which involves at least one rearrangement reaction. Name each reaction and show a mechanism for the *rearrangement* reaction.

a. $C_6H_5COC_6H_5$ $\xrightarrow{\text{Mg·Hg}}$ I $\xrightarrow{\text{H}^+}$ II

b.

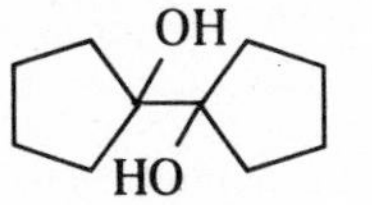

c. Oxime of (3S:5S)-3,5-dimethyl-4-heptanone $\xrightarrow{\text{PCl}_5}$

d. $(C_6H_5)_2\underset{\underset{\text{OH}}{|}}{C}CH_2OH$ $\xrightarrow{\text{H}^+}$

e. $CH_3CH_2CH(CH_3)\underset{\underset{O}{\|}}{C}-N_3$ $\xrightarrow{\Delta}$ isocyanate

                                        R isomer   Curtius rearrangement

f. (R)-2-Methyl-2-phenylbutanoic acid $\xrightarrow{\text{SOCl}_2}$ $\xrightarrow{\text{CH}_2\text{N}_2}$ I $\xrightarrow[\text{H}_2\text{O}]{\text{Ag}_2\text{O}}$ II

           (Arndt-Eistert synthesis)   A diazo   An acid
                                                       ketone

g. $CH_3\underset{\underset{O}{\|}}{C}-\underset{\underset{CH_3}{|}}{C}HCH_2CH_3$ $\xrightarrow{\text{CF}_3\text{CO}_3\text{H}}$ an ester (see Prob. 11-24*i*)

                   R isomer   Baeyer-Villiger oxidation

h.

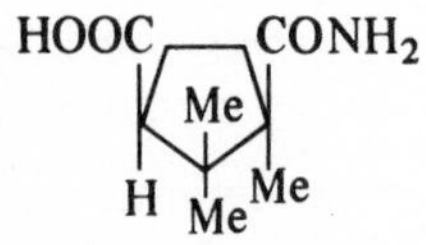

$\xrightarrow[\text{NaOH}]{\text{NaOBr}}$ I $\xrightarrow{\Delta}$ II + H_2O

13-34 Several rearrangement reactions follow the scheme:

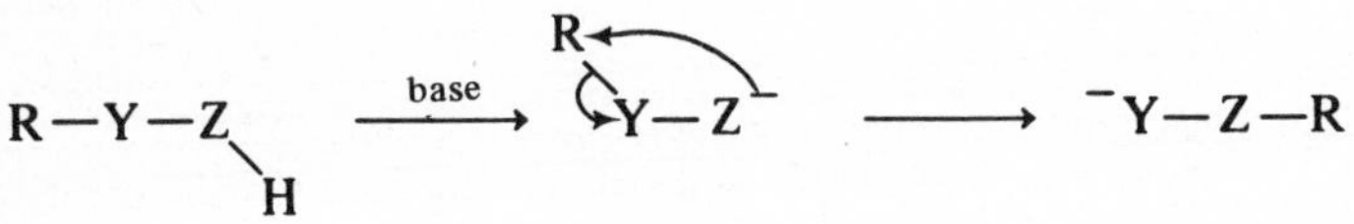

Stevens rearrangement: $R_2\overset{+}{\underset{\underset{R'}{|}}{N}}-CH_2\overset{\overset{O}{\|}}{C}R$ $\xrightarrow{\text{OH}^-}$ $R_2\underset{\underset{R'}{|}}{N}CH-\overset{\overset{O}{\|}}{C}-R$

Wittig rearrangement: $R'-O-CH_2C_6H_5$ $\xrightarrow{\text{RLi}}$ $\xrightarrow{\text{H}^+}$ $HOCH(R')C_6H_5$

Benzilic acid rearrangement: $R'-\overset{\overset{O}{\|}}{C}-\overset{\overset{O}{\|}}{C}-R$ $\xrightarrow{\text{KOH}}$ $HOOC-\underset{\underset{OH}{|}}{\overset{\overset{R'}{|}}{C}}-R$

                                α diketones

Favorskii reaction: $R'-CH_2\overset{\overset{O}{\|}}{C}-\underset{\underset{Cl}{|}}{C}H-R$ $\xrightarrow{\text{RO}^-}$ $R'-CH-\underset{\underset{\underset{O}{\|}}{C}}{}-CH-R$ $\xrightarrow{\text{RO}^-}$ $R'CH_2\underset{\underset{COOR}{|}}{C}HR$

                 α-halo ketones

Give a possible mechanism for each of the following reactions:

a.

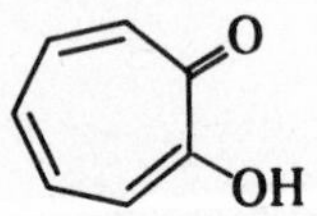

$\xrightarrow[\text{rearrangement}]{\text{benzilic acid}}$ $C_7H_6O_2$

Tropolone

b. (R)-$C_6H_5CH-\overset{+}{N}Me_2$ $\xrightarrow{OH^-}$ $C_6H_5CHCH-\underset{\underset{O}{\|}}{C}-C_6H_5$

with substituents CH_3 and $CH_2COC_6H_5$ on left, and NMe_2 above, CH_3 below on product.

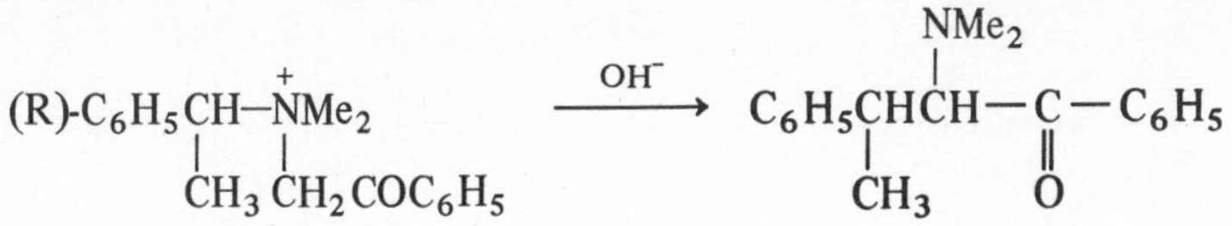

c. $CH_2=CHCH_2OCH_2CH=CH_2$ $\xrightarrow{C_6H_5Li}$ $\xrightarrow{H^+}$ $C_6H_{10}O$

d. (S)-$C_6H_5COCH_2\overset{+}{S}-CHC_6H_5$ $\xrightarrow{OH^-}$ $C_6H_5COCHCH\overset{Me}{\diagdown_{C_6H_5}}$

with $Me\,Me$ below S, and SMe below on product.

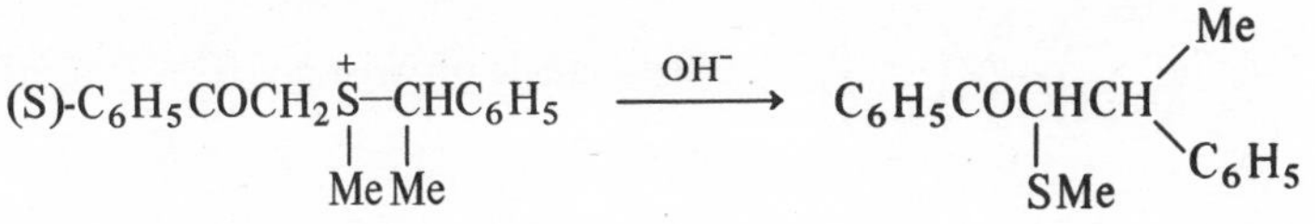

e.

ring I with COCH$_3$, ---Cl, CH$_3$, H $\xrightarrow{C_6H_5CH_2O^-}$ ring II with CH$_3$, ---COOCH$_2$C$_6$H$_5$, CH$_3$, H

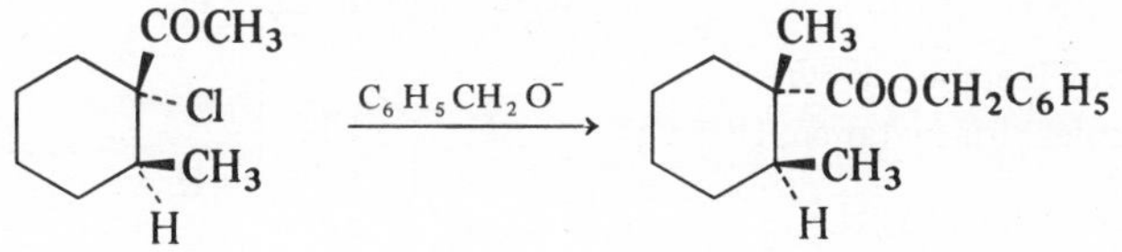

      I   II

f. (R)-Allylbenzylmethylphenylammonium iodide (I) $\xrightarrow{t\text{-}BuO^-}$

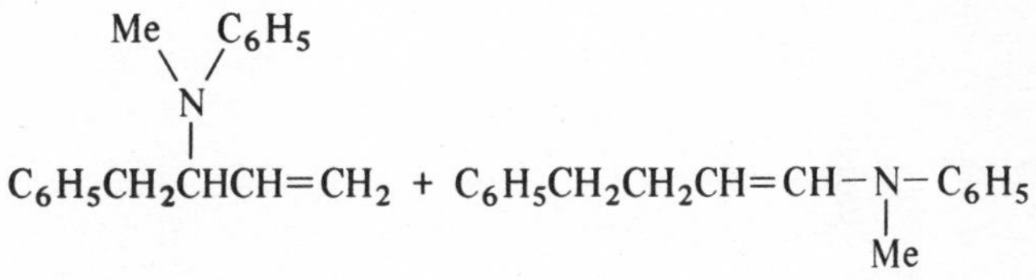

$$\underset{N}{Me\diagdown\,\diagup C_6H_5}$$

$C_6H_5CH_2CHCH=CH_2$ + $C_6H_5CH_2CH_2CH=CH-\underset{Me}{\overset{|}{N}}-C_6H_5$

                    S isomer, II   III

13-35   (R)-sec-Butyl-HgBr + racemic sec-butyl-MgBr $\longrightarrow$ sec-butyl-Hg-sec-butyl $\xrightarrow{HgBr_2}$ 2 sec-butyl-HgBr

            RS   R   RS

                                                                    A   B

Di-sec-butyl mercury (A) was prepared as shown above. If A is treated with mercuric bromide, 2 mol of
sec-butylmercuric bromide (B) are produced. If an S_E2 mechanism (see Sec. 13-6c) is operating in
A $\longrightarrow$ B, then a study of the optical activity present in the products compared to the original optical
activity present in the (R)-sec-butylmercuric bromide used to prepare A will help elucidate the stereochemistry
of the S_E2 mechanism. Assuming each sec-butyl-Hg bond in A has a 50 percent chance of breaking in the
attack by $HgBr_2$, predict the optical activity in the products if the following stereochemical possibilities
involving the configuration of the asymmetric carbon are involved.

a. Inversion   b. Retention   c. Racemization

13-36   Methane and other tetrahedral CY_4 molecules belong to a symmetry point group called Td. Identify the
number and location of C_3 and C_2 axes and planes of symmetry, σ.

13-37   Classify each of the following molecules or molecular conformations as to the symmetry point group that it
belongs to:

a. (R)-Glyceraldehyde

b.

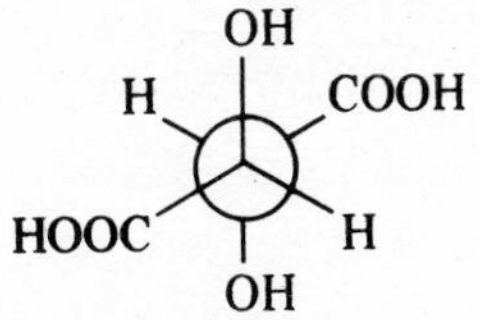

c.

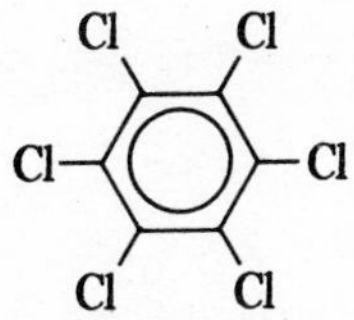

d.

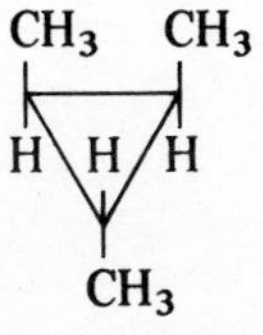

e.

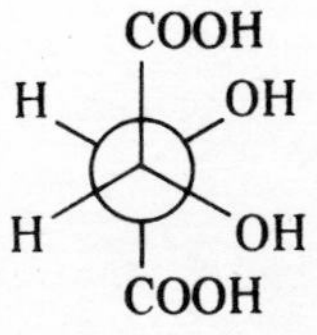

f. Eclipsed conformer
     of ethane

g.

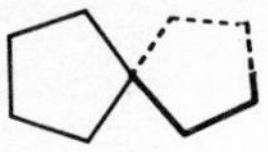

h. $Cl_2C{=}CCl_2$

i.

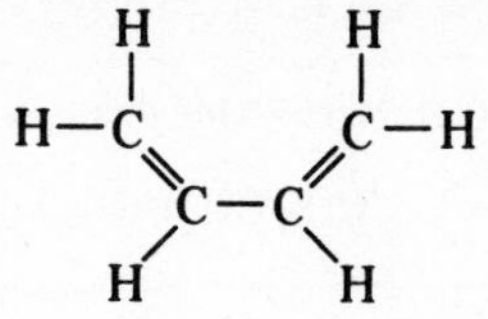

j.

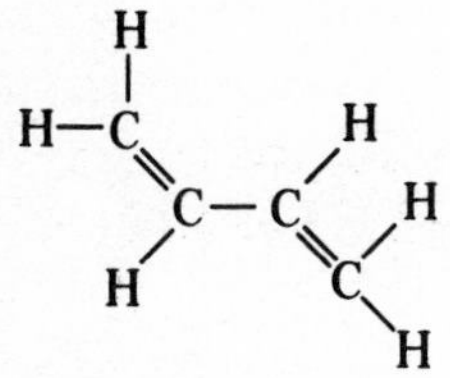

k.

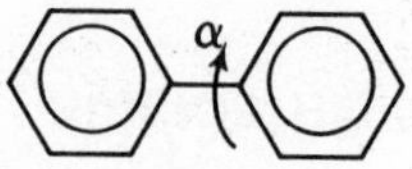

Where angle of twist α is $0° < \alpha < 90°$

l. Chair conformer of cyclohexane

m.

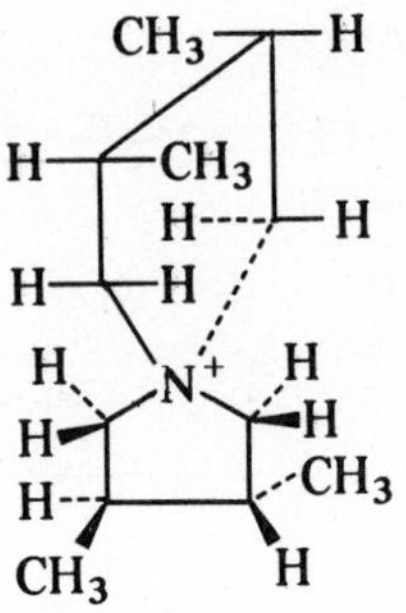

n. All heteronuclear diatomic molecules, for instance, $C{\equiv}O$

o. All linear homonuclear diatomic molecules, for instance, $HC{\equiv}CH$

14
aromatic compounds

14-1 Nomenclature

Benzene belongs to that class of cyclic unsaturated compounds called *aromatic*. These substances show unusual resistance to addition and oxidation reactions when compared to olefins. The nmr spectra of aromatic substances will have peaks at low τ (2 to 4) for protons attached to a ring which contains a closed loop of electrons (see Sec. 8-4e-3). In benzene all C–C bond lengths in benzene are equal (1.39 Å) and all C–C–C bond angles are identical (120°). Chemically, there has been observed only one monosubstitution product and three disubstitution products for benzene, leading to the conclusion that all C–H bonds are equivalent. Review Sec. 2-5 for the structure and representation of benzene.

The terms *ortho*, *meta*, and *para* (abbreviated *o*, *m*, and *p*) are used to indicate the position of substituents in *disubstituted* benzenes.

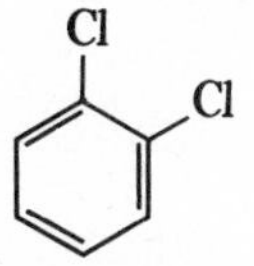

o-Dichlorobenzene 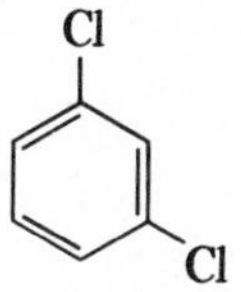   m-Dichlorobenzene 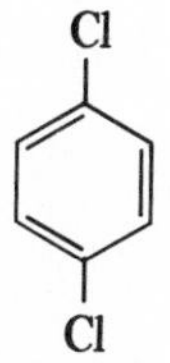   p-Dichlorobenzene

For more complicated derivatives, the positions on the ring are numbered 1 to 6.

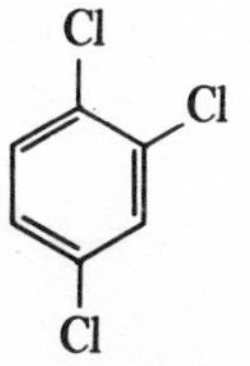

1,2,4-Trichlorobenzene   1-Bromo-2-nitro-4-fluorobenzene
(not 1,2,5 or 1,3,4)

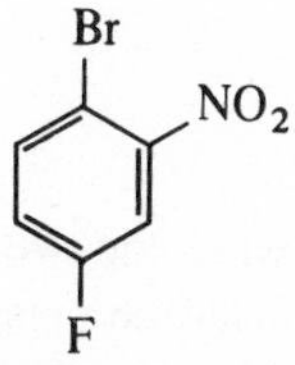

The group ⟨○⟩—, in which one hydrogen is replaced, is known as the *phenyl group* (abbreviated as C_6H_5- or ϕ or sometimes Ph).

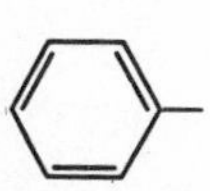

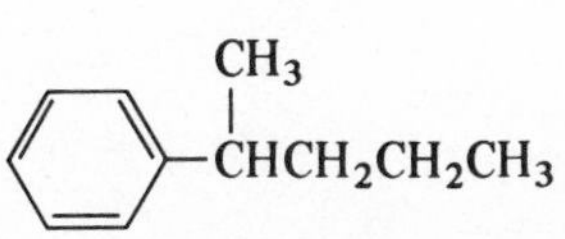

2-Phenylpentane 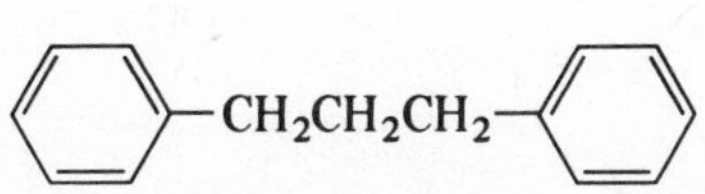   1,3-Diphenylpropane

203

The symbol Ar (aryl) is often used to stand for any aromatic substituent group in a way analogous to the use of R to mean any alkyl group. For example, aryl halides (ArX) vs. alkyl halides (RX).

Some of the common mixed aliphatic-aromatic hydrocarbons are named as follows:

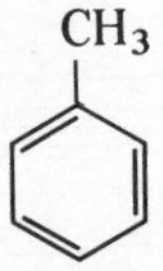

Toluene
(methylbenzene)

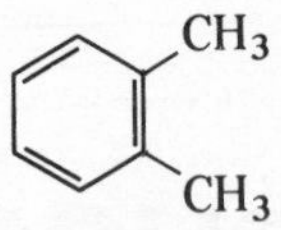

o-Xylene
(1,2-dimethylbenzene)

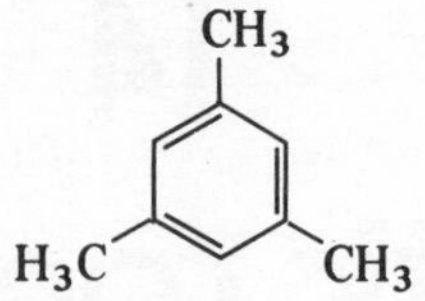

Mesitylene
(1,3,5-trimethylbenzene)

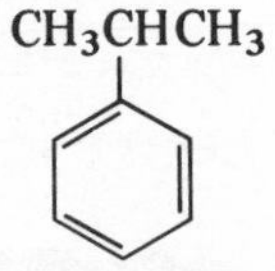

Cumene
(isopropylbenzene)

14-2 Preparation of Benzene and Benzene Derivatives

a. *From Phenylmagnesium Bromide*

$$\text{C}_6\text{H}_5-\text{MgBr} + \text{HOH} \longrightarrow \text{C}_6\text{H}_6 + \text{MgBrOH}$$

b. *From Benzoic Acid*

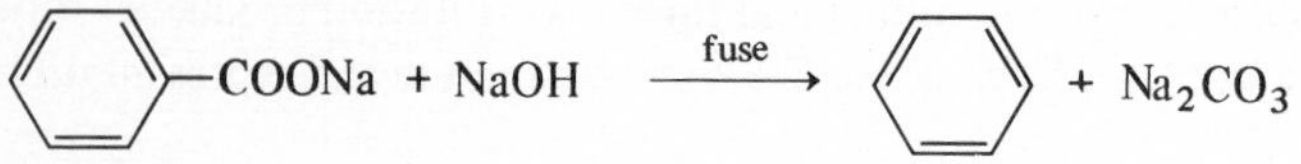

$$\text{C}_6\text{H}_5-\text{COONa} + \text{NaOH} \xrightarrow{\text{fuse}} \text{C}_6\text{H}_6 + \text{Na}_2\text{CO}_3$$

c. *Hydrolysis of Benzenesulfonic Acid*

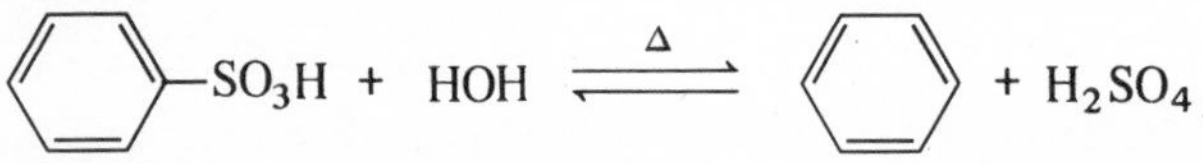

$$\text{C}_6\text{H}_5-\text{SO}_3\text{H} + \text{HOH} \underset{\Delta}{\rightleftharpoons} \text{C}_6\text{H}_6 + \text{H}_2\text{SO}_4$$

d. *Pyrolysis of Acetylenes*

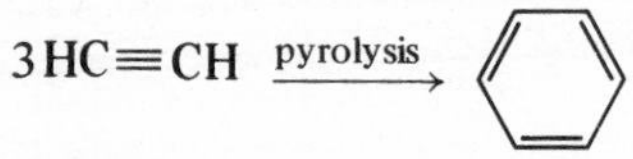

$$3\,\text{HC}\equiv\text{CH} \xrightarrow{\text{pyrolysis}} \text{C}_6\text{H}_6$$

e. *Dehydrocyclization*

$$\text{CH}_3(\text{CH}_2)_4\text{CH}_3 \xrightarrow{\text{Cr}_2\text{O}_3,\ 550°\text{C}} \text{C}_6\text{H}_6 + 4\text{H}_2$$

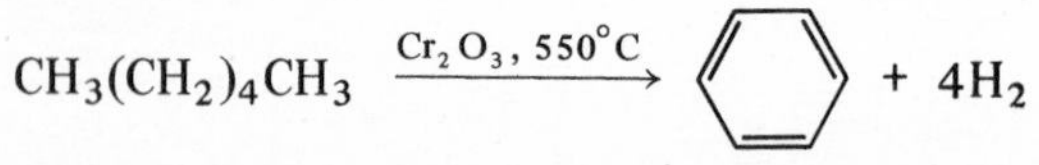

f. *Diels-Alder Reaction* A 1,3-diene will condense with an olefin when the two are warmed together to form a six-membered ring. This reaction is referred to as a [4 + 2] cycloaddition reaction in that it involves a four-π-electron system (the 1,3-diene) reacting with a two-π-electron system (the olefin or dienophile) producing a cyclic compound. The reaction proceeds best when the dienophile has electron-attracting substituents such as $-\text{NO}_2$, $-\text{CN}$, $-\text{CHO}$, and $-\text{COOR}$ on the olefinic carbons. The reaction always occurs to give cis addition. For example, butadiene and maleic anhydride react to yield *cis*-1,2,3,6-tetrahydrophthalic anhydride as shown in the second example below.

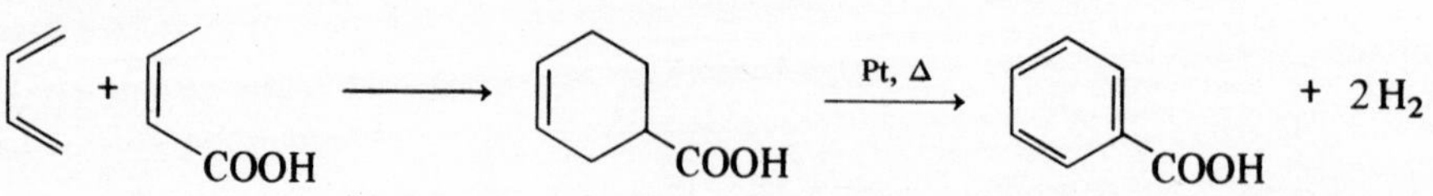

$$\text{diene} + \text{olefin(COOH)} \longrightarrow \text{cyclohexene(COOH)} \xrightarrow{\text{Pt, }\Delta} \text{Benzoic acid (COOH)} + 2\text{H}_2$$

Benzoic acid

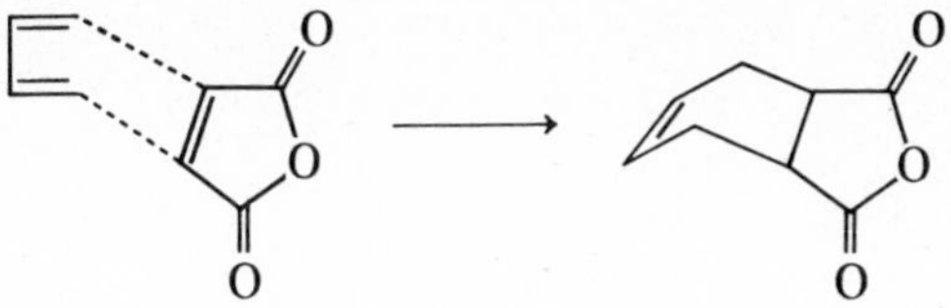

Many Diels-Alder products are readily dehydrogenated to benzene derivatives as shown above in the preparation of benzoic acid.

14-3 Reactions of Benzene and Alkylbenzenes

a. *Oxidation* Benzene does not react with alkaline $KMnO_4$ even on heating.

b. *Addition* In the presence of ultraviolet light benzene will add 3 molecules of Cl_2 or Br_2 and yield the hexahalocyclohexane.

c. *Hydrogenation* Benzene will add hydrogen, although with considerably more difficulty than olefins do. A specially prepared, highly active platinum catalyst is necessary.

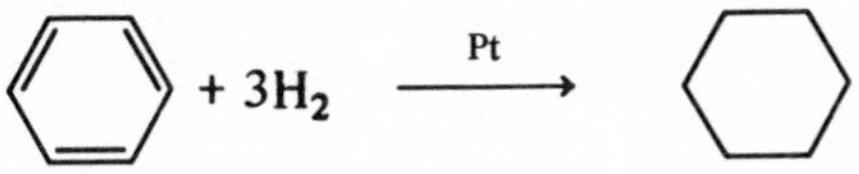

d. *Ozonolysis* (See Sec. 4-3e and Prob. 4-16.)

e. **Substitution reactions**
    1. Halogenation (Cl_2 or Br_2).

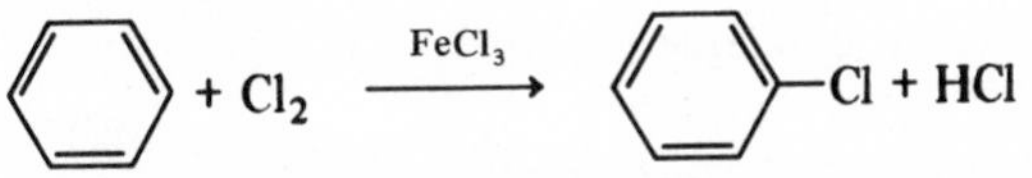

Iodobenzene can be prepared from benzene using I_2 or ICl in the presence of a catalyst such as HNO_3 or $ZnCl_2$. Most aryl iodides and fluorides are prepared from diazonium salts (see Sec. 17-5a) rather than by direct halogenation. Fluorobenzenes have been prepared by direct fluorination in some cases using XeF_2 in CCl_4 with HF as a catalyst.

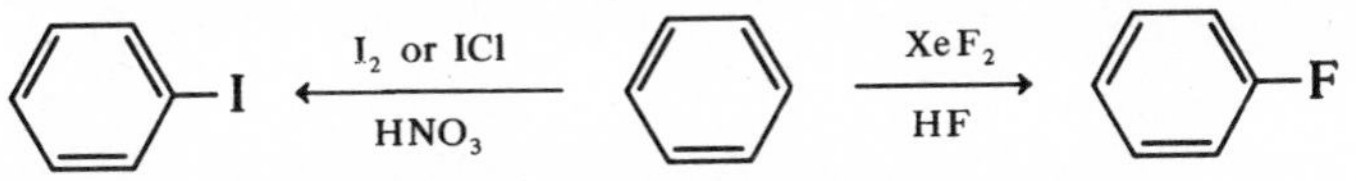

2. Nitration.

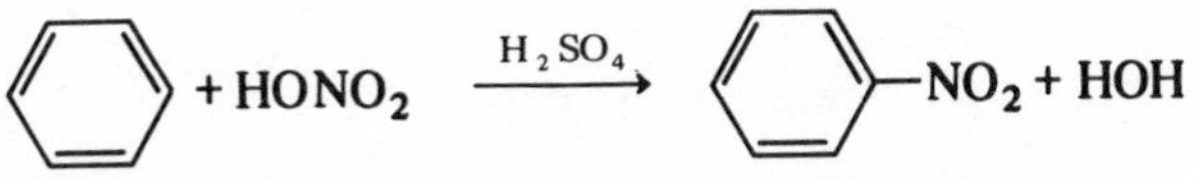

3. Sulfonation.

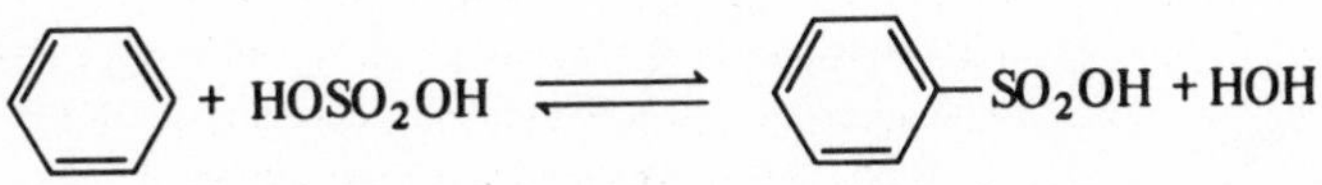

4. Alkylation (Friedel-Crafts reaction). Alkyl halides, alcohols, or olefins react with benzene in the presence of a Lewis acid such as $AlCl_3$.

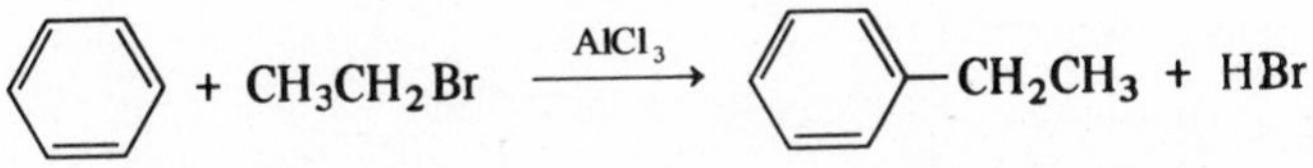

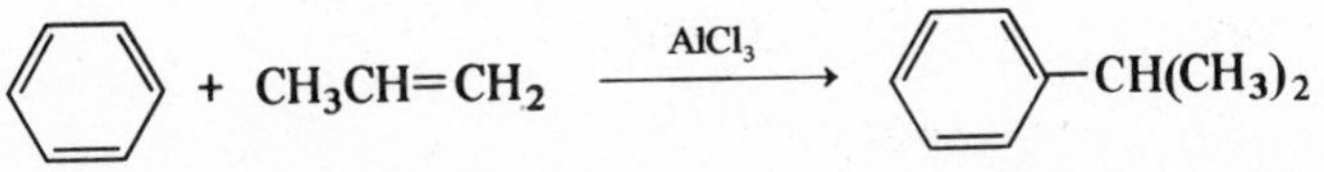

205

5. **Acylation.** An acyl halide or anhydride used in a Friedel-Crafts reaction yields an aromatic ketone (aryl ketone).

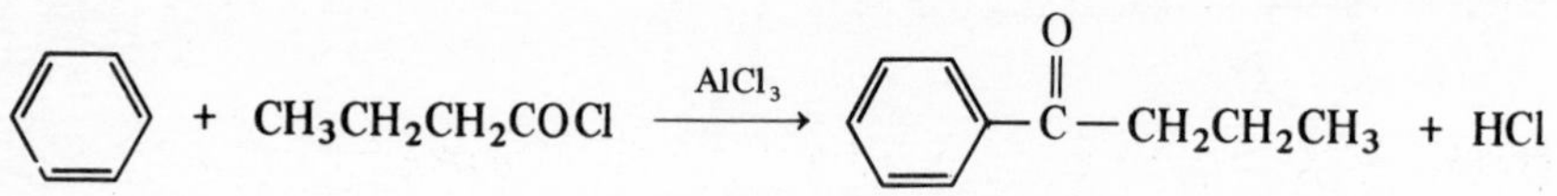

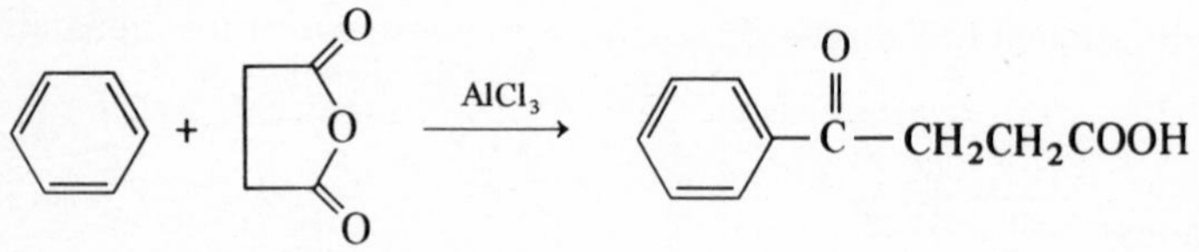

$$C_6H_5 + CH_3CH_2CH_2COCl \xrightarrow{AlCl_3} C_6H_5-\overset{\displaystyle O}{\overset{\|}{C}}-CH_2CH_2CH_3 + HCl$$

6. **Oxidation of side chains.** Note that primary and secondary side chains are oxidized to carboxyl groups but tertiary groups are unaffected.

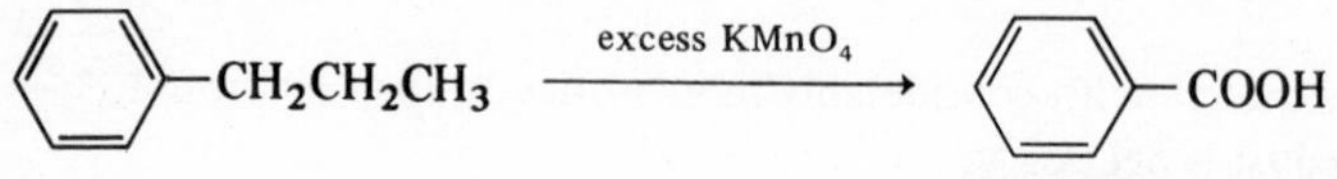

$$C_6H_5-CH_2CH_2CH_3 \xrightarrow{\text{excess } KMnO_4} C_6H_5-COOH$$

Benzoic acid

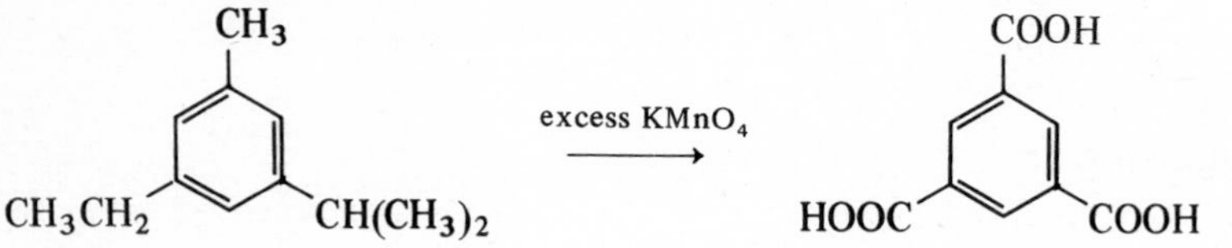

1,3,5-Benzenetricarboxylic acid

Benzene may be catalytically oxidized at 400°C to yield maleic anhydride.

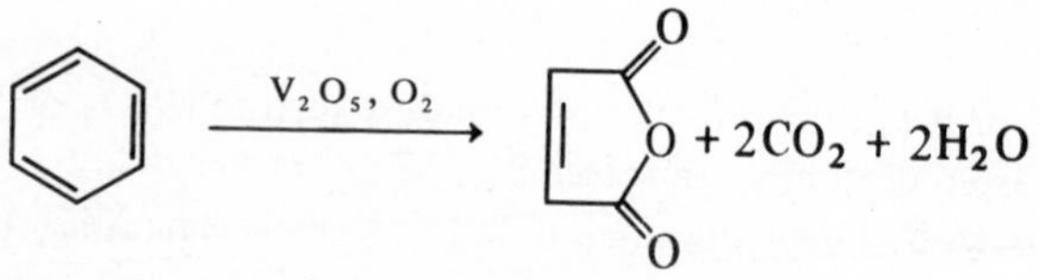

$$\text{Benzene} \xrightarrow{V_2O_5, O_2} \text{maleic anhydride} + 2CO_2 + 2H_2O$$

14-4 Synthesis of Benzene Homologs

a. *Wurtz-Fittig Reaction*

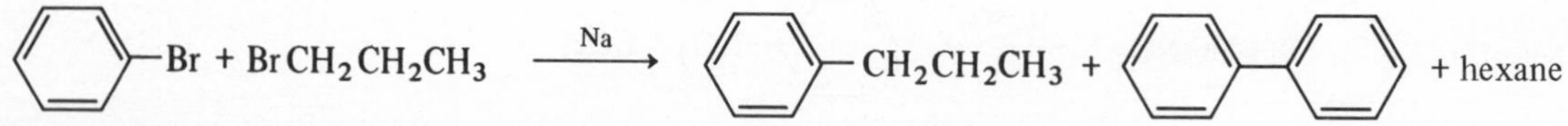

$$C_6H_5-Br + BrCH_2CH_2CH_3 \xrightarrow{Na} C_6H_5-CH_2CH_2CH_3 + \text{(biphenyl)} + \text{hexane}$$

b. *Ullmann Reaction*

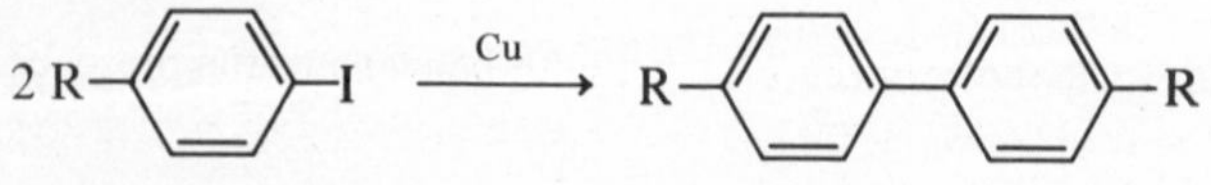

$$2\ R-C_6H_4-I \xrightarrow{Cu} R-C_6H_4-C_6H_4-R$$

c. *Clemmensen Reduction*

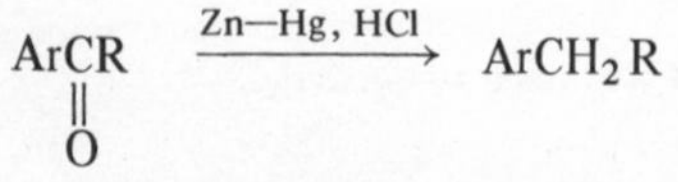

$$ArCR \xrightarrow{Zn-Hg, HCl} ArCH_2R$$
$$\ \ \|$$
$$\ \ O$$

Wolff-Kishner reduction (see Sec. 7-3*i*) works also.

d. *Alkylation by Means of Grignard Reagent* Very active halides (such as allyl and benzyl) and esters of sulfuric acid alkylate Grignard reagents.

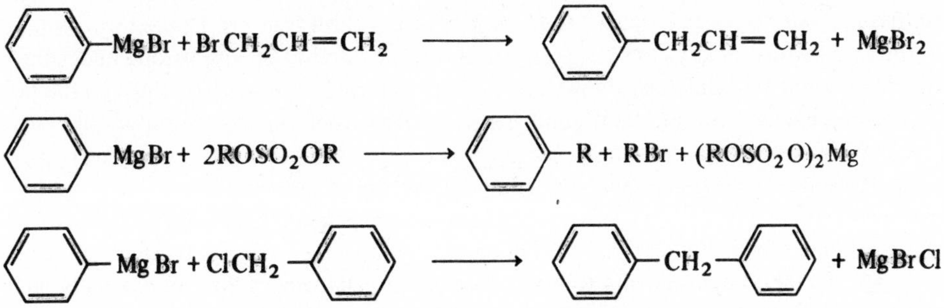

$$C_6H_5-MgBr + BrCH_2CH=CH_2 \longrightarrow C_6H_5-CH_2CH=CH_2 + MgBr_2$$

$$C_6H_5-MgBr + 2ROSO_2OR \longrightarrow C_6H_5-R + RBr + (ROSO_2O)_2Mg$$

$$C_6H_5-MgBr + ClCH_2-C_6H_5 \longrightarrow C_6H_5-CH_2-C_6H_5 + MgBrCl$$

Benzyl chloride is made by the action of light and chlorine on toluene. In this instance side-chain halogenation takes place more rapidly than ring substitution or addition.

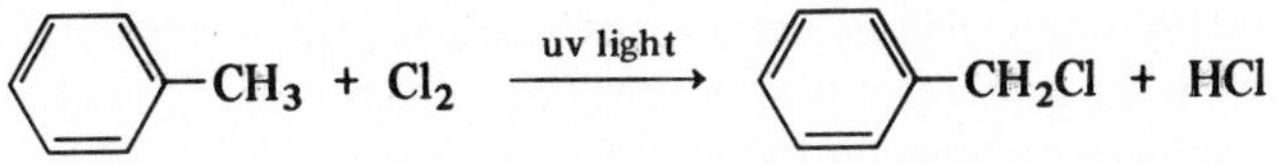

$$C_6H_5-CH_3 + Cl_2 \xrightarrow{\text{uv light}} C_6H_5-CH_2Cl + HCl$$

Benzyl chloride

Benzyl group $\equiv C_6H_5CH_2-$

e. *Friedel-Crafts Reactions* Alkyl halides, alcohols, or olefins react with aromatic compounds in the presence of Lewis acid catalysts (such as $AlCl_3$, $SnCl_4$, and $FeBr_3$) to give alkylated products. The reaction suffers from two practical disadvantages: (1) the product is a mixture of mono- and polyalkylated homologs and (2) it is difficult to prepare normal or primary alkyl derivatives above ethyl because the alkyl groups isomerize.

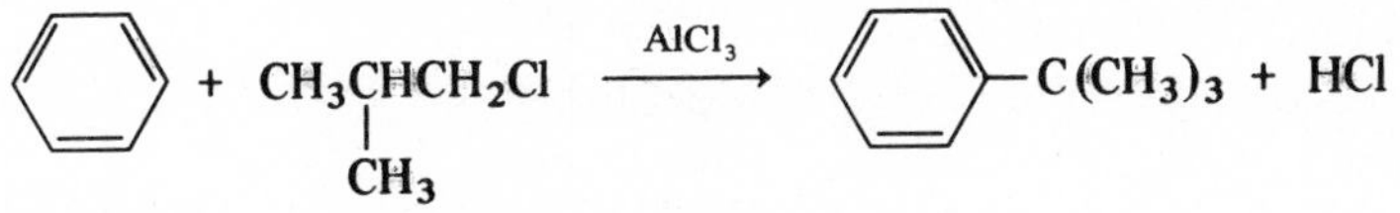

$$C_6H_6 + CH_3\underset{\underset{CH_3}{|}}{C}HCH_2Cl \xrightarrow{AlCl_3} C_6H_5-C(CH_3)_3 + HCl$$

When benzene and 3 equiv of methyl chloride are allowed to react in the presence of a large excess of $AlCl_3$ catalyst, the product is mesitylene.

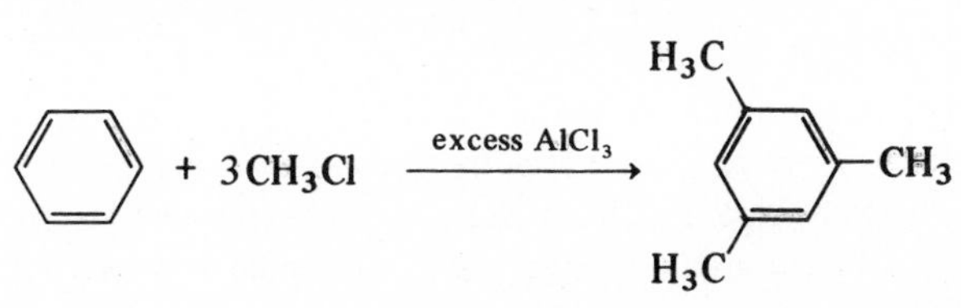

$$C_6H_6 + 3CH_3Cl \xrightarrow{\text{excess } AlCl_3} \text{mesitylene}$$

Certain limitations are recognized for the Friedel-Crafts process. The function of the catalyst is to provide an electron-deficient reagent for attack on the electron-rich ring: $RCl + AlCl_3 \longrightarrow R^+AlCl_4^-$. A similar reagent is formed from acyl halides and olefins. Thus the reagent is electron-seeking. Strong electron-attracting groups on the aromatic nucleus slow down or prevent the reaction. Nitrobenzene, for example, may be used as solvent for the reaction. Groups that have an unshared pair of electrons with which the $AlCl_3$ can coordinate (such as $-OH$ and $-NH_2$) prevent the reaction or give low yields. The most generally useful groups that may be attached to the aromatic ring are the halogens and alkyl groups.

14-5 Organic Syntheses

There are three basic problems in the synthesis of aromatic molecules: (1) formation of the aromatic ring, (2) introduction of one substituent into the ring and its possible elaboration into a complex side chain, (3) introduction of two or more substituents into the ring. Only the first two of these will be considered in this chapter.

First consider the problem of the ring. In everyday laboratory problems the solution to this is usually very easy — simply order some benzene from the nearest chemical supply house and introduce substituents into the ring by chemical means. The vast majority of aromatic syntheses are solved every day in the laboratory by this approach. However, there are instances in which this will not offer the proper solution. One of these arises occasionally when one is asked to make a molecule with a complex set of substituents in the ring. In such a case, the Diels-Alder reaction may provide a relatively easy solution, or sometimes complex condensations of the aldol type can be used. For example, heating acetone with concentrated sulfuric acid yields 1,3,5-trimethylbenzene. Methyl phenyl ketone under the same conditions gives 1,3,5-triphenylbenzene (for practice, show stepwise how this might occur). Run through the same reaction with cyclohexanone and see what you get. A different situation arises when the natural supply of benzene, toluene, etc., is not great enough and one must create more by synthesis. Until World War II, all aromatic compounds came from coal tar. Now the demand is so great that aliphatic compounds from petroleum must be converted into simple aromatics. Two approaches are open: the pyrolysis of acetylene and its simple derivatives (see Sec. 14-2d), which finds little use because of the expense of starting materials; and dehydrocyclization (see Sec. 14-2e), which is very useful in converting hexane and heptane to benzene and toluene, respectively. Dehydrocyclization is of less value for more complex derivatives of benzene because of the isomerization of dehydrogenation of alkyl groups that can occur. For example, in the dehydrogenation of octane one gets a mixture of ethylbenzene and the three possible xylenes. Furthermore, under the conditions of the reaction, some of the ethylbenzene is dehydrogenated to styrene. A third special situation arises when a need appears for benzene and its derivatives labeled with carbon isotopes in the ring. Again, reactions such as the Diels-Alder and aldol condensation will be of use, because specific carbon labeling in the starting materials will be reflected in a specific manner in the labeling of the benzene ring. Referring back to the formation of 1,3,5-trimethylbenzene from acetone, present structures for the product which would form from $(CH_3)_2{}^{14}CO$ and $({}^{14}CH_3)_2CO$.

To summarize, when you have a problem involving an aromatic ring, your first thought should be to start with benzene or a simple derivative. If this is not feasible or if the molecule is particularly complex, the Diels-Alder reaction (see Sec. 14-2f) is probably the next approach to consider. Finally, one of the three more-specialized methods, cyclic aldol condensation, pyrolysis of an acetylene, or dehydrocyclization, may turn out to be the method of choice. Consider the following examples:

I   II   III

Molecule I is an interesting molecular aggregation for contemplation because synthetic schemes can be outlined for it using each of the four methods outlined above. In fact, before reading ahead, try your hand at composing this molecule from an acetylene, a ketone (with sulfuric acid), a Diels-Alder synthesis, dehydrocyclization, and a direct process from benzene. Now let's consider the pros and cons of each method. Since it would be relatively easy (synthetically speaking) to make dimethylacetylene, pyrolysis of this compound would be an acceptable method. However, since special apparatus is needed for this reaction and since it has not been widely used, most chemists would not use it. If you discovered that the ketone necessary for condensation with sulfuric acid is methyl ethyl ketone, you would realize that this can condense in at least two ways, one giving hexamethylbenzene and the other 1,3,5-triethylbenzene. One could also get mixed methylethylbenzenes. For the Diels-Alder method one could condense 2-butene with 3,4-dimethyl-2,4-hexadiene. This would be a poor method because good yields are usually only obtained when an activating group is present in the olefin. Otherwise the diene tends to react with itself. For the dehydrocyclization reaction, one would need 3,4,5,6-tetramethyloctane. This would be quite difficult to come by synthetically and also would undergo a variety of isomerizations during the reaction. Of course the best way to prepare this compound would be with methyl chloride in the Friedel-Crafts reaction.

Compounds II and III are good examples for the diene reaction. Both have activating groups that can be included in the dienophile. Compound III is not aromatic, and it can easily be put together using 2,3-dimethyl-1,3-butadiene and 2-methyl-2-cyclopenten-1-one. Compound II requires a dehydrogenation step following the reaction of 2,3-dimethyl-1,3-butadiene with crotonaldehyde.

Now, how does one elaborate a side chain on benzene? In general, one must put some simple active group onto benzene and build up from this point. It is often harder to make a complicated piece and put it on the ring than to build it up. We shall consider two ways to get started on a benzene ring; later on we shall take up many more. Consider halogenation and the Friedel-Crafts reactions.

Chlorobenzene is too inert to form Grignard reagents easily, and iodobenzene is expensive and more difficult to prepare, so that most often one makes bromobenzene to form a Grignard reagent. With the ring now in the form of a Grignard reagent, it can be incorporated into all kinds of molecules by addition to carbonyl groups or displacement of halides from allylic halides.

Much can be done with a methyl group on benzene. It can be oxidized to benzoic acid, which can be converted into an ester for various condensations, etc. The methyl group could be halogenated to yield benzyl chloride, which will give the great variety of displacement reactions with nucleophilic reagents. In fact, almost all of the aliphatic chemistry you have learned so far can be carried out with the chloromethyl group (CH_2Cl). Relatively little interference is encountered with the ring when chemical operations are carried out on a side chain. Try working Probs. 14-4 and 14-5 for illustrations of side-chain elaboration.

Problems

14-1 Draw structures for each of the following:

a. *m*-Bromonitrobenzene
b. *o*-Ethyltoluene
c. *p*-Xylene
d. 3-Nitrobromobenzene
e. 2,4-Dibromomesitylene
f. Phenylcyclohexane
g. Biphenyl
h. Benzyl cyanide
i. Isobutylbenzene
j. *p*-Allylvinylbenzene
k. 1,4-Diphenyl-1,3-butadiene
l. *o*-Dichlorobenzene
m. *p*-Nitrocumene
n. Ethyl phenylacetoacetate
o. Benzyl γ-phenylcrotonate

14-2 Provide the missing reactant or reaction condition for the following:

a. Benzene + ? $\xrightarrow{AlCl_3}$ *t*-butylbenzene

b. Dimethyl sulfate + ? $\longrightarrow$ ethylbenzene

c. Mesitylene + ? $\longrightarrow$ [structure: benzene ring with HOOC at top, COOH at right, HOOC at bottom]

d. CH_3—[ring]—MgBr + ? $\longrightarrow$ [ring]—CH_2—[ring]—CH_3

e. Heptane + ? $\longrightarrow$ toluene
f. Benzene + ? $\xrightarrow{AlCl_3}$ hexamethylbenzene
g. Iodobenzene + ? $\longrightarrow$ biphenyl
h. $C_6H_5COCH_3$ + ? $\longrightarrow$ ethylbenzene
i. $C_6H_5COCH_3$ + ? $\longrightarrow$ $C_6H_5COCH=CHC_6H_5$
j. 1,3-Butadiene + ? $\longrightarrow$ 3-cyclohexenecarboxylic acid

14-3 Review the IUPAC rules for naming esters, amides, acid chlorides, and acid anhydrides (see Sec. 9-1 and 11-1) and then give an IUPAC name for each of the following:

a. $C_6H_5COOCH_3$
b. $C_6H_5CONH_2$
c. C_6H_5COCl
d. $(C_6H_5CO)_2O$
e. $C_6H_5COOC_6H_5$
f. $C_6H_5COOCH_2C_6H_5$

14-4 Show how to make the following conversions using any necessary reagents:

a. Benzene to ethylbenzene
b. Toluene to benzyl alcohol
c. Benzoic acid to benzonitrile ($C_6H_5C\equiv N$)
d. Styrene ($C_6H_5CH=CH_2$) to phenylacetylene
e. Benzene to biphenyl
f. Benzene to benzyl benzoate
g. Iodobenzene to benzene
h. Benzyl cyanide and ethanol to benzyl ethyl ketone
i. Toluene to 4-phenyl-1-butene
j. Ethyl benzoate to benzyl iodide
k. Benzene to triphenylcarbinol
l. Phenylacetylene to benzoic acid
m. $C_6H_5COCH_3$ to $H_2C=C(C_6H_5)C(C_6H_5)=CH_2$
n. Benzene to tartaric acid
o. Styrene to $C_6H_5CH_2CH_2OH$
p. Benzaldehyde to $C_6H_5CH=CHCH_3$

q. *p*-Xylene to HOOC—[ring with F]—COOH

r. Toluene to $C_6H_5CH_2COCHCOOEt$ with C_6H_5 below

14-5   Give structures for each of the lettered intermediates in each of the proposed syntheses:

a.   Bromobenzene + Mg $\longrightarrow$ (A) $\xrightarrow{CH_3CHO}$ $\xrightarrow{H^+}$ (B) $\xrightarrow{HBr}$ $\xrightarrow{Mg}$

(C) $\xrightarrow[\text{oxide}]{\text{ethylene}}$ $\xrightarrow{H^+}$ (D) $\xrightarrow{HBr}$ $\xrightarrow{Mg}$ $\xrightarrow{CO_2}$ $\xrightarrow{H^+}$ (E) $\xrightarrow{SOCl_2}$ (F) $\xrightarrow{NH_3}$

(G) $\xrightarrow{NaOBr}$ (H) $\xrightarrow{ketene}$ (I)

b.   Toluene $\xrightarrow[uv]{Cl_2}$ (A) $\xrightarrow{NaCN}$ (B) $\xrightarrow{EtMgI}$ $\xrightarrow{HOH}$ (C) $\xrightarrow[Zn]{BrCH_2COOEt}$ $\xrightarrow{H^+}$ (D)

c. 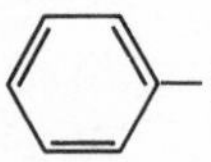   ⬡—COOCH$_3$ $\xrightarrow[CH_3ONa]{CH_3COOCH_3}$ (A) $\xrightarrow[2.\ \ CH_3I]{1.\ CH_3ONa}$ (B) $\xrightarrow[2.\ \ I_2]{1.\ CH_3ONa}$ (C) $\xrightarrow[\Delta]{H_3O^+}$ (D) $\xrightarrow[CH_3ONa]{H_2NNH_2}$ (E)

14-6   Suggest a chemical test with an easily observable change to distinguish between each of the following:
a.   Benzene and cyclohexane   b.   Benzene and cyclohexene
c.   Chlorobenzene and benzyl chloride   d.   Benzene and benzoic acid
e.   Benzonitrile and benzamide   f.   Methyl phenyl ketone and ethyl phenyl ketone

14-7   In 1874 Wilhelm Koerner pointed out that if all the positions on a benzene ring are equivalent, there should exist only three disubstituted benzenes (o, m, p). These should be distinguishable by the number of isomers each would yield when an additional substituent is introduced. Assume you have three bottles, each containing one of the three xylenes. Show how Koerner's method could be used to tell which was which. That is, how many mononitro derivatives (for example) would you expect to be formed by the nitration of each of the xylenes?

Could Koerner's method be used to distinguish between the three possible trimethylbenzenes? Between o- and m-bromochlorobenzene? Explain.

14-8   Draw the formula for the product you would expect from the Diels-Alder reaction between each of the following pairs:
a.   2,3-Dimethyl-1,3-butadiene and maleic acid   b.   2,4-Hexadiene and acrylonitrile
c.   4-Methylstyrene and maleic anhydride   d.   1,3-Cyclopentadiene and ethylene

e.   Sulfur dioxide and 1,3-butadiene   f   (furan structure) and acetylenedicarboxylic acid

14-9   What products would you expect to isolate from the ozonolysis of o-xylene after decomposition of the ozonide with Zn and dilute acetic acid?

14-10   a.   A hydrocarbon C_8H_{10}, is an alkylbenzene that gives only one ring-monobrominated derivative. Suggest a structure.
b. C_8H_{10} gives three ring-monobrominated derivatives. Suggest two structures.
c. C_9H_{12} gives one ring-monobrominated derivative. Suggest one structure.
d. C_9H_{12} gives three ring-monobrominated derivatives. Suggest two structures.
e. C_9H_{12} gives four possible ring-monobrominated derivatives. Suggest two structures.

14-11   Compound A, C_9H_{12}, showed no active hydrogen in the Zerewitinoff determination. Vigorous oxidation of A gave an acid, B. A sample of 0.10 g of B required 24.1 ml of 0.05 N NaOH for neutralization. Nitration of B gave a mixture of three mononitro compounds. Suggest a structure for A.

14-12   Neutral compound A was found to have a saponification equivalent (SE) of 178 ± 2. Distillation of the alkaline solution gave a neutral compound that gave an iodoform test and a positive Lucas test. From the residue left from the steam distillation, an acid of neutralization equivalent (NE) 136 ± 1 was isolated. Vigorous oxidation of this acid gave a new acid of NE 121 ± 1, which on fusion with NaOH gave benzene. Give a structure for A.

14-13   Suggest a chemical procedure for separating each of the following mixtures:
a.   Methanol, benzene, benzoic acid, and cyclohexanone
b.   Aniline, N-methylaniline, o-xylene, and 1,3,5-trihydroxybenzene

14-14   Under proper conditions bromobenzene is converted to phenyllithium when treated with metallic lithium.

What products would you expect from phenyllithium and each of the following?

a. Water
b. Acetone
c. Methanol
d. Phenyl isocyanate
e. Acetonitrile
f. Sulfur dioxide
g. Hydrogen sulfide
h. Benzaldehyde

14-15 Given the information that the function of the $AlCl_3$ catalyst in the Friedel-Crafts reaction is to promote the formation of carbonium ions by removal of a halogen from an alkyl halide, that tertiary carbonium ions are most stable and primary least stable, and that hydrogen or alkyl groups adjacent to a carbonium carbon may migrate to that carbon, what would you expect for the major product in the reaction between benzene and each of the following?

a. 1-Chloro-2-methylpropane
b. Isobutylene
c. Neopentyl chloride
d. 1-Chloro-2-methycyclohexane
e. Cyclopentene
f. Chloromethylcyclopentane

14-16 Given methanol, ethanol, benzene and any necessary inorganic compounds, show how you would make the following conversions:

a. Benzene to γ-phenyl-γ-ketobutyric acid
b. Cyclohexanone to 6-phenylhexanoic acid
c. 2,3-Dimethyl-1,3-butadiene to 1,2,4,5-tetracarboxybenzene
d. 2,3-Diphenyl-1,3-butadiene to 1,2-diacetyl-4,5-diphenylbenzene
e. Benzene to methyl 2-methyl-3-hydroxy-3-phenylbutanoate
f. Benzyl chloride to 1,3-diphenyl-2-methyl-2-propanol
g. Benzene to 1,6-diphenylhexane
h. Cyclohexanol to 1-phenyl-1-cyclohexene
i. Benzene to 2,4-diphenyl-2-butenal
j. $C_6H_5COCH_3$ to 3,5-diphenyl-2-cyclohexen-1-one

14-17 Refer to the Diels-Alder reaction in Sec. 14-2f. Keeping in mind that only cis addition occurs and that the position isomerism may occur when unsymmetric reagents are used, predict the number of stereoisomers that could form in each of the following examples and indicate how many would be optical isomers:

a. Fumaric acid + 1,3-butadiene
b. Maleic acid + 2-methyl-1,3-butadiene
c. 1-Phenyl-1,3-butadiene + acrolein
d. 1,3-Cyclopentadiene + maleic acid
e. Fumaric acid + 2-methyl-1,3-butadiene
f. 1,3-Cyclohexadiene + acrylonitrile

14-18 Remembering that oxidation with $KMnO_4$ gives the cis diol, that one gets cis addition in the Diels-Alder reaction, that displacement reactions occur with inversion of configuration, and that ionic additions to double bonds go trans, trace the steric course of the following reactions:

a. Cyclohexene + 1,3-butadiene $\longrightarrow$ (A) $\xrightarrow[\text{HCl}]{KMnO_4}$ (B) $\xrightarrow{CH_3CHO}$ (C)

b. 1,3-Cyclohexadiene + ethylene $\longrightarrow$ (A) $\xrightarrow{HOBr}$ (B) $\xrightarrow{\text{concd NaOH}}$ (C)

14-19 An optically active acid, A, was found to be insoluble in water and to have a NE of 125 ± 1. Heating A with Pt converted it into a new acid of NE 122 ± 1. Mild oxidation of A converted it into B, with NE 63 ± 1. When 0.126 g of A was hydrogenated using a Pd catalyst, 22.4 ml of hydrogen was absorbed. Suggest structures for A and B if B is also dissymmetric.

14-20 When carbon dioxide is added to an organolithium compound, some of the lithium salt of the acid is obtained, but a major product is a ketone. Formulate this reaction using butyllithium.

14-21 Formulate equations for the following examples illustrating the use of organolithium compounds in synthesis:

a. p-Chlorobromobenzene $\xrightarrow{Mg}$ (A) $\xrightarrow[\text{iodide}]{\text{benzyl}}$ (B) $\xrightarrow{Li}$ $\xrightarrow{CH_3CHO}$ $\xrightarrow{H^+}$ (C)

b. Benzonitrile $\xrightarrow{n\text{-}C_4H_9Li}$ $\xrightarrow{H_3O^+}$ (A) $\xrightarrow{H_2NNH_2,\ CH_3ONa}$ (B)

c. p-Chlorobenzyl chloride $\xrightarrow{Mg}$ $\xrightarrow{CO_2}$ $\xrightarrow{H^+}$ (A) $\xrightarrow{CH_2N_2}$ (B) $\xrightarrow{2C_6H_5Li}$ $\xrightarrow{H^+}$ (C)

14-22 It is possible to prepare B-alkyl and B-aryl derivatives (I) of 9-BBN (see Sec. 6-2e) in the following manner:

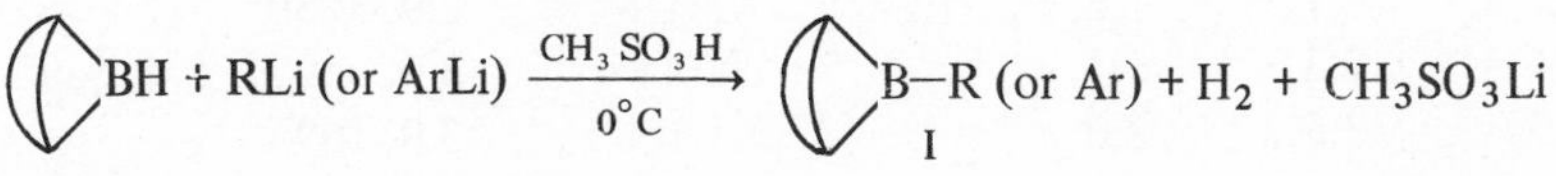

$$\text{⬡BH} + \text{RLi (or ArLi)} \xrightarrow[0°C]{CH_3SO_3H} \text{⬡B—R (or Ar)} + H_2 + CH_3SO_3Li$$
I

Using this information and reviewing Sec. 12-2e, show how to prepare the following compounds given 9-BBN, ethyl bromoacetate, cyclohexanone, methyl phenyl ketone, any organolithium compounds, and any necessary inorganic reagents:

a. 2-Phenyl-1-cyclohexanone
b. Amyl phenyl ketone
c. p-Methylphenylacetic acid

14-23   Coniine, $C_8H_{17}N$, is one of several very poisonous compounds that occur in hemlock. An extract of this material was used by the early Greeks to kill condemned men, one of whom was Socrates. Treatment of coniine with H_2 and Pt does not alter it. Coniine (1.0 g) reacts with $LiAlH_4$ to give 176.4 ml of hydrogen at STP. Dehydrogenation of coniine with sulfur or Pt converts it into conyrine, $C_8H_{11}N$, which on oxidation gives an acid, $C_6H_5NO_2$. Decarboxylation of the acid yields a basic compound, C_5H_5N. Catalytic hydrogenation of this compound, followed by repeated exhaustive methylation, yields 1,4-pentadiene. Suggest three possible structures for coniine.

14-24   *a.* When benzylmagnesium chloride is treated with formaldehyde and then aqueous acid, a compound with the formula $C_6H_{10}O$ is formed. It gives on oxidation with $KMnO_4$ a dicarboxylic acid, $C_8H_6O_4$, which readily loses water on heating. Keeping in mind the reactions of allylic compounds, suggest an appropriate structure for the Grignard product and provide a mechanistic route for the reaction.

14-25   *a.* Provide a free-radical mechanism for the conversion of toluene to benzyl chloride.
    b. On the basis of the mechanism proposed in *a*, explain why chlorination of ethylbenzene yields predominantly 1-chloro-1-phenylethane.
    c. What product would you expect from the addition of HBr to styrene? Explain.

14-26   The product formed from the reaction of an appropriate dienophile with 1,3-cyclopentadiene can be of the endo or exo configuration, depending, respectively, on whether functional groups end up cis or trans to the two-carbon bridge containing the double bond. Under usual reaction conditions the endo product is obtained. Draw three-dimensional structures for the endo and exo forms of the Diels-Alder adduct from each of the following:
    a. 1,3-Cyclopentadiene + acrylonitrile
    b. Maleic anhydride + 1,3-cyclohexadiene

14-27   Isomeric compounds X, Y, and Z (C_7H_7Br) show the following ir bands (selected): X, 12.34 μm, 810 cm^{-1}; Y, 11.76 μm, 850 cm^{-1}, and 12.99 μm, 770 cm^{-1}; Z, 13.51 μm, 740 cm^{-1}. Suggest structures for X, Y, and Z.

14-28   Compound A (MW 120) when treated with Fe and Br_2 yields compound B (MW 199) which when strongly oxidized yields compound C (NE = 96 ± 2). The nmr spectrum for A shows singlets at δ = 6.7 and 2.2 while the nmr spectrum of B shows singlets at δ = 2.2, 2.35, and 6.8. Suggest structures for A, B, and C.

14-29   Compound D contains C, H, and N (MW 116 ± 1). Its nmr spectrum shows a singlet at δ = 2.35 and a quartetlike set of peaks centered at δ = 7.3 with a respective area ratio of 1:1.33. D exhibits ir bands (selected) at 4.5 μm (2220 cm^{-1}) and 12.1 μm (826 cm^{-1}). Suggest a structure for D.

14-30   Compound E (MW 103 ± 1) shows ir bands (selected) at 3.03 μm (3300 cm^{-1}), 4.76 μm — weak band (2100 cm^{-1}), 13.33 μm (750 cm^{-1}), and 14.50 μm (690 cm^{-1}); its nmr spectrum shows a singlet at δ = 3.05 and a series of finely split peaks centered at δ = 7.3 with a respective area ratio of 1:5. Hydrolysis of E under the proper conditions gives compound F (MW 119 ± 1). The ir spectrum of F shows bands (selected) at 5.92 μm (1690 cm^{-1}), 12.98 μm (770 cm^{-1}), and 14.5 μm (690 cm^{-1}). Its nmr spectrum shows a singlet at δ = 2.6, a series of peaks centered at δ = 7.45, and another series of peaks centered at δ = 7.9 with a respective area ratio of 3:3:2. Suggest structures for E and F.

14-31   Suggest structures for compounds I through X based on the nmr spectral data and other information given below.

Compound	MW	Nmr peaks (δ, multiplicity)				Respective area ratio	Selected ir bands (μm, cm^{-1}) and other information
I	164 ± 1	7.1, *s*	3.8, *t*	2.5, *t*	2.0, *s*	2.5:1:1:1.5	Positive ferric hydroxamate test
II	150 ± 1	7.3, *s*	5.1, *s*	2.1, *s*		2.5:1:1.5	Positive ferric hydroxamate test
III	148 ± 1	7.2, *s*	3.6, *s*	2.4, *q*	1.0, *t*	2.5:1:1:1.5	5.85, 1710
IV	150 ± 1	7.2, *s*	2.7, *s*	1.6, *s*	1.2, *s*	5:2:1:6	2.9, 3450
V	211 ± 1	10.6, *s*	7.3, *s*	5.1, *s*		1:10:1	3.42, 2024; 5.9, 1695
VI	133 ± 1	7.2, *s*	3.5, *s*	2.0, *s*		2.5:1:1.5	5.9, 1695
VII	155 ± 1	7.8, *d* $J = 8$ Hz	7.3, *d* $J = 8$ Hz	2.5, *s*		1:1:1.5	5.9, 1695; contains Cl
VIII	185 ± 1	7.4, *m*	5.2, *q*	2.0, *d*		5:1:3	Contains Br
IX	200 ± 1	7.3, *s*	3.4, *t*	2.8, *t*	2.2, *p*	2.5:1:1:1	Contains Br
X	132	7.1, *m*	1.2, *s*	0.7, *m*		5:3:4	Contains C and H only

14-32 When compound M ($C_{15}H_{16}O$) is dissolved in a strong acid solution, an nmr spectrum is obtained which shows a series of finely split peaks centered at $\delta = 7.5$, a quartet at $\delta = 3.7$, and a triplet at $\delta = 1.2$ with a respective area ratio of 5:1:1.5. Suggest a structure for M and explain the observed nmr spectrum.

14-33 Compound A (MW 122) can be transformed into B (MW 138) by first a substitution reaction followed by a reduction reaction. Compound C (MW 150) can be obtained from B by first a substitution reaction and then a Grignard synthesis. Spectral data are given below for A, B, and C. Suggest structures for A, B, and C and explain the observed nmr spectrum for A.

Nmr (δ, multiplicity, number of hydrogens):

Compound A 9.75, *s*, 2H; 8.2, *d* ($J = 8$ Hz), 2H; 7.2, *d* ($J = 8$ Hz), 2H
Compound B 7.1, *d* ($J = 8$ Hz), 2H; 6.75, *d* ($J = 8$ Hz), 2H; 4.4, *s*, 2H; 3.7, *s*, 3H; 3.1, *s*, 1H
Compound C 7.2, *d* ($J = 9$ Hz), 2H; 6.8, *d* ($J = 9$ Hz), 2H; 3.75, *s*, 3H; 3.5, *s*, 2H

Selected ir bands (μm, cm^{-1}): A – 2.95, 3401; 3.4 (broad), 2941; 6.0, 1667; 11.8, 847; B – 2.95, 3401; 12.2, 820; C – 3.4 (broad), 2941; 5.9, 1695; 12.3, 813.

14-34 Compound X has MW = 106. Its nmr and ir spectra are shown. Suggest a structure for X.

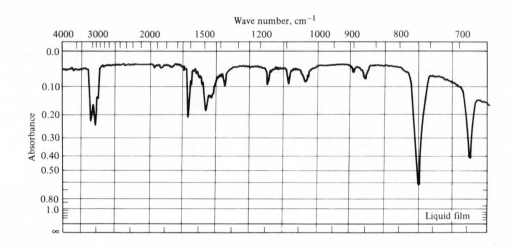

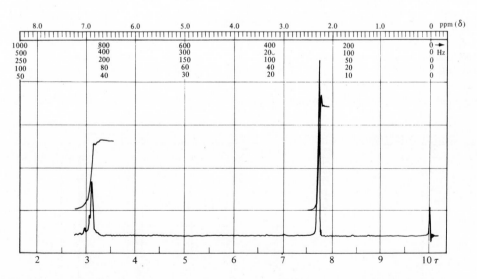

Figure P14-34. Ir and nmr spectra for Prob. 14-34. *(From R. A. Silva and R. Spenger, "The Quantitative Organic Analysis Spectra Kit," McGraw-Hill Book Company, New York, 1969. By permission of the publishers.)*

14-35   Compound Y has MW = 137 and contains N. Its nmr and ir spectra are shown. Suggest a structure for Y.

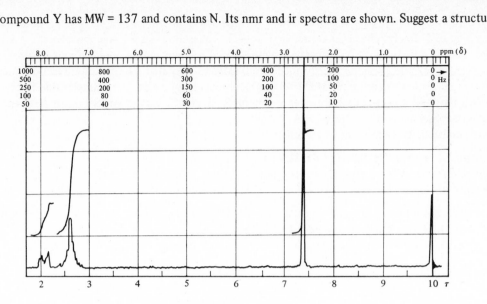

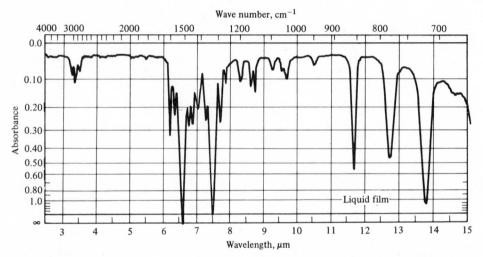

Figure P14-35. Ir and nmr spectra for Prob. 14-35. *(From R. A. Silva and R. Spenger, "The Quantitative Organic Analysis Spectra Kit," McGraw-Hill Book Company, New York,* 1969. *By permission of the publishers.)*

14-36   Compound Z has MW = 134. Its nmr and ir spectra are shown. Suggest a structure for Z.

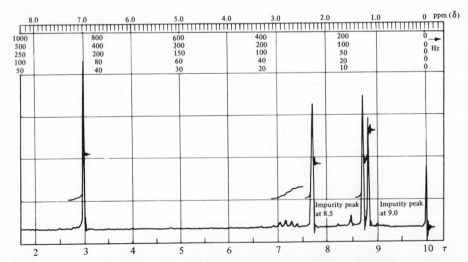

Figure P14-36. Ir and nmr spectra for Prob. 14-36. *(From R. A. Silva and R. Spenger, "The Quantitative Organic Analysis Spectra Kit," McGraw-Hill Book Company, New York,* 1969. *By permission of the publishers.)*

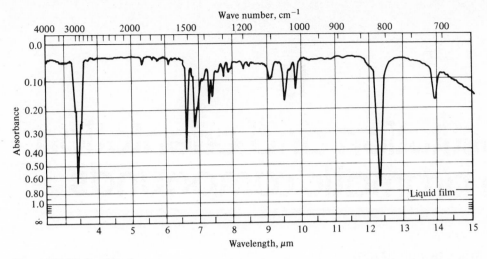

Figure P14-36. *Continued.*

15

substitution and orientation in aromatic compounds

15-1 Electrophilic Aromatic Substitution

The synthetic processes of Sec. 14-3e are typical electrophilic substitution reactions of aromatic compounds. The electrophile (E^+ in the reaction below) would be a nitronium ion, $\overset{+}{N}O_2$, in nitration and a carbonium ion, R^+, in Friedel-Crafts alkylation.

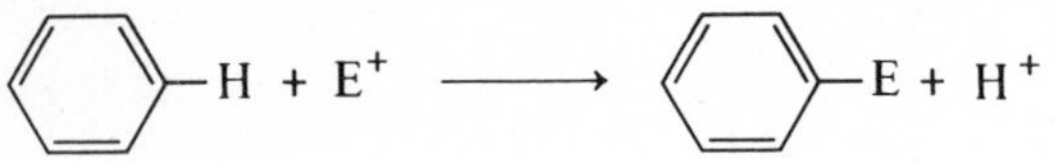

If a substituent is present in the aromatic ring undergoing electrophilic substitution, three isomers may be (and usually are) formed, but the products are in most cases predominantly ortho-para (as in toluene) or meta (as in nitrobenzene).

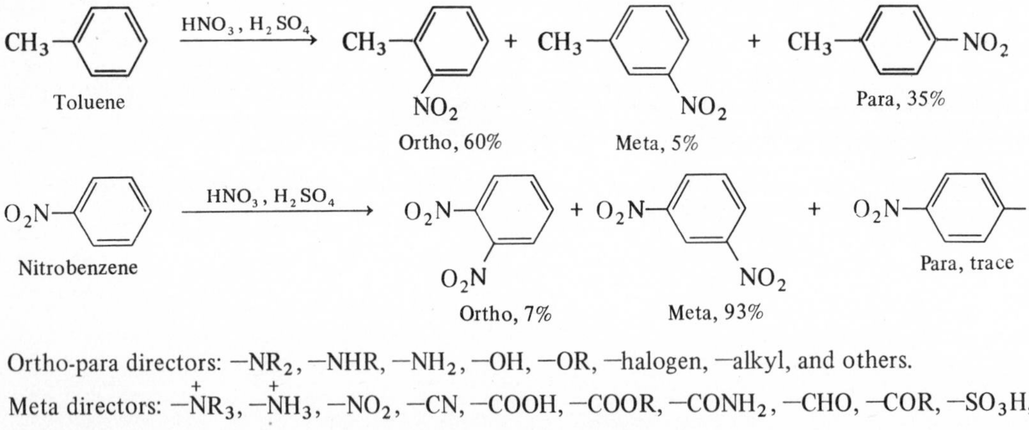

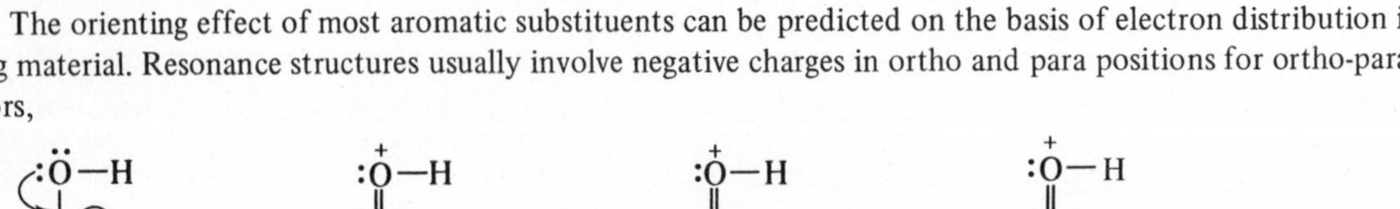

Ortho-para directors: $-NR_2$, $-NHR$, $-NH_2$, $-OH$, $-OR$, $-halogen$, $-alkyl$, and others.

Meta directors: $-\overset{+}{N}R_3$, $-\overset{+}{N}H_3$, $-NO_2$, $-CN$, $-COOH$, $-COOR$, $-CONH_2$, $-CHO$, $-COR$, $-SO_3H$, $-CX_3$, and others.

The orienting effect of most aromatic substituents can be predicted on the basis of electron distribution in the starting material. Resonance structures usually involve negative charges in ortho and para positions for ortho-para directors,

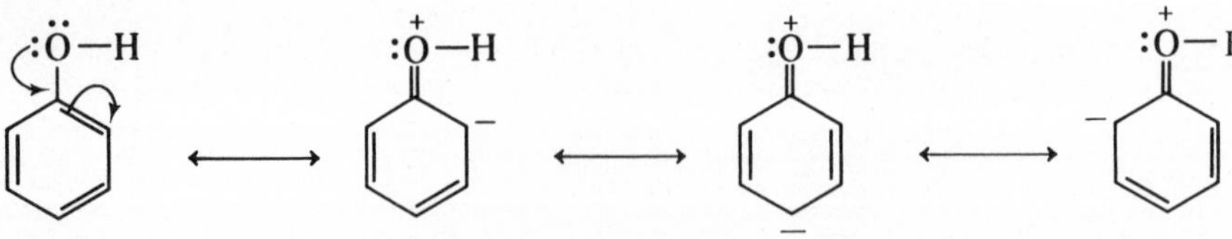

but positive charges in those positions for meta directors

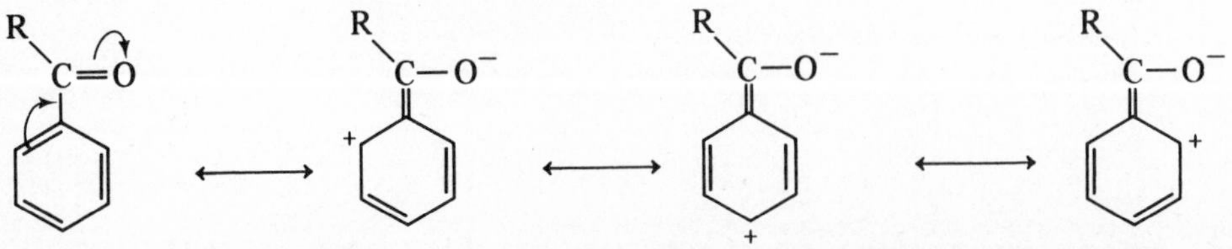

15-2 Mechanism of Electrophilic Substitution

The substitution of a proton of benzene by an electrophile, E^+, is considered to proceed via an ionic intermediate which loses the proton to a base, $B\!:$, present in the system.

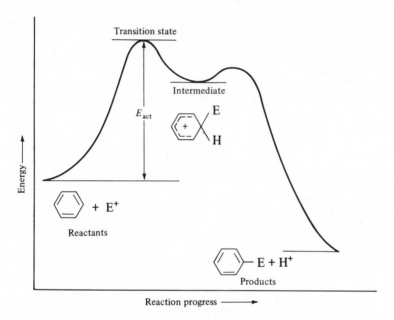

The rate of such a process is a function of the *activation energy* (E_{act} in the energy diagram, Fig. 15-1) which represents the energy necessary to get reactants to the *transition state*. The latter, in turn, is some maximum energy state at which a C—E bond is partially formed and the positive charge is beginning to be delocalized by the aromatic ring.

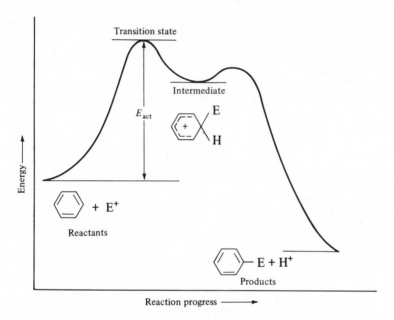

Figure 15-1. Aromatic electrophilic substitution.

A most convenient method for predicting relative rates of electrophilic substitution is to assume that those resonance effects that stabilize the ionic intermediate are also operative in the transition state. Then, the better the system is for delocalizing a positive charge, the lower will be the energy of activation and, of course, a lowered energy of activation implies that a greater proportion of reactants can pass that energy barrier at a given temperature.

Notice that the substituents listed as ortho-para and meta directors fall naturally into two other categories, depending on whether or not they will stabilize a positive charge on an adjacent atom. One group (the ortho-para) aids in delocalization, but the other does not.

217

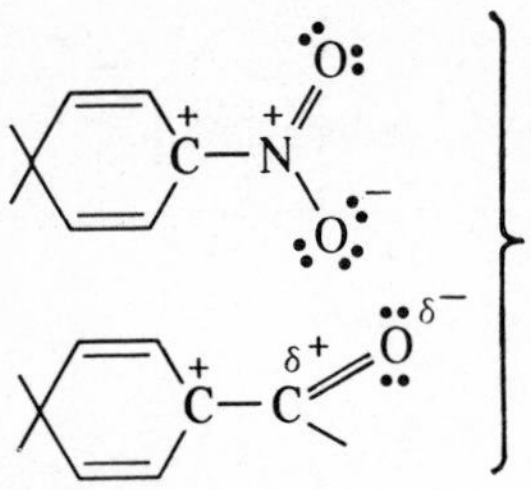

No significant delocalization, but rather an unfavorable energy situation because of adjacent centers of positive charge.

Consider the nitration of methoxybenzene at each of three different kinds of position. In the ortho and para positions the methoxyl group can help delocalize positive charge in the transition state much as it does for the intermediate, but in the meta it can be involved only via indirect means such as an inductive effect. The energy diagram is depicted in Fig. 15-2. Hence, the methoxyl group can be explained as an ortho-para director for electrophilic aromatic substitution.

$$HONO_2 + 2H_2SO_4 \rightleftharpoons \overset{+}{N}O_2 + 2HSO_4^- + H_3O^+$$

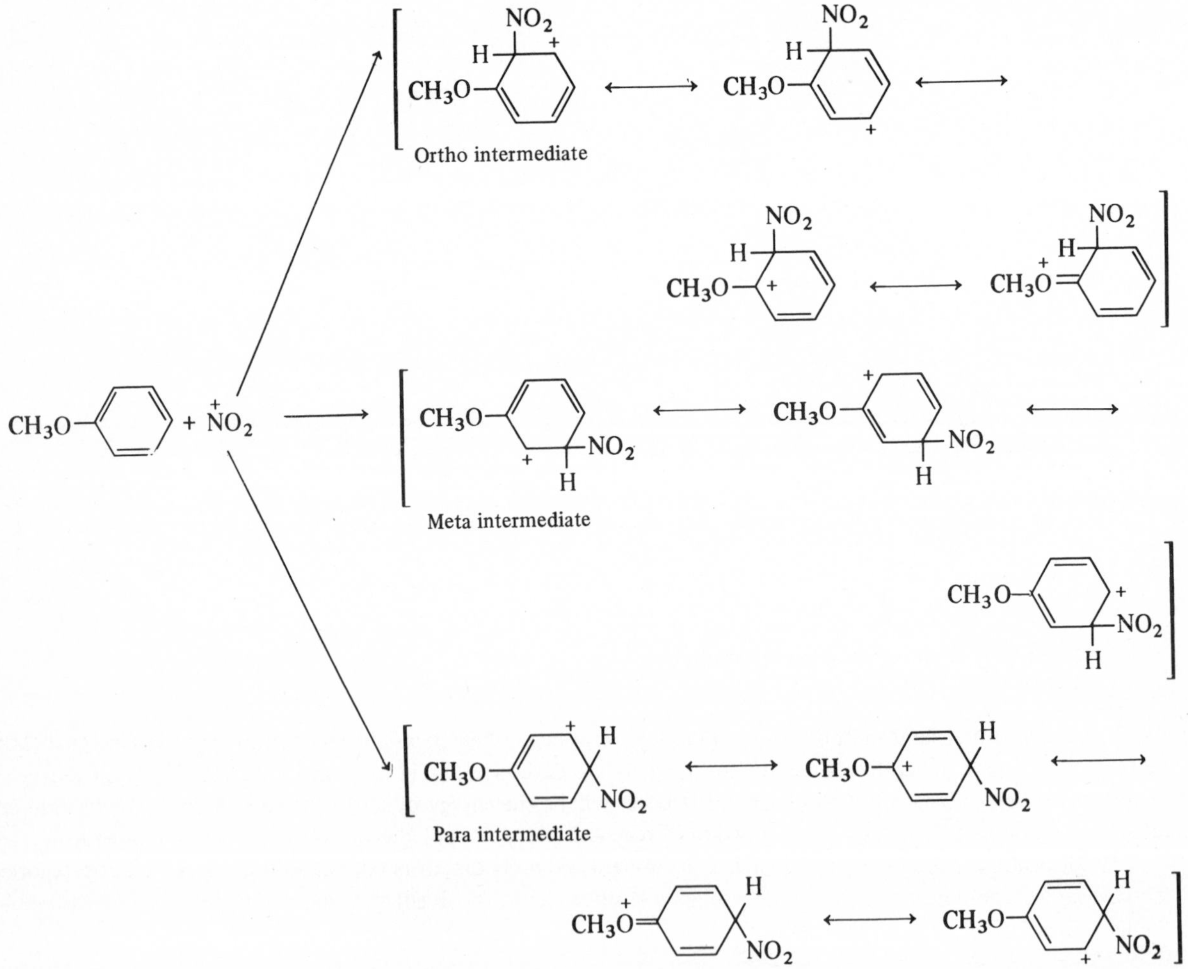

Ortho intermediate

Meta intermediate

Para intermediate

On the other hand, the nitration of nitrobenzene leads only to unfavorable intermediates (relative to benzene), but one, the meta, is more favorable than the others and that particular isomer does predominate in the product.

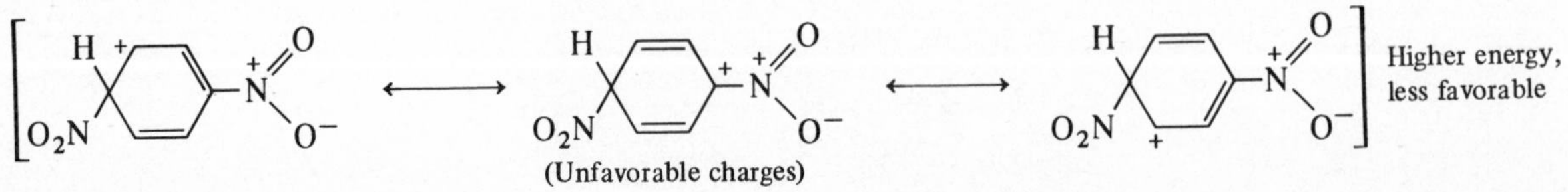

(Unfavorable charges)

Higher energy, less favorable

218

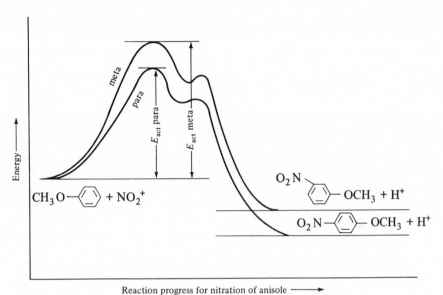

Lower energy, more favorable, because of better charge delocalization

Figure 15-2. Reaction progress for nitration of anisole.

15-3 Mechanism of Nucleophilic Substitution

Ordinarily, aromatic halides are very unreactive in S_N processes, but under forcing conditions or with proper halogen activation such substitutions can take place.

a. A process analogous to that of electrophilic substitution will occur if an activating substituent is properly situated for delocalization of *negative* charge.

b. *Benzyne* With very strong bases (NH_2^-) aromatic halides undergo an elimination reaction to yield a highly reactive intermediate benzyne:

219

When orthochlorotoluene is used in the previous reaction, two products are obtained:

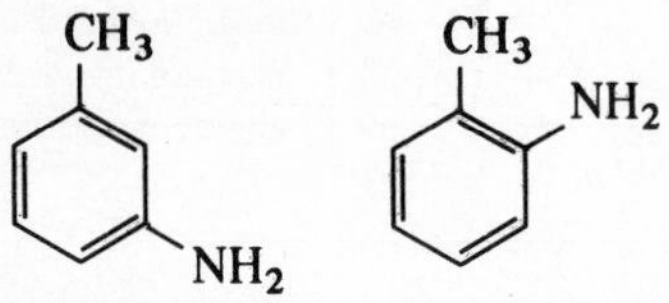

This is because NH_3 can add in two ways to the triple bond intermediate. Compounds such as 2-bromo-1,3-dimethylbenzene do not react. A hydrogen ortho to a halogen must be present for the elimination to occur.

15-4   Linear Free-Energy Relationships (LFER)

Hammett has placed in quantitative terms the long-used postulate of organic chemists that substituents have similar effects from reaction to reaction. He suggested that the ionization constants of benzoic acids (K_a) be taken as a standard for substituent effects. Thousands of examples have been found where if one plots $\log k$ (rate) or $\log K$ (equilibrium) for an aromatic reaction against $\log K_a$, a straight line is obtained. Since ionization and equilibrium constants are related to free energy ($\Delta G = RT \ln K$), these empirical linear relationships are often referred to as LFER. The general equation for such a straight line is:

$$\log k_X = \rho \log K_{aX} + c \tag{1}$$

where ρ is the slope; X identifies a particular derivative such as 4-CH_3, 3-NO_2, etc., and c is the intercept. For the special case of the parent compounds (X=H) we have:

$$\log k_H = \rho \log K_{aH} + c \tag{2}$$

Subtracting Eq. (2) from Eq. (1), we obtain:

$$\log (k_X/k_H) = \rho \log (K_{aX}/K_{aH}) \tag{3}$$

The substituent constant σ is defined as: $\sigma = \log (K_{aX}/K_{aH})$ which, when substituted into (3), yields the familiar form of the Hammett equation:

$$\log (k_X/k_H) = \rho\sigma \qquad \text{or} \qquad \log k_X = \rho\sigma + \log k_H \tag{4}$$

The σ for 4-NO_2 function is calculated as follows: K_a for benzoic acid in water at 25°C = 6.3×10^{-5}; for 4-nitrobenzoic acid, $K_a = 40 \times 10^{-5}$ (see Table 15-1)

$$\sigma_{4\text{-}NO_2} = \log \frac{40 \times 10^{-5}}{6.3 \times 10^{-5}} = \log 6.3 = 0.8$$

Table 15-1   Ionization constants of benzoic acids ($K_a \times 10^5$)

Substituent	Ortho	Meta	Para
OH	100	8.3	3.3
OCH_3	8.0	8.2	3.4
CH_3	12	5.3	4.2
Cl	120	15	10
NO_2	670	32	40

Note: with the exception of the CH_3O group, substitution in the ortho position gives the strongest acid. This is known as the "ortho effect" and is not completely understood.

220

Table 15-2 Values of substituent constants

Group	σ_{para}	σ_{meta}	σ^+_{para}	σ^+_{meta}	σ^-_{para}
N(CH$_3$)$_2$	−0.83	−0.15	−1.7	.....	
NH$_2$	−0.66	−0.16	−1.3	−0.16	.....
OH	−0.37	0.12	−0.92	......	.....
OCH$_3$	−0.27	0.12	−0.78	0.05	.....
CH$_3$	−0.17	−0.07	−0.31	−0.07	.....
C$_6$H$_5$	−0.01	0.06	−0.18	0.11	.....
H	0.00	0.00	0.00	0.00	0.00
NHCOCH$_3$	0.00	0.21	−0.60	.....	.....
F	0.06	0.34	−0.07	0.35	.....
I	0.18	0.35	0.13	0.36	.....
Cl	0.23	0.37	0.11	0.40	.....
Br	0.23	0.39	0.15	0.40	.....
CONH$_2$	0.36	0.28	.....	.....	0.63
COOH	0.45	0.37	0.42	0.32	0.73
COOCH$_3$	0.45	0.37	0.49	0.37	0.64
COCH$_3$	0.50	0.38	.....	.....	0.87
CF$_3$	0.54	0.43	0.61	0.52	.....
CN	0.66	0.56	0.66	0.56	1.00
SF$_5$	0.68	0.61	.....	.....	0.70
SO$_2$CH$_3$	0.72	0.60	.....	.....	1.05
NO$_2$	0.78	0.71	0.79	0.67	1.27
$^+$N(CH$_3$)$_3$	0.82	0.88	0.41	0.36	.....
SO$_2$CF$_3$	0.93	0.79	.....	.....	1.36

4-Nitrobenzoic acid is a stronger acid (higher K_a) than benzoic acid. It is assumed that this is because the nitro group withdraws electrons from the neighborhood of the carboxyl group, making it easier to dissociate a proton. Note that electron withdrawal is associated with positive σ values and electron release with negative σ values (see Table 15-2).

Table 15-2 contains representative σ values for meta and para substituents. The Hammett equation does not hold in general for ortho substituents. If any three of the four values k_H, k_X, ρ, and σ are known, the remaining one can be calculated. Thus, if the ionization constant for $C_6H_5NH_3^+$ is 2.4 x 10^{-5} and ρ for the reaction is 2.77, K_a for 4-methylaniline (σ for 4-CH$_3$ = −0.17) is calculated as follows:

$$\frac{\log K_a}{2.4 \times 10^{-5}} = (-0.17)(2.77)$$

$$\log K_a = (-0.17)(2.77) + \log 2.4 \times 10^{-5} = -5.09$$

$$K_a = 8.1 \times 10^{-6}$$

A positive ρ value indicates that a rate or equilibrium process is favored by electron-attracting functions, while a negative value shows that it is favored by electron-releasing groups (see Table 15-3).

Table 15-3 Values of reaction constants (ρ)

Reaction	
1. Ionization of benzoic acids in water 25°C	1.00
2. Ionization of phenols (hydroxybenzenes) in water at 25°C	2.11
3. Rate of hydrolysis of benzyl chlorides in 50% acetone at 60°C	−1.70
4. Rate of saponification of methyl benzoates in 60% acetone at 0°C	2.46

Note: See Prob. 15-17 for evaluation of ρ for a new reaction.

15-5 Interaction of Substituents with Reaction Centers

Equation (4) works best for situations where the reaction center is "insulated" by saturated groups from direct resonance interaction with substituents, e.g., X—C₆H₄—CH₂COOH. Where there is direct resonance interaction of the type shown in Sec. 15-2 and 15-3, correlations are poor. Two modified σ constants, σ^+ and σ^-, have been formulated for such situations. When a positive charge from the reaction center can be delocalized by the substituent (is conjugated with it), then σ^+ values defined by the solvolysis of 2-phenyl-2-propyl chlorides in 90% acetone yield much better correlations:

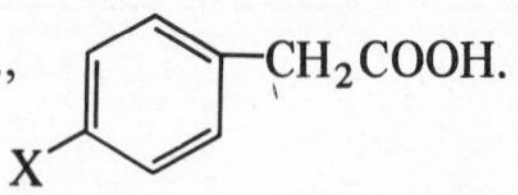

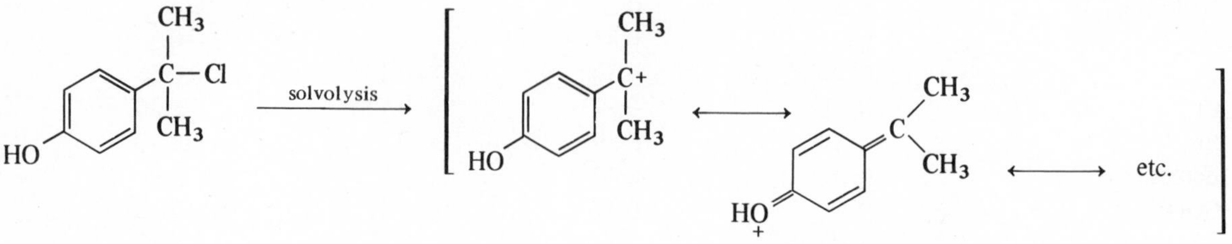

In the above example, OH illustrates the conjugation between substituent and reaction center, σ^+ is defined as:
$\sigma^+ = -(1/4.54) \log (k_X/k_H)$ where k refers to rate of solvolysis and, since ρ for this reaction is -4.54, dividing by this quantity places σ^+ values on the same scale as σ values. Note in Table 15-2 that para substituents such as OH, OCH₃, NH₂, and NHCOCH₃, which have a lone pair of electrons on the atom attached to the ring that can serve to delocalize a positive charge from the reaction center, have σ^+ values much larger than their σ values. This indicates their great activating effect in electrophilic aromatic substitution (nitration, halogenation, sulfonation, etc.). Groups which cannot accept a positive charge via resonance (i.e., have no electrons to donate) such as NO₂, CN, COOH, and COCH₃ have σ^+ values close to their σ values. Note also that σ^+ for meta substituents is about the same as σ. Since meta substituents are not conjugated with the reaction center, they show no strong resonance interaction.

A similar problem occurs when a substituent can delocalize a pair of electrons from the reaction center and the parameter σ^- has been developed for these situations.

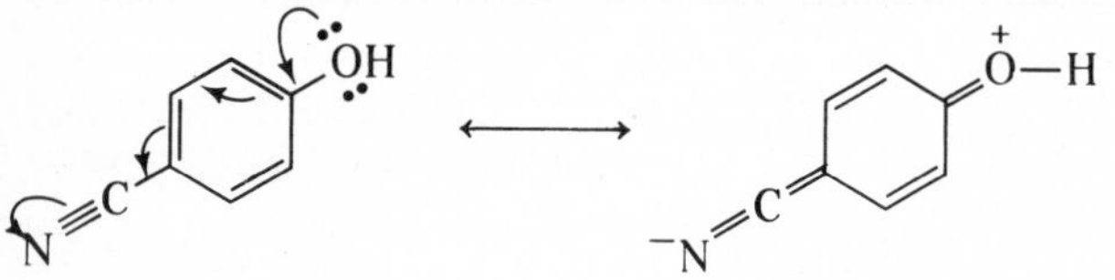

Observe in Table 15-2 that functions such as NO₂, CN, COCH₃, CONH₂, etc., which can accept a pair of electrons via resonance when conjugated with the reaction center have larger σ^- values than σ values. σ^- constants give better correlation reactions with phenols, anilines, and in nucleophilic substitutions (see Sec. 15-3).

15-6 Estimation of Resonance Energy

The stabilization of conjugated double bonds through resonance can be measured in a quantitative way by comparing them with nonconjugated bonds. An excellent method worked out by Kistiakowsky involves comparing the amounts of heat liberated on catalytic hydrogenation of the gaseous substances.

Table 15-4 Heats of hydrogenation (kcal/mol)

1-Butene	30.3	Cyclopentene	26.9	Cyclohexene	28.6
Cycloheptene	26.5	2-Methyl-2-butene	26.9	1,4-Pentadiene	60.8
1,3-Pentadiene	54.1	1,3,5-Cycloheptatriene	72.8	Divinyl ether	57.2
Benzene	49.8	Ethylbenzene	48.9	o-Xylene	47.3
Mesitylene	47.6	Styrene	77.5	Indene	69.9
Furan	36.6	1,3-Butadiene	57.1		

The hydrogenation of 1 mol of 1-butene at 82°C liberates 30.3 kcal of heat. Now one might expect that the hydrogenation of 1 mol of 1,3-pentadiene would liberate twice as much heat since it has two double bonds. However, one can see from the above table that it does not liberate 60.6 kcal, but is in a lower energy state than the olefin. This difference is known as the resonance energy. To calculate the resonance energy for mesitylene, use as a reference 2-methyl-2-butene, which has alkyl groups next to the π bonds similar to those in mesitylene. The energy of stabilization of mesitylene calculated in this fashion is 33.1 kcal/mol.

Problems

15-1 Indicate the principal product (ortho-para or meta) from the monochlorination of each of the following:
    a. Chlorobenzene   *b.* Ethyl benzoate   *c.* Cumene
    d. Diphenyl ketone   *e.* Aniline hydrochloride   *f.* Benzoyl fluoride
    g. Ethoxybenzene   *h.* Benzotrifluoride   *i.* Phenol (hydroxybenzene)

15.2 Indicate the principal isomer or isomers expected from the mononitration of each of the following:
    a. *o*-Dichlorobenzene   *b.* *p*-Bromonitrobenzene   *c.* *p*-Methylbenzoic acid (*p*-toluic acid)
    d. *m*-Nitrobenzoic acid   *e.* *m*-Ethylbenzamide   *f.* *m*-Dichlorobenzene

15-3 Draw resonance structures for each of the following involving both the ring and the functional group:
    a. Chlorobenzene   *b.* Benzamide   *c.* Nitrobenzene
    d. Benzaldehyde

15-4 The nitration of *o*-ethyltoluene yields more 4-nitro-2-ethyltoluene than 5-nitro-2-ethyltoluene. Draw resonance structures for each intermediate. Explain why one isomer should be preferred over the other.

15-5 *a.* Draw resonance structures for cinnamic acid, $C_6H_5CH=CHCOOH$ which involve both the carboxyl group and the ring.
    b. Nitration of cinnamic acid lead to ortho-para substitution. Explain this on the basis of transition-state theory.

15-6 *a.* Draw the electronic structure for nitrosobenzene, C_6H_5NO.
    b. Predict its orientation effect for electrophilic substitution.
    c. Draw electronic resonance structures for the meta and para intermediates.

15-7 When aniline, $C_6H_5NH_2$ is nitrated with HNO_3 and H_2SO_4, a large amount of *m*-nitroaniline is formed. Explain.

15-8 The nitration of naphthalene yields predominantly one monosubstitution product. Predict whether that product is α or β and explain your prediction by means of transition-state theory.

Naphthalene

15-9 It is easy to make hexamethylbenzene from methyl chloride and benzene via the Friedel-Crafts reaction, but with *tert*-butyl chloride and benzene the highest substitution product is 1,3,5-tri-*tert*-butylbenzene. Devise an explanation.

15-10 *a.* How many stereoisomers can result from the addition of 3 mol of chlorine to 1 mol of benzene in the presence of uv light?
    b. Give three-dimensional structures for all pairs of these which represent optical enantiomers.
    c. Give a conformational drawing for any isomers which have three axial chlorine atoms.
    d. Give the structure for the isomer which should undergo E_2 (trans) elimination least readily.

15-11 Point out the errors (if any) in each of the following proposed syntheses:

    a. $C_6H_6 \xrightarrow[\text{(A)}]{AlCl_3, \ CH_3CH_2CH_2Br} C_6H_5CH_2CH_2CH_3 \xrightarrow[\text{(B)}]{Cl_2, \ uv} C_6H_5CH_2CH_2CH_2Cl$

    b.

c.

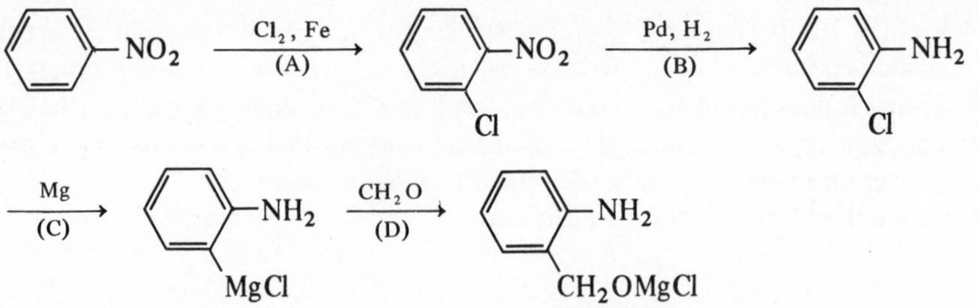

15-12 Compound A, $C_{10}H_{16}$, gave $C_{10}H_{20}$ on exhaustive hydrogenation. Mild oxidation converted A into a single product, B, an acid with NE = 101 ± 1. Suggest structures for A and B.

15-13 Compound A, C_5H_6, gave C_5H_{10} on exhaustive hydrogenation. Suggest a structure for A.

15-14 Predict the product composition for each of the following on the basis of a benzyne intermediate:

a. 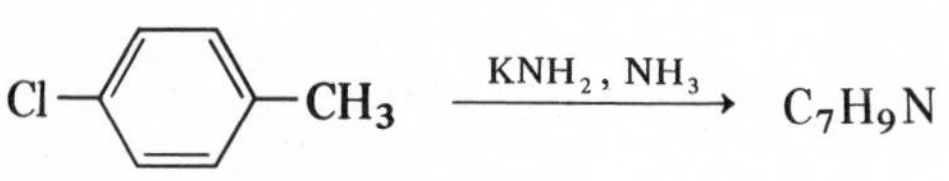 $Cl-\langle\rangle-CH_3$ $\xrightarrow{KNH_2, NH_3}$ C_7H_9N *b.* $\xrightarrow{NaNH_2,}$ $C_{15}H_{17}N$

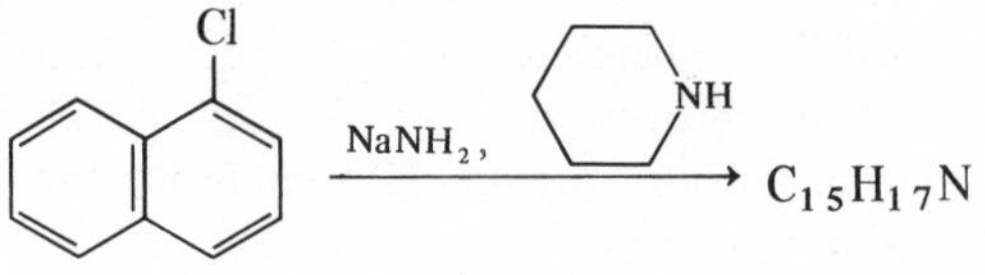

c. $Cl-\langle\rangle$ $\xrightarrow{KNH_2, NH_3}$ C_7H_9N

CH_3

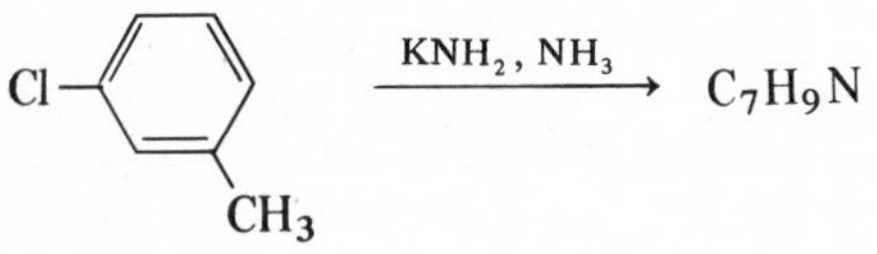

d. 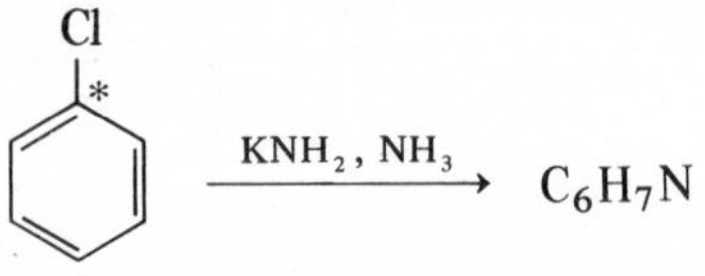 $\xrightarrow{KNH_2, NH_3}$ C_6H_7N

(Show distribution of C*)

15-15 Compound A, $C_{10}H_{16}$, gave, on exhaustive hydrogenation, $C_{10}H_{20}$. On mild oxidation, A gave a single new acid, B, of NE 66 ± 1. Suggest structures for A and B.

15-16 Compound A, C_9H_8, on exhaustive hydrogenation, gave C_9H_{16}. Vigorous oxidation of A gave an acid of NE 83 ± 1. Fusion of this acid with sodium hydroxide converted it to benzene. Suggest a structure for A.

15-17 If log K for benzoic acids is plotted vs. values of log K' for another equilibrium process (or vs. log k for a rate process) a straight line should result for *m* and *p* substituents.
a. Given the following values of K_a for benzoic acid and phenylacetic acid in water at 25°C, plot log K_a (benzoic) vs. log K_a (phenylacetic) and determine ρ for the latter:

Substituent	K_a (benzoic)	K_a (phenylacetic)
none	6.27 σ 10^{-5}	4.88 x 10^{-5}
p-Cl	10.50 x 10^{-5}	6.45 x 10^{-5}
p-Br	10.70 x 10^{-5}	6.49 x 10^{-5}
p-NO$_2$	40.00 x 10^{-5}	14.10 x 10^{-5}
p-CH$_3$	4.24 x 10^{-5}	4.24 x 10^{-5}

b. On the basis of Tables 15-2 and 15-3 make the following calculations:

1. K_a for *m*-fluorobenzoic acid in H_2O at 25°C.
2. K_a for *m*-nitrophenol in water at 25°C (K_a for phenol is 1.0 x 10^{-10}).
3. K_a for *p*-phenylbenzoic acid in H_2O at 25°C.
4. K_a for *p*-methoxyphenol in water at 25°C.

c. Predict which of the following pairs will react more rapidly under the indicated conditions and show how your predictions are related to Hammett constants:

1. Hydrolysis of p-nitrobenzyl chloride and p-methylbenzyl chloride in 50% acetone at 60°C.
2. Saponification of methyl m-bromobenzoate and methyl m-methylbenzoate in 60% acetone at 0°C.

d. Which of the reactions in Table 15-3 are favored by the introduction of a p-nitro group?

15-18 Compound A, C_6H_9N, on complete hydrogenation, gave $C_6H_{13}N$, B. When B was treated with excess methyl iodide, then silver oxide, and heated, and then this process repeated, a hydrocarbon C_6H_{10} was obtained. Hydrogenation of this gave C_6H_{14}. Ozonolysis of the C_6H_{10} compound yielded formaldehyde and a ketone $C_4H_6O_2$. Suggest a structure for A.

15-19 Compound A was found to have a molecular weight of 108. Carbon-hydrogen analysis indicated that it contained 88.89% C and 11.11% H. Complete hydrogenation gave compound B of MW 112. Oxidation of A produced an acid that had a NE of 128 ± 1. Decarboxylation of this acid gave cyclohexane. Suggest structures for A and B.

15-20 Using the data given in Sec. 15-6, calculate the resonance energy for each of the following:
   a. 1,4-Pentadiene   b. 1,3,5-Cycloheptatriene   c. Styrene
   d. Divinyl ether   e. Benzene

15-21 What product would you expect from the reaction of ICI with phenol? Why should ICI react faster than I_2?

15-22 When benzene is heated with excess DCl, perdeuterobenzene (C_6D_6) results. How can this be explained?

15-23 Arrange the following in order of ease of deuteration with DCl: benzene, toluene, anisole, chlorobenzene, nitrobenzene.

15-24 Indicate which positions (o, m, or p) in each of the following would be most likely deuterated when the compound is treated with 1 mol of DCl:
   a. Anisole   b. Nitrobenzene   c. Benzotrifluoride

15-25 For each of the following reactions plot log k values against σ, σ^+, and σ^-. Estimate the slope (ρ) for the best plot. Rationalize why the σ giving the best correlation is the most appropriate.

a. Hydrolysis of X—⬡—$OCO(CH_2)_3COO^-$.

X	log k	X	log k
4-OCH$_3$	−1.36	4-Br	−0.32
4-CH$_3$	−1.30	4-COOCH$_3$	0.68
H	−1.00	3-NO$_2$	0.80
4-Cl	−0.40	4-NO$_2$	1.72

b. Relative rate of para bromination of X—⬡.

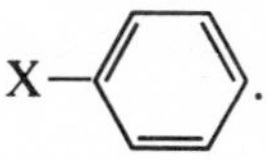

X	log k	X	log k
H	0.00	CH$_3$	3.3
N(CH$_3$)$_2$	19.5	F	0.8
OH	11.8	Cl	−0.4
OMe	9.8	Br	−0.6

c. Relative rates of the reaction:

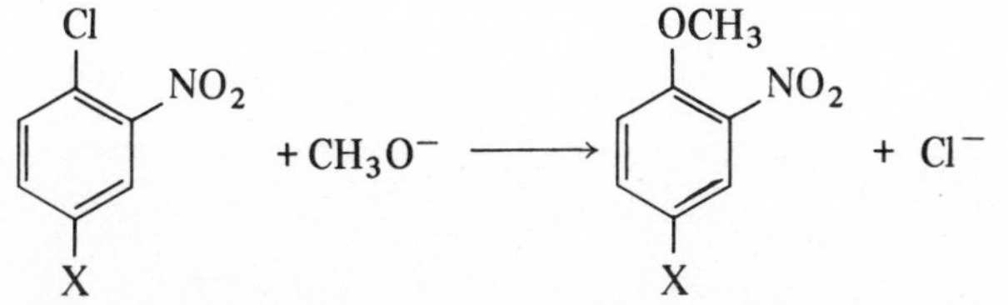

X	log k	X	log k
NO_2	-1.60	$COCH_3$	-3.51
SO_2CH_3	-2.58	Cl	-5.76
$^+N(CH_3)_3$	-3.08	H	-6.82

15-26   Explain why there is a much greater difference between σ_m and σ_p for F than there is for Cl. Explain why σ for OCH_3 has different signs for the meta and para positions. Why for $\overset{+}{N}(CH_3)_3$ is $\sigma_m > \sigma_p$ but for CN, $\sigma_p > \sigma_m$?

15-27   Which of the following substituents would aid in the delocalization of a positive charge in the electrophilic attack of $\overset{+}{E}$ on a benzene ring? Draw electronic structures to illustrate your point.

  a. –F   b. –SCH_3 c. –C_6H_5 d. –ClO_3 e. –SO_3H

15-28   Offer an explanation for each of the following observations:

  a. The three atoms of the nitro group in nitrobenzene tend to lie in the same plane as the benzene ring.

  b. Ethyl ether is much more soluble in water than ethyl vinyl ether is.

  c. Pyridine is a much stronger base than pyrrole.

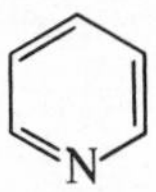

Pyridine 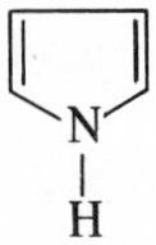   Pyrrole

  d. The pK for fluorene is 25 but that for diphenylmethane is 35.

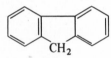

Fluorene

  e. 3,5-Dimethyl-4-nitroaniline is a stronger base than 2,6-dimethyl-4-nitroaniline.

  f. The carbon and oxygen atoms in the methoxyl group of anisole tend to lie in the same plane as that of th ring.

  g. Quinoline is a much stronger base than indole.

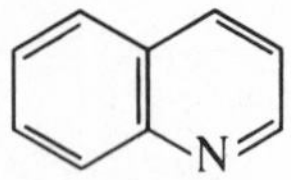

  Quinoline 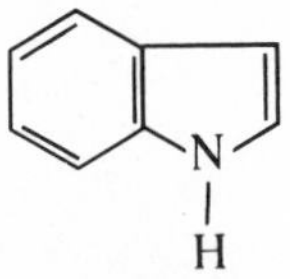   Indole

  h. The pK value for toluene is about 37 but that for indene is about 16.

  Indene

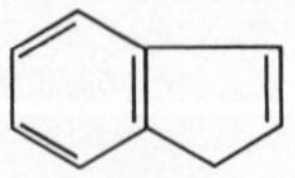

aromatic nitro
and sulfur compounds

16-1 Nomenclature

a. *Nitro Compounds*
The position of a nitro group on the aromatic ring or on its side chain is indicated in the usual fashion.

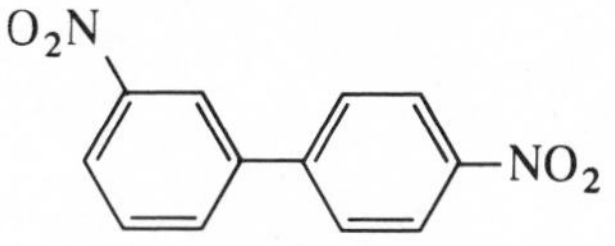

3,4′-Dinitrobiphenyl 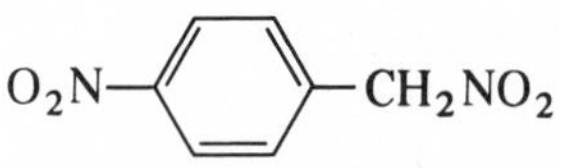   *α,p*-Dinitrotoluene 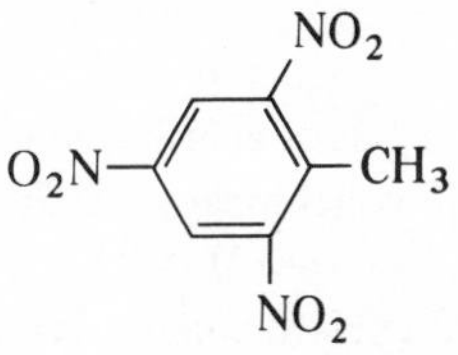   2,4,6-Trinitrotoluene (TNT)

b. *Sulfur Compounds* The following nomenclature applies to both aliphatic and aromatic sulfur compounds.
1. Divalent sulfur.

—SH   Mercapto, sulfhydryl or thiol group

—S—   Sulfide

Thioacid

—SS—   Disulfide

=S   Thial or thione

	IUPAC name	Common name
$CH_3CH_2CH_2SH$	1-Propanethiol	Propyl mercaptan
C_6H_5SH	Benzenethiol	Thiophenol (see Chap. 18)
$C_6H_5SC_6H_5$	Phenylthiobenzene	Diphenyl sulfide
$C_6H_5SSC_6H_5$	Phenyldithiobenzene	Diphenyl disulfide
$CH_3(CH_2)_3\overset{\|}{\underset{S}{C}}CH_3$	2-Hexanethione	Methyl butyl thioketone
$C_6H_5\overset{\|}{\underset{S}{C}}OH \rightleftharpoons C_6H_5\overset{\|}{\underset{O}{C}}SH$	Thiobenzoic acid	Same
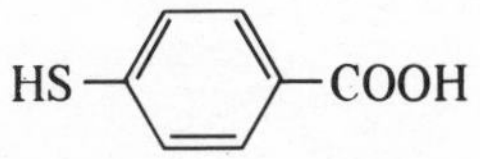 $HS{-}\langle\ \rangle{-}COOH$	*p*-Mercaptobenzoic acid	Same

2. Higher valence states.

R_2SO Sulfoxide R_2SO_2 Sulfone RSO_2H Sulfinic acid

RSO_3H Sulfonic acid $R_3S^+X^-$ Sulfonium salt

	IUPAC name	Common name
$CH_3SOCH_2CH_3$	Methylsulfinylethane	Methyl ethyl sulfoxide
$(C_6H_5)_2SO_2$	Phenylsulfonylbenzene	Diphenyl sulfone
$C_6H_5SO_2H$	Benzenesulfinic acid	Same
$C_6H_5SO_3H$	Benzenesulfonic acid	Same
$C_6H_5SO_2NH_2$	Benzenesulfonamide	Same
$C_6H_5S^+(Et)_2Cl^-$	Diethylphenylsulfonium chloride	Same

16-2 Preparation of Nitro Compounds

a. *Direct Substitution* Nitration with any of the following mixtures proceeds by way of an electrophilic attack on the aromatic ring.

Concentrated HNO_3 in glacial acetic acid (mildest conditions)
Concentrated HNO_3 in concentrated H_2SO_4
Fuming HNO_3 in concentrated H_2SO_4
Fuming HNO_3 in fuming H_2SO_4 (most vigorous conditions)

b. *Oxidation of Aromatic Amino Groups* This may be carried out with trifluoroperacetic acid in 90% hydrogen peroxide.

$$C_6H_5NH_2 + 3CF_3CO_3H \longrightarrow C_6H_5NO_2 + 3CF_3COOH + HOH$$

16-3 Preparation of Sulfur Compounds

a. *Thiols*
1. From alkyl halides.

$$RX + NaSH \longrightarrow RSH + NaX$$

Aryl halides may be used if they are activated toward nucleophilic substitution.
2. From thiourea and alkyl halides.

$$(H_2N)_2C{=}S + RX \longrightarrow [(H_2N)_2CSR]^+X^- \xrightarrow[\text{hyd}]{HO^-} RSH$$

3. Reduction of disulfides.

$$RSSR \xrightarrow{\text{Zn, HOAc}} RSH \quad (R = \text{alkyl or aryl.})$$

4. Reduction of sulfonyl chlorides.

$$RSO_2Cl \xrightarrow{\text{LiAlH}_4} RSH \quad (R = \text{alkyl or aryl})$$

5. From diazonium salts (see Sec. 17-5a).

b. *Sulfides*
1. From halides.

$$2RX + Na_2S \longrightarrow RSR + 2NaX \qquad \text{ArX may be used if it is activated toward nucleophilic substitution}$$

228

2. Alkylation of thiols.

$$RSH \xrightarrow{HO^-} RS^- \xrightarrow{R'X} RSR'$$

c. *Disulfides*
   1. From halides.

$$2RX + Na_2S_2 \longrightarrow RSSR + 2NaX$$

2. Oxidation of thiols.

$$RSH \xrightarrow{NaOI} RSSR \qquad (R = \text{alkyl or aryl})$$

d. *Sulfinic Acids* Reduction of sulfonyl chlorides.

$$2RSO_2Cl \xrightarrow{Zn} (RSO_2)_2Zn \xrightarrow{H^+} 2RSO_2H \qquad (R = \text{alkyl or aryl})$$

e. *Sulfonic Acids*
   1. Sulfonation (see Sec. 14-3*e*).
   2. Oxidation of thiols.

$$RSH \xrightarrow{HNO_3} RSO_3H \qquad (R = \text{alkyl or aryl})$$

f. *Sulfonyl Chlorides*
   1. From sulfonic acids.

$$RSO_3Na \xrightarrow{PCl_5} RSO_2Cl \qquad (R = \text{alkyl or aryl})$$

2. Chlorosulfonation.

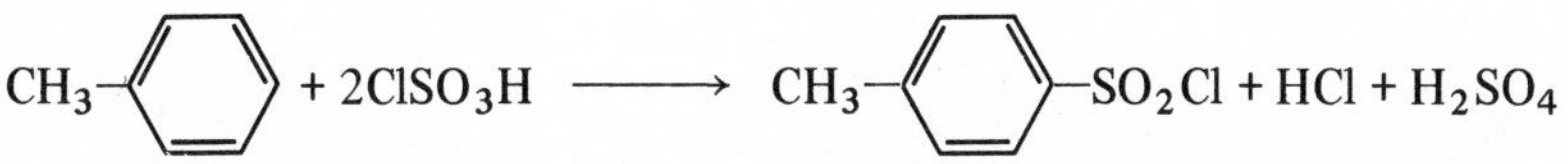

$$CH_3-\langle\text{benzene ring}\rangle + 2ClSO_3H \longrightarrow CH_3-\langle\text{benzene ring}\rangle-SO_2Cl + HCl + H_2SO_4$$

p-Toluenesulfonyl chloride or tosyl chloride (TsCl)

g. *Sulfoxides* Oxidation of sulfides.

$$R_2S \xrightarrow{1 \text{ equiv } H_2O_2} R_2SO \qquad (R = \text{alkyl or aryl})$$

h. *Sulfones*
   1. Oxidation of sulfides.

$$R_2S \xrightarrow{\text{excess } H_2O_2} R_2SO_2 \qquad (R = \text{alkyl or aryl})$$

2. Oxidation of sulfoxides.

$$R_2SO \xrightarrow{H_2O_2 \text{ or } HNO_3} R_2SO_2 \qquad (R = \text{alkyl or aryl})$$

3. Aromatic substitution.

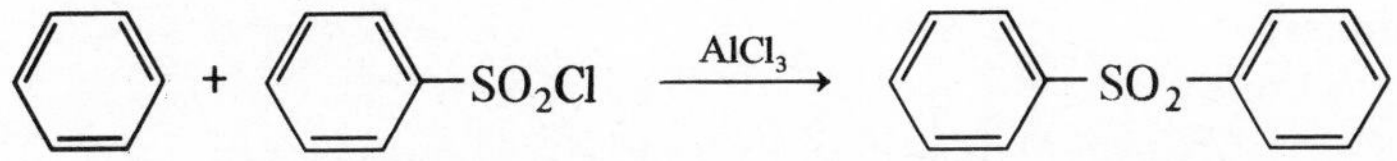

$$\langle\text{benzene}\rangle + \langle\text{benzene}\rangle-SO_2Cl \xrightarrow{AlCl_3} \langle\text{benzene}\rangle-SO_2-\langle\text{benzene}\rangle$$

i. *Sulfonium Salts* From alkyl halides and alkyl or aryl sulfides.

$$RX + R'_2S \longrightarrow R'_2\overset{+}{S}RX^-$$

16-4 Reactions of Nitro Compounds

a. *Reduction to Amines*

$$C_6H_5NO_2 + 3H_2 \xrightarrow{\text{Ni or Pd}} C_6H_5NH_2 + 2HOH$$

$$C_6H_5NO_2 \xrightarrow{\text{Sn, HCl (or SnCl}_2\text{, HCl)}} \xrightarrow{\text{NaOH}} C_6H_5NH_2$$
$$\text{Aniline}$$

b. *Intermediate Reduction Products*

$$C_6H_5NO_2 \xrightarrow{\text{Zn, NH}_4\text{Cl}} C_6H_5NHOH \xrightarrow{\text{Na}_2\text{Cr}_2\text{O}_7\text{, H}_2\text{SO}_4} C_6H_5NO$$
$$\text{Phenylhydroxylamine} \qquad\qquad \text{Nitrosobenzene}$$

$$C_6H_5NO_2 \xrightarrow{\text{MeONa, MeOH}} C_6H_5-\overset{+}{N}=N-C_6H_5$$
$$\underset{O^-}{|}$$
$$\text{Azoxybenzene}$$

$$C_6H_5NO_2 \xrightarrow{\text{Zn, NaOH (1 equiv)}} C_6H_5N=NC_6H_5$$
$$\text{Azobenzene}$$

$$C_6H_5NO_2 \xrightarrow{\text{Zn, NaOH (2 equiv)}} C_6H_5NHNHC_6H_5$$
$$\text{Hydrazobenzene}$$

Vigorous reduction of any of these intermediates leads to aniline, $C_6H_5NH_2$.

c. *Nucleophilic Aromatic Substitution* (See Sec. 15-3.)

d. Complex oxidation-reduction reactions take place between nitro compounds and Grignard reagents. Lithium aluminum hydride usually converts aromatic nitro compounds into azo compounds.

16-5 Reactions of Sulfur Compounds

a. *Thiols*
1. Oxidation to disulfides. (See Sec. 16-3c.)
2. Oxidation to sulfonic acids. (See Sec. 16-3e.)
3. Salt formation.

$$RSH + HO^- \longrightarrow RS^- + HOH$$

b. *Sulfides*
1. Oxidation to sulfoxides. (See Sec. 16-3g.)
2. Oxidation to sulfones. (See Sec. 16-3h.)

c. *Disulfides* Reduction to thiols. (See Sec. 16-3a.)

d. *Sulfinic Acids and Their Salts.*
1. Salt formation.

$$RSO_2H \xrightarrow{\text{NaOH}} RSO_2Na \qquad (R = \text{alkyl or aryl})$$

2. Alkylation to sulfinic acid esters (sulfones).

$$RSO_2Na + R'X \longrightarrow RSO_2R' + NaX \qquad (R = \text{alkyl or aryl}; R' = \text{alkyl or } S_N\text{-activated ArX}).$$

e. *Sulfonic Acids.*
1. Sulfonyl chloride formation. (See Sec. 16-3f.)

2. Replacement by OH in aromatic compounds.

$$\text{C}_6\text{H}_5\text{-SO}_3\text{H} \xrightarrow{\text{NaOH, fuse}} \xrightarrow{\text{H}^+} \text{C}_6\text{H}_5\text{-OH}$$

3. Replacement by H in aromatic compounds.

$$\text{C}_6\text{H}_5\text{-SO}_3\text{H} \xrightarrow{\text{HOH, }150°\text{C}} \text{C}_6\text{H}_5\text{-H} + \text{H}_2\text{SO}_4$$

4. Replacement by CN in aromatic compounds.

$$\text{C}_6\text{H}_5\text{-SO}_3\text{H} \xrightarrow{\text{NaCN, fuse}} \text{C}_6\text{H}_5\text{-CN}$$

5. Ionization. Sulfonic acids ionize in water to about the same extent as sulfuric acid does.

$$\text{RSO}_3\text{H} + \text{HOH} \longrightarrow \text{RSO}_3^- + \text{H}_3\text{O}^+$$

f. *Sulfonyl Chlorides*

1. Ester formation.

$$\text{RSO}_2\text{Cl} + \text{R}'\text{OH} \longrightarrow \text{RSO}_2\text{OR}' + \text{HCl} \qquad (\text{R = alkyl or aryl})$$

2. Amide formation.

$$\text{RSO}_2\text{Cl} + 2\text{R}'_2\text{NH} \longrightarrow \text{RSO}_2\text{NR}'_2 + \text{R}'_2\text{NH}_2^+\text{Cl}^- \qquad (\text{R = alkyl or aryl})$$

3. Reduction to thiols. (See Sec. 16-3a.)

g. *Sulfonate Esters* The sulfonate ion is a good leaving group in S_N reactions. Consequently sulfonate esters are effective alkylating agents.

$$\text{RSO}_2\text{O-CH}_3 + \text{NH}_3 \longrightarrow \text{RSO}_2\text{O}^- \text{CH}_3\text{NH}_3^+ \qquad (\text{R = alkyl or aryl})$$

h. *Sulfoxides* Oxidation to sulfones. (See Sec. 16-3h.)

i. *Sulfonium Salts*

1. Degradation of sulfonium hydroxides. (See Sec. 10-9.)

$$-\overset{|}{\underset{|}{\text{C}}}-\overset{|}{\underset{\underset{\text{H}}{|}}{\text{C}}}-\overset{+}{\text{S}}\text{R}_2\text{OH}^- \xrightarrow{\Delta} -\overset{|}{\text{C}}=\overset{|}{\text{C}}- + \text{R}_2\text{S} + \text{H}_2\text{O}$$

2. Sulfur ylide formation and their reactions. Treatment of certain sulfonium salts with a base yields stable sulfur ylides which may be used in Wittig-type reactions.

$$(\text{C}_6\text{H}_5)_2\overset{+}{\text{S}}\text{CH}(\text{CH}_3)_2\text{BF}_4^- \xrightarrow{(\text{CH}_3)_3\text{CLi}} (\text{C}_6\text{H}_5)_2\overset{+}{\text{S}}-\overset{-}{\text{C}}(\text{CH}_3)_2 \longleftrightarrow (\text{C}_6\text{H}_5)_2\text{S}=\text{C}(\text{CH}_3)_2$$

$$\xrightarrow{\text{R}_2\text{C}=\text{O}} \overset{}{\underset{\underset{-\text{O}\text{S}(\text{C}_6\text{H}_5)_2}{||}}{\text{R}_2\text{C}-\text{C}(\text{CH}_3)_2}} \longrightarrow \overset{}{\underset{\underset{\text{O}}{\diagdown\diagup}}{\text{R}_2\text{C}-\text{C}(\text{CH}_3)_2}} + (\text{C}_6\text{H}_5)_2\text{S}$$

16-6 Qualitative Tests

a. *Nitro Compounds*

1. Primary and secondary nitro compounds can exist in tautomeric forms and are soluble in strong bases.

$$\text{RCH}_2-\overset{+}{\text{N}}\overset{\nearrow \text{O}^-}{\diagdown\diagup}_{\text{O}} \rightleftharpoons \text{RCH}=\overset{+}{\text{N}}\overset{\nearrow \text{O}^-}{\diagdown}_{\text{OH}} \xrightarrow{\text{OH}^-} \text{RCH}=\overset{+}{\text{N}}\overset{\nearrow \text{O}^-}{\diagdown}_{\text{O}^-}$$

2. Primary and secondary nitro compounds give a positive test with Tollens' reagent after first being reduced with zinc and ammonium chloride.
3. Nitro compounds form a red-brown precipitate of $Fe(OH)_3$ when treated with $Fe(OH)_2$ (ferrous hydroxide test).

b. *Sulfur Compounds*
   1. Solubilities. Sulfonic acids are soluble in water because of a high degree of ionization.
   2. Low-molecular-weight thiols are soluble in aqueous solutions of strong bases.
   3. Sulfonic acids usually form water-insoluble calcium or barium salts.
   4. Sulfonamides with at least one N–H group are soluble in aqueous solutions of strong bases (see Hinsberg test, Sec. 10-8*b*).

16-7 Synthetic Approach

When we set out to devise a synthetic route involving an aromatic compound, we find available several principal methods of attack. Among these are electrophilic substitution, nucleophilic substitution, and interconversion of functional groups. Later free-radical substitution will be added to the list (see Sec. 17-7).

Draw pictures of benzene rings with two substituents in any of three possible orientations (ortho, meta, and para). If we start thinking of this system in terms of syntheses, what problems must be considered? The first important prerequisite is knowing all the various routes available for introducing the substituents into the ring. Any of the methods of direct substitution are important, but it might also be necessary to consider converting one group into another.

Consider the amino group. Here are some of the ways for getting an aromatic amine:

1. Reduction of a nitro group.
2. Hypobromite reaction of an amide.
3. Nucleophilic displacement of a halogen.

Notice that there is no way of introducing $-NH_2$ directly by way of *electrophilic* substitution. Nitration, however, provides a convenient indirect route by this mechanism.

Now that the methods for introducing groups are well in mind, the next logical step is to look at the orientation of the two substituents in the benzene ring. Assume two groups ortho to each other. If one is an ortho-para-directing group, it may be possible to introduce the other directly by electrophilic substitution.

Turn to Prob. 16-4*a*. There are two functional groups, Cl and NO_2, ortho to each other. Either can be introduced into the ring by a simple electrophilic substitution. (At this time it would be well to reemphasize that you would fail miserably in synthetic problem-solving of this type if you did not know these reactions intimately!) We know that chlorine is ortho-para-directing and nitro is meta-directing in such reactions. This leaves us with the necessary sequence of chlorination first and nitration second.

$$C_6H_6 \xrightarrow{Cl_2,\ Fe} Cl-C_6H_5 \xrightarrow{HNO_3,\ H_2SO_4} Cl-\!\!\!\left\langle\!\bigcirc\!\right\rangle\!\!\!-NO_2 \ + \ Cl-\!\!\!\left\langle\!\bigcirc\!\right\rangle^{\!\!O_2N}$$

Now that the equations are set up, take an introspective look at what was done. Can nitration of chlorobenzene have any effect on the halogen? Can the two isomers be separated? In this case, neither problem leads to any difficulty. We do not expect the chloride ion to be displaced by weak bases of the type present in the reaction mixture (NO_3^-, Cl^-, HSO_4^-). As for the isomers, separations are difficult only for disubstituted benzenes with the same groups ortho or para to each other.

Repeat this same process for Prob. 16-4*b*.

Problem 16-4*c* represents something a little different. We have two groups para to each other, but both of them are meta directors. Our first thought is that one of these may at some time in the synthetic route have been an electron-releasing group (ortho-para-directing for electrophilic substitution). The carboxyl group is always a good starting point because it can be made, for example, by oxidation of any alkyl group or by the carbonation of a Grignard reagent. The latter method can be excluded immediately because Grignard reactions should not be run in the presence of nitro groups. In the following oxidation, the reaction will be good because only the methyl group will react with conventional oxidizing agents.

$$O_2N-\!\!\!\left\langle\!\bigcirc\!\right\rangle\!\!\!-CH_3 \xrightarrow{\text{acid dichromate}} O_2N-\!\!\!\left\langle\!\bigcirc\!\right\rangle\!\!\!-COOH$$

232

At this point, the problem is the same as the one encountered in the first two examples. The ortho-para-directing group should be present first, and electrophilic nitration will give the proper isomer (together with ortho).

For a final example, turn to Prob. 16-4*i*. This looks similar to the first problem except that a methoxyl group cannot be introduced by electrophilic substitution. Anisole could be nitrated easily to give the desired product, but an alternative route is called for if anisole is not available.

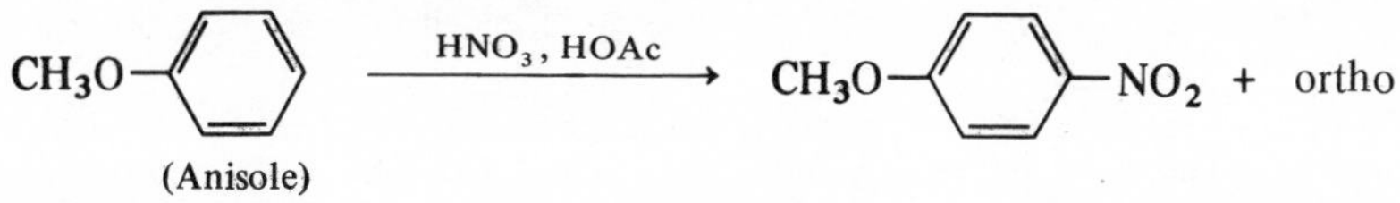

$$CH_3O-\langle\bigcirc\rangle \xrightarrow{HNO_3, HOAc} CH_3O-\langle\bigcirc\rangle-NO_2 + ortho$$

(Anisole)

Nucleophilic substitution has been considered, and it was stated that such a reaction is difficult unless a displaceable group such as halogen is ortho or para to at least one electron-accepting group such as nitro. *p*-Nitroanisole has the necessary structural features to consider synthesis via nucleophilic displacement. The methoxide ion is a strong nucleophile, the chloride is a good leaving group, and the position under consideration is para to a nitro group.

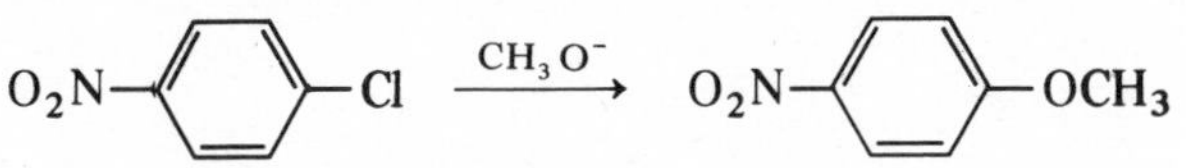

$$O_2N-\langle\bigcirc\rangle-Cl \xrightarrow{CH_3O^-} O_2N-\langle\bigcirc\rangle-OCH_3$$

The problem would be complete if *p*-nitrochlorobenzene were available, and this represents a simple synthesis as has been shown in the first synthetic sequence.

Occasionally, it is convenient or necessary to use an atom or group of atoms for a directive influence during aromatic substitution and then remove it completely (replacement by hydrogen). One group that is readily replaceable is the $-SO_3H$ group. Later, a method will be encountered for removing either a nitro or an amino group. Refer to Probs. 16-4*j* and 16-14*e* and show how to approach this synthetic process.

Consider the synthesis of methyl 4-methyl-3-nitrobenzenesulfonate. Backing up one step would require esterifying 4-methyl-3-nitrobenzenesulfonyl chloride with CH_3OH. Nitration of *p*-toluenesulfonic acid followed by treatment with PCl_5 would give the necessary sulfonyl chloride. Outline an alternate approach from *o*-nitrotoluene. Why would direct nitration of methyl *p*-methylbenzenesulfonate not be a good method?

Problems

16-1 Name the following compounds by both the IUPAC and common systems.
  a. $(CH_3)_2CHSH$
  d. $CH_3(CH_2)_5SO_2Cl$

  b. $Me_2CHSCH_2CH_3$
  e. $C_6H_5SO_2OEt$

  c. $C_6H_5SO_2CH_3$
  f. $(CH_3)_2SO$

  g. CH_2CH_2COSH

  h. $O_2N(CH_2)_3\underset{\parallel}{\overset{}{C}}H$
  S

  i.

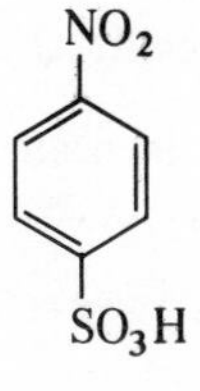

  j.

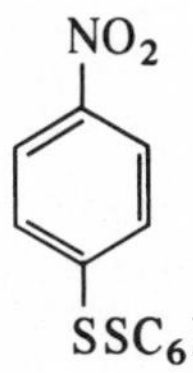

  k. $\underset{CH_3}{\overset{C_6H_5CH_2}{\diagdown}}C=S$

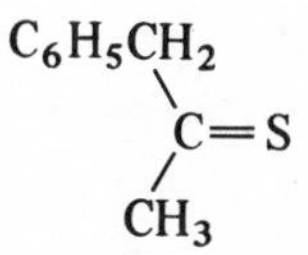

  l.

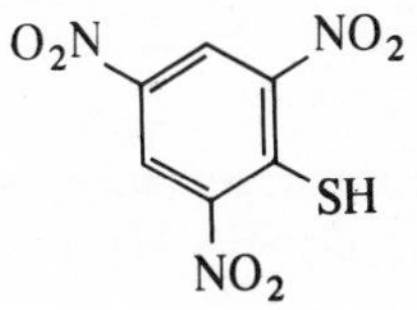

16-2 Give each of the following an acceptable name.

  a. $CH_3-\langle\bigcirc\rangle-SO_3H$

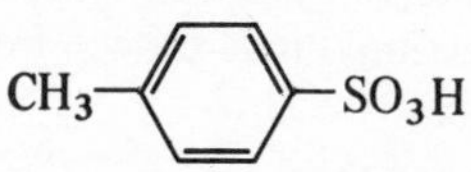

  b.

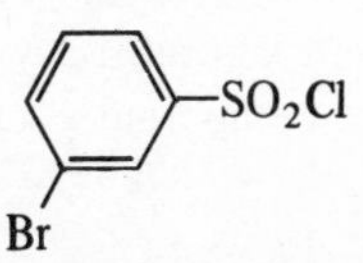

  c.

233

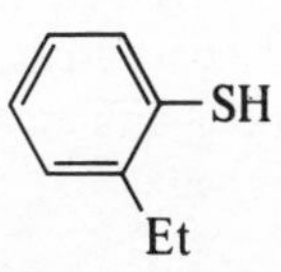

d.  O_2N—⟨benzene⟩—CH_3

e. 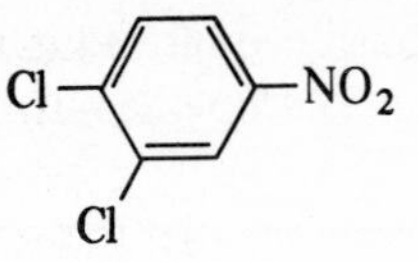 Cl—⟨benzene⟩—NO_2 with Cl

f. 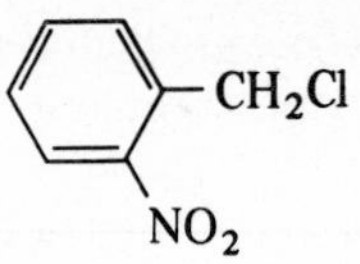 ⟨benzene⟩—CH_2Cl with NO_2

g. $C_6H_5CH_2S$ | $C_6H_5CH_2S$

h. 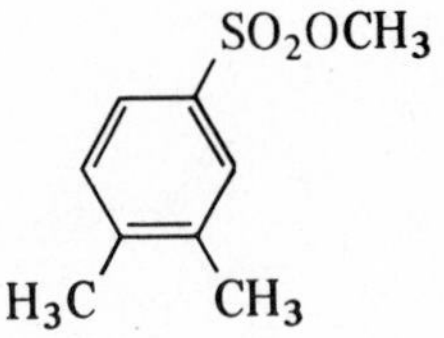 ⟨benzene⟩—SO_2OCH_3 with H_3C CH_3

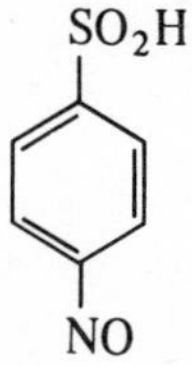

i. 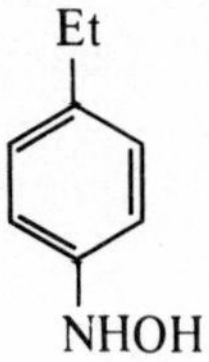 Et—⟨benzene⟩—$NHOH$

j. 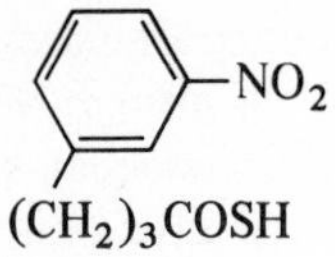 ⟨benzene⟩—NO_2 with $(CH_2)_3COSH$

k. 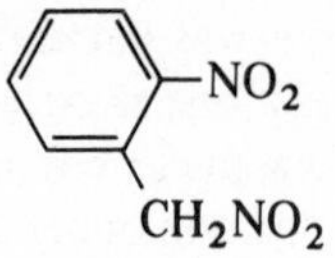 ⟨benzene⟩—NO_2 with CH_2NO_2

l. 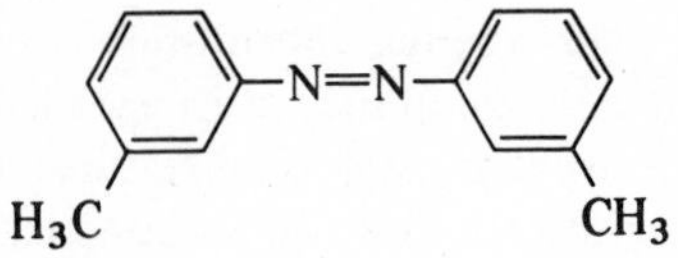 H_3C—⟨benzene⟩—$N=N$—⟨benzene⟩—CH_3

m. ⟨benzene⟩ with SO_2H and NO

n. ⟨benzene⟩ with CH_3 and $SO_2NHC_6H_5$

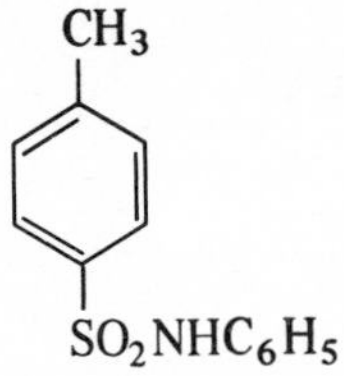

o. H_2NCNH_2 with $\parallel$ S

p. $(CH_2{=}CHCH_2)_2SO_2$

16-3 Complete the following equations. Indicate in the examples of *aromatic* substitution reactions whether it is S_N or S_E reaction.

a. $CH_3Br + CH_3CSS^-$
b. $EtBr + C_6H_5SNa$
c. (S)-2-Bromobutane + NaSH
d. $C_6H_5SH + NaOI$
e. $Et_2S + EtBr$
f. $p\text{-}CH_3C_6H_4SO_2Cl \xrightarrow{Zn} \xrightarrow{H^+}$
g. $C_6H_5SO_2Na + EtBr$
h. $Et\overset{+}{\underset{Me}{-S}}-C_6H_5 \quad OH^- \xrightarrow{\Delta}$

i. $p\text{-}ClC_6H_4NO_2 \xrightarrow{KOH, \Delta}$
j. $C_6H_5NO_2 \xrightarrow{H_2SO_4}$
k. $EtSH + EtMgBr$
l. $C_6H_5SO_2CF_3 \xrightarrow{HNO_3, H_2SO_4}$
m. $EtMgBr + SO_2$
n. $Et_2S + 1$ equiv H_2O_2
o. $(C_6H_5)_2S + $ excess H_2O_2
p. $C_6H_5SO_2Cl + CH_3NH_2$
q. (S)-α-Bromoethylbenzene $\xrightarrow{EtSK}$
r. $p\text{-}ClC_6H_4SO_3H + SOCl_2$
s. $C_6H_5SNa + (CH_3)_3CCl$
t. $p\text{-}O_2NC_6H_4Cl + C_6H_5SNa$
u. $o\text{-}ClC_6H_4NO_2 + NaOH$
v. $C_6H_5Cl \xrightarrow{NaNH_2, NH_3}$
w. 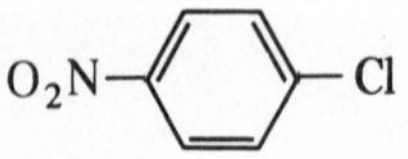 O_2N—⟨benzene⟩—$Cl + \bar{C}H(COOEt)_2$
x. $Br(CH_2)_5Br + Na_2S_2$
y. (R)-2-Bromopentane + Na_2S
z. $C_6H_5SO_2OEt + NH_3$
a'. $(NH_2)_2C{=}NH + Me_2SO_4$
b'. Thiourea + benzyl chloride
c'. (2R:3R)-2,3-Epoxybutane $\xrightarrow{NaSH}$
d'. $(C_6H_5)_2C{=}S \xrightarrow{HNO_3, H_2SO_4}$
e'. $SO_2Cl_2 + C_6H_5MgBr$
f'. $(CH_3)_2S{=}CH_2 + $ cyclohexanone

16-4 Starting with any organic compound containing four carbons or less, benzene, diphenyl sulfide, and any necessary inorganic materials, outline a convenient laboratory synthesis of the compounds shown. For those compounds marked with an asterisk, use a sulfur compound as an intermediate.

a. *o*-Chloronitrobenzene
b. *m*-Chloronitrobenzene

c. *p*-Nitrobenzoic acid
e. *s*-Butyl benzenesulfonate
g. *m*-Nitrobenzoic acid
i. *p*-Nitroanisole
k.* *p*-Hydroxytoluene (p-cresol)

d. *p*-$(CH_3)_2CHC_6H_4SO_3H$
f. *m*-$BrC_6H_4SO_2NH_2$
h. *p*-Nitrosotoluene
j.* *o*-Bromoethylbenzene
l. Sodium phenylmercaptide

m. *p*-$ClC_6H_4SO_3H$

n. 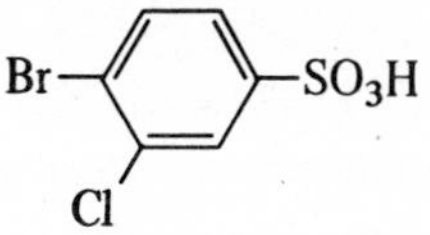 Br⎯⎯⎯SO₃H ; Cl

o.* *p*-Ethylphenyl propyl ketone
q. *p*-$O_2NC_6H_4CH_2C_6H_5$

p. *m,m′*-Dichlorohydrazobenzene
r. 2-Bromo-4-nitrobenzoic acid

s. *p*-$O_2NC_6H_4SC_6H_5$

t. 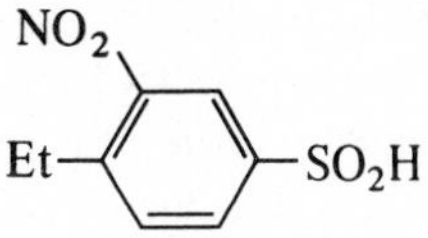 Et⎯⎯⎯SO₂H ; NO₂

u. 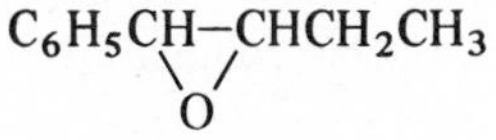 $C_6H_5CH{-}CHCH_2CH_3$ (with O bridge)

16-5 Give a chemical method for differentiating between the following pairs:
a. *p*-Nitrotoluene and α-nitrotoluene
b. Nitrobenzene and aniline
c. Benzamide and *p*-nitrosotoluene
d. Nitrobenzene and nitrohexane
e. Nitrohexane and hexyl nitrate
f. Nitrobenzene and chlorobenzene
g. *p*-Toluenesulfonic acid and sodium *p*-toluenesulfonate
h. Benzenethiol and ethyl phenyl sulfide
i. Benzenethiol and *p*-toluenesulfonic acid
j. Benzenesulfonyl chloride and benzenesulfonic acid
k. *p*-Ethylbenzenesulfonic acid and *p*-ethylbenzenesulfonamide
l. Butylamine and diethylamine

16-6 Indicate the errors in the following reactions:

a. 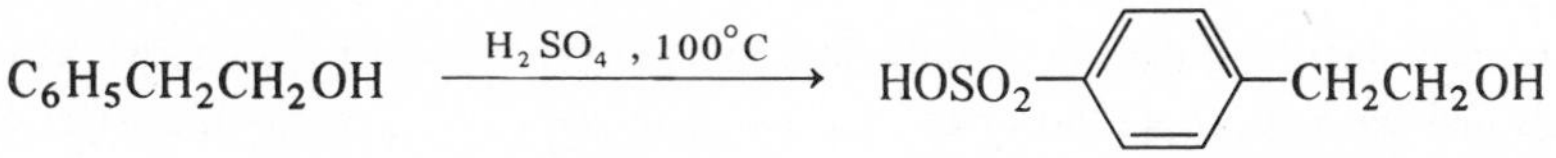 $C_6H_5CH_2CH_2OH$ $\xrightarrow{H_2SO_4, 100°C}$ HOSO₂⎯⎯⎯CH₂CH₂OH

b. Toluene $\xrightarrow[\text{(A)}]{Cl_2}$ CH₃⎯⎯⎯Cl $\xrightarrow[\text{(B)}]{NaSH}$ CH₃⎯⎯⎯SH $\xrightarrow[\text{(C)}]{HNO_3}$ CH₃⎯⎯⎯SH ; NO₂

c. C_6H_5Br $\xrightarrow[\text{(A)}]{H_2SO_4}$ O₂N⎯⎯⎯Br $\xrightarrow[\text{(B)}]{Mg}$ O₂N⎯⎯⎯MgBr $\xrightarrow{CH_3CN, H_3O^+}$ O₂N⎯⎯⎯CHCH₃ ; OH

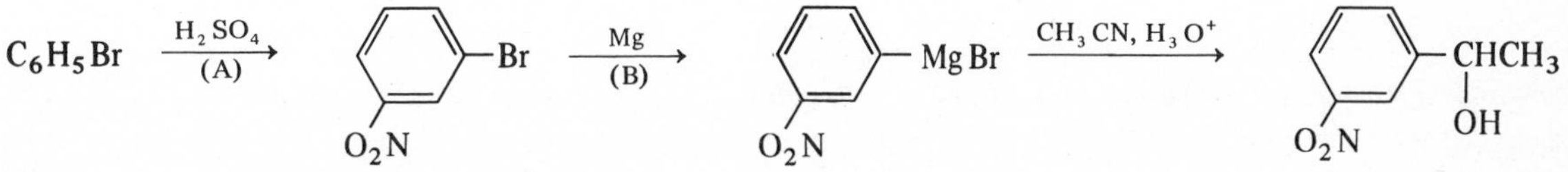

d. 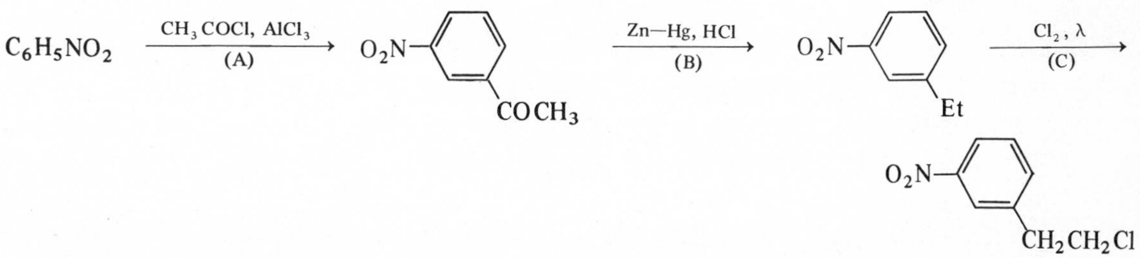 $C_6H_5NO_2$ $\xrightarrow[\text{(A)}]{CH_3COCl, AlCl_3}$ O₂N⎯⎯⎯COCH₃ $\xrightarrow[\text{(B)}]{Zn-Hg, HCl}$ O₂N⎯⎯⎯Et $\xrightarrow[\text{(C)}]{Cl_2, λ}$ O₂N⎯⎯⎯CH₂CH₂Cl

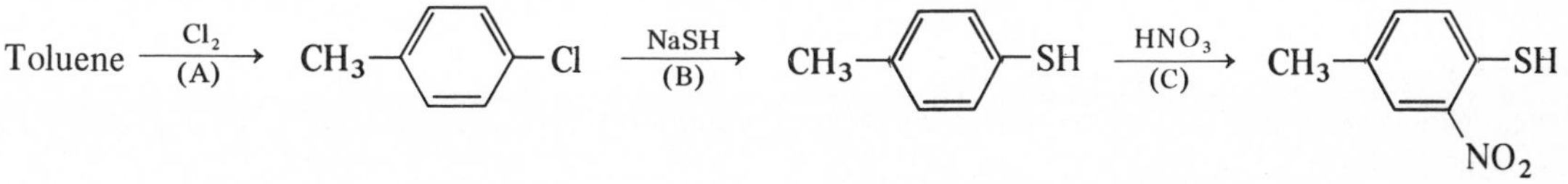

16-7 Given ethanol, benzene, and CO_2 as the only available carbon compounds, devise a convenient laboratory synthesis of the following:
a. Butyl sulfoxide
b. Crotyl ethanesulfonate
c. sec-Butyl disulfide
d. 4,4′-Diethylazobenzene
e. 2,4-Dinitropropylbenzene
f. *p*-Nitrocumene

16-8 Suggest a structure for all lettered compounds in each of the following:

a. Compound A, $C_9H_{11}O_2SCl$, was insoluble in dilute acids or bases. It gave an immediate precipitate with alcoholic $AgNO_3$. Treatment of A with Br_2 in the presence of iron yielded only a single monobromo derivative. When A was warmed for 30 min with 10% NaOH it dissolved. When this alkaline solution was strongly acidified, an acid could be isolated. Treatment of the acid with superheated steam yielded a hydrocarbon. This hydrocarbon formed a mixture of monobromo substitution products when it was treated with Br_2 and Fe.

b. Compound A, $C_{10}H_{15}O_2NS$, was soluble in dil NaOH. When a solution of 0.29 g of A in 10% NaOH was distilled, the distillate contained compound B, which required 13.7 ml of 0.10 M HCl for neutralization. B reacted with HONO to yield N_2 and compound C. C gave a positive iodoform test.

c. Compound A, $C_7H_7O_2N$, was soluble in dil NaOH. It could be oxidized by alk $KMnO_4$ to a nitrogen-free acid, NE 122.

d. Compound A, $C_7H_7O_2N$, yielded an acid, $C_7H_5O_4N$, on oxidation with $KMnO_4$. Two isomeric ring-monobrominated products could be synthesized from the acid.

e. Compound A, C_8H_9ON, was insoluble in dilute acids or bases. Vigorous oxidation with $KMnO_4$ yielded an acid, $C_8H_6O_4$, which formed a single ring-monobrominated substitution product.

f. Compound A, $C_6H_4O_4N_2$, was insoluble in dilute acids or bases. It had a zero dipole moment. Reduction yielded $C_6H_8N_2$, a compound with a zero dipole moment.

g. Compound A, $C_{12}H_{10}N_2$, existed in cis-trans forms. Hydrogenation yielded C_6H_7N, which was soluble in dilute acids and which resisted further reduction.

h. Compound A, $C_9H_{11}O_2N$, contained an asymmetric center. It was soluble in 5% NaOH but not in $NaHCO_3$. Oxidation of A yielded B, a nitrogen-free acid, NE 82 ± 1. B forms three isomeric ring-monobrominated products.

16-9 *a.* The nitration of toluene with nitric-sulfuric acids yields approximately equal portions of ortho and para compounds and a small amount of meta. Is the nitration reaction electrophilic or nucleophilic? Explain.

b. In this nitration, what is the electrophilic agent? Draw its electronic structure. What is the leaving group?

c. When HNO_3 is dissolved in H_2SO_4, the freezing-point depression of the latter is about 4 times the value expected if HNO_3 were to remain un-ionized in solution. Suggest structures for the particles that might form when 1 molecule of HNO_3 is placed in H_2SO_4 (a reaction occurs involving both acids).

d. Draw the ionic intermediate in the para nitration of bromobenzene. Draw a three-dimensional picture of the para carbon atom in this state.

16-10 *a.* Draw resonance structures for meta- and para-bromonitrobenzene. On the basis of the charge distribution that these structures indicate, which halogen would be more easily displaced by a nucleophilic attack such as by CH_3O^- on carbon?

b. In the ionic intermediate in *a* what is the spatial relationship of the atoms of the nitro group with respect to the benzene ring? Describe the hybridization of the carbon atom involved in substitution.

c. Which halogen in each of the following pairs of compounds is more easily replaced in a nucleophilic aromatic substitution? Explain.

d. Show how to convert chlorobenzene to

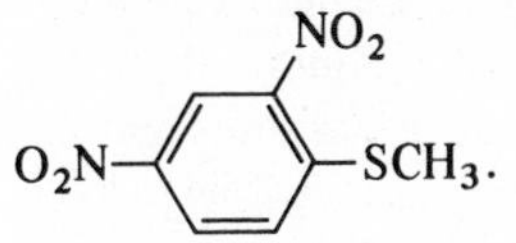

$$O_2N-\underset{\underset{SCH_3.}{}}{\overset{NO_2}{\bigcirc}}$$

16-11 Sulfinic esters have been resolved into optical antipodes; Suggest a reason why sulfinic acids have not.

16-12 Devise a method for separating the following mixtures:
a. *p*-Toluenesulfonic acid, bromobenzene, and phenylacetic acid.
b. *p*-Nitroaniline, *p*-nitrobenzoic acid, nitrobenzene, and α-nitrotoluene.

16-13 Refer to Sec. 14-3 and note that the sulfonation reaction is reversible.
a. Indicate the reaction conditions you would choose to carry out the reaction in either direction.
b. Quite often the sulfonation is carried out with fuming sulfuric acid ($SO_3 + H_2SO_4$). Would you expect this to be reversible? Explain.

16-14 Sulfonation is very subject to steric hindrance. This is illustrated by the sulfonation of bromobenzene which yields 100 percent para product.
a. Does this reaction represent electrophilic aromatic substitution? Explain.
b. Draw resonance structures for bromobenzene.
c. Draw resonance structures for the ionic intermediate in the sulfonation reaction.
d. If *p*-bromobenzenesulfonic acid were brominated with Br_2 and Fe, what isomer or isomers would result?
e. Show how the sulfonation and desulfonation reactions could be used to prepare *o*-dibromobenzene free from *p*-dibromobenzene.

16-15 Saccharin, which is about 550 times as sweet as cane sugar, has the structure shown below.
a. The first step in its synthesis involves sulfonation of toluene. What conditions would you suggest to get as much ortho isomer as possible?

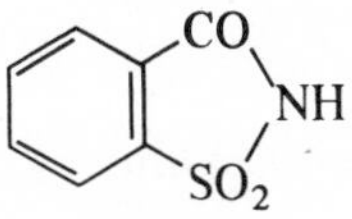

Saccharin

b. Complete the synthetic sequence from toluene.

16-16 When butane is heated with sulfur at 550°C, compound X is formed in good yield. X had a vapor density of 3.75 g/liter and an elemental composition of 57.1% carbon, 4.8% hydrogen, and 38.1% sulfur. Catalytic hydrogenation of X with an MoS_2 catalyst gave Y, which contained 54.5% carbon, 9.1% hydrogen, and 36.4% sulfur.
a. Suggest structures for X and Y.
b. The heat of combustion of X has been calculated to be 641 kcal/mol on the basis of isolated double bonds. Experiments show it to be 612 kcal/mol. How can this be rationalized? (*Hint*: see Sec. 15-6.)
c. Draw resonance structures for both ionic intermediates which could form in electrophilic substitution on thiophene.
d. Predict the position of mononitration of thiophene and justify your choice.

16-17 *a.* Draw resonance structures for nitrosobenzene.
b. Should the dipole moment of *p*-Me_2N-C_6H_4-NO be more nearly the sum or the difference of the moments of aniline and nitrosobenzene? Explain.
c. The nitroso group is ortho-para-directing for electrophilic aromatic substitution. Explain.
d. Halogen situated ortho or para to a nitroso group is subject to nucleophilic displacement. Explain.

16-18 *a.* 2,4,6-Trinitrotoluene will undergo a base-catalyzed condensation reaction with aldehydes. Draw the structure of the product that results form this aldollike reaction between TNT and benzaldehyde (C_6H_5CHO).
b. Why should TNT tend to form an anion more readily than toluene when treated with a base?
c. Draw resonance structures for this TNT anion.

16-19 Give the sign (positive or negative) for Hammett's ρ for each of the following situations and provide a specific example of a reaction in which the indicated charge conversion is taking place:
a. A rate process in which a negative charge is being generated in the transition state.

b. A rate process in which a positive charge is being generated in the transition state.

c. An equilibrium process in which a negative charge is lost in going from reactant to product.

d. An equilibrium process in which a positive charge is lost in going from reactant to product.

16-20 Explain the difference between the following σ values (Hammett) in terms of inductive and resonance effects:

a. *p-tert*-Butyl, -0.20; *p*-methyl, -0.17.

b. *p*-Chloro, $+0.23$; *m*-chloro, $+0.37$.

16-21 Starting from benzene or toluene outline the synthesis of *p*-aminobenzenesulfonamide (sulfanilamide) and *p*-$H_2NCH_2C_6H_4SO_2NH_2$ (Marfanil) which are both good bactericides.

16-22 Explain the following results:

$$C_6H_5SO_2OCH_3 \xrightarrow[H_2^{18}O]{Na^{18}OH} CH_3^{18}OH \text{ principally} \qquad CH_3COOCH_3 \xrightarrow[H_2^{18}O]{Na^{18}OH} CH_3OH \text{ principally}$$

16-23 Sulfonium salts, sulfoxides, and sulfinic esters have been resolved into optically active forms.

a. Give the structure for

$$(R)-CH_3CH_2\overset{+}{\underset{\underset{CH_3}{|}}{S}}CH_2COOH \; Br^-$$

b. Predict the number of isomers for the dimethiodide of 3,6-dithiaoctane ($EtSCH_2CH_2SEt$). Show which, if any, of the forms are optically active.

16-24 Sulfenyl chlorides add to carbon-carbon double bonds in the following way:

$$\overset{\delta^+ \; \delta^-}{RS-Cl} + \overset{\backslash}{\underset{\diagup}{C}}=\overset{\diagup}{\underset{\backslash}{C}} \longrightarrow RS-\overset{|}{\underset{|}{C}}-\overset{|}{\underset{|}{C}}-Cl$$

Give a stereochemically correct structure for the reaction product resulting from trans ionic addition of 2,4-dinitrobenzenesulfenyl chloride to each of the following:

a. 2-Methyl-2-butene   *b.* *cis*-2-Butene   *c.* *trans*-Stilbene

16-25 At room temperature the bond dissociation energy for C—D (D = deuterium) is about 1.2 kcal/mol greater than that for C—H.

a. Refer to the energy-reaction diagram in Sec. 15-2 and predict the relative rates of nitration of benzene and perdeuterobenzene (C_6D_6).

b. Propose an energy-reaction diagram such as shown in Sec. 15-2 which would be consistent with the observation that the sulfonation of C_6D_6 takes place more slowly than the sulfonation of C_6H_6.

16-26 In electrophilic substitution in each of the following aromatic molecules, reaction occurs predominantly at the position indicated. Show that this would be expected by comparing the possible intermediates and their abilities to delocalize a positive charge. Keep in mind that an aromatic sextet of electrons shows particular stability and considerable energy is required to disrupt it.

16-27 Give the total number of possible stereoisomers for each of the following compounds and indicate which compound can exist in optically active modifications:

238

6-28 Provide a mechanistic explanation for the following, based on neighboring group participation:

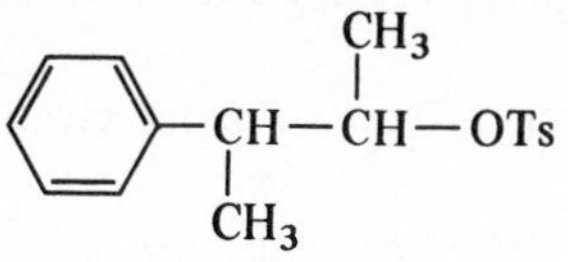

Ts = p-CH$_3$C$_6$H$_4$SO$_2^-$

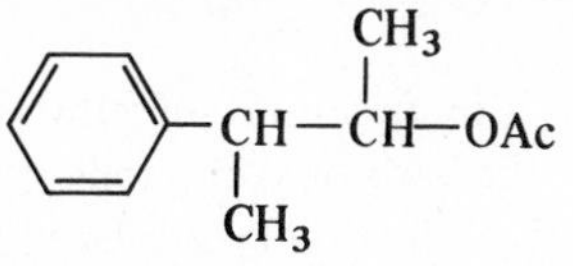

a. Optically active threo (I) reacts with NaOAc in AcOH to yield optically inactive threo (II).

b. Optically active erythro (I) reacts with NaOAc in AcOH to yield optically active erythro (II).

c. If the phenyl group of I has a p-nitro substituent, the rate of reaction is much lower than if it has a p-methoxyl substituent.

16-29 Exceptional reactions often occur in competition with normal reactions and sometimes represent the major process. This is especially true under vigorous reaction conditions or when large steric factors are involved in the normal process.

Provide mechanistic explanations for the following:

a. (CH$_3$)$_2$CH— —CH(CH$_3$)$_2$ $\xrightarrow{\text{HNO}_3}$ (CH$_3$)$_2$CH— —NO$_2$

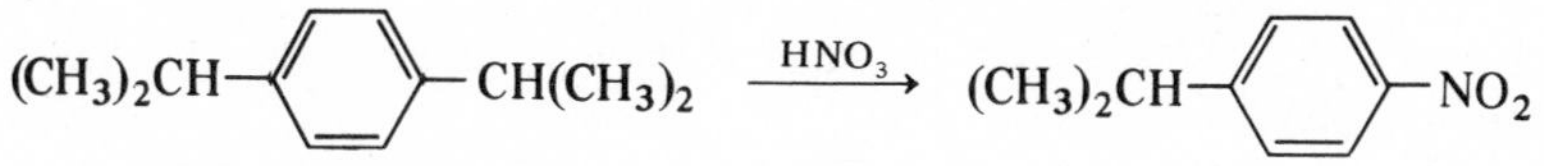

b. Br— —OCH$_3$ $\xrightarrow{\text{HNO}_3}$ O$_2$N— —OCH$_3$

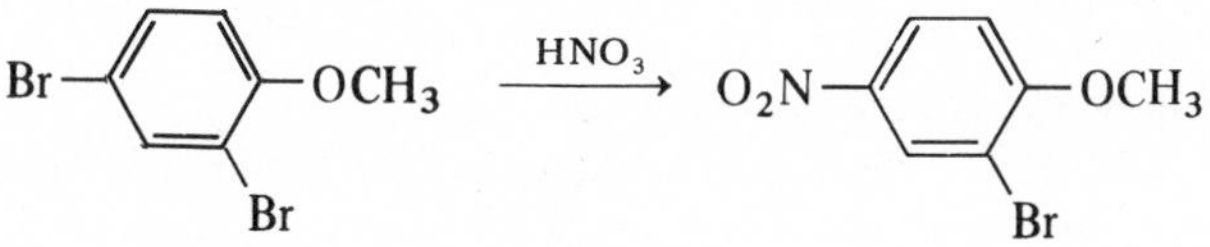

c. Benzenesulfonic acid $\xrightarrow{\text{HNO}_3,\ \text{H}_2\text{SO}_4}$ C$_{12}$H$_{10}$S$_2$ *d.* KCN + butyl sulfate $\longrightarrow$ valeronitrile

e. Benzenesulfonic acid + HOH $\xrightarrow{\Delta}$ benzene + sulfuric acid

f. EtOCOCH$_2$CH$_2$SH + B$^-$ $\longrightarrow$ + CO$_2$ + EtO$^-$ + BH

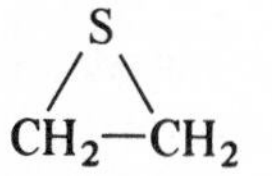

g. C$_6$H$_5$CH(OTs)CH$_3$ $\xrightarrow{\text{AcO}^-}$ C$_6$H$_5$CH(OAc)CH$_3$ but   EtCH(OTs)CH$_3$ $\xrightarrow{\text{AcO}^-}$ EtCH(OAc)CH$_3$

mostly racemization   mostly inversion

h. C$_6$H$_5$SO$_3$Na + NaCN $\longrightarrow$ C$_6$H$_5$CN + Na$_2$SO$_3$

16-30 Two isomeric oximes can be prepared from 2-chloro-5-nitrobenzaldehyde.
a. Provide structures for each oxime.
b. On the basis of the following reactions assign a specific structure to each of these (X and Y in the sequence):

(X) $\xrightarrow{\text{dil NaOH}}$ C$_7$H$_4$N$_2$O$_3$ClNa $\xrightarrow{\text{1 equiv dil HCl}}$ (X)

(Y) $\xrightarrow{\text{dil NaOH}}$ [C$_7$H$_4$N$_2$O$_3$] $\xrightarrow{\text{spontaneously}}$ 2-hydroxy-5-nitrobenzonitrile

c. Provide a mechanism for the reaction of Y with dil NaOH to yield the cyclic compound [C$_7$H$_4$N$_2$O$_3$].

16-31 *a.* Give the structure of the product that results from the reaction of aluminum chloride with pyridine (see below) and describe the geometry of the molecule.

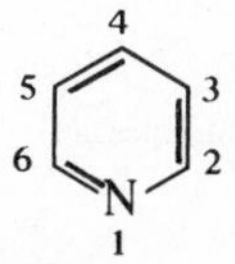

239

b. Pyridine and methylpyridines (such as 2,6-dimethylpyridine) are highly unreactive in electrophilic substitution reactions, but 2,6-di-*tert*-butylpyridine is considerably more reactive. Provide a reasonable explanation.

c. When pyridine is treated with bromine, a monobromo substitution product is formed. Predict which isomer predominates and give an explanation for the basis of your prediction.

d. When pyridine is treated with sodamide, amination takes place in the 2 position and hydrogen is formed. Provide an explanation for the orientation effect and show how hydrogen gas can be formed.

e. Phenyllithium will react with pyridine or 2-methylpyridine. One of the reactants yields benzene and the other does not. On the basis of these data, provide structures for the products of both reactions.

16-32   3,5-Dichloronitrobenzene cannot be made by direct chlorination of nitrobenzene. Explain.

16-33   Suggest structures for the compounds below based on the nmr (δ, multiplicity, and peak integration units or number of hydrogens), selected ir bonds (μm, cm^{-1}) and other data given.

   a. Contains 37.0% C, 2.8% H, 42.7% O, and 18.5% N. Nmr: 2.75, *s*, 15 units; 8.85, *s*, 10 units.

   b. $C_6H_{14}OS$. Nmr: 1.25, *d*, 30 units; 2.7, *sp*, 5 units. Ir: 9.52, 1050.

   c. MW = 76 and contains S. Nmr: 1.35, *d*, 24 units; 1.6, *d*, 4 units; 3.15, *sp*, 4 units. Ir: 3.87, 2584.

   d. MW = 108 and contains S. Nmr: 1.3, *t*, 2H; 1.9, *p*, 2H; 2.7, *q*, 4H.

   e. MW = 182 and contains S. Nmr: all singlets at 3.1, 3.85, 7.35, and 10.95 with a respective ratio of 2:2:5:1. Ir: 3.6, 2778; 5.92, 1689.

   f. MW = 152 and contains N. Nmr: 2.95, *s*, 3H; 4.65, broad singlet, 1H; 6.55, *d* (J = 9 Hz), 2H; 8.1, *d d* (J = 9 Hz), 2H. Ir: 2.9, 3448; 6.4, 1562; 7.4, 1351.

   g. $C_8H_{18}O_4S_2$. Nmr: 1.25, *s*, 15 units; 1.45, *t* 20 units; 1.8 *s*, 15 units; 2.35, *q*, 10 units; 3.5, *q*, 20 units. Ir: 7.5, 1333; 8.5, 1149.

16-34   Compound A through C are related by the following synthetic pathway:
A (MW = 190 ± 1) + B (MW = 44 ± 2) $\longrightarrow$ C (MW = 198 ± 1 and soluble in NaOH). Nmr (τ, multiplicity, and peak integration units) and selected ir data (μm, cm^{-1}) are given below. Suggest structures for A, B, and C.

Compound A Nmr: 7.5, *s*, 30 units; 2.7, *d*, 20 units; 2.2, *d*, 20 units. Ir: 7.3, 1370; 8.6, 1163.
Compound C Nmr: 8.9, *s*, 12.2 units; 7.5, *s*, 12.1 units; 6.9, *q*, 8.0 units; 4.9, *s*, 4.1 units; 2.7, *d*, 7.9 units; 2.2, *d*, 8.1 units. Ir: 7.3, 1370; 8.62, 1160.

16-35   Compounds A through D are related by the following synthetic pathway:

(A) $\xrightarrow[\text{H}_2\text{SO}_4]{\text{HNO}_3}$ (B) (MW = 157 ± 1) isomer $\xrightarrow[\text{H}_2\text{SO}_4]{\text{HNO}_3}$ (C) (MW = 202 ± 1) $\xrightarrow{\text{CH}_3\text{ONa}}$ (D) (MW = 198 ± 1)

Compound B shows nmr peaks at δ = 7.5 and 8.2 which closely resemble doublets and have equal areas. The nmr spectrum for D is sketched below in Prob. 16-36. Suggest structures for A, B, C, and D.

16-36   Compound X (MW = 106) and compound Y (MW = 74) both contain S. Their nmr spectra are sketched below. Suggest structures for X and Y.

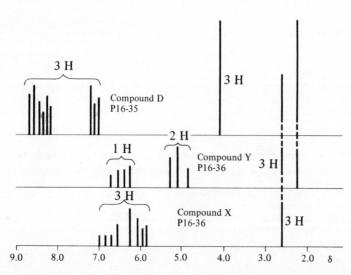

3 H

Compound D
P16-35

3 H

2 H

1 H

Compound Y
P16-36

3 H

3 H

Compound X
P16-36

3 H

9.0   8.0   7.0   6.0   5.0   4.0   3.0   2.0 δ

Figures P16-35 and 16-36. Nmr spectra for Prob. 16-35, compound D, and for Prob. 16-36, compounds X and Y.

16-37   The nmr and ir spectra for compound A (MW. = 124 ± 1 and contains N and S) and compound B (MW = 198 ± 1 and contains N) are shown. Suggest structures for A and B.

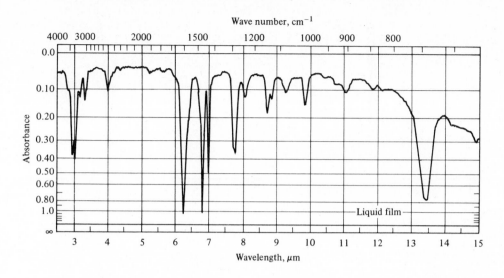

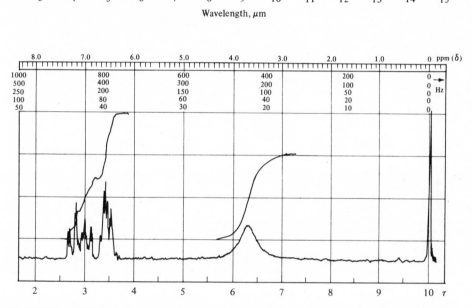

Figure P16-37 Ir and nmr spectra for Prob. 16-37, compound A. *(From R. A. Silva and R. Spenger, "The Quantitative Organic Analysis Spectra Kit," McGraw-Hill Book Company, New York, 1969. By permission of the publishers.)*

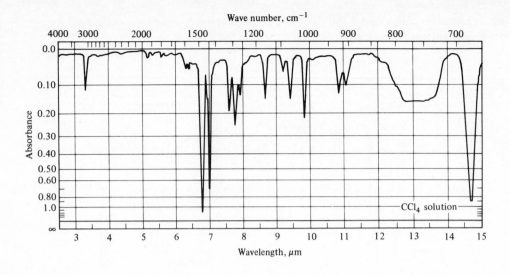

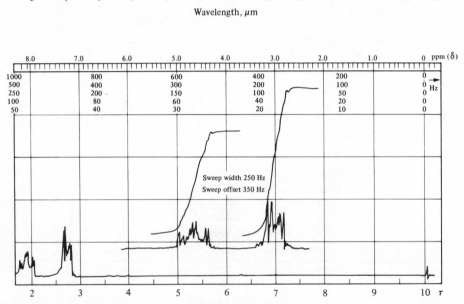

Figure P16-37 Ir and nmr spectra for Prob. 16-37, compound B. *(From R. A. Silva and R. Spenger, "The Quantitative Organic Analysis Spectra Kit," McGraw-Hill Book Company, New York, 1969. By permission of the publishers.)*

17
aromatic amines and diazonium compounds

17-1 Nomenclature

Many aromatic amines have common names, but they may also be names as substitution products of the parent aromatic compound. The prefixes *bis* (two), *tris* (three), and *tetrakis* (four) may be used to name compounds with two or more complex groups that are alike.

$C_6H_5NH_2$ $(C_6H_5)_2NH$ $H_2N-\langle\rangle-CH_3$ $H_2N-\langle\rangle-NH_2$

Aniline   Diphenylamine   *p*-Toluidine 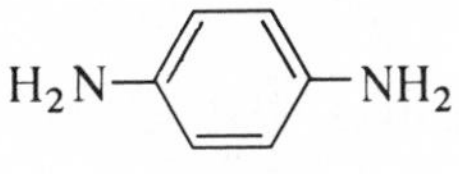   *p*-Phenylenediamine

$C_6H_5N(CH_3)_2$ $H_2N-\langle\rangle-\langle\rangle-NH_2$

N,N-Dimethylaniline   Benzidine (4,4′-diaminobiphenyl)

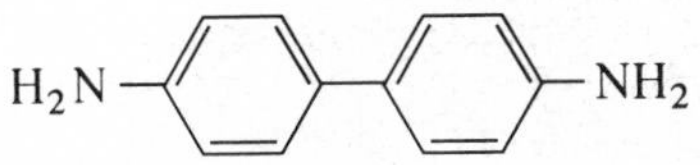

$C_6H_5\overset{+}{N}(CH_3)_3Cl^-$ $\left(Me_2N-\langle\rangle-\right)_2 CH-\langle\rangle$

Phenyltrimethylammonium chloride   *p,p′*-Bis(dimethylamino)triphenylmethane

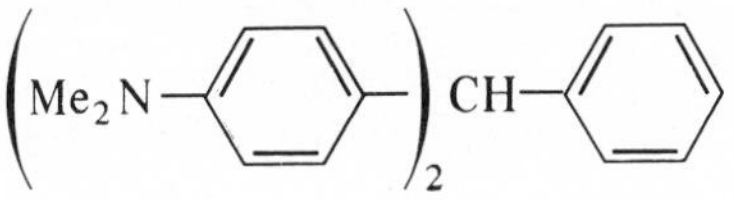

The amide composed from aniline and acetic acid, $C_6H_5NHCOCH_3$, is called acetanilide.

Compounds containing the group $-\overset{+}{N}\equiv N\ X^-$ are known as *diazonium salts.* The aromatic ring system is named first and the term *diazonium* is added. The name of the anion X^- is added as a separate word.

$CH_3CH_2-\langle\rangle-\overset{+}{N}_2Cl^-$
$\qquad\qquad\quad$ Br

$C_6H_5\overset{+}{N}_2HSO_4^-$

Benzenediazonium hydrogen sulfate   3-Bromo-4-ethylbenzenediazonium chloride

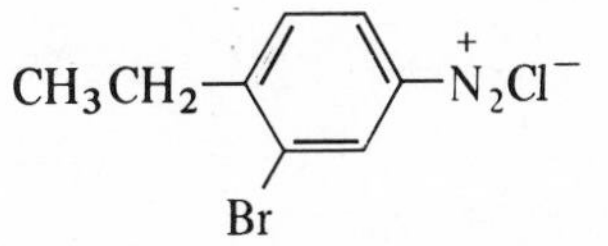

17-2 Preparation of Aromatic Amines

a. *Reduction of nitro, nitroso, hydroxylamino, azoxy, azo, or hydrazo compounds* (See Sec. 16-4a, b.)

b. *Selective Reduction of Dinitro Compounds* One nitro group of dinitro compounds can be selectively reduced using Na_2S in alcohol although it is often difficult to predict which nitro group will be reduced.

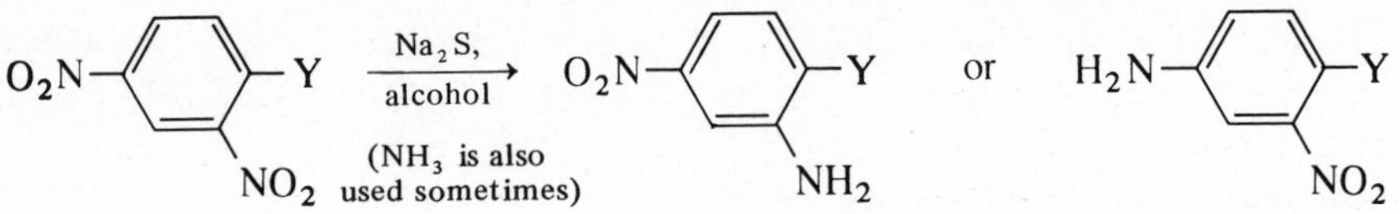

$$O_2N-\langle\text{ring}\rangle-Y \xrightarrow[\text{alcohol}]{Na_2S,} O_2N-\langle\text{ring}\rangle-Y \quad \text{or} \quad H_2N-\langle\text{ring}\rangle-Y$$

with NO_2 ortho on left ring, NH_2 ortho on middle ring (I), NO_2 ortho on right ring (II)

(NH$_3$ is also used sometimes)

I   II

I obtained when Y = OH
II obtained when Y = CH$_3$

c.   Replacement of halogen atoms ortho or para to a nitro group (See Sec. 15-3a.)

d.   *Via Benzyne Intermediates* (See Sec. 15-3b.)

e.   *Hofmann Reaction* (See Sec. 10-3d.)

f.   *Diaryl Amines*

$$C_6H_5NH_3Cl + C_6H_5NH_2 \xrightarrow{220^\circ C,\ 6\ atm} (C_6H_5)_2NH + NH_4Cl$$

g.   *Triaryl Amines*

$$(C_6H_5)_2NH \xrightarrow{Li} (C_6H_5)_2NLi \xrightarrow{C_6H_5I,\ CuI} (C_6H_5)_3N$$

17-3   Reactions of Aromatic Amines

a.   *Reactions Analogous to Those of Aliphatic Amines* Read Secs. 10-7 and 10-8 for salt formation, alkylation, acylation, and the Hinsberg reaction, and Secs. 11-2, 11-4, and 11-5 for isocyanates, ureas, and diarylthioureas.
       The basicity of diaryl amines is reduced to the extent that they fail to dissolve in aqueous solutions of strong acids.

b.   *Nitrous Acid* Primary aromatic amines form diazonium salts at $0^\circ C$.

$$C_6H_5NH_3Cl + HONO \xrightarrow{HCl,\ 0^\circ C} 2HOH + C_6H_5\overset{+}{N}\equiv N\ Cl^-$$

Benzenediazonium chloride

Secondary amines form *N*-nitroso compounds (see Sec. 10-7e). Tertiary amines are nitrosated in the para position of the ring (ortho if para is blocked).

$$C_6H_5-NR_2 + HONO \longrightarrow ON-\langle\text{ring}\rangle-NR_2$$

p-Nitroso-*N,N*-dialkylaniline

c.   *Condensation with Aldehydes or Ketones to Form "Schiff Bases"*

$$C_6H_5NH_2 + O=CHC_6H_5 \longrightarrow C_6H_5-N=CH-C_6H_5$$

Benzalaniline

d.   *Oxidation to Nitro Compounds* (See Sec. 16-2b.)

e.   *Oxidation to Quinones* The ring of aromatic amines is very sensitive to oxidizing agents and may be oxidized to a quinone or may be disrupted completely.

$$\text{Aniline} \xrightarrow{ox} O=\langle\text{ring}\rangle=NH \xrightarrow{H_2O} O=\langle\text{ring}\rangle=O$$

p-Benzoquinone

f. *Electrophilic Substutition* Aniline is halogenated so readily by aqueous chlorine or bromine that it is difficult to form any derivative but the trihalo product 2,4,6-trihaloaniline. Friedel-Crafts reactions are generally poor because of the strong Lewis acid necessary for the process.

Halogenation or nitration can be controlled, and oxidation of the ring suppressed, by converting an aromatic amine to its acetyl derivative.

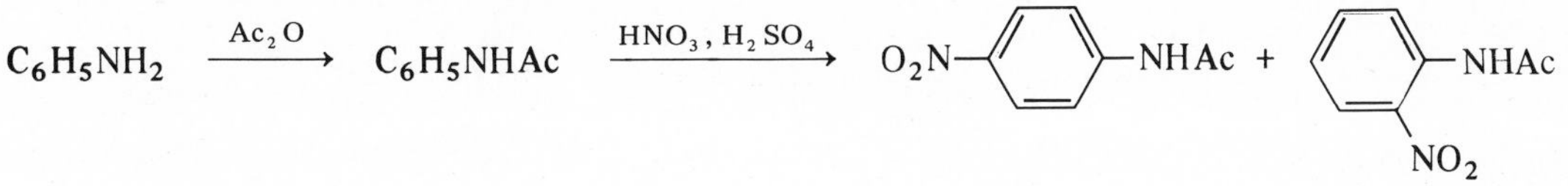

$$C_6H_5NH_2 \xrightarrow{Ac_2O} C_6H_5NHAc \xrightarrow{HNO_3, H_2SO_4} O_2N{-}\langle\ \rangle{-}NHAc + \langle\ \rangle{-}NHAc$$

with NO_2

17-4 Preparation of Aromatic Diazonium Salts

Primary aromatic amines react with cold nitrous acid in the presence of a strong mineral acid to yield a diazonium salt. The process is known as diazotization. Compare this with the reaction of aliphatic amines (see Sec. 10-7e).

$$C_6H_5NH_2 + NaNO_2 + 2HCl \longrightarrow C_6H_5\overset{+}{N}_2Cl^- + 2HOH + NaCl$$

17-5 Reactions of Aromatic Diazonium Salts

a. *Nucleophilic Displacement with Nitrogen (N_2) the Leaving Group* The anion of the diazonium salts in the following should be the same as the entering group (chloride for chloro compounds, etc.), or it should be sulfate if the first is impractical.

1. Hydroxyl.

$$C_6H_5N_2HSO_4 + HOH \xrightarrow{warm} C_6H_5OH + N_2 + H_2SO_4$$

2. Hydrogen.

$$C_6H_5N_2HSO_4 + H_3PO_2 + HOH \longrightarrow C_6H_6 + N_2 + H_3PO_3 + H_2SO_4$$

3. Iodine.

$$C_6H_5N_2HSO_4 + KI \longrightarrow C_6H_5I + KHSO_4 + N_2$$

4. Fluorine.

$$C_6H_5N_2HSO_4 + NaBF_4 \longrightarrow C_6H_5\overset{+}{N}_2BF_4^- + NaHSO_4$$

$$C_6H_5\overset{+}{N}_2BF_4^- \xrightarrow{\Delta} C_6H_5F + N_2 + BF_3$$

5. Chlorine, bromine, or the nitrile group. Cuprous salts react with diazonium salts to replace the diazonium group by Cl, Br, or CN. This is known as the *Sandmeyer reaction.*

$$C_6H_5N_2Cl \xrightarrow{CuCl, Cl^-} C_6H_5Cl + N_2$$
$$C_6H_5N_2Br \xrightarrow{CuBr, Br^-} C_6H_5Br + N_2$$
$$C_6H_5N_2HSO_4 \xrightarrow{CuCN, CN^-} C_6H_5CN + N_2 + HSO_4^-$$

6. Thiol.

$$C_6H_5N_2SO_4 + EtO{-}\overset{\overset{S}{\|}}{C}{-}SK \longrightarrow C_6H_5S{-}\overset{\overset{S}{\|}}{C}{-}OEt \xrightarrow{KOH, hyd}$$

Potassium ethyl xanthate

$$EtOH + COS + C_6H_5S^-K^+ \xrightarrow{H_3O^+} C_6H_5SH$$

245

7. Nitro.

$$C_6H_5N_2BF_4 + NaNO_2 \xrightarrow{Cu} C_6H_5NO_2 + NaBF_4 + N_2$$

8. Aryl: the Gatterman reaction.

$$2C_6H_5N_2HSO_4 \xrightarrow{Cu} C_6H_5{-}C_6H_5$$
Biphenyl

b. *Reduction to Hydrazines*

$$C_6H_5N_2Cl \xrightarrow[\text{H}_2\text{O}]{\text{Na}_2\text{SO}_3} C_6H_5NHNH_2$$

c. *Coupling*
 1. Primary or secondary amines.

$$C_6H_5{-}N_2Cl + H_2N{-}C_6H_5 \longrightarrow C_6H_5{-}N{=}N{-}NH{-}C_6H_5 + HCl$$

These products rearrange when warmed with catalytic amounts of the amine salt (compare with the rearrangements of Sec. 17-6).

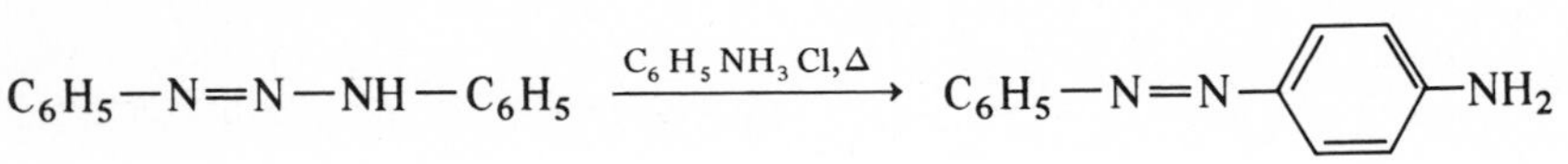

2. Aryl rings with R_2N or HO groups.

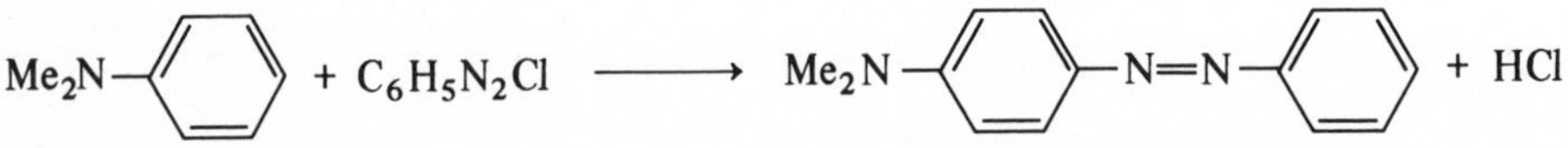

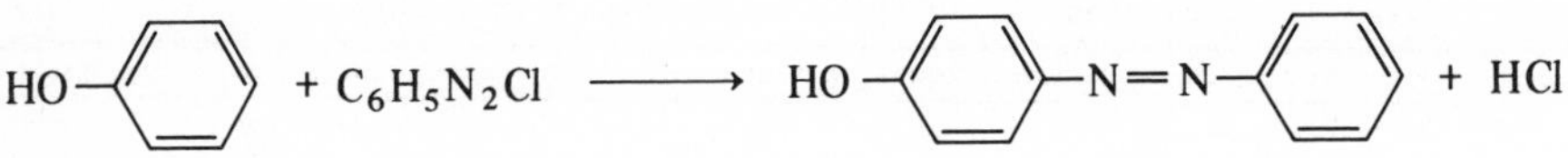

17-6 Rearrangement Reactions of *N*-Substituted Amines

Many types of *N*-substituted amines rearrange under the influence of acids or heat. The substituents migrate to the para position of the ring or to the ortho position if the other is blocked.

a. *N-Nitroso Compounds*

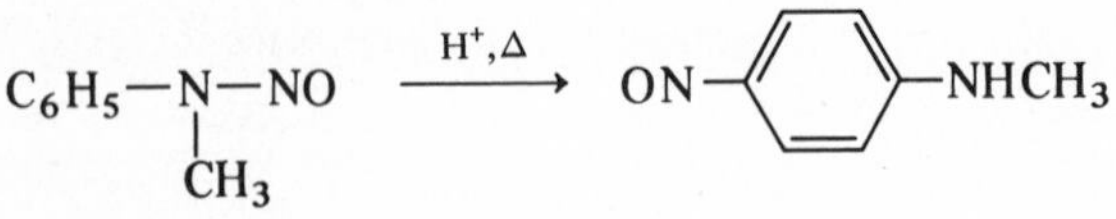

b. *Amine Sulfates*

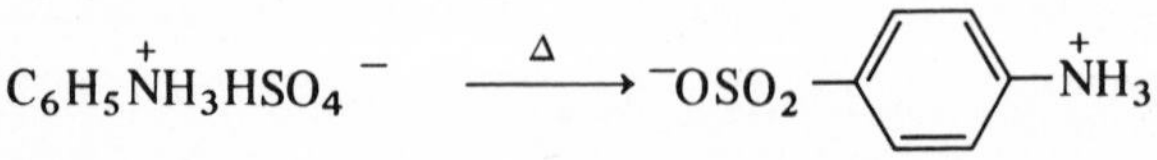

Sulfanilic acid

c. *Benzidine Rearrangement*

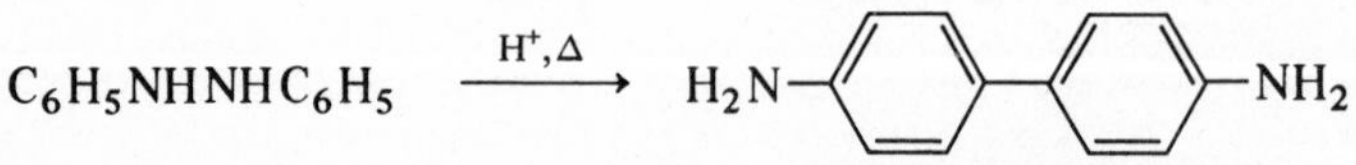

Benzidine (see Prob. 17-17)

d. *Hydroxylamines*

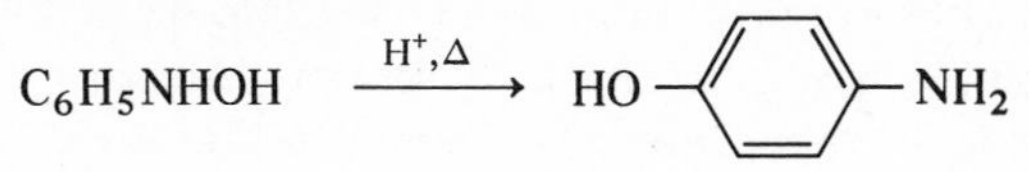

$$C_6H_5NHOH \xrightarrow{H^+, \Delta} HO-\!\!\bigcirc\!\!-NH_2$$

17-7 Free-radical Substitution

If a free radical is generated in the presence of an aromatic compound, substitution may take place

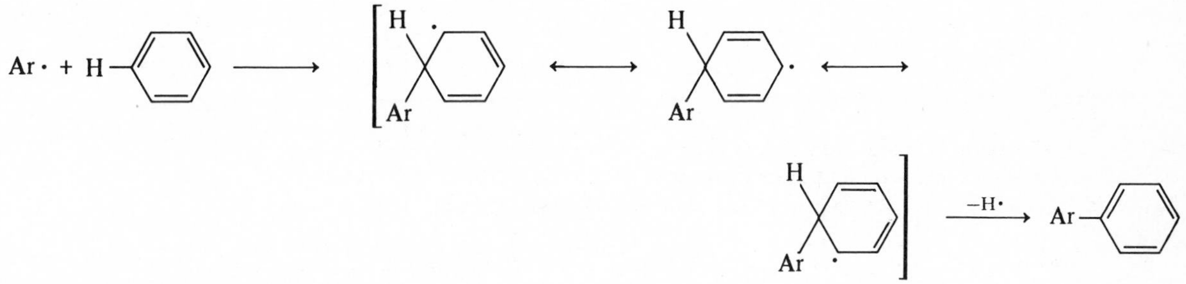

a. *Sources of Aryl Free Radicals and Synthetic Applications*

1. $ArN_2Cl \xrightarrow{OH^-} ArN=NOH \longrightarrow Ar\cdot + N_2 + HO\cdot$

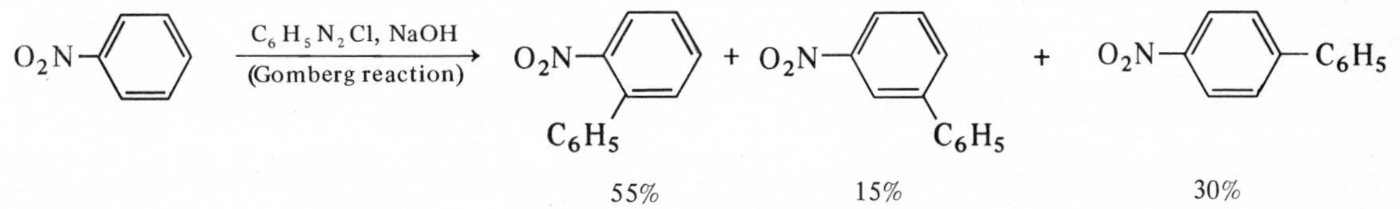

55% 15% 30%

2. $C_6H_5-CO-OO-CO-C_6H_5 \xrightarrow{\Delta} \left[C_6H_5-C\!\!\begin{smallmatrix}O\\ \\O\cdot\end{smallmatrix} \right] \longrightarrow C_6H_5\cdot + CO_2$

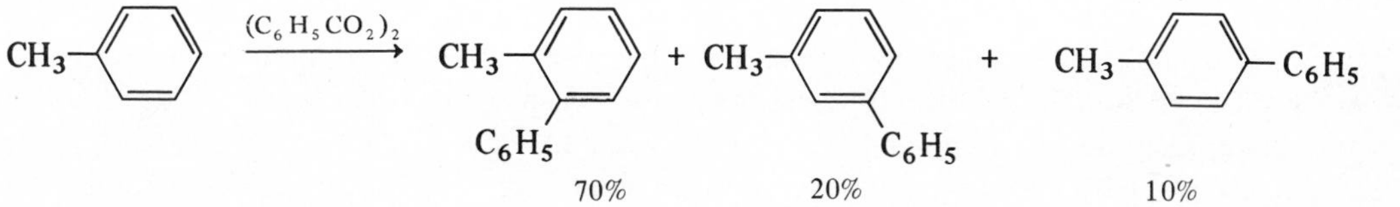

70% 20% 10%

3. $\underset{\overset{|}{NO}}{C_6H_5N}-CO-CH_3 \xrightarrow{\Delta} C_6H_5\cdot + N_2 + CH_3COO\cdot$

60% 10% 30%

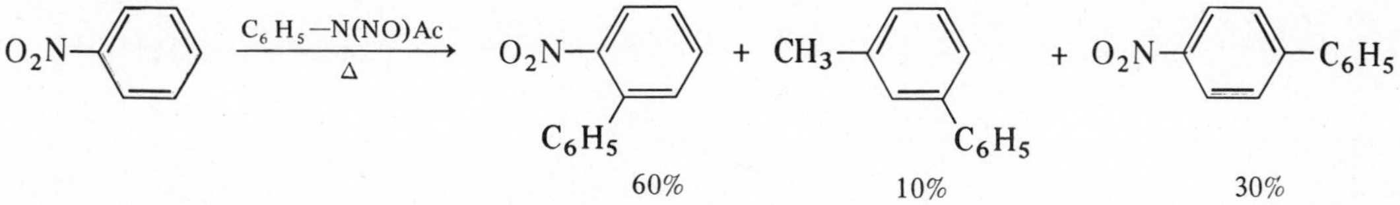

b. *Orientation in Substitution* As indicated above, the directing influence of an aromatic substituent is mainly ortho. Within experimental error the product distribution is the same regardless of the source of a given free radical. Steric factors may alter products drastically.

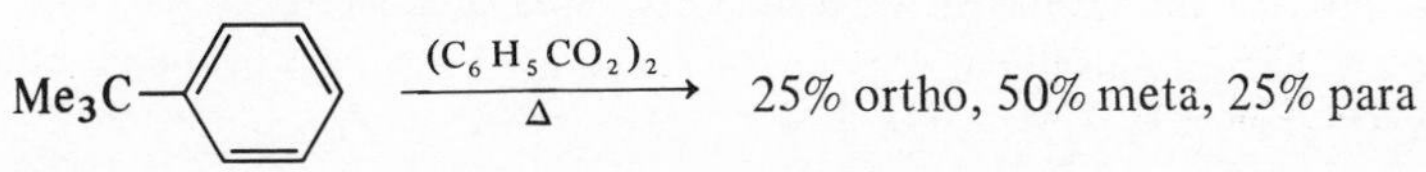

$$Me_3C-\!\!\bigcirc \xrightarrow[\Delta]{(C_6H_5CO_2)_2} \text{25\% ortho, 50\% meta, 25\% para}$$

17-8 Synthetic Routes

In the last two chapters, the problem of the introduction of two or more functional groups into the aromatic ring was approached. The principal new concept that should be added here is the ease of conversion of an aromatic amine to a wide variety of functional groups. There is now available the very convenient system that involves nitration, reduction, and diazotization. It would be well to keep a mental file of the following points:

1. The amino groups can be converted to most common functional groups by way of a diazonium salt.
2. The nitro group is an excellent meta-directing group, and it can be converted to most common functional groups by way of the amine.
3. The amino and acetylamino groups are the most convenient ortho-para-directing groups that can readily be removed.
4. Another pair of related groups are the meta-directing COR group (Friedel-Crafts acylation) and the ortho-para-directing group CH_2R.

For an example, apply these points to a synthetic procedure by illustrating the preparation of p-cyanobenzoic acid. This compound has two meta-directing groups para to each other, so direct introduction is unsatisfactory. However, the cyano group can be introduced by way of a diazonium salt.

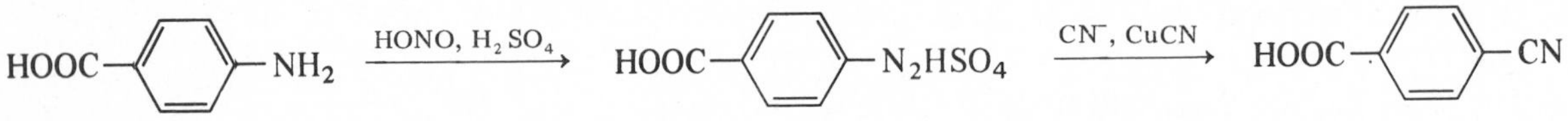

The problem has now resolved itself into the need for a combination of substitution reactions, preferably electrophilic, which can lead to p-aminobenzoic acid. First eliminate some of the possibilities. It would be impossible to start with benzoic acid because the carboxyl group is meta-directing. However, if the carboxyl groups were initially methyl, the toluene could be nitrated, oxidized, and reduced to yield the desired product.

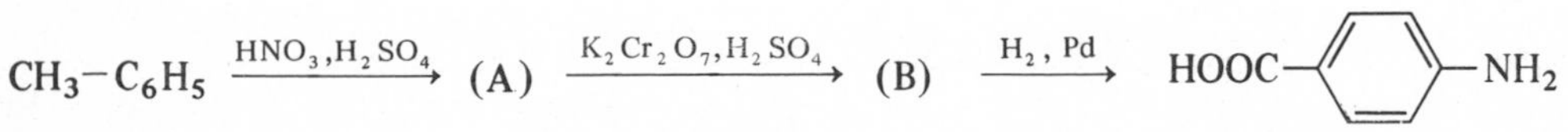

It would be possible to start with aniline but the route would be more involved. Complete the following sequence and compare it with the reaction illustrated:

Aniline ⟶ acetanilide ⟶ p-nitroacetanilide ⟶ p-nitroaniline ⟶ p-nitrobenzene-diazonium sulfate ⟶ p-nitrobenzonitrile ⟶ p-nitrobenzoic acid ⟶ p-aminobenzoic acid

Outline the more difficult synthesis of 3,5-dibromotoluene. Here all three groups are ortho-para-directing for electrophilic substitution, but they are meta to each other. At first glance, one might think of starting with a meta-directing group where the methyl group is and carrying out a dinitration. This might work, but it is a major task to reduce the original group back to methyl without altering the other groups. The second alternative is to start with a strong ortho-para-directing group, para to the methyl group, which can be removed after bromination. Such a group is the amino group.

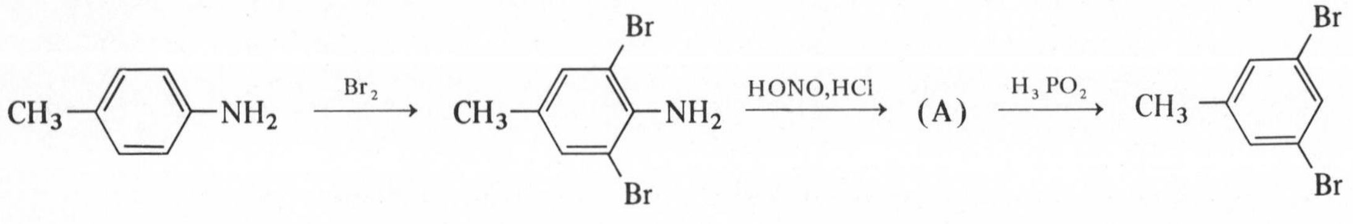

17-9 Qualitative Tests

a. Hinsberg test for amines. (See Sec. 10-8b.)
b. Diazonium salts are so seldom isolated that qualitative tests can be ignored. However, the use of nitrous acid as a reagent for primary, secondary, and tertiary amines has already been discussed (see Sec. 10-7e).
The highly colored coupling products of diazonium salts with phenols can be used as an aid in the identification of primary aromatic amines (see Sec. 17-5c).

Problems

17-1   Name the following compounds:

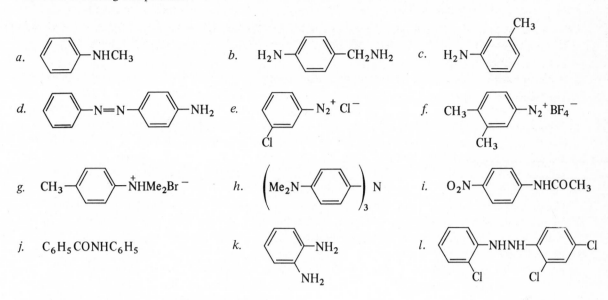

a. (phenyl)—NHCH$_3$

b. H$_2$N—(phenyl)—CH$_2$NH$_2$

c. H$_2$N—(phenyl with CH$_3$)

d. (phenyl)—N=N—(phenyl)—NH$_2$

e. (phenyl with Cl)—N$_2^+$ Cl$^-$

f. CH$_3$—(phenyl with CH$_3$)—N$_2^+$ BF$_4^-$

g. CH$_3$—(phenyl)—$\overset{+}{N}$HMe$_2$Br$^-$

h. (Me$_2$N—(phenyl)—)$_3$N

i. O$_2$N—(phenyl)—NHCOCH$_3$

j. C$_6$H$_5$CONHC$_6$H$_5$

k. (phenyl)—NH$_2$ / NH$_2$

l. (phenyl with Cl)—NHNH—(phenyl with Cl)—Cl

17-2   Predict the products of the following reactions:
  a. *p*-Nitrophenylammonium chloride + aniline in H$_2$O
  b. 2,2'-Dimethylhydrazobenzene + dil HCl, heat   *c.* *N*-Benzylaniline + dil aq HCl
  d. Diphenylamine + dil aq HCl   *e.* *p*-Toluidine + butanal
  f. Phthalimide + 1 equiv aq NaOH
  g. *o*-Toluenediazonium hydrogen sulfate + cuprous cyanide
  h. *m*-Bromobenzenediazonium fluoroborate + Δ   *i.* Benzenediazonium chloride + *N,N*-dimethylaniline
  j. *p*-Toluenediazonium chloride + aq sodium sulfite
17-3   Starting with benzene or toluene and any aliphatic compounds, give a convenient laboratory synthetic route
        for each of the following, using a diazonium reaction where applicable:
  a. Benzylamine (free from 2° and 3° RNH$_2$)   *b.* *p*-Aminobenzoic acid
  c. *o*-Chlorobenzoic acid   *d.* *p*-Phenylenediamine
  e. *o*-Nitrobenzonitrile   *f.* Benzanilide
  g. *m*-Bromoaniline   *h.* *m*-Fluorobromobenzene
  i. 1,3-Benzenedicarboxylic acid   *j.* *p*-Toluic acid
  k. *m*-Cyanobenzoic acid   *l.* 2,6-Dibromobenzoic acid
  m. *sym*-Diphenylthiourea   *n.* *N*-Nitroso-*N*-benzylaniline
  o. 3,3'-Dipropyl-4,4'-diaminobiphenyl   *p.* *p*-Aminobenzenethiol
  q. 3,3-Dinitrobiphenyl   *r.* *m*-Bromobenzenethiol
  s. *p*-Nitrophenyl ethyl ketone   *t.* *m*-Ethylaniline
  u. Tris-(*p*-tolyl)amine   *v.* 1-Fluoro-3-chloro-5-bromobenzene
  w. 1-Fluoro-2-chloro-5-bromobenzene   *x.* 1-Fluoro-2-chloro-4-bromobenzene
  y. 3-Cyano-4-fluorotoluene   *z.* 2-Amino-4'-nitrobiphenyl
17-4   List the following compounds in the order of decreasing basicity:
  a. Aniline (1), benzamide (2), benzylamine (3), *m*-chloroaniline (4)
  b. Aniline (1), *p*-nitroaniline (2), ethyl *p*-aminobenzoate (3), ethyl *m*-aminobenzoate (4)
  c. Aniline (1), hexylamine (2), *p*-toluidine (3), triphenylamine (4), diphenylamine (5), phenyltrimethyl-
      ammonium hydroxide (6)
17-5   Suggest structures for all lettered compounds in each of the following:
  a. Compounds A, B, and C (all C$_9$H$_{14}$NCl) each gave an immediate precipitate when treated with alcoholic
      AgNO$_3$. Only compound C was optically active. A was very resistant to bromination in either acid or alkaline
      solutions; it was also resistant to nitration, heat, and alkaline KMnO$_4$ oxidation. With aqueous Br$_2$,
      compound B readily formed a dibromo compound (with the formation of 2 equiv of HBr) which formed only
      a single tribromo compound on treatment with Br$_2$ and Fe; B gave a negative Hinsberg test. Compound C

reacted with $C_6H_5SO_2Cl$ and NaOH to yield an acid- and base-insoluble product; vigorous $KMnO_4$ oxidation of C gave acid D (NE 83) which formed only one monobromo substitution product.

b. Compound E, $C_{10}H_{13}ON$, was insoluble in dil HCl or NaOH. It was refluxed for 2 h with 10% NaOH, and the alkaline solution was steam-distilled. The distillate contained water-insoluble, acid-soluble organic compound F, which readily formed a tribromo derivative. The residue contained an organic acid (NE 74) that was stable to oxidation by alk $KMnO_4$. When F was treated with an excess of ethyl bromide, 2 equiv of bromide ion was formed.

c. Compound G, $C_{10}H_{15}N$, was an optically active substance soluble in dil HCl. Treatment of G with cold HONO yielded a diazonium salt, H. Reduction of H with H_3PO_2 yielded I, an optically active hydrocarbon. When H was treated with copper, compound J, $C_{20}H_{26}$, was obtained. Vigorous oxidation of J yielded an acid, K (NE 121), which gave only two monobromo substitution products when treated with Br_2 and Fe.

d. Optically active compounds L and M, both $C_{13}H_9O_3N_2Br$, were insoluble in dilute acids or bases. When treated with NaOBr, L and M yielded compounds N and O, respectively (both optically active). When N was reduced with Sn and HCl, compound P was obtained which was optically inactive and could not be resolved. Separate treatment of N and O with a cold solution of $NaNO_2$ and H_2SO_4 followed by CuBr yielded compounds Q (optically active, resolvable) and R (optically inactive, nonresolvable), respectively.

17-6 Starting with benzene, toluene, acetoacetic ester, benzoic acid-1-^{14}C (benzoic acid labeled in the 1 position with ^{14}C), and aliphatic compounds with four carbons or less, outline a convenient laboratory synthesis of each of the following:

a. N-(4-Nitrophenyl)-3-methylpentamide
b. 3-(2,4-Diaminophenyl)propanoic acid
c. 4-(4-Nitrophenyl)-2-butanone
d. m-Isobutylaniline
e. Iodobenzene-1-^{14}C
f. Fluorobenzene-3-^{14}C
g. Phenol-2-^{14}C
h. Benzenethiol-4-^{14}C
i. 3,5-Dibromoaniline-4-^{14}C
j. 2-Propoxy-5-nitroaniline
k. 2-Amine-4,5-dimethylbenzoic acid (Hint: use a Diels-Alder rection)
l. cis-2-Deuterio-1-aminocyclohexane
m. trans-2-Deuterio-1-aminocyclohexane
n. 3-(3-Bromo-4-carboxyphenyl)-2-aminopropanoic acid

17-7 Give a chemical test to differentiate between each of the following pairs of compounds:

a. p-Aminobenzoic acid and ethyl p-aminobenzoate
b. Benzamide and ammonium benzoate
c. Aniline and benzamide
d. Aniline and N-methylaniline
e. Aniline and N,N-dimethylaniline
f. Hexylamine and aniline

17-8 Indicate the errors in the following reactions sequences:

a.

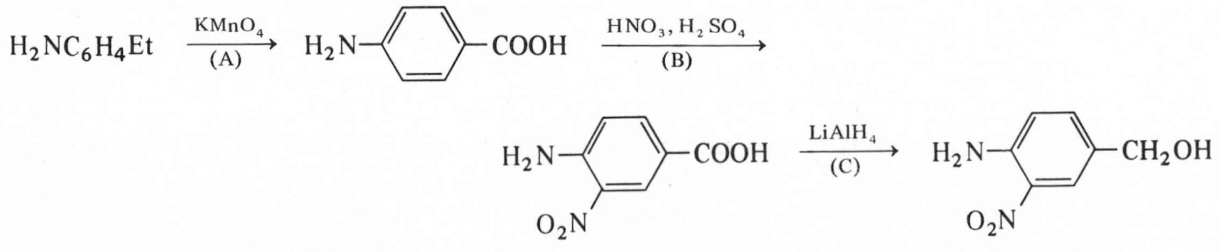

b.

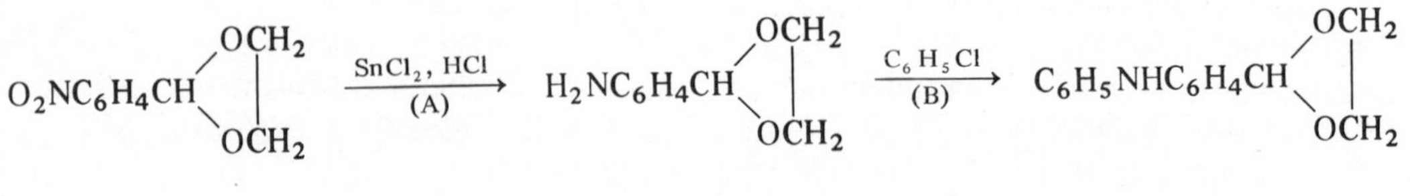

c.

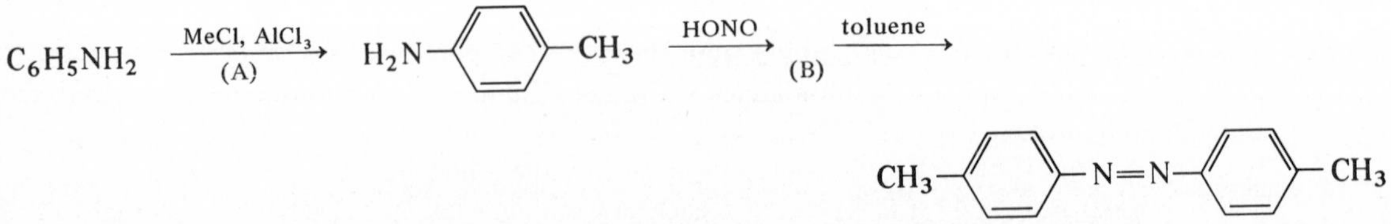

17-9 Draw electronic structures for the following, showing resonance forms involving the functional group and the ring.

a. Acetanilide
b. Benzenediazonium chloride
c. Hydrazobenzene

17-10 The following represents a synthetic route for secondary amines:

$$C_6H_5NH_2 \xrightarrow[\text{(A)}]{\text{Et X}} C_6H_5NEt_2 \xrightarrow[\text{(B)}]{\text{HONO}} ON-\!\!\!\bigcirc\!\!\!-NEt_2 \xrightarrow[\text{(C)}]{OH^-}$$

$$ON-\!\!\!\bigcirc\!\!\!-OH \;+\; NHEt_2$$

a. What types of reaction (mechanistically) are represented by A, B, and C?

b. Assume you were interested in a pure, tertiary amine. How would you separate $C_6H_5NR_2$ from $C_6H_5NH_2$, C_6H_5NHR, and $C_6H_5NR_3X$ after step A?

c. Suggest an appropriate energy-reaction diagram for step C.

17-11 Explain why the first compound in each of the following pairs is a stronger base than the second:

   a. Cyclohexylamine and aniline   *b.* *m*- and *p*-Nitroaniline

   c. Aniline and acetanilide   *d.* $C_6H_5NMe_3OH$ and NH_4OH

   e. $(H_2N)_2C=NH$ and $(H_2N)_2C=O$ *f.* Aniline and diphenylamine

17-12 Aliphatic and aromatic primary amines react quite differently with nitrous acid, the aromatic amine forming a diazonium salt that is reasonably stable at moderate temperatures.

a. Explain the difference in stability exhibited by aliphatic and aromatic products (use either resonance stabilization or the ease of formation of a carbonium ion by loss of nitrogen).

b. If treatment of 2-aminopropane with HONO resulted in the formation of a diazonium salt intermediate that immediately underwent decomposition to yield a carbonium ion, what products would you expect of the reaction? Explain.

c. If, in *b*, no free carbonium ion was formed, show how water as a base could be used to explain the formation of products.

17-13 In the coupling reaction between diazonium salts and phenols, it is necessary to maintain quite close control over the pH of the solution in order to obtain a good yield of product.

a. Phenol is a strong enough acid to form a salt with NaOH. Would excess base aid or hinder the susceptibility of phenol to electrophilic coupling?

b. Diazonium salts react with hydroxide ion to yield benzenediazoic acid.

$$C_6H_5-\overset{+}{N}\!\!\equiv\!\!N \;+\; OH^- \;\rightleftharpoons\; C_6H_5-N=N-OH \xrightleftharpoons[\text{HOH}]{\text{KOH}} C_6H_5-N=N-O^-$$

Would excess base aid or hinder the reaction from the viewpoint of the diazonium salt? Explain.

c. What stereoisomers would you expect for potassium benzenediazoate? Draw structures for these.

17-14 Diazonium salts couple with certain aliphatic compounds.

a. When benzenediazonium chloride couples with 2,4-pentanedione, the product is a phenylhydrazone, $C_{11}H_{12}O_2N_2$. Suggest a structure for the compound.

b. The phenylhydrazone from *a* is a rearrangement product of a conventional azo compound. Show this reaction.

c. Is the coupling initiated by electrophilic or nucleophilic attack on carbon? Give the structure of the attacking species.

d. Give the structure for the product that results from the coupling of benzenediazonium chloride with 1 equiv of isoprene. Justify the position of attack on the basis of the best ionic intermediate (isoprene is 2-methyl-1,3-butadiene).

17-15 The vinyl group of α-vinylpyridine $\bigcirc\!\!\!\overset{}{\underset{N}{}}\!\!\!-CH=CH_2$ added EtOH when treated with sodium ethoxide in ethanol. Under similar conditions, no addition occurs with vinylcyclohexane.

a. Predict the final reaction product and draw the ionic intermediate and its resonance forms.

b. Why should there be such a great difference in reactivity in the two vinyl groups toward nucleophilic attack?

c. Predict the sign of Hammett's ρ for the reaction in *a*.

17-16 Predict the structures for the lettered intermediates in the following:
a. p-Nitroaniline + 2ICl $\longrightarrow$ (A) $(C_6H_5N_2O_2I_2Cl)$

b. C_6H_6 $\xrightarrow{HNO_3,\ H_2SO_4}$ $\xrightarrow{H_2,\ Pd}$ $\xrightarrow{CS_2}$ (B) $(C_7H_7NS_2)$ $\xrightarrow{-H_2S}$ (C) $\xrightarrow{aniline}$ (D) $\xrightarrow{HgO\ (oxidation)}$
(E) $(C_{13}H_{10}N_2)$ $\xrightarrow{H_2O}$ (F) $\xrightarrow{hydrolysis}$ $C_6H_5NH_2 + CO_2$

c. The following synthetic scheme leads to compound I which is a difluoro analog of thyroxine $(C_{15}H_{11}NO_4I_4)$. Thyroxine is a constituent of thyroglobulin, a hormone of the thyroid gland.

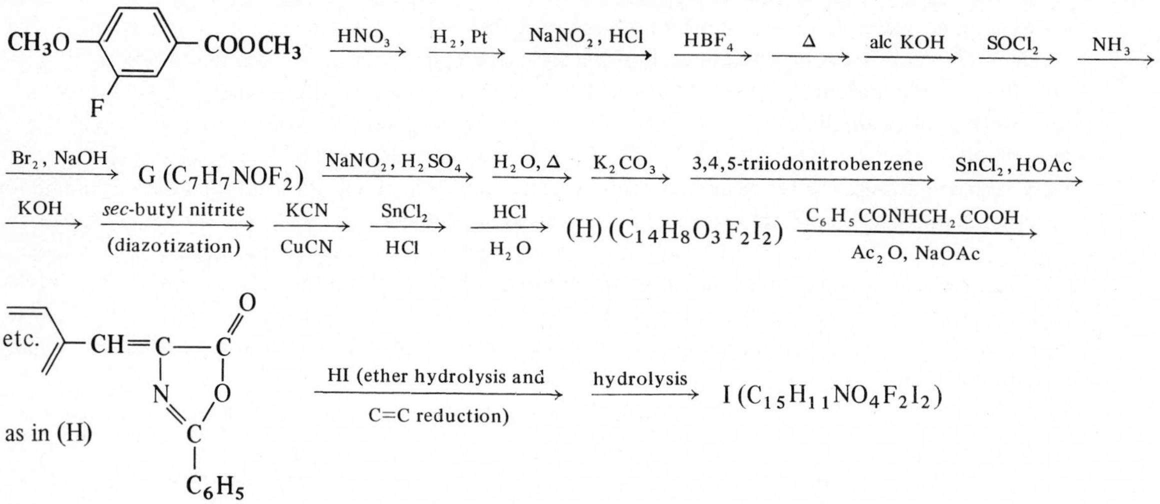

CH$_3$O— —COOCH$_3$ (with F) $\xrightarrow{HNO_3}$ $\xrightarrow{H_2,\ Pt}$ $\xrightarrow{NaNO_2,\ HCl}$ $\xrightarrow{HBF_4}$ $\xrightarrow{\Delta}$ $\xrightarrow{alc\ KOH}$ $\xrightarrow{SOCl_2}$ $\xrightarrow{NH_3}$

$\xrightarrow{Br_2,\ NaOH}$ G $(C_7H_7NOF_2)$ $\xrightarrow{NaNO_2,\ H_2SO_4}$ $\xrightarrow{H_2O,\ \Delta}$ $\xrightarrow{K_2CO_3}$ $\xrightarrow{3,4,5\text{-triiodonitrobenzene}}$ $\xrightarrow{SnCl_2,\ HOAc}$

$\xrightarrow{KOH}$ $\underset{(diazotization)}{\xrightarrow{sec\text{-butyl nitrite}}}$ $\underset{CuCN}{\xrightarrow{KCN}}$ $\underset{HCl}{\xrightarrow{SnCl_2}}$ $\underset{H_2O}{\xrightarrow{HCl}}$ (H) $(C_{14}H_8O_3F_2I_2)$ $\underset{Ac_2O,\ NaOAc}{\xrightarrow{C_6H_5CONHCH_2COOH}}$

etc. (azlactone structure) as in (H) $\underset{C=C\ reduction}{\xrightarrow{HI\ (ether\ hydrolysis\ and}}$ $\xrightarrow{hydrolysis}$ I $(C_{15}H_{11}NO_4F_2I_2)$

An azlactone

d. In the preparation of TNT by nitration of toluene, two major impurities are 2,3,4-trinitrotoluene and 2,4,5-trinitrotoluene. When the entire nitration mixture is treated with Na_2SO_3 these impurities are removed as water-soluble compounds J and K, respectively $(C_7H_5N_2O_7SNa$ isomers).

17-17 The mechanism of rearrangements such as the benzidine rearrangement is based in part on the interpretation of the nature of the products of mixed systems. Show how the following product analysis could be put to use:

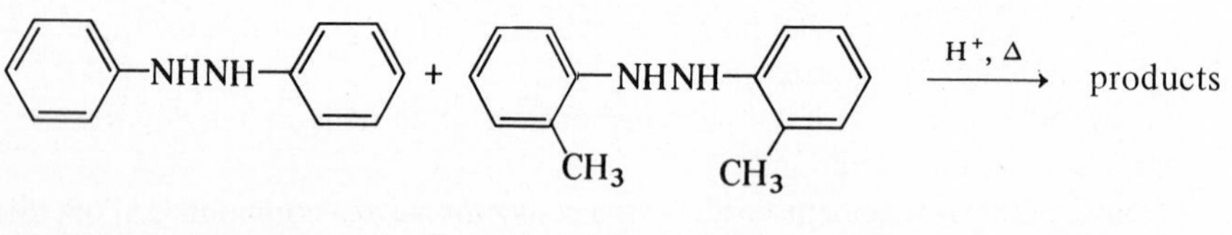

—NHNH— + —NHNH— (with CH$_3$ groups) $\xrightarrow{H^+,\ \Delta}$ products

17-18 Explain why the first compound in each of the following pairs reacts more rapidly than the second, and provide products and a mechanism for each:

a. cis-CH$_3$CH=CH— or cis-CH$_3$CH=CH— —NO$_2$ $\xrightarrow{Br_2,\ CCl_4}$ addition

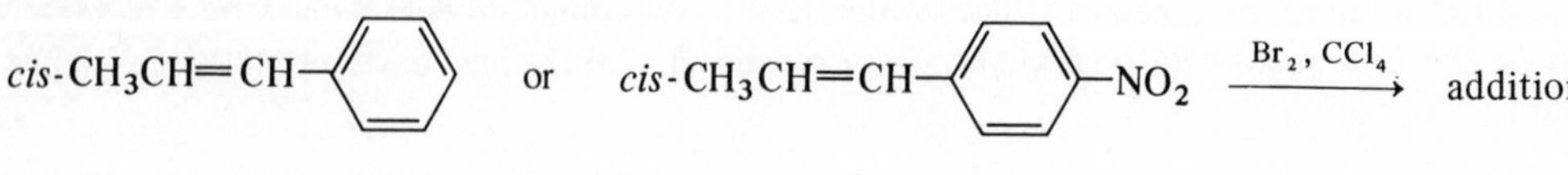

b. CH$_3$O— $\underset{OH}{\overset{C_6H_5}{\underset{|}{\overset{|}{C}}}}$ $\underset{OH}{\overset{C_6H_5}{\underset{|}{\overset{|}{C}}}}$ —OCH$_3$ or $(C_6H_5)_2\underset{OH}{\overset{}{C}}\text{—}\underset{OH}{\overset{}{C}}(C_6H_5)_2$ $\xrightarrow{pinacol\ rearrangement}$

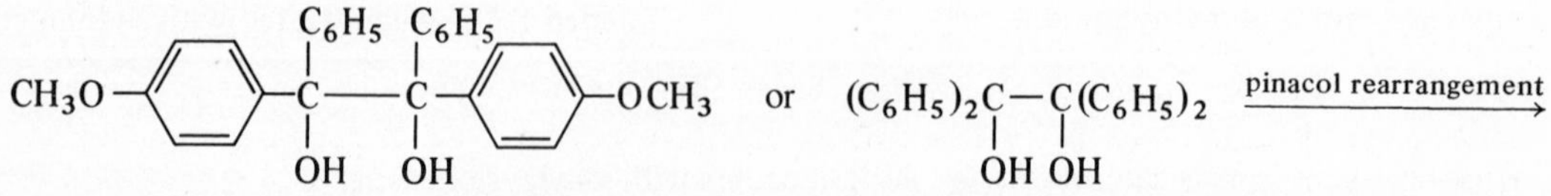

c. (2-chloropyridine) or (3-chloropyridine) $\xrightarrow{C_6H_5NH_2}$ substitution

252

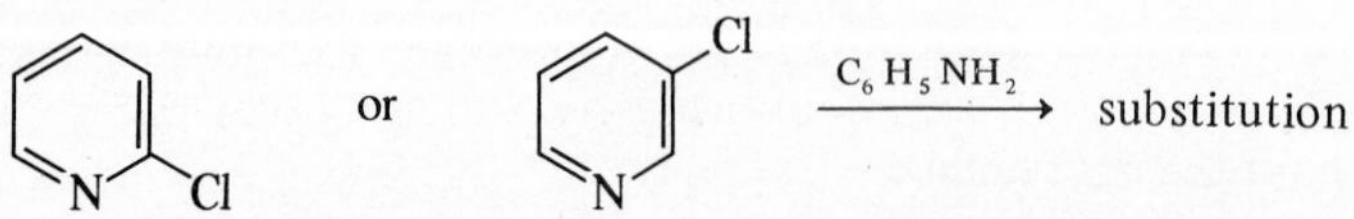

d. *cis*-1-Bromo-4-*tert*-butylcyclohexane or *trans*-1-bromo-4-*tert*-butylcyclohexane $\xrightarrow{\text{S}_\text{N}2 \text{ reaction}}$ substitution products

17-19 Four compounds with the formula $C_{12}H_{12}N_2$ were subjected to the following reaction conditions. On the basis of the experimental observations suggest structures for the unknowns:

a. Reduction with H_2, Pd
  (A) $\longrightarrow$ no reaction
  (B) $\longrightarrow$ 1 equiv NH_3
  (C) $\longrightarrow$ no reaction
  (D) $\longrightarrow$ 2 equiv C_6H_7N

b. Heat with HCl
  (A) $\longrightarrow$ salt M
  (B) $\longrightarrow$ salt N
  (C) $\longrightarrow$ salt O
  (D) $\longrightarrow$ salt M

c. Ac_2O and mononitration
  (A) $\longrightarrow$ one major isomer
  (B) $\longrightarrow$ one major isomer
  (C) $\longrightarrow$ one major isomer
  (D) $\longrightarrow$ two major isomers

17-20 Phloroglucinol (1,3,5-trihydroxybenzene) reacts with ammonia under mild conditions to yield 3,5-dihydroxyaniline. Considering the starting material as an enol which can ketonize to initiate the reaction, provide a mechanism for the transformation.

17-21 a. Provide a mechanism for the Gomberg reaction as a free-radical process proceeding via a free-radical intermediate. Start with benzenediazonium chloride and benzene and provide resonance structures for the intermediate.
  b. Rationalize the formation of more ortho than meta product in the Gomberg phenylation of nitrobenzene.
  c. Draw resonance structures for the para intermediate which is postulated for the reaction of a phenyl radical and anisole (methoxybenzene). Be careful to avoid violation of the octet rule for oxygen or carbon.
  d. Provide a mechanism for the reaction of benzenediazonium chloride with *N,N*-dimethylaniline showing the structure for the ionic intermediate and all of its significant resonance forms.

17-22 Assume that the reaction of benzenediazonium chloride with water to yield phenol proceeds by an S_N1 mechanism:

$$C_6H_5N_2^+ \xrightarrow{\text{slow}} C_6H_5^+ + N_2 \qquad C_6H_5^+ + HOH \xrightarrow{\text{fast}} C_6H_5\overset{+}{O}H_2$$

a. Draw an approximate energy diagram for the process, showing the activation energy, transition state, ionic intermediate, starting materials, and products.
  b. The rate of phenol formation is decreased by a para nitro group. Explain.
  c. The rate of phenol formation is decreased by a para methoxyl group. Explain.

17-23 a. Based on the Hammett equation and the mechanism of reaction, place the following esters in order of decreasing reactivity toward amide formation ($RCOOR' + NH_3 \longrightarrow RCONH_2 + R'OH$): methyl benzoate, methyl *p*-methylbenzoate, methyl *p*-nitrobenzoate.
  b. If the Hammett ρ constant for the rate of diazotization of aniline and related compounds is +2.31 in 0.5 M HCl (see Table 15-2 for σ constants), predict the effect of electron-releasing substituents on the reaction rate and show how to calculate the ratio of the diazotization rates of aniline and *m*-nitroaniline.

17-24 Isomeric compounds A through D (MW = 120 ± 1) are all soluble in dilute HCl. Their nmr (δ, multiplicity, and peak integration units) and selected ir (μm, cm^{-1}) data are given below. Suggest structures for A through D.

Compound A   Nmr: 7.1–6.3, series of peaks, 13.8 units; 3.3, *s*, 6.8 units; 2.4, *q*, 7.0 units; 1.15, *t*, 10.3 units. Ir: 2.85, 3509; 2.98, 3356; 3.40, 2941; 6.13, 1631; 7.83, 1277; 13.3, 750.
Compound B   Nmr: 7.25, *s*, 6.8 units; 4.05, *q*, 1.4 units; 1.3, *d*, 6.9 units (the integration for this doublet is divided into 4.8 units for the downfield peak and 2.1 units for the upfield peak). Ir: Very similar to A except additional strong peak at 14.28, 700.
Compound C   Nmr: 7.3–6.4, complex series, 8.6 units; 6.65, *s* 1.7 units; 3.31, *q*, 3.3 units; 1.11, *t*, 4.9 units. Ir: 2.92, 3430; otherwise very similar to B for $\mu m > 3.33$ (cm$^{-1} < 3000$).
Compound D   Nmr: 7.35, *s*, 8.7 units; 2.92, *t*, 3.6 units; 2.70, *t*, 3.6 units; 0.92, *s*, 3.5 units. Ir: very similar to B.

17-25 Compound X (MW = 170 ± 1) is insoluble in dilute acid or dilute base. Its nmr spectrum shows a complex series of peaks from $\tau = 2.7$ to $\tau = 3.4$ (10H total), and a singlet at $\tau = 4.45$ (1H). Compound Y is obtained when X is treated with HONO. The nmr spectrum of Y consists only of a series of peaks from $\tau = 2.6$ to $\tau = 3.2$ (10H) and its ir spectrum does not show the band at 2.9 μm (3448 cm^{-1}) that compound X does. Suggest structures for X and Y.

17-26 Suggest structures for compounds I through VII based on the nmr spectral data and other information given on p. 254.

Compound	MW	Nmr peaks (δ, multiplicity)				Respective area ratio	Selected ir bands (μm, cm^{-1}) and other information
I	128 ± 1	7.0, d	6.4, d	3.5, s		1:1:1	J for doublets = 8 Hz; contains halogen
II	120 ± 1	7.1, m	6.6, m	2.9, s		1:1.5:3	No N–H stretch bands; 13.4, 746; 14.5, 690
III	138 ± 1	7.4, two d	3.9, q	3.2, s	1.3, t	2:1:1:1.5	2.87, 3484; 2.98, 3356; 12.1, 826
IV	108 ± 1	6.9, t	6.1, d	3.5, s		1:3:4	2.85, 3509; 2.95, 3390
V	138 ± 1	7.3, m	3.9, s			2:1	2.83, 3534; 2.91, 3436; 6.5, 1538; 7.4, 1351
VI	150 ± 1	7.3, m	3.5, s	1.9, s		1.67:1:1	6.05, 1653; 13.2, 758; 14.2, 704; NE = 150 ± 1
VII	190 ± 1	9.2, s	7.4, 7.1 d, d	3.5, s	2.2, 2.3 s, s	1:2:2:2:3:3	3.15, 3175; 5.83, 1715; 6.0, 1667

Upon hydrolysis of VII a carboxylic acid is isolated with NE > 45

17-27   Suggest a structure for compound A, C_7H_9N, based on its ir spectrum shown.

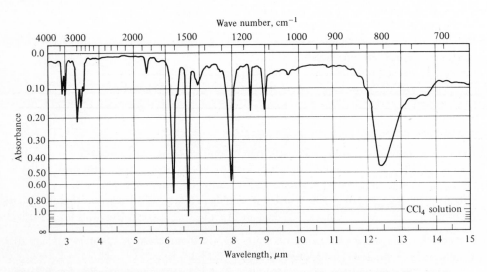

Figure P17-27. Ir spectrum for Prob. 17-27. *(From R. A. Silva and R. Spenger, "The Quantitative Organic Analysis Spectra Kit," McGraw-Hill Book Company, New York, 1969. By permission of the publishers.)*

17-28   Compound Z has MW = 138 ± 1. Its nmr and ir spectra are shown. Suggest a structure for Z.

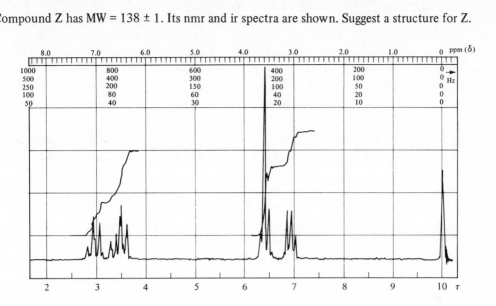

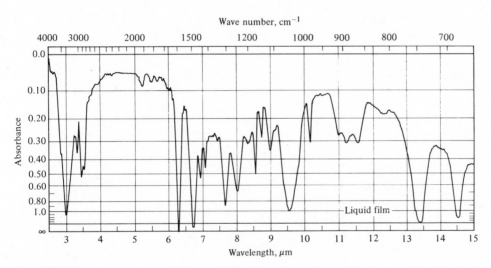

Figure P17-28. Ir and nmr spectra for Prob. 17-28. *(From R. A. Silva and R. Spenger, "The Quantitative Organic Analysis Spectra Kit," McGraw-Hill Book Company, New York, 1969. By permission of the publishers.)*

18
phenols and aromatic ethers

18-1 Nomenclature

Compounds with trivial names:

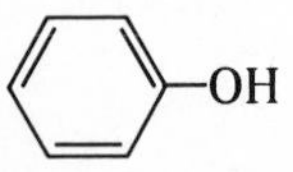

Phenol 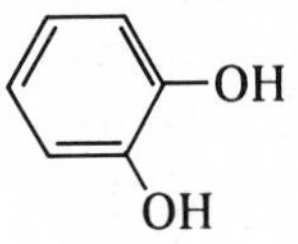   Catechol 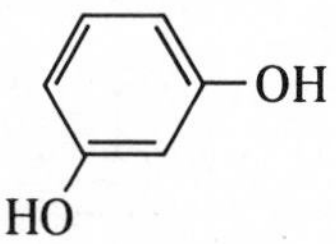   Resorcinol 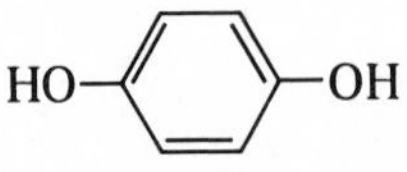   Hydroquinone

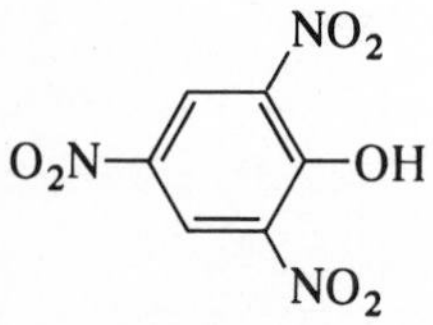

Picric acid 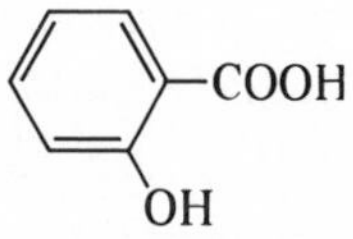   Salicylic acid 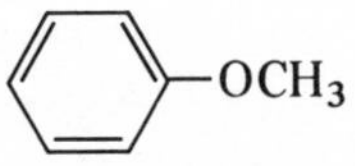   Anisole 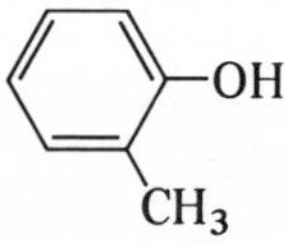   o-Cresol

p-Cresyl acetate 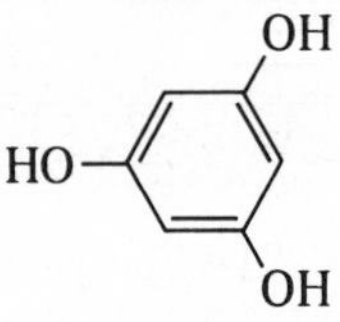   Phloroglucinol 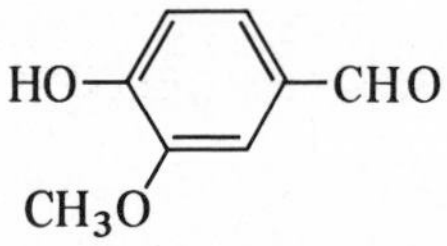   Vanillin

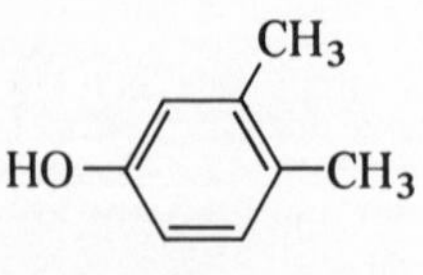

3,4-Xylenol 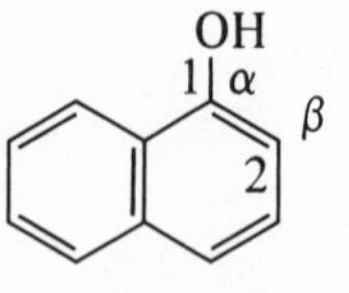   α-Naphthol   β-Naphthol
              or 1-naphthol   or 2-naphthol

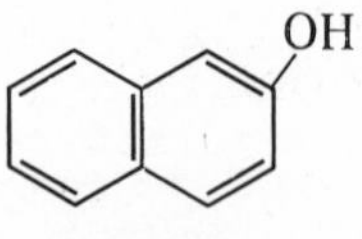

For complex structures, look for a trivial or common name which includes as much of the molecule as feasible. Thus, the parent compound for structure I might be cumene or phenol with corresponding names 3-hydroxy-5-chlorocumene or 3-chloro-5-isopropylphenol. The simplest name for II would be chlorohydroquinone and for III, 4-methoxyphthalic acid.

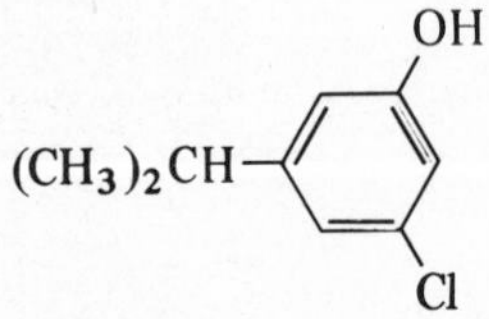

I 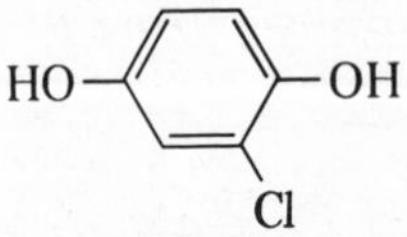   II 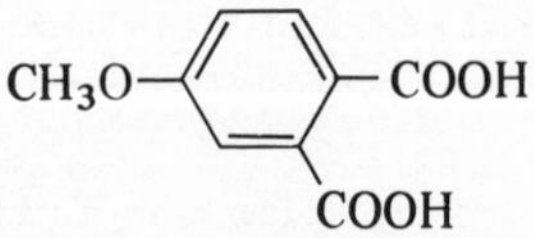   III

256

The groups RO– and ArO– as substituents are called alkoxy groups, for example, C_6H_5O- is phenoxy and $(CH_3)_3CO-$ is *tert*-butoxy. Metal salts of phenol are named as metal phenoxides.

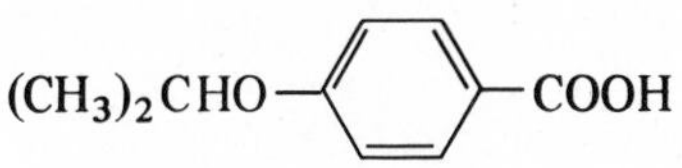

$(CH_3)_2CHO-\!\!\bigcirc\!\!-COOH$	$Cl-\!\!\bigcirc\!\!-OCH_2COOH$ (with Cl)	$\bigcirc\!\!-O^-K^+$
p-Isopropoxybenzoic acid	2,4-Dichlorophenoxyacetic acid (2,4-D)	Potassium phenoxide

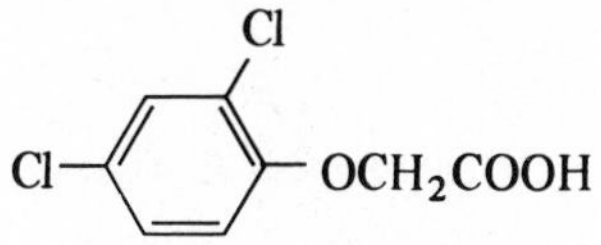

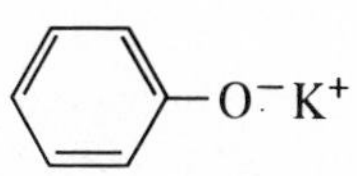

18-2 Methods of Preparation

Like the aromatic amino group, the hydroxyl group cannot be introduced directly via electrophilic substitution.

a. *Replacement of* NH_2 *by Means of a Diazonium Salt* (See Sec. 17-5a.)

b. *Nucleophilic Displacement of Halides* (See Sec. 15-3a.)

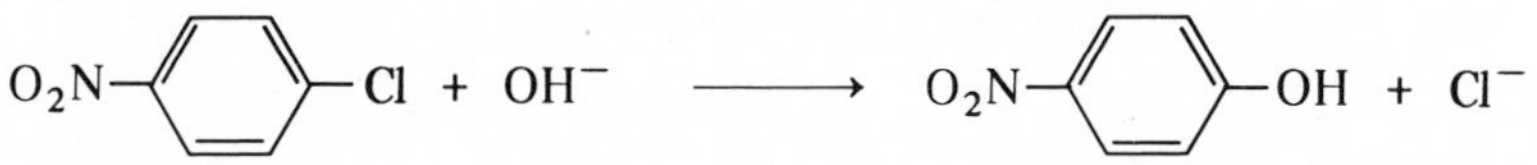

$$O_2N-\!\!\bigcirc\!\!-Cl + OH^- \longrightarrow O_2N-\!\!\bigcirc\!\!-OH + Cl^-$$

$$C_6H_5Cl + HOH \xrightarrow[\text{(commercial)}]{350°C, \text{ catalyst, high pressure}} C_6H_5OH + HCl$$

c. *Sulfonic Acid Fusion*

$$C_6H_5SO_3Na \xrightarrow{\text{NaOH, fuse}} C_6H_5ONa \xrightarrow{H_3O^+} C_6H_5OH$$

d. *Oxidation of Cumene*

$$C_6H_5-CHMe_2 \xrightarrow{O_2} C_6H_5-\underset{\underset{OOH}{|}}{C}Me_2 \xrightarrow{H_3O^+} C_6H_5OH + Me_2CO$$

Cumene hydroperoxide

e. *Oxidation of Organothallium Compounds* Treatment of aromatic hydrocarbons with thallium (III) trifluoro-acetate (TTFA) leads to ring-thallated intermediates such as A, which upon controlled oxidation with lead tetraacetate ultimately yield phenols as products.

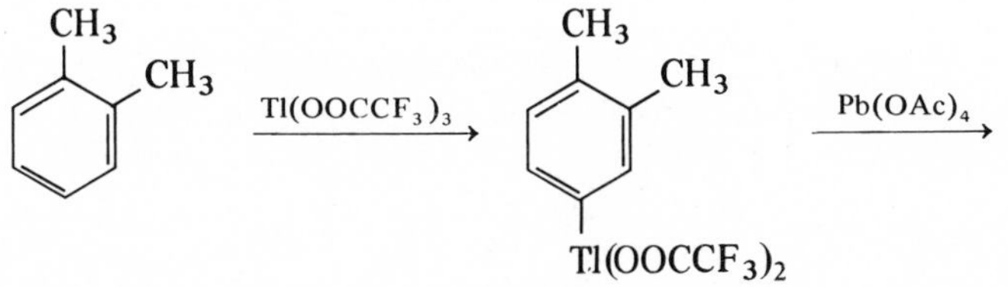

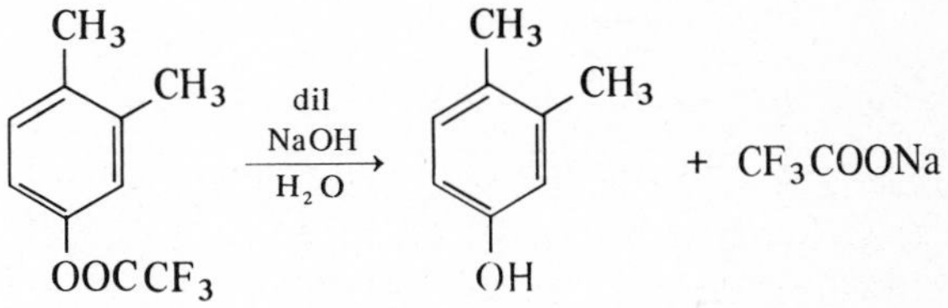

18-3 Reactions

a. *Salt Formation* Phenols are more acidic than alcohols but less acidic than carboxylic acids. The pK_a of phenol is about 1.3×10^{-10}.

$$C_6H_5OH + OH^- \longrightarrow C_6H_5O^- \xrightarrow{H_2CO_3} C_6H_5OH + HCO_3^-$$
$$C_6H_5OH + TlOEt \longrightarrow C_6H_5OTl + EtOH$$

b. *Ether Formation and Cleavage*
   1. Williamson ether synthesis.

$$C_6H_5ONa + EtI \longrightarrow C_6H_5OEt + NaI$$

$(CH_3)_2SO_4$ and CH_2N_2 are often used to make methyl ethers.
   2. Cleavage by HI or HBr. (See Sec. 6.5*a* and Prob. 6-15.)

$$ArOR + HI \xrightarrow{\Delta} ArOH + RI$$

c. *Esterification* Fischer esterification is unsatisfactory in this situation.

$$HOAc + ArOAc \xleftarrow{Ac_2O} ArOH \xrightarrow[(Ar)]{RCOCl} ArOOCR(Ar) + HCl$$

$$TlCl + ArOTs \xleftarrow{TsCl} ArOTl \xrightarrow[(Ar)]{RCOCl} ArOOCR(Ar) + TlCl$$

d. *Reduction*

$$C_6H_5OH \xrightarrow[\text{(drastic conditions)}]{Zn,\ distill} benzene$$

e. *Oxidation* Phenols, like aromatic amines (see Sec. 17-3*e*), are easily oxidized to quinones. Ring oxidation may be suppressed by acetylating the phenol.

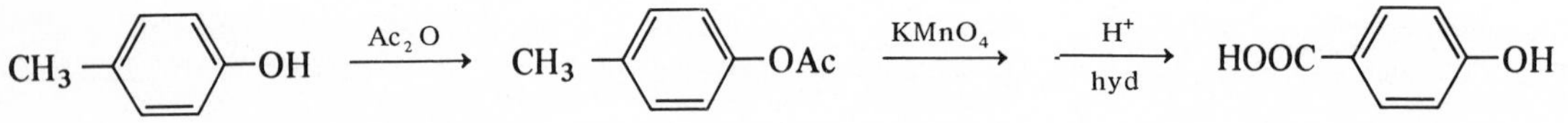

f. *Halogenation* Chlorine and bromine react so rapidly with phenols that monosubstitution requires special conditions. Iodination can be carried out with iodine monochloride.

$$Phenol + 3Br_2 \longrightarrow 2,4,6\text{-Tribromophenol}$$

g. *Nitration* Dilute nitric acid converts phenol to *o*- and *p*-nitrophenol.

h. *Sulfonation*

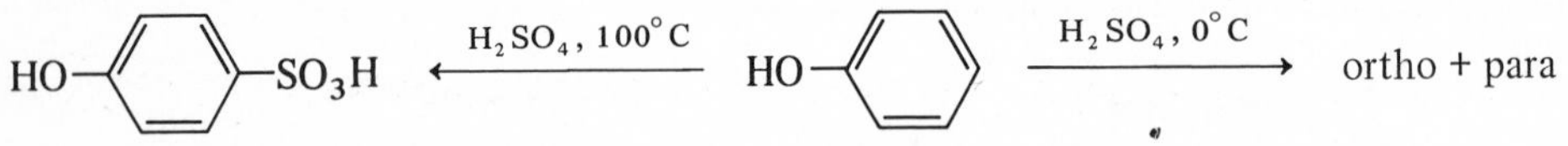

i. *Nitrosation* of phenol yields only the para isomer. Oxidation of the product gives pure *p*-nitrophenol.

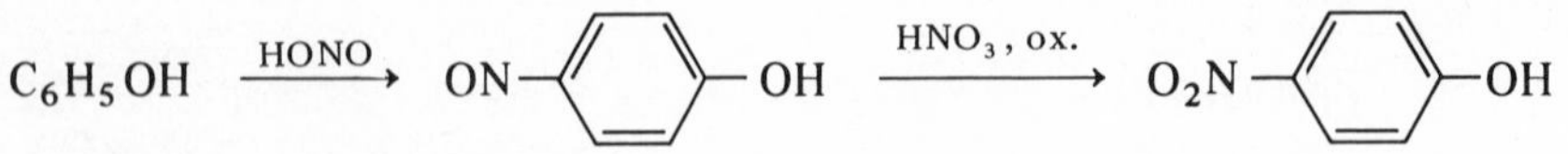

j. *Coupling with Diazonium Salts* (See Sec. 17-5*c*-2.)

k. *Introduction of* –COR *Group*
   1. Fries rearrangement. Phenols do not react smoothly in Friedal-Crafts reactions. However phenyl esters (see Sec. 18-3*c*) rearrange to give acyl derivatives.

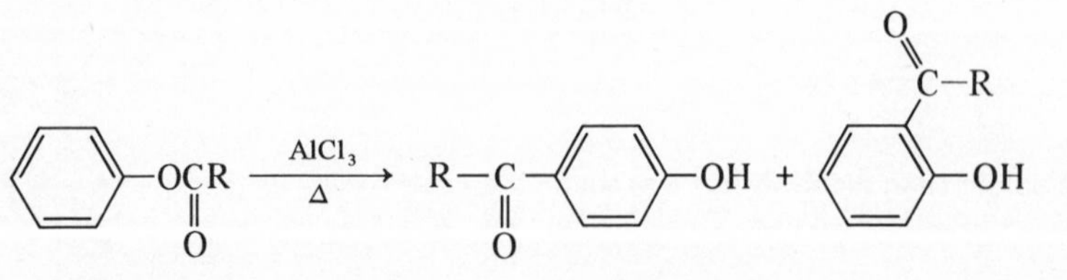

(See Sec. 18-4 for separation of products.)

2. Houben-Hoesch synthesis. This is a type of Friedel-Crafts reaction using nitriles.

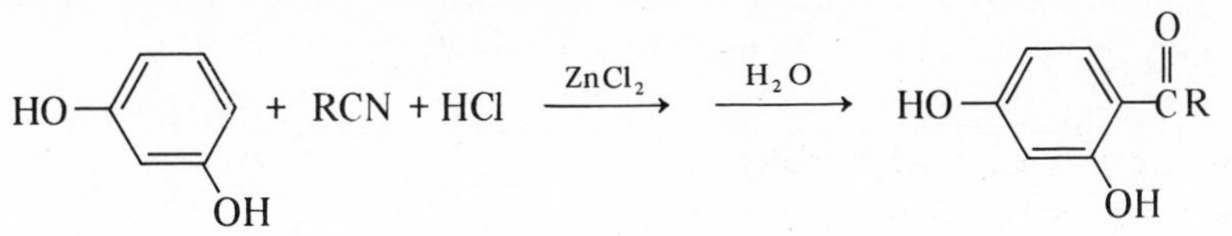

l. *Introduction of —CHO Group (Formylation)*

1. Reimer-Tiemann reaction. The ortho isomer predominates in this type of formylation, but either ortho or para is easy to isolate (see Sec. 18-4). See Prob. 18-20*a* for the mechanism.

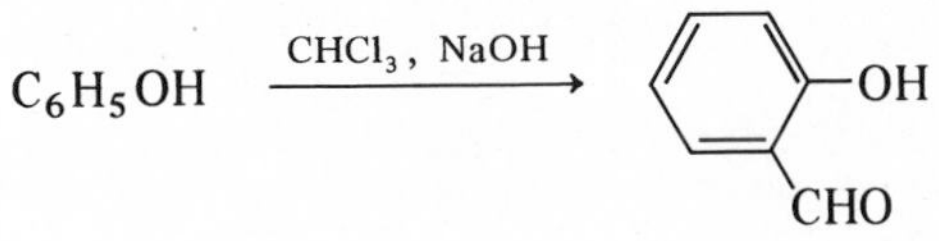

2. Gatterman reaction. This is a special case of the Houben-Hoesch reaction discussed above.

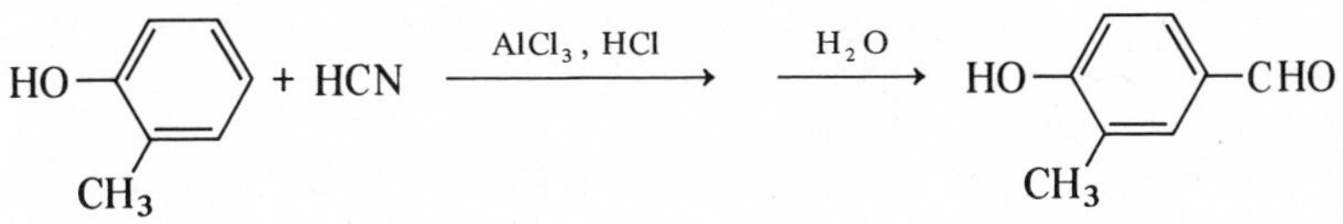

m. *Introduction of —COOH Group (Carboxylation)*

1. Kolbe synthesis.

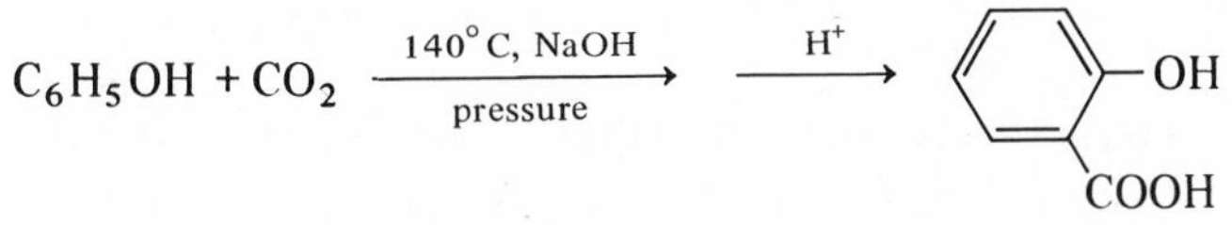

2. Use of CCl_4 in the Reimer-Tiemann reaction.

$$C_6H_5OH \xrightarrow{CCl_4,\ NaOH} \xrightarrow{H^+} \text{salicylic acid}$$

n. *Hydroxyformylation* See Prob. 18-16 for the polymerization of phenol and formaldehyde to Bakelite-type polymers.

o. *Claisen Rearrangement* Allyl ethers of phenol yield *o*-allylphenol on heating (see Prob. 18-18).

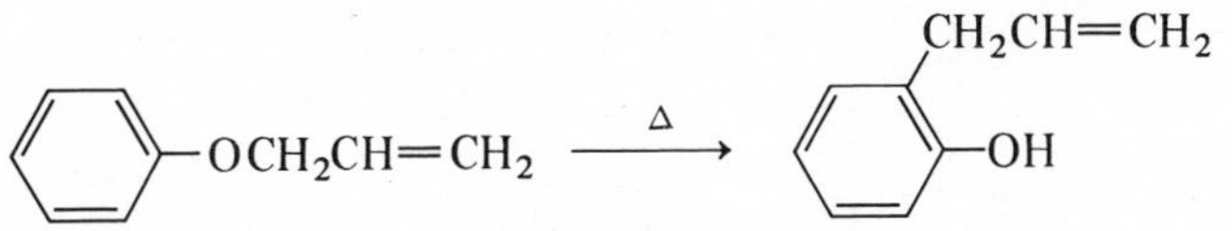

18-4 Chelate Rings

A chelate ring system forms in organic compounds whenever there is the possibility of hydrogen-bond formation that produces a five- or six-membered ring. Chelate ring formation reduces the polarity of a molecule (compared to isomers in which chelation is not possible). This has a practical synthetic application because a chelated compound such as I (*o*-nitrophenol) or II (*o*-hydroxyacetophenone) will steam distill but the meta or para isomers will not.

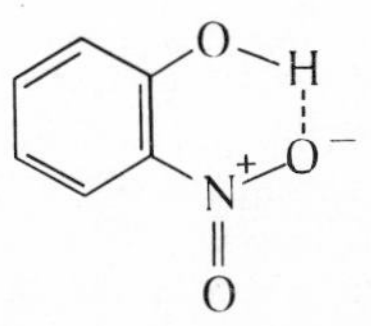

259

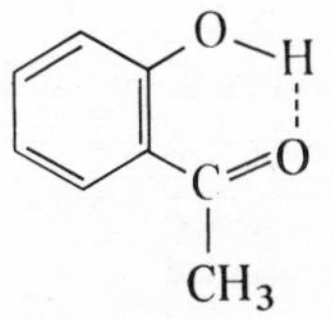

18-5 Qualitative Tests

a. Phenols are acids sufficiently strong to form salts with aqueous NaOH. Solubility in 5% NaOH but insolubility in 5% $NaHCO_3$ is useful as a test. Two ortho or para nitro groups lead to solubility in 5% $NaHCO_3$.

b. Most phenols and enols give a color (e.g., blue, green, purple, red) when treated with an aqueous solution of ferric chloride.

Problems

18-1 Name the following compounds:

a. 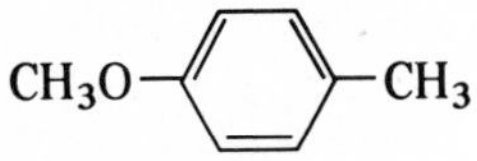 CH_3O—⟨benzene ring⟩—CH_3

b. $C_6H_5OCOCH_2CH_3$

c. 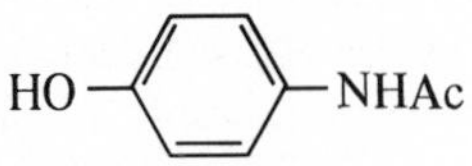 HO—⟨benzene ring⟩—NHAc

d. 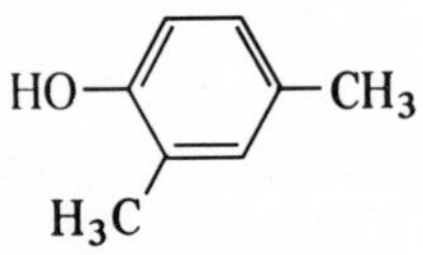 HO—⟨benzene ring with H_3C⟩—CH_3

e. HO—⟨benzene ring⟩—$CH_2CH=CH_2$

f. HO—⟨benzene ring with C_6H_5⟩—OH

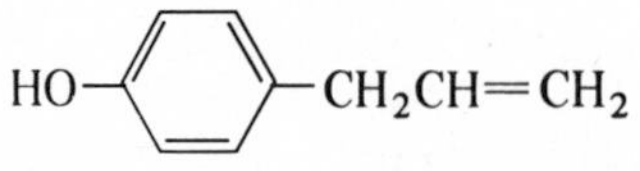

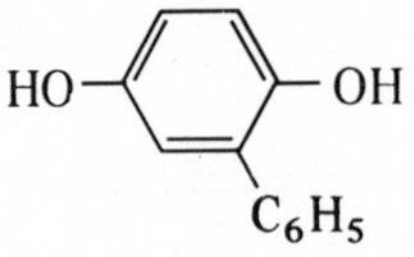

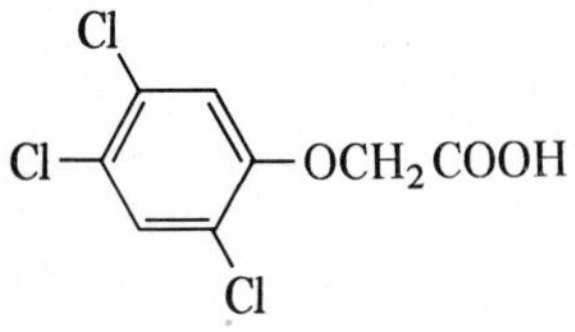

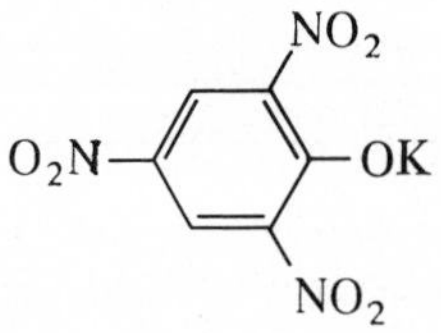

g. ⟨benzene ring with Cl, Cl, Cl⟩—OCH_2COOH

h. ⟨naphthalene⟩—OH

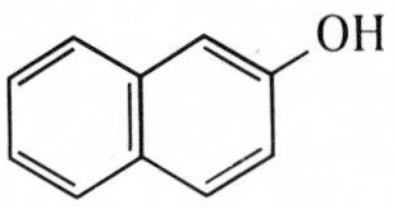

i. 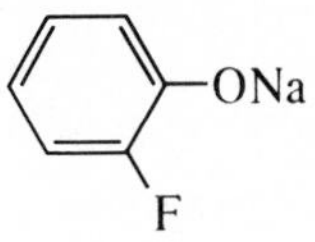 ⟨benzene ring with F⟩—ONa

j. 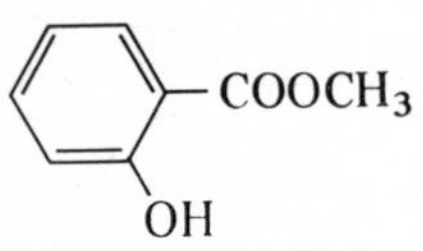 ⟨benzene ring with OH⟩—$COOCH_3$

k. 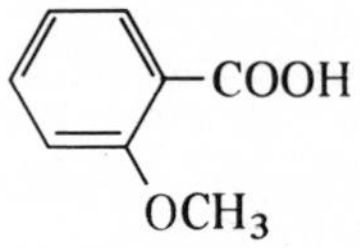 ⟨benzene ring with OCH_3⟩—COOH

l. 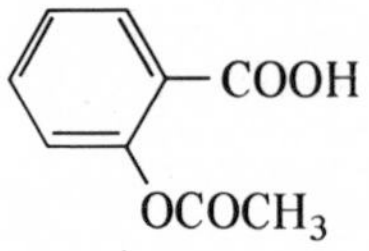 ⟨benzene ring with $OCOCH_3$⟩—COOH

m. 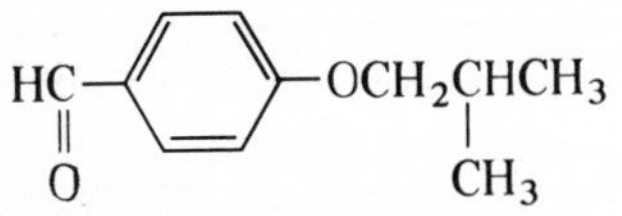 HC(=O)—⟨benzene ring⟩—OCH_2CHCH_3 (CH_3)

n. O_2N—⟨benzene ring with NO_2, NO_2⟩—OK

o. O=P(—O—⟨benzene ring⟩—CH_3)₃

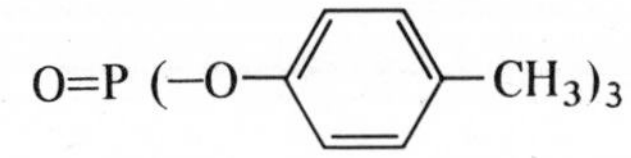

18-2 Complete and balance the following reactions:
a. C_6H_5ONa + p-ClC_6H_4OH
b. Sodium phenoxide + Me_2SO_4
c. m-Cresol + EtMgBr
d. Phenol + ketene
e. Hydroquinone + Zn + Δ
f. o-ClC_6H_4OH + $KMnO_4$
g. $C_6H_5NMe_2$ + HONO
h. Anisole + HI
i. p-Cresyl acetate + $AlCl_3$ + Δ
j. o-Cresol + KOH + $CHCl_3$
k. p-Methoxytoluene + $KMnO_4$
l. Thallium (I) phenoxide + $C_6H_5SO_2Cl$
m. o-Cresol + $C_6H_5N_2Cl$
n. Resorcinol + NaOH + CO_2 + Δ

18-3 Show how you would make the following conversions:
a. Phenyl acetate to o-ethylphenol
b. C_6H_5Cl to 2,4-dinitrophenol
c. p-Cresol and HCN to 2-hydroxy-5-methylbenzaldehyde
d. CH_3CN and resorcinol to 1,3-dihydroxy-4-ethylbenzene
e. p-Cresol to 4-methyl-1-cyclohexanone
f. p-Xylene to 2,4-xylenol
g. Phenol to acetylsalicylic acid (aspirin)
h. Aniline to salicylaldehyde
i. p-Cresol and allyl bromide to 2-allyl-4-methylphenol
j. p-Xylene to 2,5-dimethylquinone

18-4 What products would you expect from the action of nitrous acid on each of the following?
a. Benzamide
b. 4-Iodophenol
c. N,N-Diethylaniline
d. Benzylamine
e. Sulfanilamide
f. N-Methyl-p-toluidine
g. sec-Butylamine
h. 1,1-Diphenyl-2-amino-1-propanol
i. N,N-Dimethyl-4-ethylaniline

260

18-5   Suggest a chemical test with an easily observable change that would distinguish between each of the following:
   a. p-Cresol and cumene
   b. C_6H_5COONa and C_6H_5ONa
   c. p-Toluidine and p-cresol
   d. p-ClC_6H_5OH and p-ClC_6H_5COOH
   e. p-ClC_6H_4OH and p-$O_2NC_6H_5OH$
   f. Salicylic acid and ethyl salicylate
   g. o-Nitrophenol and o-aminophenol
   h. 2,4-Dinitrophenol and m-nitrophenol
   i. Cyclohexanol and phenol

18-6   Suggest an easy method for the separation of each of the following mixtures, assuming fractional distillation to be impractical;
   a. Phenol and anisole
   b. Nitrobenzene and 1-nitroheptane
   c. o-Hydroxyacetophenone and p-hydroxyacetophenone
   d. 3-Nitro-4-hydroxytoluene and 3-nitro-5-hydroxytoluene
   e. Phenol and salicylic acid
   f. Sodium benzoate and sodium phenoxide

18-7   Give the structure of the product from each of the following:

$$C_6H_5ONa \xrightarrow{ClCH_2CH=CH_2} (A) \xrightarrow{\Delta} (B) \xrightarrow{1\ mol\ O_3} (C) \xrightarrow{HOH,\ Zn} (D)+(E)$$

$$(D) \xrightarrow{OH^-} \xrightarrow{Me_2SO_4} (F) \xrightarrow{H_2NOH} (G) \xrightarrow[\Delta]{P_2O_5} (H) \xrightarrow{EtMgBr} \xrightarrow{H_3O^+} (I) \xrightarrow{H_2NNH_2,\ MeONa,\ \Delta}$$

$$(J) \xrightarrow{HI,\ \Delta} (K) \xrightarrow{Zn,\ \Delta} (L)$$

18-8   Point the errors, if any, in each of the following proposed syntheses:

   a.

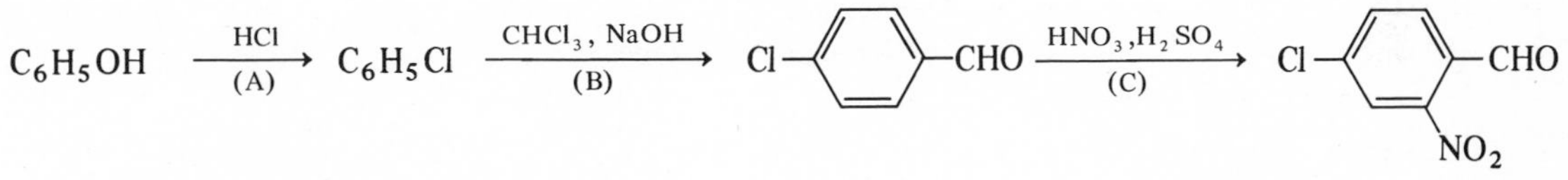

   b.

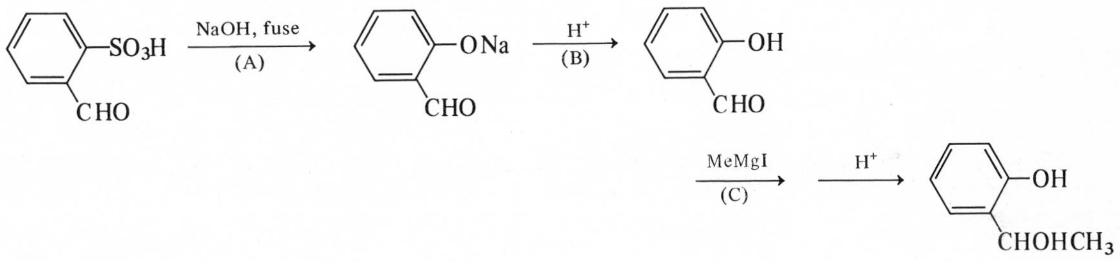

   c.

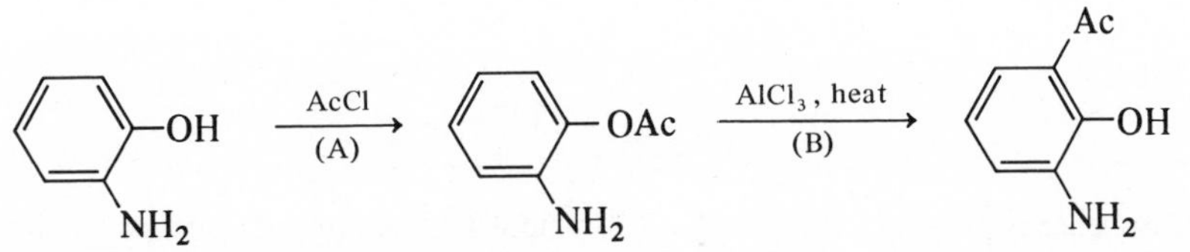

18-9   Compound A, $C_{10}H_{11}O_5N$, was found to be insoluble in dilute acid or base. When 0.40 g of A was refluxed for 4 h with 30 ml of 0.1 N NaOH, it completely dissolved. The resulting solution required 12.3 ml of 0.10 N HCl for neutralization. The neutral solution was then made alkaline and distilled.
   The aqueous distillate gave a positive iodoform test. The alkaline residue from the distillation was acidified and a nitrogen-containing acid, B, precipitated. When 0.20 g of B was titrated with 0.10 N KOH, it required 10.15 ml for neutralization. B was reduced with Sn + HCl, and the product was treated with $NaNO_2$ and HCl, then boiled, giving C. C, on heating with HI, gave D, which was found by the Zerewitinoff method to have three active hydrogen atoms. Suggest structures for A, B, C, and D.

18-10   Given toluene, benzene, methanol, ethanol, allyl bromide, triphenylphosphine, and phenol as the only organic compounds and any necessary inorganic compounds, show how you could prepare each of the following:
   a. 1-Phenylcyclohexene
   b. 2-Methoxy-6-allylphenol
   c. Phenyl cyclobutanecarboxylate
   d. 1-(4-Hydroxyphenyl)-1-aminopropane
   e. Phenyl 3,4-dichlorophenoxyacetate
   f. 4-(4-Ethoxyphenyl)-1-butene
   g. 2,6-Dinitro-4-$tert$-butyl-3-methylanisole
   h. 3,4-Diphenylphthalic acid

18-11 How many isomers would you expect from the addition of 1 mol of chlorine to each of the following?
    a. *cis*-1,2-Diphenylethylene   *b.* 1,1-Diphenylethylene   *c.* *trans*-1,2-Diphenylethylene

18-12 Formulate the intermediates in the synthesis of adrenalin:

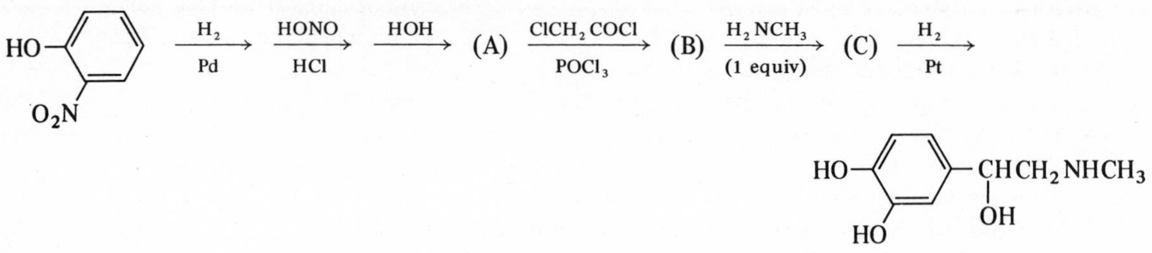

18-13 Both the nitrosation of phenol and the reaction of *p*-benzoquinone with 1 equiv of hydroxylamine can lead to the same product under proper conditions. Explain.

18-14   *a.* Draw resonance structures for phenol and the phenoxide ion.
      b. On the basis of these structures, explain why phenol is a stronger acid than cyclohexanol.
      c. From the following dipole data, determine the direction of the dipole moment of aniline and chlorobenzene.

	Debyes		Debyes
Aniline	1.5	*p*-Nitroaniline	6.3
Chlorobenzene,	1.6	*p*-Nitrochlorobenzene	2.6
Nitrobenzene	4.0		

18-15 Indicate the stronger acid in each of the following pairs and explain your choice:
    a. Phenol and *p*-nitrophenol   *b.* *p*-Nitrophenol and *m*-nitrophenol
    c. Benzyl alcohol and *m*-cresol   *d.* Toluene and TNT
    e. *p*-Nitrophenol and 3,5-dimethyl-4-nitrophenol   *f* Methyl *p*-hydroxybenzoate and
                                                      p-methoxybenzoic acid

18-16 Phenol and formaldehyde react to form polymers of the Bakelite type. Either acid or base catalysis may be used.
    a. One intermediate in the polymerization is saligenin. Show how either catalyst could lead to the formation of this product.

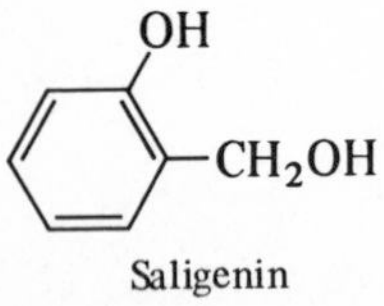

Saligenin

    b. The type of structure in Bakelite is probably that shown below. On the basis of our concept of directing groups, is the reaction electrophilic or nucleophilic aromatic substitution? Explain.

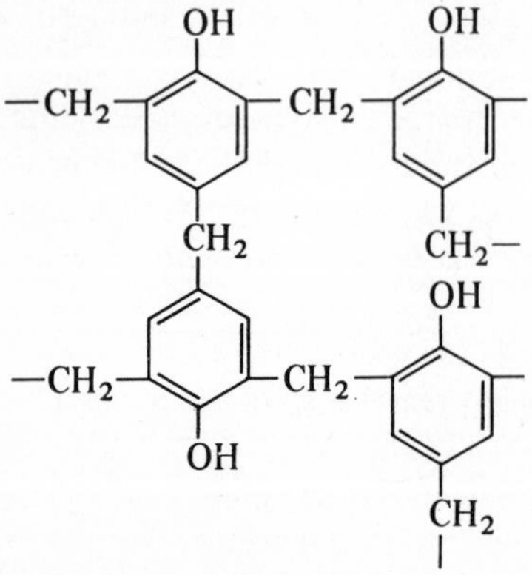

Bakelite

262

c. Show how saligenin might, under acid conditions, form a good cation that could react with phenol. Give resonance forms for the cation.

d. Show how 1 mol of acetaldehyde might react with 2 mol of *p*-cresol under acid conditions.

18-17   Aromatic aldehydes react with aromatic ketones in an aldol condensation under alkaline conditions to yield chalcones.

a. Predict the structure of chalcone from the following reaction:

$$C_6H_5-CHO + CH_3CO-C_6H_5 \xrightarrow{\text{NaOH}} \text{chalcone} + HOH$$

b. The indicated reaction can be used to prepare hesperitin (shown below), a compound of hesperidin.

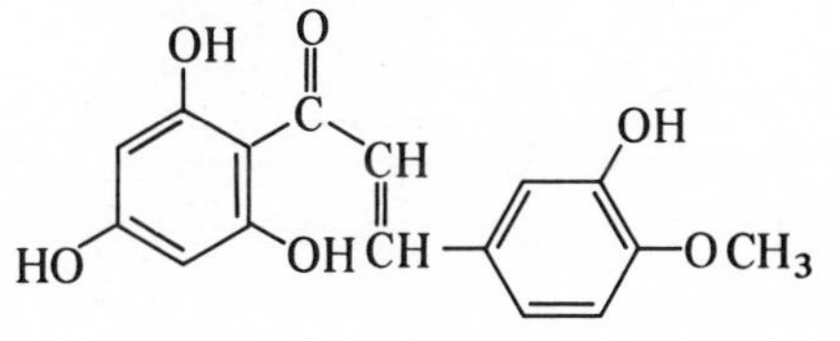

Hesperitin

Hesperidin is used to relieve capillary hemorrhaging. Follow the synthetic process by providing the intermediate products in the following sequence of reactions:

$$\text{Phloroglucinol} \xrightarrow{\text{CH}_3\text{CN, HCl, ZnCl}_2} \xrightarrow{\text{HOH}} \text{(A)} \xrightarrow[\text{(from below)}]{\text{(B)}} \xrightarrow{\text{NaOH}} \text{hesperitin}$$

$$\text{Benzene} \xrightarrow{\text{HNO}_3, \text{H}_2\text{SO}_4} \xrightarrow{\text{Pd, H}_2} \text{(M)} \xrightarrow{\text{HONO, HCl}} \text{(N)} \xrightarrow{\text{HOH}} \text{(O)} \xrightarrow{\text{Me}_2\text{SO}_4, \text{NaOH}} \text{(P)}$$

$$\xrightarrow{\text{HNO}_3, \text{HOAc}} \text{(Q) (para)} \xrightarrow{\text{H}_2, \text{Pd}} \text{(R)} \xrightarrow{\text{HONO, HCl}} \text{(S)} \xrightarrow{\text{CuCN, KCN}} \text{(T)} \xrightarrow{\text{SnCl}_2, \text{HCl}} \xrightarrow{\text{HOH}} \text{(U)}$$

$$\xrightarrow{\text{HNO}_3, \text{HOAc}} \text{(V)} \xrightarrow{\text{SnCl}_2, \text{HCl}} \xrightarrow{\text{NaOH}} \text{(W)} \xrightarrow{\text{HONO, HCl}} \xrightarrow[\Delta]{\text{HOH}} \text{(B)}$$

18-18   The Claisen rearrangement proceeds as follows:

$$C_6H_5OCH_2CH={}^{14}CH_2 \xrightarrow{\Delta}$$

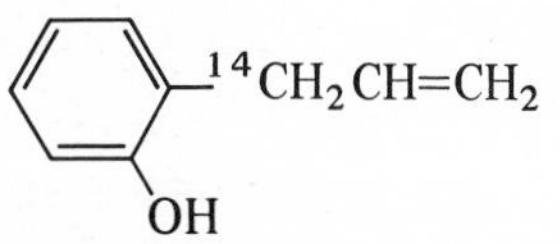

a. Predict the result of heating $C_6H_5OCH_2CH=CHC_6H_5$.

b. Predict the structure of the rearrangement product which would form from heating $CH_2=CHOCH_2CH=CH_2$.

18-19   *a.* If the Fries rearrangement involves $R-\overset{+}{C}=O$ as the migrating species, suggest an experiment which would enable one to distinguish between an intramolecular and intermolecular mechanism.

b. Arrange the *p*-nitrophenyl, *p*-cresyl, and phenyl acetates in order of increasing ease of Fries rearrangement. Explain.

18-20   *a.* Provide as complete a mechanism as possible for the Reimer-Tiemann reaction of phenol.

b. Rationalize the formation of the products A through D in the following reactions:

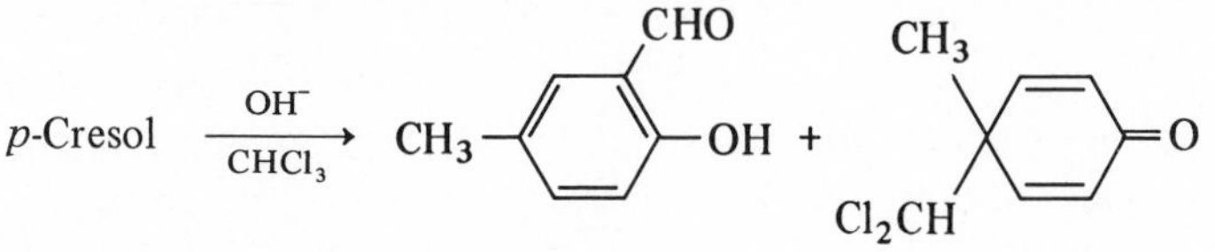

263

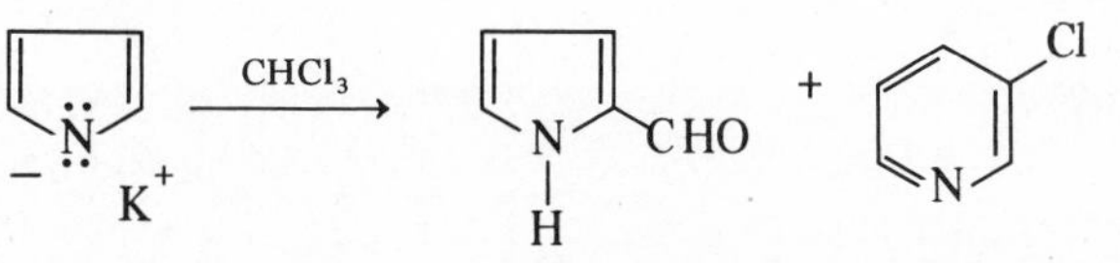

18-21 In the Houben-Hoesch synthesis using phenol, CH_3CN, HCl, and a $ZnCl_2$ catalyst, by-product A is isolated which fails to give positive ferric chloride or iodoform tests. Upon heating A with $AlCl_3$ compound B is obtained which is isomeric with A and gives positive ferric chloride and iodoform tests. Suggest structures A and B and mechanisms for their formation.

18-22 Phenols when treated with an oxidizing agent such as alkaline potassium ferricyanide, $K_3Fe(CN)_6$, undergo coupling reactions which involve a phenoxide radical as the reactive species.
a. Draw resonance structures for such a radical obtainable from phenol.
b. Explain the formation of products such as 4,4'-dihydroxybiphenyl and 2-hydroxydiphenyl ether, among others, when phenol is treated with alkaline $K_3Fe(CN)_6$.
c. Suggest a mechanism for the following:

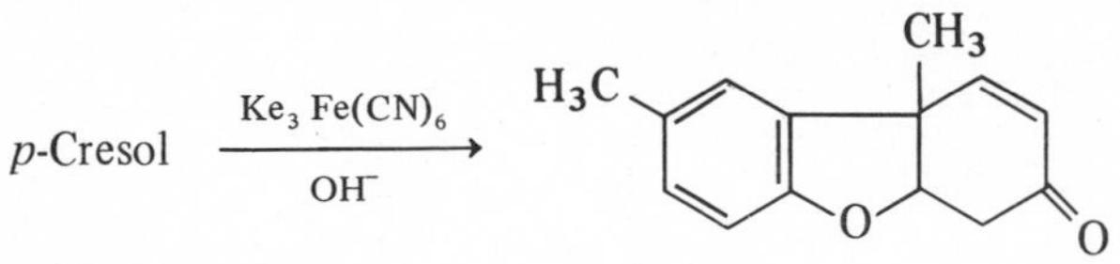

18-23 Calculate the pK_a of the following phenols under conditions where the value of Hammett's ρ is +2.364 (95% EtOH, 20°) and pK_a for phenol is 12.6:
a. o-Cresol *b.* p-Methoxyphenol *c.* m-Chlorophenol

18-24 Predict the sign of the Hammett σ constants for each of the following:
a. $m\text{-}\overset{+}{N}(CH_3)_3$ *b.* p-OAc *c.* $p\text{-}O\text{-}C_6H_5$

18-25 Predict the sign of the Hammett σ constant for each of the following and provide a reason for your choice:
a. $ArO^- + EtBr \longrightarrow ArOEt + Br^-$ *b.* Saponification of phenyl acetate

18-26 Isomeric compounds A and B (MW = 108) have the following nmr data (δ, multiplicity, peak integration units) and selected ir bands (μm, cm^{-1}). Suggest structures for A and B.

Compound A Nmr: 6.9, d, (J = 9 Hz), 4.6 units: 6.6, d (J = 9 Hz), 5.0 units; 6.0, s, 2.4 units; 2.2, s, 7.0 units. Ir: 2.7, 3704; 12.2, 820.
Compound B Nmr: 6.9, m, 7.0 units; 4.8, s, 1.8 units; 2.2, s, 5.3 units. Ir: 2.75, 3636; 13.0, 769.

18-27 Closely related compounds C, D, E, F (MW = 138, 124, 124, and 110, respectively) have the following nmr data (δ, multiplicity, peak integration units) and selected ir bands (μm, cm^{-1}). Suggest structures for C through F.

Compound C Nmr: 6.6, s, 17 units; 3.6, s, 23 units. Ir: 12.2, 820.
Compound D Nmr: 6.75, s, 15.0 units; 5.6, s, 3.8 units; 3.75, s, 11.3 units. Ir: 2.80, 3571; 12.15, 823.
Compound E Nmr: 6.7, m, 9.0 units; 6.1, s, 2.3 units; 3.55, s, 7.0 units. Ir: 2.82, 3550; 13.33, 750.
Compound F Nmr: 7.0, n, 16.2 units; 5.05, s, 8.2 units. Ir: 2.83, 3509; 13.33, 750.

18-28 Suggest structures for compounds I through V based on the nmr data and other information given below.

Compound	MW		Nmr peaks (δ, multiplicity)	Respective area ratio	Selected Ir bands (μm, cm^{-1}) and other information
I	128 ± 1	7.1, d	6.6, d; 5.5, s	2:2:1	2.75, 3636; 12.2, 820. Positive Beilstein test.
II	140 ± 1	10.5, s	Complex ArH with three groups of peaks centered at 8.0, 7.5, 6.8	1:1:1:2	3.1, 3226; 6.6, 1515; 7.6, 1316. Steam volatile.
III	122 ± 1	6.8, m	4.4, s; 2.2, s	3:1:6	2.77, 3610; 8.4, 1190.
IV	188 ± 1	7.3, d	6.7, d; 3.7, s	2:2:3	12.2, 820. Positive Beilstein test.
V	150 ± 1	7.2–6.5, 5 peaks	4.8, s; 3.2, sp; 2.3, s; and 1.25, d	3:1:1:3:6	2.85, 3509.

18-29   Compound G (MW = 195, contains N) when heated with aqueous NaOH yields compound D (see Prob. 18-27), compound H (MW = 45) and CO_2. The nmr spectrum of G shows two doublets at δ = 7.1 (2H) and 6.85 (2H), both with J = 10 Hz, and singlets at δ = 3.8 (3H) and 3.05 (6H). Its ir spectrum has characteristic bands at 5.83 μm (1715 cm^{-1}) and 12.3 μm (813 cm^{-1}). Suggest structures for G and H.

18-30   The nmr and ir spectra for compounds X (MW = 166), Y (MW = 214), and Z (MW = 180) are shown. Suggest structures for X, Y, and Z.

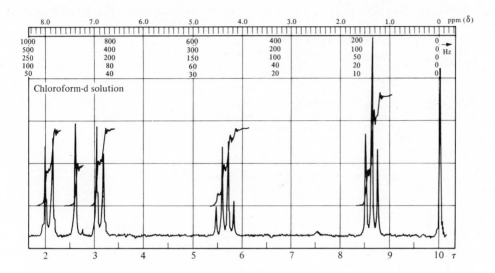

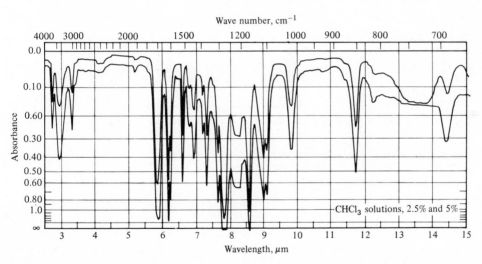

Figure P18-30. Ir and nmr spectra for Prob. 18-30, compound X. *(From R. A. Silva and R. Spenger, "The Quantitative Organic Analysis Spectra Kit," McGraw-Hill Book Company, New York, 1969. By permission of the publishers.)*

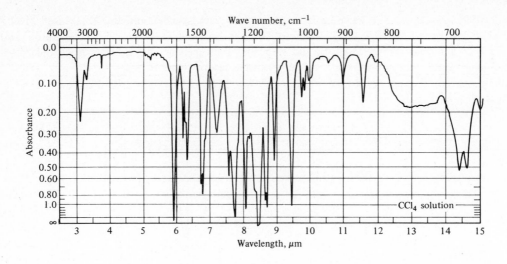

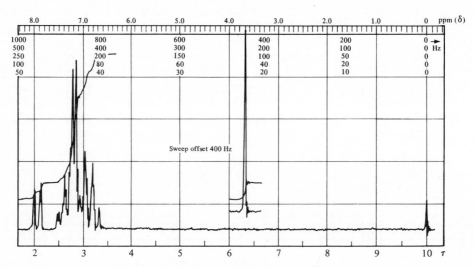

Figure P18-30. Ir and nmr spectra for Prob. 18-30, compound Y. *(From R. A. Silva and R. Spenger, "The Quantitative Organic Analysis Spectra Kit," McGraw-Hill Book Company, New York, 1969. By permission of the publishers.)*

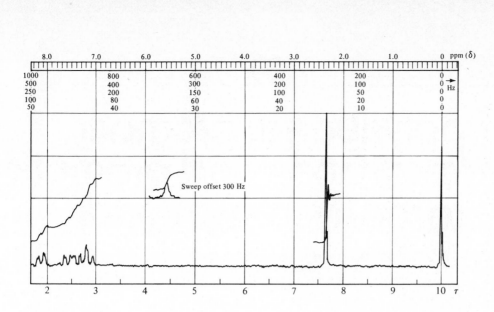

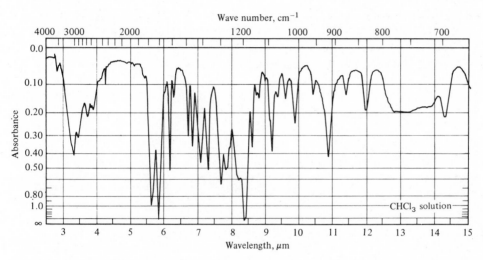

Figure P18-30. Ir and nmr spectra for Prob. 18-30, compound Z. *(From R. A. Silva and R. Spenger, "The Quantitative Organic Analysis Spectra Kit," McGraw-Hill Book Company, New York, 1969. By permission of the publishers.)*

19

polynuclear aromatic hydrocarbons and aromatic carbonyl compounds

19-1 Nomenclature

a. *Polynuclear Aromatic Hydrocarbons*

1. Naphthalene. Two different systems of designating the positions in naphthalene are used: the two possible monosubstituted naphthalenes are named primarily as in formula A while di- and polysubstituted naphthalenes are named using formula B.

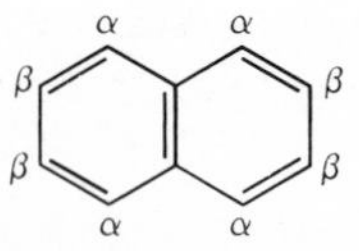

A

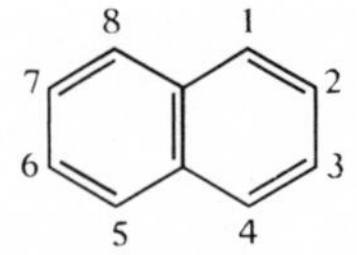

B

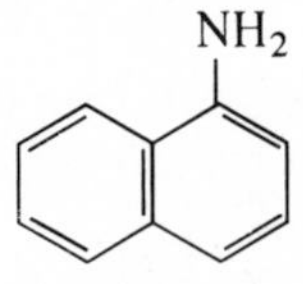

α-Naphthylamine
(1-naphthylamine)

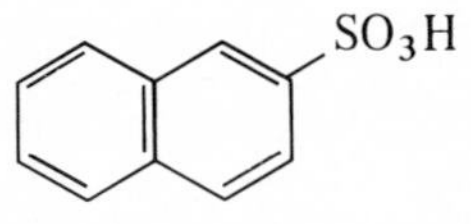

β-Naphthalenesulfonic acid
(2-naphthalenesulfonic acid)

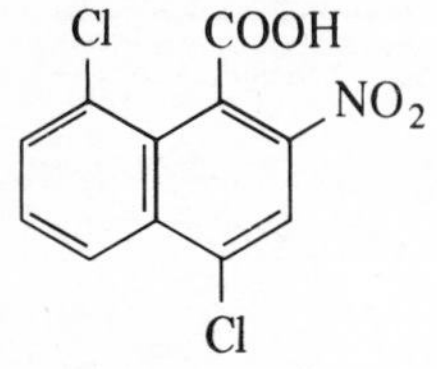

2-Nitro-4,8-dichloronaphthalene-
1-carboxylic acid

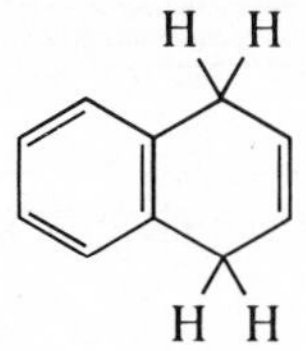

1,4-Dihydronaphthalene

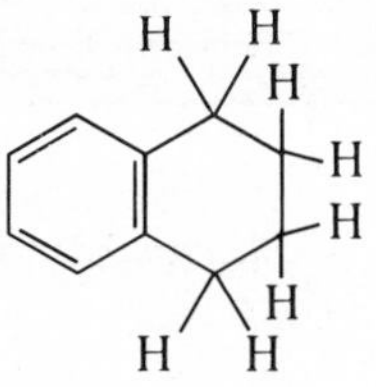

1,2,3,4-Tetrahydronaphthalene (tetralin)

2. Anthracene and phenanthrene.

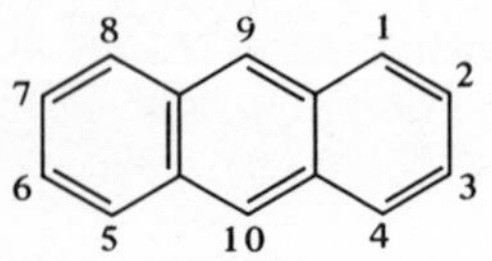

Anthracene

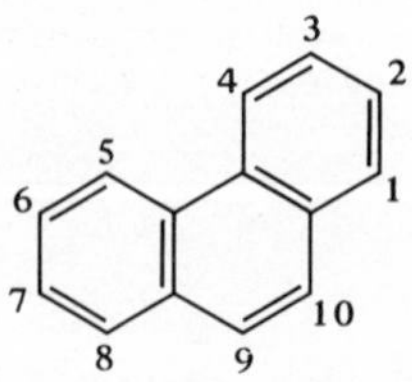

Phenanthrene

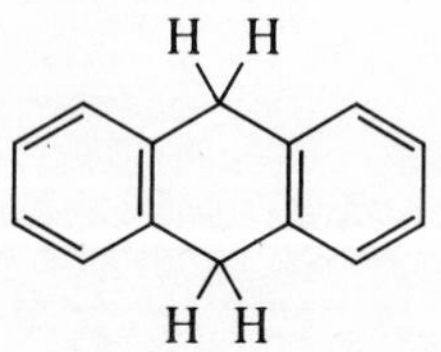

9,10-Dihydroanthracene

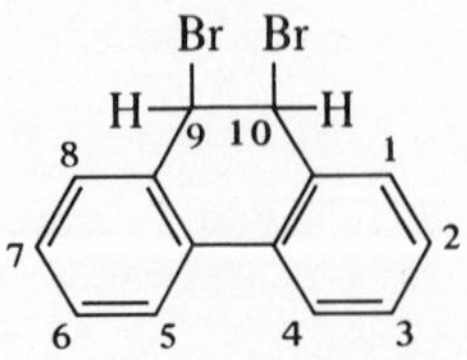

9,10-Dibromo-9,10-dihydrophenanthrene

3. Miscellaneous hydrocarbons.

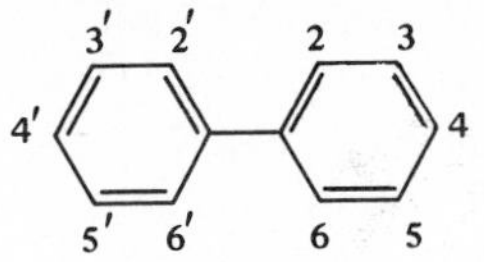

Biphenyl

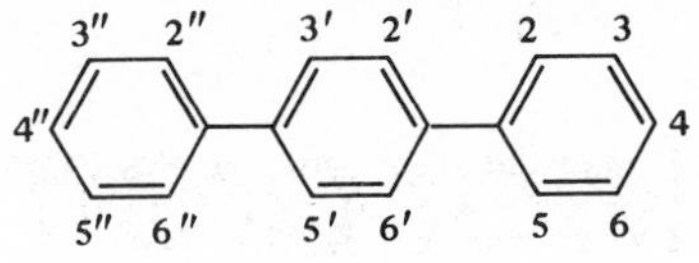

p-Terphenyl
(1,4-diphenylbenzene)

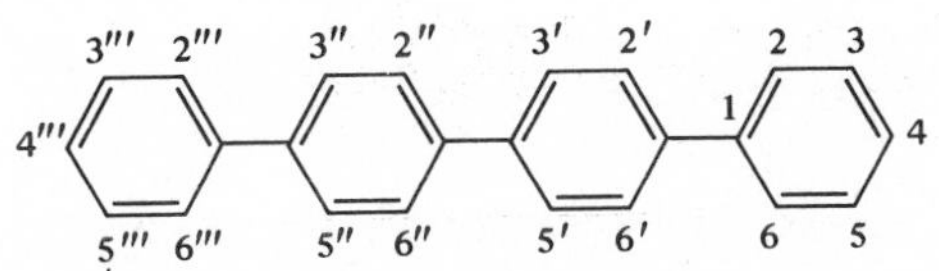

p-Quaterphenyl

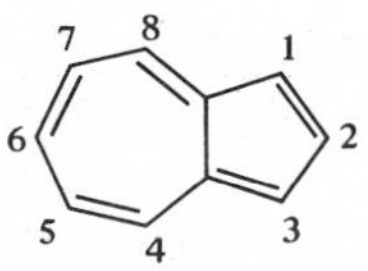

Azulene

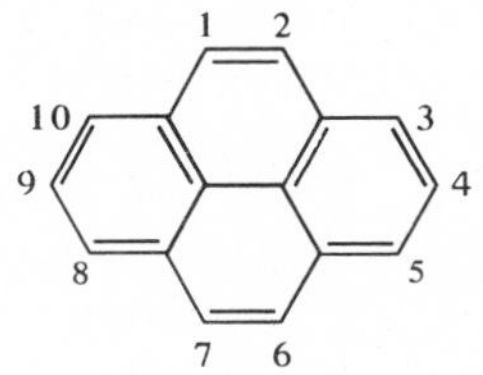

Pyrene

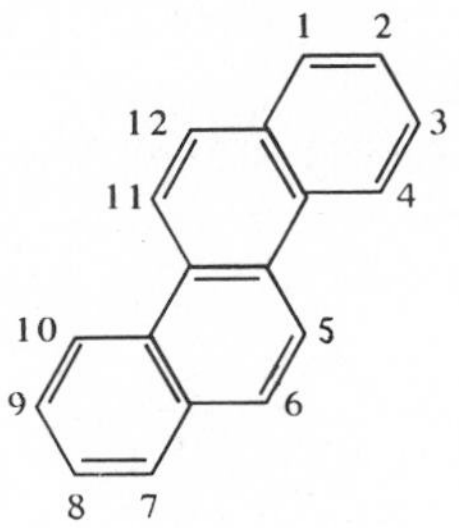

Chrysene

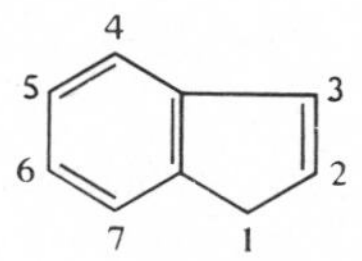

Indene

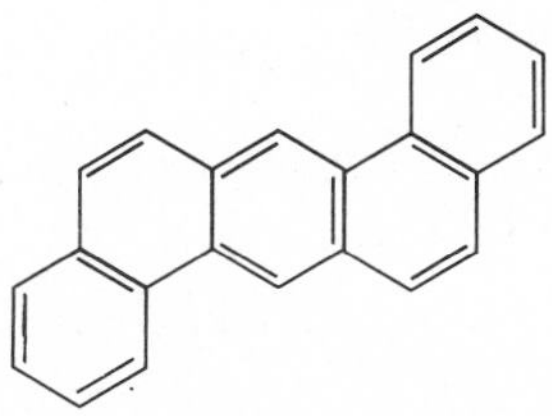

1,2,5,6-Dibenzanthracene

The term *benz* (or *benzo*) is used to indicate a benzene ring fused to a second ring. Its position is indicated by the two numbers of the positions to which it is attached, for instance, 1,2,5,6-dibenzanthracene. Or, for example, phenanthrene could be called 1,2-benznaphthalene.

b. *Carbonyl Compounds* Aromatic aldehydes, ketones, and acids may be named by methods described in the corresponding aliphatic chapters, but many have common names. In the nomenclature of aromatic ketones of the type C_6H_5COR the radical —COR is named by dropping the *ic* ending of the parent acid and adding *ophenone*. An exception is the CH_3CH_2CO— group, which is termed *propio*. Additional groups with common names are $C_6H_5COCH_2$—, the phenacyl group, and $C_6H_5\overset{\parallel}{\underset{O}{C}}$ —, the benzoyl group.

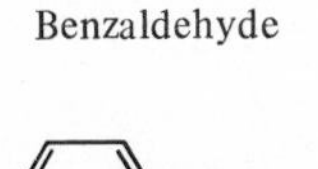

Benzaldehyde

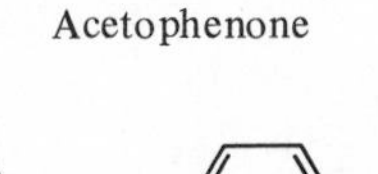

Acetophenone

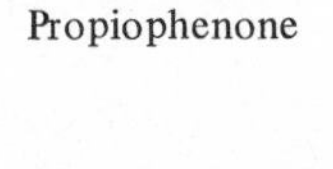

Propiophenone

Benzophenone

Cinnamaldehyde

p-Bromophenacyl bromide
(α,p-dibromoacetophenone)

269

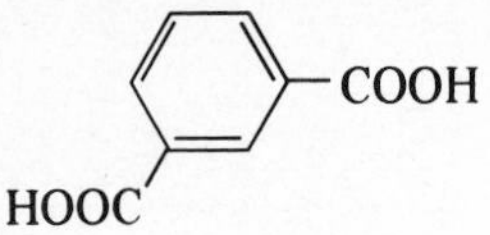

Isophthalic acid    Terephthalic acid 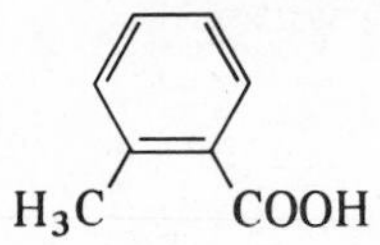   o-Toluic acid

$C_6H_5CH=CHCOC_6H_5$ $C_6H_5COC(C_6H_5)_3$ $C_6H_5COCOC_6H_5$

Chalcone
(benzalacetophenone)   Benzopinacolone   Benzil

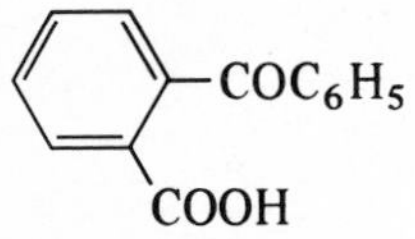

—COC_6H_5 $C_6H_5COCHC_6H_5$
|
COOH   OH

o-Benzoylbenzoic acid   Benzoin   α-Tetralone (1-tetralone) 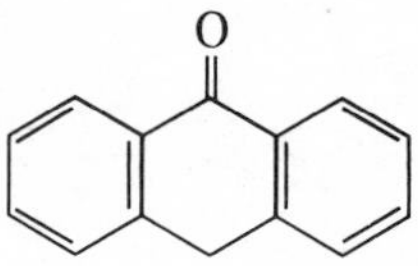   Anthrone

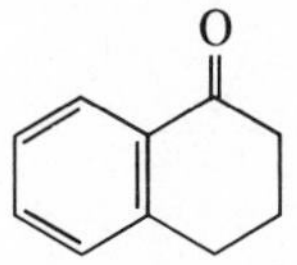

Quinones are cyclic, conjugated diketones named after the parent hydrocarbon.

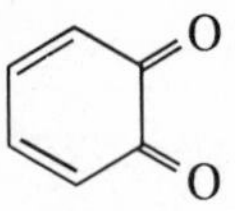

o-Benzoquinone 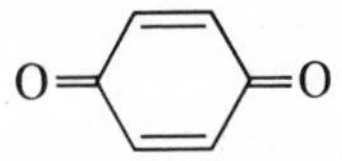   p-Benzoquinone 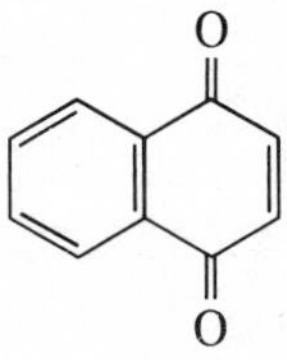   1,4-Naphthoquinone 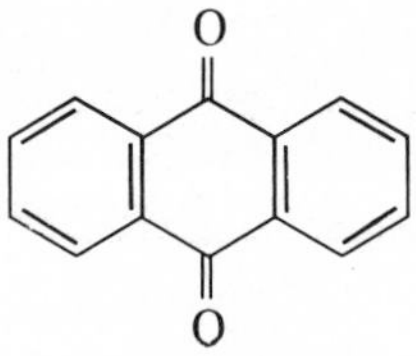   9,10-Anthraquinone

19-2 Preparation of Polynuclear Aromatic Hydrocarbons

a.   *Synthesis of Naphthalene Compounds*   Naphthalene itself is not synthesized commercially but is isolated from coal tar.
1. Via Friedel-Crafts reactions.

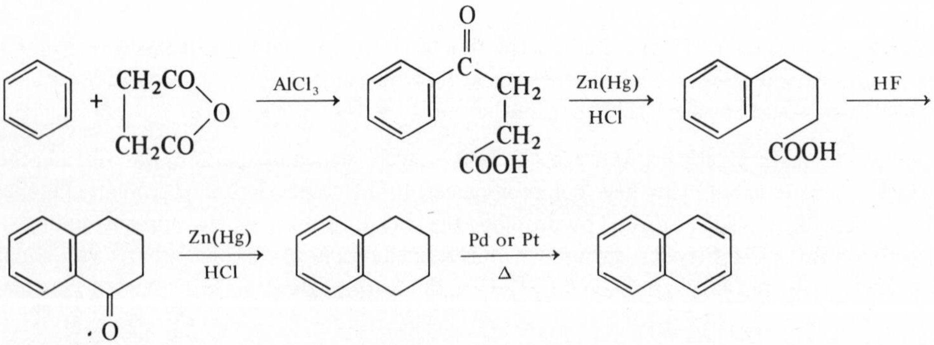

The cyclization step with HF is a general one for polycyclic molecules.
2. Via intramolecular condensation reactions.

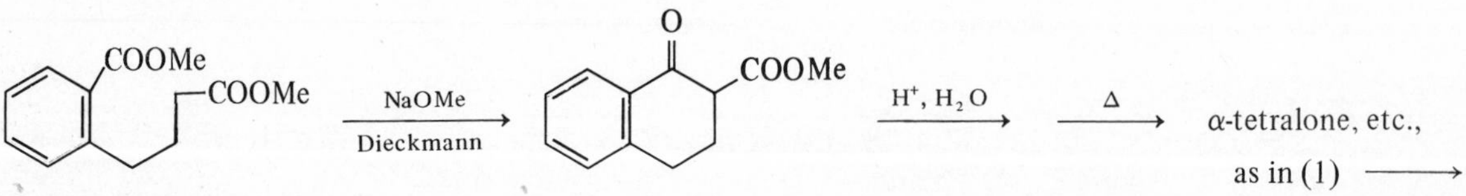

Intramolecular aldol condensations also could be used for the synthesis of naphthalene compounds.

3. Via Diels-Alder reactions

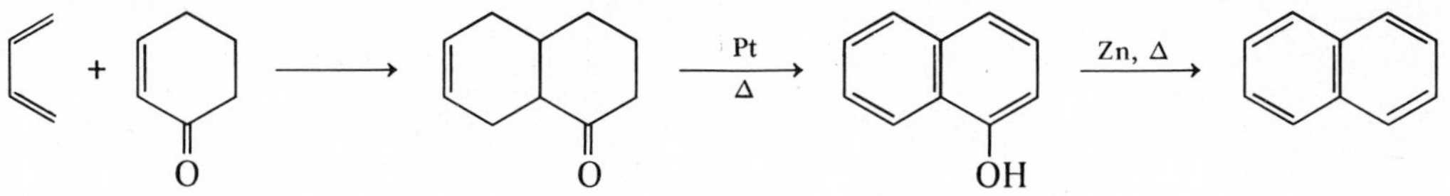

b. Synthesis of Anthracene and Phenanthrene Rings
1. Via Friedel-Crafts reactions

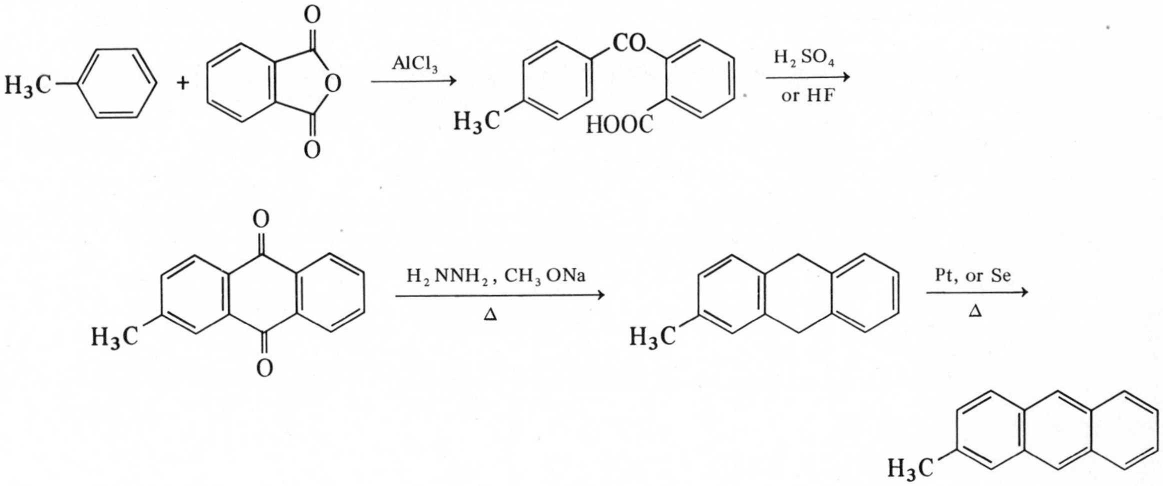

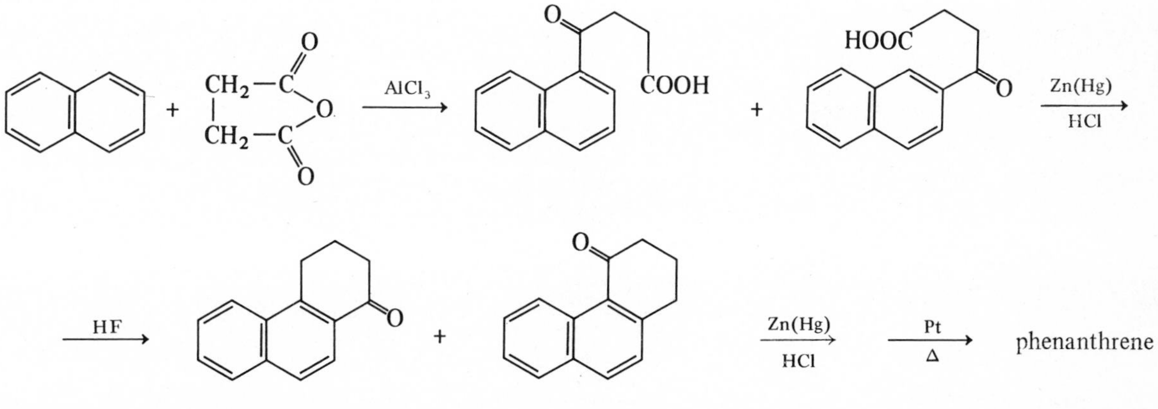

2. Via Diels-Alder reactions. See Sec. 19-4d.

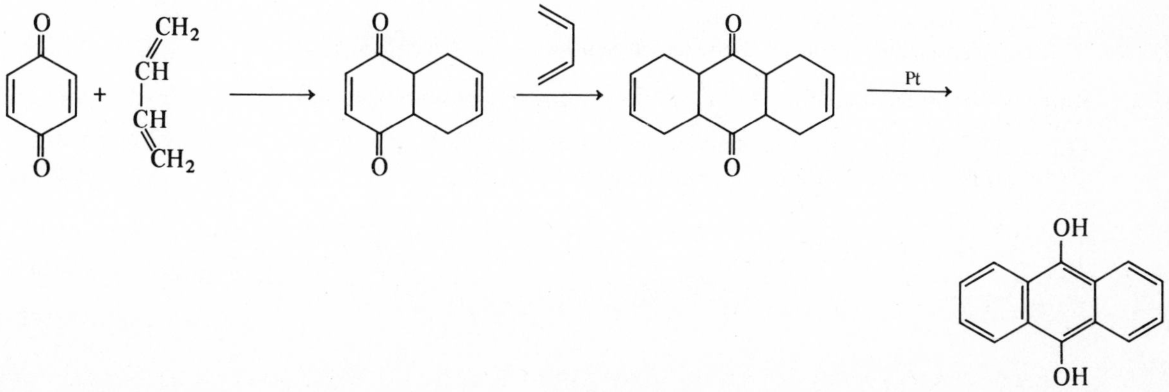

3. Phenanthrene and its derivatives may be made by means of the Pschorr ring closure.

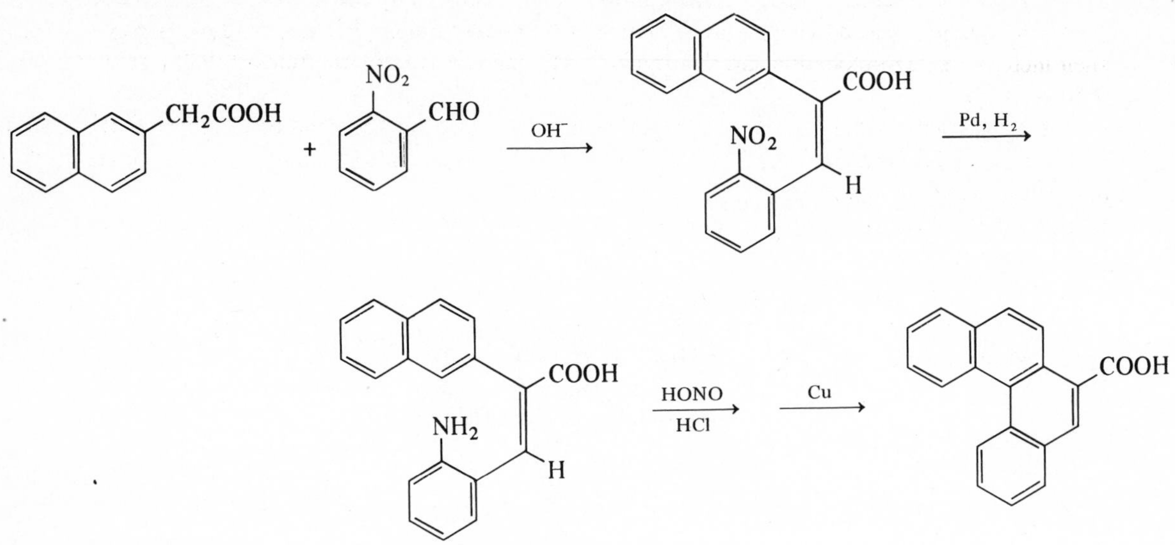

19-3 Preparation of Aromatic Carbonyl Compounds

Read Sec. 7-2 for the preparation of aldehydes and ketones. These reactions will involve the interconversion of functional groups, and they apply equally well to carbonyl compounds with or without aromatic ring systems.

   a.   *Specific Methods for Aromatic Aldehydes and Ketones*
       1. Friedel-Crafts acylation. (See Secs. 14-3*e*, 19-2*a*-1, 19-2*b*-1.)
       2. Gattermann-Koch synthesis of aldehydes. Aromatic compounds containing only alkyl groups or a single halogen atom can be formylated (–CHO, formyl group) in a reaction similar to the Friedel-Crafts reaction. Para substitution predominates.

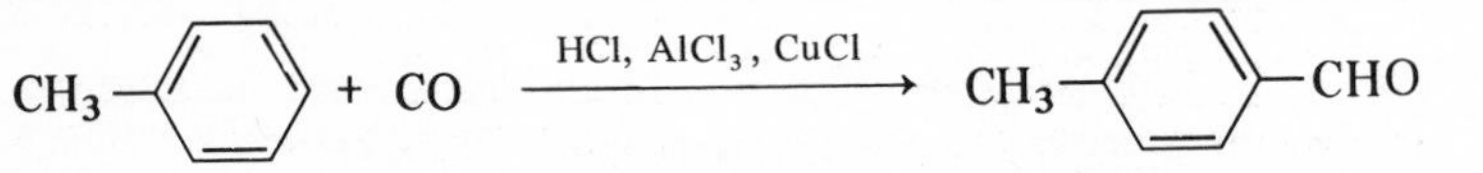

p-Tolualdehyde

       3. Gattermann synthesis of aldehydes. Formylation of phenols (see Sec. 18-3*l*-2), phenyl ethers, and some alkyl-aromatic hydrocarbons with HCN and HCl yields principally para substitution products.

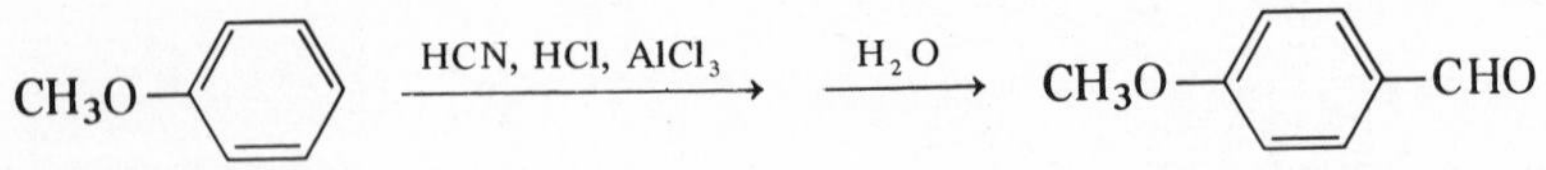

p-Anisaldehyde

       4. Reimer-Tiemann reaction. (See Sec. 18-3*l*-1.)
       5. Fries Rearrangement and Houben-Hoesch Synthesis. (See Sec. 18-3*k*-1, 2.)

   b.   *Aromatic Acids* Review Chap. 19, 12, and 14 for general reactions which lead to acids.

   c.   *Quinones* Oxidation of phenols, aminophenols, aromatic amines, and certain aromatic hydrocarbons (see Sec. 19-4*b*) leads to quinones. Common oxidizing agents are ferric chloride, acid dichromate, and Tollens' reagent.

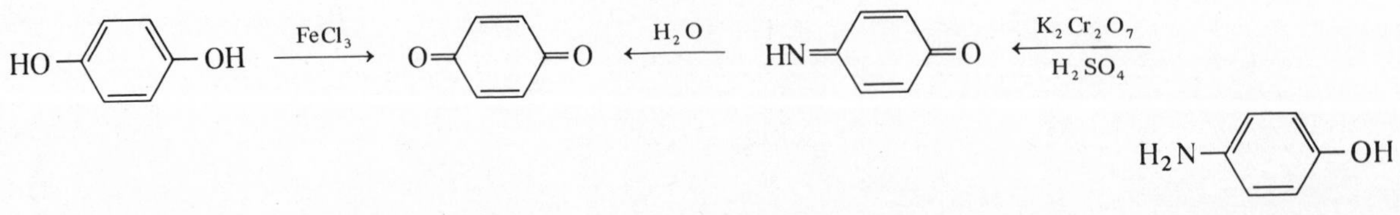

19-4 Reactions of Polynuclear Aromatic Hydrocarbons

a. *Substitution* Naphthalene undergoes electrophilic substitution readily and normally this occurs at the 1 position. Anthracene and phenanthrene undergo fewer useful electrophilic substitution reactions because of their increased ease of oxidation and their tendency to undergo addition reactions at their highly reactive 9,10 positions.

1. Nitration. Naphthalene yields α-nitronaphthalene which upon further nitration gives mainly the 1,8-dinitro and some 1,5-dinitro products. Nitric acid acts primarily as an oxidizing agent in the case of anthracene and phenanthrene.
2. Sulfonation. Because sulfonation is readily reversible (see Prob. 16-13 and Sec. 14-3e) the major substitution product will depend on the conditions.

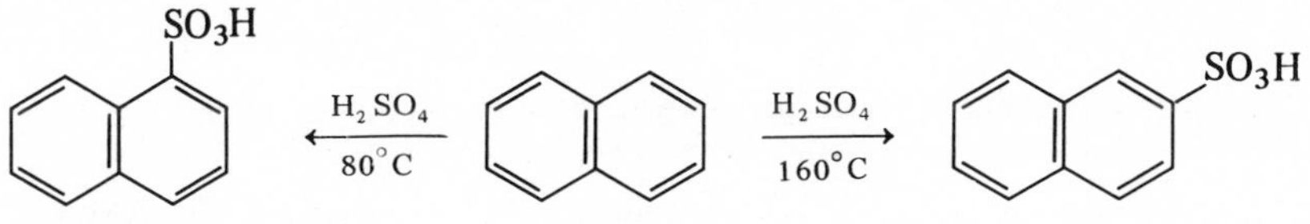

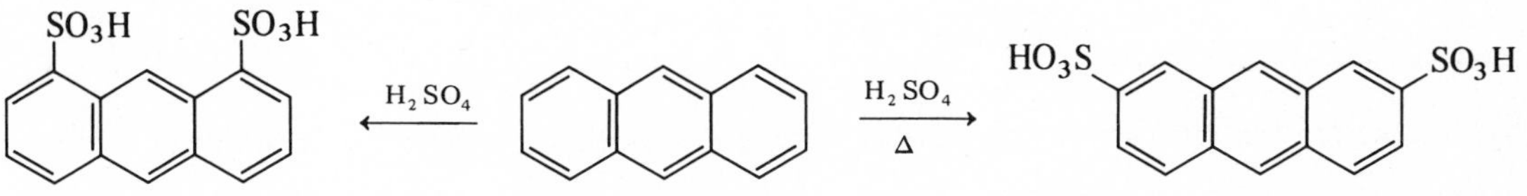

Recall that sulfonic acids when fused with NaOH lead to phenols. One useful sequence in the naphthalene series (seldom useful with benzene derivatives) is the preparation of β-naphthol and its conversion to β-naphthyl-amine with $NaHSO_3$ and NH_3 (called the *Bucherer* reaction).

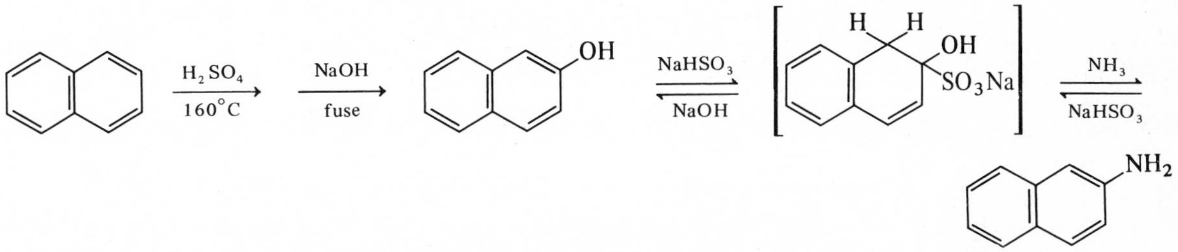

3. Friedel-Crafts reaction. Alkylation of naphthalene yields β-substituted products. With anhydrides and acyl halides, the α substitution product is the primary one formed, unless nitrobenzene is used as a solvent, in which case the β isomer predominates. The methyl halides do not react.

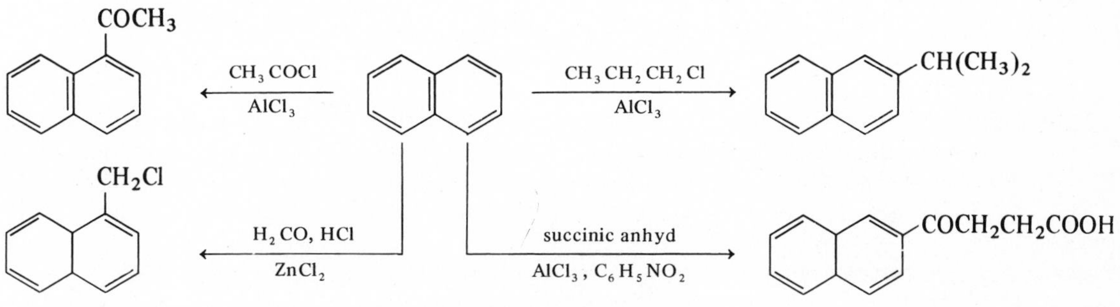

Phenanthrene gives mainly 3-substituted products while anthracene yields mainly 1-substituted products.
4. Halogenation. Naphthalene forms the same isomers as in nitration. Anthracene and phenanthrene add bromine to the 9,10 position and these compounds on heating decompose to give the 9-bromo products.

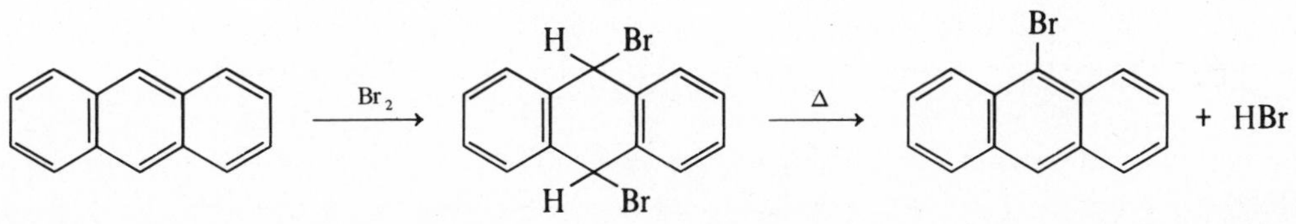

b. *Oxidation* The products observed will depend on whether mild or vigorous oxidation is used and in some cases what catalysts are used.

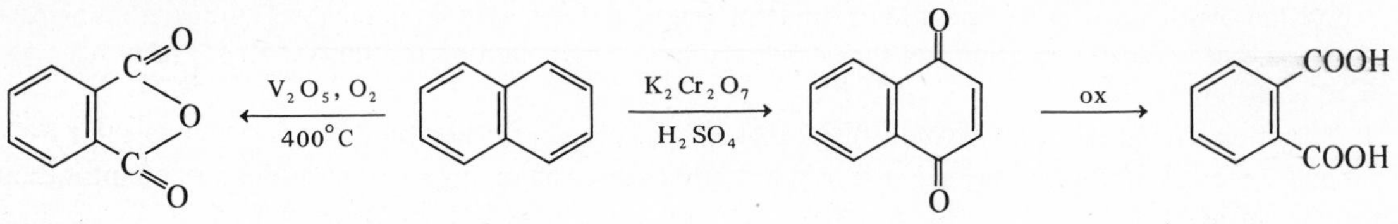

Anthracene $\xrightarrow{\text{CrO}_3}$ 9,10-anthraquinone

Phenanthrene $\xrightarrow{\text{K}_2\text{Cr}_2\text{O}_7,\ \text{H}_2\text{SO}_4}$ 9,10-phenanthrenequinone $\xrightarrow[\text{ox}]{\text{vigorous}}$ 2,2'-dicarboxybiphenyl (diphenic acid)

c. *Reduction* On reduction, naphthalene may be made to react with 1, 2, or 5 mol of hydrogen.

Naphthalene $\xrightarrow{\text{Na, EtOH}}$ 1,4-dihydronaphthalene $\xrightarrow{\text{Ni (H)}}$ tetralin $\xrightarrow{\text{Pt, H}_2}$ decalin (decahydronaphthalene)

Na(Hg), EtOH

Hydrogenation of anthracene and phenanthrene can easily be stopped at their 9,10-dihydro derivatives.

d. *Diels-Alder reactions*

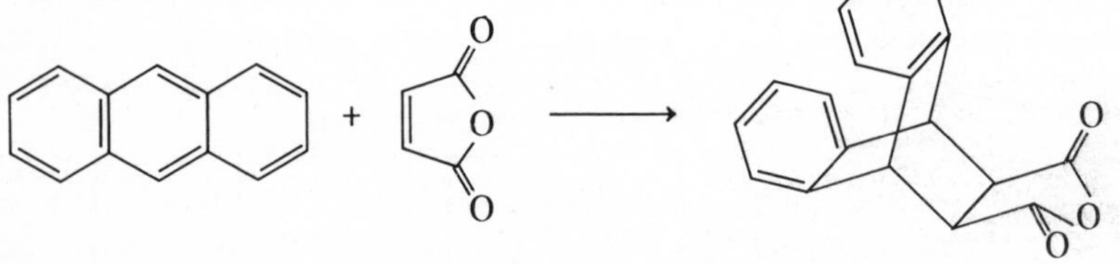

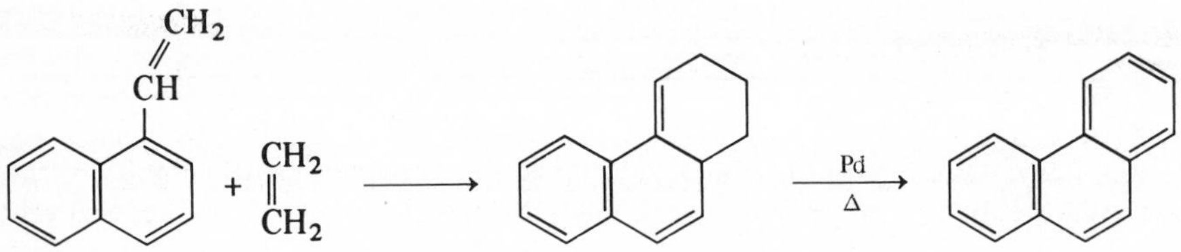

19-5 Reactions of Aromatic Carbonyl Compounds

a. *Aldehydes and Ketones* Reactions of aromatic aldehydes and ketones are comparable to those of aliphatic analogs, keeping in mind the fact that aromatic aldehydes have no α hydrogen atoms (review Secs. 7-3 and 12-2*b*-8). Ordinarily, aromatic aldehydes and ketones can be nitrated directly to give a meta product. Direct ring halogenation should be avoided because of oxidation or side-chain halogenation. The Sandmeyer reaction can be used for introduction of halogen in the ring. Side-chain halogenation can be accomplished:

$$\text{C}_6\text{H}_5\text{COCH}_3 \xrightarrow{\text{Br}_2,\ \text{AlCl}_3} \text{C}_6\text{H}_5\text{COCH}_2\text{Br} \quad \text{Phenacyl bromide}$$

A general method for reducing aldehydes and ketones not mentioned before is the use of aluminum isopropoxide in isopropyl alcohol. The reaction (Meerwein-Ponndorf-Verley reduction) is an equilibrium process which can be carried to completion by distilling the acetone as it forms in the mixture.

$$\text{R}_2\text{C=O} + [(\text{CH}_3)_2\text{CHO}]_3\text{Al} \rightleftharpoons \text{CH}_3\text{COCH}_3 + (\text{R}_2\text{CHO})_3\text{Al} \xrightarrow{\text{H}_3\text{O}^+} \text{R}_2\text{CHOH}$$

Aromatic aldehydes undergo a self-condensation catalyzed by cyanide ion called the *benzoin condensation* (see Prob. 19-29).

274

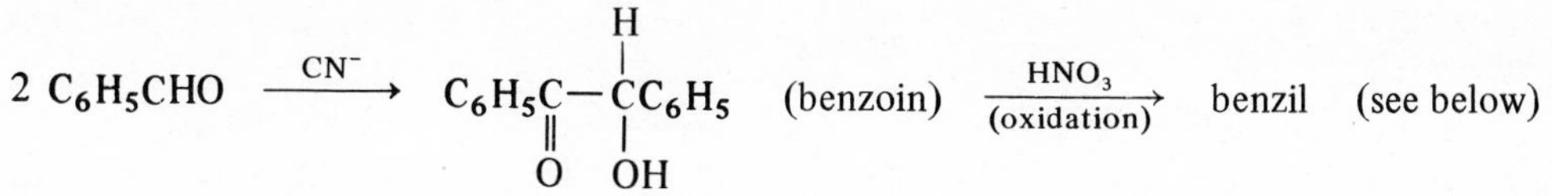

$$2\ C_6H_5CHO \xrightarrow{\ CN^-\ } \underset{\underset{\displaystyle O\ \ \ \ OH}{||\ \ \ \ |}}{C_6H_5\overset{\displaystyle H}{\underset{}{C}}-\overset{}{\underset{}{C}}C_6H_5} \quad (benzoin) \xrightarrow[\text{(oxidation)}]{HNO_3} \quad benzil \quad (see\ below)$$

α-Hydroxyketones, such as benzoin, belong to a class of compounds called *acyloins*. Aliphatic acyloins and cyclic acyloins can be prepared by a bimolecular reduction of mono or diesters, respectively, in a reaction (called the *acyloin condensation*) that is similar to the bimolecular reduction of ketones to pinacols (see Sec. 7-3c).

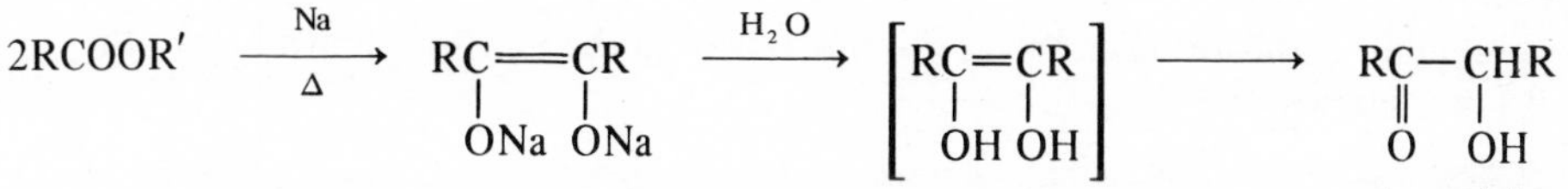

$$2RCOOR' \xrightarrow[\Delta]{Na} \underset{\underset{\displaystyle ONa\ \ ONa}{|\ \ \ \ |}}{RC=CR} \xrightarrow{H_2O} \left[\underset{\underset{\displaystyle OH\ \ OH}{|\ \ \ \ |}}{RC=CR}\right] \longrightarrow \underset{\underset{\displaystyle O\ \ \ \ OH}{||\ \ \ |}}{RC-CHR}$$

An acyloin

$$\underset{\underset{\displaystyle COOMe}{\diagdown}}{\overset{\overset{\displaystyle COOMe}{\diagup}}{(CH_2)_n}} \xrightarrow{Na} \underset{\underset{\displaystyle C-ONa}{\diagdown}}{\overset{\overset{\displaystyle C-ONa}{\diagup}}{(CH_2)_n||}} \xrightarrow{H_2O} \underset{\underset{\displaystyle CHOH}{\diagdown}}{\overset{\overset{\displaystyle C=O}{\diagup}}{(CH_2)_n}}$$

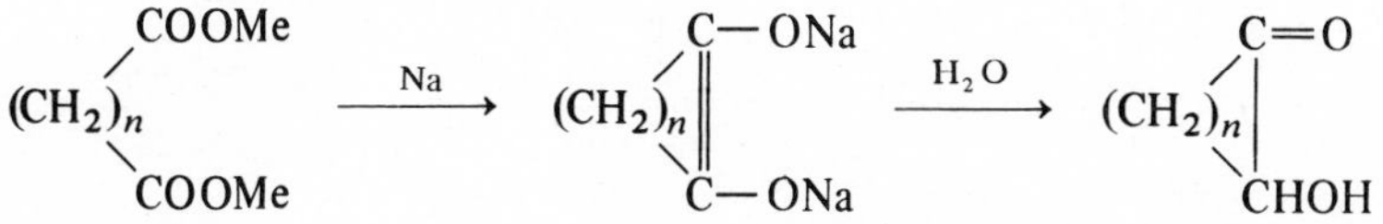

Aryl α-diketones (ArCOCOAr) such as benzil undergo a base-catalyzed reaction known as the benzil-benzilic acid rearrangement (see Prob. 13-24). α-Diketones also react with *o*-phenylenediamine to yield quinoxalines.

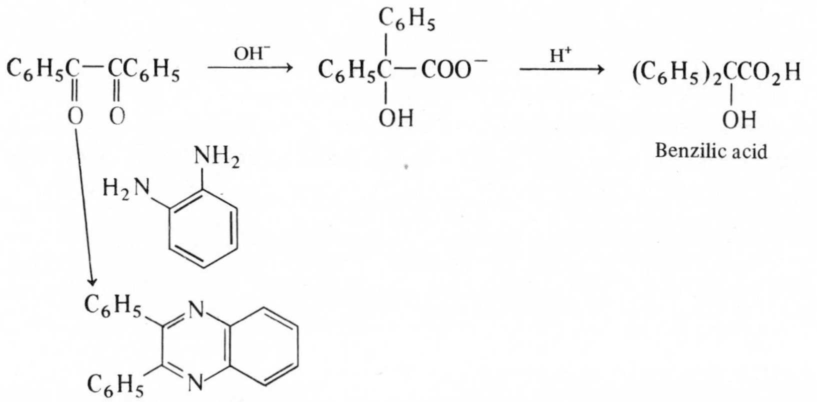

$$\underset{\underset{\displaystyle O\ \ O}{||\ \ ||}}{C_6H_5C-CC_6H_5} \xrightarrow{OH^-} \underset{\underset{\displaystyle OH}{|}}{C_6H_5\overset{\overset{\displaystyle C_6H_5}{|}}{C}-COO^-} \xrightarrow{H^+} \underset{\underset{\displaystyle OH}{|}}{(C_6H_5)_2CCO_2H}$$

Benzilic acid

2,3-Diphenylquinoxaline

b. *Acids* See Chap. 9 for general reactions of acids. Salt formation is similar for aromatic and aliphatic acids. Benzoic acid, pK_a 4.17, is only slightly stronger than acetic acid, pK_a 4.76. Ortho substituents (except amino) and electron-attracting groups in any position increase the acid strength.

c. *Esters* The reversible, acid-catalyzed esterification and hydrolysis reaction (Fischer method) is hindered by groups in the ortho positions. Hindered acids can be esterified via silver salts and alkyl halides, with diazomethane, or by the reaction of the acid chloride with alcohols. Either esterification or hydrolysis of certain hindered acids or esters may be carried out by Newman's method. This involves dissolving the acid (or ester) in concentrated H_2SO_4 and pouring the solution into an alcohol (or water).

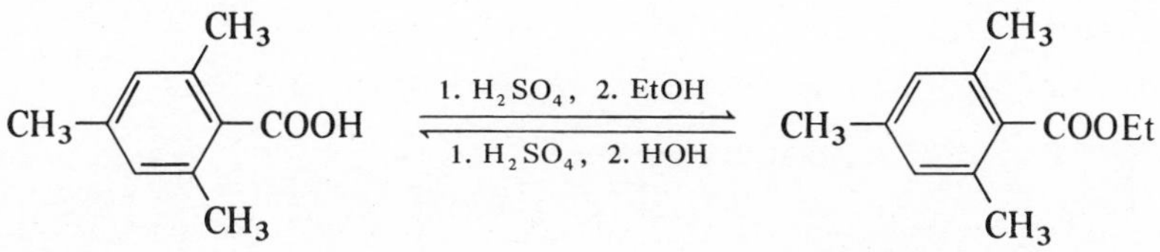

<table>
<tr><td align="center">Mesitoic acid</td><td align="center">Ethyl mesitoate</td></tr>
</table>

d. Anhydrides and Imides

Phthalic acid $\xrightarrow{\Delta}$ phthalic anhydride $\xrightarrow{NH_3,\ \Delta}$

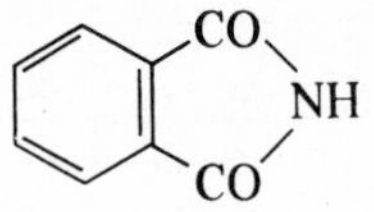

Phthalimide

e. Quinones

1. Oxidation-reduction reactions. Quinones can be reduced by reagents such as HI, H_2SO_3, $FeCl_2$, and certain hydroquinones (see Prob. 19-7d).

2. Addition of halogen; indirect substitution. When chlorine is used in the reaction shown below, the elimination of HCl is spontaneous.

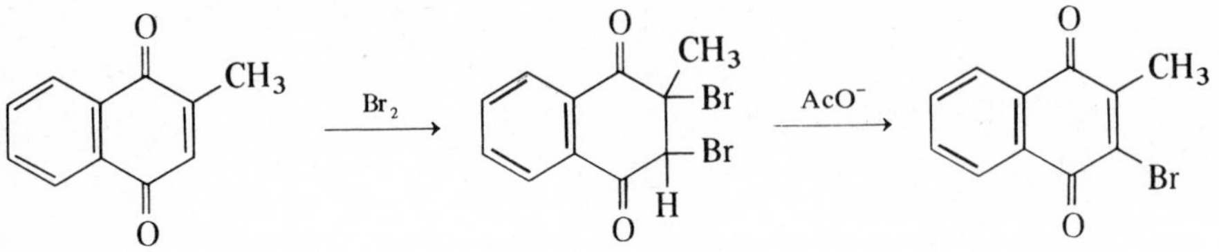

3. 1,4-additions. (See Prob. 19-30.) Common compounds which add in the 1,4 manner are alcohols (in the presence of $ZnCl_2$), amines, anhydrides, HCN, RSH, and active methylene compounds.

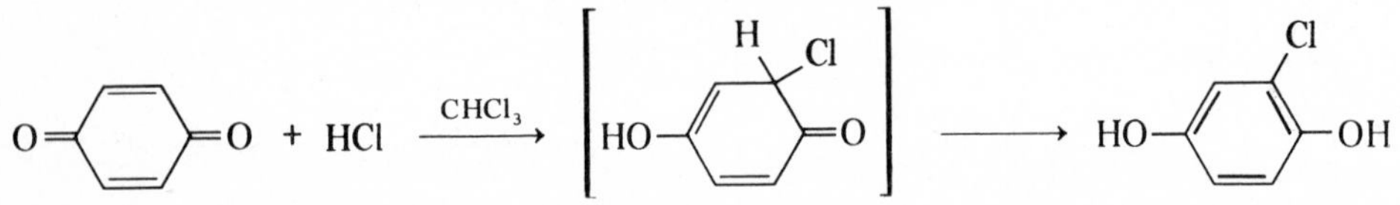

4. Substitution reactions. Direct substitution occurs through a free-radical mechanism (see the Gomberg reaction, Sec. 17-7a).

p-Benzoquinone + benzenediazonium chloride $\xrightarrow{NaOH}$

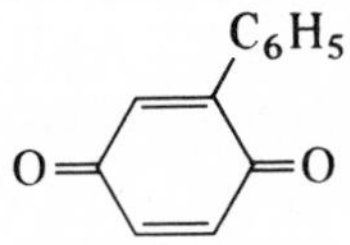

5. Quinones are convenient dienophiles for the Diels-Alder reaction (see Sec. 14-2f), and addition may occur twice in the case of *p*-benzoquinone (see Sec. 19-2b-2).

19-6 Qualitative Tests

	Tollens' reagent	Fehling's solution	Schiff's reagent	2,4-Dinitrophenylhydrazine
Aliphatic aldehydes	+	+	+	+
Aliphatic ketones	−	−	−	+
Aromatic aldehydes	+	−	+	+
Aromatic ketones	−	−	−	+
Sugars (hemiacetal form, see Chap. 21)	+	+	−	+
Quinone	+	+	−	+
Hydroquinone	+	+	−	+

Problems

19-1 Draw structures for each of the following:

    a. 9,10-Dimethylanthracene   *b.* 9,10-Phenanthrene epoxide   *c.* 1,2-Dimethylindene

    d. Indan (2,3-dihydroindene)   *e.* 2-Indanone oxime   *f.* α-Naphthoyl chloride

    g. *p*-Tolualdehyde   *h.* β-Naphthylamine   *i.* 9,10-Phenanthraquinone

    j. 1,2,3,4-Tetrahydro-5-iodonaphthalene   *k.* α-Naphthyl isocyanate

    l. Ethyl benzoylacetate   *m.* *p*-Nitrocinnamic acid   *n.* 4,4′-Dichlorobenzophenone

    o. 2-Anthracenesulfonic acid   *p.* Terephthalaldehyde   *q.* Benzal chloride

    r. Butyrophenone semicarbazone   *s.* 4,4′-Dichlorobenzoin

    t. 2,6-Naphthoquinone   *u.* 1,3,5-Trimethylazulene   *v.* 1,2,3,4-Dibenzanthracene

    w. 1,2-Dimethyl-3,4-benzpyrene   *x.* β-Benzoylpropionamide

    y. 2″,3′-Dicarbomethoxy-*p*-quaterphenyl   *z.* 1-Vinyl-12-allylchrysene

19-2 Complete the following indicating "no reaction" where appropriate:

    a. C_6H_5CHO + Tollens' reagent   *b.* C_6H_5CHO + concd NaOH

    c. C_6H_5CHO + Cl_2 *d.* *p*-Tolualdehyde + $C_6H_5NHNH_2$

    e. Phenylacetaldehyde + aq OH^- *f.* $C_6H_5CH_2CHO$ + ethylene glycol + HCl

    g. Benzophenone + $LiAlH_4$ *h.* $(C_6H_5)_2CO$ + $H_2NNHCONH_2$

    i. α-Naphthylamine + hot $KMnO_4$ *j.* α-Nitronaphthalene + hot $KMnO_4$

    k. β-Naphthyl allyl ether + Δ   *l.* α-Naphthyl acetate + $AlCl_3$ + Δ

    m. Mild oxidation of 4,4′-diaminobiphenyl   *n.* Indene + excess Cl_2 + uv light

    o. 1-Naphthaldehyde + KCN   *p.* 2-Naphthalenesulfonic acid + NaCN fusion

    q. 1-Naphthylamine sulfate + Δ   *r.* $C_6H_5CH_2CH_2COOMe$ + Na

    s. 4,4′-Diethylbenzil + OH^- *t.* Anthrone + Al(O-isoPr)$_3$

    u. $C_6H_5N=C=S$ + aniline   *v.* Benzopinacolone + $NaHSO_3$

    w. Cinnamaldehyde + HBr   *x.* Phenyl benzoate + mononitration

    y. 9,10-Dihydrophenanthrene + CH_3COCl + $AlCl_3$ *z.* C_6H_5CHO + $C_6H_5CH_2CHO$ + NaOEt

    a′. 9,10-Anthraquinone + $2H_2NOH$ *b′.* 1,4-Naphthoquinone + H_2 + Pt

    c′. *p*-Benzoquinone + HCl

    d′. C_6H_5COOMe + 4-biphenylmagnesium bromide (2 mol)

    e′. 1-Naphthaldehyde oxime + P_2O_5 (dehydrated)   *f′.* Ethyl *p*-toluate + $LiAlH_4$

19-3 Devise a satisfactory laboratory route for making the following conversions, using any necessary aliphatic and inorganic compounds:

    a. Benzonitrile to propiophenone   *b.* Benzene to butyrophenone

    c. Benzoic acid to benzophenone   *d.* Benzene to triphenylcarbinol

    e. Naphthalene to α-naphthylamine   *f.* Naphthalene to 2-naphthoic acid

    g. Naphthalene to α-naphthylacetic acid   *h.* Naphthalene to 2-butylnaphthalene

    i. Naphthalene to 5-nitro-1-aminonaphthalene   *j.* Benzene to $C_6H_5C(CH_3)=CHCOC_6H_5$

    k. Benzophenone to benzopinacolone   *l.* Ethylbenzene to *p*-ethylbenzaldehyde

    m. Hydroquinone to 6,7-dimethyl-1,4-dihydroxynaphthalene

    n. 2,6-Dimethylterephthalic acid to 2,6-dimethyl-4-carbomethoxybenzoic acid

    o. Toluene to benzaldehyde   *p.* Phenol to *o*-hydroxyacetophenone

    q. Phenanthrene to (*Hint*: use benzylic acid rearrangement)

    r. *o*-Xylene to 2-methylindene   *s.* 1-Naphthonitrile to 1-naphthaldehyde

    t. *p*-Tolualdehyde to 3-(*p*-tolyl)-2-butanol   *u.* Benzene to 1-cyclopentylnaphthalene

    v. Biphenyl to *p*-quaterphenyl   *w.* Benzene to $C_6H_5COCH_2CH(C_6H_5)CH(COOEt)_2$

    x. 1-Methylnaphthalene to 3-methyl-1,2-benzanthracene

y. Toluene to

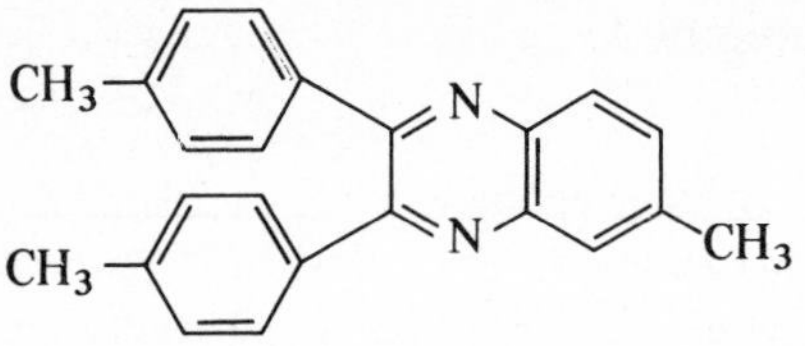

19-4 Suggest a simple test to distinguish between the following:
   a. 2-Naphthol and 2-naphthoic acid
   b. Catechol and resorcinol
   c. Benzaldehyde and phenylacetaldehyde
   d. Phenacyl chloride and acetophenone
   e. *p*-Chlorobenzaldehyde and *p*-chlorophenol
   f. 1,4-Naphthoquinone and 1,4-dihydroxynaphthalene
   g. β-Nitronaphthalene and 2-nitrodecalin
   h. Decalin and tetralin

19-5 Give a method for separating the following mixtures:
   a. Methyl benzyl ketone, propiophenone, benzoic acid, phenol, *p*-nitrophenol.
   b. α-Phenylsuccinic acid, phenyl malonate, ethyl *p*-aminobenzoate, *N*-ethyl-*p*-aminobenzoic acid.

19-6 An important factor influencing the relative acidity of a hydrogen atom in a given molecule is the ability of the anion to stabilize itself through delocalization of the negative charge. For example, cyclopentadiene will form a Grignard reagent when treated with RMgX. Write the equation for this reaction and show five resonance forms that stabilize the anion.

19-7 In each of the following, arrange the compounds listed in increasing order of the given property or characteristic.
   a. Propene, toluene, diphenylmethane, methane, 1,3-pentadiene: acidity of their most acidic hydrogen atom.
   b. Propionic acid, benzoic acid, *o*-chlorobenzoic acid, phthalimide, 2,4-dinitrobenzoic acid: acid strength.
   c. Phenylacetic acid, benzenesulfonic acid, β-phenylethanol, *p*-cresol, *p*-methylbenzenethiol: acid strength.
   d. Hydroquinones with Cl, SO_2R, CH_3, NH_2, $NHCH_3$ groups in the 2 position: ease of oxidation to the corresponding quinones.
   e. Ethyl benzoate, ethyl *p*-methylbenzoate, ethyl *o*-methylbenzoate, ethyl *p*-nitrobenzoate: ease of saponification.

19-8 Draw the indicated number of tautomeric or resonance structures in each of the following:
   a. Four tautomeric forms of compound A shown below.
   b. Five significant resonance structures for γ-pyrone, compound B shown below. Should this compound have resonance energy more nearly like benzene or dipropenyl ketone? Why does γ-pyrone fail to undergo typical carbonyl reactions such as phenylhydrazone formation?
   c. Three resonance structures for naphthalene containing no positive or negative charges. Draw four such structures for anthracene.
   d. Seven resonance structures for the ionic intermediate in electrophilic attack of Cl^+ at the 1 position in 2-methylnaphthalene and six resonance structures for such attack at the 3 position. Which substitution is more likely and why?

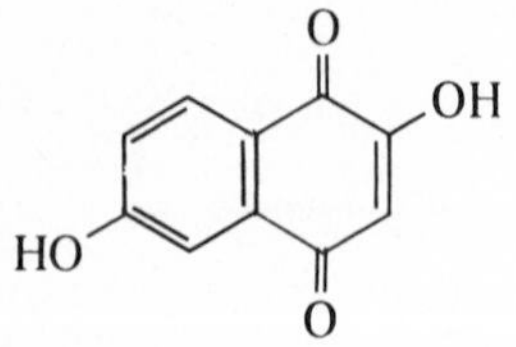

A 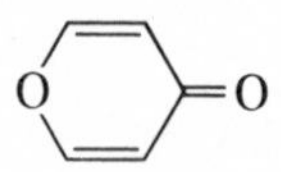   B

19-9 Explain why 2-isopropylnaphthalene gives a large amount of substitution at the 6 position when treated with succinic anhydride and $AlCl_3$. Taking advantage of this fact, show how you would convert 2-isopropyl-naphthalene into retene (1-methyl-7-isopropylphenanthrene). *Hint*: one can add a Grignard reagent to a ketone in a molecule containing a carboxyl group if one adds 2 mol of Grignard reagent. The first reacts with the active hydrogen, the second with the carbonyl group.

19-10 What would be the main product expected in the mononitration of each of the following?
   a. 1-Naphthoic acid
   b. 1-Acetylaminonaphthalene
   c. 1,4-Di-*tert*-butylnaphthalene
   d. β-Naphthol

19-11 Suggest structures for compounds, A, B, and C, all $C_{10}H_{10}O_2$, from the information given in the following table.

Compound	Tollens' test	Fehling's test	Iodoform test
A	+	+	+
B	+	−	−
C	+	−	+

Other information:

(A) $\xrightarrow{H_2, Pt}$ $C_{10}H_{14}$ $\xrightarrow{Fe, Br_2}$ only one $C_{10}H_{13}Br$

(B) $\xrightarrow{\text{vigorous ox}}$ an acid (NE 83) $\xrightarrow{\Delta}$ H_2O produced

(C) $\xrightarrow{\text{ox}}$ an acid (0.2 g neutralizes 24.1 ml of 0.1 N NaOH)

$\downarrow$ Fe, Br$_2$

$\xrightarrow{\qquad}$ only one $C_{10}H_{13}Br$

19-12 Compound A, $C_{15}H_{14}O$, when treated successively with hydroxylamine hydrochloride and PCl$_5$ in dry ether yielded B. Vigorous oxidation of A or B yielded only benzoic acid. Hydrolysis of B yielded no benzoic acid. Suggest structures for A and B.

19-13 Compound A, $C_{10}H_{10}O_4$, was soluble in 5% NaHCO$_3$. It required 32.0 ml of 0.100 N NaOH to neutralize 0.310 g of A. When A was heated with solid MnCO$_3$, compound B, C_9H_8O, was formed. B yielded a single isomer on treatment with NH$_2$OH. Suggest structures for A and B.

19-14 A hydrocarbon, $C_{12}H_{10}$, on complete hydrogenation with Pt gave $C_{12}H_{20}$. (How many sites of unsaturation?) On vigorous oxidation, the original hydrocarbon gave an acid, B, of NE 108 ± 1. Decarboxylation of B gave naphthalene. Suggest three possible structures for the hydrocarbon.

19-15 A neutral compound, $C_{17}H_{12}O$, was found to react with hydroxylamine. Heating the oxime thus formed with PCl$_5$ gave a neutral compound, B, which was hydrolyzed to give a basic compound, C, and an acid, D. Treatment of C with HONO and then heat, gave E, which on distillation with zinc dust gave benzene. D had NE 172 + 2. On vigorous oxidation, D gave a new acid, F, NE 70 ± 2. Suggest structures for A through F.

19-16 Compound A, NE 197 ± 2, contained nitrogen, was insoluble in water, but dissolved in bicarbonate solution. Treating A with SOCl$_2$, then NH$_3$, and finally NaOBr gave B. B was insoluble in water but dissolved in dilute acid. On hydrolysis with NaOH, B evolved ammonia and was converted to C, a compound soluble in acid or base, with NE 186 ± 2. Treatment of C with nitrous acid, then CuCN, gave A. Hydrolysis of A gave an acid of NE 106 ± 2. On heating, this acid evolved water and formed a neutral compound. Strong oxidation of C gave an acid D, NE 70 ± 2. Suggest structures for A through D.

19-17 Compound A, $C_{14}H_9O_3Cl$, gave a positive Tollens' test, a negative Fehling's test, and a negative test with alcoholic AgNO$_3$. It could be resolved into d and l forms. When either of these was oxidized with Tollens' reagent, B was formed, which was optically inactive and which could not be resolved. Suggest structures for A and B.

19-18 Criticize the following sequences:

a. $C_6H_5NH_2$ $\xrightarrow[\text{(A)}]{HNO_3, H_2SO_4}$ H_2N—⟨⟩—NO_2 $\xrightarrow[\text{H}_2\text{SO}_4]{HONO}$ $\xrightarrow[\text{(B)}]{CuBr}$

Br—⟨⟩—NO_2 $\xrightarrow[\text{(C)}]{CHCl_3, NaOH}$ Br—⟨⟩(CHO)—NO_2

b. O_2N—⟨⟩—CHO $\xrightarrow[\text{(A)}]{MeMgBr, H^+}$ O_2N—⟨⟩—$CHOHCH_3$ $\xrightarrow[\text{(B)}]{H_2SO_4}$ O_2N—⟨⟩(SO$_3$H)—$CHOHCH_3$

c. $Me_2CHCO-\langle\rangle-CHO$ $\xrightarrow[\text{(A)}]{\text{MeOH, H}^+}$ $Me_2CHCO-\langle\rangle-CH(OMe)_2$ $\xrightarrow[\text{(B)}]{\text{ZnHg, HCl}}$

$Me_2CHCH_2-\langle\rangle-CH(OMe)_2$

d. $\xrightarrow[\text{(A)}]{\text{Tollens' reagent}}$ $\xrightarrow[\text{(B)}]{\text{Br}_2}$

19-19 Complete the following by supplying the intermediates:

a. o-Toluidine $\xrightarrow[\text{H}_2\text{SO}_4]{\text{HONO}}$ $\xrightarrow{\text{NaBF}_4}$ (A) $\xrightarrow{\Delta}$ (B) $\xrightarrow[\text{H}_2\text{SO}_4]{\text{K}_2\text{Cr}_2\text{O}_7}$ (C)

b. p-Nitrobenzoic acid $\xrightarrow[\text{H}_2\text{SO}_4]{\text{HONO}}$ $\xrightarrow[\text{xanthate}]{\text{potassium ethyl}}$ (D) $\xrightarrow[\text{hyd}]{\text{KOH}}$ $\xrightarrow{\text{H}_3\text{O}^+}$ (E)

c. Phenol $\xrightarrow[\text{NaOH}]{\text{Et}_2\text{SO}_4}$ (F) $\xrightarrow[\text{AlCl}_3]{\text{HCN, HCl}}$ (G) $\xrightarrow{\text{aniline}}$ (H) $\xrightarrow{\text{H}_2\text{, Pd}}$ (I) $\xrightarrow[\text{HCl}]{\text{HONO}}$ (J) $\xrightarrow{\text{H}_2\text{, Pd}}$ (K)

d. C_6H_5Br $\xrightarrow[\text{AlCl}_3\text{, CuCl}]{\text{CO, HCl}}$ (L) $\xrightarrow[\Delta]{\text{20\% KOH}}$ (M) + (N)

e. Dimethyl adipate $\xrightarrow[\Delta]{\text{NaOCH}_3}$ (O) $\xrightarrow[\text{HOH}]{\text{NaOH}}$ $\xrightarrow[\Delta]{\text{H}_3\text{O}^+}$ $\xrightarrow{\text{H}_2\text{NOH}}$ $\xrightarrow{\text{PCl}_5}$ (P) $\xrightarrow{\text{NaOH, HOH}}$ $\xrightarrow{\text{HONO}}$

$\xrightarrow{\Delta}$ (Q) $C_5H_8O_2$

f. Acetophenone $\xrightarrow{\text{Mg}-\text{Hg}}$ $\xrightarrow{\text{H}^+\text{, H}_2\text{O}}$ $\xrightarrow{\text{Al}_2\text{O}_3}$ $\xrightarrow{p\text{-benzoquinone}}$ (R) $C_{22}H_{18}O_2$

g. $\xrightarrow{\text{NaOCH}_3}$ $\xrightarrow{\text{NaOH}}$ $\xrightarrow[\Delta]{\text{H}_3\text{O}^+}$ $\xrightarrow[\text{OH}^-]{\text{CH}_2\text{(COOMe)}_2}$ $\xrightarrow[\Delta]{\text{H}_3\text{O}^+}$ $\xrightarrow{\text{Pd, H}_2}$ $\xrightarrow{\text{SOCl}_2}$

$\xrightarrow{\text{NH}_3}$ $\xrightarrow{\text{NaOBr}}$ (S) $C_{11}H_{15}N$

h. Naphthalene $\xrightarrow{\text{Li, RNH}_2}$ octalin ($C_{10}H_{16}$) $\xrightarrow{\text{O}_3}$ $\xrightarrow{\text{H}_3\text{O}^+}$ diketone $\xrightarrow[(-\text{H}_2\text{O})]{\text{OH}^-}$

(T) $C_{10}H_{14}O$ $\xrightarrow{\text{Pt, H}_2}$ $\xrightarrow{\text{Pd, }\Delta}$ azulene

i. p-$ClC_6H_4CH_3$ + $ClCH_2CH_2COCl$ $\xrightarrow[\text{cyclization}]{\text{AlCl}_3}$ $\xrightarrow[\text{cyclic ketones}]{\text{H}_2\text{SO}_4}$ mixture of two $\xrightarrow[\text{reduction}]{\text{Clemmensen}}$ $\xrightarrow[\text{pyridine}]{\text{CuCN, }\Delta}$

$\xrightarrow{\text{1-naphthylMgBr}}$ $\xrightarrow{\text{H}_3\text{O}^+}$ (U) $C_{21}H_{18}O$ $\xrightarrow{\Delta\ (410^\circ\text{C}), \text{ causes loss of H}_2\text{O and cyclization}}$

(Elbs reaction)

20-Methylcholanthrene, a powerful carcinogen

j. Phthalic anhydride + acetic anhydride $\xrightarrow[\text{(Perkin-type reaction)}]{\text{KOAc}}$ (V) $C_{10}H_6O_4$ $\xrightarrow{\text{NaOCH}_3}$

$\begin{bmatrix} \text{(W) } C_{10}H_6O_4 \\ \text{unstable} \end{bmatrix}$ $\longrightarrow$ (X) $C_9H_6O_2$ $\xrightarrow[\text{2. CH}_3\text{I}]{\text{1. CH}_3\text{ONa}}$ $\xrightarrow[\text{4. CH}_3\text{I}]{\text{3. CH}_3\text{ONa}}$ (Y) $\xrightarrow[\text{reduction}]{\text{Clemmensen}}$ (Z)

19-20 Explain why the 1,4-addition product formed from the reaction of benzoquinone with HCl is energetically much more favorable than the 1,2-addition product.

19-21 The picolines (methylpyridines) often act as active methylene sources in condensation reactions of the aldol type (see Sec. 7-3*f*).

a. Will 2-picoline or 3-picoline undergo a condensation reaction more readily with benzaldehyde? Explain the relative reactivities in terms of charge distribution in the corresponding anions.

b. Give the structure for the reaction product from benzaldehyde and 2-picoline (1 equiv of H_2O is formed).

c. Give structures for the following synthetic intermediates:

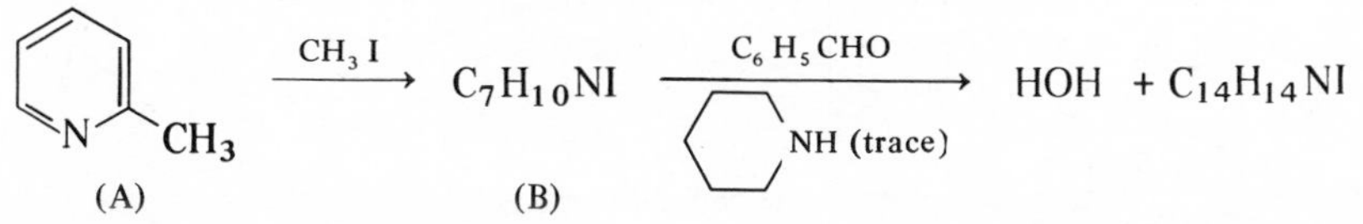

(A)   (B)

d. What is the function of the trace of piperidine in the condensation reaction of part *c*?

e. Should A or B in part *c* react more readily with benzaldehyde under the conditions indicated? Explain.

19-22 Suggest a detailed mechanism and draw an energy diagram for each of the following reactions based on the indicated experimental observations:

a. The observed reaction is:

$$C_6H_5COOCH_3 + HOH + H^+ \rightleftharpoons C_6H_5COOH + CH_3OH + H^+.$$

If $H_2{}^{18}O$ is used for hydrolysis, no ^{18}O appears in the alcohol. If the $H_2{}^{18}O$ reaction is stopped after partial hydrolysis, some heavy isotope appears in the unreacted ester.

b. The observed reaction is:

If $H_2{}^{18}O$ is used for hydrolysis the heavy isotope appears in the alcohol. The alcohol is nearly completely racemized.

c. The reaction of methyl benzoate with methanol and HCl in the absence of water gives a quantitative yield of Me_2O.

d. Hydrolysis of methyl mesitoate is carried out as indicated in Sec. 19-5*c*. Hydrolysis with aqueous HCl is very slow. When methyl benzoate is dissolved in concd H_2SO_4, the melting-point lowering indicates two particles are present in the solvent per molecule of ester introduced. For methyl mesitoate, the melting-point lowering is consistent with four particles per molecule of ester.

19-23 Indicate how many stereoisomers are possible in (*a*) decalin and (*b*) 2-methyldecalin and which ones would be optically active.

19-24 Bachmann's total synthesis of the estrogenic steroid from mare's urine included the following sequence:

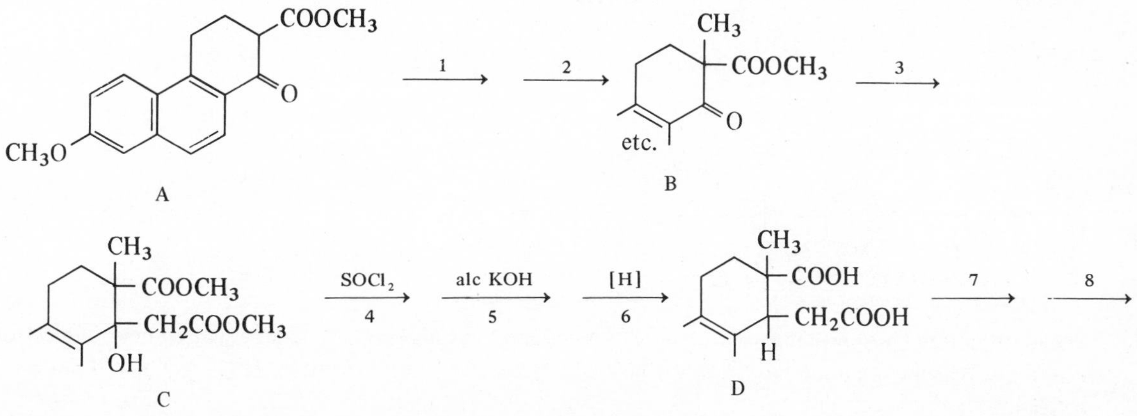

281

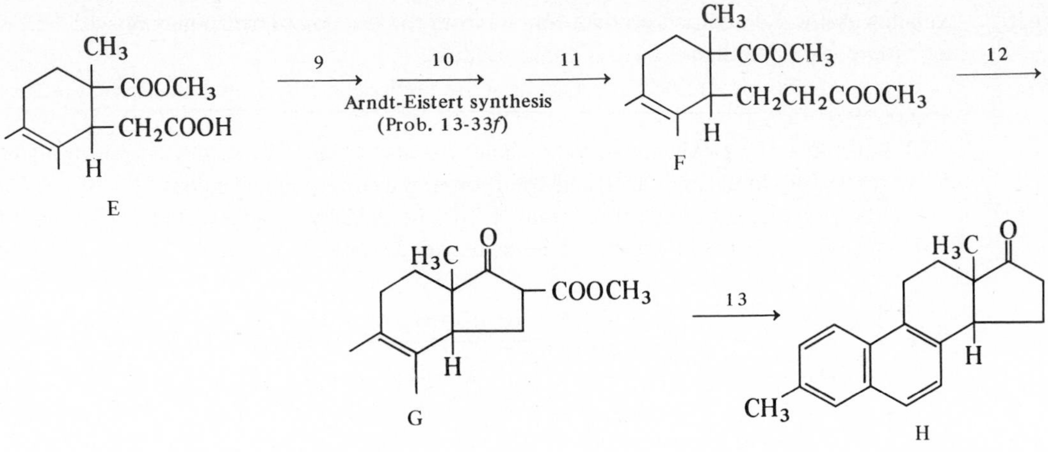

The reaction scheme shows structures E, F, G, and H with numbered steps 9, 10 (Arndt-Eistert synthesis, Prob. 13-33f), 11, 12, and 13.

a. Complete steps 1, 2, 3, 9, 10, 11, 12, and 13.

b. In step 7 both carboxyl groups are esterified and then only one is hydrolyzed in step 8. How is this possible?

c. If A were one particular stereoisomer, how many isomers of H would have been formed if no separations were made during the synthesis? At which point could simple crystallization have first been used to separate isomers in order to limit the isomers of H that are finally formed?

19-25   Predict the rearrangement product and its stereochemistry where appropriate in each of the following:

a.

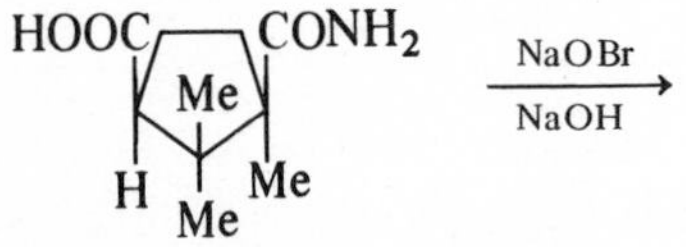

$\xrightarrow[\text{NaOH}]{\text{NaOBr}}$

b. (S)-3-Ethyl-2-heptanone oxime $\xrightarrow{\text{PCl}_5,\ \Delta}$

c.

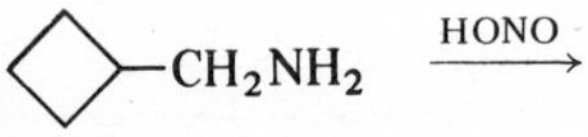

$-\text{CH}_2\text{NH}_2 \xrightarrow{\text{HONO}}$ olefin + alcohol (Demjanov rearrangement)

d.

$\text{C}_6\text{H}_5-\overset{\overset{\displaystyle \text{CH}_3}{|}}{\underset{\underset{\displaystyle \text{OH}}{|}}{\text{C}}}-\text{CH}_2\text{I} \xrightarrow{\text{AgNO}_3}$ ketone

e. $(\text{C}_6\text{H}_5)_2\underset{\underset{\displaystyle \text{OH}}{|}}{\text{C}}-\underset{\underset{\displaystyle \text{OH}}{|}}{\text{CH}_2} \xrightarrow{\text{acid catalyst}}$ aldehyde

f. $\text{CH}_3\text{CH}_2\text{C}\overset{\displaystyle O}{\underset{\displaystyle \diagdown \text{N}_3}{\diagup\!\!\!\diagup}} \xrightarrow{\Delta}$ isocyanate (Curtius rearrangement)

g. 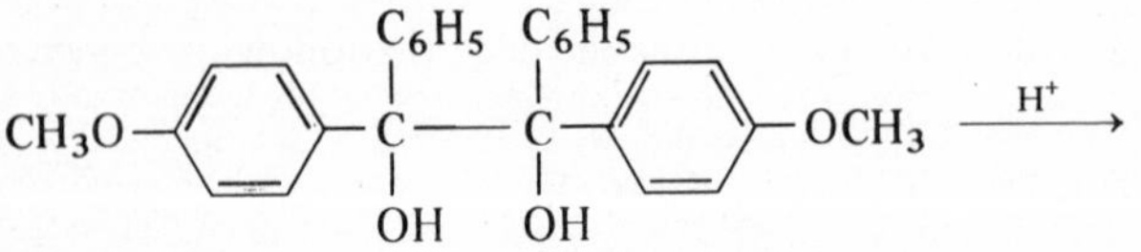 $\text{CH}_3\text{O}-$

$-\text{OCH}_3 \xrightarrow{\text{H}^+}$

h.

$\xrightarrow{\text{NaOH}}$

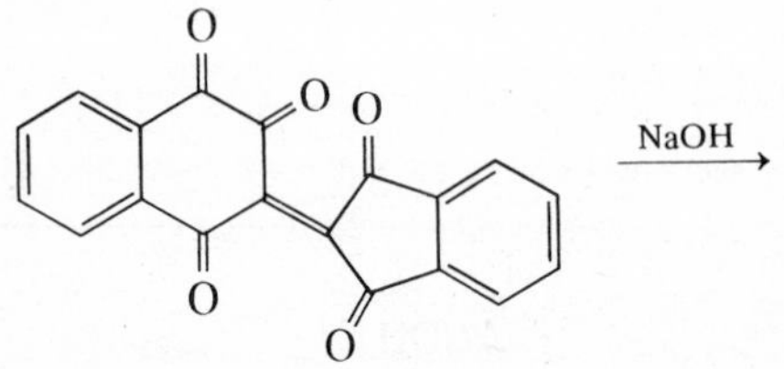

19-26   Phenanthrene and anthracene have the same number and kind of σ bonds and an equal number of electrons in π orbitals. Which compound has the lower heat of formation from the elements (that is, which is more stable)? Explain.

19-27 Give the structures for the indicated intermediates in the synthesis of lysergic acid, a polynuclear alkaloid. The diethylamide of lysergic acid is the hallucinogen called LSD.

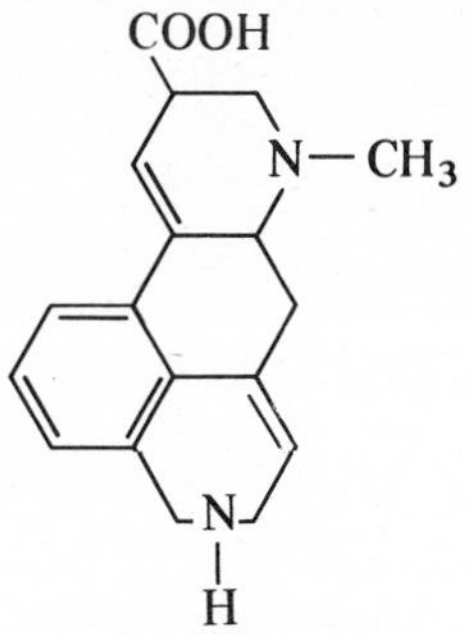

dl-Lysergic acid

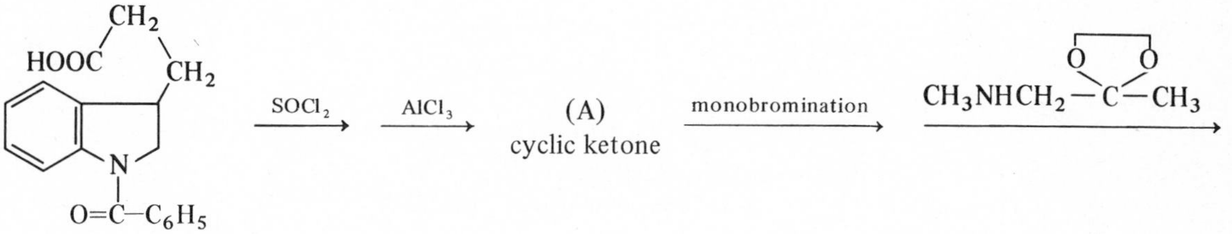

$$\underset{\substack{\text{(indoline with } O=C-C_6H_5 \text{ on N,} \\ \text{HOOC and } CH_2CH_2 \text{ substituents)}}}{}\ \xrightarrow{SOCl_2}\ \xrightarrow{AlCl_3}\ \underset{\text{cyclic ketone}}{(A)}\ \xrightarrow{\text{monobromination}}\ CH_3NHCH_2-\overset{\overset{O\diagdown\diagup O}{||}}{C}-CH_3 \longrightarrow$$

(B) (basic, bromine-free *dl* compound) $\xrightarrow{\text{acid hydrolysis}}\xrightarrow{\text{NaOEt}}$ **(C)** (four-ring compound stable to further

hydrolysis reactions) $\xrightarrow{Ac_2O}\xrightarrow[\text{reduction}]{NaBH_4\ (\text{ketone})}\xrightarrow{SOCl_2}\xrightarrow{NaCN}\xrightarrow{MeOH,\ H^+}$ **(D)** (methyl ester of acid with

three asymmetric carbons) $\xrightarrow{OH^-\ hyd}$ dihydrolysergic acid $\xrightarrow[Na_2HAsO_4]{\text{Raney Ni, }H_2O}$ *dl*-lysergic acid

19-28 After reviewing the mechanisms of the benzil-benzilic acid rearrangement (see Prob. 13-34) and pinacol formation (see Sec. 7-3c), propose mechanisms for the following reactions:

a. $C_6H_5\overset{\overset{O}{||}}{C}-\overset{\overset{O}{||}}{C}C_6H_5 \xrightarrow[CH_3OH]{NaOCH_3} (C_6H_5)_2\underset{\underset{OH}{|}}{C}-\overset{\overset{O}{||}}{C}OCH_3$

b. $CH_3OOC(CH_2)_4COOCH_3 \xrightarrow{Na}$ 2-hydroxy-1-cyclohexanone

19-29 The benzoin condensation is believed to proceed by the following mechanism:

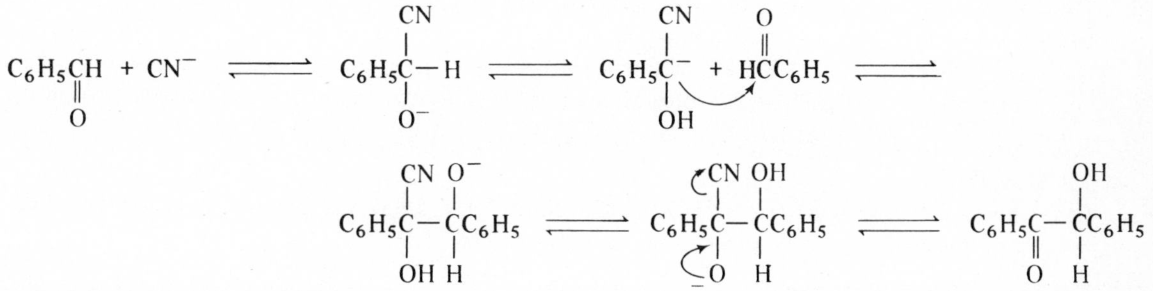

$$C_6H_5\overset{\overset{O}{||}}{C}H + CN^- \rightleftharpoons C_6H_5\underset{\underset{O^-}{|}}{\overset{\overset{CN}{|}}{C}}-H \rightleftharpoons C_6H_5\underset{\underset{OH}{|}}{\overset{\overset{CN}{|}}{C}}{}^- + H\overset{\overset{O}{||}}{C}C_6H_5 \rightleftharpoons$$

$$C_6H_5\underset{\underset{OH}{|}}{\overset{\overset{CN}{|}}{C}}-\underset{\underset{H}{|}}{\overset{\overset{O^-}{|}}{C}}C_6H_5 \rightleftharpoons C_6H_5\underset{\underset{O}{|}}{\overset{\overset{CN}{|}}{C}}-\underset{\underset{H}{|}}{\overset{\overset{OH}{|}}{C}}C_6H_5 \rightleftharpoons C_6H_5\underset{\underset{O}{||}}{\overset{}{C}}-\underset{\underset{H}{|}}{\overset{\overset{OH}{|}}{C}}C_6H_5$$

a. Discuss the role played by the cyanide ion.

b. If the step involving the *second* molecule of benzaldehyde is rate-limiting, what would the rate law be for the overall reaction?

c. Account for the following observation:

$$C_6H_5\underset{\underset{OH}{|}}{\overset{}{C}}H-\overset{\overset{O}{||}}{C}C_6H_5 + p\text{-}CH_3OC_6H_4CHO \xrightarrow{CN^-} p\text{-}CH_3OC_6H_4\underset{\underset{OH}{|}}{\overset{}{C}}H-\overset{\overset{O}{||}}{C}C_6H_5 + C_6H_5CHO$$

d. Provide a mechanism for the following reaction:

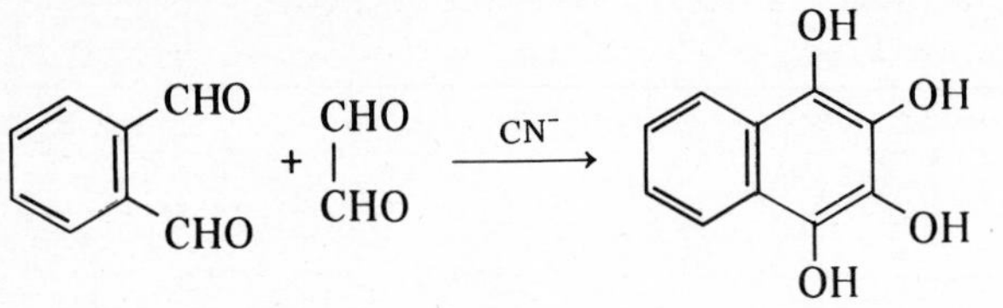

19-30 The Meerwein-Ponndorf-Verley reduction is believed to involve the following transition state:

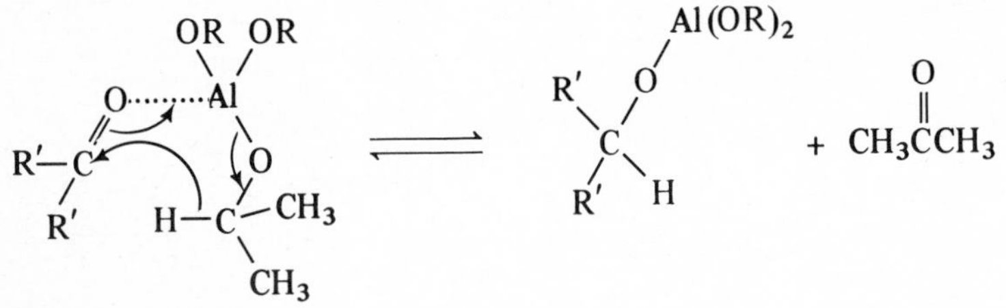

R = iso-Pr

The reverse reaction is called the Oppenauer oxidation.
a. What reaction conditions should be used in the Oppenauer oxidation?
b. Explain the mechanism of the following reaction:

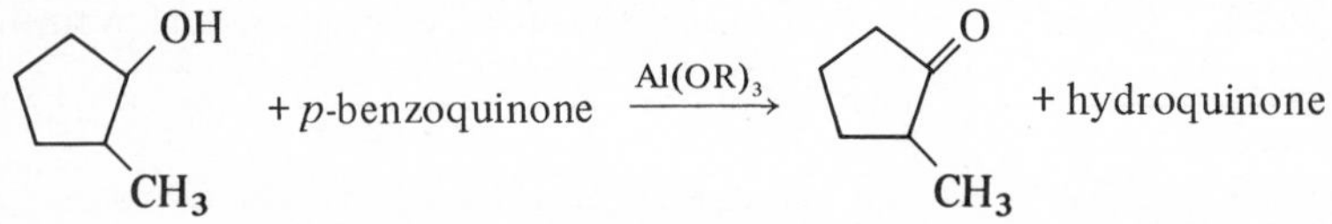

c. Explain the following:

(S)-3,3-Dimethyl-2-butanol + acetophenone $\xrightarrow{\text{Al(OR)}_3}$ (S)-α-phenylethanol + pinacolone

19-31 The nmr spectrum of compound A, $C_{10}H_{10}O$, has two singlets at $\delta = 7.1$ and 2.2 with a respective area ratio of 2:3. Upon boiling with aqueous acid A yields compound B (NE 60) and compound C which is readily oxidized to compound D, $C_6H_4O_2$. Suggest structures for A through D.

19-32 Compounds E, F, and G all contain nine carbon atoms and have molecular weights of 118, 132, and 146, respectively. Compounds F and G exhibit characteristic carbonyl stretching bands. From the nmr data (δ, multiplicity, peak integration units) given below, suggest structures for E through G. E: 7.2, s, 12.0; 2.9, t, 12.5; 2.1, p, 6.2. F: 7.2, s, 9.6; 3.5, s, 9.8. G: 7.9, m, 14.6; 3.2, s, 7.4.

19-33 Compound H, $C_8H_8O_2$, reacts with hot concentrated HBr to give compound I, $C_7H_6O_2$. When H is treated with an alcoholic solution of NaCN, compound J (MW = 352 ± 2) is produced. Upon heating with aqueous NaOH, J yields, after acidification, compound K (MW = 368 ± 2). From the spectral data (selected ir bands: μm, cm^{-1}; and nmr: δ, multiplicity, peak integration units) given below, suggest structures for H through K.

Compound H Ir: 3.53, 2833; 3.65, 2740; 5.88, 1701; 12.0, 833. Nmr: 9.7, s, 6.1; 7.7, d, 12.4; 6.9, d, 12.4; 3.8, s, 19.3.
Compound J Ir: 2.9, 3448; 6.02, 1661; 12.0, 833. Nmr: 7.9, d, 5.9; 7.3, d, 5.3; 6.8, d, 10.5; 5.8, s, 2.6; 4.4, broad singlet, 2.6; 3.7, two singlets or possibly a doublet, 15.5 (total for both peaks).

19-34 The nmr spectrum of compound L, $C_{10}H_{13}NO_3$,, shows three sets of peaks at $\delta = 7.0$ (approximate quartet), 3.8 (singlet), and 3.1 (singlet) with a respective peak area ratio of 1.33:1.0:2.0. L has characteristic ir bands at 5.82 μm (1718 cm^{-1}) and 12.2 μm (820 cm^{-1}). Heating L with aqueous NaOH followed by acidification and work-up led to the isolation or detection of the following reaction products: CO_2, dimethylamine, and compound M, $C_7H_8O_2$, which is soluble in NaOH and gives a positive $FeCl_3$ test. Suggest structures for L and M.

19-35 Suggest structures for compounds I through V based on the nmr data and other information given on p. 285.

Compound	MW	Nmr peaks (δ, multiplicity)				Respective area ratio	Other information
I	148 ± 1	7.9–7.2, m	2.8, t	1.7, sx	0.9, t	5:2:2:3	
II	180 ± 1	7.2, s	3.9, s			2:1	Easily obtained from $C_{14}H_{14}$
III	212 ± 1	8.0, m	7.3, m	5.3, s		1:4:1	Acid hydrolysis yields A (NE 122) + B (MW = 108)
IV	164 ± 1	7.3, s	4.3, t	2.9, t	2.0, s	5:2:2:3	Acid hydrolysis yields C (NE 60) + D (MW = 122)
V	179 ± 1	7.9, s	7.4 and 6.8, two d	4.0, q	2.1 and 1.4, s and t	1:2:2:2: 3:3	Acid hydrolysis yields C + F (MW = 137)

19-36 The ir and nmr spectra of compounds X (MW = 166), Y (MW = 122), and Z (MW = 158) are shown. Suggest structures for X, Y, and Z.

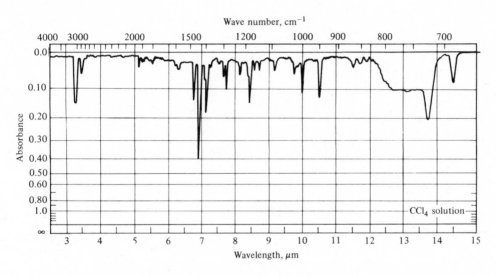

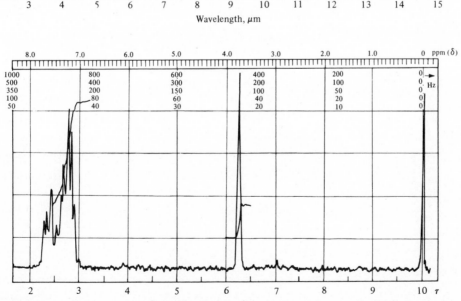

Figure P19-36. Ir and nmr spectra for Prob. 19-36, compound X. *(From R. A. Silva and R. Spenger, "The Quantitative Organic Analysis Spectra Kit," McGraw-Hill Book Company, New York, 1969. By permission of the publishers.)*

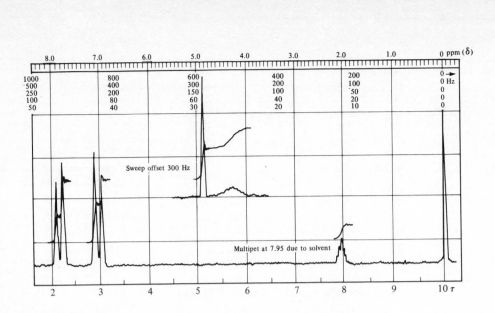

Sweep offset 300 Hz

Multipet at 7.95 due to solvent

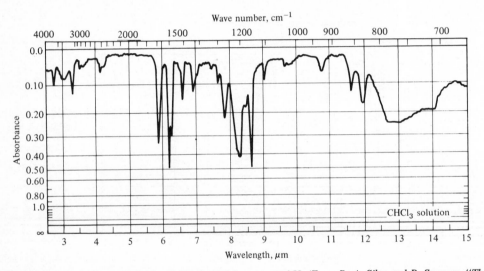

Wave number, cm⁻¹

CHCl₃ solution

Absorbance

Wavelength, μm

Figure P19-36. Ir and nmr spectra for Prob. 19-36, compound Y. *(From R. A. Silva and R. Spenger, "The Quantitative Organic Analysis Spectra Kit," McGraw-Hill Book Company, New York, 1969. By permission of the publishers.)*

286

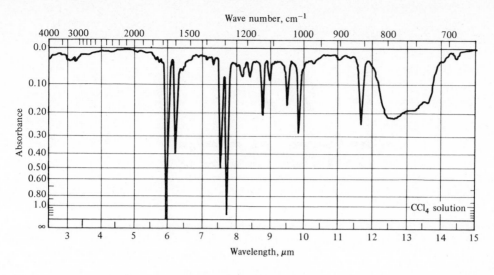

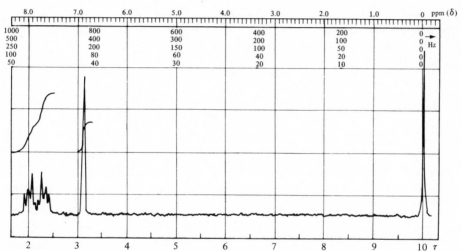

Figure P19-36. Ir and nmr spectra for Prob. 19-36, compound Z. *(From R. A. Silva and R. Spenger, "The Quantitative Organic Analysis Spectra Kit," McGraw-Hill Book Company, New York, 1969. By permission of the publishers.)*

20
amino acids

20-1 Nomenclature

The following structures are those of the important naturally occurring amino acids. The common abbreviation for each is given, but you should fill in the missing names from your text as a means of learning these.

H_2NCH_2COOH

Gly

$CH_3\underset{\underset{NH_2}{|}}{C}HCOOH$

Ala

$(CH_3)_2\underset{\underset{NH_2}{|}}{C}HCHCOOH$

Val

$(CH_3)_2CHCH_2\underset{\underset{NH_2}{|}}{C}HCOOH$

Leu

$CH_3CH_2\underset{\underset{NH_2}{|}}{\overset{\overset{CH_3}{|}}{C}}HCHCOOH$

Ileu

Phe — $CH_2\underset{\underset{NH_2}{|}}{C}HCOOH$

HO — $CH_2\underset{\underset{NH_2}{|}}{C}HCOOH$

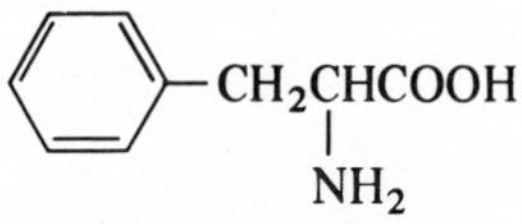

Tyr

$HOCH_2\underset{\underset{NH_2}{|}}{C}HCOOH$

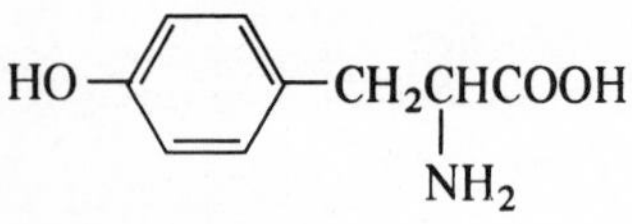

Ser

$CH_3\underset{\underset{NH_2}{|}}{\overset{\overset{OH}{|}}{C}}HCHCOOH$

Thr

$HSCH_2\underset{\underset{NH_2}{|}}{C}HCOOH$

CySH

$(-SCH_2\underset{\underset{NH_2}{|}}{C}HCOOH)_2$

CyS–SCy

$CH_3SCH_2CH_2\underset{\underset{NH_2}{|}}{C}HCOOH$

Met

Pro — COOH

Hypro — COOH

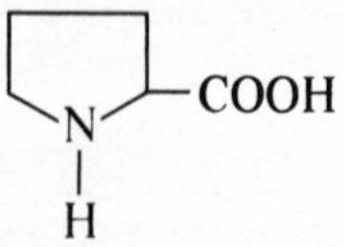

Try — $CH_2\underset{\underset{NH_2}{|}}{C}HCOOH$

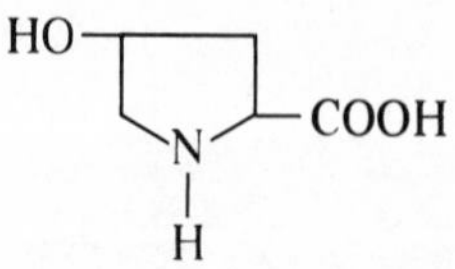

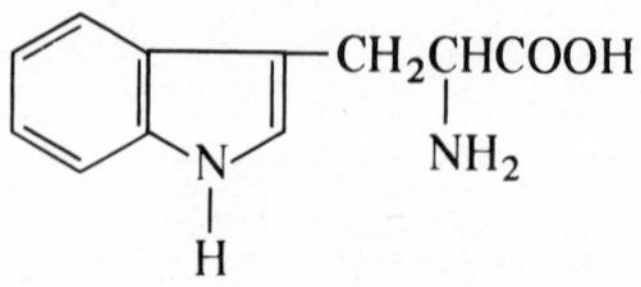

$HOOCCH_2\underset{\underset{NH_2}{|}}{C}HCOOH$

Asp

$HOOCCH_2\underset{\underset{NH_2}{|}}{C}HCOOH$

Glu

$H_2N\underset{\underset{NH}{||}}{C}NH(CH_2)_3\underset{\underset{NH_2}{|}}{C}HCOOH$

Arg

$$\underset{\underset{NH_2}{|}}{H_2N(CH_2)_4CHCOOH}$$

$$\underset{\underset{NH_2}{|}}{CH_2CHCOOH}$$ (imidazole ring structure)

Lys   His

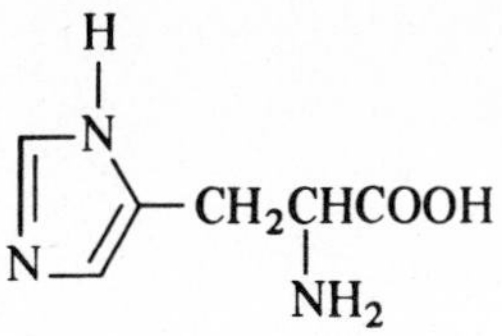

See Sec. 13-2*a* for assigning D or L configurations for the α-amino acids. Names for acid derivatives are not conventional because the trivial names do not have the *ic* ending.

$$H_2NCH_2COOEt \qquad H_2NCH_2CONH_2 \qquad CH_3CONHCH_2COOH \qquad Cl^-H_3\overset{+}{N}CH_2COOH$$

| Glycine ethyl ester | Glycylamide | Acetylglycine | Glycine hydrochloride |

The principal structural units in proteins are the α-amino acids. These are joined together through amide linkages, which are often referred to as *peptide linkages*. Compounds made of several amino acids are known as *polypeptides*. These are named from the acids from which they are derived, as follows:

$$\underset{\underset{CH_3}{|}}{H_2NCH_2CONHCHCOOH} \qquad \underset{(CH_3)_2CH \qquad \underset{SH}{\overset{|}{CH_2}}}{H_2NCH_2CONHCHCONHCHCON}\text{(ring)}COOH$$

Glycylalanine (Gly·Ala)

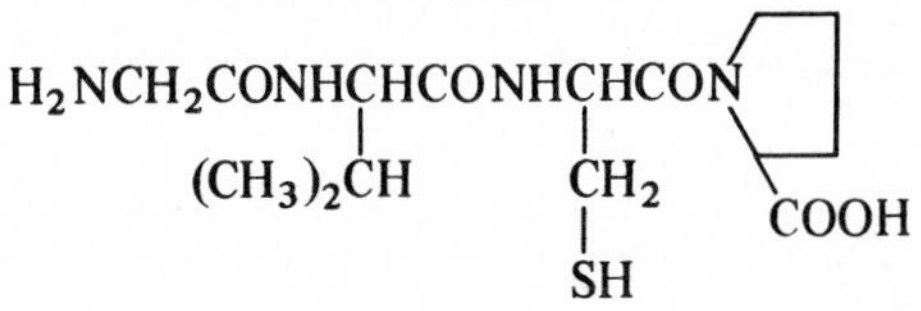

Glycylvalylcysteinylproline (Gly·Val·CySH·Pro)

20-2 Preparation of Amino Acids

a.   *Ammonolysis of Halo Acids*

$$\underset{\underset{Br}{|}}{CH_3CHCOOH} \quad \xrightarrow{NH_3 \text{ (excess)}} \quad \underset{\underset{NH_2}{|}}{CH_3CHCOOH}$$

b.   *Gabriel Synthesis* (See Sec. 10-3*e*.)

c.   *Strecker Synthesis* This reaction is a modification of the cyanohydrin synthesis (see Sec. 7-3*a*-1).

$$RCHO + NH_3 + HCN \rightleftharpoons \underset{\underset{NH_2}{|}}{RCHCN} + HOH \xrightarrow{\text{acid hydrolysis}} \underset{\underset{NH_2}{|}}{RCHCOOH}$$

d.   *Acetylaminomalonate Synthesis* The key compound in this preparation is ethyl acetylaminomalonate. It possesses an active hydrogen atom and can be used in typical malonic ester syntheses.

$$\underset{\underset{NHAc}{|}}{EtOOCCHCOOEt} \xrightarrow{NaOEt} \xrightarrow{EtBr} \underset{Et \quad NHAc}{EtOOCCCOOEt} \xrightarrow{H_3O^+, \Delta} \underset{\underset{NH_2}{|}}{EtCHCOOH}$$

The starting material can be made from malonic ester.

$$CH_2(COOEt)_2 \xrightarrow{HONO} ON-CH(COOEt)_2 \xrightarrow{H_2, Ni} H_2NCH(COOEt)_2 \xrightarrow{Ac_2O} AcNHCH(COOEt)_2$$

e. *Reductive Amination of Keto Acids*

$$\underset{\text{Pyruvic acid}}{CH_3COCOOH} \xrightarrow{NH_3,\ H_2,\ Pt} \underset{\underset{NH_2}{|}}{CH_3CHCOOH}$$

20-3 Preparation of Polypeptides

When two or more amino acids are joined together in head-to-tail fashion by amide linkages, such molecules are called polypeptides or simply peptides. Proteins are the naturally occurring polymers of amino acids joined by such linkages.

a. *Chloroacyl Chloride Method*

$$ClCH_2COCl + \underset{\underset{NH_2}{|}}{CH_3CHCOOCH_3} \longrightarrow \underset{\underset{CH_3}{|}}{ClCH_2CONHCHCOOCH_3} \xrightarrow{NH_3}$$

$$\underset{\underset{CH_3}{|}}{H_2NCH_2CONHCHCOOCH_3}$$

At this point, the ester can be hydrolyzed to give glycylalanine, or the ester can be treated with another chloroacyl chloride so as to give a tripeptide, etc.

b. *Carbobenzoxy Synthesis*

$$C_6H_5CH_2OH + ClCOCl\ \text{(phosgene)} \longrightarrow \underset{\text{Carbobenzoxy chloride}}{C_6H_5CH_2OCOCl + HCl}$$

In this method, the carbobenzoxy group is used to protect the amino group of an amino acid. It is removed on completion of the synthesis by hydrogenolysis (cleavage of the bond by hydrogen).

$$C_6H_5CH_2OCOCl + H_2NCH_2COONa \longrightarrow C_6H_5CH_2OCONHCH_2COOH + NaCl \xrightarrow{SOCl_2} \text{etc., } -COCl$$

$$\xrightarrow{RCH(NH_2)COONa} \underset{\underset{R}{|}}{C_6H_5CH_2OCONHCH_2CONHCHCOOH}$$

At this point, the product can be treated with $SOCl_2$ again and the peptide can be increased by an amino acid unit, or treatment with hydrogen and Pd will yield the following:

$$C_6H_5CH_3\ \text{(toluene)} + CO_2 + \underset{\underset{R}{|}}{H_2NCH_2CONHCHCOOH}$$

c. *Phthalyl Synthesis* In this method the amino group is covered by use of phthalyl chloride, the desired number of peptide bonds are formed by means of acid chloride intermediates, and the phthalyl group is removed by means of hydrazine.

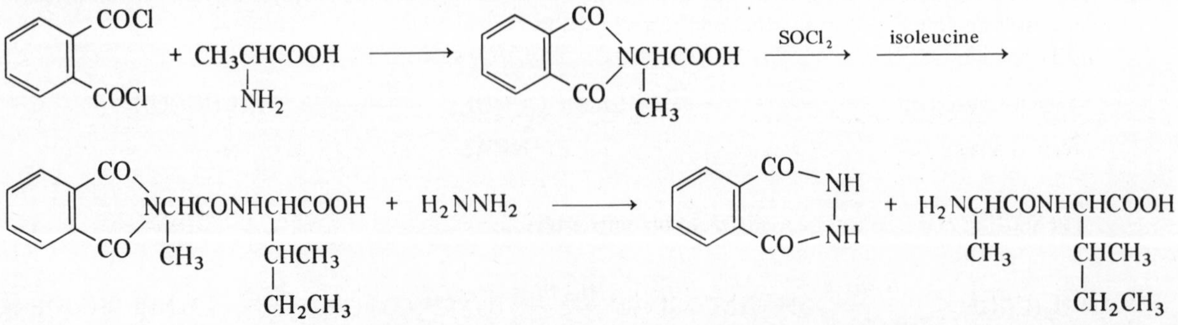

290

d. *Solid-phase Peptide Synthesis* In this procedure an amino acid with its amino group protected by a *t*-butoxy-carbonyl group, is attached via a benzyl ester linkage to an insoluble polymer such as polystyrene and then hydrolyzed to free the amino group. Dicyclohexylcarbodiimide (DCC) is added to activate the carboxyl group of a second amino-protected acid and to promote formation of a new amide linkage. The sequence of acid hydrolysis followed by addition of DCC and another amino-protected acid is repeated as many times as desired. The final step is removed of the benzyl-P group by treatment with HBr.

$$t\text{-BuOCONHCHCOO}^- + \text{ClCH}_2\text{C}_6\text{H}_4\text{—P} \longrightarrow t\text{-BuOCONHCHCOOCH}_2\text{C}_6\text{H}_4\text{P}$$

with R substituents below each CHCOO, and

P = polymer

$$\xrightarrow{\text{H}^+} \text{H}_2\text{NCHCOOCH}_2\text{C}_6\text{H}_4\text{—P}$$

(R below)

carbodiimide structure:

cyclohexyl—N=C=N—cyclohexyl (DCC)

$+ \text{HOOCCHNHOCO—}t\text{-Bu}$ (R′ above)

$$t\text{-BuOCONHCHCONHCHCOOCH}_2\text{C}_6\text{H}_4\text{—P} \xrightarrow{\text{H}^+}$$

(R′ and R below)

$$\text{DCC} + \text{HOOCCHNHOCO—}t\text{-Bu}$$

(R″ above)

(last two steps repeated many times)

$$\xrightarrow[\text{(final step)}]{\text{HBr}} \text{H}_2\text{N—CHCO} \sim\!\sim\!\sim \text{NHCHCONHCHCONHCHCOOH} + \text{BrCH}_2\text{C}_6\text{H}_4\text{P}$$

(R^L, R″, R′, R below)

Last amino acid
residue added

20-4 Reactions

a. *Acid-Base Equilibria* Amino acids are amphoteric compounds because of the presence of both the acidic carboxyl group and the basic amino group. The molecule exists as an inner salt or dipolar ion, II, which can be neutralized as follows:

$$\text{CH}_3\text{CHCOO}^- \underset{\text{OH}^-}{\overset{\text{H}^+}{\rightleftharpoons}} \text{CH}_3\text{CHCOO}^- \underset{\text{OH}^-}{\overset{\text{H}^+}{\rightleftharpoons}} \text{CH}_3\text{CHCOOH}$$

(NH$_2$ below, I) (NH$_3^+$ below, II) (NH$_3^+$ below, III)

If an electrical potential is applied to an aqueous solution of alanine, a migration of ions may occur. In an acidic solution, alanine exists as a cation, III, and will migrate to the cathode. In an alkaline solution, the anion, I, will migrate to the anode. There is also a pH at which the molecule exists as the simple inner salt, II, which shows no tendency to migrate toward either electrode. This pH is known as the *isoelectric point* of the amino acid.

b. *Acylation* The free amino group will react readily with either anhydrides or acyl halides. Base is usually added to expedite this reaction.

$$\overset{+}{\text{H}_3}\text{NCH}_2\text{COO}^- \xrightarrow{^-\text{OH}} \text{H}_2\text{NCH}_2\text{COO}^- \xrightarrow{\text{RCOCl}} \text{RCONHCH}_2\text{COOH}$$

c. *Esterification* Acyl halide formation or amide formation at the carboxyl group proceeds normally if the interfering lone pair of electrons on nitrogen is removed by acylation or salt formation.

1. $\overset{+}{\text{H}_3}\text{NCHRCOO}^- \xrightarrow[\text{CH}_3\text{COCl}]{\text{NaOH}} \text{CH}_3\text{CONHCHRCOOH} \xrightarrow{\text{SOCl}_2} \xrightarrow{\text{ROH}} \text{CH}_3\text{CONHCHRCOOR}$

2. $\overset{+}{\text{H}_3}\text{NCH}_2\text{COO}^- \xrightarrow{\text{HCl}} \overset{+}{\text{H}_3}\text{NCH}_2\text{COOH} \xrightarrow[\text{HCl}]{\text{CH}_3\text{OH}} \overset{+}{\text{H}_3}\text{NCH}_2\text{COOCH}_3$

d. *Cyclic Amides* These reactions are analogous to lactide and lactone formation of hydroxy acids.

1. α-Amino acids form diketopiperazines.

$$2\ H_2NCH_2COOH \xrightarrow{\Delta} \quad \underset{\text{CO—CH}_2}{\overset{\text{CH}_2\text{—CO}}{HN \qquad NH}} + 2\,HOH$$

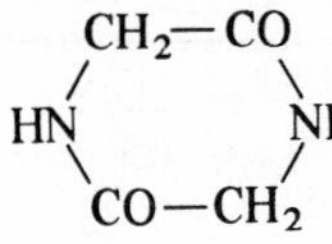

2. β-Amino acids lose ammonia readily to form unsaturated acids.

$$H_2NCH_2CH_2COOH \longrightarrow CH_2{=}CHCOOH + NH_3$$

3. γ-Amino acids and δ-amino acids form lactams; the former do so more readily.

$$H_2NCH_2CH_2CH_2COOH \longrightarrow \underset{\lfloor\text{—NH—}\rfloor}{CH_2CH_2CH_2C{=}O} + HOH$$

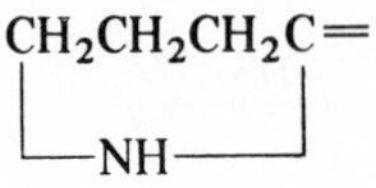

γ-Butyrolactam

20-5 Qualitative Tests

α-Amino acids give a characteristic blue-to-violet color when treated with ninhydrin (triketohydrindene hydrate),

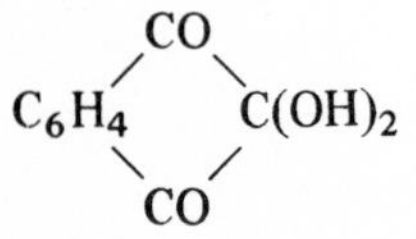

$$\underset{\text{CO}}{\overset{\text{CO}}{C_6H_4}}\;C(OH)_2$$

20-6 Structure Determination of Peptides and Proteins

To specify the gross structure of a protein one must know its molecular weight, the kinds and relative numbers of amino acids present, and their order of linkage in the protein chain. The following example illustrates how this is accomplished.

An unknown polypeptide was subjected to acid hydrolysis and 1 g of the resulting pure amino acids analyzed. The following percentages were obtained: Gly, 27.2; Phe, 39.9; Val, 14.2; and His, 18.8. The molecular weight was found to be about 1,400. From this information we can calculate an empirical formula and then a molecular formula using amino acids instead of atoms as one usually does for simple compounds.

$$\frac{27.2}{75\ (\text{MW Gly})} = 0.363 \div 0.121 = 3 \qquad \frac{14.2}{117\ (\text{MW Val})} = 0.121 \div 0.121 = 1$$

$$\frac{39.9}{165\ (\text{MW Phe})} = 0.242 \div 0.121 = 2 \qquad \frac{18.8}{155\ (\text{MW His})} = 0.121 \div 0.121 = 1$$

Empirical formula = $Gly_3Phe_2ValHis$.

However, from the molecular weights of the individual acid residues (remember that in forming an amide bond between two amino acids a molecule of water is lost) we see that this peptide would have a MW of only 719. So the molecular formula would be $(Gly_3Phe_2ValHis)_2$. This of course tells us nothing of the order in which the 14 amino acids occur in the chain.

To determine the end amino acid, use is made of 2,4-dinitrofluorobenzene (DNFB), O_2N—⟨ ⟩—F. This reagent reacts rapidly via an S_N2 reaction with the free amino group on the end of the peptide. NO_2

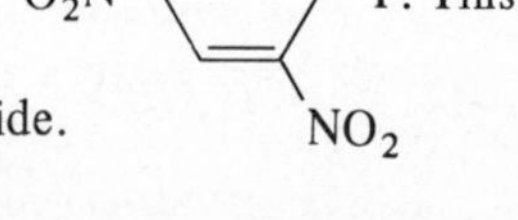

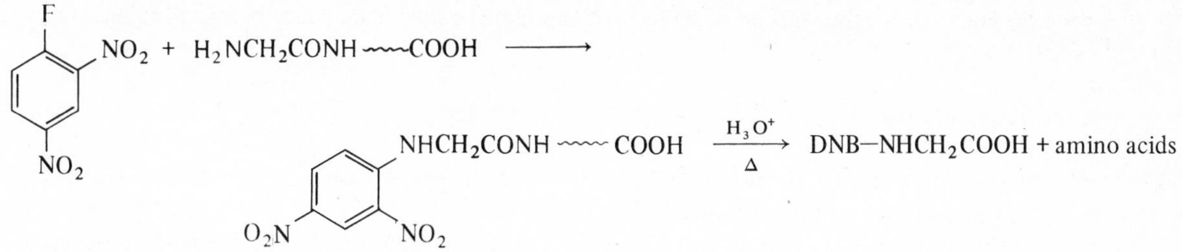

In the case of our unknown the end amino acid was glycine and hydrolysis gives DNB-Gly which can be identified by paper chromatography. The amino acid at the other end of the chain can often be determined by using the enzyme carboxypeptidase. This enzyme cleaves amide linkages adjacent to free α carboxyl groups. Such hydrolysis with our unknown followed by paper chromatography shows valine to be at other end of the peptide. So we may write Gly·X_{12}·Val, where X represents the 12 amino acids whose order we do not know. At this point a sample of the unknown is treated with DNFB and then the peptide is partially hydrolyzed with HCl. The resulting polypeptides are separated by paper chromatography and their structures determined by end-group analysis and partial hydrolysis. In our case the following structures were established:

1. DNB·Gly·Phe·Val   2. Gly·His·Gly
3. Phe·His·Gly   4. Phe·Phe·Gly
5. Phe·Gly·Gly·Val   6. Val·Gly·His
7. His·Gly·Phe·Phe

Now we must put these pieces together using jigsaw-puzzle tactics. Our end-group analysis tells us that peptides 1 and 5 constitute the tail and head, i.e.,

Gly·Phe·Val   Phe·Gly·Gly·Val

Since we know from our molecular formula that there are only two Val in the molecule, 6 must be a continuation of 1. Since there are only two His and one is preceded by Gly then 2 must follow 6. That is, Gly·Phe·Val·Gly·His·Gly·Phe·Gly·Gly·Val. We now have the position of two of the four Phe groups. Inspection of 4 and 7 shows that 5 must be a continuation of 7. This gives the order of 12 of the 14 and the Phe·His part of 3 must fit in to give Gly·Phe·Val·Gly·His·Gly·Phe·His·Gly·Phe·Phe·Gly·Gly·Val.

Problems

20-1 Draw structures for the following peptides:
   a. Lysylalanylcysteine   b. Glutamylleucylthreonine   c. Glycylalanylalanylamide
   d. Serylmethionylproline   e. Isoleucyllysyltyrosine   f. Lysylglycylglycylamide
20-2 Place the amino acids given under Nomenclature, Sec. 20-1, into the following categories:
   a. No asymmetric carbon atoms   b. Two unlike asymmetric carbon atoms
   c. Two like asymmetric carbon atoms   d. One asymmetric carbon atom
20-3 How many optical isomers are possible for each of the following:
   a. Alanylalanine   b. Alanylglycylproline   c. Hypro·Try·Lys·Gly
   d. Tetraglycylglycine   e. Threonylthreonine   f. Gly·Phe·Hypro·Ala
20-4 Predict the relative migratory tendencies of alanine, lysine, and aspartic acid in distilled water during electrophoresis.
20-5 Write balanced equations for the following:
   a. Valine + HCl   b. Leucine + NaOH
   c. Glycine hydrochloride + $SOCl_2$ d. Alanine + $NaNO_2$ + HCl
   e. Ammonium cyanide + ethanal   f. Malonic ester + nitrous acid
   g. Potassium phthalimide + ethyl chloroacetate h. Acetylphenylalanyl chloride + CH_3NH_2
   i. Ethyl α-acetylaminomalonate + NaOEt j. Alanine + heat
   k. γ-Aminobutyric acid + heat   l. Leu·Met·Asp + boiling HCl
20-6 Amino acids react much more slowly than simple amines with acetic anhydride or acetyl chloride when the two are mixed. Why? How can the reaction be speeded up?

20-7 Write the equation for the reaction of alanine with each of the following. Where necessary assume the presence of NaOH.

    a. Ketene (2 mol)   *b.* $CH_3OH + HCl$ *c.* $CH_3N=C=S$

    d. $(CH_3O)_2SO_2$ (excess)   *e.* Carbobenzoxy chloride   *f.* $CH_3N=C=O$

    g. $(CH_3)_3COCOCl$ *h.* Phthalyl chloride   *i.* Heat

20-8 In alkaline solution glycine, $H_2NCH_2COO^-$, has two basic groups. What product would form on the addition of 1 equiv of protons? 2 equiv? 3 equiv?

20-9 In acid solution glycine, $H_3\overset{+}{N}CH_2COOH$, has protons of several degrees of acidity. Show the product of reaction with 1 equiv of base, 2 equiv, and 3 equiv, one of which is a very strong base.

20-10 The name for threonine was chosen because of its stereo relationship to threose. Most naturally occurring amino acids belong to the L series. Make a Fischer-type projection of L-threonine and indicate whether the asymmetric centers are R or S.

20-11 Make a Fischer projection for naturally occurring lysine. Indicate whether the asymmetric center is R or S.

20-12 It has been estimated that 1 mg of diphtheria toxin, *Clostridium botulinum*, will kill 30 million mice. The molecular weight of this protein is 900,000. How many molecules of this protein are needed to kill one mouse?

20-13 Why does alkaline hydrolysis of proteins cause racemization of the amino acids while acid hydrolysis does not?

20-14 The chloroacetyl chloride method of peptide synthesis depends on the fact that, in the hydrolysis to remove the ester end group, the amide linkages are not hydrolyzed. Why would you expect the amide linkages to be more resistant to hydrolysis than the ester linkage?

20-15 How could you distinguish chemically between the following pairs of compounds?

    a. Aspartic acid and malic acid   *b.* Phenylalanine and acetylphenylalanine

    c. Serine and threonine   *d.* Glycine ethyl ester and valine

20-16 Given carbobenzoxy chloride, phthalyl chloride, methanol, ethanol, 1-propanol, dicyclohexylcarbodiimide, and any necessary inorganic reagents, show how you would make each of the following:

    a. Valylalanine, using phthalyl chloride

    b. Leucylglycylvaline, using carbobenzoxy chloride

    c. Alanylglycine, using the haloacyl halide method

    d. Gly·Ala·Leu·Val, using the solid-phase peptide method

20-17 Draw the electronic structure for the product of the reaction between arginine and 2 equiv of HCl. The proper position for the proton on the terminal group is that which leaves the greatest possibility for distribution of the positive charge by resonance. Draw these resonance structures.

20-18 Given methanol, ethanol, and malonic ester, show how you would make the following conversions:

    a. Ethyl iodoacetate to glycine via Gabriel's prep   *b.* Ethyl bromide to *dl*-2-aminobutyric acid

    c. L-Alanine to L-lactic acid   *d.* Benzoic to hippuric acid

    e. Glutaric anhydride to γ-aminobutyric acid   *f.* Isobutyl bromide to *dl*-leucine

    g. $C_6H_5CH_2Cl$ to *dl*-aspartic acid   *h.* 1-Propanol to *dl*-α-aminobutyric acid

    i. Ethyl acetate to *dl*-aspartic acid   *j.* Acrolein to *dl*-methionine

    j. Acroelin to *dl*-methionine

20-19 Formulate the following synthesis of D L-threonine.

$$\text{\textit{trans}-Crotonic acid} \xrightarrow[Br_2]{CH_3OH} \underset{(A)}{C_5H_9O_3Br} \xrightarrow{SOCl_2} (B) \xrightarrow{2Et_2NH} C_9H_{18}O_2NBr \xrightarrow{NH_3} \xrightarrow{\text{hydrolysis}} \underset{\text{chiefly D L-threonine}}{}$$

If compound A is treated with ammonia, one obtains the diastereoisomeric D L-allothreonine instead of threonine. How can this be explained?

20-20 The configuration of naturally occurring serine can be related to L-lactic acid as follows. Formulate with three-dimensional drawings.

$$\text{Serine} \xrightarrow[HCl]{CH_3OH} \xrightarrow{PCl_5} \xrightarrow{H_3O^+} \xrightarrow[H_2]{Pt} \xrightarrow{HONO} \text{L-lactic acid}$$

If we assume that serine has the same configuration as L-lactic acid, what must we assume about the five reactions used in this conversion?

20-21 A naturally occurring amino acid formed a monoacetyl derivative with excess acetic anhydride, but treatment with nitrous acid yielded no nitrogen. Suggest a structure for the compound.

20-22 Compound A, an amino acid, was optically active. When 0.445 g of A was treated with nitrous acid, 112 ml of nitrogen gas (STP) was evolved. Suggest a structure for A.

20-23 Compound A, an amino acid, was optically active. It has the empirical formula $C_5H_{11}O_2N$. It formed a salt with nitrous acid, and the original compound could be regenerated by neutralization with dil NaOH. Suggest a structure for A.

20-24 A tripeptide gave glycine and alanine on hydrolysis. When the tripeptide was first treated with nitrous acid and then hydrolyzed, it yielded glycolic acid in addition to glycine and alanine. Suggest two structures for the peptide.

20-25 The protein hemoglobin was found to contain 0.34% Fe. Assuming one atom of iron per molecule, what is the molecular weight of hemoglobin?

20-26 The enzyme cytochrome C contains 0.43% Fe and 1.48% S. What is the lowest molecular weight possible? How many atoms of sulfur would this molecule contain?

20-27 Hydrolysis of a polypeptide was found to yield equal amounts of alanine, leucine, serine, and phenylalanine. When it was treated with DNFB and then subjected to partial hydrolysis with dil HCl, the amino acids alanine, leucine, and serine could be detected. In addition there were two dipeptides that contained (serine + alanine) and (serine + leucine), and a dipeptide containing (DNB-phenylalanine + leucine). Formulate a structure for the unknown peptide, assuming a MW of 436.

20-28 Often carboxyl groups in proteins are not free but are found as amides. If the molecular weight of the protein is known, then from NH_3 obtained on hydrolysis one can calculate the number of terminal amide groups. From the data given, calculate the number of amide groups in each of the following:
   a. Oxytocin, MW = 994. Hydrolysis of 1 g yields 0.051 g of NH_3.
   b. β-Lactoglobulin, MW = 42,000. Hydrolysis of 0.3 g yields 0.00393 g NH_3.

20-29 Oxytocin, the hormone from the posterior pituitary gland, has a MW of 994 and is found by hydrolysis to contain the following amino acids in equivalent amounts: CySSCy, Ileu, Tyr, Pro, Leu, Gly, Glu, and Asp. As shown in Prob. 20-28*a*, certain of the carboxyl groups are present in the peptide as amides. The reagent performic acid cleaves cystine, oxidizing the sulfur atoms to sulfonic acids (CySSCy $\longrightarrow$ $2CySO_3H$). When oxytocin is treated with performic acid, a single peptide containing the same amino acids is isolated. Partial hydrolysis of the oxidized material yields the following peptides:

Asp·CySO$_3$H   Ileu·Glu   CySO$_3$H·Pro·Leu

CySO$_3$H·Tyr   Leu·Gly   Tyr·Ileu·Glu

Enzymatic hydrolysis of oxidized oxytocin yields glycine amide and two tetrapeptides: CySO$_3$·(Glu, Ileu, Tyr) and Asp·(CySO$_3$H, Leu, Pro). In the two tetrapeptides end-group analysis with DNFB indicates that CySO$_3$H and Asp are end residues; however, the position of the other residues as indicated by commas instead of dots shows that their sequence is not known. From the above information construct first the oxidized oxytocin molecule and then oxytocin. Show which carboxyls in oxytocin are present as $-CONH_2$.

20-30 The structure of the ACTH hormone β-corticotropin from the hog pituitary has been determined. This protein has a MW of about 4,500. This plus amino acid analysis indicates the following empirical formula of 39 amino acid units: Ala$_3$Arg$_3$Asp$_2$Glu$_5$HisLeu$_2$Lys$_4$MetPhe$_3$Pro$_4$Ser$_2$TryTyr$_2$Val$_3$Gly$_3$. Hydrolysis of a 10-mg sample yields 0.0374 mg NH_3. Treatment with DNFB indicates a terminal Ser. Carboxypeptidase hydrolysis yielded a tripeptide whose structure was shown to be Leu·Glu·Phe. Enzymatic hydrolysis of β-corticotropin with the enzyme trypsin gave four peptides whose structures were shown to be

   1. Ser·Tyr·Ser·Met·Glu·His·Phe·Arg
   2. Try·Gly·Lys·Pro·Val·Gly·Lys
   3. Val·Tyr·Pro·Asp·Gly·Ala·Glu·Asp·Glu·Leu·Ala·Glu·Ala·Phe·Pro·Leu·Glu·Phe
   4. Lys·Arg·Arg·Pro·Val·Lys

Hydrolysis with the enzyme chymotrypsin yielded six peptides:

   5. Ser·Tyr
   6. Ser·Met·Glu·His·Phe
   7. Arg·Try
   8. Gly·Lys·Pro·Val·Gly·Lys
   9. Glu·Phe
   10. Lys·Arg·Arg·Pro·Val·Lys·Val·Tyr·Pro·Asp·Gly·Ala·Glu·Asp·Glu·Leu·Ala·Glu·Ala·Phe·Pro·Leu.

The structures of these 10 peptides were determined by end-group analysis and by partially hydrolyzing each of these and then fitting the pieces together. Ammonia was shown to associate with peptides 3 and 10. With what residues in the protein might the ammonia be associated? Construct a formula for β-corticotropin.

20-31 The first protein structure to be determined was that of beef insulin. For this the English chemist Sanger was awarded the Nobel Prize. From the following information work out the structure of insulin. Performic acid cleaves insulin into two peptide chains termed A and B. Amino acid analysis of A yields empirical formula Ala, Gly, Ileu, Val_2, Leu_2, Asp_2, Ser_2, Tyr_2, Glu_4, $CySO_3H_4$. B has the empirical formula Pro, Arg, Lys, Asp, Ser, Thr, Ala_2, His_2, Tyr_2, $CySO_3H_2$, Gly_3, Val_3, Phe_3, Glu_3, Leu_4. Treatment of B with DNFB and hydrolysis gave DNB-Phe·Val while Ala was isolated with the carboxypeptidase. Partial acid hydrolysis gave the following peptides:

Phe·Val·Asp·Glu·His   Ser·His·Leu·Val·Glu
Tyr·Leu·Val·$CySO_3$H   Thr·Pro·Lys·Ala
Leu·Val·CySO H·Gly   Gly·Glu·Arg·Gly

Hydrolysis with pepsin gave

Leu·Val·$CySO_3$H·Gly·Glu·Arg·Gly·Phe
Tyr·Thr·Pro·Lys·Ala
Val·Glu·Ala·Leu
Phe·Val·Asp·Glu·His·Leu·$CySO_3$H·Gly·Ser·His·Leu

Hydrolysis with the enzyme trypsin gave Gly·Phe·Phe·Tyr·Thr·Pro·Lys. Construct the sequence for the B chain. The sequence for the oxidized A chain is Gly·Ileu·Val·Glu·Glu·$CySO_3$H·$CySO_3$H·Ala·Ser·Val·$CySO_3$H·Ser·Leu·Tyr·Glu·Leu·Glu·Asp·Tyr·$CySO_3$H·Asp. Chain A has an internal 20-membered ring involving a disulfide linkage of cystine, Chain A is attached to chain B through two such disulfide linkages. In addition there are six amide groups on residues 5, 15, 18, and 21 of chain A and 3 and 4 of chain B.

21
carbohydrates

21-1 Nomenclature

Most simple sugars are hydroxy aldehydes, $HOCH_2(CHOH)_n CHO$, or hydroxy ketones, $HOCH_2 CO(CHOH)_n CH_2 OH$, known as aldoses and ketoses, respectively. Naturally occurring sugars are usually pentoses (five-carbon sugars) or hexoses (six-carbon sugars) related stereochemically to D-glyceraldehyde by virtue of the configuration of the highest numbered asymmetric carbon atom (see Sec. 13-2a). A few have the L-configuration.

Monosaccharide: A simple sugar. Hydrolytic processes cause no further division of the molecule.

Disaccharide: A molecule composed of two monosaccharide units.

Polysaccharide: A molecule composed of several monosaccharide units.

Epimers: Aldoses which are identical except for the configuration of the C-2 carbon atom (the following aldoses are shown successively as epimeric pairs).

Aldopentoses:

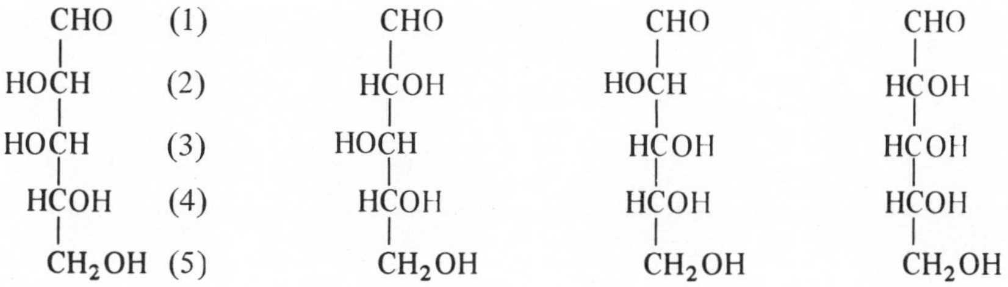

$$
\begin{array}{cc}
CHO & (1) \\
| & \\
HOCH & (2) \\
| & \\
HOCH & (3) \\
| & \\
HCOH & (4) \\
| & \\
CH_2OH & (5)
\end{array}
\qquad
\begin{array}{c}
CHO \\
| \\
HCOH \\
| \\
HOCH \\
| \\
HCOH \\
| \\
CH_2OH
\end{array}
\qquad
\begin{array}{c}
CHO \\
| \\
HOCH \\
| \\
HCOH \\
| \\
HCOH \\
| \\
CH_2OH
\end{array}
\qquad
\begin{array}{c}
CHO \\
| \\
HCOH \\
| \\
HCOH \\
| \\
HCOH \\
| \\
CH_2OH
\end{array}
$$

    D-Lyxose   D-Xylose   D-Arabinose   D-Ribose

D-*Aldohexoses* (the horizontal lines show positions of hydroxyl groups):

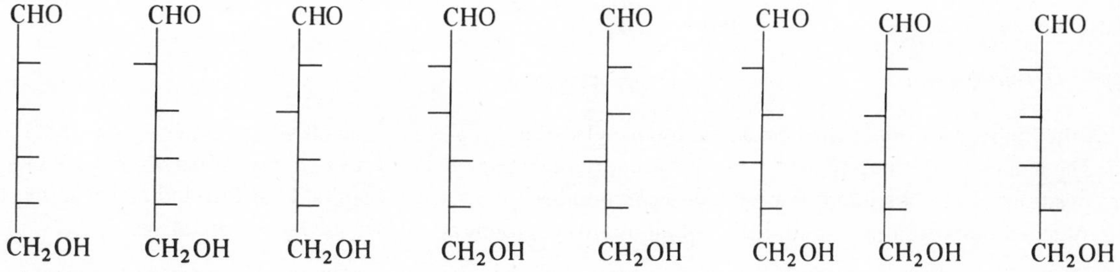

CHO	CHO	CHO	CHO	CHO	CHO	CHO	CHO
CH₂OH	CH₂OH	CH₂OH	CH₂OH	CH₂OH	CH₂OH	CH₂OH	CH₂OH
Allose	Altrose	Glucose	Mannose	Gulose	Idose	Galactose	Talose

Ketohexoses: The most common ketose is fructose, a 2-ketohexose whose 3, 4, and 5 carbon atoms have the same configuration as the same atoms in glucose.

21-2 Structure and Derivatives

Simple sugars exist in a cyclic hemiacetal (or hemiketal in the case of ketoses) form having five or six atoms in the ring. The carbonyl carbon now becomes asymmetric, giving rise to two compounds called α or β anomers which differ only in the configuration at the number 1 carbon (or number 2 in the case of ketoses). In Fischer projection formulas shown below, the OH group at C-1 is on the right in the α anomer and on the left in the β anomer. If the Haworth formulas are used, the α anomer always has the OH group at C-1 pointing down. Note that whether Haworth or Fischer formulas are used, the α anomer has the S configuration and the β anomer has the R configuration at the anomeric carbon. The nomenclature of these cyclic structures is dependent on the ring size. If the ring is five-membered, the sugar is known as a *furanose*, and if six-membered, as a *pyranose*.

Fischer formula

Haworth formula

α-D-Glucopyranose

Pyran

β-D-Glucofuranose

Furan

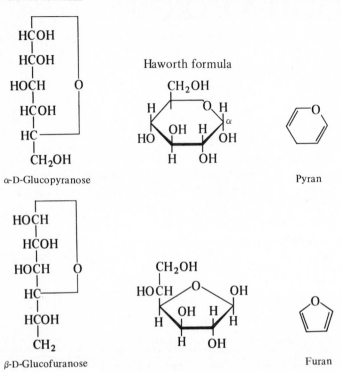

When drawn in their true chair conformations the most stable conformation is that which has more of the bulky substituents (especially the CH_2OH group) in equatorial positions.

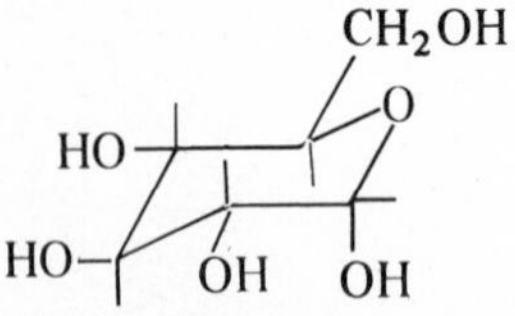

α-D-Glucopyranose

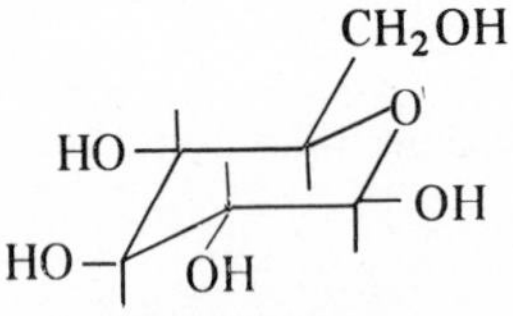

β-D-Glucopyranose

If the hydrogen atom of the hemiacetal hydroxyl group is replaced by an alkyl (or related) group, a *glycoside* is formed. The name of the group is given first, followed by the name of the sugar with the ending *ose* replaced by *oside*. In other positions, each substituent is given a position number followed by a capital O or C to indicate whether the group is attached to oxygen or carbon. Esters of all hydroxyl groups are named as O-acyl derivatives.

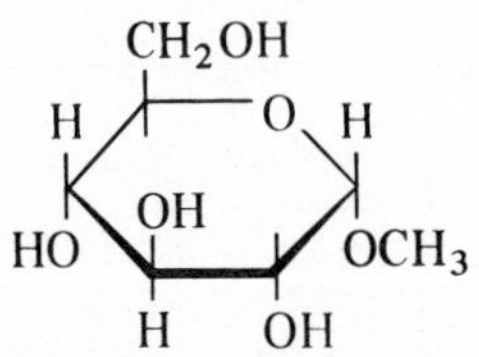

Methyl α-D-glucopyranoside

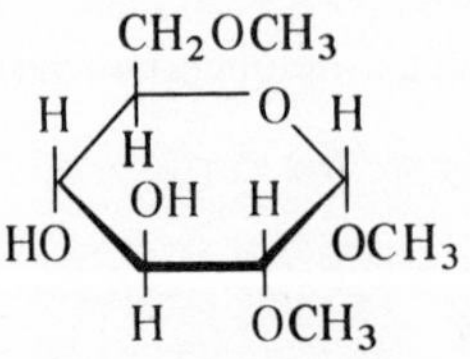

Methyl 2,6-di-O-methyl-α-D-glucopyranoside

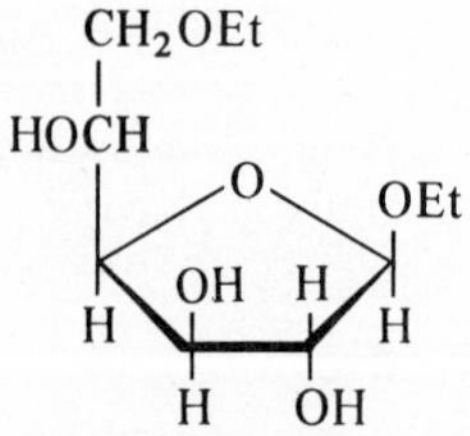

Ethyl 6-O-ethyl-β-D-glucofuranoside

1,2,3,4-Tetra-O-acetyl-α-D-allopyranose

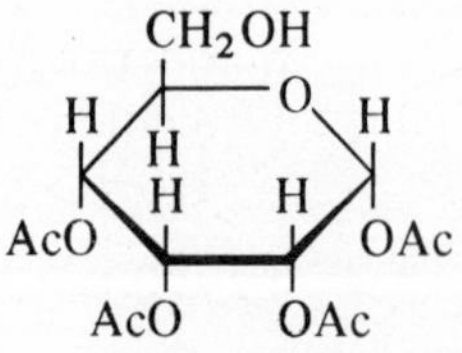

If the number 1 carbon atom of aldoses is oxidized to a carboxyl group, an *aldonic acid* is formed. The *ose* ending of the sugar is changed to *onic*, and the word *acid* is added. In the lactone form, the *ose* ending becomes *ono* and this is followed by -γ-lactone or -δ-lactone, depending on the ring size. If the end-carbon atoms of aldoses are both converted to carboxyl groups, an *aldaric acid* is formed. The *ose* ending of the aldose is changed to *aric* and the word *acid* is added. If the number 1 carbon atom of aldoses is converted to a CH_2OH group, an *alditol* is formed. The *ose* ending of the sugar is changed to *itol*.

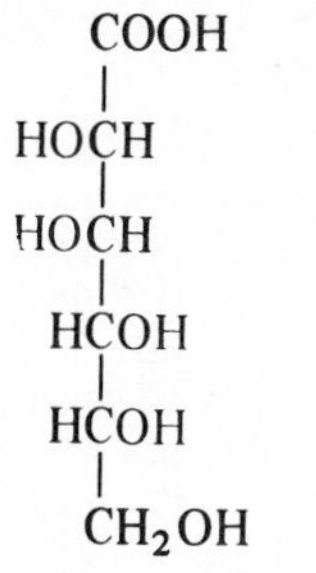

D-Mannaric acid 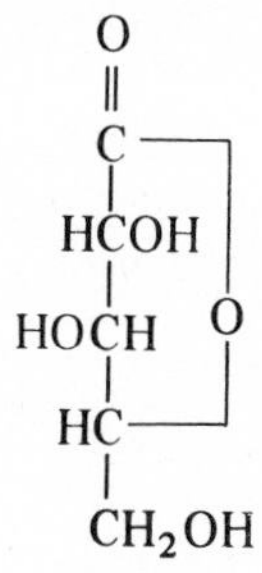   D-Xylono-γ-lactone 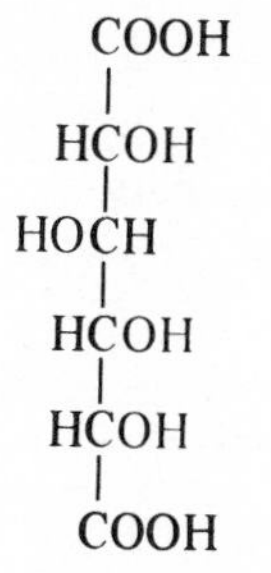   D-Glucaric acid 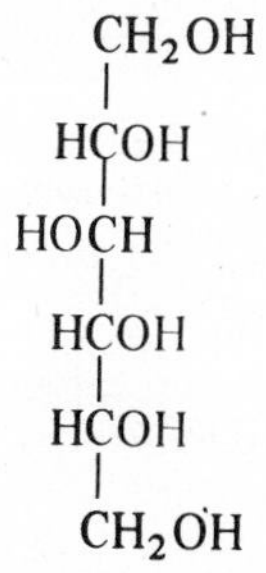   D-Glucitol (sorbitol)

In naming the cyclic sugar group minus the hemiacetal hydrogen atom, the ending *ose* is changed to *osyl*.

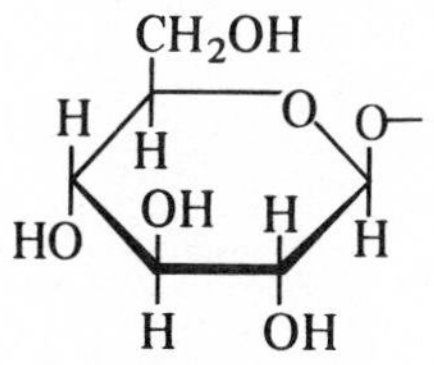

β-D-Glucopyranosyl group

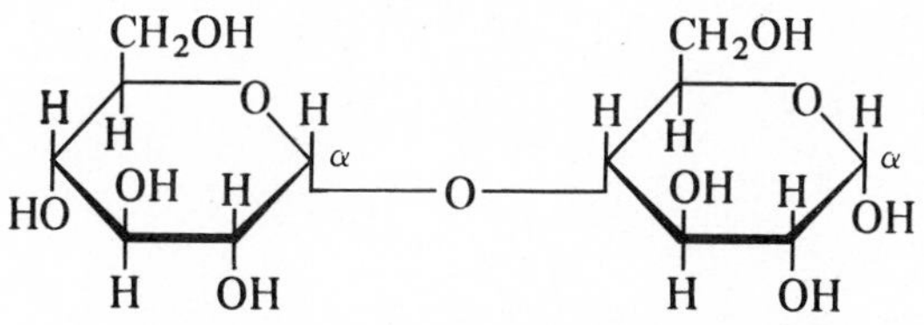

4-O-α-D-Glucopyranosyl-α-D-glucopyranose (maltose)

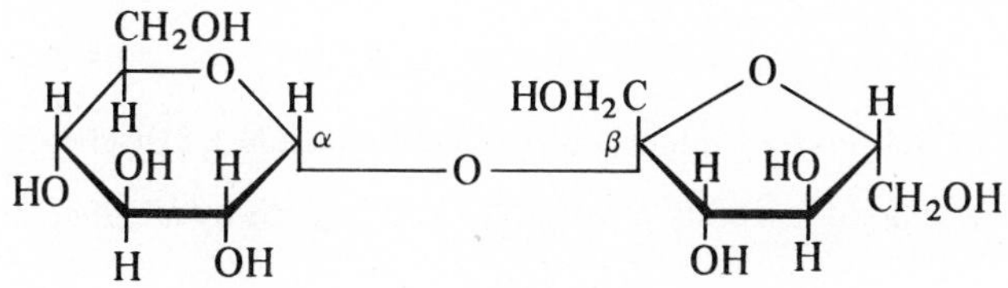

β-D-Fructofuranosyl-α-D-glucopyranoside (sucrose)

In addition to maltose and sucrose (shown above), two other disaccharides are very common.

   Lactose: 4-O-β-D-Galactopyranosyl-β-D-glucopyranose.

   Cellobiose (the repeating unit of cellulose): 4-O-β-D-Glucopyranosyl-β-D-glucopyranose.

21-3.   Reactions – The Carbonyl Group

a.   *Reduction*

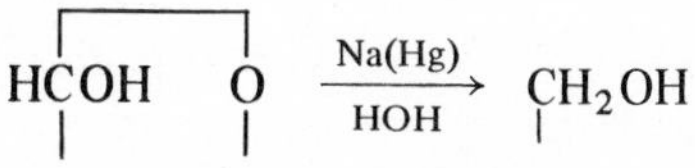

glucose $\xrightarrow[\text{HOH}]{\text{Na(Hg)}}$ sorbitol

b.   *Oxidation to Aldonic Acids*

D-Glucose $\xrightarrow{Br_2, CaCO_3}$ $\xrightarrow{HCl}$ D-glucono-δ-lactone

c.   *Oxidation to Aldaric Acids*

D-Glucose $\xrightarrow{HNO_3}$ D-glucaric acid

d. *Cyanohydrin Synthesis; Lengthening the Carbon Chain (Kiliani-Fischer Synthesis)*

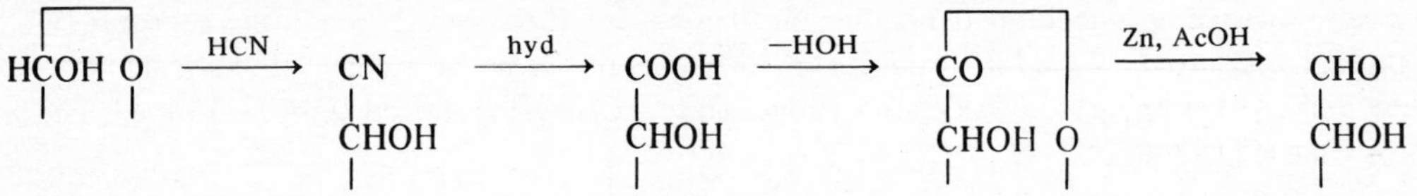

$$\begin{array}{c} \overline{\text{HCOH}} \ \overline{\text{O}} \\ | \end{array} \xrightarrow{\text{HCN}} \begin{array}{c} \text{CN} \\ | \\ \text{CHOH} \\ | \end{array} \xrightarrow{\text{hyd}} \begin{array}{c} \text{COOH} \\ | \\ \text{CHOH} \\ | \end{array} \xrightarrow{-\text{HOH}} \begin{array}{c} \text{CO} \\ | \\ \overline{\text{CHOH}} \ \overline{\text{O}} \\ | \end{array} \xrightarrow{\text{Zn, AcOH}} \begin{array}{c} \text{CHO} \\ | \\ \text{CHOH} \\ | \end{array}$$

Aldose, C_n Aldose, C_{n+1}

e. *Ruff or Wohl Degradations; Shortening the Carbon Chain*

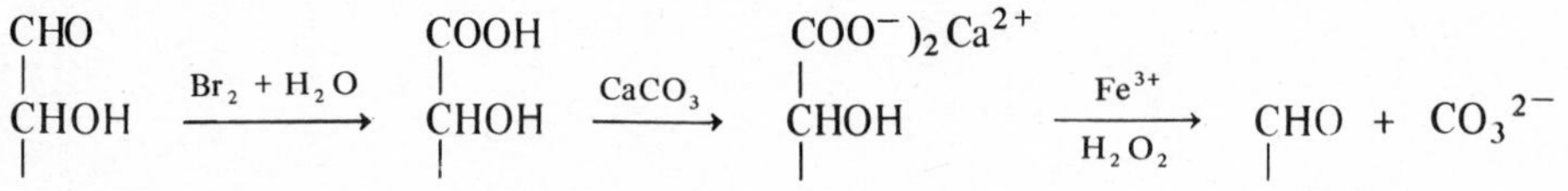

$$\begin{array}{c} \text{CHO} \\ | \\ \text{CHOH} \\ | \end{array} \xrightarrow{\text{Br}_2 + \text{H}_2\text{O}} \begin{array}{c} \text{COOH} \\ | \\ \text{CHOH} \\ | \end{array} \xrightarrow{\text{CaCO}_3} \begin{array}{c} \text{COO}^-)_2\text{Ca}^{2+} \\ | \\ \text{CHOH} \\ | \end{array} \xrightarrow[\text{H}_2\text{O}_2]{\text{Fe}^{3+}} \begin{array}{c} \text{CHO} \\ | \end{array} + \text{CO}_3^{2-}$$

Aldose, C_n Aldonic acid, C_n Aldose, C_{n-1}

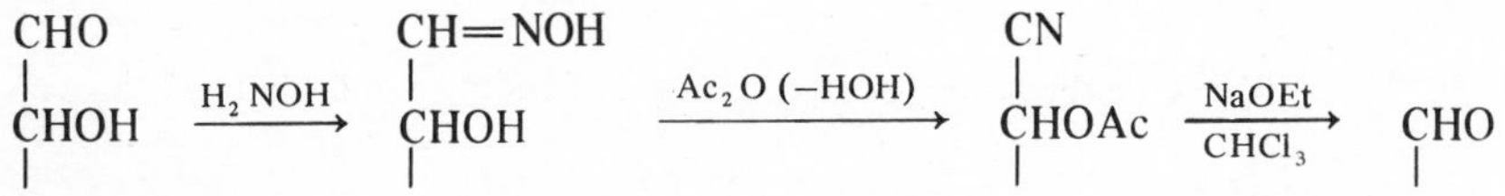

$$\begin{array}{c} \text{CHO} \\ | \\ \text{CHOH} \\ | \end{array} \xrightarrow{\text{H}_2\text{NOH}} \begin{array}{c} \text{CH}{=}\text{NOH} \\ | \\ \text{CHOH} \\ | \end{array} \xrightarrow{\text{Ac}_2\text{O} (-\text{HOH})} \begin{array}{c} \text{CN} \\ | \\ \text{CHOAc} \\ | \end{array} \xrightarrow[\text{CHCl}_3]{\text{NaOEt}} \begin{array}{c} \text{CHO} \\ | \end{array} \quad \text{(Wohl degradation)}$$

Aldose, C_n Aldose, C_{n-1}

f. *Glycosides*

α-D-Glucopyranose $\xrightarrow{\text{MeOH, H}^+}$ methyl α-D-glucopyranoside

21-4 Reactions – The Hydroxyl Groups

a. *Ether Formation*

α-D-Glucose + 5(CH₃)₂SO₄ + 5NaOH → methyl 2,3,4,6-tetra-O-methyl-α-D-glucopyranoside + 5HOH + 5CH₃SO₄Na

(Note: The methyl group associated with the acetal bond is readily replaced by acid but not by alkaline hydrolysis.)

b. *Esterification*

D-Glucopyranose $\xrightarrow{\text{Ac}_2\text{O}}$ 1,2,3,4,5-penta-O-acetyl-D-glucose

c. *Reduction*

An aldonic acid $\xrightarrow{\text{HI, P, }\Delta}$ caproic acid

d. *Isopropylidene Derivatives (Acetonides)* If hydroxyl groups on adjacent carbon atoms are cis, they will react with acetone to form cyclic ketals.

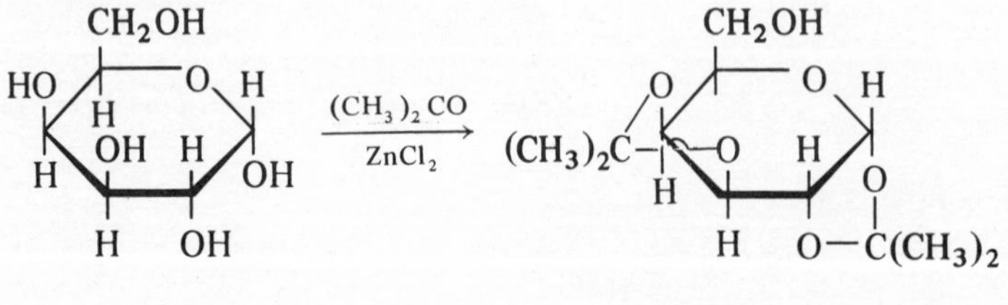

α-D-Galactopyranose   1,2,3,4-Di-O-isopropylidene-
                                                         D-galactopyranose

e. *Cleavage of Glycols* A useful tool in the determination of sugar structure is the reaction of glycols with periodic acid. The carbon-carbon bond between two CHOH groups is cleaved with the formation of aldehydes.

$$CH_3CH-CHCH_3 + HIO_4 \longrightarrow 2\,CH_3CHO + HIO_3 + HOH$$
$$||$$
$$OHOH$$

If a CHOH group is joined to two other such groups, it is converted to formic acid.

$$HOCH_2CHCH_2OH + 2\,HIO_4 \longrightarrow 2\,CH_2O + HCOOH + 2\,HIO_3 + HOH$$
$$|$$
$$OH$$

Glyoxylic   D-Glyceric acid
acid

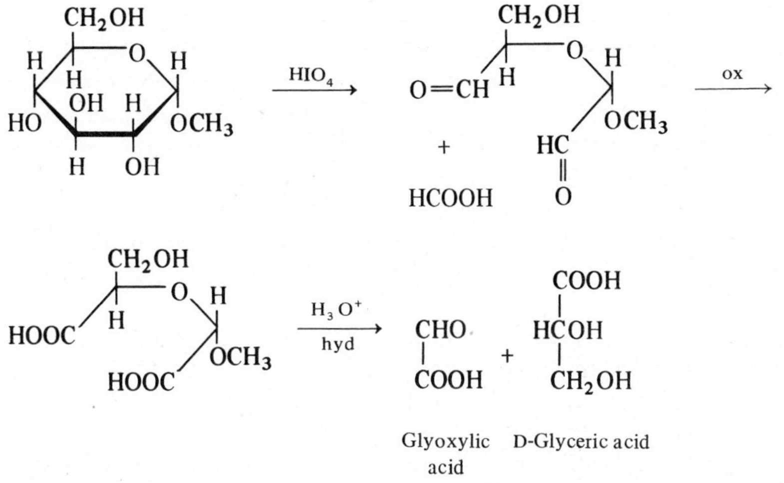

21-5 Qualitative Tests

a. Both aldoses and ketoses are oxidized by Fehling's solution (alk cupric tartrate) or Tollens' reagent (ammoniacal $AgNO_3$). Glycosides do not give positive reactions with these reagents because the carbonyl group must be free or in the form of a hemiacetal. Thus, sucrose is not a reducing sugar but maltose is.

    Sugars do not give a Schiff test and can thus be differentiated from common aliphatic aldehydes.

b. Aldoses and ketoses react with phenylhydrazine to give phenylhydrazones. In the presence of an acidic solution of excess phenylhydrazine, an adjacent CHOH group is oxidized to a carbonyl group and this reacts with one more molecule of phenylhydrazine to give an *osazone*. The rate of osazone formation and osazone crystal structure are useful in identification of a sugar.

    In ketoses, the C-1 carbon atom is oxidized to a carbonyl group. Consequently, pairs of compounds such as glucose and fructose yield identical osazones, demonstrating the stereochemical similarity of the C-3, C-4, C-5, and C-6 carbon atoms.

    Osazones are formed by any compound having the structure $R-C-CH-R$.
$$\||$$
$$OOH$$

Problems

21-1   *a.* Draw the structures for lactose and cellobiose.

     b. Which of these will reduce Tollens' reagent?

21-2   Show how to make the following conversions. Draw complete structures for reactants and products and indicate precise conditions for each reaction.

   a. D-Glucopyranose to methyl D-glucopyranoside
   b. D-Galactopyranose to methyl 2,3,4,6-tetra-O-methyl-D-galactopyranoside
   c. D-Glucose to D-arabinose   *d.* D-Fructofuranose to glucosazone
   e. D-Arabinose to D-glucose   *f.* D-Galactose to D-galactonic acid
   g. D-Galactose to galactaric acid   *h.* Methyl α-L-arabinofuranoside to L-glyceric acid
   i. α-D-Glucofuranose to its diisopropylidene derivative
   j. D-Ribofuranose to butyl 2,3,5-tri-O-ethyl-D-ribofuranoside
   k. D-Galactopyranose to 2,3,4,6-tetra-O-methyl-D-galactopyranose
   l. D-Mannopyranose to penta-O-propionyl-D-mannopyranose
   m. D-Xylofuranose to D-xylose phenylhydrazone   *n.* L-Gulose to L-xylose
   o. D-Arabinose to D-mannose   *p.* L-Glucose to L-glucaric acid
   q. D-Gulose to D-gulonic acid   *r.* α-D-Ribopyranose to a diisopropylidene derivative

21-3   In Prob. 21-2*g*, the DL designation was omitted from galactaric acid but not from D-glucaric acid in Sec. 21-3. Why would D or L be redundant for the former?

21-4   Name all the sugars that will yield the same osazone as D-glucopyranose.

21-5   How many different stereoisomeric aldohexofuranoses are possible?

21-6   The polysaccharide glycogen was completely methylated and then hydrolyzed to monosaccharides. It yielded mainly 2,3,6-tri-O-methyl-D-glucopyranose but also 2,3,4,6-tetra-O-methyl-D-glucopyranose and 2,3-di-O-methyl-D-glucopyranose. How could this help establish the structure of glycogen?

21-7   The naturally occurring disaccharide $C_{12}H_{22}O_{11}$ reduced Fehling's solution. It was hydrolyzed by the enzyme emulsin (hydrolyzes only β-glucosides) to 2 molecules of D-glucopyranose. When this disaccharide was methylated and hydrolyzed, equal parts of 2,3,4,6-tetra-O-methyl-D-glucopyranose and 2,3,4-tri-O-methyl-D-glucopyranose were formed. Suggest a structure for the compound.

21-8   Compound A, $C_6H_{12}O_6$, reduced Fehling's solution and was optically active. On the basis of this and the following reactions, suggest a structure for A.

   (A) $\xrightarrow{\text{HCN}}$ $\xrightarrow{\text{hydrolysis}}$ $\xrightarrow{\text{HI, P, }\Delta}$ 2-methylhexanoic acid

   (A) $\xrightarrow{\text{C}_6\text{H}_5\text{NHNH}_2 \text{ (excess)}}$ D-glucosazone

   (A) $\xrightarrow{\text{CH}_3\text{OH, H}^+}$ $\xrightarrow{\text{HIO}_4}$ no formic acid in products

21-9   Compound A, $C_6H_{12}O_6$, was optically active and reduced Fehling's solution. On the basis of this and the following reactions, suggest a structure for A.

   (A) $\xrightarrow{\text{HCN}}$ $\xrightarrow{\text{hydrolysis}}$ $\xrightarrow{\text{HI, P, }\Delta}$ heptanoic acid

   (A) $\xrightarrow{\text{Na(Hg), HOH}}$ $C_6H_{14}O_6$, optically inactive

   (A) $\xrightarrow{\text{Wohl degradation}}$ an aldose $\xrightarrow{\text{Na(Hg), HOH}}$ $C_5H_{12}O_5$, optically inactive

   (A) $\xrightarrow{\text{MeOH, H}^+}$ $\xrightarrow{\text{HIO}_4}$ 1 equiv formic acid plus B

   (B) $\xrightarrow{\text{oxidation, hydrolysis}}$ D-glyceric acid

21-10   The following sugars were isolated from natural products: sugar A from corn cobs; sugar B from Barbados aloes; sugar C from cell nuclei; sugar D from talose.
   a. Each sugar yielded caproic acid on reaction with HCN followed by hydrolysis and reduction with HI.
   b. On treatment with HIO_4, none of the methyl glycosides of these sugars yielded formic acid or formaldehyde. Each reaction product was identical and yielded equivalent amounts of D-glyceric acid and glyoxylic acid after oxidation and hydrolysis.
   c. Each was reduced with Na(Hg) to yield a pentahydroxy compound. The reduction products of sugars A and C were optically inactive, and those of B and D were optically active.
   d. B and C formed the same osazone. A and D yielded the same osazone, different from the osazone of B and C.

e. B could be converted synthetically into two epimeric aldohexoses, both of which formed glucosazone.

   On the basis of these observations, give the structures of the four sugars. Which of the pentahydroxy compounds in *c* are identical?

21-11   Give a convenient method for distinguishing between the following:

a. Methyl D-glucopyranoside and 2-O-methyl-D-glucopyranose

b. 2,3-Pentanediol and 2,4-pentanediol

c. Water solutions of pentanal and mannose

d. D-Glucose and sorbitol

e. Sucrose and maltose

f. Sucrose and propionaldehyde

g. Sorbitol and lyxose

h. Xylono-γ-lactone and ribopyranose

i. Glucaric acid and methyl D-glucopyranoside

j. D-Glucose and L-glucose

21-12   Draw the most favorable chair form for α-D-glucopyranose. Identify each substituent on the ring carbon atoms as axial (*a*) or equatorial (*e*).

21-13   Name an aldohexopyranose which must have two axial hydroxyl groups.

21-14   The bulky –CH$_2$OH group is conformation-controlling in the pyranose ring in the same manner as the *tert*-butyl group in cyclohexane (see Sec. 12-3*b*-2). Draw conformationally correct structures for the following:

a. β-D-Ribopyranose

b. β-D-Fructopyranose

c. β-L-Glucopyranose

d. α-D-Idopyranose

e. Methyl α-D-mannopyranoside

f. D-Galactono-δ-lactone

21-15   A cis diol increases the conductivity of a solution of boric acid by the following trans formation:

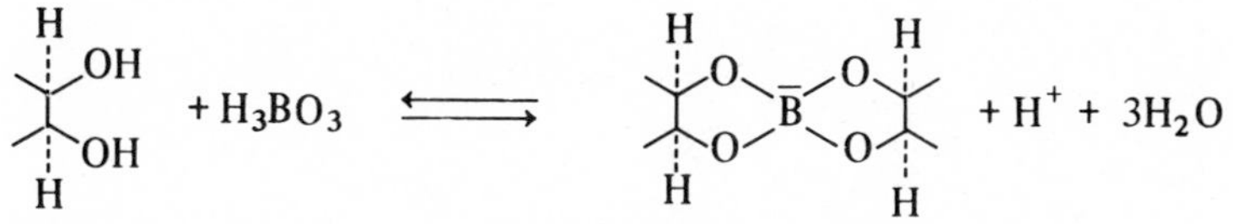

a. Will the conductivity of such a solution be increased more by the diol from the reaction of water with cyclohexene oxide or the diol from the oxidation of cyclohexene with dilute permanganate? Explain.

b. Give the structure for all methyl D-aldopentopyranosides which will not increase the conductivity.

c. Explain the following: The conductivity of a solution of α-D-glucopyranose in aqueous boric acid gradually decreases with time and approaches an equilibrium value.

d. Explain the following: The conductivity of a solution of β-D-glucopyranose in aqueous boric acid gradually increases with time and approaches an equilibrium value.

21-16   Using chair-form pyranose rings, provide an appropriate drawing for the following:

a. Cellobiose, the repeating unit of cellulose

b. Maltose, the repeating unit of starch

21-17   Provide structures for the synthetic intermediates in the following reaction sequence, which is used to convert D-glucose to L-ascorbic acid (vitamin C).

a. D-glucose $\xrightarrow[\text{hydrogenation}]{\text{catalytic}}$ (A) $\xrightarrow[\textit{Acetobacter suboxydans}]{\text{oxidation by}}$ (B)   Compound B is a ketohexose which gives the same osazone as L-gulose or L-idose.

b. (B) $\xrightarrow{\text{spontaneous}}$ (C) (hemiacetal) $\xrightarrow[\text{HCl}]{\text{MeOH}}$ acetal $\xrightarrow{\text{HIO}_4}$ no CH$_2$O or HCOOH

c. (C) $\xrightarrow[\text{ZnCl}_2]{\text{acetone}}$ (D) (C$_{12}$H$_{20}$O$_6$)

d. (D) $\xrightarrow{\text{KMnO}_4}$ (E) (C$_{12}$H$_{18}$O$_7$) $\xrightarrow{\text{H}_3\text{O}^+}$ $\xrightarrow{\text{MeOH, HCl}}$ (F) (C$_7$H$_{12}$O$_7$)

e. (F) $\xrightarrow{\text{MeO}}$ (G) (sodium salt of an *enediol*, C$_7$H$_{11}$O$_7$Na)

f. (G) $\xrightarrow{\text{HCl}}$ L-ascorbic acid (an enediol and γ-lactone)

21-18   *a.* Give structures for all stereoisomers of inositol (1,2,3,4,5,6-hexahydroxycyclohexane) and indicate which of these are optically active.

b. One inositol isomer is an essential biological factor for certain microorganisms. It is an optically inactive form which is identical in configuration with Phytin, a salt of the hexaphosphate ester of inositol. Enzymatic hydrolysis of Phytin yielded an optically active tetraphosphate, an optically active diphosphate, and an optically inactive monophosphate. Which possible structures are ruled out by these observations?

c. *Acetobacter suboxydans* is an interesting microorganism which can oxidize an axial hydroxyl group to a carbonyl function. Equatorial hydroxyl groups are not altered. Suggest a structure for inositol on the basis of its oxidation by *A. suboxydans* to $C_6H_{10}O_6$.

answers

Chapter 1

1-1   *a.* $:\!\overset{\cdot\cdot}{\underset{\cdot\cdot}{Cl}}\!:\!\overset{\cdot\cdot}{\underset{\cdot\cdot}{Cl}}\!:$

b. $:\!\overset{\cdot\cdot}{O}\!:\!:\!C\!:\!:\!\overset{\cdot\cdot}{O}\!:$

c.
$$\begin{array}{c} H \\ H\!:\!\overset{H}{\underset{H}{C}}\!:\!H \end{array}$$
Tetrahedral

d.
$$\overset{:\overset{\cdot\cdot}{Cl}:}{:\!\overset{\cdot\cdot}{Cl}\!:\!C\!:\!\overset{\cdot\cdot}{Cl}\!:}$$
:Cl:
Tetrahedral

e. $H\!:\!C\!:\!:\!:\!N\!:$

f.
$$\overset{:\overset{\cdot}{O}\cdot}{\underset{H:\overset{\cdot\cdot}{O}\cdot\overset{C}{\cdot}\overset{\cdot\cdot}{O}:H}{}}$$

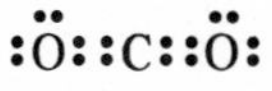

g.
$$\begin{array}{c} H \\ H\!:\!\overset{H}{\underset{H}{C}}\!:\!Mg\!:\!\overset{\cdot\cdot}{\underset{\cdot\cdot}{I}}\!: \end{array}$$

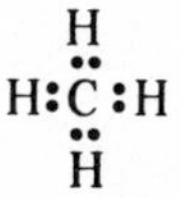

h. $H \quad \overset{\cdot\cdot}{\overset{\cdot}{O}}\quad \overset{\cdot\cdot}{\underset{\cdot\cdot}{Cl}}$

i. $\overset{\cdot\cdot}{O}\quad N\!:\!N\quad \overset{\cdot\cdot}{O}$ (with O atoms)

j. $H\!:\!C\!:\!:\!:\!C\!:\!H$

k. $:\!\overset{\cdot\cdot}{\underset{\cdot\cdot}{I}}\!:\!\overset{\cdot}{\underset{\cdot\cdot}{Cl}}\!:$

l.
$$\begin{array}{cc} H & H \\ C\!:\!:\!C \\ H & H \end{array}$$

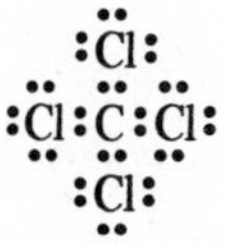

m.
$$\begin{array}{c} :\!\overset{\cdot\cdot}{Cl}\quad \overset{\cdot\cdot}{Cl}\cdot \\ C \\ :\overset{\cdot\cdot}{O}. \end{array}$$

n.
$$\begin{array}{c} :\!O:\;:O: \\ N\quad N \\ O\quad O\quad O \end{array}$$

o.
$$\begin{array}{c} H \\ :\!\overset{\cdot\cdot}{Cl}\!:\!\overset{\cdot\cdot}{C}\!:\!\overset{\cdot\cdot}{Cl}\!: \\ :\overset{\cdot\cdot}{Cl}: \end{array}$$
Tetrahedral

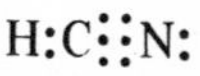

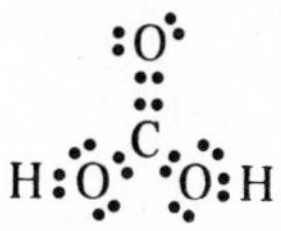

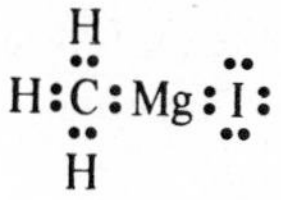

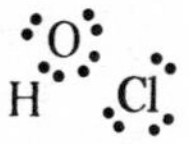

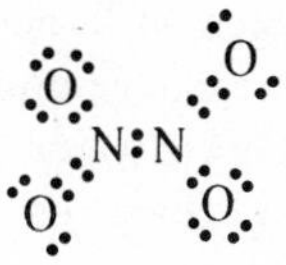

1-2   *a.*
$$\begin{array}{c} \overset{\cdot\cdot}{S} \\ \overset{\cdot\cdot}{O}\quad \overset{\cdot\cdot}{O} \end{array}$$

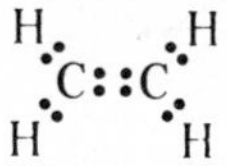

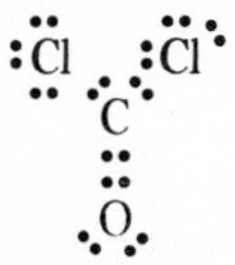

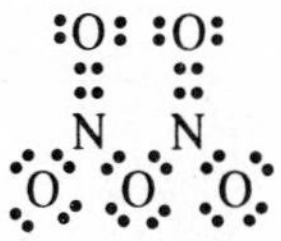

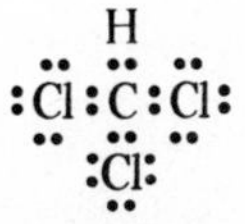

b.
$$\begin{array}{c} H \\ :\!\overset{\cdot\cdot}{O}\!: \\ :\!O\!:\!S\!:\!O\!: \\ :\!\overset{\cdot\cdot}{O}\!: \\ H \end{array}$$

c.
$$\begin{array}{c} :\!\overset{\cdot\cdot}{F}\!: \\ :\!\overset{\cdot\cdot}{F}\!: \\ :\!\overset{\cdot\cdot}{F}\!:\!S\!:\!\overset{\cdot\cdot}{F}\!: \\ :\!\overset{\cdot\cdot}{F}\!: \\ :\!\overset{\cdot\cdot}{F}\!: \end{array}$$

d. $\overset{\cdot\cdot}{\underset{\cdot\cdot}{O}}\!:\!:\!C\!:\!:\!C\!:\!:\!C\!:\!:\!\overset{\cdot\cdot}{\underset{\cdot\cdot}{O}}$

e.
$$\begin{array}{c} \overset{\cdot\cdot}{O} \\ \overset{\cdot\cdot}{F}\quad \overset{\cdot\cdot}{F} \end{array}$$

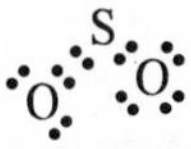

f.
$$\left[\begin{array}{c} :\!\overset{\cdot\cdot}{O}\!: \\ :\!O\!:\!\overset{\cdot\cdot}{Cl}\!:\!\overset{\cdot\cdot}{O}\!: \\ :\!\overset{\cdot\cdot}{O}\!: \end{array}\right]^{-}$$

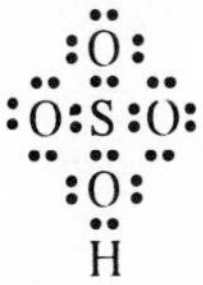

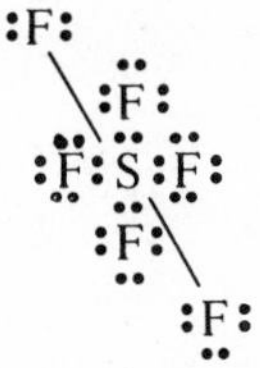

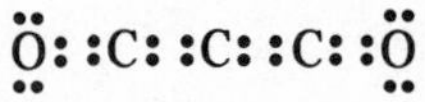

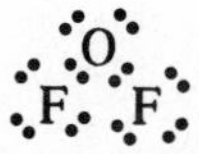

305

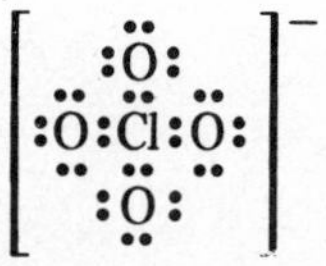

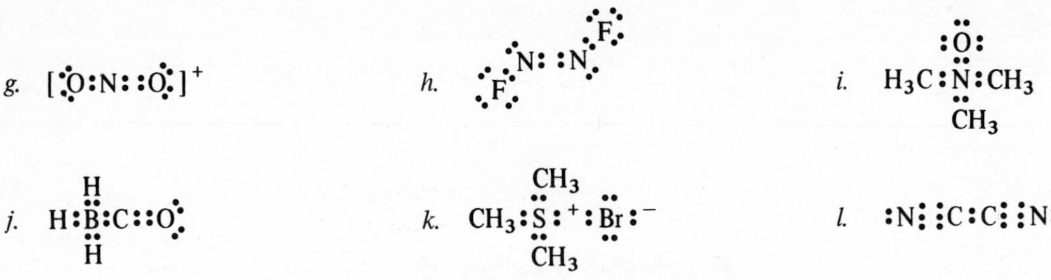

g. $[\ddot{\text{O}}\!:\!\text{N}\!:\!:\ddot{\text{O}}\!:]^+$ h. (Lewis structure of N₂F₂)   i. $H_3C\!:\!\overset{..}{\underset{CH_3}{N}}\!:\!CH_3$ with $:\overset{..}{O}:$ above N

j. $H\!:\!\overset{H}{\underset{H}{B}}\!:\!C\!:\!:\overset{..}{\underset{..}{O}}$ k. $CH_3\!:\!\overset{CH_3}{\underset{CH_3}{S}}\!:^+\!:\!\overset{..}{\underset{..}{Br}}\!:^-$ l. $:N\!:\!:C\!:\!C\!:\!:N:$

1-3 a. II   b. IV   c. IV   d. II   e. III   f. III

1-4 a. $\overset{+}{O}=\overset{..}{S}-\overset{..}{\underset{..}{O}}:^- \longleftrightarrow {}^-:\overset{..}{\underset{..}{O}}-\overset{..}{\overset{+}{S}}=\overset{..}{O}:$

b. $CH_2=CH\overset{+}{C}H_2 \longleftrightarrow \overset{+}{C}H_2CH=CH_2$ c. ${}^-:\overset{..}{\underset{..}{O}}-\overset{..}{N}=O \longleftrightarrow O=\overset{..}{N}-\overset{..}{\underset{..}{O}}:^-$

d. $I^+I^- \longleftrightarrow I^-I^+$ e. ${}^-CH_2-CH=CH-\overset{+}{C}H_2 \longleftrightarrow \overset{+}{C}H_2-CH=CH-\overset{-}{C}H_2$

f. (resonance structures of cyclic system shown)

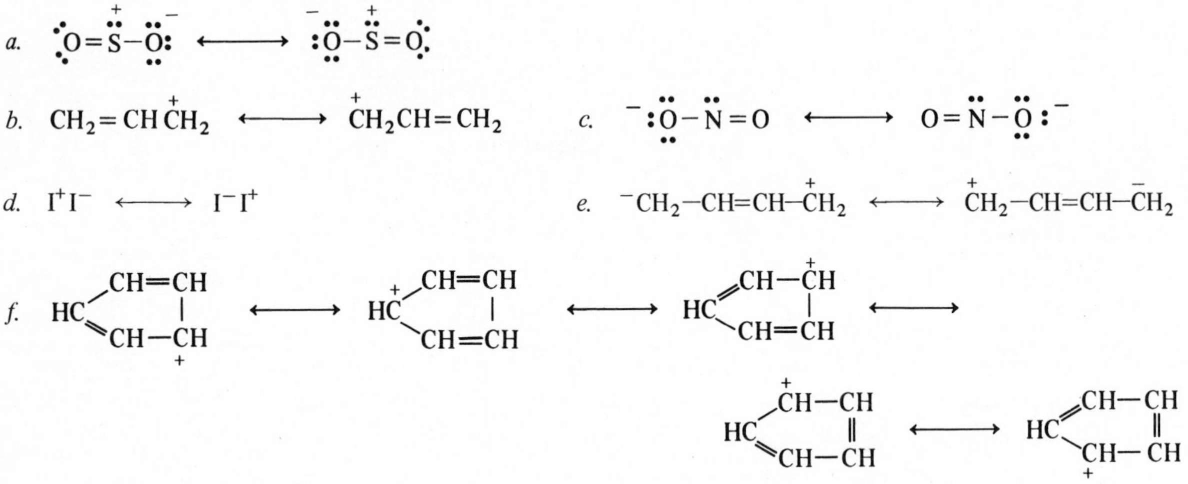

1-5 a. Negative on Se, negative ion   b. No charge on O, negative charge on S, negative ion
     c. No charge on N, neutral compound   d. Positive charge on O, positive ion
     e. Negative charge on N, negative ion   f. No charge on Cl, positive charge on I, positive ion
     g. Negative charge on Al, negative ion   h. Positive charge on N, neutral compound
     i. Negative charge on B, negative ion

1-6 a. C, 74.8%; H, 25.13%   b. C, 52.14%; H, 13.13%; O, 34.73%
     c. C, 10.06%; H, 0.84%; Cl, 89.10%   d. C, 68.27%; H, 6.28%; N, 3.79%; O, 21.66%
     e. C, 41.61%; H, 4.08%; N, 8.09%; O, 27.71%; S, 18.51%
     f. C, 47.4%; H, 2.8%; Cl, 50.0%

1-7 A, CH_3O; B, CH; C, $C_7H_8N_2O$; D, C_3H_5
1-8 A, $C_2H_6O_2$; B, C_2H_2; C, $C_7H_8N_2O$; D, C_6H_{10}
1-9 $C_7H_8N_2O$ 1-10 C_6H_6 1-11 136; one
1-12 Two   1-13 136, $C_8H_8O_2$ 1-14 83
1-15 254 ml

1-16 a. $\xrightarrow{\quad}$ $C-O$ b. $\xleftarrow{\quad}$ $C-Si$ c. $\xrightarrow{\quad}$ CH_3-CF_3

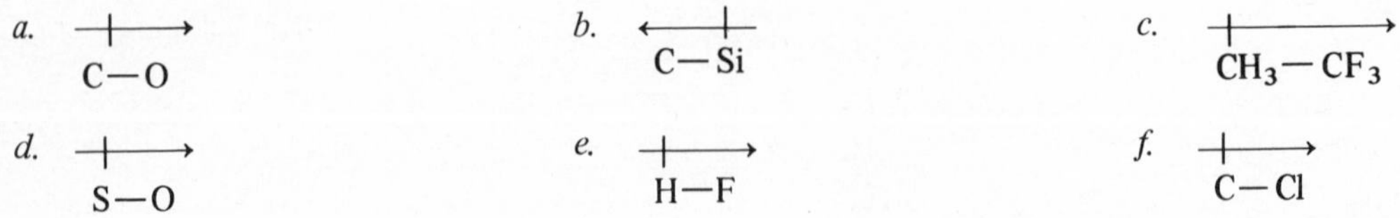

d. $\xrightarrow{\quad}$ $S-O$ e. $\xrightarrow{\quad}$ $H-F$ f. $\xrightarrow{\quad}$ $C-Cl$

1-17 a. $(CH_3)_2C=\overset{..}{\underset{..}{O}}: \longleftrightarrow (CH_3)_2\overset{+}{C}-\overset{..}{\underset{..}{O}}:^-$ b. $CH_3\overset{..}{N}=\overset{..}{\underset{..}{O}}: \longleftrightarrow CH_3\overset{+}{\overset{..}{N}}-\overset{..}{\underset{..}{O}}:^-$

c. $:\overset{..}{\underset{..}{S}}=C=\overset{..}{\underset{..}{S}}: \longleftrightarrow {}^-:\overset{..}{\underset{..}{S}}-\overset{+}{C}=\overset{..}{\underset{..}{S}}: \longleftrightarrow :\overset{..}{\underset{..}{S}}=\overset{+}{C}-\overset{..}{\underset{..}{S}}:^-$

d. (resonance structures of $CH_2=CH-N$ nitro group shown)

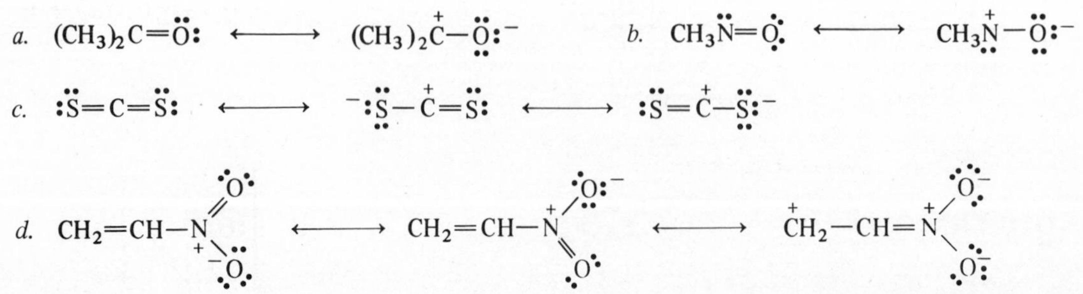

$e.$

$$CH_3\overset{\overset{\displaystyle :O:}{\|}}{C}-\ddot{\underset{\cdot\cdot}{C}}\ddot{l}: \quad\longleftrightarrow\quad CH_3\overset{\overset{\displaystyle :\ddot{O}:^-}{|}}{\underset{+}{C}}-\ddot{\underset{\cdot\cdot}{C}}\ddot{l}: \quad\longleftrightarrow\quad CH_3\overset{\overset{\displaystyle :\ddot{O}:^-}{|}}{C}=\ddot{\underset{\cdot\cdot}{C}}l^+$$

$f.$

$$CH_3\overset{\displaystyle}{\underset{\underset{\displaystyle :\ddot{O}CH_3}{|}}{\ddot{O}}}-B-\ddot{\underset{\cdot\cdot}{O}}CH_3 \longleftrightarrow CH_3\overset{+}{\underset{\underset{\displaystyle :\ddot{O}CH_3}{|}}{\ddot{O}}}=\bar{B}-\ddot{\underset{\cdot\cdot}{O}}CH_3 \longleftrightarrow CH_3\overset{\displaystyle}{\underset{\underset{\displaystyle +\ddot{O}CH_3}{\|}}{\ddot{O}}}-\bar{B}-\ddot{\underset{\cdot\cdot}{O}}CH_3 \longleftrightarrow CH_3\overset{\displaystyle}{\underset{\underset{\displaystyle :\ddot{O}CH_3}{|}}{\ddot{O}}}-\bar{B}=\overset{+}{\ddot{O}}CH_3$$

$g.$

$$CH_3C\equiv\ddot{N} \quad\longleftrightarrow\quad CH_3\overset{+}{C}=\ddot{N}:^-$$

$h.$

$$H_2C=CH-\ddot{\underset{\cdot\cdot}{O}}CH_3 \quad\longleftrightarrow\quad H_2\bar{C}-CH=\overset{+}{O}CH_3 \quad\longleftrightarrow\quad H_2\bar{C}-\overset{+}{C}H-\ddot{\underset{\cdot\cdot}{O}}CH_3$$

$i.$

$$HC\equiv C-C\equiv CH \quad\longleftrightarrow\quad H\bar{\underset{\cdot\cdot}{C}}=C=C=\overset{+}{C}H \quad\longleftrightarrow\quad H\overset{+}{C}=C=C=\bar{\underset{\cdot\cdot}{C}}H \quad\longleftrightarrow$$

$$HC\equiv C-\overset{+}{C}=\bar{\underset{\cdot\cdot}{C}}H \quad\longleftrightarrow\quad H\bar{\underset{\cdot\cdot}{C}}=\overset{+}{C}-C\equiv CH \quad\longleftrightarrow\quad H\bar{\underset{\cdot\cdot}{C}}=\overset{+}{C}-\bar{\underset{\cdot\cdot}{C}}=\overset{+}{C}H$$

1-18 $a.$ $\longrightarrow$ $2NH_3$ $b.$ $\longrightarrow$ $(CH_3)_2\overset{+}{O}\bar{B}F_3$ $c.$ $\longrightarrow$ $CH_3\overset{+}{O}H_2Cl^-$

$d.$ $\longrightarrow$ $_+HOSO_2Cl_-$ $e.$ $\longrightarrow$ $(CH_3)_2\overset{+}{S}\bar{A}lCl_3$ $f.$ $\longrightarrow$ $CH_3OH + H_2O$

$g.$ $\longrightarrow CH_2-CH_2-\bar{B}F_3$ $h.$ $\longrightarrow$ $Ag(NH_3)_2^+$ $i.$ $\longrightarrow$ $CH_3\overset{+}{O}H_2 + F^-$

1-19 $a.$ More to the left. The negative charge, spread out more on the larger S atom, will not bind the proton as effectively.

$b.$ More to the right. The negative charge on the OH^- is localized on oxygen and hence better able to hold a proton. The two negative charges on the carbonate anion are spread between three oxygens by resonance.

$c.$ More toward the left. Reasoning similar to a.

$d.$ More toward the left. C—H bond more covalent than N—H.

$e.$ More to the right. Cl better able to accommodate negative charge, less able to accept proton compared to smaller O.

$f.$ More to the left. Reasoning similar to d.

$g.$ More to the right. Reasoning similar to a.

$h.$ More to the left. Cl^- is more electronegative than inert gas Ne.

Chapter 2

2-1 $a.$ 8 $b.$ 18 $c.$ 6 $d.$ 10 $e.$ 2 $f.$ 0. No $2d$ orbitals.

2-2 $a.$ 2 $b.$ 6 $c.$ 2 $d.$ 18 $e.$ 1 $f.$ 32

2-3 $a.$ $1s^2 2s^2 2p^6 3s$ or $1s^2 2p_x{}^2 2p_y{}^2 2p_z{}^2 3s$ $b.$ $1s^2 2s^2 2p^6$ $c.$ $1s^2 2s^2 2p$

$d.$ $1s^2 2s^2 2p^3$ $e.$ $1s^2 2s^2 2p^5$ $f.$ $1s^2 2s^2 2p^6 3s^2 3p^5$

$g.$ $1s^2 2s^2 2p^6 3s^2 3p^6 4s^2 3d^{10} 4p^6 5s^2 4d^{10} 5p^5$ $h.$ $1s^2 2s^2 2p^6$

2-4 $a.$ sp^3, tetrahedral $b.$ sp^3, tetrahedral $c.$ sp^3, about $105°$

$d.$ sp^3, tetrahedral $e.$ sp^3, tetrahedral $f.$ sp^2, planar trigonal

$g.$ p, about $90°$ $h.$ sp^3, tetrahedral

2-5 $a.$ All carbon atoms have sp hybridization; linear.

$b.$ Entire molecule is planar; about $120°$ bond angles for the sp^2 carbons of C=C. Linear arrangement for the sp carbons of C≡C.

$c.$ Tetrahedral nitrogen atoms; H—N—H bond angles about like those in ammonia.

$d.$

$$\overset{\displaystyle H}{\underset{\diagdown}{}}N=\overset{+}{N}=\bar{N}$$

Central N is sp. The H—N—N bond angle is about $120°$ because of sp^2 hybridization.

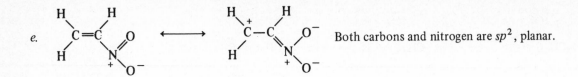

e.   Both carbons and nitrogen are sp^2, planar.

f. $:C\equiv C:^-$ *sp* hybridization.   *g.* All carbons *sp*, linear molecule.
h. CH_3 is tetrahedral; other carbons are sp^2. Planar molecule except for CH_3.

i. 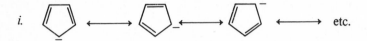   etc.

All carbons equivalent and sp^2. Planar molecule.

2-6   *a* and *c* can be detected. *b* is too unstable.

2-7

2-8   *a.* Yes, $n = 0$.
      f. Yes, $n = 2$; *b, c, d,* and *e* do not contain $(4n + 2)$ π electrons.

2-9   *a.* 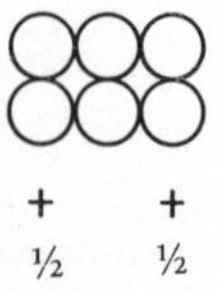   *b.*

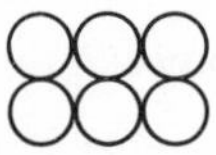

       +   +
       ½   ½

In the cation the plus charge is delocalized between the two end carbons. In the radical each orbital contains one electron.

2-10   *a.* $CH_2=CH-\overset{+}{C}H_2 \longleftrightarrow \overset{+}{C}H_2CH=CH_2$ Such resonance is not possible in $CH_3CH_2{}^+$.

       b. 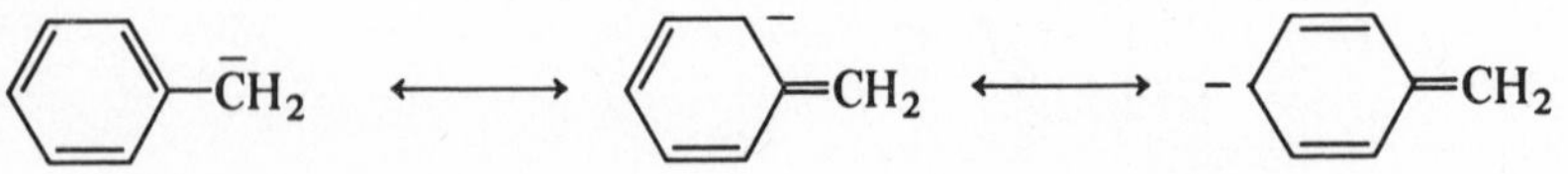   etc.

In the second molecule the charge is insulated from resonance with the ring electrons by the CH_2 unit.

       c. 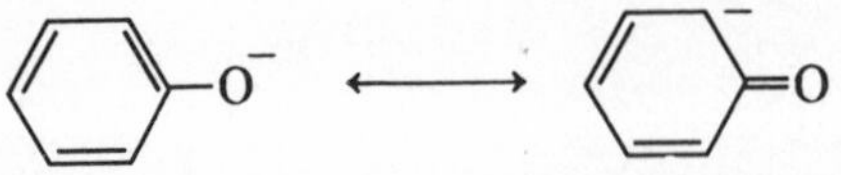   etc.

       d. Hückel's rule applies to    but not to the bicyclic molecule.

       e. 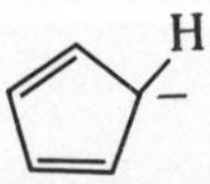   Fits $4n + 2$ rule, other ion does not.

f. 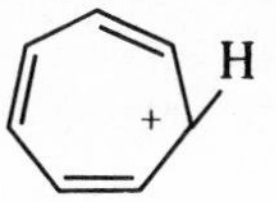 Fits $4n + 2$ rule, other ion does not.

2-11

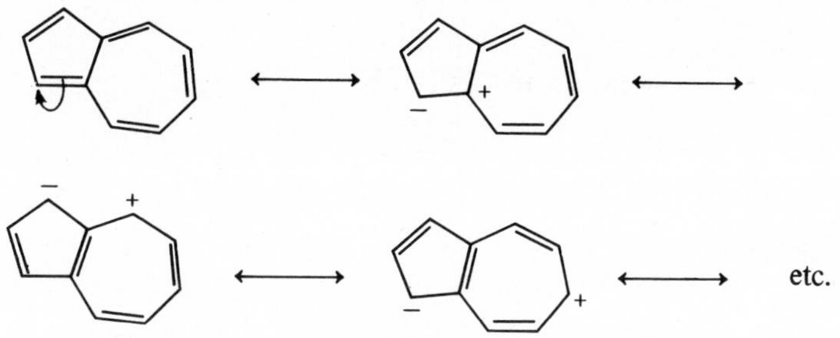

etc.

The above type of resonance with the positive charge in the seven-membered ring and the negative charge in the five-membered ring allows both rings to conform to $4n + 2$ rule.

2-12

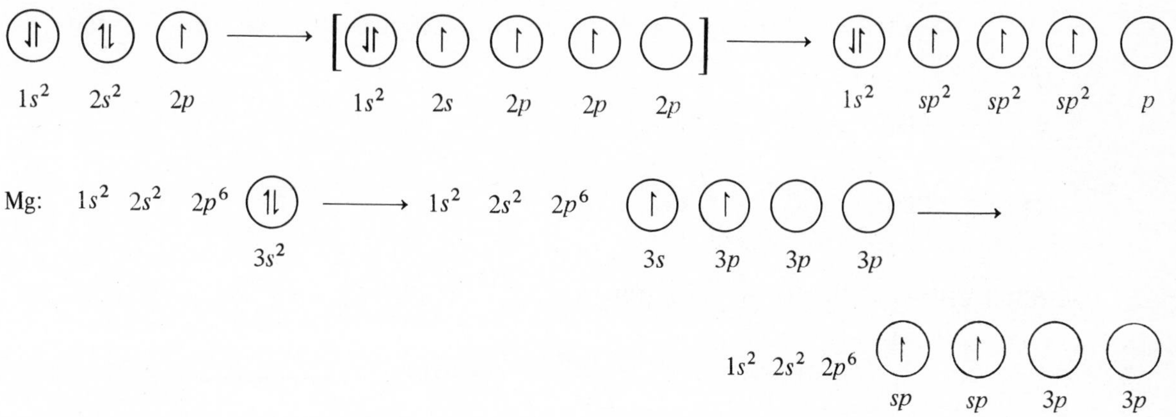

2-13   In linear form carbon has *sp* hybridization with one unpaired electron in each *p* orbital. In the other form carbon appears to have a structure intermediate between the unhybridized form (90° angles) and sp^2 hybridization (120° angles).

2-14   *a.*

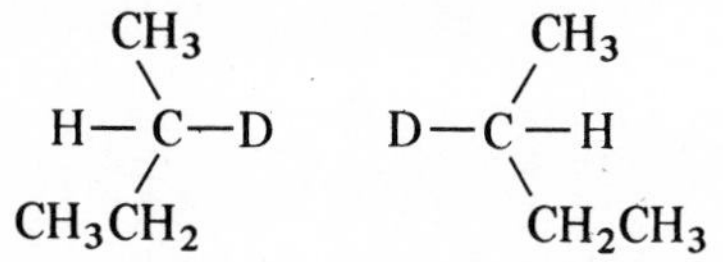

Both optically active

b.

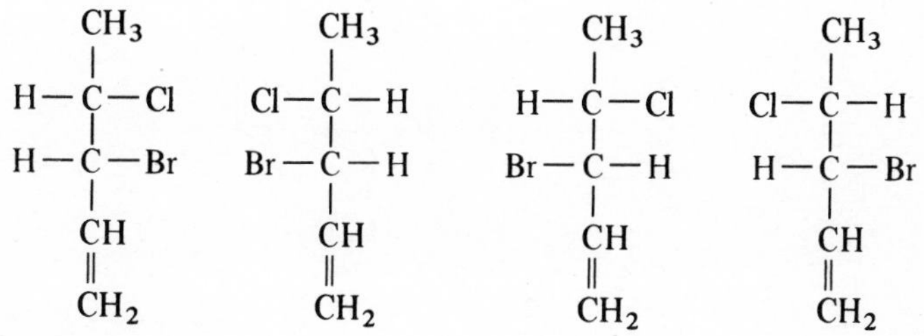

All optically active

c. No isomers.

d.

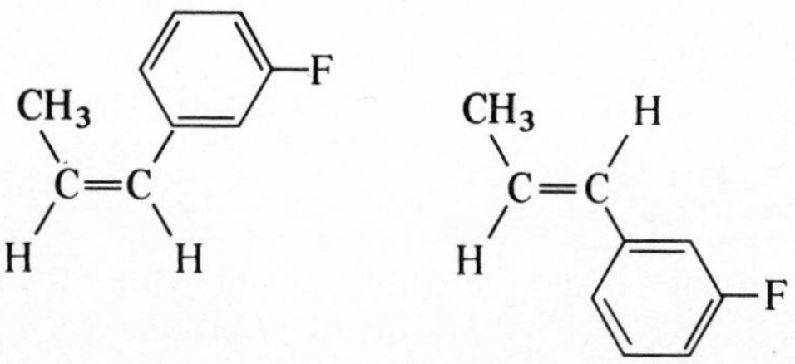

e.

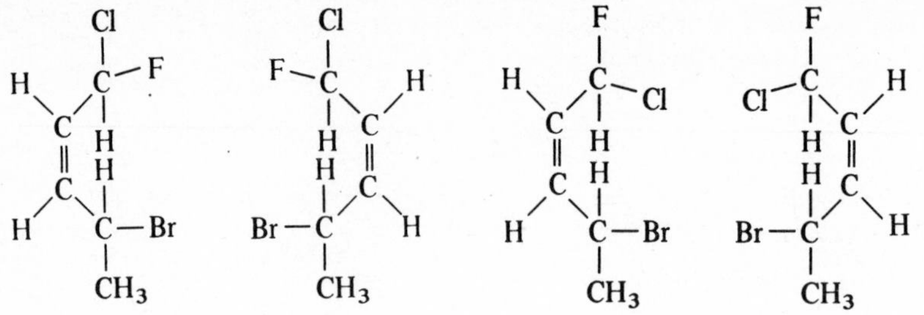

Plus a set of trans isomers (all optically active).

f. $2^4 = 16$ isomers.

2-15 a.

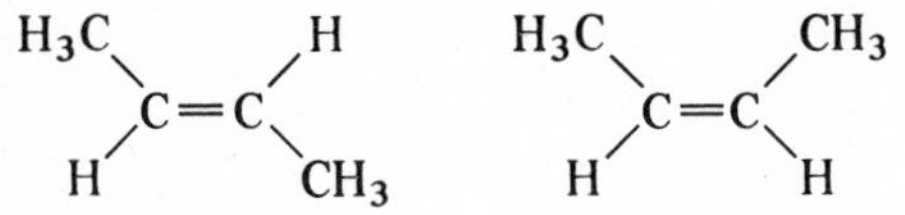

E Z

b.

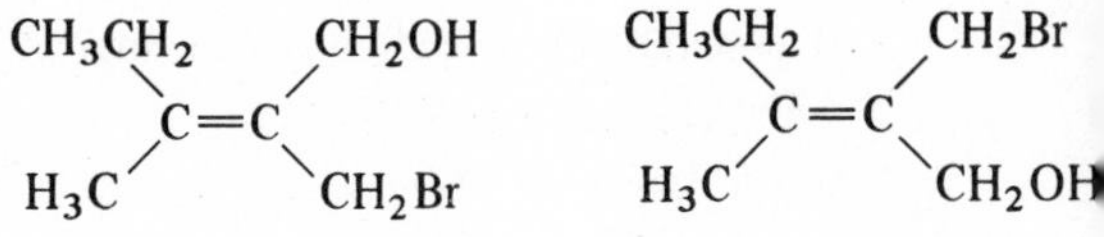

E Z

c.

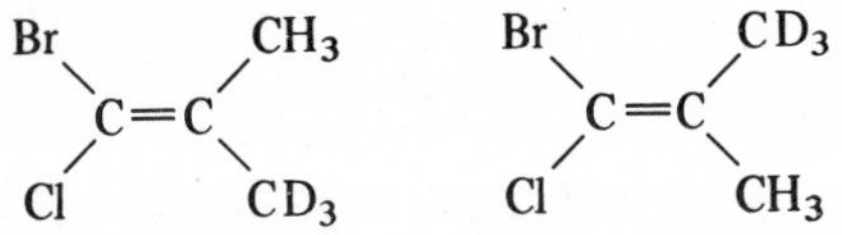

E Z

d.

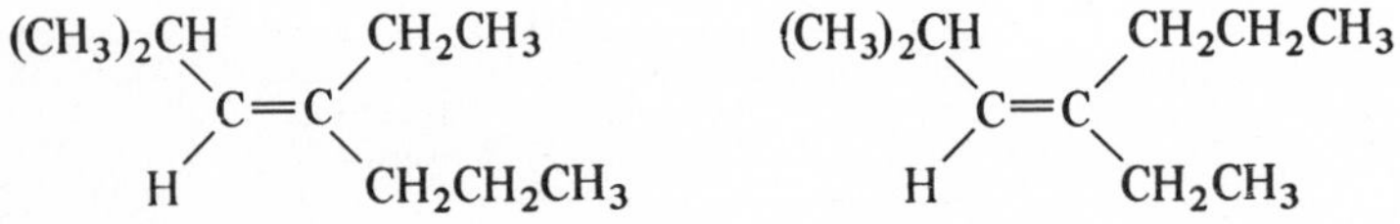

E Z

2-16 $4n + 2$ rule obeyed; 14 overlapping π electrons, $n = 3$.

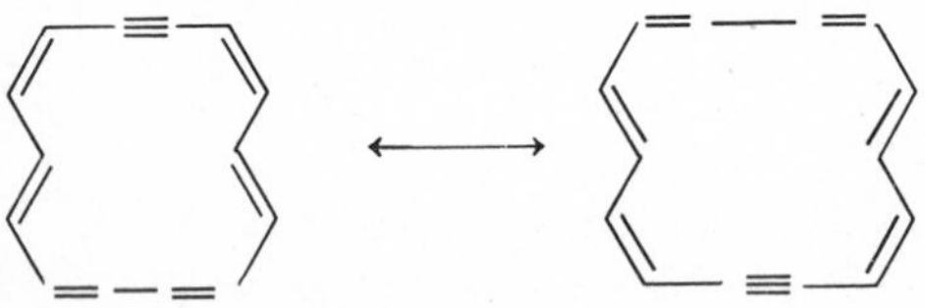

2-17

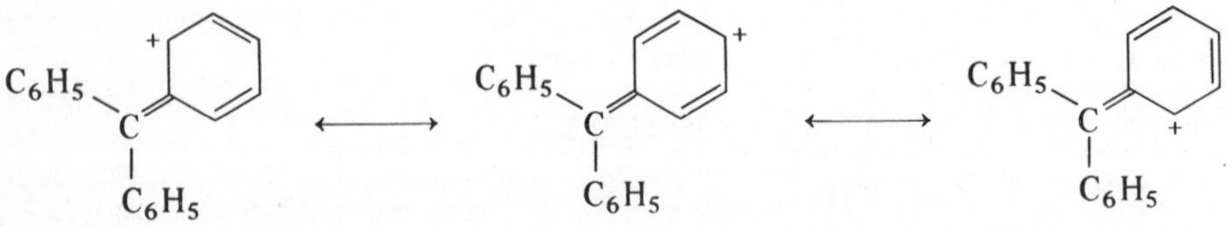

plus the equivalent forms for the other two rings.

2-18   Central C has sp hybridization.

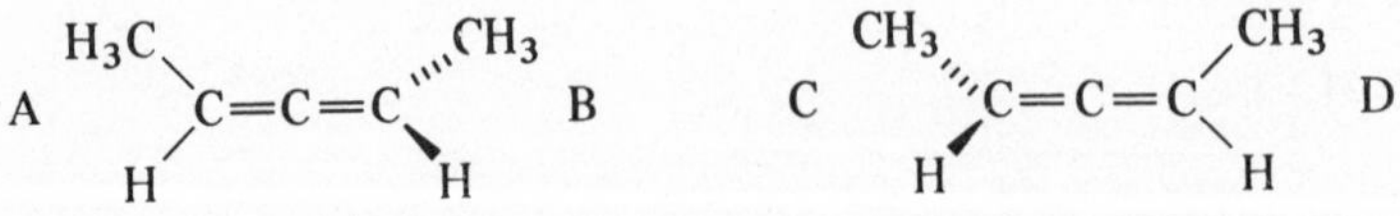

Substituents A + D are in the plane of the page while substituents B and C are in front and behind the plane of the page. Hence the two forms are not superimposable.

310

Chapter 3

3-1

	Common name	IUPAC name
a.	Propane	Propane
b.	Isobutane	2-Methylpropane
c.	*n*-Nonane	Nonane
d.	Isopentane	2-Methylbutane
e.	Isononane	2-Methyloctane
f.	Neopentyl chloride	1-Chloro-2,2-dimethylpropane

3-2   *a.*   1-Chlorobutane   *b.*   2,3-Dimethylbutane

        c.   1,2-Dimethylcyclobutane   *d.*   2,2-Dimethylbutane

        e.   1-Bromo-2-methylpropane   *f.*   1-Nitrobutane

        g.   4-Propyl-4-*tert*-butyldecane   *h.*   3,4-Diethylhexane

        i.   4-Cyclopropyl-5-ethyl-2,3,3-trimethyl-5-nitroheptane

3-3   *a.* 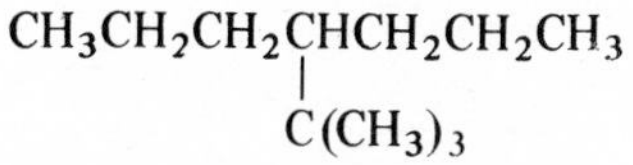 $CH_3CH_2CH_2CHCH_2CH_2CH_3$

                              $C(CH_3)_3$

b. 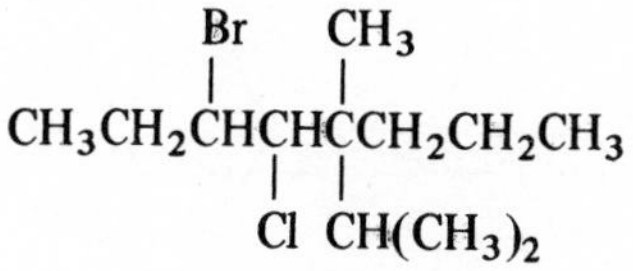   Br CH_3

          $CH_3CH_2CHCHCCH_2CH_2CH_3$

                   Cl $CH(CH_3)_2$

c. $ClCH_2CH_2CH(CH_2CH_2CH_2Cl)_2$

d. 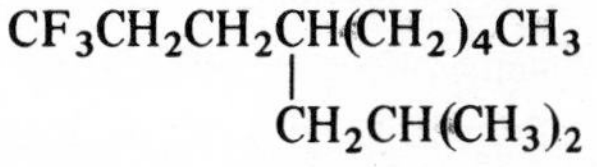 $CF_3CH_2CH_2CH(CH_2)_4CH_3$

                  $CH_2CH(CH_3)_2$

e. 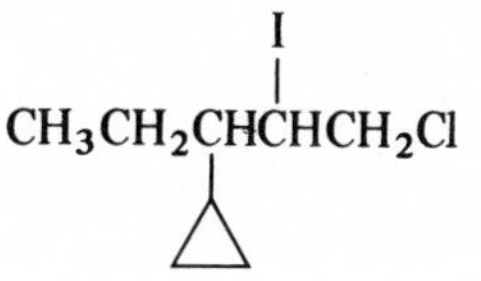   I

     $CH_3CH_2CHCHCH_2Cl$

f. 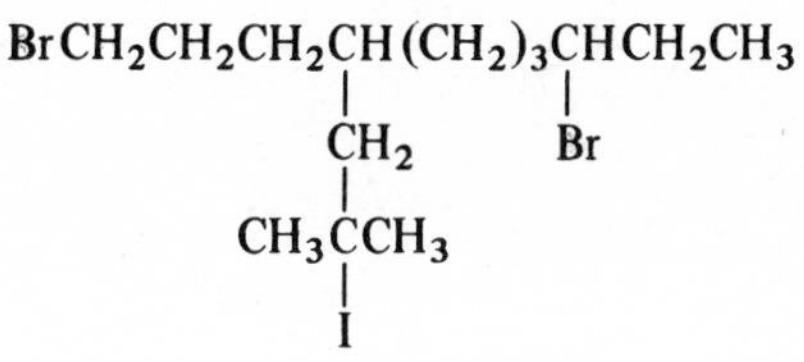 $BrCH_2CH_2CH_2CH(CH_2)_3CHCH_2CH_3$

                     CH_2 Br

              CH_3CCH_3

                   I

3-4 $CH_3CH_2CH_2CH_2CH_2CH_2CH_3$, heptane; $CH_3CH(CH_3)CH_2CH_2CH_2CH_3$, 2-methylhexane; $CH_3CH_2CH(CH_3)CH_2CH_2CH_3$, 3-methylhexane; $(CH_3)_2CHCH_2CH(CH_3)_2$, 2,4-dimethylpentane; $(CH_3)_2CHCH(CH_3)CH_2CH_3$, 2,3-dimethylpentane; $(CH_3)_3CCH_2CH_2CH_3$, 2,2-dimethylpentane; $CH_3CH_2C(CH_3)_2CH_2CH_3$, 3,3-dimethylpentane; $CH_3CH_2CH(CH_2CH_3)CH_2CH_3$, 3-ethylpentane; $(CH_3)_2CHC(CH_3)_3$, 2,2,3-trimethylbutane

3-5   *a.*   Only one isomer   *b.* $CH_3CH_2CH_2Cl$, $CH_3CHClCH_3$ *c.* CH_3CHCl_2, $ClCH_2CH_2Cl$

        d. $CH_3CH_2CHClBr$, $CH_3CClBrCH_3$, $CH_3CHClCH_2Br$, $CH_3CHBrCH_2Cl$, $ClCH_2CH_2CH_2Br$

        e. $ClCH_2CHBrCH_2F$, $BrCH_2CHClCH_2F$, $ClCH_2CHFCH_2Br$, $ClBrCHCHFCH_3$, $ClBrCHCH_2CH_2F$, $BrFCHCHClCH_3$, $BrFCHCH_2CH_2Cl$, $ClFCHCHBrCH_3$, $ClFCHCH_2CH_2Br$, $CH_3CClBrCH_2F$, $CH_3CClFCH_2Br$, $CH_3CFBrCH_2Cl$, $BrFClCCH_2CH_3$

3-6   *a.* $2CH_3(CH_2)_4CH_3 + 19O_2 \longrightarrow 12CO_2 + 14H_2O$

        b. $2CH_3CHBrCH_3 + 2Na \longrightarrow (CH_3)_2CHCH(CH_3)_2 + 2NaBr$

        c. $4(CH_3)_2CHCH_2CH_3 + 4CH_2N_2 \overset{\lambda}{\longrightarrow} (CH_3)_2CHCH_2CH_2CH_3 + (CH_3)_3CCH_2CH_3$
             $+ (CH_3)_2CHCH(CH_3)_2 + (CH_3CH_2)_2CHCH_3 + 4N_2$

        d. $CH_3CH_2MgI + H_2O \longrightarrow CH_3CH_3 + MgIOH$

        e. $(CH_3)_3CCH_2Cl + H_2 \overset{Pt}{\longrightarrow} (CH_3)_4C + HCl$

        f. $CH_3CH_2CH_2CH_2NO_2$, $CH_3CH_2CH(NO_2)CH_3$, $CH_3CH_2CH_2NO_2$, $CH_3CH(NO_2)CH_3$, $CH_3CH_2NO_2$, CH_3NO_2

3-7 $^{5.0}\!/_{156}$ = 0.032 mol ethyl iodide. Each mole of ethyl iodide yields $\frac{1}{2}$ mol butane, so yield of butane in moles would be 0.016. But only 60 percent yield was obtained, so 0.016 x 0.60 = 0.0096 mol butane. 0.0096 x 58 = 0.56 g of butane.

3-8 $CH_4 + 4Cl_2 \longrightarrow CCl_4 + 4HCl.$ $^{10.0}\!/_{22.4}$ = 0.45 mol CH_4. Each mole of CH_4 yields 1 mol CCl_4; hence one should get 0.45 x 154 = 69.3 g (154 = molecular weight CCl_4).

3-9 $CH_4 \longrightarrow CH_3Cl \longrightarrow CH_3CH_3 \longrightarrow CH_3CH_2Br.$ $^{100}\!/_{16}$ = 6.25 mol CH_4; this would give the same number of moles of CH_3Cl except yield is only 40 percent. 6.25 x 0.40 = 2.5. Each 2 mol CH_3Cl gives 1 mol of CH_3CH_3 but in 50 percent yield, so 1.25 x 0.50 = 0.625 mol ethane. Each mole ethane gives 1 mol of CH_3CH_2Br but only in 60 percent yield. 0.625 x 0.60 = 0.375 mol CH_3CH_2Br. 0.375 x 109 = 40.9 g ethyl bromide.

3-10   The Wurtz reaction would also lead to *n*-hexane and *n*-butane, thus lowering the yield of *n*-pentane.

3-11   *a.* $CH_3CHICH_3 \xrightarrow{Mg} (CH_3)_2CHMgI + H_2O \longrightarrow CH_3CH_2CH_3 + MgIOH$

     b. $2(CH_3)_2CHCH_2Br + 2Na \longrightarrow (CH_3)_2CHCH_2CH_2CH(CH_3)_2 + 2NaBr$

     c. $CH_3CH_3 + Br_2 \xrightarrow{\lambda} CH_3CH_2Br + HBr$ $2CH_3CH_2Br + 2Na \longrightarrow CH_3CH_2CH_2CH_3 + 2NaBr$

     d. $CH_4 \xrightarrow[\text{light}]{CH_2N_2} CH_3CH_3 + N_2$ $CH_3CH_3 \xrightarrow{Cl_2} CH_3CH_2Cl + HCl$

     e. $(CH_3)_4C \xrightarrow[\text{light}]{CH_2N_2} CH_3CH_2C(CH_3)_3$

     f.

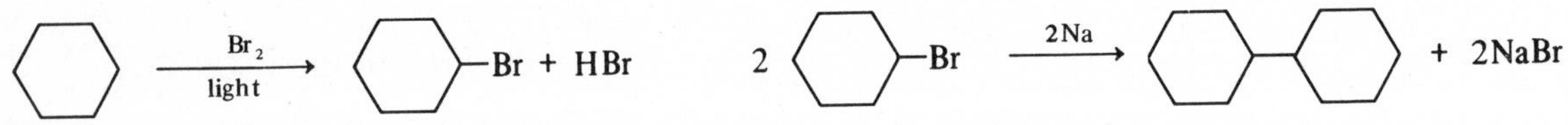

3-12   The chlorination of hydrocarbons occurs in a fairly random fashion. Therefore, the chlorination of ethane would be most practical because there is only one monochloro substitution product.

3-13   Nine 1° hydrogens $\longrightarrow$ 90% isopentane; one 3° hydrogen $\longrightarrow$ 10% neopentane. Thus 3° C—H is more reactive than 1° C—H bond in this reaction.

3-14

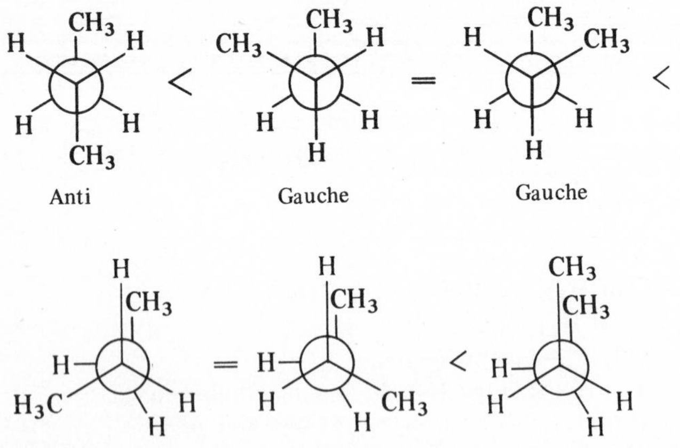

    Anti   Gauche   Gauche

3-15   *a,*

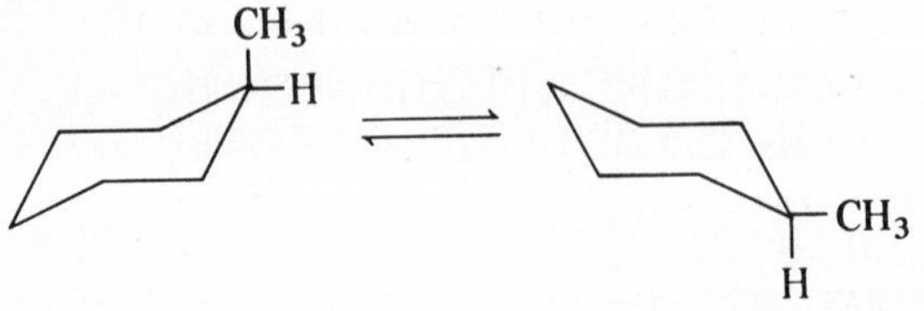

    b.

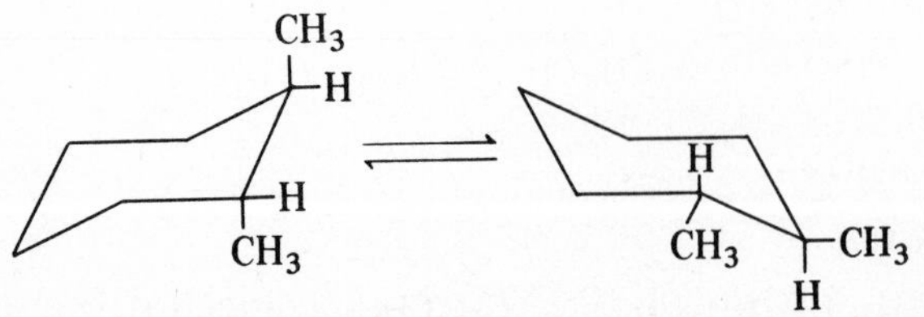

        trans-1,2-Dimethylcyclohexane 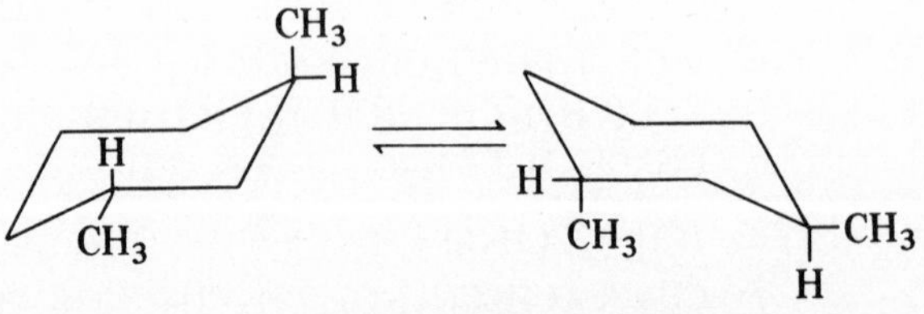   *trans*-1,3-Dimethylcyclohexane

c.

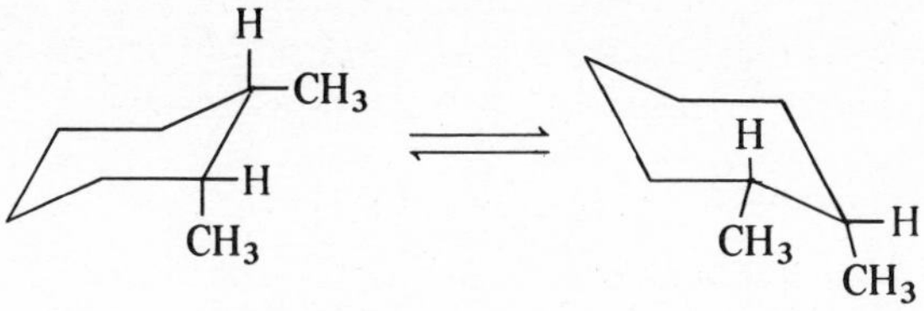

cis-1,2-Dimethylcyclohexane

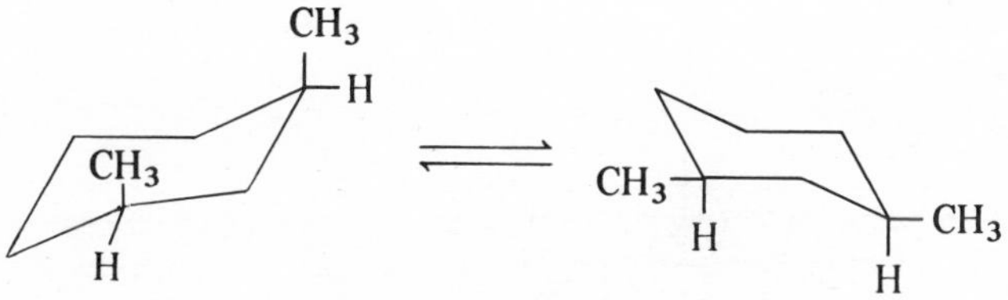

cis-1,3-Dimethylcyclohexane

3-16

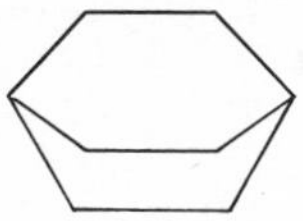

Molecule exists with all three cyclohexane rings in the boat conformation. Chair conformations would involve extreme strain.

3-17 *a.*

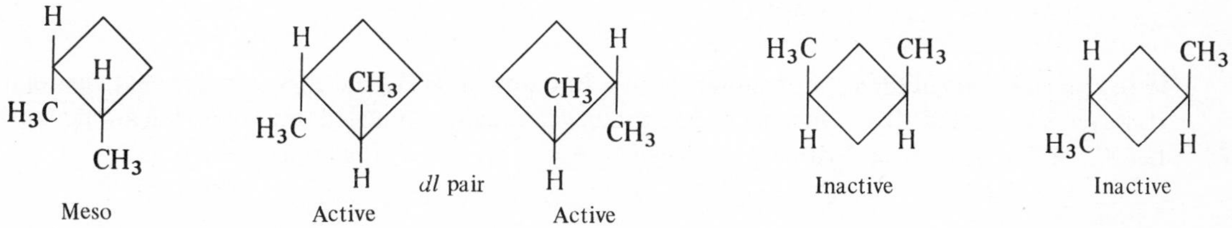

Meso   Active   *dl* pair   Active   Inactive   Inactive

b.

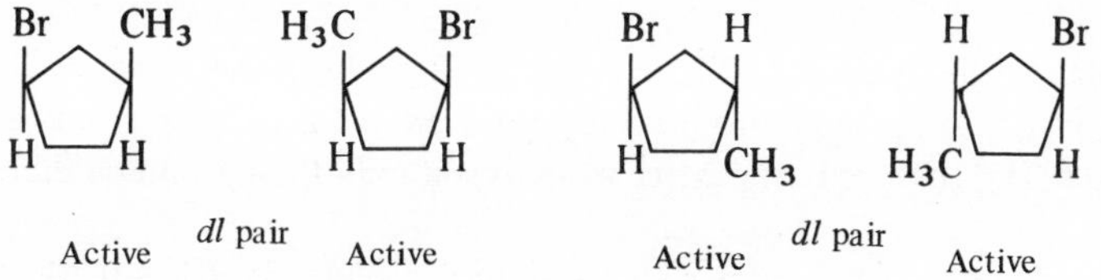

Active   *dl* pair   Active   Active   *dl* pair   Active

3-18

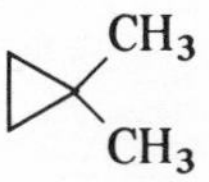

3-19   Volume at STP = 158 ml. MW = $22,400/158$ × 1 = 142 less 127 for I = 15. CH_3 = 15; hence compound is CH_3I.

3-20   Volume at STP = 286 ml. MW = $22,400/286$ × 1 = 78.3 less 35.5 for Cl = 42.8. C_3H_7 = 43; hence $CH_3CH(Cl)CH_3$ or $CH_3CH_2CH_2Cl$ would fit data.

3-21 C_7H_{16}, 83.89% C, 16.11% H; C_8H_{18}, 84.19% C, 15.81% H; no.

3-22

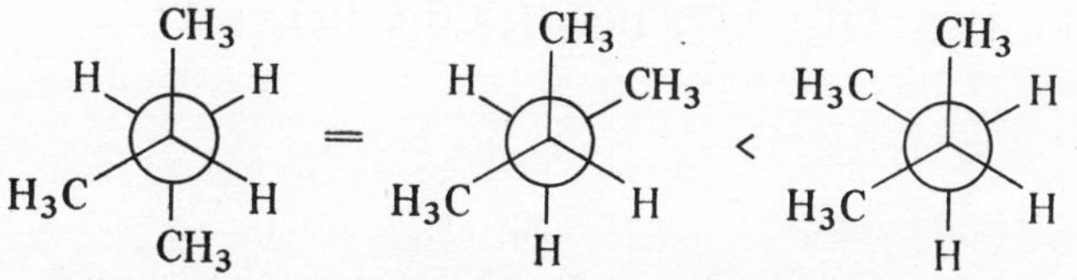

3-23

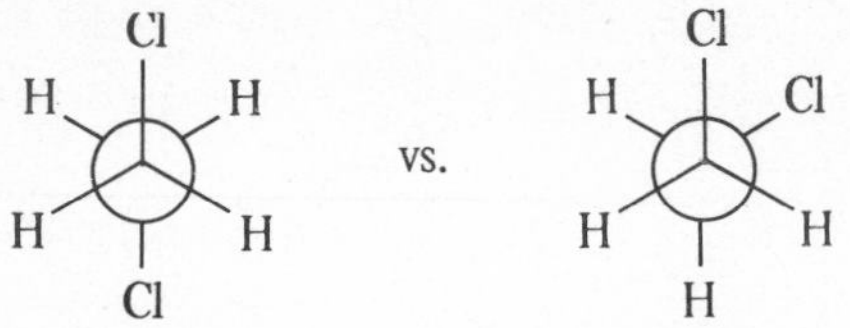

Cl   Cl

(Newman projection, anti)   vs.   (Newman projection, gauche)

$\mu = 0$ (anti) $\mu \neq 0$ (gauche)

At higher temperatures the population of the gauche conformations is increased.

3-24

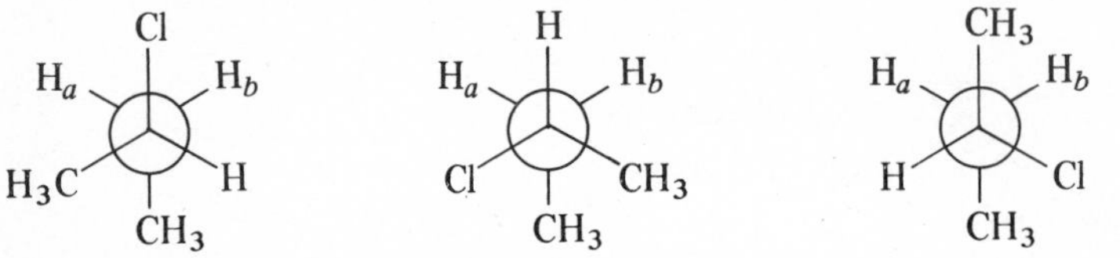

(three Newman projections with H_a, H_b, Cl, CH_3, H substituents)

Here H_a and H_b are in a different environment in each of the three conformations (none of which are of equal energy).

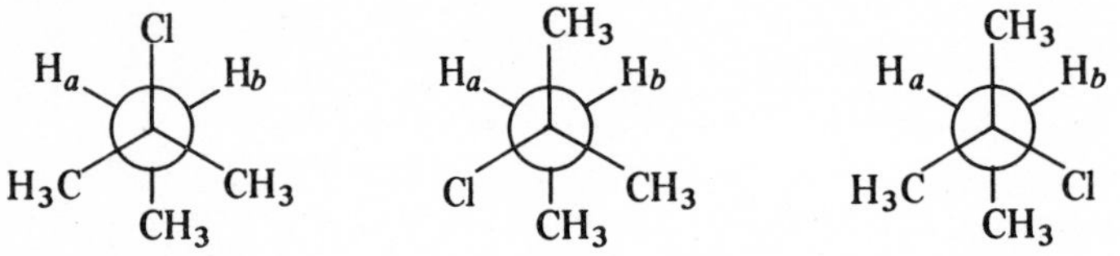

(three Newman projections labeled I, II, III)

I   II   III

Here H_a and H_b are in the same environment in I. In II and in III H_a and H_b are in different environments. However, since II and III are of equal energy and hence equally populated, the environments are averaged out.

3-25   Let X = MW of the gaseous hydrocarbon. Then 80 + X − 1 = MW of the bromo compound

$$\frac{1.37}{79 + X} = \frac{0.580}{X} \qquad \text{Solving,} \quad X = 58$$

If the alkane is C_nH_{2n+2}, its molecular weight is given by $12n + 2n + 2 = 58$; and n, the number of carbon atoms, is equal to 4. The alkane is C_4H_{10} and the monobromo compound is C_4H_9Br. Of the two possible hydrocarbons, $CH_3CH_2CH_2CH_3$ and $(CH_3)_3CH$, only the latter would yield only three isomeric dibromo derivatives.

3-26   Let X = unknown MW. Then, if only one atom of Fe (at wt 55.9) is in the molecule, 55.9/X = 0.30, X = 186 = MW.

3-27   12,571 (see Prob. 3-26).

3-28 C_2H_5OCl

3-29 $2CH_3CH_2 \cdot \longrightarrow CH_3CH_2CH_2CH_3$, $CH_3CH_2 \cdot + Cl \cdot \longrightarrow CH_3CH_2Cl$, $2Cl \cdot \longrightarrow Cl_2$

3-30   *a.* $\dfrac{\text{Butyl chloride}}{\textit{sec}\text{-Butyl chloride}} = \dfrac{6(1^\circ H)}{4(2^\circ H)}\dfrac{1.0}{3.8} = \dfrac{6}{15.2}$ or $\dfrac{28.3\%}{71.7\%}$

     b. $\dfrac{\text{Isobutyl chloride}}{\textit{tert}\text{-Butyl chloride}} = \dfrac{9(1^\circ H)}{1(3^\circ H)}\dfrac{1.0}{5} = \dfrac{9}{5}$ or $\dfrac{64.3\%}{35.7\%}$

     c.   Three products: $ClCH_2CH_2CH(C_2H_5)CH(C_2H_5)_2$, $CH_3CH(Cl)CH(C_2H_5)CH(C_2H_5)_2$, and $CH_3CH_2CCl(C_2H_5)CH(C_2H_5)_2$

$$\frac{\text{1-Chloro}}{\text{2-Chloro}} = \frac{12(1^\circ H)}{8(2^\circ H)}\frac{1.0}{3.8} = \frac{12}{30.4}$$

$$\frac{\text{2-Chloro}}{\text{3-Chloro}} = \frac{8(2°\text{H})}{2(3°\text{H})}\frac{3.8}{5.0} = \frac{30.4}{10}$$

1-Cl : 2-Cl : 3-Cl = 12 : 30.4 : 10 or 23% : 58% : 19%

d. Four products: 1, 2, or 3-chloro-2,3,4-trimethylpentane and 3-chloromethyl-2,4-dimethylpentane

$$\frac{\text{1-Chloro}}{\text{2-Chloro}} = \frac{12(1°\text{H})}{2(3°\text{H})}\frac{1.0}{5.0} = \frac{12}{10}$$

$$\frac{\text{3-Chloro}}{\text{3-Chloromethyl}} = \frac{1(3°\text{H})}{3(1°\text{H})}\frac{5}{1.0} = \frac{5}{3}$$

1-Cl : 2-Cl : 3-Cl : 3-CH$_2$Cl = 12 : 10 : 5 : 3 or 40% : 33.3% : 16.7% : 10%

3-31   *a.* H$_2$(104) + F$_2$(37) $\longrightarrow$ 2HF(2 x 135)
   ΔH = 104 + 37 − 270 = −129 kcal/mol
   b. CH$_3$—H(102) + I$_2$(36) $\longrightarrow$ CH$_3$I(53) + HI(71)
   ΔH = 102 + 36 − (53 + 71) = +14 kcal/mol
   c. (CH$_3$)$_3$C—H(91) + Cl(58) $\longrightarrow$ (CH$_3$)$_3$C—Cl(75) + HCl(103)
   ΔH = 91 + 58 − (75 + 103) = −29 kcal/mol
   d. C$_6$H$_5$CH$_2$—H(78) + Br$_2$(46) $\longrightarrow$ C$_6$H$_5$CH$_2$—Br(51) + HBr(87)
   ΔH = 78 + 46 − (51 + 87) = −14 kcal/mol
   e. C$_6$H$_5$—H(102) + Cl$_2$(58) $\longrightarrow$ C$_6$H$_5$—Cl(86) + HCl(103)
   ΔH = 102 + 58 − (86 + 103) = 29 kcal/mol
   f. CH$_3$—CH$_3$(84) + Br$_2$(46) $\longrightarrow$ 2CH$_3$Br(2 x 67)
   ΔH = 84 + 46 − (134) = −4 kcal/mol
3-32   See Fig. A3-32.
3-33   See Fig. A3-33.

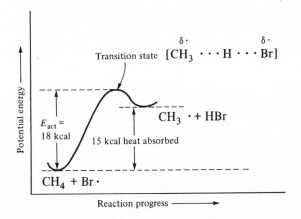

Figure A3-32. Bromination of methane−energy diagram.

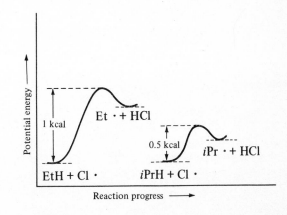

Figure A3-33. Chlorination of ethane vs. propane−energy diagram.

3-34   SO$_2$Cl$_2$ $\longrightarrow$ Cl· + SO$_2$Cl· $\longrightarrow$ 2Cl· + SO$_2$
   Cl· + RH $\longrightarrow$ HCl + R·   R· + SO$_2$Cl· $\longrightarrow$ RSO$_2$Cl
3-35   *a.* The nitrogen and oxygen atoms are probably both in the *sp*^{2} state. This allows for good *p* overlap in the system. The σ bonds between N and O are formed by *sp*^{2} bonds overlapping. Carbon is in the *sp*^{3} state.
   b. Magnesium is in the *sp* state in this linear compound.
   c. Boron is in the *sp*^{2} state. It forms three bonds to *sp*^{3} carbon, all in the same place. Boron has a vacant *p* orbital at right angles to this plane which means it can accept a pair of electrons easily. This explains its strong acid character.

Chapter 4

4-1 *a.* Propylene or methylethylene
 c. Trimethylethylene
 e. Isoamylene, or isopropylethylene
 g. Isopropenyl chloride
 i. Propargyl chloride
 k. Propynyl bromide

 b. Ethylacetylene
 d. Diethylacetylene
 f. Allyl chloride
 h. Propenyl chloride
 j. *t*-Butylacetylene
 l. Vinylacetylene

4-2 *a.* 2,3-Dimethyl-2-pentene
 c. 1,4-Dichloro-1-butyne
 e. 1,3,5-Heptatriyne
 g. 1-Allyl-2-methyl-1,3-cyclohexadiene
 i. 1-(1-Cyclohexenyl)-4-cyclohexyl-2-butyne
 k. 4-Cyclopentyl-1-buten-3-yne

 b. 2-Chloro-3-hexene
 d. 3-Methyl-2,4-hexadiene
 f. 3-Cyclopropyl-1,4-pentadiyne
 h. 3,6-Dimethylcyclooctyne
 j. 3-Isopropyl-1-pentene
 l. 1-(3-cyclohexenyl)-3,4,4-trimethyl-2-hexene

4-3 *a.* (*E*)-2-Chloro-3-methyl-2-pentene
 c. (*Z*)-1-chloro-2-fluoro-2-methoxy-1-nitroethene

 b. (*E*)-3-Trifluoromethyl-4-methyl-2-pentene

4-4 *a.* $CH_3CHBrCH_2Br$ *b.*

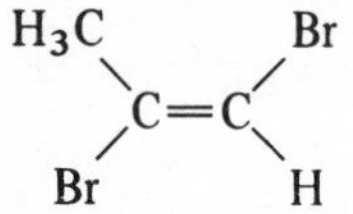

$$\underset{Br}{\overset{H_3C}{}}C=C\underset{H}{\overset{Br}{}}$$

c. ⬡—OH

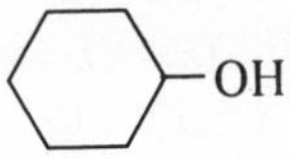

d. 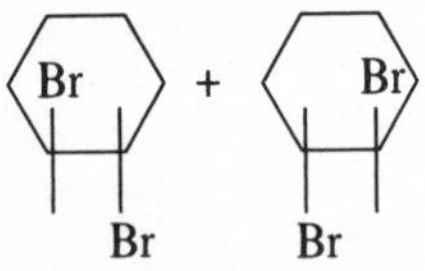 ⬡(Br) + ⬡(Br) (Br below) *e.* 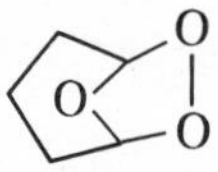 (bicyclic peroxide structure) *f.* 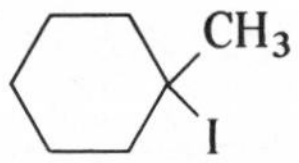 ⬡ with CH_3 and I

g. $CH_3CH_2CHCH_2Cl$ with OH *h.* $[(CH_3)_2CHCH_2CH_2]_3B$ *i.* $CH_3CH_2CHCH_3$ with OH

j. $CH_3CH_2CBr_2CH_3$ *k.* $CH_3CBr_2CHBr_2$ *l.* $CH_3CH_2CH_2C=CH_2$ with CH_3COO

m. 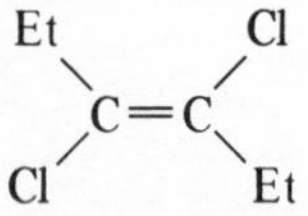 $\underset{Cl}{\overset{Et}{}}C=C\underset{Et}{\overset{Cl}{}}$ *n.* $CH_3(CH_2)_3CHBrCH_3$ *o.* $CH_3(CH_2)_3CH_2CH_2Br$

p. $EtCCl_2CCl_2Et$ *q.* $CH_3(CH_2)_3CHICH_3$ *r.* $CH_3(CH_2)_3CHICH_3$

s. 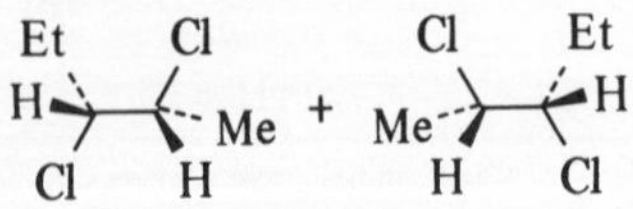 (two Fischer projections) *dl* threo *t.* 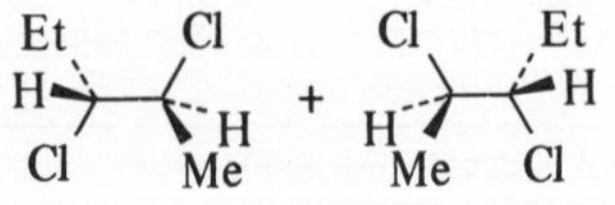 (two Fischer projections) *dl* erythro

u. $CH_3CH_2CH_2CCH_2CH_3$ with O (double bond)

v. 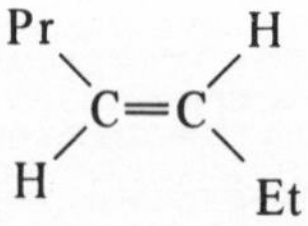 $\underset{H}{\overset{Pr}{}}C=C\underset{Et}{\overset{H}{}}$ *w.* 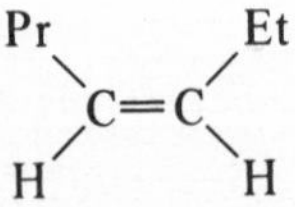 $\underset{H}{\overset{Pr}{}}C=C\underset{H}{\overset{Et}{}}$

x. 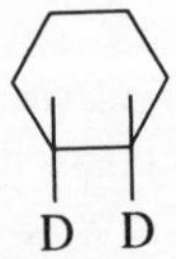 ⬡ with D D

316

4-5 a. Propene   b. Propene   c. Propyne

d. Propyne   e. Propyne   f. 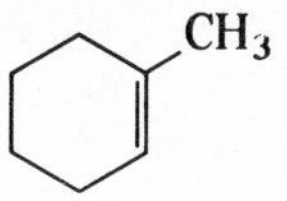 (cyclohexene with CH₃ — structure)

$f.$ shows a cyclohexene ring with a CH$_3$ group

g. 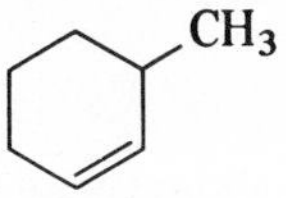 (cyclohexene ring with CH$_3$)   h. 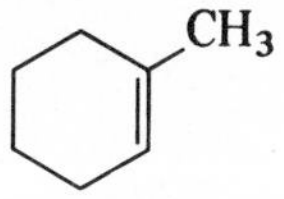 (cyclohexene ring with CH$_3$)   i. 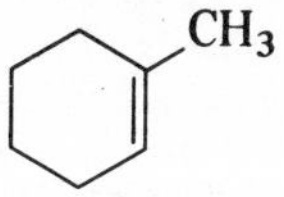 (cyclohexene ring with CH$_3$)

j. 2-Methyl-2-butene   k. Mostly *trans*-2-butene   l. 1-Hexene
m. $(CH_3)_2C=C=O$ n. $CH_2=C=O$ o. Isobutylene

4-6 a. $CH_2=CH_2 + 4KMnO_4 \longrightarrow 2K_2CO_3 + 4MnO_2 + 2HOH$
b. $C_8H_{14} + 6KMnO_4 \longrightarrow 2K_2CO_3 + C_6H_9O_3K + 6MnO_2 + KOH + 2HOH$
c. $C_7H_{12} + K_2Cr_2O_7 + 4H_2SO_4 \longrightarrow C_7H_{12}O_3 + Cr_2(SO_4)_3 + K_2SO_4 + 4HOH$
d. $3Me_2C=CMe_2 + 2KMnO_4 + 4HOH \longrightarrow 3Me_2C(OH)C(OH)Me_2 + 2MnO_2 + 2KOH$

4-7 a. $CH_3CH=CH_2 \xrightarrow{HI} CH_3CHICH_3$ b. $CH_3CH=CHCH_3 \xrightarrow{KMnO_4} 2CH_3COOK$

c. $CH_3CH=CH_2 \xrightarrow[HBr]{peroxide} CH_3CH_2CH_2Br$

d. $CH_3CH_2CHClCH_3 \xrightarrow[KOH]{alc} CH_3CH=CHCH_3 \xrightarrow[KMnO_4]{cold, dil} CH_3\underset{\underset{OH\,OH}{|\;\;|}}{CHCHCH_3}$

e. $CH_3CH_2CH_2CH_2Br \xrightarrow[KOH]{alc} CH_3CH_2CH=CH_2 \xrightarrow{HBr} CH_3CH_2CHBrCH_3$

f. $CH_3CH(OH)CH_3 \xrightarrow{HBr} CH_3CHBrCH_3 \xrightarrow{Na} (CH_3)_2CHCH(CH_3)_2$

g. $CH_3CH_2C≡CH \xrightarrow{H_2,\,Pd} CH_3CH_2CH=CH_2 \xrightarrow[HBr]{peroxide} CH_3CH_2CH_2CH_2Br$

h. $CH_3CH_2CH_2CH_2Br$ (from g) $\xrightarrow{Na} CH_3(CH_2)_6CH_3$

i. $CH_3CH_2CH_2CH_2Br \xrightarrow{hot,\,alc\,KOH} CH_3CH_2CH=CH_2 \xrightarrow{Br_2} CH_3CH_2CHBrCH_2Br$
$\xrightarrow{NaNH_2,\,heat} CH_3CH_2C≡CH$

j. $CH_3CH_2CH_2Br \xrightarrow{hot,\,alc\,KOH} CH_3CH=CH_2 \xrightarrow{Br_2} CH_3CHBrCH_2Br \xrightarrow{hot,\,alc\,KOH} CH_3CH=CHBr$

$\left(E2 \text{ via} \right.$ Newman projection with Br, Br, H, H₃C, H, H $\left. \right)$

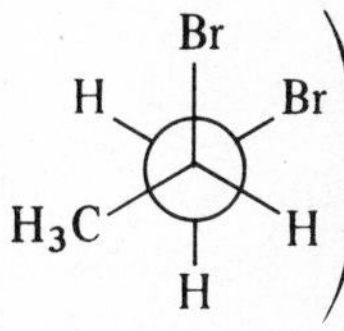

k. 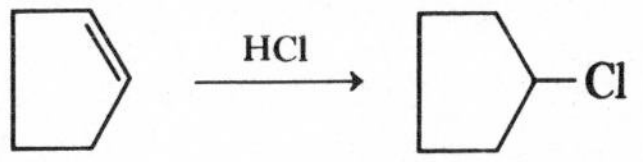 (cyclopentene) $\xrightarrow{HCl}$ (cyclopentyl chloride with —Cl)

l. 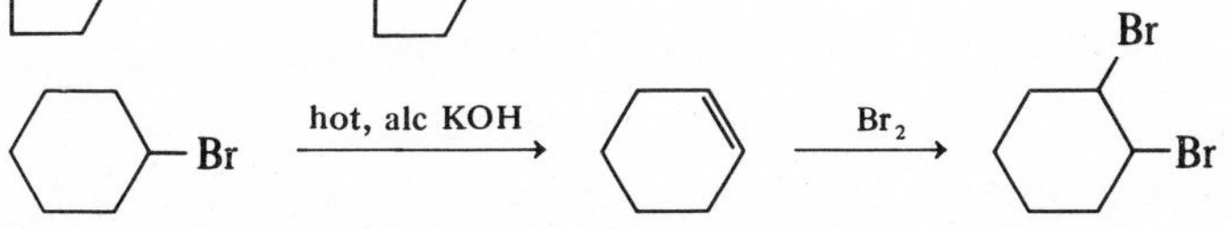 (cyclohexyl bromide —Br) $\xrightarrow{hot,\,alc\,KOH}$ (cyclohexene) $\xrightarrow{Br_2}$ (cyclohexane with Br, —Br)

m. $CH_3CH_2CH=CHCH_2CH_3 \xrightarrow{O_3} \xrightarrow{Zn,\,H_2O} CH_3CH_2CHO$

n. $(CH_3)_2C=CH_2 \xrightarrow[HBr]{peroxide} (CH_3)_2CHCH_2Br$

o. 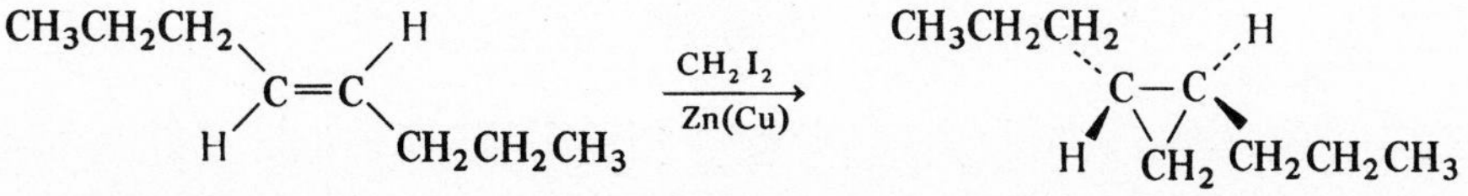 $\underset{H}{\overset{CH_3CH_2CH_2}{>}}C=C\underset{CH_2CH_2CH_3}{\overset{H}{<}} \xrightarrow[Zn(Cu)]{CH_2I_2}$ (cyclopropane product with CH₃CH₂CH₂, H, CH₂, CH₂CH₂CH₃)

317

p. 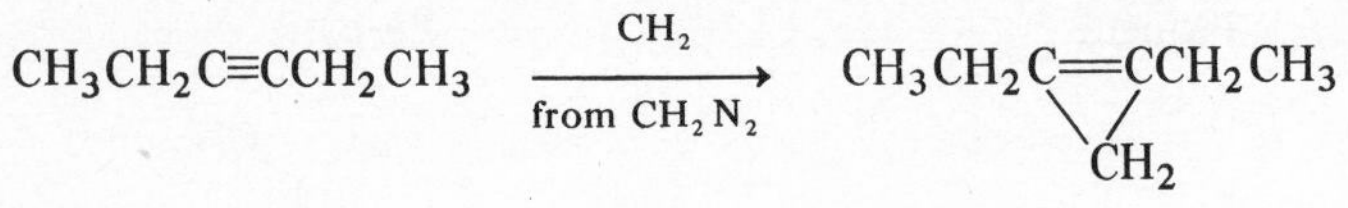 $CH_3CH_2C{\equiv}CCH_2CH_3 \xrightarrow[\text{from } CH_2N_2]{CH_2} CH_3CH_2C{=}CCH_2CH_3$ with CH_2 below

4-8   *a.* $2CH_2O + OHCCHO$ *b.* 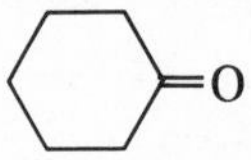   (cyclohexanone) $={}O + CH_2O$ *c.* $CH_3C(CH_2)_4CCH_3$ with O, O below

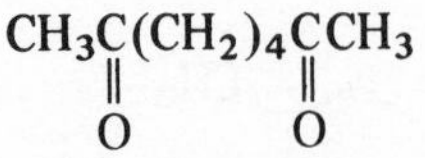

     d. $CH_3CH_2CH_2COOH + CH_3COOH$ *e.* $CH_2O + OHCCH_2COOH + HCOOH$
     f. $2HCOOH + HOOCCH_2CH_2COOH$ *g.* $CH_2O + OHCCHO + CH_3CHO$
     h. $CH_3CHO + OHCCOOH$ *i.* $2CH_2O + 2OHCCOOH$

4-9   2,3-Dimethyl-2-pentene

4-10   A, $CH_3CH_2CH{=}CHCH_2CH_3$; B, CH_3CH_2COOK

4-11   A, 2-butene; B, 2-bromobutane

4-12 $^{29}/_{22,400} = 0.0129$ mol of acetylene. $^{10}/_{346} = 0.0289$ mol of bromide. Since each mole of bromide should yield 1 mol of acetylene, $^{0.0129}/_{0.0289} \times 100 = 44.6$ percent yield.

4-13 $107.9/x = 0.668$; $x = 161 = $ MW if only one Ag atom is present in the acetylide. $161 - 107.9 = 53 = $ MW of carbon-hydrogen part of the molecule. $C{\equiv}C{-} = 24$; $53 - 24 = 29$; the sum of the atomic weights of $CH_3CH_2 = 29$; hence, $CH_3CH_2C{\equiv}CAg = $ silver salt and $CH_3CH_2C{\equiv}CH = $ unknown acetylene.

4-14   The carbon skeleton is established by hydrogenation as

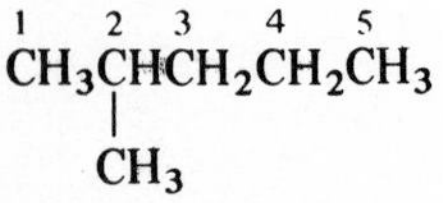

$$\overset{1}{C}H_3\overset{2}{C}H\overset{3}{C}H_2\overset{4}{C}H_2\overset{5}{C}H_3$$
$$\text{with } CH_3 \text{ below carbon 2}$$

Hydration indicates the presence of a triple bond that could be in only two positions, between 3 and 4 or between 4 and 5. The 4—5 position is ruled out because a terminal triple bond would react with sodium or ammoniacal CuCl. Hence, the structure must be $(CH_3)_2CHC{\equiv}CCH_3$.

4-15 $4.643 \times 22.4 = 104 = $ MW. $0.2/172 = x/22,400$; $x = 26 = $ equivalent weight in terms of H_2. $^{104}/_{26} = 4 = $ number of sites of double or triple bonds. As only one product, OHCCHO, was obtained on ozonolysis, the molecule

must be symmetric and cyclic. The only possible structure would be

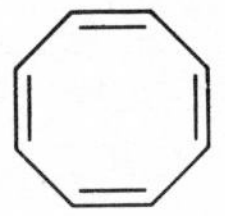

4-16   *a.*  (benzene ring)   *b.* 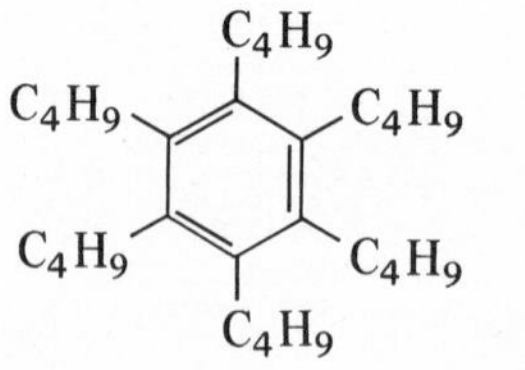 (benzene ring with substituents C_4H_9 at all positions)   *c.* (benzene ring with C_2H_5, C_2H_5, C_2H_5)

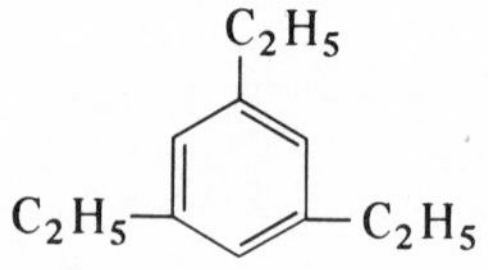

4-17   *a.*   No isomerism

     b. 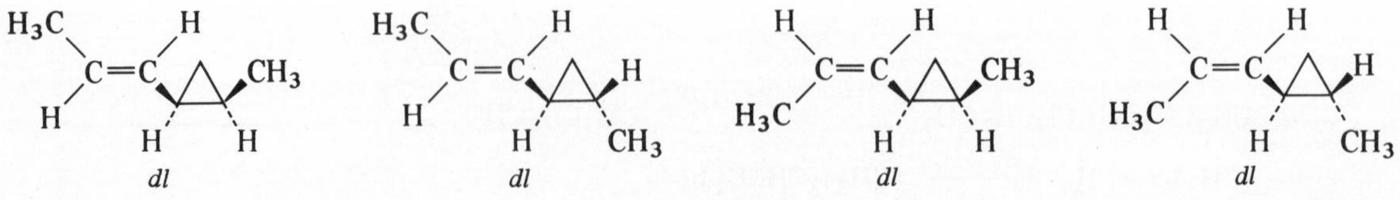   (four structures of methylcyclopropane/alkene, each labeled *dl*)

     c. 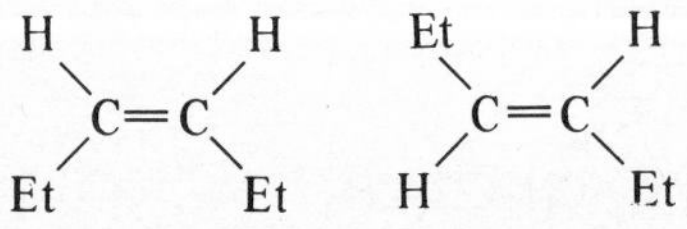   (two alkene structures with Et and H groups)

318

d. (structural formulas - four chlorocyclohexane with two CH₃ groups, each labeled)

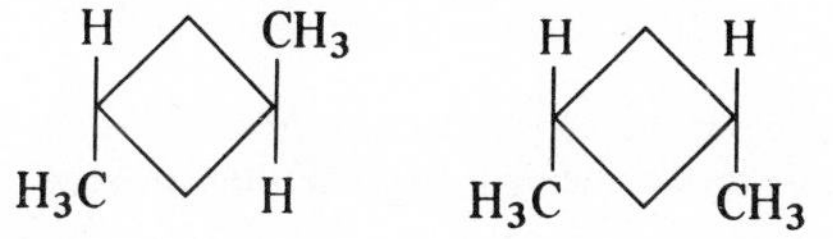

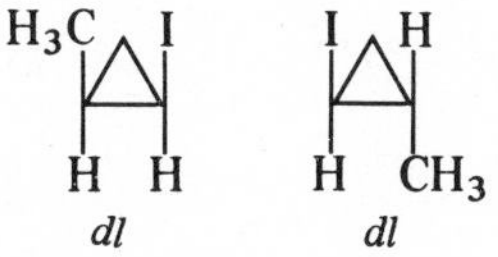

dl   dl   dl   dl

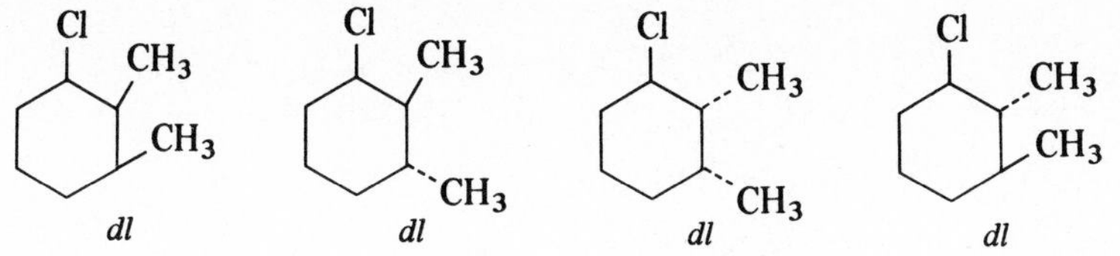

e. (three diene structural formulas)

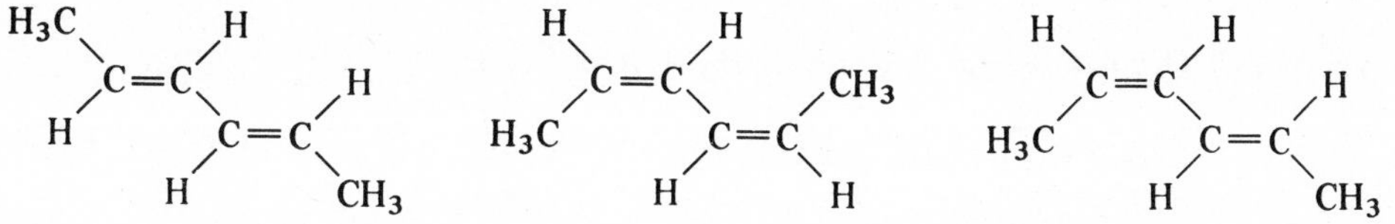

f. (two cyclobutane structures with CH₃ groups)   **g.** No isomerism

h. (four fatty acid structural formulas with C_5H_{11} and $(CH_2)_7COOH$)

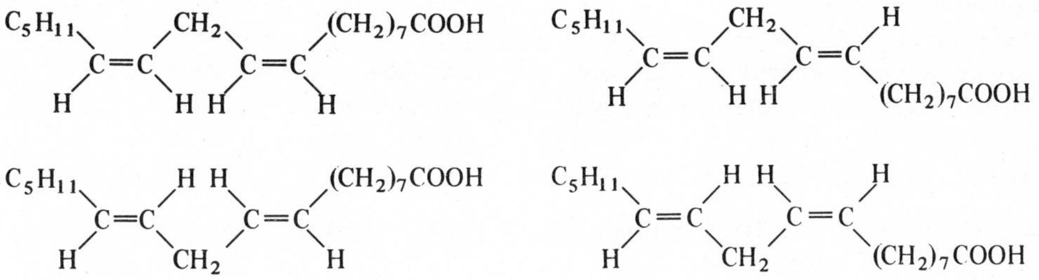

i. (two cyclopropane structures)   **j.** (two alkene structures with Cl and CH₃)

dl   dl

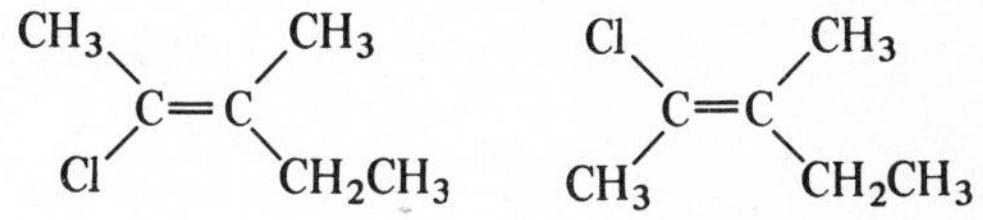

k. No isomerism

4-18 $CH_3CH_2C(CH_3)=CH_2$

4-19   (cyclobutyl)$-C\equiv CCu$ (methylcyclopropyl)$-C\equiv CCu$ cis or trans   (methylcyclopropyl)$-C\equiv CCu$

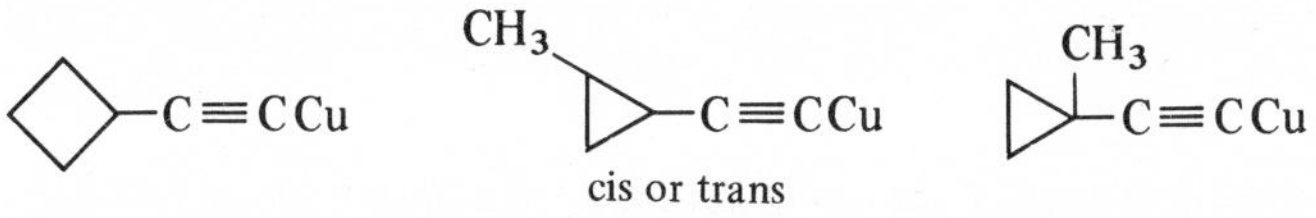

4-20   *a.* 2   *b.* 3   *c.* 4   *d.* 1   *e.* 6   *f.* 3

4-21   *a.* 2   *b.* 1   *c.* 2   *d.* 2

4-22   *a.* $HC\equiv CH \xrightarrow{NaNH_2} HC\equiv CNa \xrightarrow{CH_3I} CH_3C\equiv CH$

b. $CH_3C\equiv CH$ (from *a*) $\xrightarrow{NaNH_2} CH_3C\equiv CNa \xrightarrow{CH_3I} CH_3C\equiv CCH_3$

c. $CH_3C\equiv CH$ (from *a*) $\xrightarrow{H_2, Pd} CH_3CH=CH_2 \xrightarrow{Cl_2} CH_3CHClCH_2Cl$

d. $CH_2=CH_2 \xrightarrow{HI} CH_3CH_2I \xrightarrow{HC\equiv CNa} \text{(from } a) CH_3CH_2C\equiv CH \xrightarrow{H_2, Pd} CH_3CH_2CH=CH_2$

e. $CH_3CH_2CH=CH_2$ (from *d*) + HI $\longrightarrow CH_3CH_2CHICH_3$

f. $CH_3CH=CH_2$ (from *c*) + HBr $\xrightarrow{peroxides} CH_3CH_2CH_2Br$

g. $HC\equiv CNa$ (from *a*) + $CH_3CH_2CH_2Br$ (from *f*) $\longrightarrow CH_3CH_2CH_2C\equiv CH \xrightarrow{HOH, Hg^{2+}, H^+} CH_3CH_2CH_2CCH_3$ (with $\overset{\|}{O}$)

h. $CH_3C\equiv CCH_3$ (from *b*) $\xrightarrow{H_2, Pd}$ (cis-2-butene structure)

319

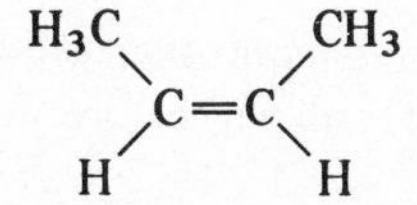

i. $CH_3C{\equiv}CCH_3$ (from *b*) $\xrightarrow{\text{Na, NH}_3}$

(cis-2-butene structure: H_3C and H on left carbon, H and CH_3 on right carbon, C=C)

j. $CH_3C{\equiv}CH$ (from *a*) $\xrightarrow[\text{H}_2\text{O}]{\text{Hg}^{2+}, \text{H}^+}$ $CH_3\overset{\displaystyle O}{\overset{\|}{C}}CH_3$

k. $CH_3CH{=}CH_2$ (from *c*) $\xrightarrow{\text{ICl}}$ $CH_3CHClCH_2I$

l. $CH_3CH_2CH{=}CH_2$ (from *d*) + HBr $\xrightarrow{\text{peroxide}}$ $CH_3CH_2CH_2CH_2Br$ $\xrightarrow[\text{(from 1-Butyne + Na)}]{CH_3CH_2C{\equiv}CNa}$

$$CH_3CH_2CH_2C{\equiv}CCH_2CH_2CH_3$$

$$\Big\downarrow \text{H}_2\text{, Pt}$$

cis-4-octene $\longleftarrow$

4-23   *a.* 1-Hexene will decolorize a solution of Br_2 in CCl_4.

     b. Heptane will not dissolve in an authentic sample of water; or, pour the two together and observe that the top phase will be the lighter heptane.

     c. Sodium chloride is a solid.

     d. Ethyl alcohol will dissolve in water; or, 1-octene will rapidly decolorize Br_2 in CCl_4.

     e. 1-Butyne will form a precipitate with ammoniacal silver nitrate.

     f. 1-Butyne will form a precipitate with ammoniacal silver nitrate.

4-24   *a.* (resonance structures of ethyl cation / protonated species)

     b. $H_3C{-}\overset{+}{\underset{\displaystyle CH_3}{C}}{-}CH_2Cl$ Delocalization of the positive charge by the eight C—H bonds via hyperconjugation as in *a*.

     c. (cyclobutyl cation with I)   Delocalization of positive charge by the three adjacent C—H bonds.

     d. $Cl{-}\overset{+}{\underset{\displaystyle H}{C}}CH_2Br$ $\longleftrightarrow$ $\overset{+}{Cl}{=}CHCH_2Br$ + two hyperconjugation possibilities

     e. $CH_3\overset{+}{\underset{\displaystyle H}{C}}{-}OCH_3$ $\longleftrightarrow$ $CH_3\underset{\displaystyle H}{C}{=}\overset{+}{O}CH_3$ + three hyperconjugation possibilities

     f. $CH_3CH_2\overset{+}{\underset{\displaystyle F}{C}}{-}F$ $\longleftrightarrow$ $CH_3CH_2\underset{\displaystyle F}{C}{=}F^+$ $\longleftrightarrow$ $CH_3CH_2\overset{F^+}{\overset{\|}{C}}{-}F$ + two hyperconjugation possibilities

4-25 $Br_2 + CH_2{=}CH{-}CH{=}CH_2 \longrightarrow$ $[Br{-}CH_2{-}\overset{+}{C}H{-}CH{=}CH_2 \longleftrightarrow Br{-}CH_2{-}CH{=}CH{-}\overset{+}{C}H_2]\,Br^-$

     In the intermediate, the positive charge is delocalized over two carbon atoms. Each of these is a potential site for bonding with the bromide ion. If bonding takes place at the terminal carbonium carbon, the product will be $BrCH_2CH{=}CHCH_2Br$.

4-26   The first step of the reaction is protonation of the diene to yield the best carbonium ion. In the second step, the chloride ion can bond at one of two different sites, yielding the indicated products.

$$CH_2{=}CH{-}CH{=}CH_2 + HCl \longrightarrow [CH_3{-}\overset{+}{C}H{-}CH{=}CH_2 \longleftrightarrow CH_3{-}CH{=}CH{-}\overset{+}{C}H_2] \xrightarrow{Cl^-}$$

$$CH_3{-}CHCl{-}CH{=}CH_2 + CH_3{-}CH{=}CH{-}CH_2Cl$$

     1,2-Addition product is formed faster (kinetically controlled product) while 1,4-addition product is the more stable (thermodynamically controlled product) and will be favored at higher temperatures and longer reaction times.

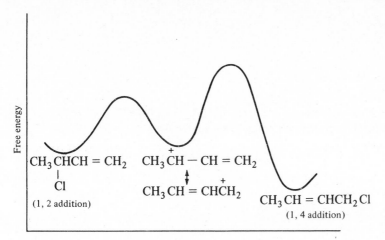

$CH_3CHCH=CH_2$ (with Cl below, 1,2 addition) ... $CH_3\overset{+}{C}H-CH=CH_2 \updownarrow CH_3CH=CH\overset{+}{C}H_2$ ... $CH_3CH=CHCH_2Cl$ (1,4 addition)

Figure A4-26. Energy diagram for 1,2 vs. 1,4 addition to 1,3-butadiene.

4-27 $BrCH_2CH_2Br$ $BrCH_2CH_2Cl$ $BrCH_2CH_2OH$

For $ClCH_2CH_2Cl$ to form there must be initial attack by Cl^+ which is not possible with only NaCl present.

$$Br_2 + H_2O \rightleftharpoons HO^-Br^+ + CH_2=CH_2 \longrightarrow HOCH_2CH_2Br$$

4-28   When HCl adds to acrolein, one of two potential carbonium ions can form:

$$HCl + CH_2=CH-CH=O \longrightarrow CH_3-\overset{+}{C}H-CH=O \text{ or } \overset{+}{C}H_2-CH_2-CH=O$$

In the first of these, there is an unfavorable interaction between the charge on the carbonium carbon and the partial positive charge on the carbon containing the double-bond oxygen. The unfavorable interaction between like charges is less severe when the two sites of low electron density are isolated by a methylene group. Consequently the relative stability of positive charges will favor the anti-Markovnikov addition.

4-29   *a.* Since a methyl group is electron-releasing relative to hydrogen, the carbon-carbon double bond will have a higher electron density in 2-butene than in 1-butene. Hence the former will react more rapidly with electrophilic reagents.

b. The butadiene is more reactive toward an electrophilic attack because of the strong electron-attracting power that deactivates the double bond in acrolein (see the answer to Prob. 4-28). The activation energy for the addition of HBr to the diene is very low because the positive charge is highly delocalized (see the answer to Prob. 4-25).

c. In the given solvent system at a high temperature, the reaction will be $E1$ for a tertiary halide. Consequently the basicity of the reagent (HO^- or H_2N^-) will have no influence on the rate of reaction.

d. With a primary halide, even in an ionizing solvent, the carbonium ion mechanism is relatively slow compared with the $E2$ process. In this case the reaction rate will be a function of the basicity of the reagent. Amide ion will react faster.

4-30   *a.* $CH_3OCH=CH_2 \longleftrightarrow CH_3\overset{+}{O}=CH\overset{-}{C}H_2$

b. $CH_2=CHC\equiv CH \longrightarrow \overset{+}{C}H_2-\overset{-}{C}HC\equiv CH \longleftrightarrow \overset{-}{C}H_2-\overset{+}{C}HC\equiv CH \longleftrightarrow CH_2=CH\overset{+}{C}=\overset{-}{C}H \longleftrightarrow$
$CH_2=CH\overset{-}{C}=\overset{+}{C}H \longleftrightarrow \overset{+}{C}H_2-CH=C=\overset{-}{C}H \longleftrightarrow \overset{-}{C}H_2-CH=C=\overset{+}{C}H$

c. $HC\equiv C-F \longleftrightarrow H\overset{-}{C}=\overset{+}{C}-F \longleftrightarrow H\overset{+}{C}=\overset{-}{C}-F \longleftrightarrow H\overset{-}{C}=C=\overset{+}{F}$

d. $CH_2=CHC\equiv N \longleftrightarrow \overset{+}{C}H_2\overset{-}{C}HC\equiv N \longleftrightarrow \overset{-}{C}H_2CH=C=\overset{+}{N} \longleftrightarrow CH_2=CH\overset{+}{C}=\overset{-}{N}$

e. Same as *d* + $\overset{-}{C}l=CHCH=C=\overset{-}{N}$ *f.* $(CH_3)_2NCH=CH_2 \longleftrightarrow (CH_3)_2\overset{+}{N}=CH\overset{-}{C}H_2$

4-31   *a.* $CH_3\underset{+}{C}=CH_2 \longleftrightarrow H-\overset{H^+}{\underset{H}{C}}=C=CH_2 \longleftrightarrow H-\overset{H}{\underset{H^+}{C}}=C=CH_2 \longleftrightarrow H^+ \overset{H}{\underset{H}{C}}=C=CH_2$

b. $ICH_2\underset{+}{C}F \longleftrightarrow ICH_2\overset{H}{C}=\overset{+}{F}$ + two hyperconjugation structures of type shown in *a*

321

c. $CH_3CH=\overset{}{\underset{+}{C}}-F \longleftrightarrow CH_3CH=C=\overset{+}{F}$

d. $CH_3\overset{+}{C}H\overset{\underset{|}{Cl}}{C}HCF_3$ + four hyperconjugation structures

e. $\overset{+}{C}H_2CH_2CF_3$ This intermediate would form rather than $CH_3\overset{+}{C}HCF_3$ because the strong electron-attracting CF_3 group makes the adjacent CH group electron-deficient.

f. $CH_3\overset{+}{O}CHCH_3 \longleftrightarrow CH_3\overset{+}{O}=CHCH_3$

4-32   a. HI is stronger. The negative charge in I^- is spread over a larger ion than in Cl^-. Hence, it binds a proton less tightly.

b. CH_3OH is stronger because O is more electronegative than N.

c. The C—H bond in acetylene is more polar than that in ethane.

d. $\overset{+}{N}H_4$ is the stronger acid.

e. CH_3SH is stronger for reason given in a

f.

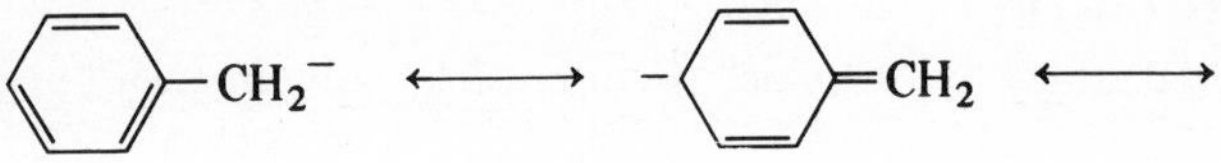

Resonance spreads negative charge over large area, making this a weak anion.

g.

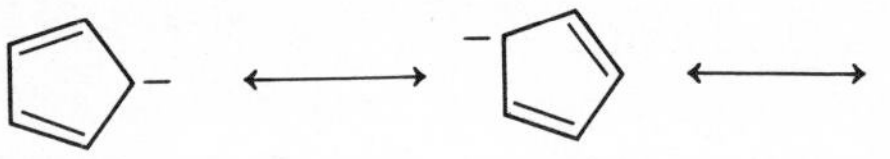

More effective dispersion of the negative charge by resonance means that this anion is more stable than $CH_2=CHCH_2^-$.

h. Cl_3CH is stronger because of stronger inductive pull of electrons by three chlorine atoms.

i. $CH_2(CN)_2$ is stronger because of the electronegativity of N compared to carbon.

j.

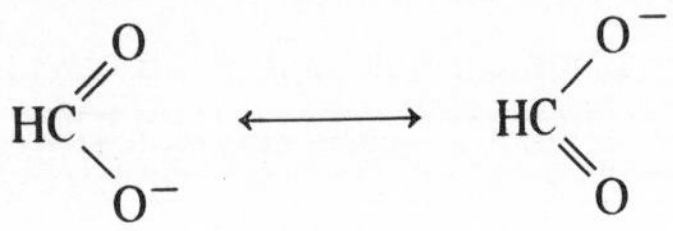

Stabilization of the anion by resonance makes this the stronger acid.

k. Both anions are stabilized by resonance $^-CH_2CH=O \longleftrightarrow CH_2=CH-O^-$; however, oxygen is more electronegative than carbon; hence, this is the stronger acid.

l. $OH^- \longrightarrow O^{2-} + H^+$. The O^{2-} ion with two negative charges on the same atom would be much less stable than ^-OH. Thus, water is the stronger acid.

4.33

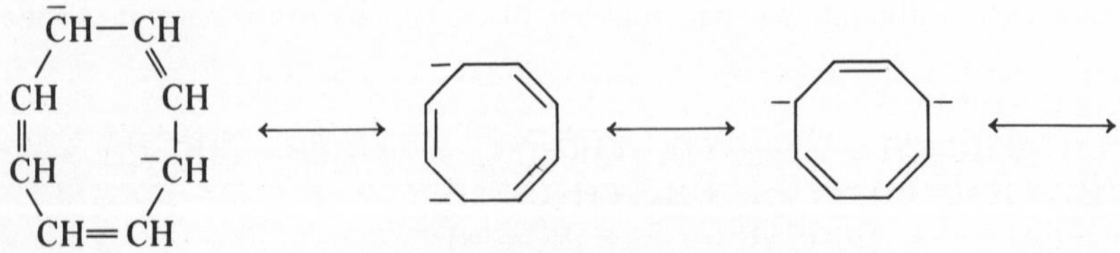

When two electrons are added to the π-electron system in cyclooctatetraene, the total is 10 instead of 8. The $4n + 2$ rule is satisfied and a stable resonance system results. The stability gained through resonance more than compensates that lost by going from the tub to the planar form.

4-34   a. Carbon and nitrogen are in the sp^2 state. Oxygen is close to sp^3. Bond angles around C and N would be close to $120°$.

b. Carbon and nitrogen would be in the sp state. Two p orbitals from C and two from N would overlap to complete the triple bond. A fluorine p orbital would overlap with a C sp orbital to form this σ bond. Nitrogen would have a lone pair of electrons in an sp orbital.

c. The two acetylenic carbons would be in the sp state while the two olefinic would be in the sp^2 state. The σ bond joining the two central carbons would be the result of overlap of an sp and an sp^2 orbital.

322

d. The electronic structure of peroxide is uncertain; *p* orbitals may be used to form the $O-O$ bond or may be in the sp^3 state.

4-35   The $4n + 2$ rule will be satisfied if a pair of electrons is moved into the ring from the diazo group as follows:

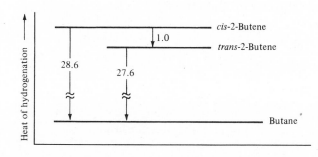

4-36   If 1 liter weighs 3.57 g, then 22.4 liters weigh 3.57 × 22.4 = 80 g = MW of hydrocarbon. Br_2 solution contains 0.2 mol/liter, so 0.025 liter × 0.2 mol/liter = 0.005 mol of Br_2. If 0.2 g hydrocarbon reacts with 0.005 mol, then $\frac{80}{0.2}$ × 0.005 = 2 gives the number of moles of Br_2 that react with 1 mol (80 g) of hydrocarbon. Since each double bond reacts with 1 mol, there are two double bonds present.

4-37   The molecular weight of the compound is about 368 and there are 12 double bonds (or 6 triple bonds; or a combination of the two). In the range of molecular weights that might be included within experimental error of 368 there are many possible structures, such as $CH_3CH_2(CH=CH)_{12}CH_2CH_3$ (MW 370). If this compound contained one ring, the MW would be exactly 368.

4-38 $CH_2{=}CH{-}\ddot{\underset{\cdot\cdot}{Cl}}:$ ⟷ $\bar{C}H_2{-}CH{=}\overset{+}{\underset{\cdot\cdot}{Cl}}:$ This type of resonance is in opposition to normal electron displacement in $C \longmapsto Cl$

4-39   *a.*

Br
H ╱╲ CH₃
H │ CH₃
H
I

Br
H₃C ╱╲ H
H │ CH₃
H
II

Br
H ╱╲ H
H │ CH₃
CH₃
III

b. trans-2-Butene via the more favorable conformation II in part *a.*

c. Fig. P4-39*c*

Heat of hydrogenation —

cis-2-Butene

1.0

trans-2-Butene

28.6

27.6

Butane

Figure A4-39c. Heat of hydrogenation, *cis*- vs. *trans*-2-butene.

d. In the energy diagram shown in *c* above, 1-butene would have an energy level above that of both the other butenes. Since it is at higher energy than *trans*-2-butene it is *less* stable. Consequently an elimination reaction that favors the more-stable product would favor *trans*-2-butene over 1-butene.

Chapter 5

5-1 $CH_3CH_2CH_2CHCl_2$ $CH_3CHClCHClCH_3$ (3)
$CH_3CH_2CHClCH_2Cl$ (2) $CH_3CHClCH_2CH_2Cl$ (2)
$Cl(CH_2)_4Cl$ $CH_3CH_2CCl_2CH_3$
$(CH_3)_2CHCHCl_2$ $(CH_3)_2CClCH_2Cl$
$CH_3CH(CH_2Cl)_2$

5-2   *a.*   2-Iodopropane
        Isopropyl iodide

b.   3-Chloropropene
      Allyl chloride

c.   2-Fluoropropane
      Isopropyl fluoride

d.   Bromoethene
      Vinyl bromide

e.   1,2-Dibromoethane
      Ethylene bromide

f.   Dichloromethane
      Methylene chloride

g. 1-Chloropropene
Propenyl chloride

h. 3-Bromopropyne
Propargyl bromide

i. Trichloromethane
Chloroform

j. 1-Chloro-2-methylpropane
Isobutyl chloride

k. Iodocyclohexane
Cyclohexyl iodide

l. 2-Iodobutane
sec-Butyl iodide

m. Tribromomethane
Bromoform

n. Tetrafluoromethane
Carbon tetrafluoride

o. 1,2-Dichloropropane
Propylene chloride

5-3 *a.* 2-Fluoro-3-chloro-4-methyl-1-pentanol   *b.* 3-Chloro-6-bromo-2-methyl-1,4-hexadiene
c. 1-Chloro-3-(2-chloroethyl)-5-methylhexane   *d.* 2-Chloro-3-(1-chloroethyl)-4-fluoro-4-isopropylheptane

5-4 *a.* $CH_3CH_2CH_2OH$ *b.* $(CH_3)_2CHOCH_3$ *c.* CH_3OH

d. CH_3SCH_3 *e.* cyclohexyl—SH   *f.* $N{\equiv}C(CH_2)_5C{\equiv}N$

g. EtSSEt   *h.* $HC{\equiv}CEt$ *i.* $ClCH{=}CHCH_2OOCCH_3$

j. $CH_3OOCCOOCH_3 + 2I^-$ *k.* $CH_3C{\equiv}CCH_3 + 2NaI$ *l.* $CH_3(CH_2)_5CN + I^-$

5-5 *a.* $MeI + HO^-$ *b.* $EtI + HS^-$

c. $CH_3S^- + CH_3I$; or, $2CH_3I + S^{2-}$ *d.* $HC{\equiv}CNa + ICH_2CH_2CH_3$

e. $EtC(=O)O^- + EtI$ *f.* $CH_3I + SCN^-$

g. $CH_3I + NH_3$ (forms the protonated form, $CH_3\overset{+}{N}H_3$, but excess NH_3 can remove H^+)

h. $(CH_3)_2PH + CH_3I$; or, $2CH_3I + CH_3PH_2$; or, $3CH_3I + PH_3$

i. $CH_3C(=S)S^- + CH_3I$ *j.* $3EtI + NH_3$

k. $EtI + EtO^-$ *l.* Large excess $EtI + NH_3$

5-6 *a.* $:C{\equiv}N:$ *b.* $CH_3{-}C{\equiv}\overset{+}{N}:$ and $CH_3{-}\overset{+}{N}{\equiv}\overset{-}{C}:$

c. Same as the second structure in *b*, because nitrogen is more electronegative than carbon.

d. A = $:CCl_2$ B = $R\overset{+}{N}H_2\ddot{C}Cl_2$ C = $R\overset{+}{N}{\equiv}C:$

e.  $CH_3{-}O{-}N{=}O$ and $CH_3{-}\overset{+}{N}(=O)O^-$

f. $CH_3CH_2CH_2CH_2{-}\overset{+}{N}(=O)O^-$

5-7 *a.* 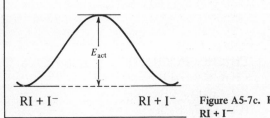   *b.* Same as the structure in *a*.

c. Since the energy levels for reactants and products are identical the activation energies will be identical.

![graph] E_{act} RI + I⁻   RI + I⁻

Figure A5-7c. Reaction progress for RI + I⁻

d. The product gives equal parts of (R)- and (S)-2-iodobutane. Since iodide ion is both a good nucleophile and leaving group, iodide ion displaces iodide from iodobutane with inversion so that eventually equal amounts of the two forms remain. Initially there is more R than S so that the rate of formation of S is greater than that of R. But as the concentration of the two approach each other their rates of formation will become equal.

5-8 *a.*

$$\left[\begin{array}{c} CH_2 \\ \parallel \\ CH \\ Y\text{-----}C\text{-----}Cl \\ H \quad CH_3 \end{array} \right]^{-}$$

b. $[\,Y\text{---}CH_2 = CH = CH\text{---}Cl\,]^{-}$ with CH_3 below the central CH

c,d. The S_N1 and S_N1' transition states are identical. The transition state involves ionization of the carbon-chlorine bond.

$$\overset{\delta^+}{CH_2} = CH = \overset{\delta^+}{CH_2} \text{---} \overset{\delta^-}{Cl}$$

e.

$$CH_2 = CH - CH = CH - CH_2 - Cl$$
$$Y^{-}$$

5-9 *a.* $EtI + CH_3OK \longrightarrow EtOCH_3 + KI \qquad (S_N2)$ *b.* $CH_3Cl + KCN \longrightarrow CH_3CN + KCl \qquad (S_N2)$

c. Secondary alkyl halides often react by a mixed S_N1-S_N2 mechanism. Isopropyl alcohol would be formed by both mechanisms.

d. $Mg + ICH_2CH_2I \longrightarrow MgI_2 + CH_2 = CH_2 \qquad (E2)$

e. $4C_4H_9Br + LiAlH_4 \longrightarrow 4C_4H_{10} + LiBr + AlBr_3 \qquad (S_N2)$

f. $(CH_3)_3CI + NaOH \longrightarrow (CH_3)_2C = CH_2 + NaI + H_2O \qquad (E1)$

g. $CH_3I + KSCN \longrightarrow CH_3SCN + KI \qquad (S_N2)$

h. $CH_2 = CHCH_2Br + CH_3COONa \longrightarrow CH_2 = CHCH_2OCOCH_3 + NaBr \qquad (S_N1)$

i. $2CH_3CH_2C(CH_3)_2I + Ag_2O + H_2O \longrightarrow 2CH_3CH = C(CH_3)_2 + 2AgI + 2H_2O \qquad (E1)$

j. $(CH_3)_2CHCI(CH_3)_2 + NaSH \longrightarrow (CH_3)_2C = C(CH_3)_2 + NaI + H_2S \qquad (E1)$

k. $CH_3I + NaNO_2 \longrightarrow CH_3NO_2 + NaI \qquad (S_N2)$

l. $CH_3Br + NaC \equiv CH \longrightarrow CH_3C \equiv CH + NaBr \qquad (S_N2)$

m. $CH_3CH = CHCl + NaNH_2 \longrightarrow CH_3C \equiv CH + NaCl + NH_3 \qquad (E2)$

n. $4CH_3MgI + SnCl_4 \longrightarrow (CH_3)_4Sn + 4MgICl \qquad (S_N2)$

o. $EtI + NH_3 \longrightarrow CH_3CH_2\overset{+}{N}H_3I^{-} \qquad (S_N2)$

p. $CH_3CCl_2CH_3 \xrightarrow[\text{NaOH}]{H_2O} (CH_3)_2C\big\langle{}^{Cl}_{OH} \quad (S_N1\text{-}S_N2) \xrightarrow{-HCl} (CH_3)_2C = O$

Unstable

q. $(CH_3)_2CCl_2 \xrightarrow[\text{KOH}]{\text{alc}} CH_3C \equiv CH \qquad (E2)$

r. $C_6H_5CH_2Br + HO^{-} \longrightarrow C_6H_5CH_2OH + Br^{-} \qquad (S_N1)$

5-10 *a.* $n\text{-}C_4H_9Br + HO^{-} \longrightarrow n\text{-}C_4H_9OH + Br^{-}$

b. $n\text{-}C_4H_9Br \xrightarrow{\text{hot, alc KOH}} CH_3CH_2CH = CH_2 \xrightarrow{HBr} CH_3CH_2CHBrCH_3$

c. $n\text{-}C_3H_7Br \xrightarrow{\text{hot, alc KOH}} CH_3CH = CH_2 \xrightarrow{NBS} BrCH_2CH = CH_2$

d. $CH_3CHBrCH_3 \xrightarrow{\text{hot, alc KOH}} CH_3CH = CH_2 \xrightarrow[\text{peroxide}]{HBr} CH_3CH_2CH_2Br$

e. $CH_3CH_2CH = CH_2 \xrightarrow{\text{HBr, peroxide}} \xrightarrow{CN^{-}} CH_3(CH_2)_3CN$

f. $CH_3CH = CH_2 \xrightarrow{Br_2} \xrightarrow{KNH_2} CH_3C \equiv CH \xrightarrow{Br_2} CH_3CBr_2CHBr_2$

g. $CH_3C \equiv CCH_3 \xrightarrow{H_2,\ Pd} CH_3CH = CHCH_3 \xrightarrow{HBr} \xrightarrow{HS^{-}} CH_3CH_2CHSH$ with CH_3 below the CHSH

h.

i. $n\text{-}C_3H_7Br \xrightarrow{\text{hot, alc KOH}} CH_3CH=CH_2 \xrightarrow{Br_2} \xrightarrow{KNH_2} CH_3C\equiv CH$

j. $CH_3C\equiv CH \text{ (from } i) \xrightarrow{Na} CH_3C\equiv CNa \xrightarrow{CH_3CH_2CH_2Br} CH_3C\equiv CCH_2CH_2CH_3$

k. Bromocyclohexane $\xrightarrow{\text{hot, alc KOH}}$ cyclohexene $\xrightarrow{Br_2}$ 1,2-dibromocyclohexane

5-11 *a.* In $CH_3CH=CHCH_2Cl$ the stability of the intermediate is greater because of the delocalization of the positive charge $CH_3CH=CH\overset{+}{C}H_2 \longleftrightarrow CH_3\overset{+}{C}HCH=CH_2$. Thus Cl^- is more easily removed since as the positive charge forms through removal of Cl^-, its influence is mitigated.

b. The carbonium ion is flat. Such a flat ion cannot form at the ring juncture in this molecule; hence an S_N1 process goes only with great difficulty.

c. Same type situation prevails as in *a.* In the α-halo ether the positive charge which results in an S_N1 process can be delocalized as it forms, allowing Cl^- to be more easily removed.

$$R\overset{\cdot\cdot}{\underset{\cdot\cdot}{O}}\underset{|}{C}HCH_3 \longrightarrow [R\overset{\cdot\cdot}{\underset{\cdot\cdot}{O}}-\overset{+}{C}HCH_3 \longleftrightarrow R\overset{\cdot\cdot}{\underset{+}{O}}=CHCH_3] + Cl^-$$
(with Cl below)

d. $CH_3CH=CH-Cl \longleftrightarrow CH_3\overset{-}{C}H-CH=\overset{+}{C}l$ Such resonance means that carbon-chlorine bond has considerable double-bond character making chlorine more difficult to displace.

e. The transition state for the process resembles a carbonium ion. Therefore the phenyl compound is more reactive because the aromatic ring can aid in delocalizing charge in the transition state and in the carbonium ion intermediate.

f. Since the first step is ionization of the carbon-chlorine bond, both starting materials yield the same intermediate ion. Any subsequent reaction would have to yield the same products under identical reaction conditions. The intermediate ion is $[CH_3CH=CH\overset{+}{C}H_2 \longleftrightarrow CH_3\overset{+}{C}H-CH=CH_2]$.

5-12 *a.* In the S_N2 process the attacking reagent approaches the C—Cl bond from the back side. With neopentyl chloride the approach of the nucleophilic group is blocked by the large *tert*-butyl group.

b. In addition to normal S_N2 displacement the reaction may occur as follows:

$$CH_3CH=CH-CH_2-Cl \longrightarrow \underset{CN}{CH_3CHCH=CH_2} + Cl^-$$
(with CN^- and curved arrows shown)

c. As in *a*, back-side approach of the reagent is prevented.

d. The carbon-iodine bond is not as strong as the carbon-chlorine bond. Also I^- is a more stable ion than Cl^- because the negative charge is spread over a larger ion.

e. In the formation of a carbonium ion intermediate, the two end carbon atoms become equivalent because of resonance:

$$\overset{+}{\underset{*}{C}H_2}-CH=CH_2 \longleftrightarrow \underset{*}{CH_2}=CH-\overset{+}{C}H_2$$

The reaction with a nucleophile will give equal amounts of $N-\overset{*}{C}H_2-CH=CH_2$ and $\overset{*}{C}H_2=CH-CH_2-N$ which upon ozonolysis will give $\overset{*}{C}H_2O$ and CH_2O.

5-13 *a.* In A, a vicinal dihalide yields an alkene with Mg. In B, no reaction occurs between a Grignard reagent and a secondary halide.

b. In A, a very poor yield of the indicated product would result because chlorination is too random; also, separation of isomers would be very difficult. In B, the sodium acetylide is too basic to give a substitution reaction; only elimination would occur.

c. HCl adds in the indicated way whether peroxide is present or not. Only elimination will take place in B.

d. In A the wrong halohydrin is shown; the hydroxyl group ends up where the carbonium ion carbon appears in the intermediate. In B the hydroxyl group is acidic enough to react with a Grignard reagent.

e. Step A is satisfactory but step B will not occur because displacement of vinyl halides by nucleophiles is too difficult.

f. Fluorine reacts violently with hydrocarbons, yielding carbon tetrafluoride with HF. In the second step, alkyl fluorides are unreactive toward nucleophilic attack.

5-14

$$\text{ClCH}_2\text{CHCH}_3 \quad \xrightarrow{\text{HO}^-} \quad \text{Cl}-\text{CH}_2-\text{CH}-\text{CH}_3 \quad \longrightarrow \quad \text{CH}_2-\text{CH}-\text{CH}_3 \quad + \text{Cl}^-$$

with HSCH_2 on the first structure, $\text{S}-\text{CH}_2$ on the middle, and $\text{S}-\text{CH}_2$ on the last.

5-15 Hydrolysis: $\text{Me}_2\text{CHBr} + \text{HOH}$
- $\xrightarrow{S}$ $\text{Me}_2\text{CHOH} + \text{HBr}$
- $\xrightarrow{E}$ $\text{CH}_3\text{CH=CH}_2 + \text{H}_2\text{O} + \text{HBr}$

Ammonolysis: $\text{Me}_2\text{CHBr} + \text{NH}_3$
- $\xrightarrow{S}$ $\text{Me}_2\text{CH}\overset{+}{\text{N}}\text{H}_3 \ \text{Br}^-$
- $\xrightarrow{E}$ $\text{CH}_3\text{CH=CH}_2 + \text{NH}_4\text{Br}$

Ethanolysis: $\text{Me}_2\text{CHBr} + \text{CH}_3\text{CH}_2\text{OH}$
- $\xrightarrow{S}$ $\text{Me}_2\text{CHOCH}_2\text{CH}_3 + \text{HBr}$
- $\xrightarrow{E}$ $\text{CH}_3\text{CH=CH}_2 + \text{CH}_3\text{CH}_2\text{OH} + \text{HBr}$

Acetolysis: $\text{Me}_2\text{CHBr} + \text{CH}_3\text{COOH}$
- $\xrightarrow{S}$ $\text{Me}_2\text{CHOOCCH}_3 + \text{HBr}$
- $\xrightarrow{E}$ $\text{CH}_3\text{CH=CH}_2 + \text{CH}_3\text{COOH} + \text{HBr}$

5-16 With tertiary halides (*tert*-butyl halides, 2-halo-2-methylbutane, etc.) any reasonably good base will lead to elimination. Bases such as HO^-, RO^-, RS^-, $\text{HC}\equiv\text{C}^-$, H_2N^-, RCOO^-, H_3N, and H_2O will work. With secondary halides, use the stronger bases (the first five in the sequence) at elevated temperature.

5-17 $\text{HC}\equiv\text{CCH}_2\text{Cl}$

5-18 X, $\text{CH}_3\text{CH}_2\text{CH}_2\text{CH}_2\text{Cl}$; Y, $\text{CH}_3\text{CH}_2\text{CH=CH}_2$

5-19 *a.* Ethanol is soluble in water.

b. Allyl chloride will react rapidly with alcoholic AgNO_3 to yield a precipitate of silver chloride.

c. Butyl chloride will give a positive test with NaI in acetone.

d. Butyl chloride will give a positive test with NaI in acetone.

e. 1-Bromo-2-butene will react rapidly with alcoholic AgNO_3.

f. Ethyl iodide will yield silver iodide when treated with alcoholic AgNO_3. The other compound is very unreactive.

5-20 The 3-methyl-1-butene and 2-methyl-1-butene result from simple hydrogenation of one or the other of the double bonds. The 2-methyl-2-butene can be formed from a 1,4-addition of hydrogen to the conjugated system, which can be shown symbolically as:

$$\overset{\text{CH}_3}{\underset{|}{\text{CH}_2=\text{C}-\text{CH}=\text{CH}_2}}$$

with H atoms adding at the ends.

The 2-methylbutane will form through further hydrogenation of any one of the previous monoalkenes, and the starting material remains because some of the molecules have used up two equivalents of hydrogen.

5-21 3,4-Dibromo-1-butene, 1,4-dibromo-2-butene (from 1,4 addition), 1,2,3,4-tetrabromobutane (from further addition of bromine to one of the first two products mentioned), and some unreacted diene.

5-22

5-23 The aluminum is tetrahedral, with bonding to each hydrogen, and it has a formal negative charge. Lithium is present as a cation. Boron is more electronegative than aluminum so the hydrogens attached to it in the BH_4^- ion would have less hydride character. Furthermore, B–H bonding would be better than Al–H bonding because the orbitals are more nearly the same size.

5-24   a. Linear with Be in *sp* state.   b. Linear with both carbon and Mg in *sp* state.
       c. Planar.   d. Planar, trigonal.

5-25   a. In the transition state the negative charge is more diffuse. Therefore there will be better solvation of starting materials than of transition state with a more-polar solvent, and the energy of activation will be greater. The reaction is slower in a more-polar solvent.

b. Charge is being developed in the transition state v. the neutral reactants; thus, a less-polar solvent will decrease the rate.

c. The substrate carbon atom that is undergoing substitution becomes more negative as we go from reactants to transition state. Therefore, adjacent groups that are electron-withdrawing (such as the carbonyl group) will lower the energy of the transition state and speed up the substitution process.

d. Strong inductive effect of CF_3 groups will weaken C—Cl bond and delocalize the positive charge in the intermediate carbonium ion, thus increasing the rate.

e. As a generalization, less-basic leaving groups are more reactive in S_N2 reactions than are more-basic leaving groups. Since acetate ion is a much weaker base than is methoxide ion (CH_3O^-) it will increase the reaction rate.

f. $(CH_3)_2C=O$, being a less-polar solvent and less able to solvate the carbonium ion formed, will slow down the reaction.

g. No effect in changing bases because the base is not involved in the rate-determining step of an $E1$ reaction.

h. The elimination would be slowed down. Acetate is a much weaker base than methoxide and an $E2$ elimination responds to base strength.

5-26   a. $(CH_3)_3CCH_2Br$, $(CH_3)_2C=CHCH_3$, $(CH_3)_2CBrCH_2CH_3$
       b. $CH_3CH=CHCH_2Cl$, $CH_3CHClCH=CH_2$ c. $(CH_3)_3C-OH$, $(CH_3)_2C=CH_2$
       d. $CH_3CH_2CH=C(CH_3)CH_2CH_3$, $CH_3CH_2CH_2\overset{\underset{\displaystyle CH_3}{|}}{C}=CHCH_3$

       e. $(CH_3)_2C=C(CH_3)_2$, $(CH_3)_3CCH=CH_2$ f. $(CH_3)_2CHOCH_3$, $CH_3CH=CH_2$
       g. $(CH_3)_3CCH_2OCOCH_3$, $(CH_3)_2\overset{\underset{\displaystyle OCOCH_3}{|}}{C}CH_2CH_3$ h. $(CH_3)_2C=CH_2$, $(CH_3)_3COEt$

5-27   a. This is a trans elimination because the conformation necessary to yield the indicated product has the bromine atoms trans to each other (as shown below). If the bromine atoms had to be in the cis arrangement, the methyl groups would also be cis and would yield *cis*-2-butene.

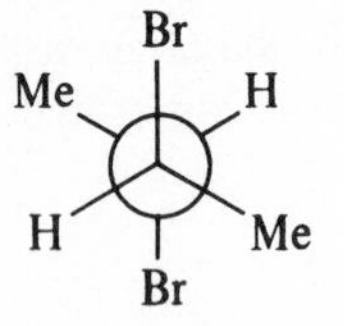

Meso

The mechanism would have to involve iodide ion in the transition state because it appears in the rate expression. Therefore the process could be pictured as attack by iodide ion on one of the bromines with simultaneous loss of the other bromine as bromide ion.

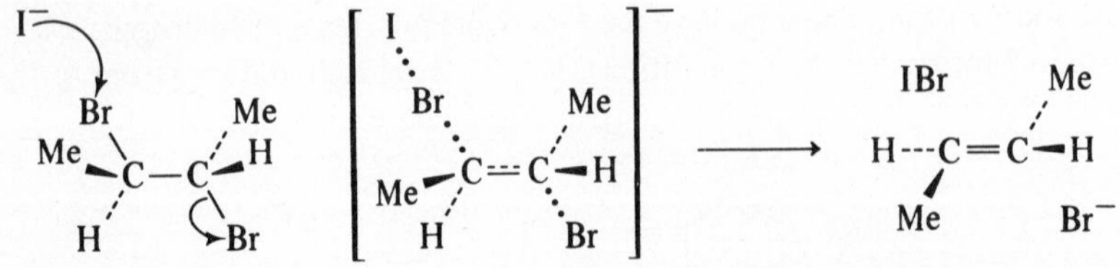

b.

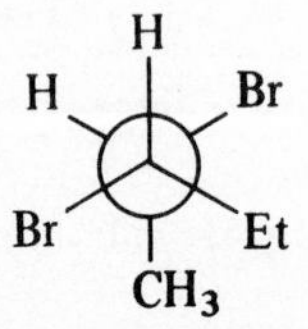

c.

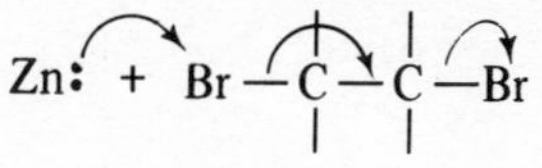

$$Zn\colon + Br-\overset{|}{\underset{|}{C}}-\overset{|}{\underset{|}{C}}-Br$$

d.

$$\underset{Br}{\overset{CH_2}{\diagdown}}CH=CH-CH_2 \quad \xrightarrow[\text{(one step)}]{\text{concerted}} \quad CH_2=CH-CH=CH_2$$

$$\overset{\uparrow}{I^-} \text{ (or } Mg\colon\text{)}$$

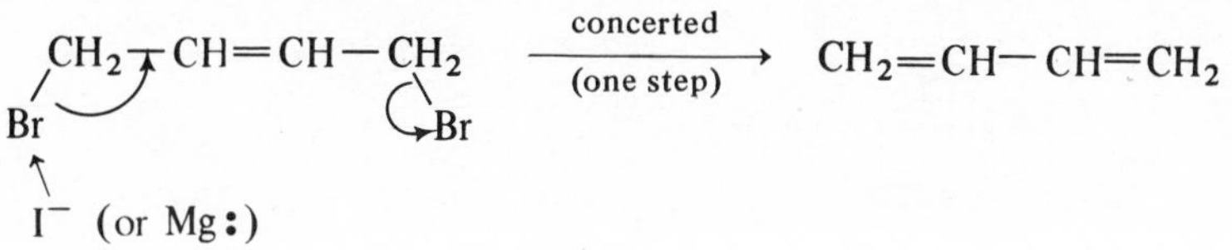

e. The first thing that must happen when the acid is brought into the presence of a base is the transfer of a proton. Subsequently CO_2 and bromide ion could be lost in a concerted process as shown:

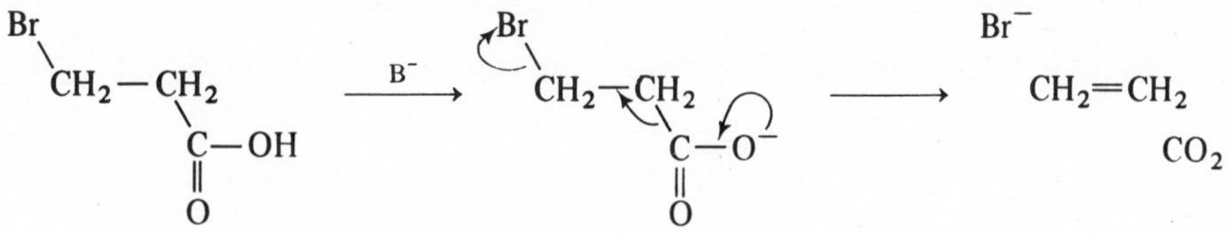

$$\underset{\overset{|}{\underset{O}{C}}-OH}{\overset{Br}{\diagdown}} \quad \xrightarrow{B^-} \quad \underset{\overset{|}{\underset{O}{C}}-O^-}{\overset{Br}{\diagup}} \quad \longrightarrow \quad \begin{array}{c} Br^- \\ CH_2=CH_2 \\ CO_2 \end{array}$$

f.

$$\equiv \quad H_3C\overset{Br}{\underset{H}{-}C}-\overset{CH_3}{\underset{COOH}{-}C}H$$

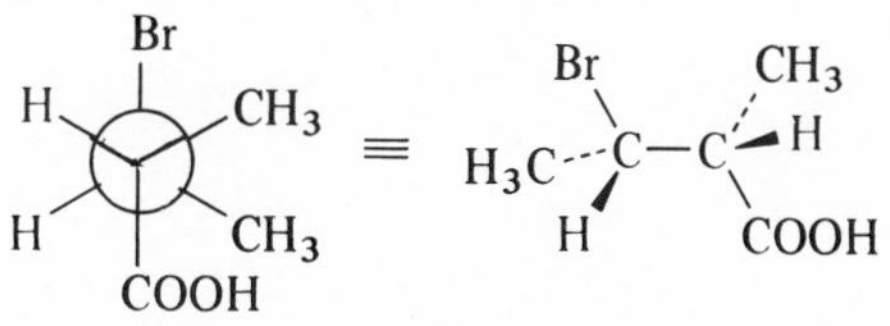

g.

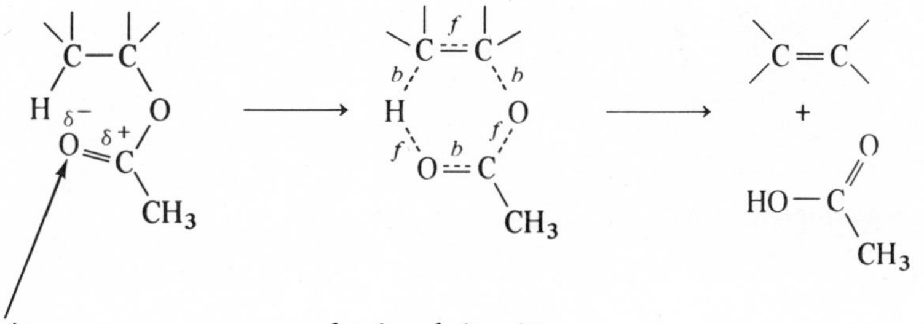

Basic oxygen atom

b = bonds breaking
f = bonds forming

h. $CH_3CH=CH_2$; $CH_3CH_2CH=CH_2$

i. Since the process is a cis elimination a conformation will have to be attained that will have the acetate group cis to either the D or the H on the front carbon atom. If the bond is rotated so that it is cis to the D, the benzene rings will be eclipsed. This is a much higher energy conformation than that in which the acetate group is cis to the H because the benzene rings interact only with very small atoms. Consequently CH_3COOH will be eliminated in preference to CH_3COOD, and the production will be *trans*-$C_6H_5CH=CDC_6H_5$.

5-28 *a.*

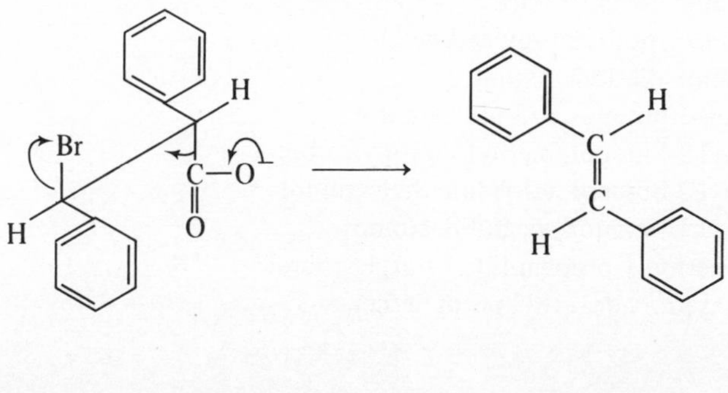

329

b.

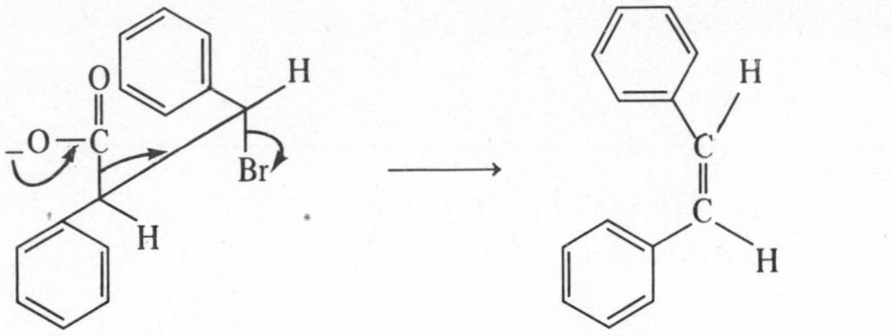

Conformation in *a* is less hindered; thus reaction will go faster.

5-29 *a.* 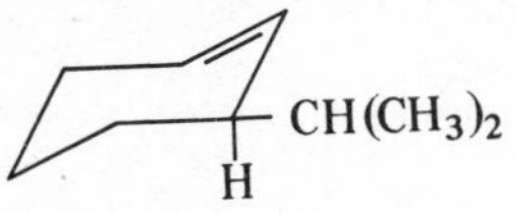 $\diagup$ CH(CH$_3$)$_2$ and $\diagup$ CH(CH$_3$)$_3$

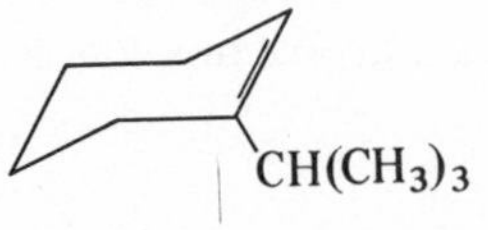

There are hydrogens trans diaxial to Cl on both adjacent carbons; thus, two modes of elimination are possible.

b.

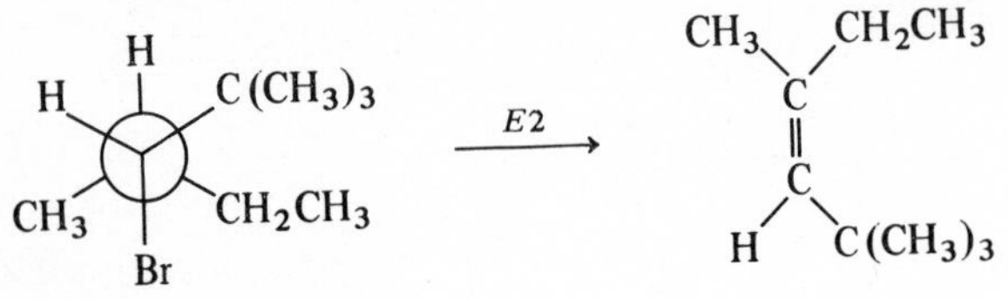

H   H   C(CH$_3$)$_3$
CH$_3$ CH$_2$CH$_3$
Br

$\xrightarrow{E2}$

CH$_3$ CH$_2$CH$_3$
C
‖
C
H   C(CH$_3$)$_3$

Necessary conformation

c.

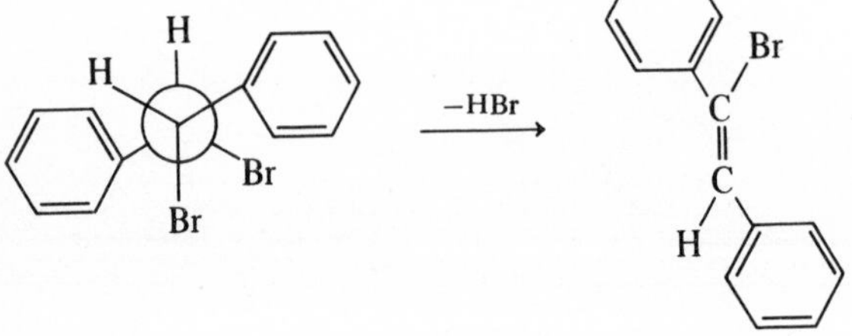

$\xrightarrow{-HBr}$

Necessary conformation — no crowding of rings

d.

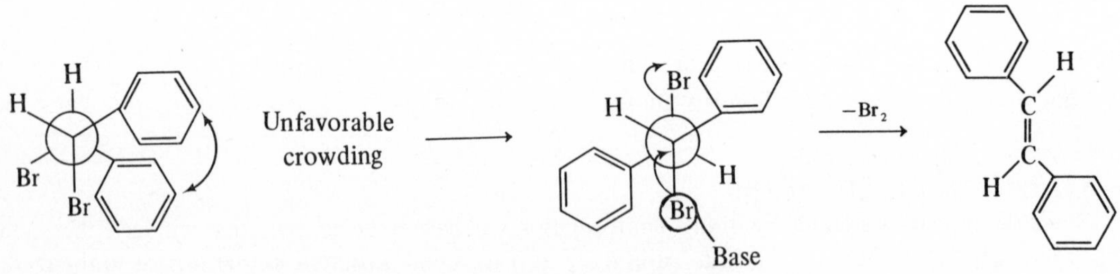

Unfavorable crowding $\longrightarrow$ $\xrightarrow{-Br_2}$

Base

Necessary conformation for HBr elimination

Chapter 6

6-1 *a.* CH$_3$CH$_2$CH$_2$CH$_2$CH$_2$OH   1-Pentanol, butylcarbinol
    b. CH$_3$CH$_2$CH$_2$CH(OH)CH$_3$ 2-Pentanol, methylpropylcarbinol
    c. CH$_3$CH$_2$CH(OH)CH$_2$CH$_3$ 3-Pentanol, diethylcarbinol
    d. (CH$_3$)$_2$CHCH$_2$CH$_2$OH   3-Methyl-1-butanol, isobutylcarbinol
    e. (CH$_3$)$_2$CHCH(OH)CH$_3$ 3-Methyl-2-butanol, methylisopropylcarbinol
    f. (CH$_3$)$_2$C(OH)CH$_2$CH$_3$ 2-Methyl-2-butanol, ethyldimethylcarbinol
    g. CH$_3$CH$_2$CH(CH$_3$)CH$_2$OH   2-Methyl-1-butanol, *sec*-butylcarbinol
    h. (CH$_3$)$_3$CCH$_2$OH   2,2-Dimethyl-1-propanol, *tert*-butylcarbinol
    i. CH$_3$CH$_2$OCH$_2$CH$_2$CH$_3$ 1-Ethoxypropane, ethyl propyl ether

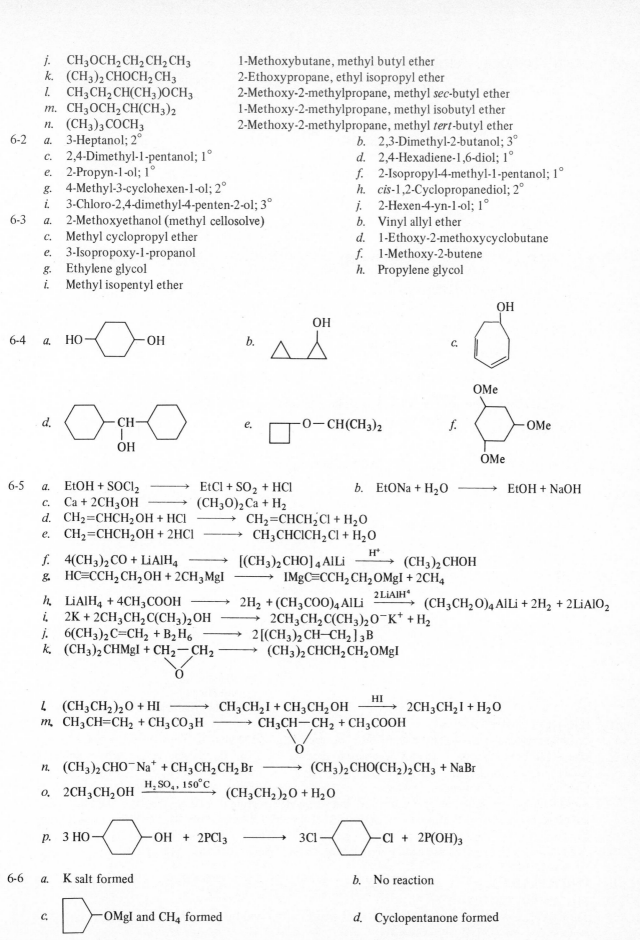

j. $CH_3OCH_2CH_2CH_2CH_3$ 1-Methoxybutane, methyl butyl ether
k. $(CH_3)_2CHOCH_2CH_3$ 2-Ethoxypropane, ethyl isopropyl ether
l. $CH_3CH_2CH(CH_3)OCH_3$ 2-Methoxy-2-methylpropane, methyl sec-butyl ether
m. $CH_3OCH_2CH(CH_3)_2$ 1-Methoxy-2-methylpropane, methyl isobutyl ether
n. $(CH_3)_3COCH_3$ 2-Methoxy-2-methylpropane, methyl tert-butyl ether

6-2 a.   3-Heptanol; 2°   b.   2,3-Dimethyl-2-butanol; 3°
     c.   2,4-Dimethyl-1-pentanol; 1°   d.   2,4-Hexadiene-1,6-diol; 1°
     e.   2-Propyn-1-ol; 1°   f.   2-Isopropyl-4-methyl-1-pentanol; 1°
     g.   4-Methyl-3-cyclohexen-1-ol; 2°   h.   cis-1,2-Cyclopropanediol; 2°
     i.   3-Chloro-2,4-dimethyl-4-penten-2-ol; 3°   j.   2-Hexen-4-yn-1-ol; 1°

6-3 a.   2-Methoxyethanol (methyl cellosolve)   b.   Vinyl allyl ether
     c.   Methyl cyclopropyl ether   d.   1-Ethoxy-2-methoxycyclobutane
     e.   3-Isopropoxy-1-propanol   f.   1-Methoxy-2-butene
     g.   Ethylene glycol   h.   Propylene glycol
     i.   Methyl isopentyl ether

6-4 a.   b.   c.

     d.   e.   f.

6-5 a. $EtOH + SOCl_2 \longrightarrow EtCl + SO_2 + HCl$ b. $EtONa + H_2O \longrightarrow EtOH + NaOH$
     c. $Ca + 2CH_3OH \longrightarrow (CH_3O)_2Ca + H_2$
     d. $CH_2=CHCH_2OH + HCl \longrightarrow CH_2=CHCH_2Cl + H_2O$
     e. $CH_2=CHCH_2OH + 2HCl \longrightarrow CH_3CHClCH_2Cl + H_2O$

     f. $4(CH_3)_2CO + LiAlH_4 \longrightarrow [(CH_3)_2CHO]_4AlLi \xrightarrow{H^+} (CH_3)_2CHOH$
     g. $HC\equiv CCH_2CH_2OH + 2CH_3MgI \longrightarrow IMgC\equiv CCH_2CH_2OMgI + 2CH_4$

     h. $LiAlH_4 + 4CH_3COOH \longrightarrow 2H_2 + (CH_3COO)_4AlLi \xrightarrow{2LiAlH_4} (CH_3CH_2O)_4AlLi + 2H_2 + 2LiAlO_2$
     i. $2K + 2CH_3CH_2C(CH_3)_2OH \longrightarrow 2CH_3CH_2C(CH_3)_2O^-K^+ + H_2$
     j. $6(CH_3)_2C=CH_2 + B_2H_6 \longrightarrow 2[(CH_3)_2CH-CH_2]_3B$
     k. $(CH_3)_2CHMgI + CH_2-CH_2 \longrightarrow (CH_3)_2CHCH_2CH_2OMgI$
          $\qquad\qquad\qquad\qquad O$

     l. $(CH_3CH_2)_2O + HI \longrightarrow CH_3CH_2I + CH_3CH_2OH \xrightarrow{HI} 2CH_3CH_2I + H_2O$
     m. $CH_3CH=CH_2 + CH_3CO_3H \longrightarrow CH_3CH-CH_2 + CH_3COOH$
          $\qquad\qquad\qquad\qquad\qquad\qquad O$

     n. $(CH_3)_2CHO^-Na^+ + CH_3CH_2CH_2Br \longrightarrow (CH_3)_2CHO(CH_2)_2CH_3 + NaBr$

     o. $2CH_3CH_2OH \xrightarrow{H_2SO_4, 150°C} (CH_3CH_2)_2O + H_2O$

     p. $3\ HO-\!\!\!\bigcirc\!\!\!-OH + 2PCl_3 \longrightarrow 3Cl-\!\!\!\bigcirc\!\!\!-Cl + 2P(OH)_3$

6-6 a.   K salt formed   b.   No reaction

     c. $\square$—OMgI and CH_4 formed   d.   Cyclopentanone formed

331

6-7 $CH_3CH_3 < CH_3OCH_3 < CH_3C\equiv CH < CH_3CH_2OH < CH_3COOH$

6-8   *a.* $(CH_3)_2CHOH \xrightarrow[325°C]{Al_2O_3} CH_3CH=CH_2 \xrightarrow[peroxide]{HBr} CH_3CH_2CH_2Br \xrightarrow{NaOH} C_3H_7OH$ or hydroboration-

   oxidation of propene

   b. $CH_3CH=CH_2 \xrightarrow{H_2SO_4} (CH_3)_2CHOSO_2OH \xrightarrow{H_2O} (CH_3)_2CHOH \xrightarrow[325°C]{Cu} (CH_3)_2CO$

   c. $CH_3CH_2OH \xrightarrow{HBr} \xrightarrow{Mg} CH_3CH_2MgBr \xrightarrow{CH_2O} CH_3CH_2CH_2OMgBr \xrightarrow{H^+} CH_3CH_2CH_2OH$

   d. $CH_3CH_2MgBr + \underset{O}{CH_2\backslash\!\!/CH_2} \longrightarrow CH_3(CH_2)_3OMgBr \xrightarrow{H^+} CH_3(CH_2)_3OH$

   e. $CH_3(CH_2)_3OH \xrightarrow[Al_2O_3]{350°C} CH_3CH_2CH=CH_2 \xrightarrow{H_2SO_4} \xrightarrow{H_2O} \xrightarrow[350°C]{Cu} \xrightarrow{CH_3MgX} \xrightarrow{H^+} ans$

   f. $CH_3CH_2CH_2COOH \xrightarrow{LiAlH_4} \xrightarrow{H^+} CH_3(CH_2)_3OH \xrightarrow[350°C]{Al_2O_3} \xrightarrow{HI} CH_3CH_2CHICH_3$

   g. $CH_3CH_2MgBr + CH_3CHO \longrightarrow CH_3CH_2CH(OMgBr)CH_3 \xrightarrow{H^+}$

   $CH_3CH(OH)CH_2CH_3 \xrightarrow{Na} \xrightarrow{CH_3Br} ans$

   h. $CH_2=CH-Cl \xrightarrow{Cl_2} ClCH_2CHCl_2 \xrightarrow{CH_3O^-Na^+} CH_3OCH_2CH(OCH_3)_2$

   i. $CH_3CH=CH_2 \xrightarrow{HBr, peroxide} CH_3CH_2CH_2Br \xrightarrow{CH\equiv C^-Na^+} CH_3CH_2CH_2C\equiv CH \xrightarrow[Pt]{1 \text{ mol } H_2}$

   $CH_3CH_2CH_2CH=CH_2 \xrightarrow{CH_3COOOH} ans$

   j. $CH_3CH=CH_2 \xrightarrow{HBr} \xrightarrow{Mg} CH_3CH(MgBr)CH_3 \xrightarrow{CH_3CHO} \xrightarrow{H^+}$

   $(CH_3)_2CHCH(OH)CH_3 \xrightarrow[H_2SO_4]{K_2Cr_2O_7} \xrightarrow{(CH_3)_2CHMgBr} \underset{OH}{(CH_3)_2CHC(CH_3)CH(CH_3)_2} \xrightarrow{Al_2O_3} \xrightarrow[Pt]{H_2} ans$

   k. $(CH_3)_3COH \xrightarrow{Al_2O_3, 350°} (CH_3)_2C=CH_2 \xrightarrow{cold, dil KMnO_4} (CH_3)_2C(OH)CH_2OH$

   l. $\underset{O}{CH_2\backslash\!\!/CH_2} \xrightarrow{CH_3CH_2MgBr} \xrightarrow{H^+} CH_3CH_2CH_2CH_2OH \xrightarrow{Al_2O_3, 350°} CH_3CH_2CH=CH_2 \xrightarrow{Br_2}$

   $\xrightarrow[NH_3]{2NaNH_2} CH_3CH_2C\equiv CH \xrightarrow[1 \text{ mol}]{HBr} \underset{Br}{CH_3CH_2C=CH_2}$

6-9   *a.*   Propene   *b.*   1-Butene
      c.   Mixture of 2- and 3-hexenes   *d.*   1,3-Butadiene
      e.   2,5-Dimethyl-2-hexene   *f.*   1,3-Cyclohexadiene

6-10 $(CH_3)_2CHCH(OH)CH_3$

6-11   To obtain the equivalent weight of A: 0.88 g of A reacts to give 224 ml CH_4. Thus, 0.88/224 = X/22,400, where X is the number of grams of A necessary to produce 1 mol of CH_4. Solving, we get X = 88, or 88 g of X would produce 1 mol of methane if X had one active hydrogen (if A had two or three active hydrogens, the molecular weight of A would be 2 x 88 = 176 or 3 x 88 = 264, respectively). First assume that there is only one active hydrogen per molecule of A and try to find a suitable structure that fits *all* the data. The Lucas test indicates that we have a 3° alcohol. The simplest 3° alcohol is *tert*-butyl (MW 74). The difference between this and our molecular weight for one active hydrogen (88) is 14. This is the weight of one $-CH_2-$. Adding this unit to *tert*-butyl alcohol gives a five-carbon *tert*-alcohol, i.e., *tert*-pentyl, $CH_3CH_2C(CH_3)_2OH$.

6-12 $HC\equiv CCH_2CH_2CH_2OH$

6-13 $CH_3CH_2OCH=CH_2$

6-14 $CH_3CH_2OCH_2CH_2OH$

6-15   Three OCH_3 groups present since the equivalent weight is 44.7 and the molecular weight is 134. A possible structure for A is $(CH_3OCH_2)_2CHOCH_3$.

6-16   *a.* Vinyl alcohol does not exist; hence A is impossible. If one did have vinyl chloride, step B is not possible. Heating olefins with Na causes them to polymerize.

b. Step A is correct. Step B is bad because hydrogenation with Pt would reduce halide to the saturated hydrocarbon.

c. It is not possible to reduce a COOH group with H_2 + Pt. In A the C=C would be reduced. In step B, allyl chloride would result.

d. A is wrong because of strong oxidizing action of Cl_2. Also attack by Cl_2 would not favor C-3. B is wrong because this Grignard would react with itself.

e. 1,2-Diiodoethane is not very stable under ordinary conditions. Compounds with iodine on adjacent carbons tend to decompose to give olefins and iodine.

f. *tert*-Alkyl halides undergo elimination; product will be isobutylene.

g. *tert*-Alcohols undergo dehydration to olefins rather than to ethers.

h. This method of ether synthesis will produce a mixture of three ethers. The Williamson method should be used.

i. Hydroboration of isobutylene will lead to triisobutylborane not tri-*tert*-butylborane.

j. Step A is incorrect because only the dialkylborane is formed, $[(CH_3)_2CHC(CH_3)]_2BH$. Step B is wrong because in hydroboration-carbonylation reactions leading to $2°$ alcohols an equimolar quantity of H_2O must be added with CO. The use of $LiBH_4$ in B would lead to a $1°$ alcohol.

k. It is not possible to get free-radical addition of HI to a double bond in the presence of peroxide (HBr only does this). B is correct.

l. *tert*-Butyl bromide in presence of a base such as HC≡CNa will undergo elimination to give isobutylene. B is also wrong because *tert*-butyl methyl ketone would be obtained instead.

m. Step A is bad because a mixture of three ethers would be obtained. B is poor because a variety of monochlorinated products would be obtained.

n. Step A is correct. Step B is wrong because a base like NaCN will cause elimination of HBr and formation of 1-methylcyclohexene as the major product.

6-17   *a.* 1-Hexanol will evolve H_2 with Na or heat with acetyl chloride. Hexane is inert to both reagents.

b. 1-Butanol is soluble in water; the hydrocarbon is not. The acetylene will give a precipitate with ammoniacal Ag^+.

c. 2-Propanol will give a positive Lucas test.

d. Methanol is water-soluble.

e. Ethyl allyl ether will reduce permanganate or decolorize bromine soln.

f. Dibutyl ether will form an oxonium salt with concentrated sulfuric acid and dissolve. *n*-Octane will not.

g. 1-Butanol will react with sodium to evolve hydrogen; the ether will not. 1-Butanol is very soluble in water. Ether is only slightly soluble.

h. 1-Butyne will evolve hydrogen with sodium or give a precipitate with ammoniacal silver nitrate.

i. The iodide will give a Beilstein test. The iodide does not form an onium salt with sulfuric acid; hence, it is insoluble in this reagent. The ether is soluble.

j. $(CH_3CH_2)_2S$ will dissolve in cold, concentrated H_2SO_4 whereas the bromide will not, but the bromide will give a positive Beilstein test.

k. 2-Propanol is water-soluble, 2-octanol is not.

l. The chloro compound will give a positive Beilstein test whereas the fluoro compound will not.

6-18   *a.* $CH_3CH_2CH_2OH \xrightarrow[\Delta]{H^+} \xrightarrow[\text{peroxide}]{HBr} \xrightarrow{CH_3O^-} CH_3CH_2CH_2OCH_3$

b. $CH_3(CH_2)_4OCH_3 \xrightarrow[\Delta]{HBr} \xrightarrow[\text{KOH}]{\text{alc}} \xrightarrow{Br_2} \xrightarrow{NaNH_2} CH_3CH_2CH_2C≡CH$

c. $CH_2{=}CH_2 \xrightarrow{HOCl} \xrightarrow[\Delta]{H_2SO_4} \xrightarrow{NaNH_2} CH_2{=}CHOCH{=}CH_2$

d. $\xrightarrow[\text{KOH}]{\text{alc}} \xrightarrow{CH_3COOOH}$

e. $BrCH_2CH_2Br \xrightarrow{NaNH_2} \xrightarrow{CH_3MgI} \xrightarrow{(CH_3)_2C=O} \xrightarrow{H^+} (CH_3)_2\underset{\underset{OH}{|}}{C}C≡CH$

333

f. $CH_3C{\equiv}CH \xrightarrow{2HBr} \xrightarrow{2CH_3O} (CH_3)_2C(OCH_3)_2$

g. $\xrightarrow{CH_3COOOH} \xrightarrow{HC{\equiv}CMgX} \xrightarrow{H^+} \xrightarrow{H_2,\,Pd}$

h. $HC{\equiv}CH \xrightarrow{2CH_3MgX} \xrightarrow{2(CH_3)_2CO} \xrightarrow{H^+} (CH_3)_2CC{\equiv}CC(CH_3)_2$
$\phantom{HC{\equiv}CH \xrightarrow{2CH_3MgX} \xrightarrow{2(CH_3)_2CO} \xrightarrow{H^+} (CH_3)_2}$ OH OH

i. $-OH \xrightarrow[\Delta]{Al_2O_3} \xrightarrow{HOCl} \xrightarrow{KCN}$

j. $CH_3CH{-}CHCH_3 \xrightarrow{CH_3MgX} \xrightarrow{H^+} \xrightarrow[\Delta]{Al_2O_3} (CH_3)_2C{=}CHCH_3$
$\phantom{CH_3CH{-}CH}O$

k. $CH_3CH{=}CH_2 \xrightarrow{Cl_2,\,500^\circ C} \xrightarrow{HOCl} \xrightarrow{OH^-} HOCH_2CH(OH)CH_2OH$

6-19 *a.* Convert propene to 1-bromopropane with HBr and peroxide. Make sodium salt of acetylene and react this with 1-bromopropane; then hydrogenate triple bond to a double bond. Oxidize this olefin with peracetic acid or add HOCl and then treat with concentrated NaOH.

b. Treat propene with Cl_2 at $500^\circ C$ to get allyl chloride. With Wurtz reaction this gives 1,5-hexadiene to which add chlorine and then treat with CH_3ONa. Allyl halides give good yields in the Wurtz reaction whereas other unsaturated halides polymerize.

c. Hydrogenate acetylene to ethylene and then oxidize to ethylene oxide. Make Grignard reagent from acetylene and add this to ethylene oxide.

d. Oxidize methanol to formaldehyde. Make isopropyl bromide from propene and HBr and form Grignard reagent. React this with formaldehyde and dehydrate alcohol to isobutylene. Hydrobromination of isobutylene, formation of *t*-butyl Grignard reagent and treatment of this with ethylene oxide will give 3,3-dimethyl-1-butanol which can then be dehydrated and hydrogenated.

e. Treat propene with HOBr. Make sodium salt of alcohol and treat with methyl iodide. Treat 2-methoxy-1-bromopropane with NaCN.

f. Hydroboration-carbonylation of 2-butene will give 3,5-dimethyl-4-*sec*-butyl-4-heptanol. Treat this with sodium metal and then with CH_3I to give the methyl ether.

6-20 *a.* $CH_3CH_2OH \xrightarrow[\Delta]{H^+} CH_2{=}CH_2 \xrightarrow{Br_2} \xrightarrow[KOH]{alc} \xrightarrow[Pt]{D_2} CHD_2CHD_2$

b. $HC{\equiv}CH \xrightarrow[H_3O^+]{HgSO_4} CH_3CHO \xrightarrow{LiAlD_4} \xrightarrow{H_2O} CH_3CHDOH$

c. $CH_4 \xrightarrow[Cl_2]{\lambda} CCl_4 \xrightarrow[Pt]{D_2} CD_4 + 4DCl$ 　　 *d.* $(CH_3)_2C{=}O \xrightarrow{LiAlD_4} \xrightarrow{D_3O^+} (CH_3)_2CDOD$

e. $CaC_2 \xrightarrow{2D_2O} DC{\equiv}CD \xrightarrow{D_2,\,Pt} CD_3CD_3$ 　　 *f.* $CH_3COOH \xrightarrow{LiAlD_4} \xrightarrow{H^+} CH_3CD_2OH$

6-21 *a.* S_N2 　 *b.* S_N1 　 *c.* S_N2 or S_N1 　 *d.* S_N2 　 *e.* S_N1 　 *f.* S_N1

6-22 $H_2C{=}CH{-}O{-}\textcircled{H} \longrightarrow CH_3CH{=}O$

6-23 $CH_2{=}CH\overset{+}{C}H_2 \longleftrightarrow \overset{+}{C}H_2CH{=}CH_2$, $CH_2{=}CH\overset{-}{C}H_2 \longleftrightarrow \overset{-}{C}H_2CH{=}CH_2$, $CH_3CH{=}CHCH_2Cl$ and $CH_3CHClCH{=}CH_2$

6-24 $\overset{-}{>}C{-}\overset{+}{O}$ because oxygen is more electronegative than carbon.

6-25 *a.* $HOCH_2CH_2OH$ *b.* $HC\equiv CCH_2CH_2OMgX$ *c.* $CH_3OCH_2CH_2OH$

    d. $H_2NCH_2CH_2OH$ *e.* $CH_3NHCH_2CH_2OH$ *f.* $HOCH_2CH_2C\equiv N$

    g. $HSCH_2CH_2OH$ *h.* $BrCH_2CH_2OH$

6-26 *a.* $(CH_3)_2CClCH_2OH$ *b.* $(CH_3)_2C(OH)CH_2OCH_3$ *c.* $(CH_3)_2C(CN)CH_2OH$

6-27 Molecular weight of X as calculated from camphor data is 102. The empirical formula is $C_6H_{14}O$. Reaction of X with acetyl chloride and Na indicates an alcohol which upon dehydration to an alkene and then oxidation with permanganate yields a ketone and an acid. Subtracting 45(COOH) from the NE (74) leaves a residue of 29 or C_2H_5. The acid is then CH_3CH_2COOH obtained from a $CH_3CH_2CH=$ linkage in Y. The ketone has to be acetone. Thus a structure for X would be $(CH_3)_2C(OH)CH_2CH_2CH_3$.

6-28 *a.* $(CH_3)_2BCH=CH_2 \longleftrightarrow (CH_3)_2\bar{B}=CH\overset{+}{C}H_2$

    b. $CH_3OCH=CHCH=O \longleftrightarrow CH_3OCH=CH\bar{C}H-O^- \longleftrightarrow CH_3O\overset{+}{C}HCH=CH-O^- \longleftrightarrow$

          $CH_3\overset{+}{O}=CHCH=CH-O^-$

    c. $(CH_3)_2NCH=CH\overset{+}{N}\begin{smallmatrix} O^- \\ \\ O \end{smallmatrix} \longleftrightarrow (CH_3)_2\overset{+}{N}=CH\bar{C}H\overset{+}{N}\begin{smallmatrix} O^- \\ \\ O \end{smallmatrix} \longleftrightarrow$

          $(CH_3)_2\overset{+}{N}=CHCH=\overset{+}{N}\begin{smallmatrix} O^- \\ \\ O^- \end{smallmatrix} \longleftrightarrow (CH_3)_2NCH=CH-\overset{++}{N}\begin{smallmatrix} O^- \\ \\ O^- \end{smallmatrix}$

    d.

6-29 *a.*

    b.

    c.

6-30 1-Methylcyclopentene $\xrightarrow{B_2H_6}$ $\xrightarrow{H_2O_2,\ OH^-}$ *trans*-2-methyl-1-cyclopentanol $\xrightarrow{Na}$ $\xrightarrow{CH_3I}$ ans

6-31 *a.* $(CH_2=CH)_2O < CF_3OH < CH_3OH < (CH_3)_2O$

    b. $CH_3\underset{\underset{O}{\parallel}}{C}OH \xrightarrow{H^+} \left[CH_3\underset{\underset{+OH}{\parallel}}{C}OH \longleftrightarrow CH_3\underset{\underset{OH}{\mid}}{\overset{+}{C}}OH \longleftrightarrow CH_3\underset{\underset{OH}{\mid}}{C}=\overset{+}{O}H \right]$

6-32 *a.* Add HCN to ethylene oxide and dehydrate.

    b. Add ethylene oxide to methanol, convert this product to sodium salt with Na, and react this with ethyl iodide.

    c. HCN can be added to the product made in *a*. $CH_2=CH-C\equiv N \longleftrightarrow \overset{+}{C}H_2CH=C=N^-$. The course of the addition is determined by polarization due to resonance.

    d. Add ethylene oxide to $CH_2=CHMgX$. Then add Br_2 or HOCl to the resulting olefin and hydrolyze.

    e. Add $CH_2=CHMgX$ to ethylene oxide to get $CH_2=CHCH_2CH_2OH$; convert this to the Grignard reagent and add to ethylene oxide to get $CH_2=CHCH_2CH_2CH_2CH_2OH$; add Br_2 and dehydrohalogenate to $HC\equiv C(CH_2)_4OH$. Convert to the bromide (PBr_3) and treat with ^-SH.

    f. Add HI to ethylene oxide. Convert this product to the sodium salt and add to another mole of ethylene oxide. Treat resulting product with PI_3.

6-33 A, $BrCH_2CHBrCH_2Br$; B, $BrCH=CHCH_2Br$; C, $BrCH=CHCH_2C(CH_3)_3$

6-34   *a.* $CH_3CH_2CH_2CHO$ and $(CH_3)_2CHCHO$; $CH_3CH_2CH_2CH_2OH$ and $(CH_3)_2CHCH_2OH$

b. 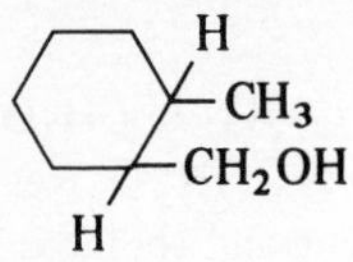 (cis and trans) and

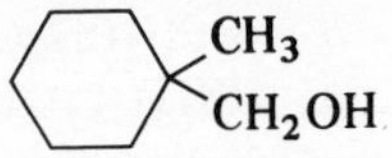

6-35   Per mole of A there are two active hydrogens (OH) and 2 mol of hydrogen are absorbed. Two possible structures are $HOCH_2C\equiv CCH_2OH$ or $HOCH=CHCH=CHOH$. The diene does not exist as such, but exists in the form $OHCCH_2CH_2CHO$ (see Sec. 4-3a−2).

6-36   In compound A the chlorine has only one carbon adjacent to it that has a hydrogen in the necessary trans arrangement ($180°$) to the chlorine. Thus elimination of HCl from A gives only one alkene. In compound B both carbons adjacent to the chlorine have trans hydrogens suitable for elimination and thus two alkenes are possible.

6-37   *a.*

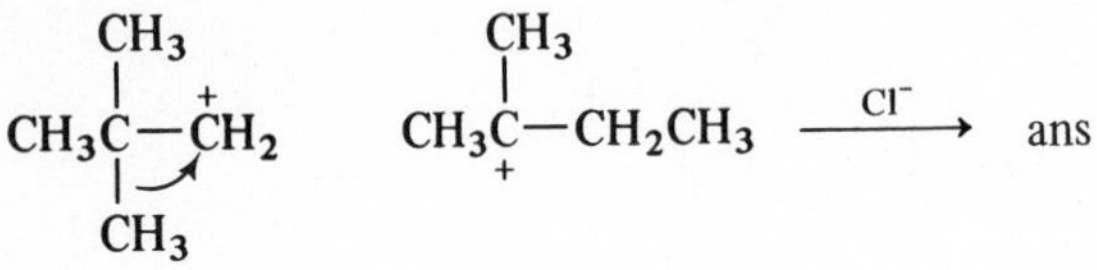

b.

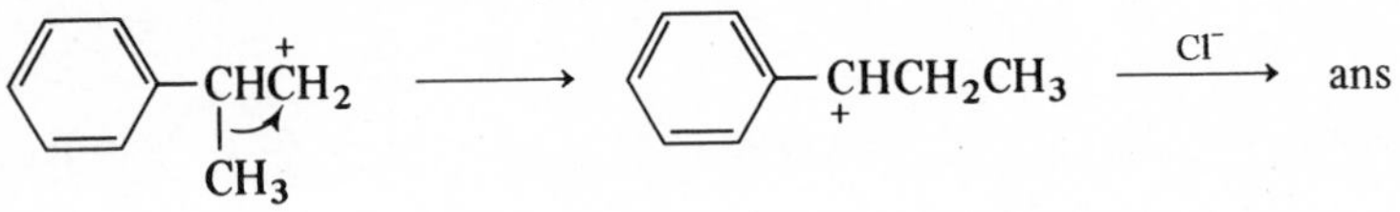

c.

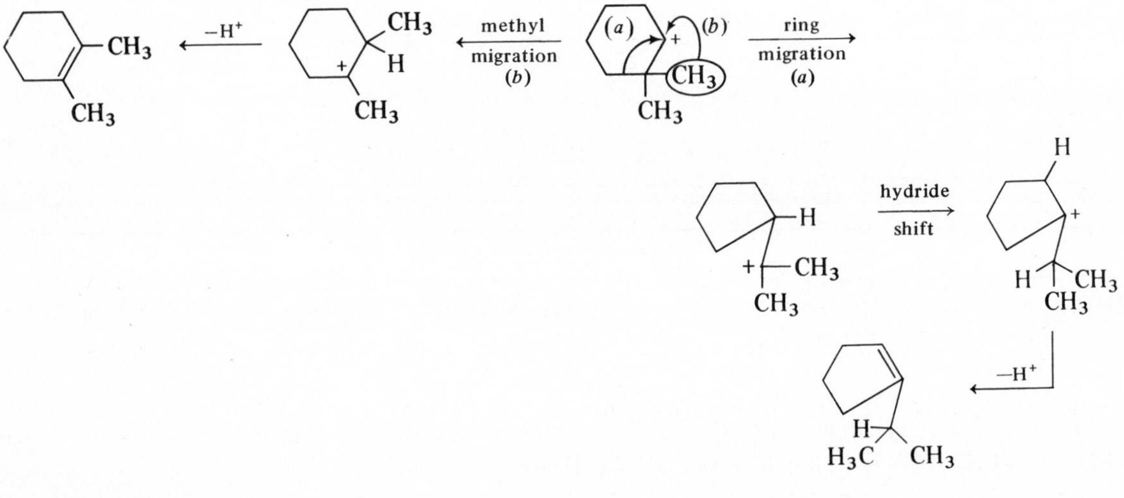

d.

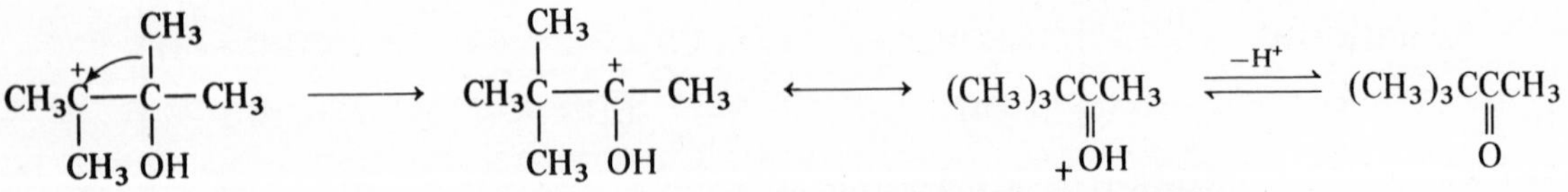

e.

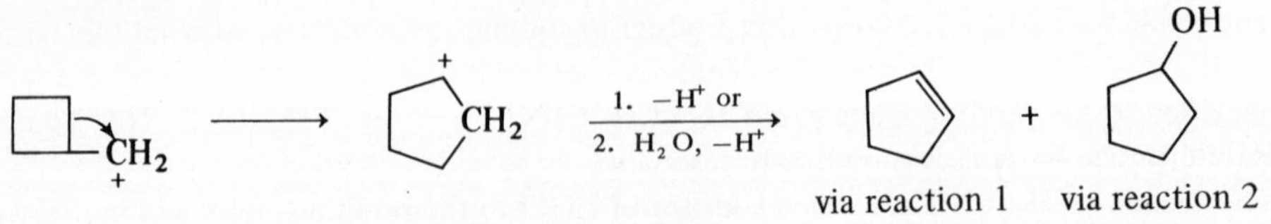

via reaction 1   via reaction 2

f.

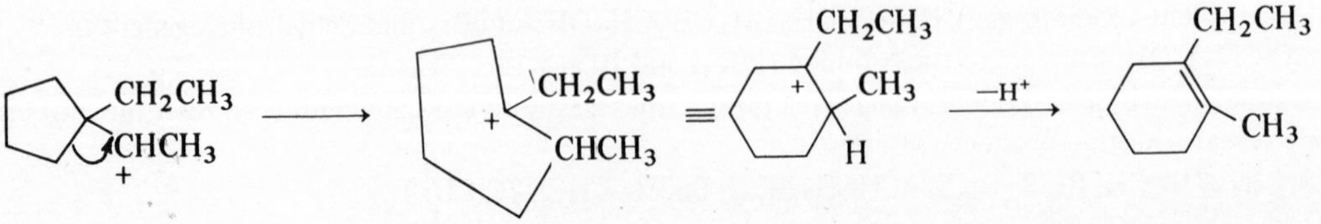

336

6-38

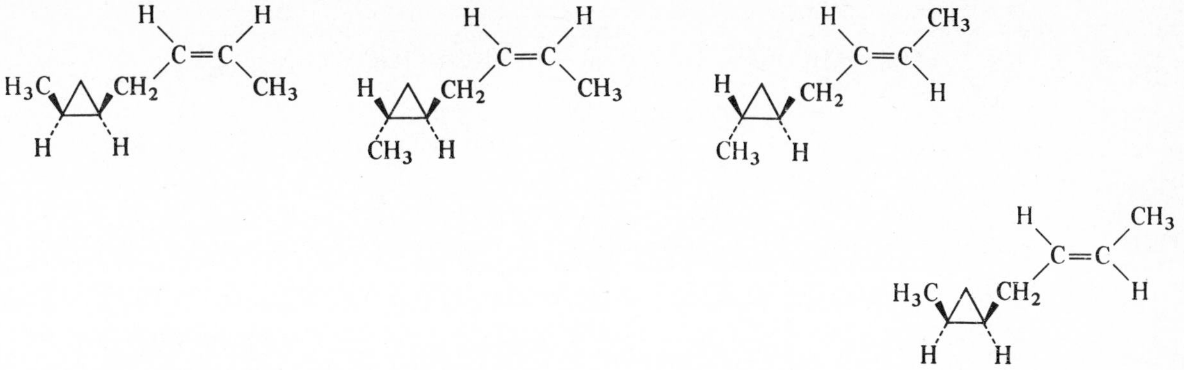

6-39 The anion is particularly stable since it forms a cyclic resonance system of $4n + 2$ electrons, e.g.,

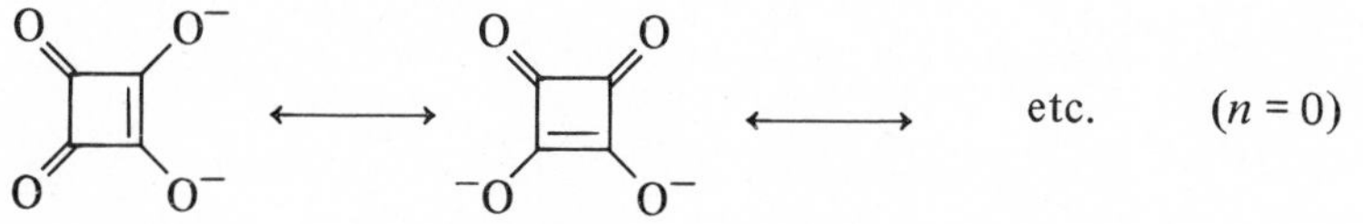

etc. $(n = 0)$

6-40 The phenoxide anion is stabilized by resonance whereas this is not possible in the case of the anion derived from 2,4-cyclopentadien-1-ol.

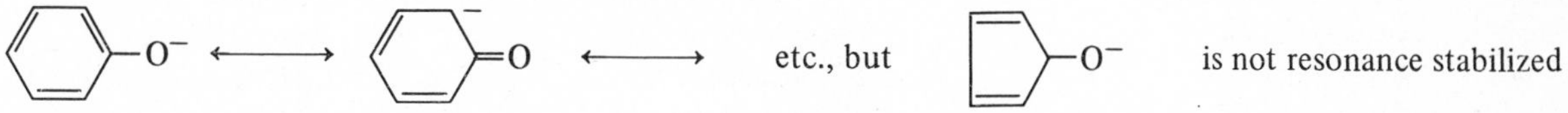

etc., but

is not resonance stabilized

6-41 If A and B each contain only one atom of I (at wt = 127) then $127/X = 0.894$ gives the MW of iodide A and $127/Y = 0.690$ gives the MW of iodide B. MW of A = 142; MW of B = 184. 142 less 127 = 15; 184 less 127 = 57. Thus, A = CH_3I and B = $(CH_3)_3CI$, which would undergo elimination with KOH to yield gaseous isobutylene. The ether is $CH_3OC(CH_3)_3$.

6-42 In the unsaturated ether the lone-pair electrons on oxygen are delocalized by resonance. This makes them much less available for hydrogen bonding with water.

6-43

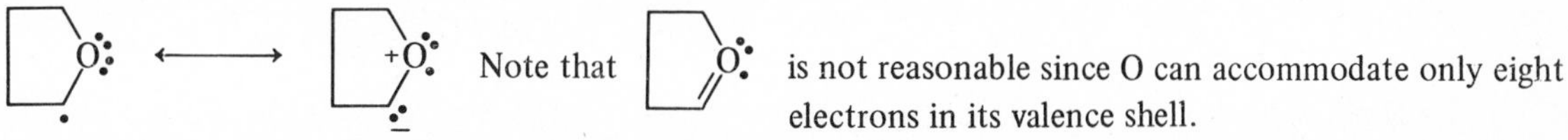

Note that

is not reasonable since O can accommodate only eight electrons in its valence shell.

6-44 The trans isomer can exist in a conformation where the S with its lone-pair electrons is trans diaxial to the Cl. Thus the neighboring S assists the leaving of the Cl by formation of a three-membered ring which is rapidly reopened by attack by H_2O to give the alcohol. The leaving of the Cl is not assisted by S in the cis isomer since the S and the Cl cannot assume the necessary trans diaxial arrangement.

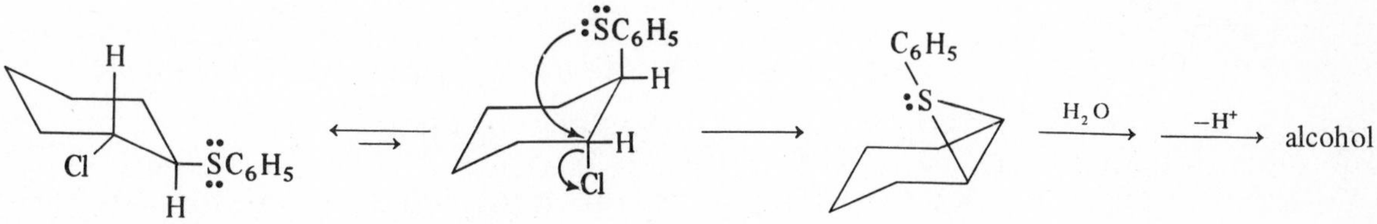

6-45 *a.*

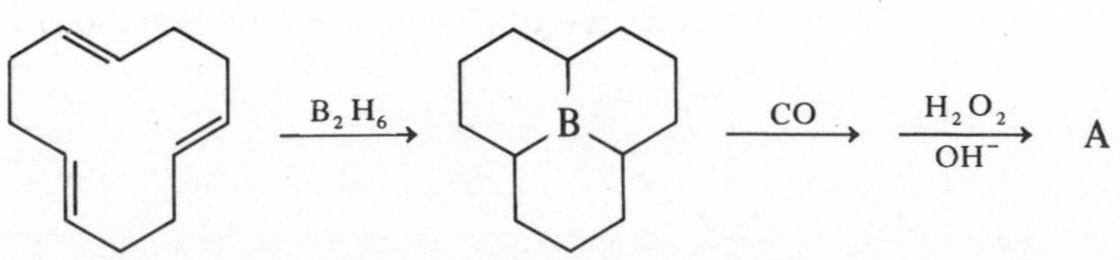

337

b.

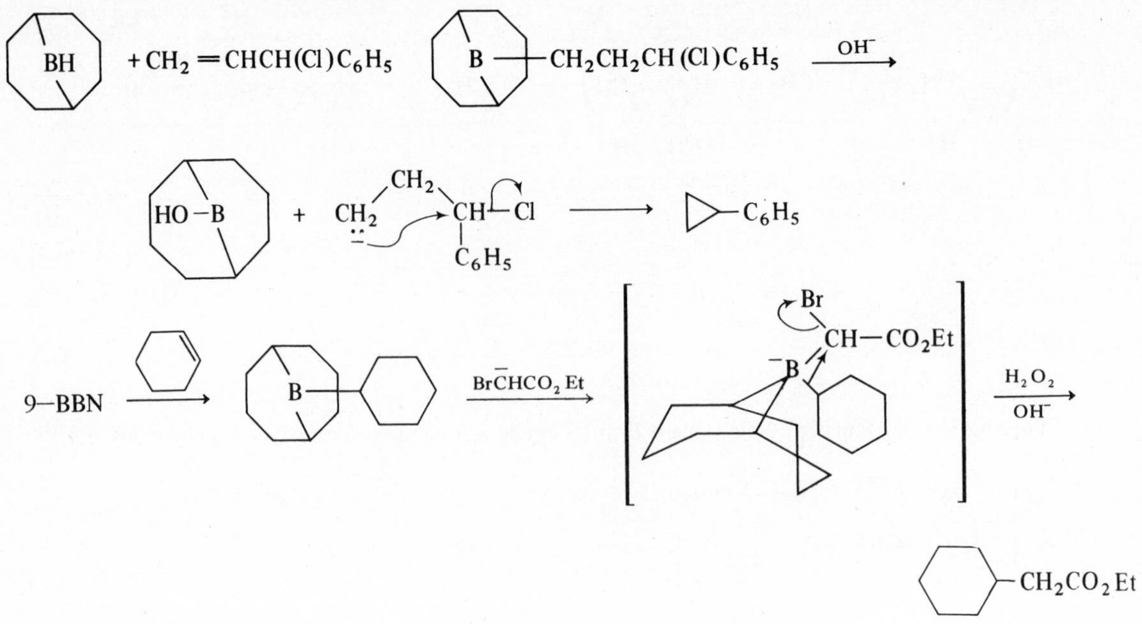

Chapter 7

7-1   *a.*   2-Methylpropanal   *b.*   4-Methyl-2-pentanone   *c.*   Propenal
        Isobutyraldehyde   Methyl isobutyl ketone   Acrolein
      d.   Ethanedial   *e.*   2,3-Butanedione   *f.*   3-Hydroxybutanal
        Glyoxal   Biacetyl (diacetyl)   Aldol
      g.   3-Methyl-2-cyclobuten-1-one   *h.*   Cyclopentanecarbaldehyde
        Same   Cyclopentylformaldehyde
      i.   2,4-pentadienal   *j.*   4-Methyl-3-penten-2-one
        Same   Mesityl oxide

7-2   *a* $CH_3COCH=CHC{\equiv}CCH_3$ *b.* $CH_3CH_2CH(CH_3)COCH(CH_3)_2$ *c.* $CH_3COCH=CHCOCH_3$

d.

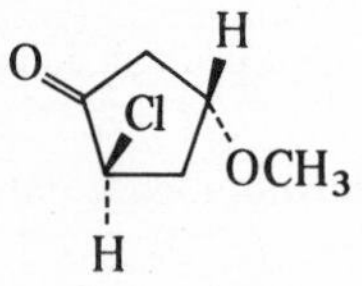

e. $HOCH_2CH_2\overset{\overset{\displaystyle NO_2}{|}}{C}=CHCHCHO$

f.   *g.* $CH_3COCH_2CH=CHCH_2CHO$

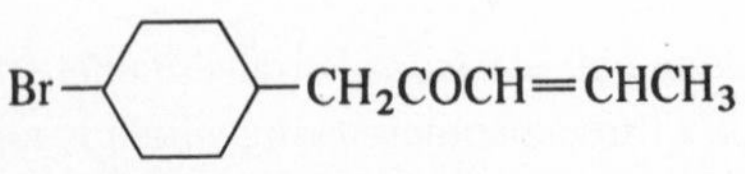

h.

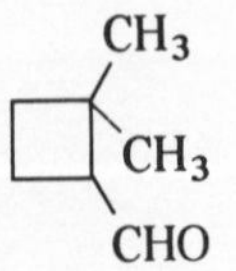

7-3   *a.* $(CH_3)_2CO + H_2NNHCONH_2 \longrightarrow (CH_3)_2C=NNHCONH_2 + HOH$
      b. $CH_3(CH_2)_3CHO + PCl_5 \longrightarrow CH_3(CH_2)_3CHCl_2 + POCl_3$
      c. $2(CH_3)_3CCHO + OH^- + \Delta \longrightarrow (CH_3)_3CCH_2OH + (CH_3)_3CCO_2^-$
      d. $CH_3CH_2OH + 6NaOH + 4Br_2 \longrightarrow HCBr_3 + HCOONa + 5NaBr + 5H_2O$
      e. $(CH_3CH_2\underset{\underset{\displaystyle CH_3}{|}}{CH}-)_3B + 4H_2CrO_4 \longrightarrow 3CH_3CH_2COCH_3 + 2Cr_2O_3 + 4H_2O + B(OH)_3$

      f. $=O + H_2NOH \longrightarrow$ $=NOH + H_2O$

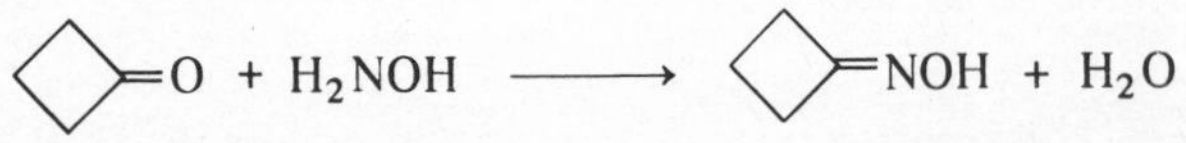

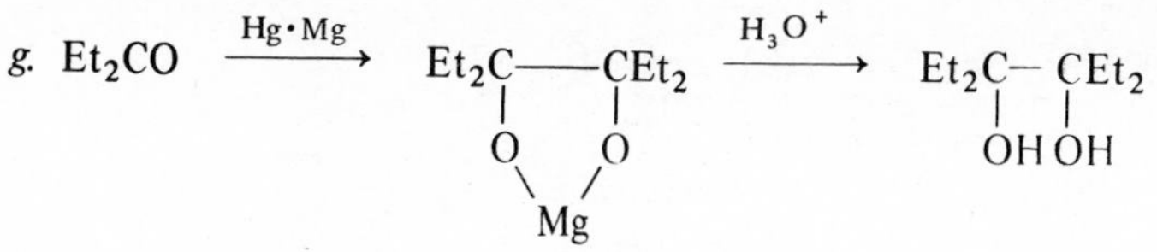

g. Et_2CO $\xrightarrow{Hg\cdot Mg}$ $Et_2C\!-\!\!-\!CEt_2$ (with O, O bridged to Mg below) $\xrightarrow{H_3O^+}$ $Et_2C\!-\!CEt_2$ (OH OH)

$$Et_2CO \xrightarrow{Hg\cdot Mg} \underset{\substack{O \;\; O \\ \diagdown\; \diagup \\ Mg}}{Et_2C\!-\!CEt_2} \xrightarrow{H_3O^+} \underset{OH\; OH}{Et_2C\!-\!CEt_2}$$

h. $CH_3CH_2CH_2CN + SnCl_2 \xrightarrow{H_3O^+} CH_3CH_2CH_2CH{=}NH\cdot H_2SnCl_6 \xrightarrow{H_3O^+} CH_3CH_2CH_2CHO$

i. $CH_3CH_2CH_2CHO + (C_6H_5)_3\overset{+}{P}{-}\overset{-}{C}(CH_3)_2 \longrightarrow CH_3CH_2CH_2CH{=}C(CH_3)_2 + (C_6H_5)_3PO$

j. $(CH_3CO_2^{-})_2Ca^{2+} \xrightarrow{\Delta} (CH_3)_2C{=}O + CaCO_3$

k. $CH_3C{\equiv}CH + HOH \xrightarrow{Hg^{2+},\,H^+} (CH_3)_2C{=}O$

l. $(CH_3)_2CHC(CH_3)_2BH_2 + 2CH_2{=}CHCH_2CO_2Et \longrightarrow (CH_3)_2CHC(CH_3)_2B(CH_2CH_2CH_2CO_2Et)_2$

m. $(CH_3)_2C{=}O + HC(OEt)_3 \longrightarrow (CH_3)_2C(OEt)_2 + HCOOEt$

n. $CH_3CH_2COCH_3 + NaOCl \longrightarrow CH_3CH_2COONa + HCCl_3$

o. $(C_6H_5)_3P + CH_3CH(Br)CH_2CH_3 \longrightarrow (C_6H_5)_3P^+{-}CH(CH_3)CH_2CH_3\ Br^-$

p. $\underset{OH}{(CH_3)_2CCH_2OH} \xrightarrow{H^+} (CH_3)_2CHCHO + H_2O$

q. $CH_3COCH_2CH{=}CH_2 \xrightarrow{LiAlH_4} \xrightarrow{H^+} CH_3CH(OH)CH_2CH{=}CH_2$

r. $[(CH_3)_3C]_2CO \xrightarrow{NaHSO_3}$ no reaction

s. 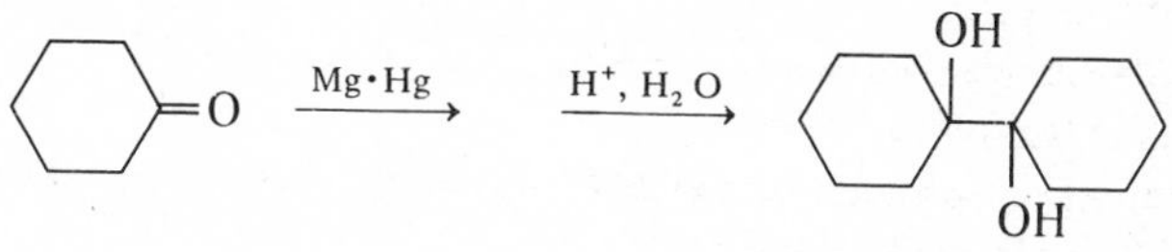 (cyclohexanone) $\xrightarrow{Mg\cdot Hg}$ $\xrightarrow{H^+,\,H_2O}$ (1,1'-bi(cyclohexyl)-1,1'-diol, with OH groups)

t. $CH_3CH(OH)CH_3 + KOH + I_2 \longrightarrow CH_3COOK + CHI_3$

u. 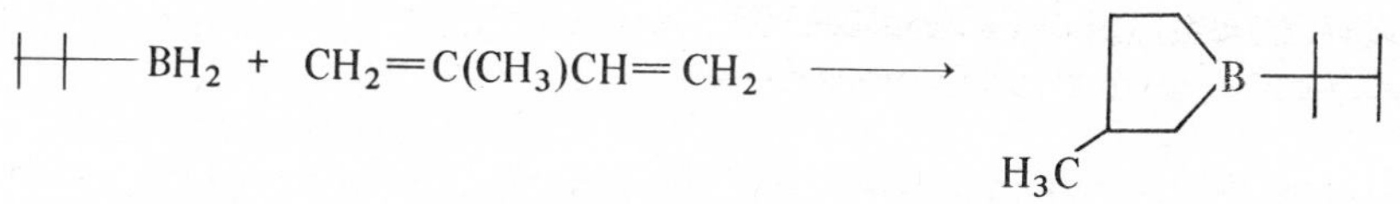 (structure—BH_2) $+\ CH_2{=}C(CH_3)CH{=}CH_2 \longrightarrow$ (cyclic boron ring with H_3C and B—structure)

v. $CH_2{=}CHCOCl + H_2 \xrightarrow[NH_3]{Pd\cdot BaSO_4} CH_3CH_2CHO + HCl$

w. $CH_2{=}CHCHO + Ag^+ \longrightarrow CH_2{=}CHCOOH + Ag$

x. $(CH_3)_2CHBr + (EtO)_3P \longrightarrow (CH_3)_2CHPO(OEt)_2 + EtO^- + Br^-$

y. $HC{\equiv}CCHO + (C_6H_5)_3P{=}CHCH{=}CH_2 \longrightarrow HC{\equiv}CCH{=}CHCH{=}CH_2 + (C_6H_5)_3PO$

z. $CH_2O + (CH_3)_3CCHO + N{:}OH \longrightarrow HCOONa + (CH_3)_3CCH_2OH.$

7-4

a. $CH_3CH_2COOH \xrightarrow{ThO_2} CH_3CH_2COCH_2CH_3 \xrightarrow{H_2NNH_2,\,CH_3ONa}$ pentane

b. $CH_3CH_2COOH \xrightarrow{SOCl_2} CH_3CH_2COCl \xrightarrow{Pd\cdot BaSO_4,\,H_2} CH_3CH_2CHO$

c. $CH_3CHO \xrightarrow{NaOI} HCI_3 \xrightarrow{EtONa} HC(OEt)_3$

d. $CH_3CH_2CHO \xrightarrow{PCl_5} CH_3CH_2CHCl_2 \xrightarrow{NaNH_2} CH_3C{\equiv}CH$

e. $CH_3CHO \xrightarrow{OH^-} CH_3CH(OH)CH_2CHO \xrightarrow{\Delta} CH_3CH{=}CHCHO$

f. $CH_3CH_2CH_2CHO \xrightarrow{OH^-} \xrightarrow{Pt,\,H_2} \xrightarrow{Al_2O_3,\,350^\circ} \underset{Et}{CH_3CH_2CH{=}CHC{=}CH_2}$

g. 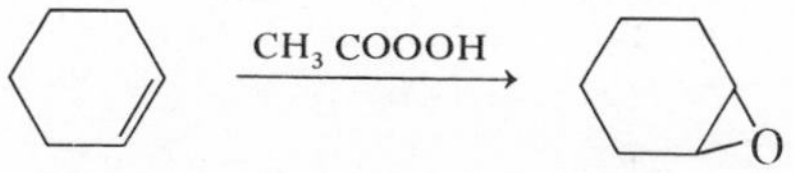 (cyclohexene) $\xrightarrow{CH_3COOOH}$ (cyclohexene oxide, with O)

h. $CH_3COCH_3 \xrightarrow{HCN} (CH_3)_2C(OH)C{\equiv}N \xrightarrow{Al_2O_3,\,\Delta} CH_2{=}C(CH_3)C{\equiv}N$

i. $CH_3CH_2CHO \xrightarrow{OH^-} \xrightarrow{Pt, H_2} CH_3CH_2CH(OH)CH(CH_3)CH_2OH$

j. (cyclohexanone) $\xrightarrow{(C_6H_5)_3P=CH_2}$ (methylenecyclohexane)

k. $CH_3CH_2COOH \xrightarrow{SOCl_2} CH_3CH_2COCl \xrightarrow{Pd \cdot BaSO_4, H_2} \xrightarrow{EtMgBr} \xrightarrow{H^+}$

$CH_3CH_2CH(OH)CH_2CH_3 \xrightarrow{Al_2O_3, \Delta} CH_3CH=CHCH_2CH_3$

l. $CH_3CH_2CH_2CHO \xrightarrow{OH^-, \Delta} CH_3CH_2CH_2CH=C(Et)CHO \xrightarrow{H_2, Pt} CH_3(CH_2)_3CH(Et)CH_2OH$

m. $CH_3COCH_3 \xrightarrow{OH^-} CH_3COCH_2\overset{\displaystyle OH}{\underset{\displaystyle |}{C}}(CH_3)_2 \xrightarrow{LiAlH_4} \xrightarrow{H^+} \xrightarrow[\Delta]{Al_2O_3} CH_2=CHCH=C(CH_3)_2$

n. $CH_3COCH_3 \xrightarrow{Mg \cdot Hg} \xrightarrow{H^+, H_2O} (CH_3)_2\overset{\displaystyle OH}{\underset{\displaystyle |}{C}}-\overset{\displaystyle OH}{\underset{\displaystyle |}{C}}(CH_3)_2 \xrightarrow[\Delta]{Al_2O_3} CH_2=C(CH_3)C(CH_3)=CH_2$

o. $CH_3\overset{\displaystyle C}{\underset{\displaystyle ||}{}}-CHCHO \xrightarrow{2(C_6H_5)_3\overset{+}{P}-\overset{-}{C}H_2} CH_3\overset{\displaystyle C}{\underset{\displaystyle ||}{}}CH(CH_3)CH=CH_2 \xrightarrow{thexylborane} \xrightarrow[1,000\ lb/in^2]{CO}$
$\underset{\displaystyle O}{} \ \underset{\displaystyle CH_3}{} \underset{\displaystyle CH_2}{}$

$\xrightarrow[OH^-]{H_2O_2} CH_3-$ (4-methylcyclohexanone with CH₃) $=O \xrightarrow[CH_3ONa]{H_2NNH_2} CH_3-$ (methylcyclohexane with CH₃)

p. $(CH_3)_2C=CH_2 \xrightarrow{B_2H_6} \xrightarrow{CO} \xrightarrow{H_2O, OH^-} [(CH_3)_2CHCH_2]_3COH \xrightarrow[\Delta]{Al_2O_3} \xrightarrow{Pt, H_2}$

$[(CH_3)_2CHCH_2]_3CH$

7-5 Reaction of the Grignard reagent with carbonyl compounds, ethylene oxide, cyanides, and CO_2; sodium cyanide plus alkyl halide; pyrolysis of acids over ThO_2; aldol condensation; addition of HCN to carbonyl group; Wurtz reaction; cadmium alkyls plus acid halides; pinacol reduction; hydroboration-carbonylation-oxidation of alkenes to alcohols and to ketones (using thexylborane); the Wittig reaction of alkenes and ylides; addition of CN^- to epoxides; alkyl halides plus sodium acetylides; carbene insertion and addition reactions.

7-6 *a.* No reaction. Ethers inert to alkali.
 b. *E*2 reaction yielding $(CH_3)_2C=CH_2$.
 c. No reaction.
 d. No reaction. Steric hindrance by *tert*-butyl group prevents backside attack by OH^-.
 e. Essentially no reaction. RONa is almost 100 percent hydrolyzed in aqueous solution.
 f. No reaction.
 g. No reaction.
 h. Aldol condensation yielding $(CH_3)_2C=CHC(CH_3)_2CHO$.

7-7 *a.* Acetic acid will turn litmus paper or cause carbonate to fizz. Propanal will give a Schiff test.
 b. Only acetaldehyde will give an iodoform test.
 c. Only 2-pentanone will give an iodoform test or react with $NaHSO_3$.
 d. Butanal will give a Schiff test or yield a phenylhydrazone.
 e. Only the ether will form a soluble oxonium salt with H_2SO_4.
 f. Only ethanol will give an iodoform test.
 g. Only ether is soluble in concentrated sulfuric acid.
 h. Only 1-chloro-1-pentene will decolorize permanganate.
 i. Only acetaldehyde will give a Schiff test.
 j. Only 4-pentyn-2-one would form a precipitate when treated with $Ag(NH_3)_2^+$.

7-8 You could first test the liquids with alcoholic silver nitrate; only the iodide would give a precipitate. Alternatively, you could wet a copper wire with each liquid and heat the wire in a flame. Halogen compounds, except fluorine, give a green flame. Next you could test the liquids with phenylhydrazine or semicarbazide; the two carbonyl compounds could then be tested with NaOI; only acetone would give a precipitate of HCI_3.

7-9 CH_3CHO

7-10 $CH_3CH_2OCH_2CH_3$, $CH_3OCH_2CH_2CH_3$, $CH_3OCH(CH_3)_2$

7-11 $CH_2=CH-CH=O \longleftrightarrow CH_2=CH-\overset{+}{C}H-O^- \longleftrightarrow \overset{+}{C}H_2-CH=CH-O^- \longleftrightarrow \overset{+}{C}H_2-\overset{-}{C}H-CH=O$

$\overset{+}{C}H_2-CH=CH-O^- \xrightarrow{H^+Cl^-} ClCH_2CH=CH-OH \rightleftharpoons ClCH_2CH_2CHO$

$CH_2=CHCHO \xrightarrow{CH_3MgX} \xrightarrow{H^+} CH_3CH_2CH_2CHO$

$CH_2=CHCHO \xrightarrow{CH_3SH} CH_3SCH_2CH_2CHO$; $CH_2=CHCHO \xrightarrow{NH_3} H_2NCH_2CH_2CHO$

$CH_2=CHCHO \xrightarrow{HCN} NCCH_2CH_2CHO$ and $CH_2=CHCH(OH)CN$

7-12 $(CH_3)_3CCH(OH)CH_3 \xrightarrow{H^+, \Delta} (CH_3)_3C\overset{+}{C}HCH_3 \longrightarrow (CH_3)_2\overset{+}{C}CH(CH_3)_2 \longrightarrow (CH_3)_2C=C(CH_3)_2 + H^+$

7-13 $CH_2=CHC\equiv N \longleftrightarrow CH_2=CH\overset{+}{C}=\overset{-}{N} \longleftrightarrow \overset{+}{C}H_2-CH=C=N^- \xrightarrow{RMgX}$

$RCH_2CH=C=NMgX$ or $CH_2=CHC(R)=NMgX$

7-14 *a.* In A, dichromate would also oxidize the double bond. B is all right, but in C, Br would be removed by hydrogenolysis.

b. In A it is not possible to have ketone and Grignard functions in the same molecule.

c. Step A gives elimination instead of S_N reaction. Step C is wrong; NaOI cleaves only methyl ketones.

d. Wurtz reaction causes polymerization of olefins; also, the reaction is only useful for symmetric molecules. Compounds with I on adjacent carbons are unstable.

e. In A the OH group will not survive at $450°C$. Elimination will occur. In B only methyl ketones will react with $NaHSO_3$.

f. In A, $BrCH_2CH_2OH + CH_3O^- \longrightarrow BrCH_2CH_2O^-$ occurs. The latter can then react with itself to give ethylene oxide.

g. In A it is not possible to hydrogenate Grignard reagents.

h. In A presence of the acid HBr will cause a pinacol rearrangement of the diol. In compound B, Mg, like Zn, will cause dehalogenation of the vicinal dihalide giving 2,3-dimethyl-2-butene.

i. In A the presence of a CHO group in the ylide would not be suitable because the ylide would react with itself. B is all right.

j. In A hydroboration would result in the boron of thexylborane attacking the least substituted olefinic carbon atom yielding bis-1-(2-methylcyclohexyl)ketone as the product. B is all right.

k. A is all right. In B, unless $Pd \cdot BaSO_4$ is used as the catalyst, it is not possible to stop the reduction of the acid chloride at the aldehyde but rather it will proceed on to the alcohol. In C the crossed-Cannizzaro reaction will not work because 2-methylpropanal has an α hydrogen and will undergo an aldol condensation with itself instead.

7-15 The main product would be that from acetaldehyde condensing with itself, i.e., aldol. Aldehydes are more reactive than ketones. Some acetone would condense with itself and some with acetaldehyde. Write equations for the two ways in which acetaldehyde and acetone can condense with each other.

7-16 *a.* Convert acetylene to acetaldehyde using Hg^{2+} catalyst. Run aldol condensation and dehydrate to get 2-butenal. Oxidize this with Tollens' reagent and hydrogenate. Or, hydrogenate the 2-butenal to 1-butanol and oxidize.

b. HCN can be added catalytically to acetylene, or add HCN to CH_3CHO made in *a* and dehydrate.

c. Convert acetylene to 2-butyne with CH_3I. Hydrogenate with Pd to 2-butene. Oxidize to diol with cold, dil $KMnO_4$ and then dehydrogenate to dione.

d. Add 2 mol of HCN to acrolein; the first adds across the double bond, the second to the CHO group.

e. Starting with 2-butene (part *c* above) hydroboration followed by treatment with CO and H_2O would give 3,5-dimethyl-4-heptanone which can then be reduced with hydrazine and CH_3ONa.

f. Convert acetylene to propyne with CH_3I. Hydrogenate with Pd to propylene. Hydroborate propylene followed by carbonylation and then alkaline peroxide.

7-17   a. Make Grignard reagent from 1-butyne and add CO_2. Make acid chloride of resulting acid and add 2 mol of ethylmagnesium iodide.

b. Take the alcohol made in the previous problem and hydrogenate the triple bond to a double bond and then dehydrate the alcohol.

c. Add CO_2 to $HC \equiv CMgI$ and hydrogenate to CH_3CH_2COOH. Pass this acid over hot thoria to get 3-pentanone. Pinacol reduction gives the diol that can be dehydrated to diene.

d. Convert acetylene to CH_3COOH via acetaldehyde. Convert this to acetone using CH_3MgX and run a pinacol reduction. Then pinacol rearrangement gives $(CH_3)_3CCOCH_3$, which, with NaOCl, gives Na salt of desired product.

e. Make Grignard reagent from propyne and add 2 mol of this to glyoxal, CHOCHO.

f. Pinacol reduction of acetone followed by a pinacol rearrangement would give $(CH_3)_3CCOCH_3$.

7-18   a. Oxidize cyclopentanol to cyclopentanone and add HCN. Dehydrate the cyanohydrin over Al_2O_3.

b. Reduce acetylene to ethylene and hydroborate. Carbonylation of triethylborane followed by alkaline H_2O_2 oxidation will give 3-ethyl-3-pentanol. Convert the alcohol to a Grignard reagent and add CH_2O. Convert the primary alcohol to the chloride with $SOCl_2$ and convert this to the hydrocarbon via the Grignard reagent or catalytic hydrogenation. $SOCl_2$ will result in a minimum of rearrangement while HCl would cause mostly rearrangement.

c. Oxidation of cyclopentanol with HNO_3 gives $HOOC(CH_2)_3COOH$. Reduce this with $LiAlH_4$, convert to di-Grignard reagent, and treat with CO_2. Pass this acid over hot thoria to get cyclohexanone. Alternatively, 1,5-pentanediol could be dehydrated to 1,4-pentadiene with Al_2O_3. Then perform hydroboration with thexylborane (obtained from hydroboration of 2,3-dimethyl-2-butene) followed by carbonylation and alkaline H_2O_2, yielding cyclohexanone. Reduce to cyclohexanol, dehydrate, and oxidize with performic acid to the epoxide. To this add cyclopentylmagnesium bromide and oxidize, resulting in the desired ketone.

d. Make butanal from acetylene via acetaldehyde and aldol condensation or from 1-propanol. Condense butanal in aldol condensation and dehydrate. Add EtMgX to this unsaturated aldehyde and then hydrogenate to the saturated alcohol.

e. Condense 2 mol of CH_2O with 1 mol of CH_3CH_2CHO in aldol condensation.

f. Make nitromethane from CH_3I using $NaNO_2$ and dimethylformamide. Condense nitromethane with acetone.

g. Treat $CH_2=CHCOCl$ with $(CH_3CH_2CH_2)_2Cd$. Convert 1-propanol to propene, chlorinate at $500°C$, and hydrolyze to get $CH_2=CHCH_2OH$. Dehydrogenate to the aldehyde and oxidize with Tollens' reagent to the acid; convert to the acid chloride with $SOCl_2$. Alternatively, $CH_3CH_2CH_2COCH_3$ can be condensed with CH_2O to give $CH_3CH_2CH_2COCH_2CH_2OH$ via aldol condensation. The αH on the methyl is much more active than that on propyl (why?), so a good yield is obtained. Dehydration gives the product.

h. Aldol condensation of acetone and dehydration gives $(CH_3)_2C=CHCOCH_3$. Hydrogenation gives the saturated alcohol. Oxidize this to the ketone and add HCN.

i. Dehydrate cyclopentanol to cyclopentene. Peroxidation using CH_3COOOH would yield the epoxide. Open the epoxide ring with CN^- to give the hydroxynitrile.

j. Dehydrate cyclopentanol and 1-propanol to cyclopentene and propylene, respectively. Convert propylene to chloride and treat with CN^- to give $CH_2=CHCH_2CN$. Hydroborate 2,3-dimethyl-2-butene to thexylborane and treat this with 1 mol of cyclopentene and then 1 mol of $CH_2=CHCH_2CN$. Carbonylation (1,000 lb/in^2 of CO) of the resulting trialkylborane followed by alkaline H_2O_2 oxidation yields the desired compound.

k. Equilibrate acetone with D_2O in presence of acid or base. Run pinacol reduction then rearrangement with D_2SO_4 to avoid introduction of H. Run haloform cleavage with NaOI.

l. Treat CaC_2 with D_2O to get $DC \equiv CD$. Hydrate with $HgSO_4$ and D_2O to D_3CCDO and add CH_3MgX followed by H^+.

m. Make 4-octyne from $CH_3CH_2CH_2C \equiv CNa$ and $CH_3CH_2CH_2Br$. Hydroborate 4-octyne and treat with CH_3COOD (made by oxidizing acetaldehyde to acetic acid, then D_2O) to get product.

n. Hydroboration of 4-octyne (part m) followed by H_2O_2 to get 4-octanone. The hydroborating agent should be $[(CH_3)_2CHCH(CH_3)]_2BH$ which can be made from B_2H_6 and 2-methyl-2-butene.

7-19   a. $(CH_3)_2C(OEt)_2 \xrightarrow{H_3O^+} (CH_3)_2CO + 2EtOH$

b. No reaction except protonation of ether oxygen

c. $(CH_3)_2C(OH)CH(OH)CH_3 \xrightarrow{H^+} (CH_3)_3CCHO + H_2O$

342

d. $(CH_3)_2C\overset{\displaystyle\diagdown}{\underset{\displaystyle O}{}}\!\!\!-\!\!\!\overset{\displaystyle\diagup}{}C(CH_3)_2 \xrightarrow{H^+} (CH_3)_2\overset{+}{C}C(CH_3)_2 \longrightarrow (CH_3)CCOCH_3$
$\qquad\qquad\qquad\qquad\qquad\qquad\qquad\qquad \underset{\displaystyle OH}{|}$

e. $CH_2\!-\!CH_2 \xrightarrow{H_3O^+} CH_3CHO \;+\; HOCH_2CH_2OH$
$\qquad \overset{\diagdown}{\underset{O}{}}\!\!\diagup$

f. $(CH_3)_3CCH(OH)CH_3 \xrightarrow{H^+} (CH_3)_2C\!=\!C(CH_3)_2$

7-20 $\quad ^-CH_2CH\!=\!CHCH\!=\!O \longleftrightarrow CH_2\!=\!CH\!-\!CH\!=\!CH\!-\!O^-$
      a. $\;CH_3CH\!=\!CHCH\!=\!CHCH\!=\!CHCHO$ *b.* $\;(HOCH_2)_3CCH\!=\!CHCHO$

7-21   *a.* $\;CH_3COCH\!=\!CH_2 < CH_3CH\!=\!CHCHO < CH_3COCH_3 < CH_3CHO < F_3CCHO$
      b. $\;CH_2\!=\!CHCHO < CH_3CH_2CHO < BrCH_2CHO < ClCH_2CHO < CH_3CF_2CHO$
      c. $\;HSCH_2CH_2CHO < CH_3SeCH_2CHO < CH_3SCH_2CHO < CH_3OCH_2CHO < NCCH_2CHO$

7-22 $CH_2\!=\!CHCHO \xrightarrow[CH_3OH]{H^+} CH_2\!=\!CHCH(OMe)_2 \xrightarrow{Cl_2,\,H_2O} ClCH_2CH(OH)CH(OMe)_2 \xrightarrow{NaOH}$

  $\xrightarrow{H_3O^+} HOCH_2CH(OH)CHO$

7-23 $CH_2\!=\!CH\!-\!\overset{\displaystyle O}{\overset{\displaystyle \|}{C}}\!-\!OH \;\longleftrightarrow\; \overset{+}{C}H_2\!-\!CH\!=\!\overset{\displaystyle O^-}{\overset{\displaystyle |}{C}}\!-\!OH \xrightarrow{HCN} NCCH_2CH\!=\!\overset{\displaystyle O^-}{\overset{\displaystyle |}{C}}\!-\!OH \longleftrightarrow$

                                              $NCCH_2\overset{-}{C}H\!-\!\overset{\displaystyle O}{\overset{\displaystyle \|}{C}}\!-\!OH \xrightarrow{H^+} NCCH_2CH_2\overset{\displaystyle O}{\overset{\displaystyle \|}{C}}\!-\!OH$

Such resonance stabilization is not possible in RCH=CHR.

7-24 $\underset{\displaystyle |}{\underset{\displaystyle H}{RCHCHO}} \xrightarrow{base} \underset{\displaystyle -}{RCHCHO} \longleftrightarrow RCH\!=\!CH\!-\!O^-$ Only carbanions resulting from the loss

of an α hydrogen can be stabilized by resonance.

7-25 $-C\!\equiv\!N$

7-26   *a.* $\left[CH_2\!=\!CH\!-\!\overset{\displaystyle O}{\overset{\displaystyle \|}{C}}CH_3 \longleftrightarrow \overset{+}{C}H_2CH\!=\!\overset{\displaystyle O^-}{\overset{\displaystyle |}{C}}CH_3 \right] \xrightarrow{^-C(NO_2)_3}$

              $\left[(O_2N)_3CCH_2CH\!=\!\overset{\displaystyle O^-}{\overset{\displaystyle |}{C}}CH_3 \longleftrightarrow (O_2N)_3CCH_2\overset{-}{C}H\overset{\displaystyle O}{\overset{\displaystyle \|}{C}}CH_3 \right] \xrightarrow{H^+} ans$

    b. $(CH_3)_2C(OH)CH_2I \xrightarrow{HgO} [(CH_3)_2C(OH)CH_2{}^+] \longrightarrow CH_3\underset{\displaystyle |}{\underset{\displaystyle OH}{\overset{+}{C}CH_2CH_3}} \longrightarrow$

                                        $CH_3\underset{\displaystyle O}{\overset{\displaystyle \|}{C}}CH_2CH_3 \;+\; H^+$

    c. $-CH_2OH \xrightarrow{H^+,\,\Delta} \left[\text{}-\overset{+}{C}H_2 \longrightarrow \text{} \right] \longrightarrow \text{} + H^+$

    d. $CH_3COCH_3 \xrightarrow{OH^-} CH_3COCH_2{}^- + [CH_2\!=\!CH\!-\!C\!\equiv\!N \longleftrightarrow \overset{+}{C}H_2CH\!=\!C\!=\!N^-] \longrightarrow \xrightarrow{H^+} ans$

343

e. $CH_2(CN)_2 \xrightarrow{OH^-} (CN)_2CH^- + [CH_3CH=CHCH=CHCO_2CH_3 \longleftrightarrow$

$$CH_3\overset{+}{C}H-CH=CH-CH=\overset{\overset{O^-}{|}}{C}OCH_3] \xrightarrow{\;H^+\;} \text{ans}$$

f. $(C_6H_5)_3\overset{+}{P}-CH_2(CH_2)_3\underset{\underset{O}{\parallel}}{C}C_6H_5\; Cl^- \xrightarrow{CH_3Li} (C_6H_5)_3\overset{+}{P}-\overset{-}{C}H$, $(CH_2)_3$, $O=C$, $C_6H_5 \longrightarrow$

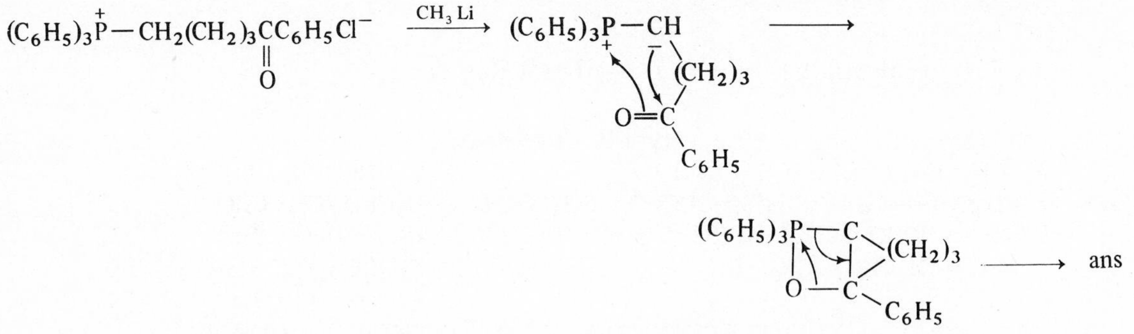

$$(C_6H_5)_3P-C , (CH_2)_3 , O-C , C_6H_5 \longrightarrow \text{ans}$$

g. $H-\underset{\underset{O}{\parallel}}{C}-\underset{\underset{O}{\parallel}}{C}-H \xrightarrow{OH^-} H-\underset{\underset{O}{\parallel}}{C}-\overset{\overset{H}{|}}{\underset{\underset{O^-}{|}}{C}}-OH \longrightarrow H-\overset{\overset{H}{|}}{\underset{\underset{OH}{|}}{C}}-\underset{\underset{O}{\parallel}}{C}-OH \xrightarrow{OH^-} HOCH_2CO_2^-$

h. [cyclohexane ring with OH and $\underset{\underset{OH}{|}}{\overset{\overset{CH_3}{|}}{C}}-CH_3$] $\xrightarrow[\Delta]{H^+}$ [cyclohexane ring with OH and $\overset{+}{C}$, CH_3, $-CH_3$] $\longrightarrow$

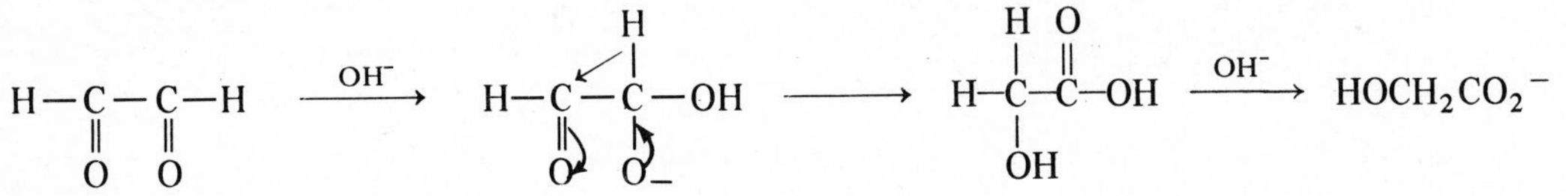

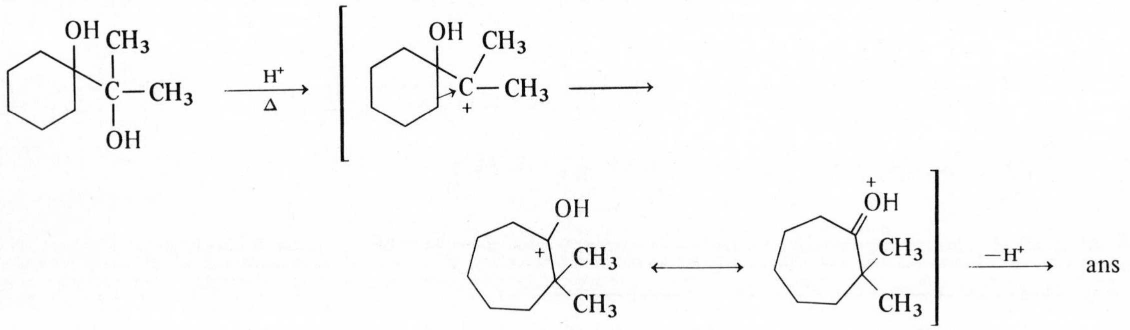

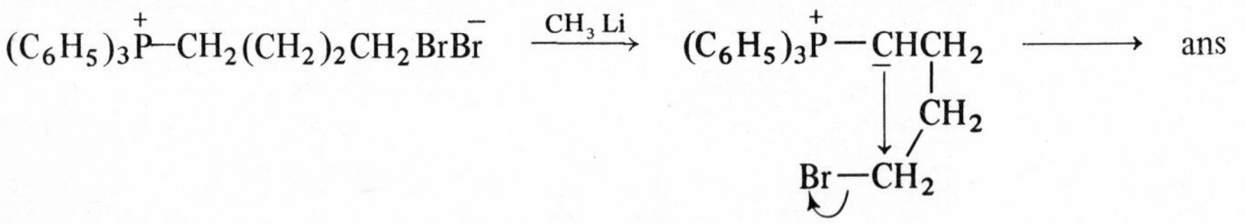

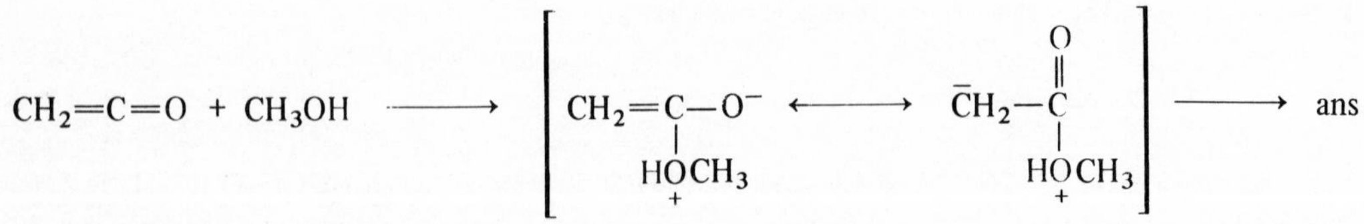

[cycloheptane ring with OH, $+$, CH_3 $\longleftrightarrow$ cycloheptane ring with $\overset{+}{O}H$, CH_3, CH_3] $\xrightarrow{-H^+}$ ans

i. $(C_6H_5)_3\overset{+}{P}-CH_2(CH_2)_2CH_2Br\; \overset{-}{Br} \xrightarrow{CH_3Li} (C_6H_5)_3\overset{+}{P}-\overset{-}{C}H\underset{\underset{\underset{Br-CH_2}{|}}{CH_2}}{CH_2} \longrightarrow$ ans

7-27 a. $CH_2=C=O + CH_3OH \longrightarrow \left[CH_2=\underset{\underset{H\overset{+}{O}CH_3}{|}}{C}-O^- \longleftrightarrow \overset{-}{C}H_2-\underset{\underset{H\overset{+}{O}CH_3}{|}}{\underset{\underset{O}{\parallel}}{C}} \right] \longrightarrow$ ans

b. $CH_3\underset{\underset{O}{\parallel}}{C}NHCH_3$ and $CH_3\underset{\underset{O}{\parallel}}{C}O\underset{\underset{O}{\parallel}}{C}CH_3$

c. $CH_3COCH_3 \xrightarrow{\Delta} CH_3\underset{\underset{O}{\parallel}}{C}\!\cdot + CH_3\cdot$ Initiation

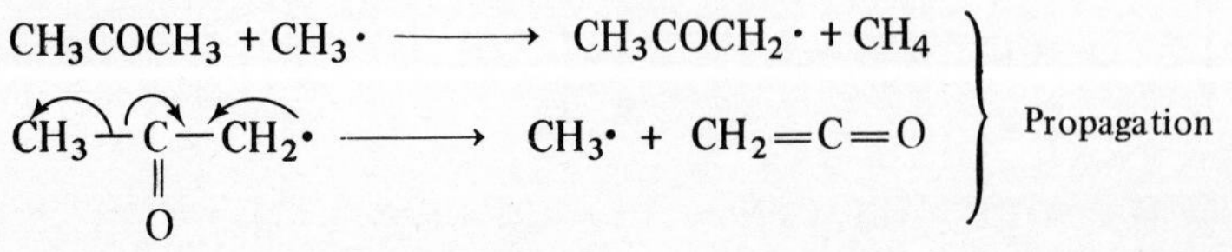

$CH_3COCH_3 + CH_3\cdot \longrightarrow CH_3COCH_2\cdot + CH_4$

$CH_3-\underset{\underset{O}{\parallel}}{C}-CH_2\cdot \longrightarrow CH_3\cdot + CH_2=C=O$ } Propagation

344

7-28 *a.* CH_3COCHO $\xrightarrow{HOCH_2CH_2OH,H^+}$ $CH_3C-CH{\begin{smallmatrix}O-CH_2\\|\\O-CH_2\end{smallmatrix}}$ A
$\qquad\qquad\qquad\qquad\qquad\qquad\quad\overset{\|}{O}$

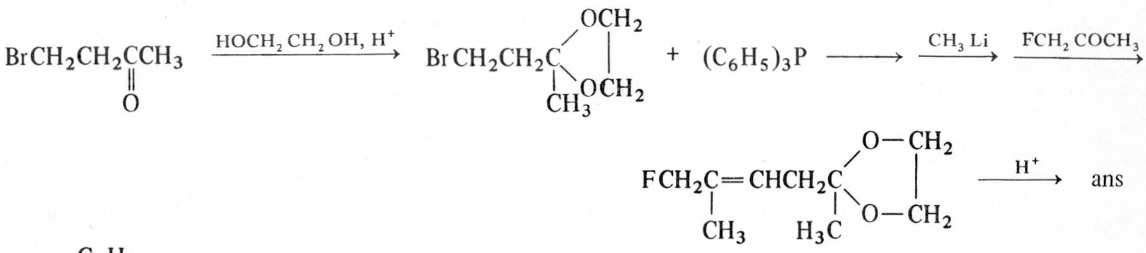

$(C_6H_5)_3P + (CH_3)_2C=CHCH_2Br \longrightarrow \xrightarrow{CH_3Li} (C_6H_5)_3P=CH-CH=C(CH_3)_2 \xrightarrow{A}$

$\qquad\qquad\qquad (CH_3)_2C=CHCH=C(CH_3)CH{\begin{smallmatrix}OCH_2\\|\\OCH_2\end{smallmatrix}} \xrightarrow{H^+} ans$

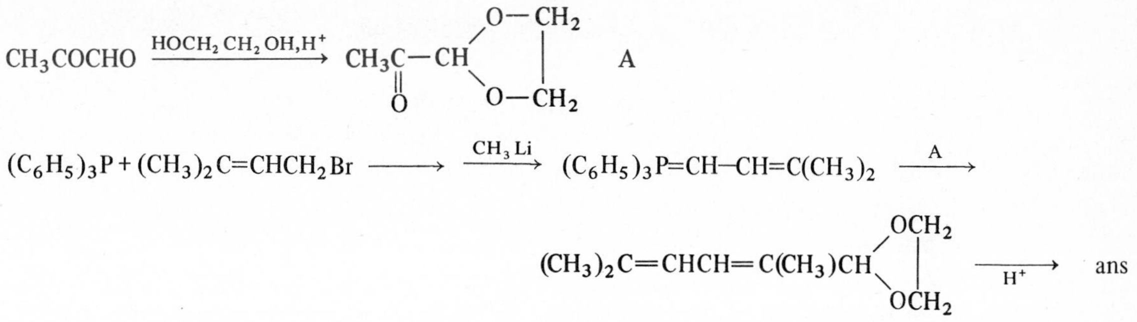

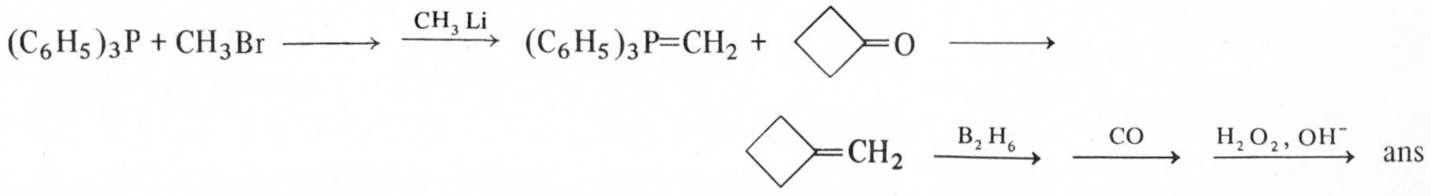

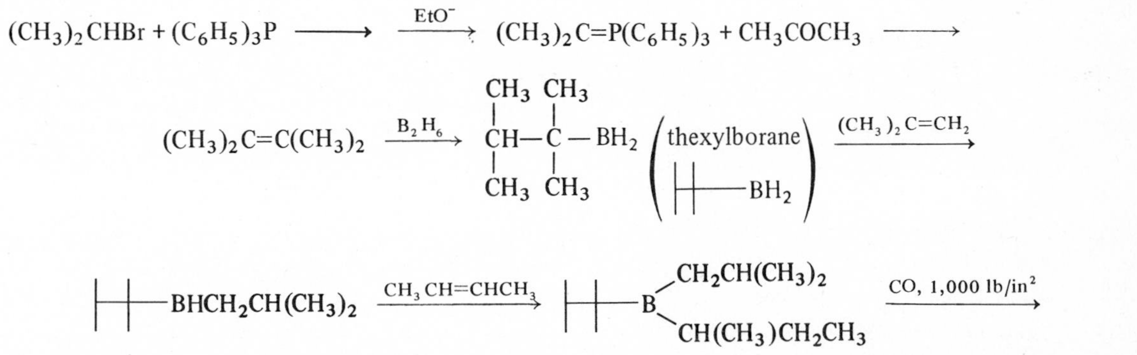

b. $BrCH_2CH_2CCH_3 \xrightarrow{HOCH_2CH_2OH,\ H^+} BrCH_2CH_2C{\begin{smallmatrix}OCH_2\\|\\OCH_2\end{smallmatrix}} + (C_6H_5)_3P \longrightarrow \xrightarrow{CH_3Li} \xrightarrow{FCH_2COCH_3}$
$\qquad\quad\overset{\|}{O}\qquad\qquad\qquad\qquad\qquad\qquad\qquad\underset{CH_3}{}$

$\qquad\qquad\qquad\qquad\qquad FCH_2C=CHCH_2C{\begin{smallmatrix}O-CH_2\\|\\O-CH_2\end{smallmatrix}} \xrightarrow{H^+} ans$
$\qquad\qquad\qquad\qquad\qquad\quad\underset{CH_3}{|}\qquad\underset{H_3C}{|}$

7-29 $CH_3{-}\overset{\overset{C_6H_5}{\vdots}}{\underset{\underset{O}{\|}}{P}}{-}CH_2CH_3$

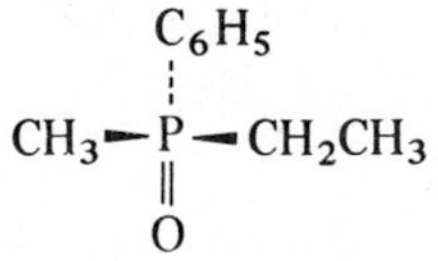

7-30 *a.* $(C_6H_5)_3P + CH_3Br \longrightarrow \xrightarrow{CH_3Li} (C_6H_5)_3P=CH_2 + \square{=}O \longrightarrow$

$\qquad\qquad\qquad\qquad \square{=}CH_2 \xrightarrow{B_2H_6} \xrightarrow{CO} \xrightarrow{H_2O_2,\ OH^-} ans$

b. $(CH_3)_2CHBr + (C_6H_5)_3P \longrightarrow \xrightarrow{EtO^-} (CH_3)_2C=P(C_6H_5)_3 + CH_3COCH_3 \longrightarrow$

$\qquad (CH_3)_2C=C(CH_3)_2 \xrightarrow{B_2H_6} \underset{\underset{CH_3}{|}}{\overset{\overset{CH_3}{|}}{CH}}{-}\underset{\underset{CH_3}{|}}{\overset{\overset{CH_3}{|}}{C}}{-}BH_2 \left(\text{thexylborane} \; \mapsto{-}BH_2\right) \xrightarrow{(CH_3)_2C=CH_2}$

$\qquad \mapsto{-}BHCH_2CH(CH_3)_2 \xrightarrow{CH_3CH=CHCH_3} \mapsto{-}B{\begin{smallmatrix}CH_2CH(CH_3)_2\\ \\CH(CH_3)CH_2CH_3\end{smallmatrix}} \xrightarrow{CO,\ 1,000\ lb/in^2}$

$\qquad\qquad\qquad\qquad\qquad\qquad \xrightarrow{H_2O_2,\ OH^-} \xrightarrow{NH_2OH} ans$

c. $CH_3COCOCH_3 + 2CH_2=P(C_6H_5)_3\ (\text{part } a) \longrightarrow$

$\qquad \underset{CH_3C=CH_2}{CH_3C=CH_2} \xrightarrow[(\text{part } b)]{\mapsto{-}BH_2} \underset{H_3C}{\overset{H_3C}{}}{\bigcirc}{B}\mapsto \xrightarrow{CO\ 1,000\ lb/in^2} \underset{H_3C}{\overset{H_3C}{}}{\bigcirc}{=}O\quad A$

$\qquad BrCH_2CH_2CH_2Br + 2(C_6H_5)_3P \longrightarrow \xrightarrow{base} (C_6H_5)_3P=CHCH_2CH=P(C_6H_5)_3 \xrightarrow{2\ mol\ A}\xrightarrow{Pt,\ H_2} ans$

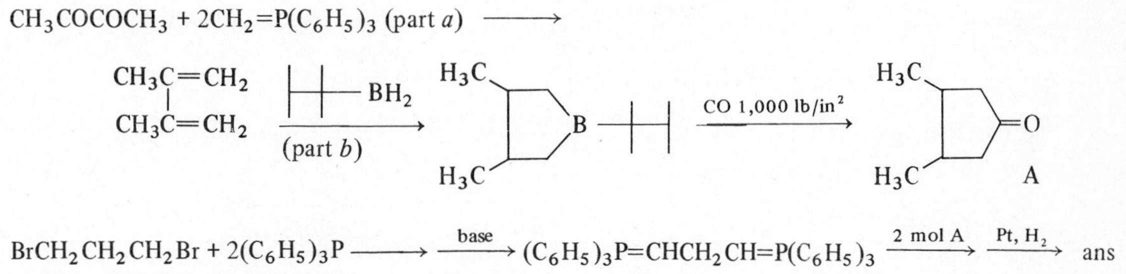

d. $OHCCH_2CH_2CHO + 2(CH_3)_2C=P(C_6H_5)_3\ (\text{part } b) \longrightarrow$

$\qquad\qquad\qquad (CH_3)_2C=CHCH_2CH_2CH=C(CH_3)_2 \xrightarrow{HCOOOH} ans$

345

e. $CH_3(CH_2)_2CHO + CH_2=P(C_6H_5)_3 \longrightarrow CH_3(CH_2)_2CH=CH_2 \xrightarrow{B_2H_6} \xrightarrow{CO,\ LiBH_4} \xrightarrow{H_2O_2,\ OH^-}$

$CH_3(CH_2)_4CHO \xrightarrow{OH^-,\ aldol} CH_3(CH_2)_4CH=C(CHO)(CH_2)_3CH_3 + CH_2=P(C_6H_5)_3 \longrightarrow$ ans

f. Convert $CH_3OCH_2CH_2CHO$ to $CH_3OCH_2CH_2CH=CH_2$ via Wittig reaction.

BH_2 (part b) $\xrightarrow{CH_3OCH_2CH_2CH=CH_2} \xrightarrow{CH_2=C(CH_3)_2} \xrightarrow{CO,\ 1{,}000\ lb/in^2}$

$CH_3OCH_2(CH_2)_3COCH_2CH(CH_3)_2 \xrightarrow{CH_2=P(C_6H_5)_3} \xrightarrow{B_2H_6} \xrightarrow{CO,\ H_2O} \xrightarrow{H_2O,\ OH^-}$

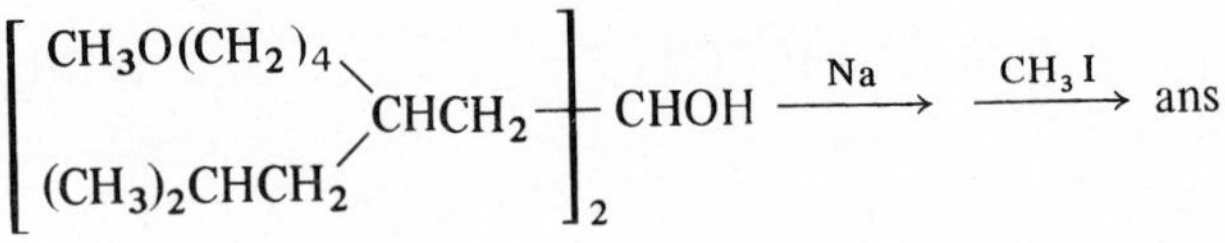

$$\left[\begin{array}{c} CH_3O(CH_2)_4 \\ \\ (CH_3)_2CHCH_2 \end{array} CHCH_2 \right]_2 CHOH \xrightarrow{Na} \xrightarrow{CH_3I} \text{ans}$$

7-31 a.

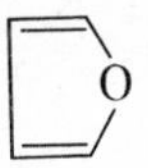

b.

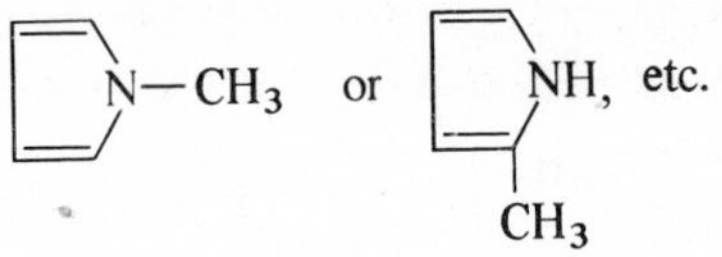

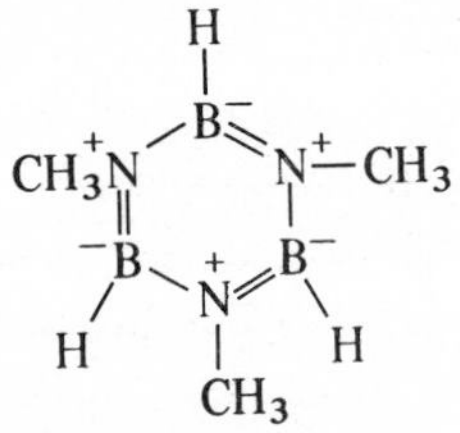

7-32

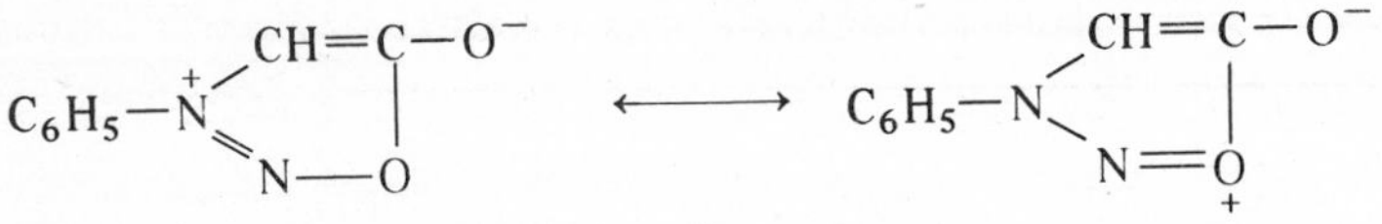

The $4n + 2$ rule holds with one p electron on each N and C atom and two from oxygen. All of the atoms in the ring probably have sp^2 hybridization.

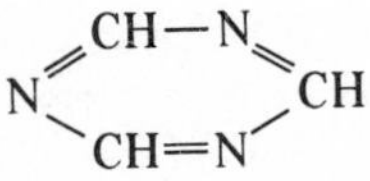

7-33 Tropone $\xrightarrow{H^+}$

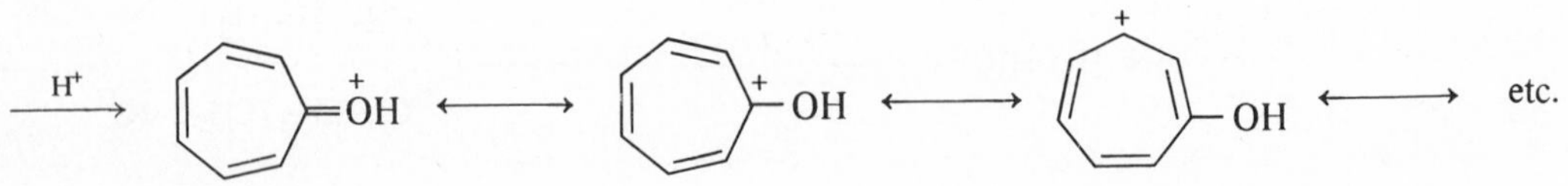

7-34 $CH_3CHO \xrightarrow{OH^-} {}^-CH_2CHO \xrightarrow{H_2C=O} {}^-OCH_2CH_2CHO \xrightarrow{H_2O} HOCH_2CH_2CHO \xrightarrow{OH^-}$

$HOCH_2\overset{-}{C}HCHO \xrightarrow{H_2C=O} \xrightarrow{H_2O} \xrightarrow{OH^-} \xrightarrow{H_2C=O} \xrightarrow{H_2O} (HOCH_2)_3CCHO$ (A)

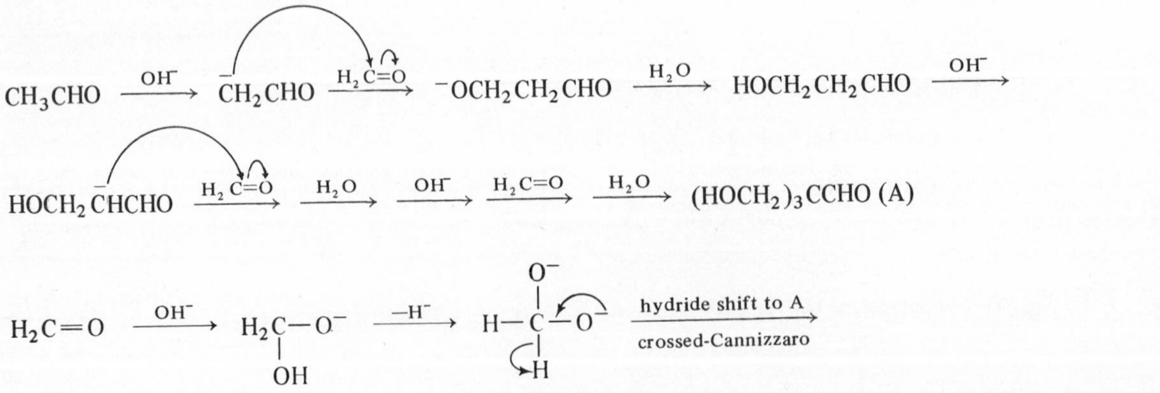

$(HOCH_2)_3CCH_2O^- + HCOO^- \xrightarrow{H_2O} (HOCH_2)_4C + HCOO^-$

346

7-35

$$RC{=}CHR \xleftarrow{R_2'BH} RC{\equiv}CR \xrightarrow{B_2H_6} \left(RCH{=}C{-}B\right)_3$$

with $R_2'B$ on the left structure and R below on the right structure.

$$CH_3C{\equiv}CC(CH_3)_3 \xrightarrow{R_2'BH} CH_3C{=}CHC(CH_3)_3 \xrightarrow[H_2O]{H_2O_2} CH_3C{=}CHC(CH_3)_3 \rightleftharpoons$$

with $R_2'B$ below the middle structure and OH below the right structure.

Enol

$$CH_3CCH_2C(CH_3)_3$$

with $\overset{\|}{O}$

Chapter 8

8-1
a.	225 nm	*b.*	227 nm	*c.*	237 nm	*d.*	222 nm
e.	~360 nm	*f.*	258 nm	*g.*	234 nm	*h.*	259 nm
i.	283 nm	*j.*	355 nm	*k.*	323 nm		

l. 283 nm. Although there are three C=C in this structure the base value is that for a homodiene system with no extension of the conjugation. This is because the third C=C is "cross-conjugated" to the homodiene system rather than "through-conjugated" as in sample *k*.

m. 244 nm. Another example of "cross-conjugation."

n. 383 nm   *o.* 298 nm

8-2 A; 284 nm (calc 280 nm)   B; 278 nm (calc 278 nm)   C; 317 nm (calc 317 nm)
D; 258 nm (calc 254 nm)   E; 293 nm (calc 293 nm)

8-3 21,000

8-4
a. H_2O, three modes: symmetric and asymmetric stretching, and one in-plane bending mode.
CO_2, four modes: symmetric and asymmetric stretching, and two in-plane bending modes.
b. H_2 and the C–C symmetric stretching in acetylene.
c. 51.7×10^{12} c/s or 1724 cm^{-1}; $\% T = 15.9$.

8-5
a. 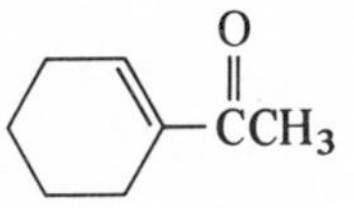 (cyclohexenyl)$-CCH_3$ with $\overset{O}{\|}$ above the C

b. $CH_3CH(OH)CH_3$

c. $CH_3CH_2C{\equiv}CCH_2CH_3$

d. $CH_3(CH_2)_3C{\equiv}CH$

e. 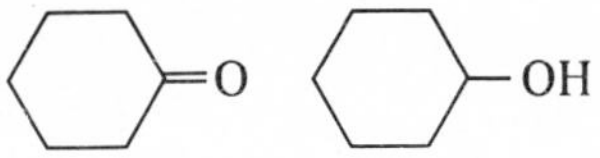 (cyclohexanone) $=O$ E   (cyclohexanol) $-OH$ F

f. $CH_3CH_2CH_2CHO$

g. $CH_2{=}CHCH_2CH_2CH_2CHO$

h. $(CH_3)_3CCCH_3$ with $\overset{\|}{O}$ — I $(CH_3)_3CCHCH_3$ with OH above CH — J

$$\underset{H_3C}{\overset{H_3C}{>}}C{=}C\underset{CH_3}{\overset{CH_3}{<}}$$ K   Via carbonium ion rearrangement of 10

i. $(CH_3)_2C{=}CH{-}C(CH_3)_2$ with CHO below the right C

j.  (cyclohexene ring drawing)

k. $CH_3CH_2CH{=}C$ with CHO and CH_3 branches

l. $CH_3CH_2CH_2OCH_2CH_2CH_3$

8-6
a. $\delta = \dfrac{366}{60 \times 10^6}\,10^6 = 6.1$; $\tau = 10 - \delta = 3.9$

347

b. 100-Hz sweep width: $\delta = 0$ to 1.67; $\tau = 10$ to 8.33
250-Hz sweep width: $\delta = 0$ to 4.17; $\tau = 10$ to 5.83
1,000-Hz sweep width: $\delta = 0$ to 16.67; $\tau = 10$ to -6.67
Chemical shift is independent of sweep width.

8-7 a. CH_3CH_3, $CH_3CH_2CH_3$, $(CH_3)_3CH$
b. CH_3CH_3, CH_3CH_2I, CH_3CH_2OH, CH_3CHI_2, $CH_3CH(OH)_2$
c. $(CH_3)_4Si$, $(CH_3)_3P$, $(CH_3)_2S$, CH_3Cl, CH_3F

8-8 a.

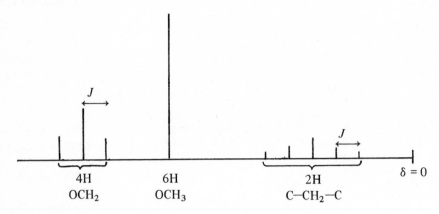

b. $-OCH_2-$ triplet is in 1:2:1 ratio while $C-CH_2-C$ quintet is in 1:4:6:4:1 ratio.

c. For OCH_2: ↑↓ ↑↑ ↓↑   For CCH_2C:
$$\begin{array}{ccccc} & & \uparrow\uparrow\uparrow\uparrow & & \\ \uparrow\uparrow\uparrow\downarrow & \uparrow\uparrow\downarrow\uparrow & \uparrow\downarrow\uparrow\uparrow & \downarrow\uparrow\uparrow\uparrow & \\ \uparrow\uparrow\downarrow\downarrow & \uparrow\downarrow\downarrow\uparrow & \uparrow\downarrow\uparrow\downarrow & \downarrow\downarrow\uparrow\uparrow & \downarrow\uparrow\uparrow\downarrow & \downarrow\uparrow\downarrow\uparrow \\ \uparrow\downarrow\downarrow\downarrow & \downarrow\uparrow\downarrow\downarrow & \downarrow\downarrow\uparrow\downarrow & \downarrow\downarrow\downarrow\uparrow & \\ & & \downarrow\downarrow\downarrow\downarrow & & \end{array}$$

d. $J \cong 6$ to 8 Hz, only one coupling constant $(-OCH_2CH_2CH_2O-)$

8-9 a. $J_{ortho} = 7$ to 10 Hz

b. $(CH_3)_2CHC\equiv CH$
    $J = 5$ to 7 Hz $J = 2$ to 4 Hz

c. $CH_3CH_2OCH_2\overset{\overset{\displaystyle O}{\|}}{C}H$
    $J = 6$ to 8 Hz $J = 1$ to 3 Hz

d.
$J = 0.5$ to 3 Hz

H_3C .... H

$J = 4$ to 10 Hz $C\!=\!C$ $J = 1$ to 3 Hz $J = 12$ to 18 Hz

H ← $J = 6$ to 14 Hz → H

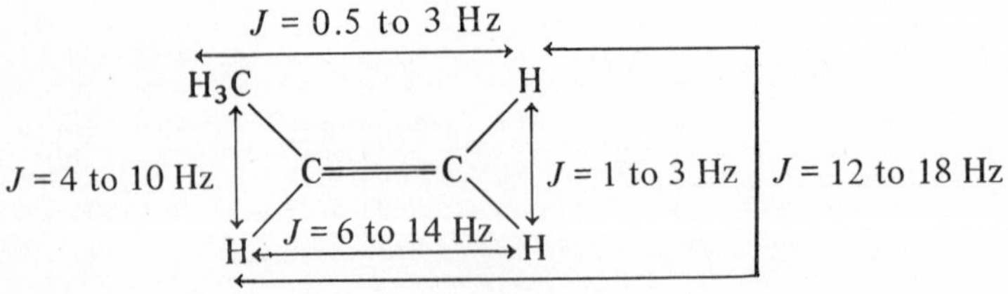

8-10 Abbreviations used for multiplicity designation: s = singlet, d = doublet, t = triplet, q = quartet, p = quintet, sx = sextet, sp = septet, m = multiplet.

a. $\overset{a}{C}H_3\overset{b}{C}H_2\overset{c}{C}H_2Br$ a, t; b, sx; c, t

b. $Cl_2\overset{a}{C}H\overset{b}{C}H_2Cl$ a, t; b, d

c. $\overset{a}{C}H_3\overset{b}{C}H$ a, d; b, q
    $\quad\quad\|$
    $\quad\quad O$

d. $Cl\overset{a}{C}H_2\overset{b}{C}H_2\overset{a}{C}H_2Cl$ a, t; b, p

348

e.

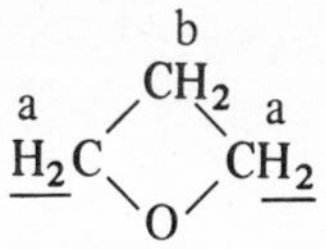

$$\overset{a}{H_2C}\underset{O}{\overset{\overset{b}{CH_2}}{\diagdown}}\overset{a}{CH_2}$$

a, t; b, p

f. $(CH_3)_2\overset{a}{C}\overset{b}{HNO_2}$ a, d; b, sp

g.

$$\underset{\underset{H}{b}}{\overset{a}{H}}C=C\underset{Cl}{\overset{\overset{c}{CH_2Cl}}{\diagup}}$$

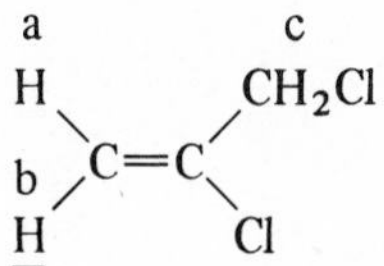

a, d (proton a is strongly coupled only to proton b; $J_{ac} \cong 0$); b, sx (proton b is split by protons c into a triplet each of which is further split by proton a into a doublet, resulting in a sextet overall); c, d (protons c are strongly coupled only to proton b).

h. $\overset{a}{CH_3}\overset{b}{CH_2}\overset{c}{CH_2}\overset{d}{OH}$ a, t; b, sx; c, t; d, s (broad peak, unresolved).

i. $H\overset{a}{C}\equiv C\overset{b}{CH_2}\overset{c}{OH}$ a, t; b, d; c, s (hydroxyl protons usually appear as singlets due to rapid chemical exchange). See Sec. 8-4f and Figs. 8-6 and 8-9.

j.

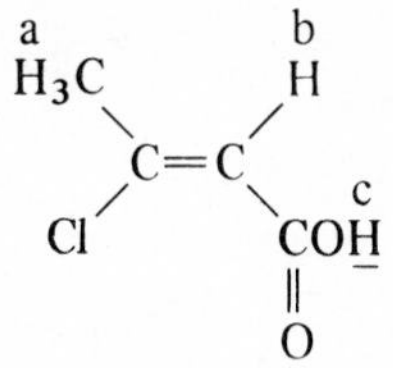

$$\underset{Cl}{\overset{a}{H_3C}}C=C\overset{\overset{b}{H}}{\underset{\underset{O}{\overset{||}{C}OH}}{\diagdown}}$$

a, d; b, q (a doublet and a quartet are expected in theory, but under normal resolution these signals would more resemble singlets since $J_{ab} \cong 0$); c, s (carboxylic acid protons are usually broad singlets due to rapid exchange).

k. $\overset{a}{CH_3}\overset{b}{CH_2}\overset{c}{CH}(Br)\overset{d}{COOH}$ a, t; b, p; c, t; d, s

l.

$$\underset{\overset{b}{H_3C}}{\overset{a}{H_3C}}C=C\underset{Cl}{\overset{\overset{c}{H}}{\diagup}}$$

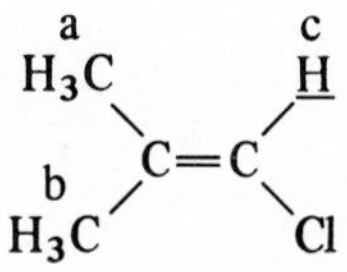

a, d; b, d; c, m (again $J_{ac} \cong 0$; see part j)

m. 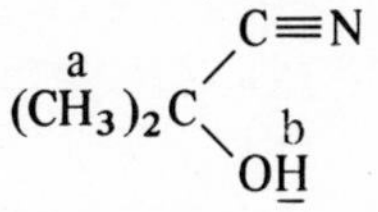 $(CH_3)_2\overset{a}{C}\underset{\underset{OH}{b}}{\overset{\diagup C\equiv N}{\diagdown}}$ a, s; b, s (most likely broad peak)

n. 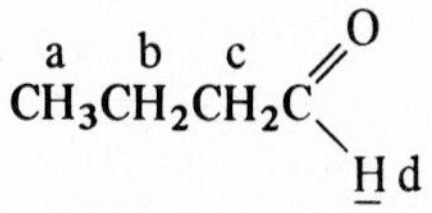 $\overset{a}{CH_3}\overset{b}{CH_2}\overset{c}{CH_2}C\underset{\underset{H\,d}{\diagdown}}{\overset{\diagup O}{}}$

a, t; b, sx; c, sx (protons c are strongly coupled with protons b resulting in a triplet which is further split into a sextet by the rather weak coupling of protons c with proton d).

o. $\overset{a}{CH_3}\overset{\overset{O}{||}}{C}O\overset{b}{CH_2}\overset{c}{CH_3}$ a, s; b, q; c, t

p. $(CH_3)_3\overset{a}{C}\overset{b}{NH_2}$ a, s; b, s

349

$q.$

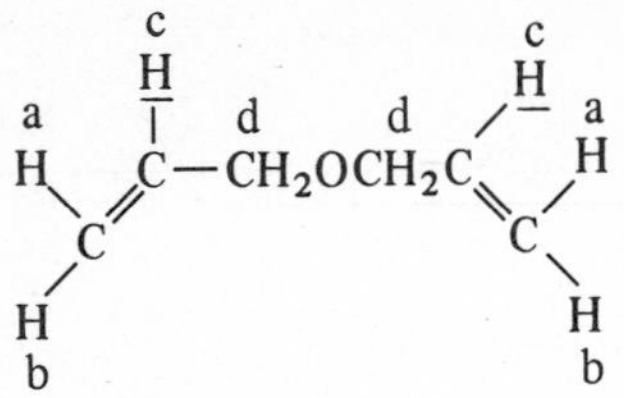

$a, m;$ b, $m;$ c, $m;$ d, $m.$ Complex spectrum due to the various long-range couplings, for example, $J_{bc}, J_{ba},$ and $J_{bd}.$

$r.$

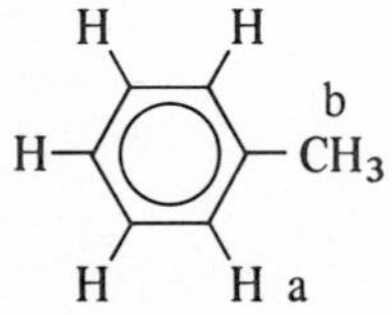

Ring protons = a

a, s (ring protons in simple monosubstituted benzenes frequently appear as an unresolved singlet at $\delta = 7$ to 7.5); b, $s.$

$s.$

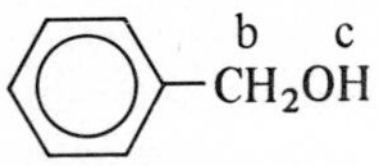

$a, s;$ b, $s;$ c, s

Ring protons = a

$t.$ $(C_6H_5)_3SiH$ $\quad$ $a, m;$ b, s $\qquad$ u

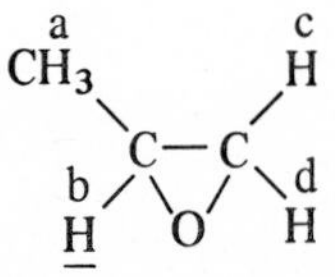

$a, d;$ b, $m;$ c, $q;$ d, q

8-11 $\quad$ A, $(CH_3)_2CHCH(OH)CH_3$; $\quad$ B, $(CH_3)_2CHCCH_3$ (with $\overset{O}{\overset{\|}{}}$); $\quad$ C, $(CH_3)_2C(OH)CH_2CH_3$; $\quad$ D, $(CH_3)_3COCH_3$; E, $(CH_3CH_2)_2C=O$; $\quad$ F, $(CH_3)_3CCHO$; $\quad$ G, $(CH_3)_2CHOCH_3$; $\quad$ H, $(CH_3)_3COH$;

I, $(CH_3)_2CHCH_2CCH_3$ (with $\overset{\|}{O}$); $\quad$ J, $(CH_3CH_2CH_2)_2O$; $\quad$ K, $C_6H_5CH_2OH$; $\quad$ L, CH_3CHO; $\quad$ M, $(CH_3)_2CHOH$;

N, $CH_3OCH_2CH_2OCH_3$; $\quad$ O, $CH_3CH_2OCH_2CH_3$; $\quad$ P, $CH_3CH(OCH_3)_2$;

Q,

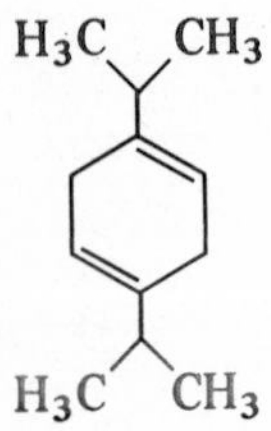

or

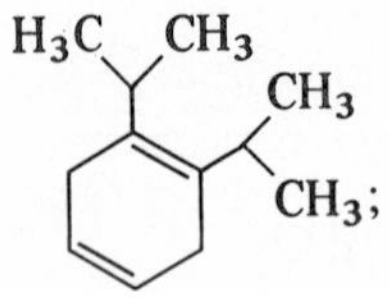

$;$ $\quad$ Q$'$, $(CH_3)_2CHCCH_2COOH$ (with $\overset{\|}{O}$); $\quad$ R, $HOC(CH_3)_2C\equiv CH$;

S, $(CH_3)_2C(NO_2)CH_2OH$; $\quad$ T, $(CH_3)_2C=C(CH_3)_2$;

U, $CH_3CH_2C{=}CCH_2CH_3$ (with $\underset{CH_3\ CH_3}{\big|\ \big|}$); $\quad$ V.

$;$

W, $(CH_3)_2C(Br)CH_2Br$; $\quad$ X, $CH_3C(Cl)_2CH_2Cl$; $\quad$ Y, $CH_3CF_2CH_2Cl$; $\quad$ Z, $ClCH_2CF_2Cl$

8-12 $\quad$ A, $(CH_3)_2CHCCH_2CH_3$ (with $\overset{\|}{O}$); $\quad$ B, $(CH_3)_2CHCHCH_2CH_3$ (with $\underset{OH}{\big|}$); $\quad$ C, $(CH_3)_2C=CHCH_2CH_3$

350

8-13   A, $(CH_3)_2CHCHO$;   B, $HOCH_2CH_2OH$;   C, $(CH_3)_2CHCH\begin{matrix} O-CH_2 \\ | \\ O-CH_2 \end{matrix}$

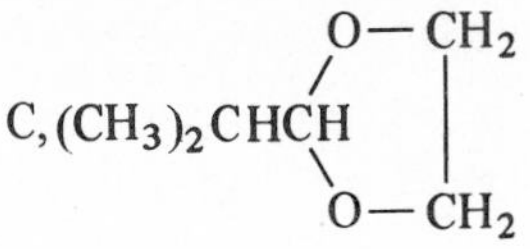

8-14 $-OH$ (alcohols, phenols), $-NH_2$ (amines), $-SH$ (thiols), $-C=N-OH$ (oximes), $-COOH$ (only in polar solvents). Upfield.

8-15   a. $CH_4 \xrightarrow{Cl_2,\ \lambda} CCl_4 + D_2 \text{ (1 mol)} \xrightarrow{Pt} CDCl_3 + DCl$

b. $CCl_4 \xrightarrow{D_2,\ Pt} CD_4 \xrightarrow{Cl_2,\ \lambda} CD_3Cl \xrightarrow{Mg} CD_3MgCl(A) \qquad CaC_2 \xrightarrow{D_2O} DC\equiv CD \xrightarrow{Hg^{2+},\ D_2O}$

$CD_3CDO + A \longrightarrow CD_3CD(OH)CD_3 \xrightarrow{[O]} CD_3COCD_3$

c. $3CaC_2 + 3D_2O \longrightarrow 3DC\equiv CD \xrightarrow[TiCl_4]{Al(CH_2CH_3)_3} C_6D_6$

8-16   a. Convert CaC_2 to $HC\equiv CH \xrightarrow{Hg^{2+},\ H^+} CH_3CHO$.

$CH_3COCH_3 \xrightarrow{D_2,\ Pt} CH_3CD(OD)CH_3 \xrightarrow{PBr_3} \xrightarrow{Mg} CH_3CD(MgBr)CH_3$

$\xrightarrow{CH_3CHO} \xrightarrow{H^+,\ H_2O} (CH_3)_2CDCH(OH)CH_3 \xrightarrow{PBr_3} (CH_3)_2CDCHBrCH_3$

b. Use double-resonance technique where one radiofrequency beam corresponds to the resonance frequency of proton B and another stronger radiofrequency beam corresponds to the resonance frequency of the methine proton on C-3. This will decouple proton B from all protons except the A protons.

8-17   a. (I) $\xrightarrow[B]{aldol\ cond}$ (II)

b. (I) $\xrightarrow[B]{(C_6H_5)_3P=CHCH=CH_2}$ (II)

c. (I) $\xrightarrow[B]{alc\ KOH}$ (II) $\xrightarrow[H]{H_2,\ Ni}$ $\xrightarrow{Cu,\ \Delta}$ (III) $\xrightarrow[H]{H_2NNH_2,\ KOH}$ (IV)

d. (I) $\xrightarrow[B]{Al_2O_3}$ (II) $\xrightarrow[H]{(C_6H_5)_3P=CH_2}$ $\xrightarrow{H_2,\ Pt}$ (III)

e. 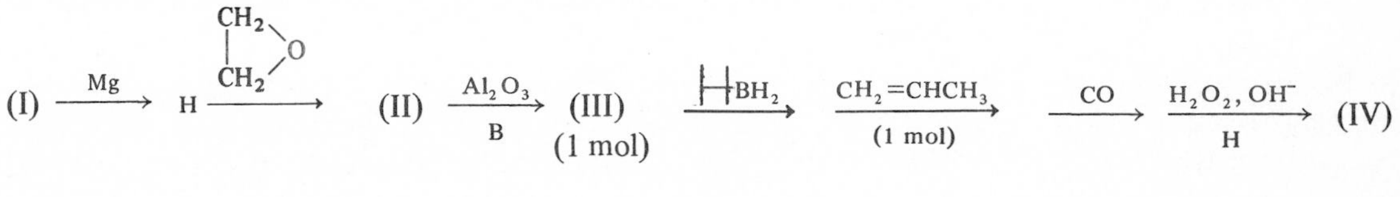 (I) $\xrightarrow{Mg}$ $H\underset{\begin{matrix}|\\CH_2\end{matrix}}{\overset{\begin{matrix}CH_2\\|\end{matrix}}{}}\!\!\diagdown O$ (II) $\xrightarrow[B\ (1\ mol)]{Al_2O_3}$ (III) $\xrightarrow{\rceil BH_2}$ $\xrightarrow[(1\ mol)]{CH_2=CHCH_3}$ $\xrightarrow{CO}$ $\xrightarrow[H]{H_2O_2,\ OH^-}$ (IV)

8-18   a. $C_6H_5CH_2\underset{\overset{\|}{O}}{C}CH_2Br$

b. $CH_3CH_2OCH\underset{\begin{matrix}|\\CH_3\end{matrix}}{\overset{\begin{matrix}CH_3\\|\end{matrix}}{C}}=CH_2$

c. $CH_3C\equiv CC(CH_3)_2CH_2OH$

d. $CH_3C\equiv CCH_2\underset{\begin{matrix}|\\OH\end{matrix}}{C}(CH_3)_2$

e. $ClCH_2CH_2CH_2Cl$

f. 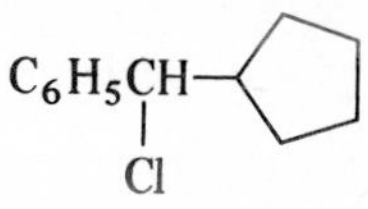 $C_6H_5\underset{\begin{matrix}|\\Cl\end{matrix}}{CH}\!\!-\!\!\bigcirc$

g. $(CH_3)_2\underset{\begin{matrix}|\\OH\end{matrix}}{\overset{\begin{matrix}OH\\|\end{matrix}}{C}}C\equiv N$

h. $(CH_3)_3COCH(CH_3)_2$

i. $HC\equiv CCH_2Br$

j. 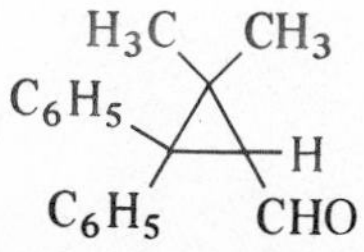 $\begin{matrix} H_3C \quad CH_3 \\ C_6H_5\diagdown\!\!\triangle\!\!\diagup H \\ C_6H_5 \quad CHO \end{matrix}$

8-19 A, $(C_6H_5)_2\underset{\underset{OH}{|}}{C}CH_2C(CH_3)_2CH_2\underset{\underset{OH}{|}}{C}(C_6H_5)_2$

B, $(C_6H_5)_2C=CHC(CH_3)_2CH=C(C_6H_5)_2$

C, $(CH_3)_2C(COOH)_2$

D, $C_6H_5\underset{\underset{O}{\|}}{C}C_6H_5$

8-20 At $-75°C$ the rate of change from one conformer to another is slow enough that a signal for each of the two possible conformers is observed: $\delta = 4.64$ for the conformer with Br axial and $\delta = 3.97$ for the equatorial Br conformer which predominates $[4.6/(1.0 + 4.6) = 82$ percent$]$. At room temperature the rate of exchange between conformers is given by $\pi J/\sqrt{2}$ where $J = 40.2$ Hz $[(4.64 - 3.97)60]$ and is equal to 89.3 exchanges per second.

8-21 $4.49 = 4.63$ Ne $+ 3.83$ Na or $4.49 = 4.63$ Ne $+ 3.83$ $(1 - $ Ne$)$. Ne $= 0.825$. Thus 82.5 percent has Br axial.

8-22 B, because the protons would be held in a relatively rigid situation above the benzene ring and thus are shielded by the π-electron-ring current.

8-23

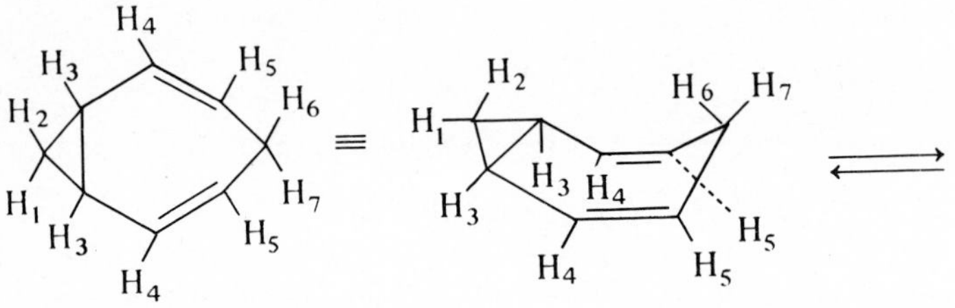

$-50°C$	$25°C$	$180°C$	
τ For isomer I	τ	τ I, II	
4.3 4 vinyl H's (H_4, H_5)	4.2 4 vinyl H's	4.3 $H_4\equiv H_4$	
7.2 H_6, H_7	7–10 other H's	6.4 $2H_3\equiv 2H_5$	
8.9 H_3		8.2 $H_2\equiv H_6$	
9.7 H_1, H_2		8.9 $H_1\equiv H_7$	

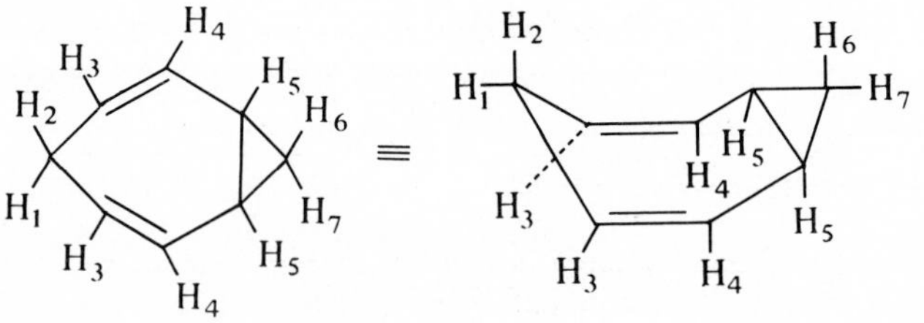

At $-50°C$ fluxionalism has slowed to the point where only one isomer is present (either I or II); at $25°C$ the fluxionalism is rapid enough so that the nmr spectrometer sees a time averaged equilibrium mixture of I and II; by $180°C$ the rearrangement is so rapid on the nmr time scale that only four sharp bonds occur.

8-24 $ClCH_2CH_2COOH$

8-25 $(CH_3)_2CHCH_2CH_2OH$

8-26 $CH_3O-\!\!\left\langle\;\right\rangle\!\!-OCH_3$

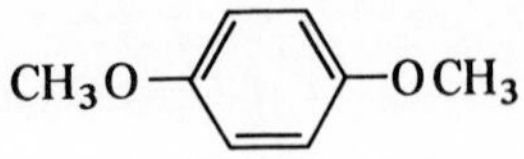

8-27 $(CH_3)_2CHCH_2CH_2Br$

8-28 $N\equiv C(CH_2)_4C\equiv N$

8-29

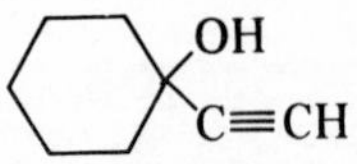

8-30

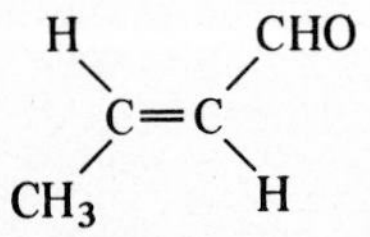

8-31

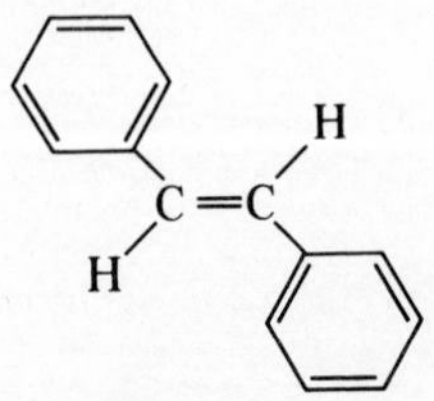

8-32

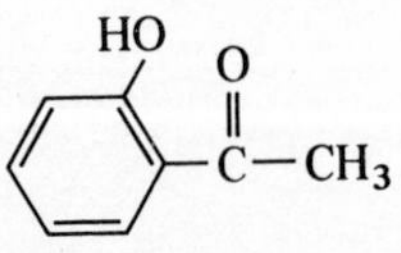

Chapter 9

	IUPAC name	Common name
$CH_3CH_2CH_2COCl$	Butanoyl chloride	Butyryl chloride
$(CH_3CH_2CH_2CO)_2O$	Butanoic anhydride	Butyric anhydride
$(CH_3)_2CHCOCl$	2-Methylpropanoyl chloride	Isobutyryl chloride
$[(CH_3)_2CHCO]_2O$	2-Methylpropanoic anhydride	Isobutyric anhydride
$CH_3(CH_2)_2COOCOCH(CH_3)_2$	2-Methylpropanoic butanoic anhydride	Isobutyric butyric anhydride

9-2
a. 2,2-Dibromopropanoic acid   α,α-Dibromopropionic acid
b. Ethanoyl chloride   Acetyl chloride
c. Trifluoroethanoic acid   Trifluoroacetic acid
d. Ethanoic anhydride   Acetic anhydride
e. Ethanoic 2-methylpropanoic anhydride   Acetic isobutyric anhydride
f. 2-Butenoyl chloride   Crotonyl chloride
g. 2-Cyclopropylpropanoic acid   Methylcyclopropylacetic acid
h. 2,2-Dimethylpropanoic anhydride   Trimethylacetic anhydride

9-3
a. 2-Butynoyl bromide   b. 2-Ethylbutanedioic acid
c. Pentanedioyl chloride   d. 2-Hepten-5-ynoic acid
e. 2,2-Diethylbutanoic anhydride   f. 2,4-Pentadienoic acid
g. 2-Cyclohexene-1-carbonyl chloride   h. 1,4-Cyclohexanedicarboxylic acid
i. 4-Oxo-2-pentenoic acid

9-4
a. $CH_2=CH-CH(CH_3)COOH$ b. $CH_3(CH_2)_5CHBrCHBrCHBrCOOH$ c. $(FCH_2CHFCH_2CH_2CO)_2O$

d. *trans*-$HOOCCH=CHCOOH$ e. 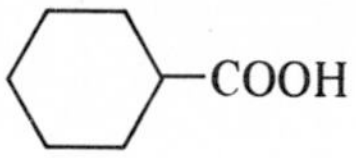 ⬡—COOH   f. $CH_3(CH_2)_{12}COCl$

g. $[(CH_3)_2CHCH_2]_3CCOBr$ h. CH_3COCH_2COOH i. 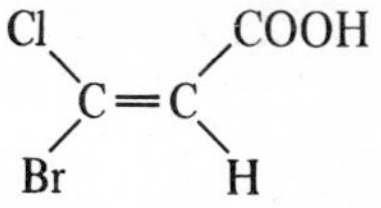
$$\underset{Br}{\overset{Cl}{}}C=C\underset{H}{\overset{COOH}{}}$$

j. 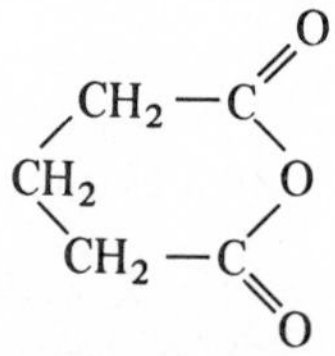
$$\begin{array}{c} CH_2-C{\overset{O}{\diagdown}} \\ CH_2 \quad\quad O \\ CH_2-C{\underset{O}{\diagup}} \end{array}$$

9-5
a. Methyl vinyl ketone   b. Methylmagnesium iodide   c. Correct
d. 2-Butanol   e. 2-Chloro-4-bromoheptane   f. 2-Methyloctanoic acid

9-6
a. $CH_3CH_2CH_2COOH + NaCl, S_E$ b. $CH_3COOEt + AgI, S_N$
c. $CH_3(CH_2)_5COOMgI, A_N$ d. $CH_3CH_2CH_2COOMgBr + CH_3CH_3, S_E$
e. $BrCH_2CH(OH)CH_2CH_3, A_E$ f. $ClCH_2CH_2CHO, A_E$
g. $CH_3C\equiv CCH_3 + 2HBr, E$ h. $(CH_3)_2C(OH)CN, A_N$
i. CH_3CH_2COBr, S_N j. $LiAlO(OCH_2CH_2CH_2CH_3)_2, A_N$
k. $(CH_3)_2C=CH_2 + NaBr + H_2O, E$ l. $CH_3CONH_2 + NH_4Cl, S_N$

9-7
a. $CH_2=CHCOOH + HBr \longrightarrow BrCH_2CH_2COOH$
b. $CH_3CHBrCH_2COOH + 2NaOH \longrightarrow CH_3CH=CHCOONa + NaBr + 2H_2O$
c. $CH_3CH=CHCOOH + HCOOOH \longrightarrow CH_3\underset{\underset{O}{\diagdown\diagup}}{CH}CHCOOH + HCOOH$

d. $CH_3COCHO + 2AgOH + NaOH \longrightarrow CH_3COCOONa + 2Ag + 2H_2O$
e. $OHCCHO + 4AgOH + 2NaOH \longrightarrow NaOOCCOONa + 4Ag + 4H_2O$

f. $HC \equiv CNa + (CH_3)_2CO \longrightarrow (CH_3)_2C(ONa)C \equiv CH$

g. $CH_2 = C = O + ClCH_2COOH \longrightarrow ClCH_2COOCOCH_3$

h. $O_2NCH_2COOH + SOCl_2 \dashrightarrow O_2NCH_2COCl + SO_2 + HCl$

i. $HOOCCOOH + 2PCl_5 \longrightarrow ClOCCOCl + 2POCl_3 + 2HCl$

j. $HOCH_2COOH + 2PCl_5 \longrightarrow ClCH_2COCl + 2POCl_3 + 2HCl$

k. $CH_3MgI + CS_2 \longrightarrow CH_3\underset{\underset{S}{\|}}{C}-SMgI$ *l.* $CH_3C \equiv N + HCl \longrightarrow CH_3\underset{\underset{Cl}{|}}{C}=NH$

m. $CH_3COCl + AgCN \longrightarrow CH_3COCN + AgCl$

n. $ClOCCH_2CH_2COCl + (CH_3)_2Cd \longrightarrow CH_3COCH_2CH_2COCH_3 + CdCl_2$

o. Succinic anhydride $+ EtOH \longrightarrow EtOOCCH_2CH_2COOH$

p. $2CH_2 = C = O + HOCH_2CH_2OH \longrightarrow CH_3COOCH_2CH_2OOCCH_3$

q. $CH_3CH = CHCOOH + HBr \longrightarrow CH_3CH(Br)CH_2COOH$

r. $HC \equiv CNa + (CH_3)_3CCl \longrightarrow (CH_3)_2C = CH_2 + HC \equiv CH + NaCl$

s. $2CH_3(CH_2)_{14}COOTl + 3Br_2 \longrightarrow 2CH_3(CH_2)_{13}CH_2Br + 2CO_2 + Tl_2Br_4$

t. $ClCOCH_2CH_2COCl + 2CuCN \longrightarrow NCCOCH_2CH_2COCN + 2CuCl$

u. $HC \equiv CCOCl + 2NH_3 \longrightarrow HC \equiv CCONH_2 + NH_4Cl$

v. $CH_3CH_2COCl + H_2 \xrightarrow{\text{Pd} \cdot \text{BaSO}_4} CH_3CH_2CHO + HCl$

w. $CH_3COOCOC_3H_7 + H_2O \longrightarrow CH_3COOH + CH_3CH_2CH_2COOH$

x. $HCOOH + SOCl_2 \longrightarrow CO + 2HCl + SO_2$

y. $2(CH_3)_3CCOOTl + SOCl_2 \longrightarrow [(CH_3)_3CCO]_2O + SO_2 + 2TlCl$

z. $3CH_3CH_2CH(OH)CH_2CH_2COOH + 2PBr_3 \longrightarrow 3CH_3CH_2CHBrCH_2CH_2COBr + 2H_3PO_3$

9-8   *a.* Fuse the acid with sodium hydroxide to cause decarboxylation.

b. Use the Hunsdiecker reaction to obtain pentyl bromide. Run a Wurtz reaction on this.

c. Oxidize the alcohol to propanoic acid and treat this with PCl_3.

d. Dehydrate the alcohol to propene; add HBr to get isopropyl bromide; make Grignard reagent and treat with CO_2 or treat bromide with NaCN followed by hydrolysis of the resulting cyanide.

e. Treat acetone with NaOCl to get sodium acetate. Convert this to acetyl chloride using $SOCl_2$ and react with sodium acetate.

f. Hell-Volhard-Zelinsky (HVZ) reaction yields $BrCH_2COOH$; treat this with $SOCl_2$ or PCl_3.

g. Convert acetic acid to acid chloride with $SOCl_2$; Rosenmund reduction of this yields CH_3CHO; react this with NH_2OH.

h. Chlorinate CH_3COOH to Cl_3COOH. Treat this with SbF_3 and then with PBr_3.

i. Treat with HCN and dehydrate.

j. HVZ reaction yields $CH_3CH_2CHBrCOOH$; then dehydrohalogenate with alcoholic KOH.

k. Convert acetic acid to acetone by passing over hot thoria; then add HCN to get cyanohydrin followed by hydrolysis to give product.

l. Run aldol condensation with acetaldehyde and oxidize with Tollens' reagent.

m. Reduce with $LiAlH_4$. Dehydrate the alcohol to propylene followed by vapor-phase oxidation or add HOCl and treat with concd KOH.

n. Reduce the anhydride to 1-butanol with $LiAlH_4$. Dehydrate over hot Al_2O_3.

o. Oxidize the alcohol to the acid; convert to the acid chloride; treat with SbF_3 or anhydrous HF and NaF.

p. Treat acetyl chloride with excess I^- (for example, LiI).

q. Dehydrate lactic acid over hot Al_2O_3 to get acrylic acid; add Br_2 to this and dehydrohalogenate $BrCH_2CHBrCOOH$ to propiolic acid with $NaNH_2$.

r. Catalytic hydrogenation.

s. Wolff-Kishner reduction.

t. Treat $HC \equiv CH$ with Hg^{2+} and H^+ to obtain CH_3CHO; oxidize this to acetic acid which then is brominated under HVZ conditions and hydrolyzed.

u. Haloform reaction yields CH_3CH_2COONa; convert this to the acid chloride with PCl_3.

v. Reduce the acid bromide to the aldehyde via Rosenmund reaction; treat the aldehyde with PBr_5 followed by dehydrohalogenation of the $CH_3CH_2CH_2CHBr_2$ with $NaNH_2$.

w. Make the acid chloride and run Rosenmund reduction to get CH_3CHO; treat this with ethanol using H^+ catalyst to get the acetal.

x. Convert the acid to diethyl ketone with thoria. Treat the ketone with PCl_5 to get the dichloride and then dehydrohalogenate with sodamide.

y. Dehydration of malic acid using concentrated H_2SO_4 followed by gentle oxidation of the resulting HOOCCH=CHCOOH with neutral $KMnO_4$ will yield tartaric acid.

z. Prepare Grignard reagent from the iodide; add to CO_2 and hydrolyze. Or treat the iodide with NaCN and hydrolyze the resulting cyanide to valeric acid.

a′. Reduce the acid chloride to ethanol and convert this successively to ethyl bromide and to diethylcadmium. React this with CH_3COCl to get $CH_3COCH_2CH_3$. Pinacol reduction of this gives product.

b′. Add HBr to the alkene, convert the bromide to the Grignard reagent; carbonation of this yields product.

c′. Convert to the acid chloride and reduce this to butanal; aldol condensation of butanal and dehydrate to get product.

d′. Oxidize (HNO_3) to adipic acid. HVZ reaction on this (2 mol Br_2) followed by dehydrohalogenation with alcoholic KOH gives the product.

e′. Reduce acid with H_2 and Pt to get β-hydroxybutyric acid; convert this to the bromo acid and treat with triphenylphosphine and then CH_3Li to obtain the ylide. React the ylide with acetoacetic acid to obtain the product.

f′. Hydroboration of the $CH_2=CHCH_2CN$ (2 mol) with thexylborane; then carbonylation-oxidation will yield $(NCCH_2CH_2CH_2)_2CO$. Hydrolysis of this to the diacid followed by Wolff-Kishner reduction of the keto group give the product.

9-9   A, $HO(CH_2)_3COOH$; B, $HOOCCH_2CH_2COOH$

9-10 CH_3OCH_2COCl 9-11 $CH_3COCH(CH_3)COOH$

9-12   A,

CH_2—$CHCH_3$	

B, $\underset{HOOC\quad COOEt}{CH_2CHCH_3}$ C, $\underset{EtOOC\quad COOH}{CH_2-CHCH_3}$ D, $\underset{EtOOC\quad COOEt}{CH_2CHCH_3}$

9-13 *a.* Only formic acid will reduce $KMnO_4$.

b. Crotonic acid will reduce $KMnO_4$ or decolorize Br_2 in CCl_4.

c. Hexanoic acid is soluble in dilute alkali.

d. 1-Propanol has less than five carbons and is soluble in water.

e. Levulinic acid will react with semicarbazide or give a positive iodoform test.

f. Lactic acid will give a positive iodoform test.

g. Beilstein test for halogen; or, evolution of heat on addition of acetyl chloride to water.

h. Acetic acid is water-soluble.

i. Caproic anhydride will dissolve slowly in dil NaOH as it hydrolyzes.

j. Ketene will react with water to form acetic acid which will be water-soluble.

9-14 *a.* First extract methanol with water; then hexanoic acid with dilute sodium hydroxide. Separate the remaining aldehyde from the hydrocarbon with bisulfite.

b. Extract formic acid with water; then extract triethylacetic acid with dil NaOH. 2-Hexanone will form a solid bisulfite addition product while the other ketone remains soluble.

c. Extract the succinic acid with water. Then extract the suberic acid with dil NaOH. Next extract the aldehyde with bisulfite solution.

d. Extract hexanoic acid with dil NaOH. Then extract hexanal with bisulfite solution. Next form the silver salt of 1-hexyne by extracting with ammoniacal silver nitrate and filter this off leaving the hexane. 1-Hexyne may be regenerated from the Ag salt by treatment with acid.

9-15 *a.* Nitroacetic, 1.7; fluoroacetic, 2.7; α-chloropropionic, 2.8; β-chloropropionic, 4.1; acetic, 4.8; propionic, 4.9.

b. Trifluoroacetic, 0.23; iodoacetic, 3.1; CH_3SCH_2COOH, 3.7; formic, 3.8; acetic, 4.8; isobutyric, 4.9.

c. Trichloroacetic, 0.7; dichloroacetic, 1.3; chloroacetic, 2.8; bromoacetic, 2.9; iodoacetic, 3.1; acetic, 4.8.

d. Oxalic, 1.2; malic, 3.4; lactic, 3.9; succinic, 4.2.

e. Pyruvic, 2.5; acetoacetic, 3.3; formic, 3.8; levulinic, 4.4.

9-16 $CH_3COOH + CH_3OH \rightleftharpoons CH_3COO^- + CH_3\overset{+}{O}H_2$

$CH_3COOH + CH_3C\equiv N \rightleftharpoons CH_3COO^- + [CH_3C\equiv\overset{+}{N}H \longleftrightarrow CH_3\overset{+}{C}=NH]$

$K_e = \dfrac{[CH_3COO^-]\,[CH_3\overset{+}{O}H_2]}{[CH_3COOH]\,[CH_3OH]}$ $K_e = \dfrac{[CH_3COO^-]\,[CH_3C\equiv\overset{+}{N}H]}{[CH_3COOH]\,[CH_3C\equiv N]}$

9-17 *a.* Isovaleryl chloride, 5-chloro-3-penten-1-ol, 5-chloro-2-pentanone, 1-chloro-1-penten-3-ol.

b. Butanoyl chloride, allyl chloride, 2-methyl-2-chloropropane, 2-iodopentane, 2-chlorobutane, CCl_4.

9-18 *a.* Isobutylene > propene > ethylene > vinyl iodide

   b. Allylacetic > crotonic > acrylic > fumaric acid

9-19 *a.* Acetyl chloride *b.* Ethyl iodide *c.* Propionyl chloride

   d. Allyl chloride *e.* Acetyl chloride *f.* Acetyl fluoride

   g. Acetyl chloride *h.* Butanoyl chloride

9-20 *a.* Prepare sodium formate by treating CO with NaOH at high pressure. Fusion of sodium formate with NaOH will give oxalic acid.

   b. Prepare maleic anhydride from benzene as described in the problem. Hydrolyze the anhydride to maleic acid which upon catalytic hydrogenation will yield succinic acid.

   c. HVZ reaction on acetic acid to give bromoacetic acid; treat this with NaCN to get cyanoacetic acid which can be hydrolyzed to malonic acid.

   d. Reduce succinic acid with $LiAlH_4$ to the diol; convert this to the dibromide, then react it with NaCN, and hydrolyze to adipic acid.

   e. Use approach similar to that used in *d.*

   f. Oxidize cyclohexanone to adipic acid with HNO_3. Make the disilver salt of adipic acid and treat with bromine. The resulting 1,4-dibromobutane can be hydrolyzed to the diol and then oxidized to succinic acid.

9-21 *a.* Convert 1-propanol to propene and add HBr, then make the Grignard reagent. Add ethylene oxide (made from ethanol) to this and convert the resulting isopentyl alcohol to the bromide with PBr_3. Convert this bromide to the desired acid via the Grignard reagent and CO_2 or by means of NaCN and hydrolysis.

   b. Reduce acid made in *a* to the alcohol with $LiAlH_4$; dehydrate to the olefin and add HBr. Convert bromide to the acid as in *a.*

   c. Treat isopentyl bromide from *a* with Na to give the product via the Wurtz reaction.

   d. Dehydrate 1-propanol to propene, add HBr, then treat with NaCN and hydrolyze. Convert isobutyric acid to its thallium salt with thallium ethoxide. Treat this with $SOCl_2$ to obtain product. Or, convert half of the isobutyric acid to the acyl halide with $SOCl_2$ and the other half to the sodium salt. React these two together.

   e. Oxidize ethanol to acetic acid, treat with $SOCl_2$, and treat the acetyl chloride with the thallium or sodium salt of isobutyric acid made in *d*. Or, isobutyric acid and ketene will react to give the desired anhydride.

   f. Oxidize 1-propanol to propanal and run the aldol condensation, then dehydrate and oxidize with Tollens' reagent. Convert this acid to acid chloride with $SOCl_2$.

   g. Convert acid from (*a*) to acid chloride and reduce to the aldehyde via the Rosenmund reaction. A Grignard reaction with the aldehyde and isopropylmagnesium bromide followed by oxidation will give the desired ketone.

   h. Convert ethanol or 1-propanol to butanal. Run aldol condensation and dehydrate to get 2-ethyl-2-hexenal. Hydrogenate this to the alcohol, and convert this to the acid as done in *a.*

   i. Oxidize 1-propanol to propanoic acid and pass over hot thoria to get 3-pentanone. Run the pinacol reduction on this ketone and then use pinacolone rearrangement. Cleave the resulting methyl ketone with NaOI to triethylacetic acid and convert the acid to the acid bromide with PBr_3.

   j. Oxidize ethanol to acetic acid; convert to bromacetic acid by HVZ reaction. Treat this with $NaNO_2$ in dimethylformamide to get nitroacetic acid. Treat acid with $SOCl_2$ to get acid chloride and treat this with NaF and HF to get the acid fluoride.

   k. Oxidize isopentyl alcohol made in *a* to the acid; convert this under HVZ conditions to the α-bromo acid. Treat this with $NaNH_2$ to get the α-amino acid.

   l. Make pinacol from acetone and rearrange to pinacolone. Haloform reaction with this followed by $SOCl_2$ gives product.

9-22 *a.* $CH_2=CHCH=CH_2$ $\xrightarrow[\text{pressure}]{\text{H-BH}_2 \quad CO}$ $\xrightarrow{H_2O_2,\ OH^-}$ cyclopentanone $\xrightarrow{CH_2=P(C_6H_5)_3}$ $\xrightarrow{B_2H_6}$

   $\xrightarrow{H_2O_2,\ OH^-}$ cyclopentylcarbinol $\xrightarrow{K_2Cr_2O_7}$ $\xrightarrow{SOCl_2}$ product

   b. $BrCH_2COOCH_3 + (C_6H_5)_3P$ $\longrightarrow$ $\xrightarrow{CH_3Li}$ $(C_6H_5)_3P=CHCOOCH_3$ $\xrightarrow{\text{cyclopentanone from } a}$ product

   c. $CH_2=CHCH_2COCl$ $\xrightarrow{CH_3OH}$ $CH_2=CHCH_2COOCH_3$ $\xrightarrow{\text{H-BH}_2}$ $\xrightarrow{CO,\ \text{pressure}}$ $\xrightarrow{H_2O_2,\ OH^-}$ product

   d. $OHCCH_2CHO + 2(C_6H_5)_3P=CH_2$ $\longrightarrow$ 1,4-pentadiene $\xrightarrow{\text{H-BH}_2}$ $\xrightarrow{CO,\ \text{pressure}}$ $\xrightarrow{H_2O_2,\ OH^-}$

   cyclohexanone (A) $Br(CH_2)_4Br + 2(C_6H_5)_3P$ $\longrightarrow$ $\xrightarrow{CH_3Li}$ $(C_6H_5)_3P=CHCH_2CH_2CH=P(C_6H_5)_3$

   $\xrightarrow{\text{2 mol A}}$ $\xrightarrow{H_2,\ Pt}$ product

e. $(CH_3)_2CHCHO \xrightarrow{OH^-} \xrightarrow{\text{dehydrate}} (CH_3)_2C=CHC(CH_3)_2CHO \xrightarrow{(C_6H_5)_3P=C(CH_3)_2} \text{product}$

f. Cyclohexanone from d $\xrightarrow{(C_6H_5)_3P=CH_2} \xrightarrow{B_2H_6} \xrightarrow{H_2O_2,\ OH^-}$ cyclohexylcarbinol $\xrightarrow{K_2Cr_2O_7} \xrightarrow{CH_3CH_2OTl}$

$\xrightarrow{SOCl_2}$ product

9-23 A, $CH_2=CHCH_2C(CH_3)=NMgX$;   B, $CH_2=CHCH_2\overset{\overset{\displaystyle CH_3}{|}}{C}=O$;   C, $CH_2=CHCH_2CH_2CH_3$;

D, $ClCH_2CH(OH)CH_2CH_2CH_3$;   E, $\underset{\displaystyle O}{CH_2-CHCH_2CH_2CH_3}$;   F, 3-hexanol

9-24 a. $\left[\underset{\displaystyle HC-OH}{\overset{\displaystyle +OH}{\overset{\|}{}}} \longleftrightarrow \underset{\displaystyle HC-OH}{\overset{\displaystyle OH}{\overset{|}{}}} \longleftrightarrow \underset{\displaystyle HC=OH}{\overset{\displaystyle OH}{\overset{|}{}}} \right]$ Delocalization of charge makes this the more stable form.

b. $H_2\overset{+}{O}Cl$

c. $\left[\underset{\displaystyle H_2N-C-NH_2}{\overset{\displaystyle +NH_2}{\overset{\|}{}}} \longleftrightarrow \underset{\displaystyle H_2N-C-NH_2}{\overset{\displaystyle NH_2}{\overset{|}{}}} \longleftrightarrow \underset{\displaystyle H_2N=C-NH_2}{\overset{\displaystyle NH_2}{\overset{|}{}}} \longleftrightarrow \underset{\displaystyle H_2N-C=NH_2}{\overset{\displaystyle NH_2}{\overset{|}{}}} \right]$

Same reason as in a.

d. $HO\overset{+}{N}H_3$ Nitrogen is less electronegative than oxygen.

e. $^-OCH_2C\equiv CH$ Carbon is less electronegative than oxygen.

f. $CH_3C\equiv\overset{+}{N}H \longleftrightarrow CH_3\overset{+}{C}=NH$ Lone pair of electrons on N most available for protonation.

g. $HOCH_2CH_2\underset{+}{\overset{\overset{\displaystyle H}{|}}{O}}CH_3$ Higher electron density on ether because electron-releasing ability of CH_3 is greater than H.

h. $HOCH_2CH_2S^-$ Negative charge on O is on smaller atom and therefore more localized for protonation.

9-25 a. $CH_3CH_2^-$, because $HC\equiv C^-$ is a weaker base as shown by acidic nature of acetylenic hydrogen.

b. $^-CH_2CN$ is a stronger base since negative charge is delocalized over only one CN group.

c. $CH_3CH_2^-$ is a stronger base; no delocalization of negative charge possible.

d. NH_3 is a stronger base; N is less electronegative than O.

e. $C_4H_9^-$ is a stronger base than the more-electronegative nitrogen compound.

f. CH_3O^- is a stronger base because the negative charge is localized.

g. $CH_3CH_2^-$ is a stronger base because the charge is localized.

h. $C_6H_5CH_2^-$ is a stronger base because the charge is delocalized over only one benzene-ring system.

i. CH_3NH_2, because nitrogen is less electronegative than oxygen.

9-26 a. $\left[\underset{\displaystyle CH_3C-Cl}{\overset{\displaystyle O}{\overset{\|}{}}} \longleftrightarrow \underset{\displaystyle CH_3\underset{+}{C}-Cl}{\overset{\displaystyle O^-}{\overset{|}{}}} \longleftrightarrow \underset{\displaystyle CH_3C=Cl^+}{\overset{\displaystyle O^-}{\overset{|}{}}} \right]$

b. $\left[CH_3\overset{\overset{\displaystyle O}{\|}}{C}-O-\overset{\overset{\displaystyle O}{\|}}{C}CH_3 \longleftrightarrow CH_3\overset{\overset{\displaystyle O^-}{|}}{\underset{+}{C}}-O-\overset{\overset{\displaystyle O}{\|}}{C}CH_3 \longleftrightarrow CH_3\overset{\overset{\displaystyle O^-}{|}}{C}=\underset{+}{O}-\overset{\overset{\displaystyle O}{\|}}{C}CH_3 \longleftrightarrow \right.$

$\left. CH_3\overset{\overset{\displaystyle O}{\|}}{C}-O-\underset{+}{\overset{\overset{\displaystyle O^-}{|}}{C}}CH_3 \longleftrightarrow CH_3\overset{\overset{\displaystyle O}{\|}}{C}-\underset{+}{O}=\overset{\overset{\displaystyle O^-}{|}}{C}CH_3 \right]$

c. $\left[CH_2=CH-\underset{\underset{\displaystyle O}{\|}}{C}-OH \longleftrightarrow CH_2=CH-\overset{+}{\underset{\underset{\displaystyle O^-}{|}}{C}}-OH \longleftrightarrow CH_2=CH-\underset{\underset{\displaystyle O^-}{|}}{C}=\overset{+}{O}H \longleftrightarrow \right.$

$\left. \overset{+}{C}H_2-CH=\underset{\underset{\displaystyle O^-}{|}}{C}-OH \longleftrightarrow \overset{+}{C}H_2-\overset{-}{C}H-\underset{\underset{\displaystyle O}{\|}}{C}-OH \right]$

357

d. $\left[\begin{array}{c} \end{array} \right.$ CH$_3$OCH=CH–C–OH $\longleftrightarrow$ CH$_3$OCH=CH–$\overset{+}{\text{C}}$–OH $\longleftrightarrow$ CH$_3$OCH=CH–C=$\overset{+}{\text{O}}$H $\longleftrightarrow$
$\quad\quad\quad\quad\quad$ || $\quad\quad\quad\quad\quad\quad\quad\quad\quad$ | $\quad\quad\quad\quad\quad\quad\quad\quad\quad\quad$ |
$\quad\quad\quad\quad\quad$ O $\quad\quad\quad\quad\quad\quad\quad\quad\quad$ –O $\quad\quad\quad\quad\quad\quad\quad\quad\quad\quad$ –O

CH$_3$$\overset{+}{\text{O}}$=CH–$\overset{-}{\text{CH}}$–C–OH $\longleftrightarrow$ CH$_3$$\overset{+}{\text{O}}$=CH–CH=C–OH $\longleftrightarrow$ CH$_3$O$\overset{+}{\text{CH}}$–CH=C–OH $\longleftrightarrow$
$\quad\quad\quad\quad\quad\quad$ || $\quad\quad\quad\quad\quad\quad\quad\quad\quad$ | $\quad\quad\quad\quad\quad\quad\quad\quad\quad\quad$ |
$\quad\quad\quad\quad\quad\quad$ O $\quad\quad\quad\quad\quad\quad\quad\quad\quad$ –O $\quad\quad\quad\quad\quad\quad\quad\quad\quad\quad$ –O

$\quad\quad\quad\quad\quad\quad\quad\quad\quad\quad\quad\quad\quad\quad\quad\quad\quad$ CH$_3$O$\overset{+}{\text{CH}}$–$\overset{-}{\text{CH}}$C–OH $\left. \begin{array}{c} \end{array} \right]$
$\quad\quad\quad\quad\quad\quad\quad\quad\quad\quad\quad\quad\quad\quad\quad\quad\quad\quad\quad\quad\quad\quad\quad$ ||
$\quad\quad\quad\quad\quad\quad\quad\quad\quad\quad\quad\quad\quad\quad\quad\quad\quad\quad\quad\quad\quad\quad\quad$ O

9-27   *a.* (See Fig. P9-27*a.*)   *b.* (See Fig. P9-27*b.*)

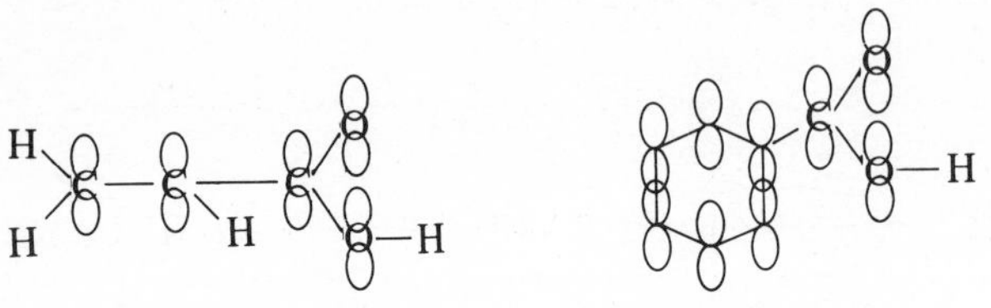

         a.   *b.*

Figure A9-27 a,b. All of the carbon and oxygen atoms are in the sp^2 state.

9-28   *a.*   CH$_3$COONa + H$_2$O $\rightleftharpoons$ CH$_3$COOH + NaOH

       b.   (CH$_3$)$_3$CI $\underset{\text{CH}_3\text{OH}}{\rightleftharpoons}$ (CH$_3$)$_2$C=CH$_2$ + HI

       c.   HCOOH + CH$_3$OH $\rightleftharpoons$ HCOOCH$_3$ + H$_2$O

       d.   CH$_3$CH(OH)CH$_2$CHO $\underset{\text{HCOOH}}{\rightleftharpoons}$ CH$_3$CH=CHCHO + H$_2$O

       e.   CH$_3$COOCH$_3$ + H$_3$O$^+$ $\rightleftharpoons$ CH$_3$COOH + CH$_3$$\overset{+}{\text{O}}H_2$

       f.   CH$_3$CHClOCH$_3$ + H$_2$O $\rightleftharpoons$ CH$_3$CHO + HCl + CH$_3$OH

9-29   Solvolysis occurs with sulfuric acid acting as a proton acceptor: RCHICH$_3$ + HOSO$_2$OH $\longrightarrow$ H$_2$$\overset{+}{\text{O}}SO_3$H + I$^-$ + RCH=CH$_2$.

9-30   *a.*   BF$_3$, HClO$_4$(Al$_2$O$_3$ at high temperatures)   *b.* $^-$OH, EtO$^-$, CH$_3$NH$_2$

9-31   NH$_3$ + C$_4$H$_9$C–O–C C$_4$H$_9$ $\rightleftharpoons$ C$_4$H$_9$–C–O–C–C$_4$H$_9$ $\longrightarrow$
$\quad\quad\quad\quad\quad$ || $\quad\quad\quad$ || $\quad\quad\quad\quad\quad\quad\quad\quad$ |
$\quad\quad\quad\quad\quad$ O $\quad\quad\quad$ O $\quad\quad\quad\quad\quad\quad\quad\quad$ $^+$NH$_3$

$\quad\quad\quad\quad\quad\quad\quad\quad\quad$ O $\quad\quad\quad\quad\quad$ O $\quad\quad\quad\quad\quad\quad\quad\quad\quad\quad\quad$ O $\quad\quad\quad\quad\quad\quad$ O
$\quad\quad\quad\quad\quad\quad\quad\quad\quad$ || $\quad\quad\quad\quad\quad$ || $\quad\quad\quad\quad\quad\quad\quad\quad\quad\quad\quad$ || $\quad\quad\quad\quad\quad\quad$ ||
$\quad\quad\quad\quad\quad\quad\quad\quad$ C$_4$H$_9$C–$\overset{+}{\text{N}}$H$_3$ + $^-$O–CC$_4$H$_9$ $\overset{\text{NH}_3}{\longrightarrow}$ C$_4$H$_9$CNH$_2$ + C$_4$H$_9$C–O$^-$ + NH$_4^+$

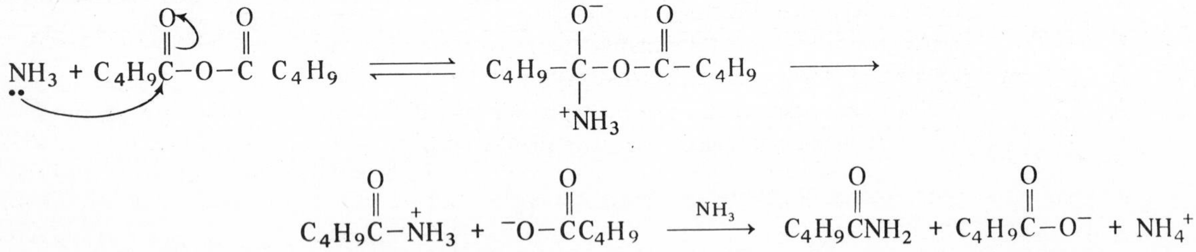

9-32   CH$_3$COAc $\overset{\text{H}^+}{\rightleftharpoons}$ CH$_3$C–OAc $\overset{\text{C}_4\text{H}_9\text{COOH}}{\rightleftharpoons}$ HOAc + CH$_3$C–OCC$_4$H$_9$ $\overset{\text{C}_4\text{H}_9\text{COOH}}{\rightleftharpoons}$
$\quad\quad\quad\quad\quad$ || $\quad\quad\quad\quad\quad\quad\quad$ | $\quad\quad\quad\quad\quad\quad\quad\quad\quad\quad\quad\quad\quad\quad\quad\quad\quad\quad$ || $\quad$ ||
$\quad\quad\quad\quad\quad$ O $\quad\quad\quad\quad\quad\quad\quad$ H$\overset{+}{}$ $\quad\quad\quad\quad\quad\quad\quad\quad\quad\quad\quad\quad\quad\quad\quad\quad$ O$\overset{+}{}$ O

$\quad\quad\quad\quad\quad\quad\quad\quad$ H $\quad\quad\quad\quad\quad\quad\quad\quad\quad\quad\quad$ $-$H$^+$
$\quad\quad\quad\quad\quad\quad\quad\quad$ | $\quad\quad\quad\quad\quad\quad\quad\quad\quad\quad\quad\quad\quad$
$\quad\quad\quad\quad$ CH$_3$COH + C$_4$H$_9$C–O–CC$_4$H$_9$ $\rightleftharpoons$ C$_4$H$_9$COCC$_4$H$_9$
$\quad\quad\quad\quad\quad\quad$ || $\quad\quad\quad\quad$ || $\quad$ || $\quad\quad\quad\quad\quad\quad\quad\quad$ || ||
$\quad\quad\quad\quad\quad\quad$ O $\quad\quad\quad\quad$ O$\overset{+}{}$ O $\quad\quad\quad\quad\quad\quad\quad\quad$ O O

     Use excess acetic anhydride or distill off acetic acid as it forms.

9-33   *a.* One   *b.* One   *c.* None   *d.* Four   *e.* One   *f.* Two

9-34   X, ;   Y,   CH$_3$COCH$_2$CH(COOH)CH$_2$COCH$_3$

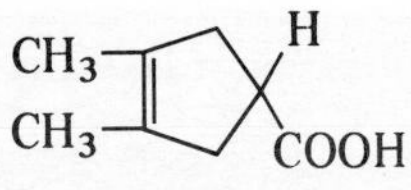

9-35 Moles of AgCl formed = 0.9/143.3. Therefore equivalent weight of A (based on AgCl formed) can be calculated: 0.482/X = 0.9/143.3; X = 76.7; 221(273/303)(740/760) = 194 ml of H_2 at STP. Thus, 1.0/Y = 193/22,400; Y = 116 (molecular weight of B). If B has only one double bond then A must have 2 Cl. One double bond (CH=CH) = 26. One COOH = 45; 45 + 26 = 71. Difference between 116 and 71 is 45 which represents 1 COOH. A, ClOCCH=CHCOCl; B, HOOCCH=CHCOOH; C, maleic anhydride.

9-36 A, $HOCH_2C(CH_3)_2CHO$; B, $(CH_3)_2C(COOH)_2$ **9-37** A, $(CH_3)_2C=C=O$; B, $(CH_3)_2CHCOOH$

9-38 Z,

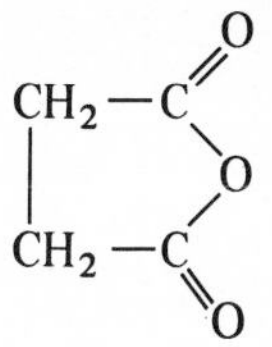

9-39 M,

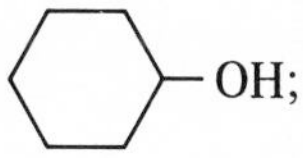

—OH;   N, CH_3—

=O   O, $HOOC(CH_2)_4COOH$;

P, $HOOCCH_2CH(CH_3)CH_2CH_2COOH$

9-40 A,

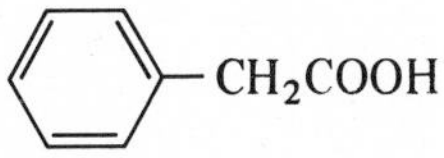

—CH_2COOH **9-41** B, butanoic acid

9-42 C,

; D,

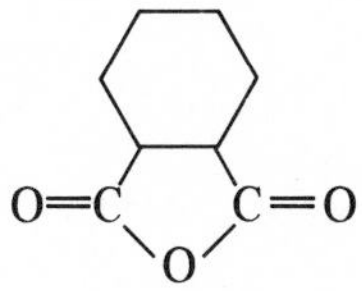

Chapter 10

10-1

	IUPAC name	Common name	Degree
a.	N-Methylisopropylamine	Methylisopropylamine	Secondary
b.	1-Amino-2-propene	Allylamine	Primary
c.	2-Butanamine	sec-Butylamine	Primary
d.	1,3-Propanediamine	Trimethylenediamine	Primary
e.	N,N-Dimethylisopropylamine	Dimethylisopropylamine	Tertiary
f.	2-Aminoethanol	Ethanolamine	Primary
g.	Aminoethanoic acid	Aminoacetic acid (glycine)	Primary
h.	2-Methyl-2-propanamine	tert-Butylamine	Primary
i.	N,N-Dimethylvinylamine	Dimethylvinylamine	Tertiary
j.	4-Diethylaminobutanal	4-Diethylaminobutyraldehyde	Tertiary
k.	4-Aminocyclohexylethanoic acid	4-Aminocyclohexylacetic acid	Primary
l.	1,5-Diamino-3-pentanone	Di-2-ethylamino ketone	Primary

10-2

a. $(CH_3)_2NCH_2CH_2CHO$

b. $(CH_3)_2NCH_2CH_2CH_2N(Et)_2$

c. $CH_3CHCH=CHCOOH$
     |
     $CH_3NCH_2CH_3$

d. $CH_3COCHCH_2CH_2CH_3$
     |
     $N(CH_2CH=CH_2)_2$

e. $(CH_3)_2NCH_2COOH$

f. $H_2NCH_2C\equiv CCH_2CH_3$

g. $[(CH_3)_2CHCH_2]_4N^+NO_3^-$

h.

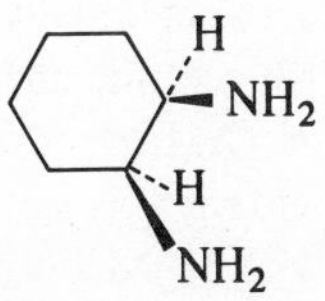

$i.$ $CH_3(CH_2)_3\underset{\overset{|}{CH=CH_2}}{N}CH(CH_3)CH_2CH_2CH_2OH$

$j.$ $\underset{\underset{H}{|}}{CH_3}C=\underset{\underset{H}{|}}{C}\overset{CH_2NH_2}{}$

10-3
$a.$ $CH_3COOH + H_2NEt \longrightarrow CH_3CO\bar{O} + H_3\overset{+}{N}Et$

$b.$ $F_3B + :NH_2C(CH_3)_3 \longrightarrow F_3\bar{B}:\overset{+}{N}H_2C(CH_3)_3$

$c.$ $HONO + CH_3NH_2 \longrightarrow CH_3OH + N_2 + H_2O$

$d.$ $2(CH_3CO)_2O + HOCH_2CH_2NH_2 \longrightarrow CH_3COOCH_2CH_2NHCOCH_3 + 2CH_3COOH$

$e.$ $2(Et)_4\overset{+}{N}I^- + Ag_2SO_4 \longrightarrow 2AgI + [(Et)_4N^+]_2SO_4{}^{2-}$

$f.$ $(Et)_2C=NOH + 2H_2 \xrightarrow{Pt} (Et)_2CHNH_2 + H_2O$

$g.$ $Na_2NC\equiv N + 2EtI \longrightarrow Et_2NCN + 2NaI$

$h.$ $OCHCH_2CH_2CHO + 2CH_3NH_2 + 2H_2 \xrightarrow{Pt} CH_3NH(CH_2)_4NHCH_3 + 2H_2O$

$i.$ See Sec. 10-3e.

$j.$ $EtCONH_2 + NaOBr + 2NaOH \longrightarrow EtNH_2 + Na_2CO_3 + H_2O + NaBr$

$k.$ $(CH_3)_3N: + HONO \longrightarrow (CH_3)_3\overset{+}{N}H\ ONO^-$ $l.$ $NCCH_2CH_2CN + 4H_2 \xrightarrow{Pt} H_2N(CH_2)_4NH_2$

$m.$ $CH_2=C=O + CH_3NH_2 \longrightarrow CH_3CONHCH_3$ $n.$ $CH_3NH_2 + 2CH_3MgI \longrightarrow 2CH_4 + CH_3N(MgI)_2$

$o.$ $[CH_3(CH_2)_4CH_2]_3B + 2H_2NOSO_3H \xrightarrow{4NaOH} 2CH_3(CH_2)_4CH_2NH_2 + 2\ Na_2SO_4$
$+\ n\text{-hexyl-}B(OH)_2 + 2H_2O$

$p.$ $2(CH_3CO)_2O + HSCH_2CH_2NHCH_3 \longrightarrow CH_3COSCH_2CH_2N(CH_3)COCH_3 + 2CH_3COOH$

$q.$ $Et_2NC\equiv N + 2H_2O \xrightarrow[\Delta]{OH} Et_2NCOOH + NH_3 \longrightarrow Et_2NH + CO_2$

$r.$ $CH_2=CHCN + 3H_2 \xrightarrow{Pt} CH_3CH_2CH_2NH_2$

$s.$ $(CH_3CH_2CH_2)_4N^+OH^- \xrightarrow{\Delta} (CH_3CH_2CH_2)_3N + CH_3CH=CH_2 + H_2O$

$t.$ $(CH_2=CHCH_2)_2NH + C_6H_5SO_2Cl \xrightarrow[+]{NaOH} C_6H_5SO_2N(CH_2CH=CH_2)_2 + NaCl + H_2O$

$u.$ $(CH_3)_3B + NH_3 \longrightarrow (CH_3)_2\bar{B}:\overset{+}{N}H_3$

$v.$ $2(CH_3)_2CO + H_2NCH_2(CH_2)_4CH_2NH_2 \xrightarrow{2H_2,\ Pt} (CH_3)_2CHNHCH_2(CH_2)_4CH_2NHCH(CH_3)_2 + H_2O$

10-4
$a.$ Convert acetic acid to the chloride with PCl_3, and treat this with NH_3. Convert the resulting acetamide to methylamine with NaOBr.

$b.$ Use direct reductive alkylation with hydrogen and ammonia or convert acetone to the oxime with NH_2OH and then hydrogenate.

$c.$ Convert methyl chloride to acetonitrile with NaCN. Hydrogenate to ethylamine.

$d.$ Add HBr, then treat product with NaCN and hydrogenate.

$e.$ Add HCN to acetone and hydrogenate.

$f.$ Ethane reacts at $500°C$ with nitric acid to yield nitroethane. Hydrogenate this to the amine.

$g.$ Make ethyl bromide and react this with sodium cyanamide. Hydrolysis gives diethylamine which on treatment with HONO yields the product.

$h.$ Make 1,2-dibromoethane and treat it with NaCN. Catalytic hydrogenation gives diamine.

$i.$ Convert diethylamine with EtI and sodium hydroxide to tetraethylammonium iodide. On treatment with silver hydroxide the quaternary hydroxide forms.

$j.$ Passing amines over hot alumina causes the elimination of ammonia just as alcohols eliminate water. Alternatively the amine could be converted to the quaternary ammonium hydroxide with methyl iodide and then silver hydroxide. Heating the hydroxide gives smooth conversion to the olefin.

$k.$ Reductive alkylation with formaldehyde gives trimethylamine which on oxidation with hydrogen peroxide yields product.

$l.$ Use same technique as in a.

$m.$ Run aldol condensation with acetaldehyde and subject product to reductive alkylation with ammonia.

$n.$ Run pinacol reduction with acetone and then pinacolone rearrangement. Make oxime of resulting ketone and catalytically hydrogenate. Or, run reductive alkylation with NH_3.

$o.$ Hydroboration of cyclopentene (cis addition of H–B across the C=C) followed by treating the intermediate organoborane with H_2NOSO_3H and then sodium hydroxide will give the $trans$-2-methylcyclo-pentylamine.

p. Make nitroethane as in *f* and condense this with 2 mol of formaldehyde in aldol-type condensation. Then catalytic hydrogenation gives product.

q. Hydroboration-carbonylation-oxidation of isobutylene would give 2,6-dimethyl-4-heptanone. Reductive alkylation of methylamine with this ketone gives the product.

r. Run an aldol condensation on isobutyraldehyde and dehydrate the aldol product. Convert isobutyraldehyde to isobutylamine via reductive alkylation with NH_3. Combine this amine with the dehydrated aldol product in another reductive alkylation step reducing the C=C at the same time.

s. Use same technique as in *r*.

t. Hydroboration of divinylacetic acid using thexylborane followed by carbonylation and oxidation with alkaline peroxide would give 4-oxo-1-cyclohexanecarboxylic acid which could be reduced ($LiAlH_4$) to product.

10-5 *a.* Hinsberg test. Dimethylamine forms a product with benzenesulfonyl chloride that is insoluble in dil NaOH.

b. Triethylamine does not react with benzenesulfonyl chloride. Butylamine reacts to give a solid product soluble in dil NaOH.

c. Ammonium chloride reacts with nitrous acid to give off nitrogen. Dimethylammonium chloride gives nitrosodimethylamine.

d. Dibutylamine is soluble in dil HCl. The neutral cyanide is not.

e. Diethylamine is soluble in water.

f. Hexylamine will dissolve in dilute acid; 1-hexanol will not.

g. Dibutylamine will react with benzenesulfonyl chloride to give an alkali-insoluble product. The tertiary amine will not.

h. Ethylammonium chloride is soluble in water.

10-6 *a.* First extract the propylene glycol with water. Next extract the tributylamine with dilute acid. The 2-octanone can then be separated from the heptane as the bisulfite addition product.

b. Extract the trimethylamine with water and the heptylamine with dilute acid. The 2-hexanone can be separated as a bisulfite complex.

c. Extract the 2-octanamine with dilute acid. Next extract the octanoic acid with dil NaOH.

10-7 *a.* 101 *b.* 89 *c.* 30 *d.* 44 *e.* 119

10-8 *a.* $CH_2=CH_2 + (CH_3)_2NEt$
  b. $CH_2=CH_2 + CH_3SC_3H_7$
  c. $(CH_3)_2NCH_2CH_2CH(CH_3)_2 + CH_2=CH_2$
  d. $(CH_3)_3N$ + cyclopentene
  e. $CH_2=CH_2 + EtSC_3H_7$
  f. $CH_3CH=CH_2 + (CH_3)_2NCH_2CH(CH_3)_2$

10-9 In this reaction a proton is removed by OH from the carbon atom β to the nitrogen. The more electron-deficient this carbon atom is, the more easily the proton is removed. Two methyl groups attached to this carbon atom increase the density on it more than one isopropyl group. The chlorine-containing group is most rapidly eliminated since the strong inductive pull of this atom decreases the electron density.

10-10 *a* $ClNH_2, NH_3, CH_3NH_2, (CH_3)_2NH$
  b. $(CH_3)_2NCH=CHCHO, (CH_3)_2NCH=CH_2, (CH_3)_3N$
  c. $HONH_2, H_2NNH_2, NH_3, CH_3CH_2NH_2$
  d. $(C_6H_5)_2NH, C_6H_5NH_2, C_6H_5NHCH_3, C_6N_{11}NH_2$

10-11 Bulky alkyl groups around a base center tend to prevent the approach of an acid. The bigger the acid the more pronounced this effect becomes.

10-12 $(CH_3)_2NCH_2CH_2CH_3, (CH_3)_2NCH(CH_3)_2, (CH_3CH_2)_2NCH_3$

10-13 *a.* Isobutylene
  b. 2-Pentene
  c. 1-Butene and propene
  d. Cyclopentene
  e. 1,4-Pentadiene and 1,3-pentadiene (due to rearrangement of 1,4 diene to the conjugated, more stable 1,3 diene)
  f. Cyclopentene and 1,3- and 1,4-pentadiene (see part *e*).

10-14 *a.* 2,2,4-Trimethyl-3-pentanamine
  b. Cyclohexylamine
  c. Butylamine
  d. 3-Propyl-1,5-pentanediamine

e. 3-Pentanamine

f.

10-15

10-16 $(CH_3)_2CHCH_2CH_2\overset{}{\underset{}{CH_2}} \xrightarrow{\Delta} (CH_3)_2NOH$ and $(CH_3)_2CHCH_2CH=CH_2$

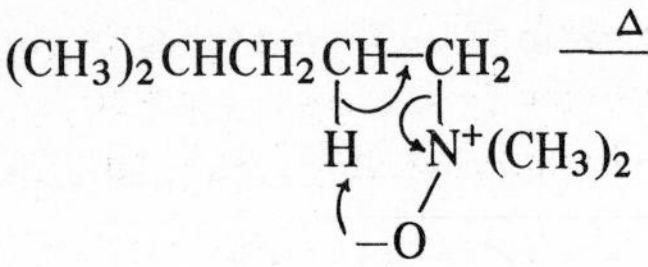

(structure shows H and $\overset{+}{N}(CH_3)_2$ with $-O$)

10-17 As soon as one methyl group is attached, this increases the electron density on this nitrogen atom. The more electron-rich nitrogen is, the more effective it is in an S_N reaction.

10-18 *a.* In step B a bad side reaction would occur between halogen and copper at the high temperature. Halogen in organic compounds reacts with most metals at $350°C$ and above. Step C is wrong if halogen is present.
b. In step B elimination would occur. Amine oxides (step C) cannot be formed from primary amines.
c. HCN does not add to simple olefins. Step C would give a very poor yield of the indicated product. The primary carbonium ion formed would tend to rearrange to a tertiary.
d. In step A amines are very easily oxidized to a variety of products. In step C amine C—N bonds are not hydrolyzed.

10-19 A, $CH_3OCH_2CH_2NH_2$; B, $CH_3OCH_2CH_2NHCOCH_3$

10-20 $CH_3NHCH_2CH_2OH$

10-21 A, $(CH_3)_2NCH_2CH_2C\equiv N$; B, $(CH_3)_2NCH_2CH_2CH_2NH_2$; C, $(CH_3)_2\overset{+}{N}HCH_2CH_2COOH$ $\overset{-}{Cl}$

10-22 A, $(CH_3)_2NCH_2CH(CH_3)CH(CH_3)CH(CH_3)CH_2N(CH_3)_2$;
B, $CH_2=C(CH_3)CH(CH_3)C(CH_3)=CH_2$; C, $CH_3COCH(CH_3)COCH_3$;
D, $HOOCCH(CH_3)CQOH$

10-23 A, $CH_3CH_2CH_2CH_2\overset{+}{N}\equiv\overset{-}{C}$; B, $CH_3CH_2CH_2CH_2NHCH_3$; C, $CH_3CH_2CH_2CH_2N=C=O$; D, $CH_3CH_2CH_2COOH$

10-24 ⟨N—CH₃⟩ $CH_2=CHCH_2CH=CH_2$ (Some 1,3-pentadiene might be formed from rearrangement of 1,4-pentadiene)

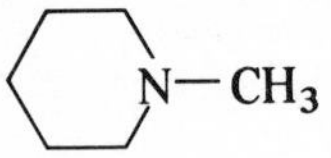

10-25 A, (pyridine ring)—$C\equiv N$ B, (pyridine ring)

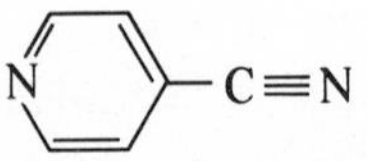

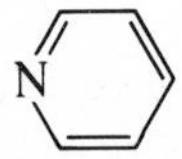

10-26 *a.* Convert acetone to isopropylamine using reductive alkylation. Add amine to ethylene oxide.
b. Oxidize CH_3CHO to acetic acid and pass over hot thoria to get acetone. Convert acetone to mesityl oxide via aldol condensation. Run reductive alkylation with mesityl oxide and methylamine.
c. Add HCN to CH_3CHO and hydrogenate or condense CH_3CHO with CH_3NO_2 and hydrogenate.
d. Run aldol condensation with CH_3CHO, add HCN to product and hydrogenate.
e. Treat $BrCH_2CH_2Br$ with NaCN and add isopropylmagnesium bromide to this dinitrile. Hydrolysis gives a diketone which can be converted to the desired product via conversion to the dioxime and hydrogenation, or via reductive alkylation of the diketone.
f. Add $CH_2=CHMgX$ to $CH_2=CHCHO$. Convert the resulting secondary alcohol to the halide with PBr_3 and convert this to the amine using Gabriel's method.
g. Treat 2-butene with peracetic acid to get 2,3-epoxybutane. Treat this with dimethylamine to get desired product.
h. Condense $HOCH_2CH_2NO_2$ with butanal. Hydrogenate the NO_2 to an amino group.
i. Convert acetone to trimethylacetyl chloride via pinacol formation, pinacol-pinacolone rearrangement, and the iodoform reaction followed by $SOCl_2$ treatment. Convert the acid chloride to the amide with NH_3.
j. Convert the alcohol to the Grignard reagent and treat this with ethylene oxide. Oxidize the resulting alcohol to the acid. Convert the acid to the amide and treat with NaOBr to get neopentylamine.
k. Dehydrate alcohol to isobutylene. A hydroboration-carbonylation-oxidation sequence performed on isobutylene (using thexylborane) followed by reductive alkylation (using NH_3) of the resulting 2,6-dimethyl-4-heptanone will give the product.
l. Oxidize cyclohexanol to cyclohexanone. Add HCN to the ketone and hydrogenate.
m. Condense CH_3CHO with 3 mol of CH_2O. Reductive alkylation of the resulting $(HOCH_2)_3CCHO$ with ammonia will give product.
n. Reductive alkylation of $(CH_3)_2CHCHO$ with $H_2NCH_2CH_2OH$ will give product.
o. Convert allyl bromide to allylamine with $NaNH_2$. Treatment of the amine with thexylborane followed by carbonylation and oxidation yields $(H_2NCH_2CH_2CH_2)_2C=O$(A). Prepare $CH_2=CHCH=P(C_6H_5)_3$ from allyl bromide using $(C_6H_5)_3P$ followed by CH_3Li. Use this in a Wittig reaction on A to give the product.

362

p. Prepare cyclopentanone from 1,3-butadiene using the thexylborane hydroboration-carbonylation-oxidation technique. React cyclopentanone with CH_3MgBr and dehydrate to 1-methylcyclopentene, followed by hydroboration and treatment with H_2NOSO_3H to give *trans*-2-methylcyclopentylamine. Treatment of this with acetyl chloride will give the product.

10-27 *a.* $(CH_3)_3COH$ *b.* $CH_3COC(CH_3)_3$
       c. Cyclobutanol and cyclobutene   *d.* Cycloheptanone
       e. $(CH_3)_3CCHO$ *f.* *tert*-Pentyl alcohol and $(CH_3)_2C=CHCH_3$

10-28 $CH_3\ddot{N}=\overset{+}{N}=\overset{-}{\underset{\cdot\cdot}{N}}\colon \longleftrightarrow CH_3\overset{+}{\underset{\cdot\cdot}{N}}-\overset{+}{N}\equiv N\colon$

10-29 $H_2NCH_2CH_2OH + H_2SO_4 \xrightarrow{\text{NaOH}} H_3\overset{+}{N}CH_2CH_2OSO_3^- \longrightarrow$

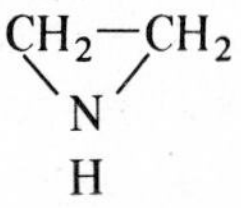

$$\underset{\underset{H}{N}}{CH_2-CH_2}$$

One could also use HCl and displace Cl^- instead of $SO_4{}^{2-}$.

a. $\xrightarrow{H_2S} H_2NCH_2CH_2SH$

b. $\xrightarrow{CH_2=CHC\equiv N}$ $\underset{CH_2}{\overset{CH_2}{\big|}}NCH_2CH_2C\equiv N$

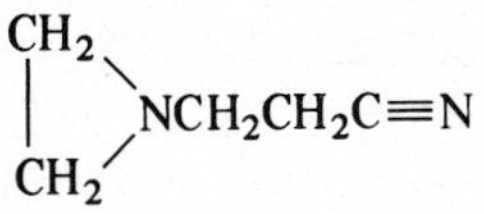

c. $\xrightarrow{CH_2=C=O}$ $\underset{CH_2}{\overset{CH_2}{\big|}}NCOCH_3$

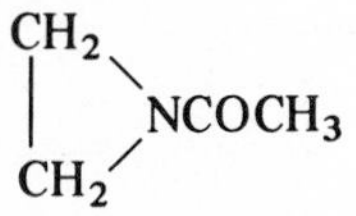

d. Alkali neutralizes the HCl liberated; if present, the HCl would open the imine ring to give $RCONHCH_2CH_2Cl$.

10-30 *a.* $H_2NCH_2CH_2OH$ *b.* 3-Dimethylamino-2-butanol
       c. $CH_3CH_2CH(OH)CH_2SCH_3$ *d.* 2-Methoxy-1-butanol
       e. $HOCH_2CH_2OC(CH_3)CH_2OH$ *f.* 1-Amino-2-methyl-2-propanol
                        $\underset{Et}{\big|}$

       g. $(CH_3)_2CCH_2SCH_3$ *h.* 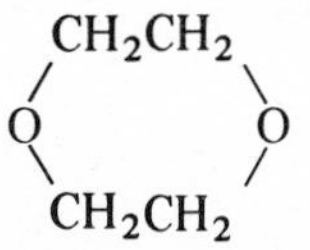 $\underset{CH_2CH_2}{\overset{CH_2CH_2}{O\diagdown\diagup O}}$
                  $\underset{SH}{\big|}$

10-31 Nitrogen uses sp^3 hybridization bonding to the four different alkyl groups at the corners of a tetrahedron. Optical activity is possible because the compound can exist as two nonsuperimposable mirror images, i.e., a *dl* pair:

$$CH_3CH_2-\overset{\overset{\displaystyle CH(CH_3)_2}{|}}{\underset{\underset{\displaystyle CH_2CH_2CH_3}{|}}{N^+}}-CH_3 \quad \overset{-}{Cl} \qquad\qquad CH_3-\overset{\overset{\displaystyle CH(CH_3)_2}{|}}{\underset{\underset{\displaystyle CH_2CH_2CH_3}{|}}{N^+}}-CH_2CH_3 \quad \overset{-}{Cl}$$

10-32

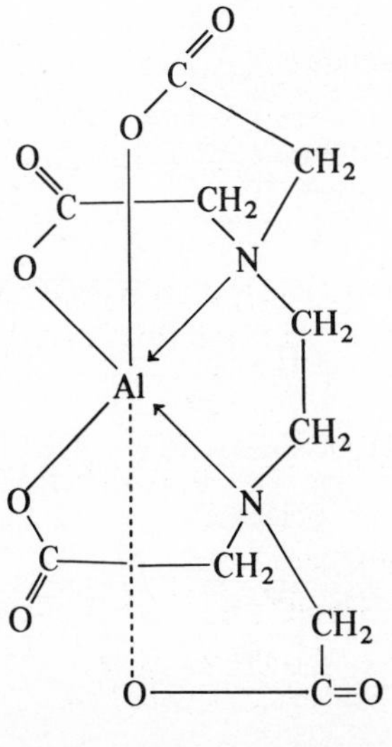

10-33   A, $(CH_3O)_2CCH_2CH_2C\equiv N$;   B, $(CH_3O)_2CCH_2CH_2COOH$;   C, $CH_3CCH_2CH_2COOH$

A below: CH_3 ; B below: CH_3 ; C below: O

10-34   X, $(CH_3)_2NNH_2$ 10-35   A, $(CH_3)_3CCH_2NH_2$
10-36   Y, $(CH_3)_2NCH_2CH_2OH$;   Z, $(CH_3)_2NCH_2CH_2OCOCH_3$

10-37   A, ⬡—$CH_2C\equiv N$;   B, ⬡—$CH_2CH_2NH_2$;   C, ⬡—CH_2COOH;

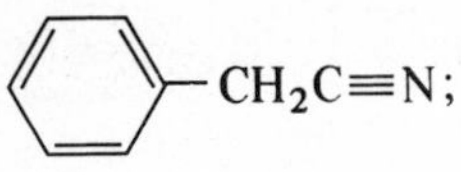

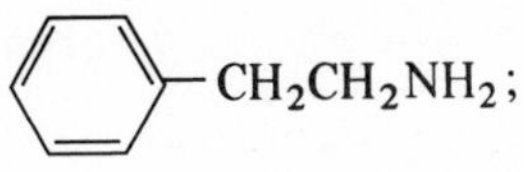

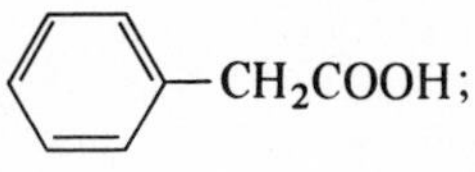

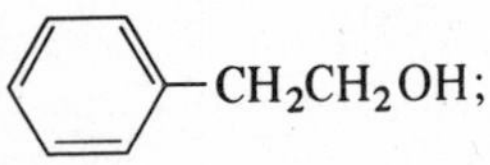

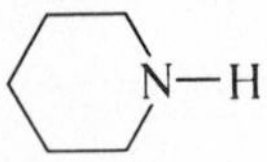

D, ⬡—CH_2CH_2OH;   E, ⬡—$CH_2\overset{O}{\overset{\|}{C}}CH_2$—⬡

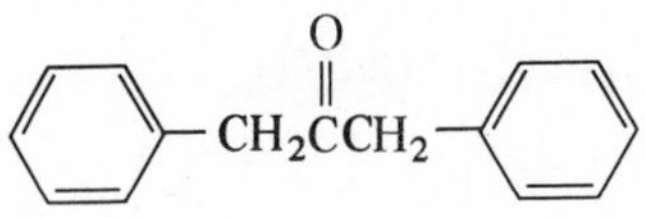

10-38   A, [piperidine]N—H   B, [piperidine]N—N=O

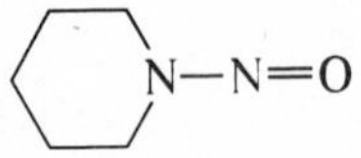

10-39   X, O⬡NH;   Y, $(CH_2=CH)_2O$;   Z, $CH_3CH_2OCH_2CH_3$

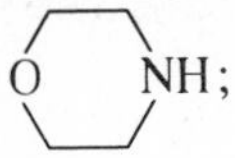

10-40   ⬡—$NHCH_2CH_2OH$

Chapter 11

11-1

IUPAC name	Common name
a. Pentanamide	Valeramide
b. Methyl butanoate	Methyl butyrate
c. Ethyl propanoate	Ethyl propionate
d. 2-Butenamide	Crotonamide
e. Trimethoxymethane	Trimethyl orthoformate
f. tert-Butyl propenoate	tert-Butyl acrylate
g. N-Bromopropanamide	N-Bromopropionamide
h. Dimethyl butanedioate	Dimethyl succinate
i. N,N'-Dimethylpropanediamide	N,N'-Dimethylmalonamide
j. Propynamide	Propiolamide
k. Vinyl 4-hydroxybutanoate	Vinyl γ-hydroxybutyrate
l. N-Methyl-N-ethyl-2-oxopropanamide	N-Methyl-N-ethylpyruvamide
m. N-Methylcyclohexanecarboxamide	N-Methylcyclohexylformamide
n. 3-Propanolide	β-Propiolactone
o. Cyclopropyl cyclopentanecarboxylate	Cyclopropyl cyclopentylformate

11-2   a. $CH_3CH_2CH_2C\equiv N$ b. $CH_2=CHCH_2N=C=O$ c. $CH_3COOCH_2CHCH_2OOCCH_3$
   c below: CH_3COO

d. $CH_3CH=CHCOOCH(CH_3)_2$ e. $EtNHCONHEt$ f. $CH_3OOCCH_2CHCH_2COOCH_3$
   f below: CH_3OOC

g. $CH_3CH_2CH_2ONO$ h. $C(OCH_3)_4$ i. 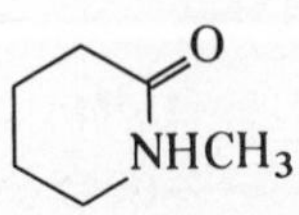 [ring with O and NHCH_3]

364

j.

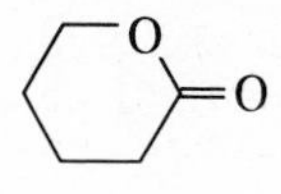

k. $(Et)_2NCNH_2$ with S (i.e. $(Et)_2N\overset{\|}{C}NH_2$, $C=S$)

l. $CH_3(CH_2)_7CH=CH(CH_2)_7$
$CH_3(CH_2)_{14}CH_2OC=O$

m.

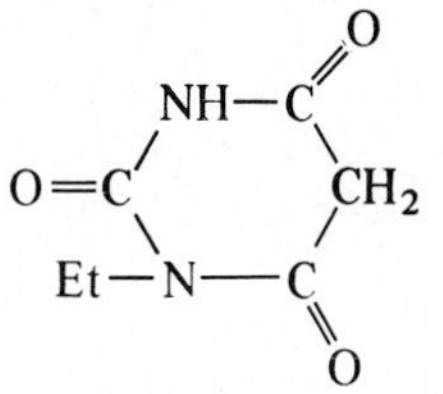

n. $CH_3CNHCNHCCH_3$ (each C double-bonded to O)

o.

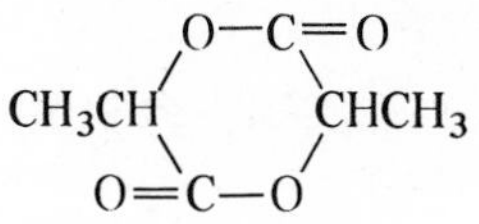

p. $(CH_2=CH)_2CO_3$

q. $(Et)_2NCONEt$ with $N=O$

r. $OCHCH_2CHCH_2CHO$
$\quad\quad\quad COOCH_2CH_3$

11-3 *a.* $C_3H_7COCl + HOCH_2CH(CH_3)_2 \longrightarrow C_3H_7COOCH_2CH(CH_3)_2 + HCl$

b. $(CH_3CO)_2O + (CH_3)_2CHOH \longrightarrow CH_3COOCH(CH_3)_2 + CH_3COOH$

c. $CH_3CH_2COCl + CH_3CH(NH_2)CH_2CH_3 \longrightarrow CH_3CH_2CONHCH(CH_3)CH_2CH_3 + HCl$

d. $(CH_3CH_2CH_2CO)_2O + CH_2=CHCH_2NH_2 \longrightarrow CH_2=CHCH_2NHCOCH_2CH_2CH_3 + C_3H_7COOH$

e. $H_2NOCCH_2CH_2CONH_2 + 2P_2O_5 \longrightarrow NCCH_2CH_2CN + 4HPO_3$

f. $(CH_3)_2CHCH_2CONH_2 + Br_2 + 4NaOH \longrightarrow (CH_3)_2CHCH_2NH_2 + Na_2CO_3 + 2NaBr + 2H_2O$

g. $2CH_2N_2 + CH_2(COOH)_2 \longrightarrow CH_3OOCCH_2COOCH_3 + 2N_2$

h. $2CH_2=C=O + HO(CH_2)_3OH \longrightarrow CH_3COOCH_2CH_2CH_2OOCCH_3$

i. $CH_3COCH_2CH_3 + (EtO)_3CH \longrightarrow CH_3C(OEt)_2CH_2CH_3 + HCOOEt$

j. $CH_3(CH_2)_3COOCH_3 + 2CH_3MgI \longrightarrow CH_3CH_2CH_2CH_2C(CH_3)_2OMgI + CH_3OMgI$

k. $H_2NCSNH_2 + 2HONO \longrightarrow 2N_2 + COS + 3H_2O$

l. $HCOOCH_3 + CH_3NH_2 \longrightarrow HCONHCH_3 + CH_3OH$

m. $H_2NOC(CH_2)_3CONH_2 \longrightarrow$ 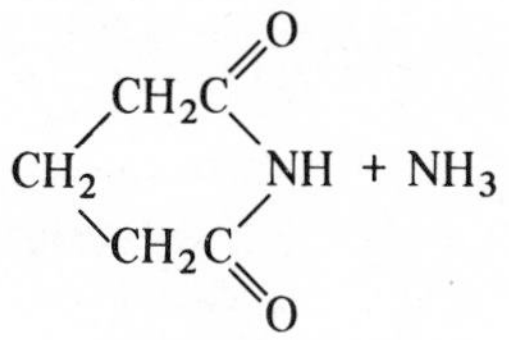 $+ NH_3$

n. $(CH_3)_2NCONH_2 + 2NaOH \longrightarrow (CH_3)_2NH + Na_2CO_3 + NH_3$

o. $C_3H_7COOCH_2CH(OOCC_3H_7)CH_2OOCC_3H_7 + H_2 \xrightarrow{cat} glycerol + 3CH_3CH_2CH_2CH_2OH$

p. $3CH_3ONa + Cl_3CCH_3 \longrightarrow (CH_3O)_3CCH_3 + 3NaCl$

q. $CH_3CH_2COOAg + EtI \longrightarrow CH_3CH_2COOEt + AgI$

r. $CH_3(CH_2)_4COOCH_3 + 4(H) \longrightarrow CH_3(CH_2)_5OH + CH_3OH$

s. $CH_3(CH_2)_{10}COONH_4 \longrightarrow CH_3(CH_2)_{10}CONH_2 + H_2O$

t. $2CH_3CH_2CH_2CONH_2 + 2K \longrightarrow 2CH_3CH_2CH_2CONHK + H_2$

u. $3C_3H_7COOCH_2CH=CH_2 + 3NH_2OH + Fe^{3+} \longrightarrow (C_3H_7CONHO)_3Fe + 3CH_2=CHCH_2OH + 3H^+$

v. $CH_2=C=O + CH_2=CHCH_2NH_2 \longrightarrow CH_3CONHCH_2CH=CH_2$

w. 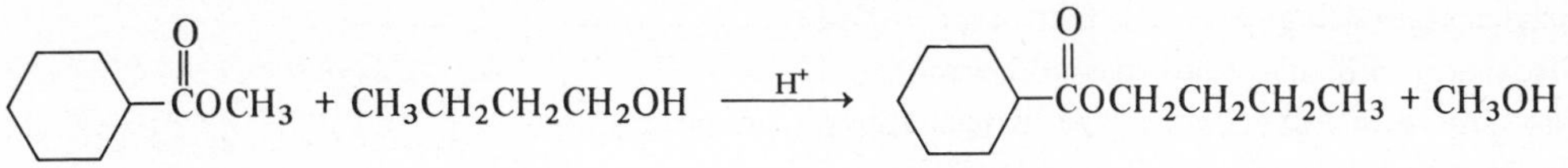 $+ CH_3CH_2CH_2CH_2OH \xrightarrow{H^+}$ $+ CH_3OH$

x. 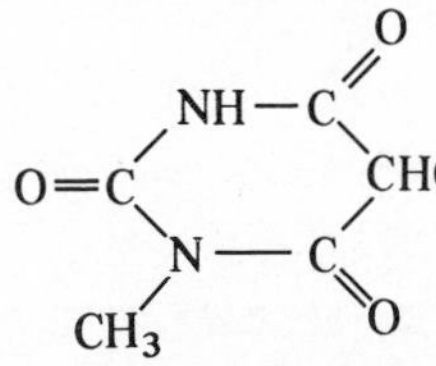 $+ 4NaOH \longrightarrow CH_3CH(COONa)_2 + NH_3 + CH_3NH_2 + Na_2CO_3$

y. $(CH_3CO)_2NH + NaOBr \longrightarrow (CH_3CO)_2NBr + NaOH$

z. $CH_3CH(OH)CONH_2 + 3CH_3MgI \longrightarrow CH_3CH(OMgI)CON(MgI)_2$

a'. $EtOSOOEt + 2NaOH \longrightarrow Na_2SO_3 + 2EtOH$

b'. $CH_3COCl + (CH_3)_3COH \longrightarrow CH_3COOH + (CH_3)_3CCl$ Unless special conditions are observed one does not obtain esters when tertiary alcohols are treated with acyl halides.

c'.

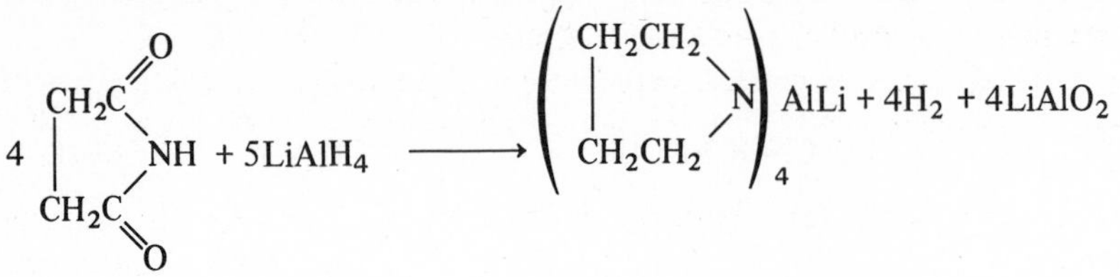

d'. $CH_3CH_2CH_2C \equiv N + 2H_2O_2 \xrightarrow{NaOH} CH_3CH_2CH_2\overset{\overset{\displaystyle O}{\|}}{C}NH_2 + H_2O + O_2$

e'.

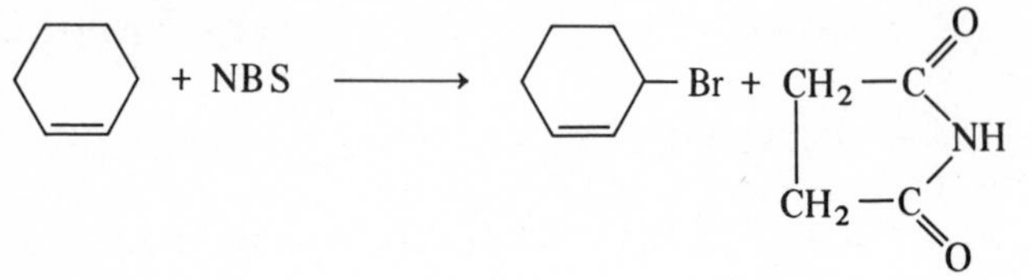

f'. $(H_2N)_2C=NH + 2NaOH + H_2O \longrightarrow 3NH_3 + Na_2CO_3$

g'.

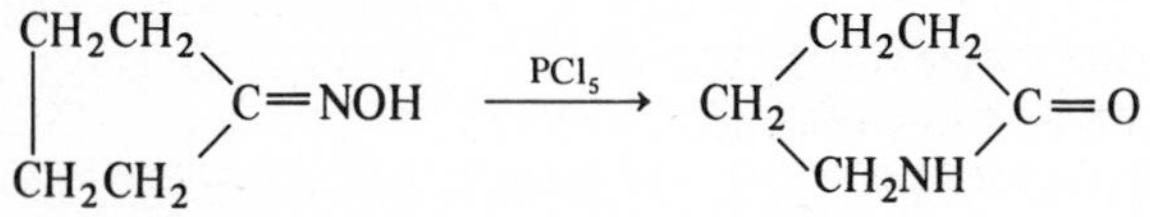

h'. $(CH_3NH)_2C=S + 2NaOH \longrightarrow 2CH_3NH_2 + Na_2CO_2S$

i'. $HOCH_2CH_2CH_2COOH \xrightarrow{\Delta}$
$+ H_2O$

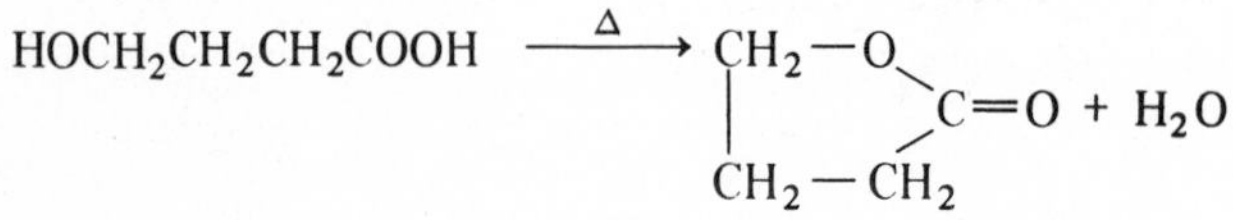

j'.

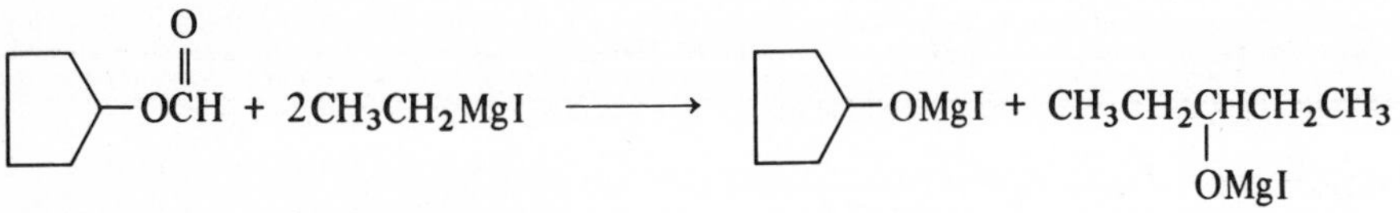

11-4 a. Caproic acid is soluble in dil $NaHCO_3$.

      b. Butyl bromide will give a Beilstein test.

      c. Upon treatment with HONO only urea will evolve N_2.

      d. Upon treatment with HONO only the amide will evolve N_2.

      e. 2-Propanol is soluble in water.

      f. Hexanal will react with Tollens' reagent.

      g. The urea is soluble in water while the amide is not.

      h. Semicarbazide forms insoluble semicarbazones with carbonyl compounds.

      i. Use litmus since the hydroxide is a strong base.

      j. The amine is soluble in dil HCl.

      k. Butyl acetate will give a ferric hydroxamate test.

      l. 2-Hexanone will give a positive iodoform test.

      m. Vinyl butyrate will decolorize permanganate or bromine solution.

      n. The ammonium hexanoate is soluble in water.

      o. Acetamide is soluble in water.

      p. The diamide is soluble in dilute alkali.

      q. The carbonate evolves NH_3 when heated with sodium hydroxide solution.

      r. The barbituric acid is soluble in dilute alkali.

11-5   Ozonolysis of double and triple bonds, oxidation of double and triple bonds, action of NaOX on amides, Beckmann rearrangement, action of NaOX on methyl ketones and secondary alcohols of the type $RCH(OH)CH_3$, action of bromine on the silver or thallium salt of an acid, decarboxylation of acids by fusion with sodium hydroxide, and the isonitrile reaction.

11-6 *a.* Hydrolyze to propionamide and treat with NaOCl.

b. Allylic bromination using *N*-bromosuccinimide and then alcoholic KOH.

c. Convert the ester to the amide with ammonia and then dehydrate with P_2O_5 to the nitrile.

d. Add HCN to propanal; then convert the cyanohydrin to the amide with H_2O_2.

e. Convert acetone to its oxime and run a Beckmann rearrangement.

f. Convert cyclohexene to cyclohexanone by hydroboration followed by oxidation with chromic acid. Convert cyclohexanone to the oxime and run the Beckmann rearrangement to get the cyclic amide. Hydrolyze this to the amino acid product.

g. Convert ethene to succinonitrile by treatment with Br_2 and then NaCN. Convert this to the amide with H_2O_2 and then to the imide by heating. Hofmann reaction with NaOCl on this imide gives the product.

h. Reduce ethyl acetate to ethanol with $LiAlH_4$. Convert ethanol to ethylmagnesium bromide. React this with more ethyl acetate to obtain product.

i. Hydrolyze the nitrile to the acid and reduce to 1-butanol. Esterify butyric acid with 1-butanol.

j. Convert cyclobutene to cyclobutanone via hydroboration-oxidation. Convert the ketone to its oxime and run Beckmann rearrangement to obtain product.

k. Prepare ketene by pyrolysis of acetone. Prepare isopropyl alcohol by $LiAlH_4$ reduction of acetone. React ketene with isopropyl alcohol to obtain product.

l. Convert the ester to the amide with NH_3. Reduce the amide with $LiAlH_4$ to get product.

m. Add Br_2, then NaCN. Convert this to the amide with H_2O_2 and proceed as in *g* above.

n. Convert the acid to the acid chloride and treat this with NH_3 to get the amide which is treated with NaOCl to get *tert*-butylamine. React this with phosgene to get *tert*-BuNHCOCl which upon heating will give product.

o. Hydroboration followed by oxidation with alkaline peroxide yields methyl 4-hydroxybutanoate. Saponification of this followed by heating the sodium salt of the hydroxyacid with an acid catalyst will give the lactone.

p. Convert propionamide to ethylamine with NaOCl. Treat 2 mol of ethylamine with 1 mol of phosgene.

q. Convert 1-butene to 1-butanol via hydroboration-oxidation with alkaline H_2O_2. Treat this with phosgene to get $CH_3CH_2CH_2CH_2OCOCl$ which upon treatment with NH_3 gives product.

r. Aldol condensation of ethanal followed by dehydration gives crotonaldehyde. Convert this to its acetal with ethanol and H^+.

11-7 *a.* SE 88; 2.20 g   *b.* SE 130; 3.25 g   *c.* SE 87; 2.18 g

d. SE 87; 2.18 g   *e.* SE 51; 1.28 g   *f.* SE 87; 2.18 g

g. SE 79; 1.98 g   *h.* SE 89; 2.23 g   *i.* SE 66; 1.65 g

11-8 *a.* 200   *b* 30.7   *c.* 27   *d.* 269

11-9 *a.*

$$CH_3O\overset{\overset{O}{\|}}{C}-\bigcirc-\overset{\overset{O}{\|}}{C}OCH_3$$

b. $H_2NCOCOOCH_3$

c. $CH_3CH_2COOCH_2CH_2CH_3$

d. (β-propiolactone structure)

11-10 *a.* LiI (N) + CH_3COOCH_3 ⟶ CH_3I + CH_3COOLi

b. BF_3 (E) + CH_3COOCH_3 ⟶ $CH_3C\overset{\overset{\bar{O}BF_3}{\|}}{\underset{+OCH_3}{}}$

c. CH_3NH_2 (N) + CH_3COOCH_3 ⟶ $CH_3CONHCH_3$ + CH_3OH

d. $AlCl_3$ (E) + CH_3COOCH_3 ⟶ $CH_3C\overset{\overset{\bar{O}AlCl_3}{\|}}{\underset{+OCH_3}{}}$

e. CH_3SH (N) + CH_3COOCH_3 ⟶ CH_3COSCH_3 + CH_3OH

11-11 *a.* CH_3CN *b.* $CH_3CH_2CH_3$ *c.* CH_4 *d.* $CH_3C{\equiv}CCH_3$ *e.* CH_3I

11-12 SE 116; isopropyl propionate (A) and dimethyl decanedioate (B) both fit SE. These could be distinguished by distillation of the hydrolyzed esters giving isopropyl alcohol (which gives a positive iodoform test) in the case of A. Hydrolysis of B gives methyl alcohol which does not give an iodoform test.

11-13 $CH_3COOCH(CH_3)_2$

11-14 $CH_3CH{=}CHCONH_2$, $CH_2{=}CHCH_2CONH_2$, $CH_2{=}C(CH_3)CONH_2$

367

11-15   *a.* 152   *b.* 224

11-16 $CH_3OCH_2CH_2C{\equiv}N$

11-17
$$\begin{array}{l} CH_3CHCONH_2 \\ \quad | \\ CH_3CHCONH_2 \end{array}$$

$$\begin{array}{l} CH_3CHCO \\ \quad | \qquad\qquad NH \\ CH_3CHCO \end{array}$$

$$\begin{array}{l} \qquad\quad CH_3 \\ \qquad\quad | \\ CH_3CHCHCOOH \\ \qquad\qquad | \\ \qquad\qquad Y \end{array}$$

          (A)   (B)   (C) Y = NH_2; (D) Y = OH

$$CH_3CH(COOH)_2 \longrightarrow CH_3CH_2COOH + CO_2$$
           (E)

11-18   A, $CH_3NHCH_2CH(OH)CH_3$; B, $CH_3CON(CH_3)CH_2CH(CH_3)OOCCH_3$

11-19   A, $(CH_3CH_2)_2C{=}NOH$; B, $CH_3CH_2CONHCH_2CH_3$; C, CH_3NH_2; D, CH_3CH_2COOH

11-20   X, $EtOOCCH_2COOEt$; Y, $HOOCCH_2COOH$; Z, CH_3CH_2OH

11-21   M, $CH_3OOCCH_2C{\equiv}N$; N, CH_3COOH

11-22   *a.* Sodium hydroxide would cause displacement of Cl too. In B, elimination would occur with the *tert*-halide yielding isobutylene.

    b. In A, H^+ is needed to catalyze reaction. In step B it would be better to introduce Br in the acid rather than in the ester. In C, $LiAlH_4$ would cleave the carbon-bromine bond also.

    c. Formyl chloride does not exist.

    d. Step C is wrong. The amide nitrogen is not electron-rich enough to give an S_N reaction.

    e. Step A is wrong. An acid chloride of an amino acid would react with itself.

    f. Step A is wrong. The C—N bond in amines is not hydrolyzed by acid or alkali.

11-23   *a.* $CO + NaOH \xrightarrow[\Delta]{pressure} HCOONa \xrightarrow{H^+} HCOOH \xrightarrow{EtOH,\ H^+} ans$

    b. Convert 1-propanol to methylacetylene and this in turn to its sodium salt. Treat with CO_2 and then acid to get $CH_3C{\equiv}CCOOH$. Catalytic hydrogenation with Pd gives the *cis*-2-butenoic acid. Esterify with diazomethane. Esterification of this cis acid by refluxing with methanol and a mineral acid would result in the acid-catalyzed conversion of the cis acid to trans.

    c. Treat 1-propanol with HBr to get propyl bromide. Convert to Grignard reagent and add CO_2 to get butyric acid. Convert 1-propanol to 2-propanol by dehydration to the olefin and then treatment with H_2SO_4 followed by hydrolysis. Esterify by refluxing the alcohol and acid together with a small amount of H_2SO_4 catalyst.

    d. Convert ethanol to ethylene oxide and this to ethylene glycol. Esterify this with ketene, acetyl chloride, or acetic acid to get the final diester.

    e. Oxidize ethanol to acetaldehyde. Treat this with ethylmagnesium bromide and hydrolyze to 2-butanol. Convert this to the Grignard reagent and carbonate with CO_2. Esterify by refluxing with ethanol and an acid catalyst.

    f. Oxidize 2-propanol from *c* to acetone. Pinacol reduction of acetone and rearrangement with H_2SO_4 gives $(CH_3)_3CCOCH_3$. Haloform reaction then gives trimethylacetic acid. Convert this to the Ag salt with AgOH and treat this with 2-bromobutane made in *e*.

    g. Run an aldol condensation with acetaldehyde and dehydrate product to crotonaldehyde. Oxidize this to the acid with Tollens' reagent and treat with $SOCl_2$. Treat this acid chloride with NH_3.

    h. Treat propene with NBS to get allyl bromide. Add HBr to this in presence of peroxide; then convert this dibromide to trimethylenediamine via Gabriel's method. Heat the diamine with phosgene.

    i. Convert ethanol to acetamide. Hofmann reaction with this gives methylamine which gives methyl isocyanate when heated with phosgene. Add methyl isocyanate to 2-aminoethanol made from the action of NH_3 on ethylene oxide.

    j. Treat acetone with $(CH_3)_2C{=}P(C_6H_5)_3$ prepared from triphenylphosphine and isopropyl bromide and CH_3Li. Treat the resulting 2,3-dimethyl-2-butene with 4 equiv of NBS.

    k. Prepare isobutylene via a Wittig reaction of acetone and $CH_2{=}P(C_6H_5)_3$. Treat isobutylene with 2 equiv of NBS. Treat resulting dibromide with NaCN.

    l. Reaction of ethylene oxide with CH_3NH_2 gives $CH_3NHCH_2CH_2OH$. When this is treated with 1 mol acetyl chloride, the more-nucleophilic nitrogen reacts faster than the OH group to give the desired monoacetyl derivative.

m. Heating phosphoric acid with ethanol or treating $POCl_3$ with ethanol gives triethyl phosphate.

n. Add HOBr to ethylene and treat resulting bromohydrin with CH_3COCl or $CH_2=C=O$.

o. Treat chloroform with sodium ethoxide.

p. Prepare allyl bromide as in *h.* Convert this to the Grignard reagent and add this to ethyl formate made in *a.*

q. Treat $CH_3C{\equiv}CNa$ made in *b* with CH_3MgI to get the Grignard reagent. React with $H_2C=O$ and dehydrogenate to get $CH_3C{\equiv}CCHO$. Convert to the acetal with CH_3OH; then hydrogenate over Pd.

r. Hydrolysis of NBS gives succinimide which with NaOH gives sodium salt. React this with EtBr.

s. Add HCN to acetone, dehydrate, and hydrolyze to get α-methylacrylic acid. Convert this to the acid chloride and treat with methylamine.

t. Add $HC{\equiv}CMgI$ to acetone. Conver the triple bond in the resulting compound to a ketone by hydration using Hg^{2+} catalyst.

u. Hydrolyze allyl bromide from *h* to allyl alcohol and dehydrogenate to the aldehyde. Condense this with 2-methyl-1-nitropropane in an aldol-type condensation. Treat isobutyl iodide with silver nitrate to make the nitro compound.

v. First reduce the carbonyl group to an OH and dehydrate. Make the quaternary ammonium salt by treatment with excess CH_3I and NaOH. Convert this to the hydroxide with silver hydroxide and heat to cause elimination to diolefin. Repeat this process to get 1,3,5-cyclooctatriene. One mole of Br_2 adds 1,6 to this triolefin to give 1,6-dibromo-2,4-cyclooctadiene. Treat this with excess trimethylamine to get the diquaternary bromide. Convert this to the hydroxide with silver hydroxide and heat to cause double elimination to the tetraene.

w. Treat allyl bromide from *h* with NaCN to get $CH_2=CHCH_2CN(A)$. Hydroborate 2,3-dimethyl-2-butene

made in *j* to get thexylborane $\left(\left|\!\!-\!\!\!+\!\!\!-\!\!\right|\!-BH_2\right)$. Add 1 mol of isobutylene and then 1 mol of A to thexylborane,

followed by carbonylation and alkaline H_2O_2 yielding $(CH_3)_2CHCH_2COCH_2CH_2CH_2CN$. Hydrolysis of this would give the acid followed by reduction of the ketone group to the alcohol using catalytic hydrogenation will give the δ-hydroxy acid. Heating the sodium salt of the acid results in the lactone formation.

11-24 *a.*

$$EtC^{18}OEt \underset{O}{\overset{H^+}{\rightleftharpoons}} Et\overset{+}{C}^{18}OEt \underset{OH}{\overset{H_2O}{\rightleftharpoons}} EtC^{18}OEt \underset{OH}{\rightleftharpoons} EtC\overset{H}{\underset{HO}{-^{18}\overset{+}{O}Et}} \overset{-Et^{18}OH}{\rightleftharpoons}$$

$$Et\overset{+}{C}-OH + Et^{18}OH \overset{-H^+}{\rightleftharpoons} EtCOH$$
$$\underset{OH}{} \qquad\qquad \underset{O}{}$$

b.

$$CH_3\underset{O}{C}OC(CH_3)_3 \overset{H^+}{\rightleftharpoons} CH_3C\underset{O}{-}\overset{H}{\underset{+}{O}}-C(CH_3)_3 \rightleftharpoons CH_3\underset{O}{C}OH + (CH_3)_3C^+ \overset{H_2^{18}O}{\rightleftharpoons}$$
$$\qquad\qquad\qquad\qquad\qquad\qquad\qquad \text{Stable 3° carbonium ion}$$

$$CH_3COOH + (CH_3)_3C^{18}\overset{+}{O}H_2 \overset{-H^+}{\rightleftharpoons} (CH_3)_3C^{18}OH$$

c.

$$CH_3COCH_3 \overset{H^+}{\rightleftharpoons} CH_3\overset{+}{C}OCH_3 \overset{(CH_3)_2CHCH_2CH_2OH}{\rightleftharpoons} CH_3\underset{OH}{C}OCH_3 \rightleftharpoons$$
$$\underset{O}{} \qquad\qquad \underset{OH}{} \qquad\qquad\qquad\qquad\qquad \overset{\overset{+}{H}OCH_2CH_2CH(CH_3)_2}{}$$

$$\underset{CH_3\underset{\underset{OH}{H}}{\overset{|\,+}{C}}OCH_3}{\overset{OCH_2CH_2CH(CH_3)_2}{}} \rightleftharpoons \underset{CH_3\overset{+}{C}}{\overset{OCH_2CH_2CH(CH_3)_2}{\underset{OH}{}}} + CH_3OH \overset{-H^+}{\rightleftharpoons} CH_3COCH_2CH_2CH(CH_3)_2$$
$$\qquad\qquad\qquad\qquad\qquad\qquad\qquad\qquad\qquad\qquad \underset{O}{}$$

d. $(CH_3)_2C=C=O \longleftrightarrow (CH_3)_2C=\overset{+}{C}-\overset{-}{O} \xrightarrow{CH_3OH} (CH_3)_2C=C-O^- \longleftrightarrow$

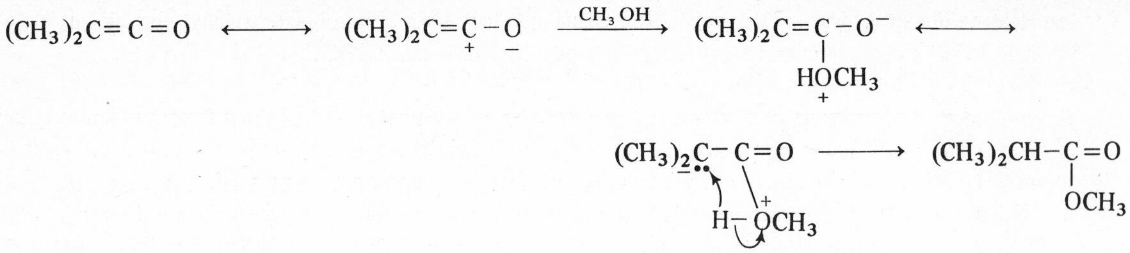

$\overset{|}{\underset{+}{H}OCH_3}$

$(CH_3)_2\overset{\cdot\cdot}{\underset{-}{C}}-C=O \longrightarrow (CH_3)_2CH-C=O$

$H-\overset{+}{O}CH_3 \qquad\qquad OCH_3$

e. (cyclohexanone) $=O + (CH_2-\overset{+}{N}\equiv N \longleftrightarrow CH_2=\overset{+}{N}=\overset{-}{N}) \longrightarrow$ (cyclohexane with) O^-, $CH_2-\overset{+}{N}\equiv N \longrightarrow$

(epoxide) O $CH_2 + N_2$

f. (1-methylcyclobutene) CH_3 ... $H \quad + \quad BH_3 \xrightarrow[\;1\;]{\substack{cis\ add\\B-H}}$ (cyclobutane) CH_3, H, H, BH_2 $\xrightarrow[\substack{repeat\ reaction\ 1\\twice\ more}]{2\ (1\text{-methylcyclobutene})\ CH_3}$ (cyclobutane) CH_3, H, H, BR_2 where R is (cyclobutane) CH_3, H, H

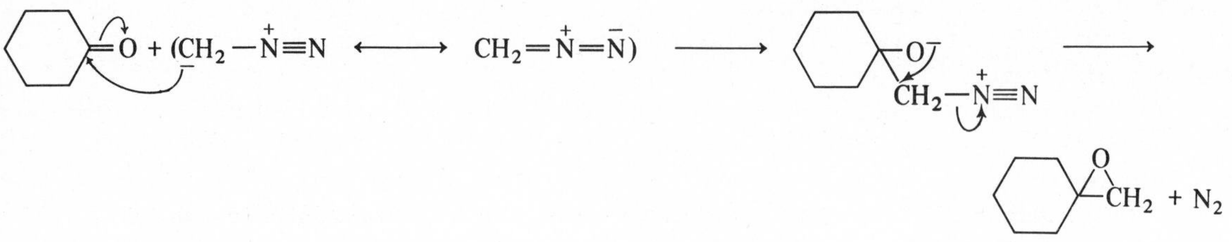

$\xrightarrow[\;2\;]{NH_2OSO_3H}$ (cyclobutane) CH_3, H, H, $\overset{|}{\underset{\underset{^+NH_2-OSO_3H}{|}}{BR_2}}$ $\xrightarrow{\;3\;}$ (cyclobutane) CH_3, H, H, $\overset{+}{NH_2BR_2}$ $\xrightarrow[NaOH]{H_2O}$ (cyclobutane) CH_3, H, H, NH_2

g. $\underset{C_6H_5}{\overset{CH_3}{C}}=N-OH \xrightarrow{\;H^+\;} \underset{C_6H_5}{\overset{CH_3}{C}}=N-\overset{+}{O}H_2 \longrightarrow CH_3\overset{+}{C}=N-C_6H_5 + H_2O \xrightarrow{H_2O}$

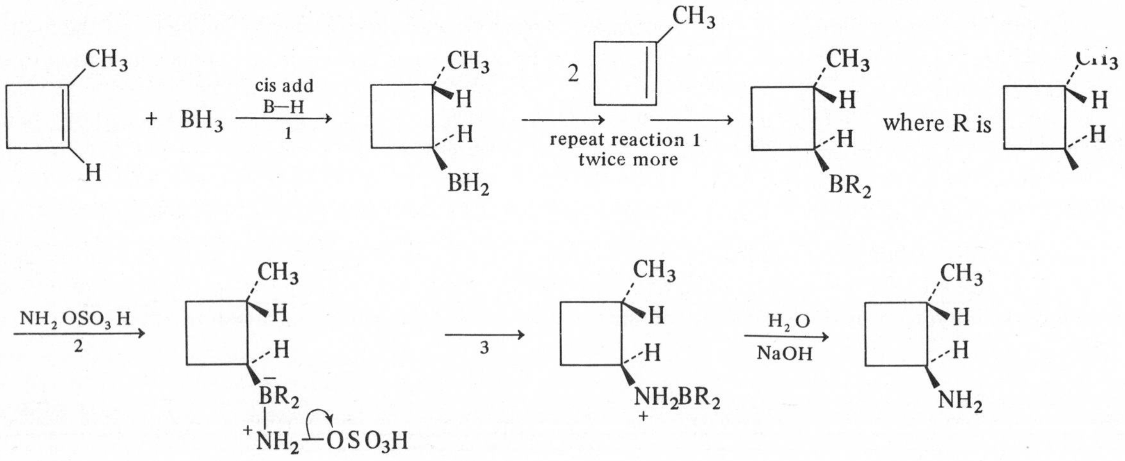

$\underset{^+OH_2}{CH_3C=NC_6H_5} \xrightarrow{-H^+} \underset{OH}{CH_3C=NC_6H_5} \rightleftharpoons \underset{\overset{||}{O}}{CH_3CNHC_6H_5}$

h. (phthalimide: benzene fused ring with) $C=O$, NH, $C=O$ $\xrightarrow[H_2O]{OH^-}$ (benzene with) COH ($C=O$) and CNH_2 ($=O$) $\xrightarrow{NaOBr}$ (benzene with) COH ($C=O$) and $CNHBr$ ($=O$) $\xrightarrow{OH^-}$

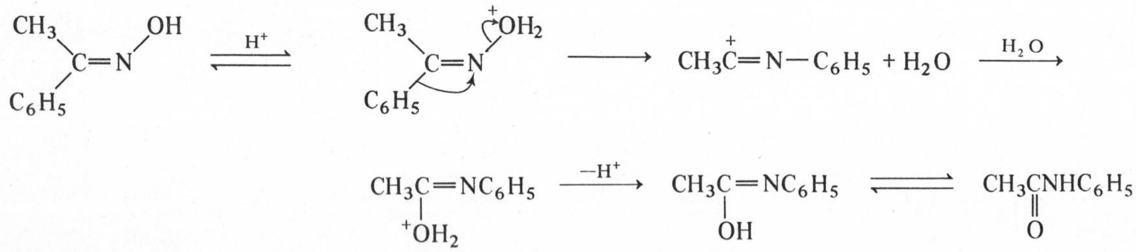

(benzene with) $COOH$ and $C-N-Br$ ($\overset{||}{O}$) $\longrightarrow$ (benzene with) $COOH$ and $N=C=O$ $\xrightarrow{H_2O}$ (benzene with) $COOH$ and NH_2 $+ CO_2$

370

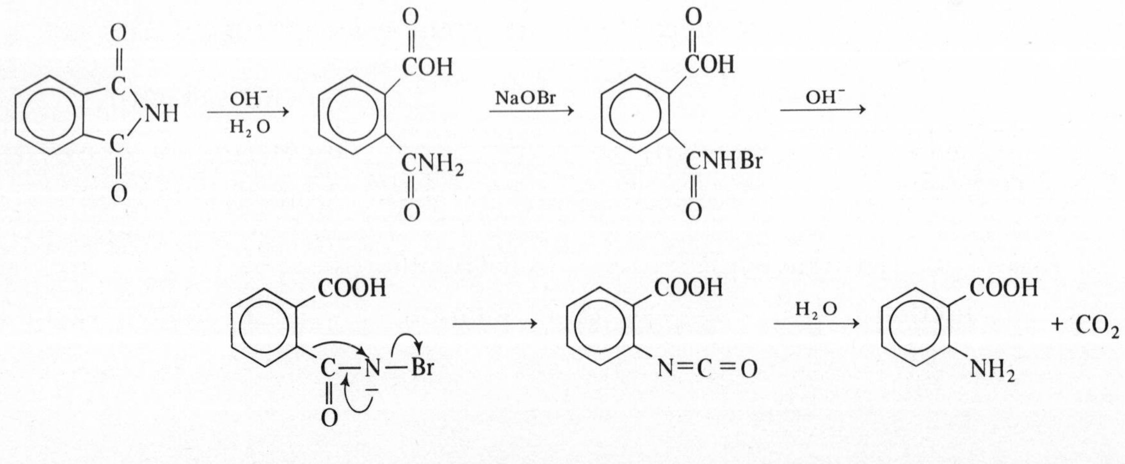

i. $C_6H_5\overset{\displaystyle O}{\underset{\displaystyle \|}{C}}-CH_3 + CF_3\overset{\displaystyle O}{\underset{\displaystyle \|}{C}}\overset{\displaystyle ..}{O}H \longrightarrow CH_3\overset{\displaystyle C_6H_5}{\underset{\displaystyle \underset{\displaystyle :OH}{|}}{C}}-O-\overset{\displaystyle O}{\underset{\displaystyle \|}{O}}CCF_3 \quad \xrightarrow{\;H^+\;}$

$$CH_3\overset{\displaystyle +}{\underset{\displaystyle \overset{\displaystyle \|}{OH}}{C}}-OC_6H_5 + CF_3\overset{\displaystyle O}{\underset{\displaystyle \|}{C}}O^- \rightleftharpoons CH_3\overset{\displaystyle O}{\underset{\displaystyle \|}{C}}OC_6H_5 + CF_3\overset{\displaystyle O}{\underset{\displaystyle \|}{C}}OH$$

11-25

In the trans isomer the hydroxyl group is not able to get near enough to the carbonyl carbon in order to attack.

11-26   In an aldehyde the positive charge on the carbon of the carbonyl group cannot be effectively delocalized by resonance; hence, this atom is more active toward electron-rich reagents. In esters the positive charge can be delocalized more by means of the noncarbonyl oxygen atom.

11-27   Formic, 112; acetic, 104; propionic, 91.9; butyric, 41.9; isobutyric, 33.6; crotonic, 1.3; 2,3-dimethyl-butyric, 1.2; 2-methylpropenoic, 0.8.

11-28   The β carbon is also electron-deficient through resonance and therefore a possible reaction site for an electron-rich group.

11-29   An oil contains many more sites of unsaturation due to carbon-carbon double bonds which make the oil molecule more rigid and decrease the number of ways neighboring molecules can get close enough and have the forces of attraction operate between them. Because these intermolecular attractive forces are weakened the oil molecule has a greater degree of freedom to move about and is a liquid at room temperature.

11-30   *a.* $CH_3O-\overset{\displaystyle O}{\underset{\displaystyle \|}{C}}-OCH_3 \longleftrightarrow CH_3O-\overset{\displaystyle O^-}{\underset{\displaystyle +}{C}}-OCH_3 \longleftrightarrow CH_3\overset{+}{O}=\overset{\displaystyle O^-}{C}-OCH_3 \longleftrightarrow$

$$CH_3O-\overset{\displaystyle O^-}{C}=\overset{+}{O}CH_3$$

b. $CH_2=CH-O-\overset{\displaystyle O}{\underset{\displaystyle \|}{C}}CH_3 \longleftrightarrow CH_2=CH-O-\overset{\displaystyle O^-}{\underset{\displaystyle +}{C}}CH_3 \longleftrightarrow CH_2=CH-\overset{+}{O}=\overset{\displaystyle O^-}{C}CH_3 \longleftrightarrow$

$$\overset{-}{C}H_2-CH=\overset{+}{O}-\overset{\displaystyle O}{\underset{\displaystyle \|}{C}}CH_3 \longleftrightarrow \text{etc.}$$

c. $H_2N-\overset{\displaystyle NH}{\underset{\displaystyle \|}{C}}-NH_2 \longleftrightarrow H_2N-\overset{+}{\underset{\displaystyle \underset{\displaystyle NH^-}{|}}{C}}-NH_2 \longleftrightarrow H_2\overset{+}{N}=\overset{\displaystyle }{\underset{\displaystyle \underset{\displaystyle NH^-}{|}}{C}}-NH_2 \longleftrightarrow H_2N-\overset{\displaystyle }{\underset{\displaystyle \underset{\displaystyle NH^-}{|}}{C}}=\overset{+}{N}H_2$

d. $Br-C\equiv N \longleftrightarrow Br-\overset{+}{C}=N^- \longleftrightarrow \overset{+}{B}r=C=N^-$

e. $CH_3N=C=O \longleftrightarrow CH_3N=\overset{+}{C}-O^- \longleftrightarrow CH_3\overset{+}{N}\equiv C-O^-$

f. $Cl-\overset{\displaystyle O}{\underset{\displaystyle \|}{C}}-Cl \longleftrightarrow Cl-\overset{+}{\underset{\displaystyle \underset{\displaystyle O^-}{|}}{C}}-Cl \longleftrightarrow {}^+Cl=\overset{\displaystyle }{\underset{\displaystyle \underset{\displaystyle O^-}{|}}{C}}-Cl \longleftrightarrow Cl-\overset{\displaystyle }{\underset{\displaystyle \underset{\displaystyle O^-}{|}}{C}}=Cl^+$

g. $CH_3O\overset{\displaystyle O}{\underset{\displaystyle \|}{C}}NH_2 \longleftrightarrow CH_3O-\overset{+}{\underset{\displaystyle \underset{\displaystyle ^-O}{|}}{C}}-NH_2 \longleftrightarrow CH_3\overset{+}{O}=\overset{\displaystyle }{\underset{\displaystyle \underset{\displaystyle ^-O}{|}}{C}}-NH_2 \longleftrightarrow CH_3O-\overset{\displaystyle }{\underset{\displaystyle \underset{\displaystyle ^-O}{|}}{C}}=\overset{+}{N}H_2$

h. $\bar{C}H_2-\overset{+}{N}\equiv N \longleftrightarrow CH_2=\overset{+}{N}=\bar{N}$

i. 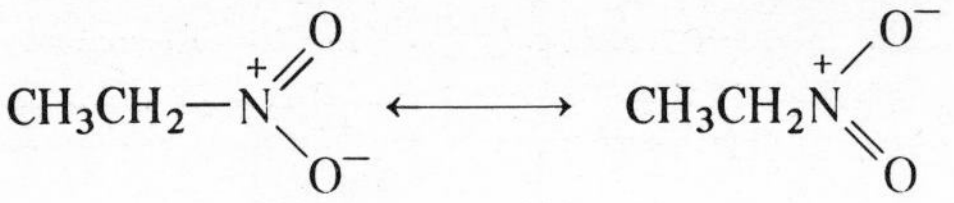 $CH_3CH_2-\overset{+}{N}\big(\!\!=\!O\big)O^- \longleftrightarrow CH_3CH_2\overset{+}{N}\big(O^-\big)\!\!=\!O$

j. 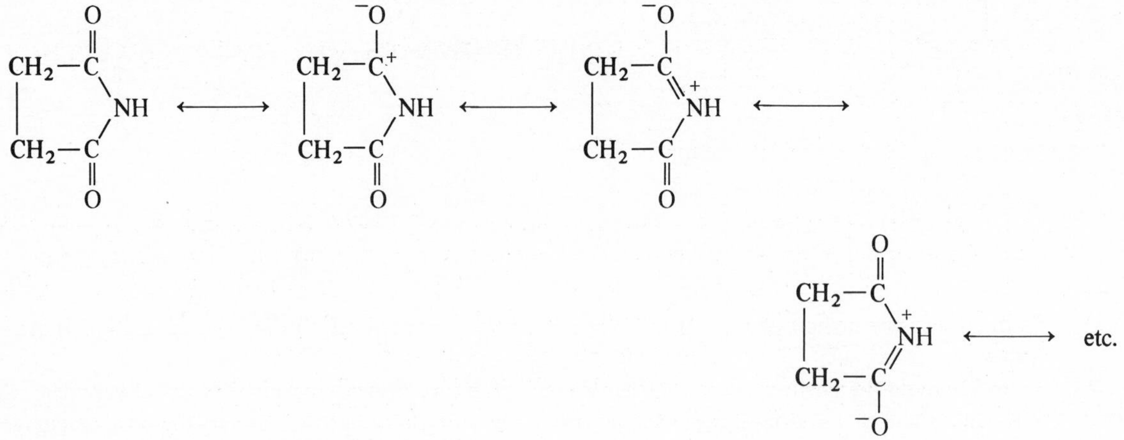 (resonance structures of a cyclic imide)

$$\text{(succinimide)} \longleftrightarrow \longleftrightarrow \longleftrightarrow \longleftrightarrow \text{etc.}$$

k. 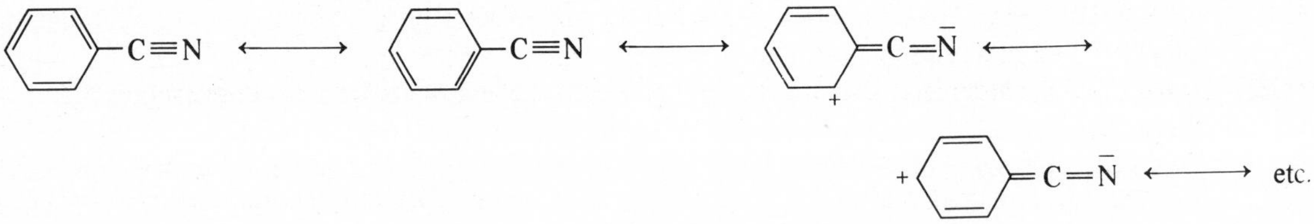 $C_6H_5-C\equiv N \longleftrightarrow C_6H_5-C\equiv N \longleftrightarrow =C=\bar{N} \longleftrightarrow =C=\bar{N} \longleftrightarrow \text{etc.}$

l. 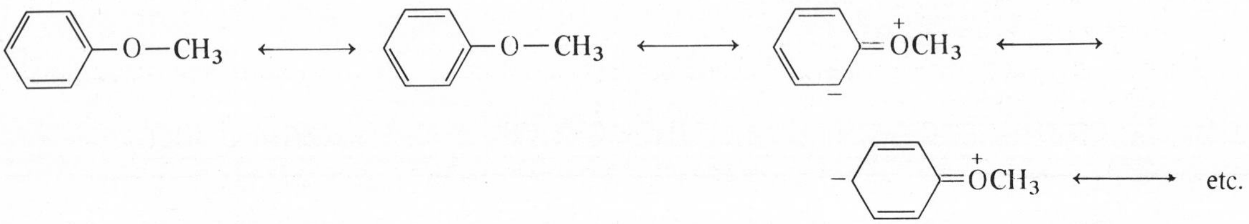 $C_6H_5-O-CH_3 \longleftrightarrow C_6H_5-O-CH_3 \longleftrightarrow =\overset{+}{O}CH_3 \longleftrightarrow =\overset{+}{O}CH_3 \longleftrightarrow \text{etc.}$

11-31 a. Carbon in the methyl group is sp^3, nitrogen is sp^2 and the central carbon is sp. Oxygen could use pure p orbitals or it might be hybridized.

b. All of the carbon atoms are in the sp^2 state. Whether oxygen and nitrogen use p orbitals or hybridized orbitals is uncertain.

c. $N=N=CCOOCH_3$ (with $-$ on first N, $+$ on second N, and an H on the central C)

The central N atom is sp, the carbon in the methyl group is sp^3, the other two carbons are sp^2. Whether oxygen uses p or sp^2 orbitals is uncertain.

d. Carbon is sp^2. Nitrogen and oxygen may use p orbitals or they may be in the sp^2 state.

e. The carbon atoms of the methyl groups are sp^3. Nitrogen and oxygen are also probably sp^3.

11-32 a. In methyl carbonate, resonance in the carbonyl group results in this carbon being quite electron-deficient, hence readily attacked by nucleophilic OH. Also the planar structure of the carbonate offers no hindrance to attack. In the orthocarbonate no partially vacant orbital is present with which the OH can react. Also steric effects hinder the reaction.

b. The nitrogen of the amide is not nucleophilic enough because of the delocalization of its lone pair of electrons to displace the OEt of the ester. The strong base removes a proton from the nitrogen, producing a good nucleophile.

c. Addition of a proton to the =NH portion of guanidine gives a symmetric ion stabilized by three equivalent resonance forms. This is not possible in urea. Also in urea the more-electron-attracting oxygen is less basic per se than nitrogen.

d. Bromine first adds to the double bond. Then nitrogen causes an S_N reaction to give a cyclic quaternary salt. One Br is then in the ionic form and one in the covalent.

e.

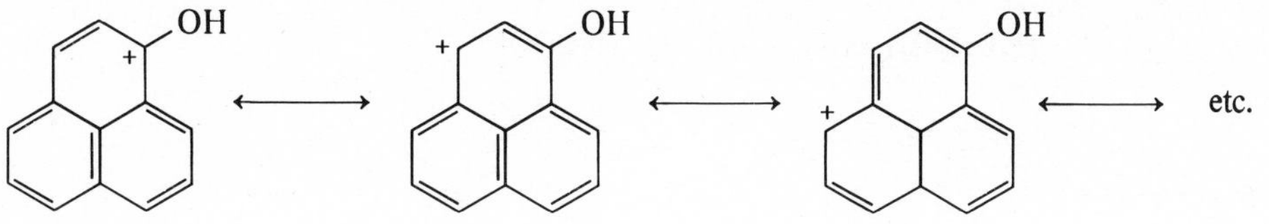

Resulting salt is resonance stabilized

f. The lone pair of electrons in the amine is in an sp^3 orbital, that of the imine in an sp^2 orbital, and that of the nitrile in an sp orbital. The more s character the orbital has, the weaker its bonding strength is.

11-33   *a.* Both   *b.* Both   *c.* Acid   *d.* Both   *e.* Both   *f.* Both   *g.* Acid   *h.* Both

11-34   *a.* Acetamide, *N,N*-dimethylacetamide, ammonia, methylamine, ethylamine.

       b. Acetic acid, succinimide, acetamide, *N,N*-dimethylpropanamide, ammonia.

       c. Dimethyl ether, ammonia, trimethylamine, tetramethylammonium hydroxide.

11-35   *a.* $(CH_3)_2NMgI + CH_3CH(OMgI)CH_3$ *b.* $CH_3N{=}C(OMgI)CH_3$

       c. $(CH_3)_3COMgI$ *d.* $(CH_3)_3COMgI$

11-36   *a.* $CH_3NH_2 + COS$ *b.* $(CH_3NH)_2C{=}S$

       c. $CH_3NHCSSCH_3$ *d* $CH_3NHCSOCH_2CH{=}CH_2$

11-37   *a.* $R{-}\underset{\underset{NH_2}{|}}{C}{=}NH$, amidines   *b.* $R{-}\underset{\underset{OR}{|}}{C}{=}NH$, imido esters

11-38   *a.* Oxidize cyclohexene to adipic acid. Heating the ammonium salt of adipic acid gives the diamide which on dehydration at high temperature gives adiponitrile. Catalytic hydrogenation of the nitrile gives hexamethylenediamine. This diamine plus adipic acid gives a polymeric salt which on heating loses water to yield the polymeric amide nylon.

       b. Reduce adipic acid from *a* to 1,6-hexanediol. Convert hexamethylenediamine to its diisocyanate which when reacted with the diol will give the polymer polyurethane.

11-39 $\underset{\underset{O}{\diagdown\diagup}}{CH_2{-}CH_2} \xrightarrow{CH_3COOH} HOCH_2CH_2OOCCH_3 \xrightarrow{PBr_3} BrCH_2CH_2OOCCH_3 \xrightarrow{(CH_3)_3N}$

$(CH_3)_3\overset{+}{N}CH_2CH_2OOCCH_3 \xrightarrow{dry HCl} ans$

11-40   *a.* $\underset{\underset{C_6H_5}{}}{\overset{\overset{CH_3}{}}{C}}(COOH)_2 \xrightarrow[EtOH]{H^+} \underset{\underset{C_6H_5}{}}{\overset{\overset{CH_3}{}}{C}}(COOEt)_2 \xrightarrow{\overset{O}{\overset{\|}{NH_2CNH_2}}} ans$

       b. $NH_2CH_2COOH \xrightarrow{\Delta}$

           $+$ $O{=}C\underset{\diagdown NHCH_2}{\overset{\diagup CH_2NH}{\diagdown}}C{=}O$

       $HOOCCH_2NH_2 \xleftarrow{OH^-}$

11-41 $CH_3CH_2CH_2OOCCH_2CH_2COCl$

11-42 $O{=}C\underset{\diagdown OCH_2CH_2CH_2CH_2O}{\overset{\diagup CH_2CH{=}CHCH_2}{\diagdown}}C{=}O$ B, $HOCH_2CH_2CH_2CH_2OH$;   C, $HOOCCH_2CH{=}CHCH_2COOH$

11-43

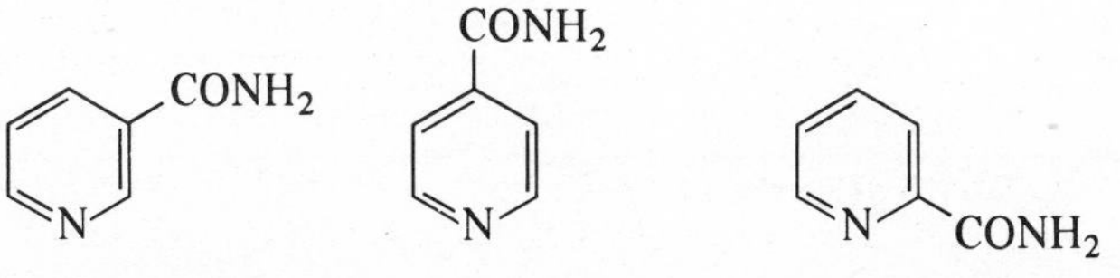

Nicotinamide

11-44 $(H_2N)_2C=\overset{+}{O}HNO_3^-$

11-45 X, $(CH_3)_2CHOOCC\equiv CCOOCH(CH_3)_2$;   Y, $(CH_3)_2CHOH$;   Z, $HOOCC\equiv CCOOH$

11-46 X, $HC\equiv CCH_2COCH_2CH_2OH$

11-47 A, $CH_3CH_2CONH_2$;   B, $CH_3CH_2CH_2NH_2$;   C, CH_3CH_2COOH

11-48 M,

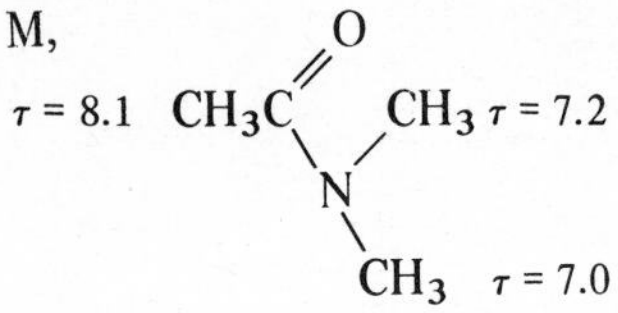

$\tau = 8.1$ CH_3C CH_3 $\tau = 7.2$

CH_3 $\tau = 7.0$

Hindered rotation about the $\overset{O}{\overset{\|}{C}}$–N bond makes the two N-methyl groups nonequivalent and the methyl protons on the methyl group cis to the carbonyl group absorb farther upfield.

11-49 X, 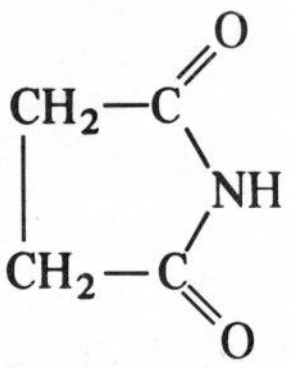   Y, 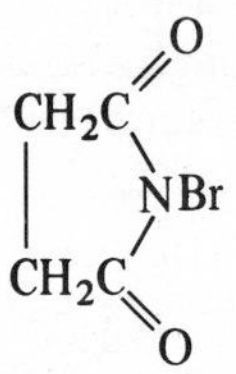   Z, $H_2NCH_2CH_2COOH$

When Z is treated with D_2O the amine and carboxylic hydrogens are exchanged to give $D_2NCH_2CH_2COOD$, which would only show two triplets in its nmr spectrum due to the nonequivalent methylene protons.

11-50 A. 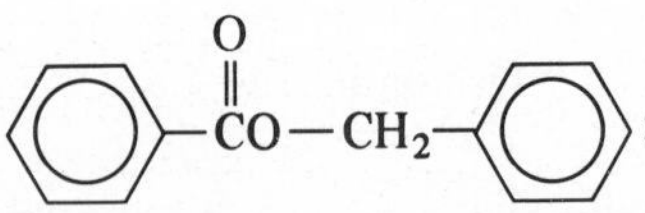 ; B.

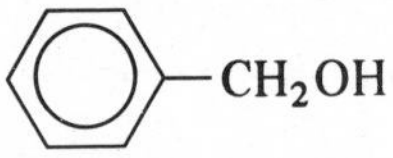

11-51 C, $CH_3CH_2CH_2COOCH_2CH_3$

11-52 D, $CH_3CH(OH)COOCH_2CH_3$

11-53 E, 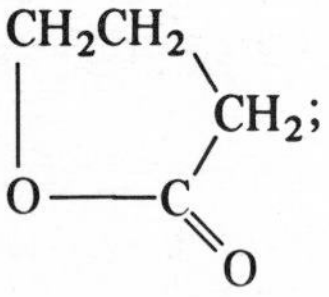 ;   F, $HO(CH_2)_4OH$

11-54 G, *trans*-EtOOCCH=CHCOOEt;   H, *cis*-EtOOCCH=CHCOOEt

Chapter 12

12.1 a. 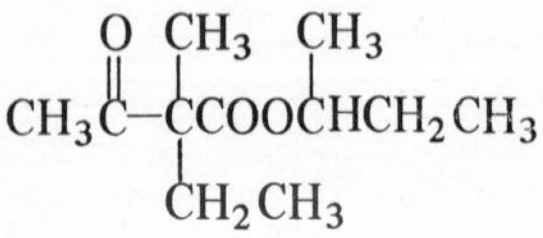   b. $EtCOCH_2CH_2COOH$ c. $N\equiv CCH_2COOH$

d. $CH_3OOCC(CH_3)_2COOCH_3$ e. 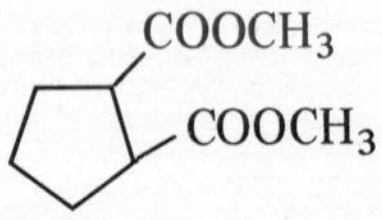   f. $HOOCCH_2COCH_2COOH$

g. $(CH_3)_2NCH(COOCH_3)_2$ h.

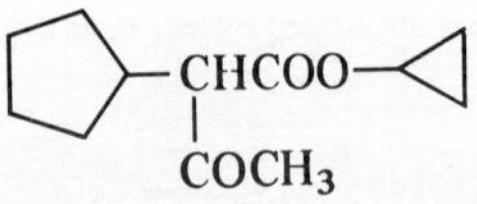

12-2   Acid hydrolysis $\xleftarrow{\text{H}^+,\ \text{H}_2\text{O}}$ S_N reaction product $\xrightarrow[\text{NaOH}]{\text{concd}}$ alkaline hydrolysis products
products

a. $CH_3COCH_2CH_2CH=CH_2$

     $CH_3COCHCOOCH_3$
            |
          $CH_2CH=CH_2$

    $CH_2=CHCH_2CH_2COONa$ and
    $CH_3CH_2CH=CHCOONa$†

b. $CH_3COCH_2CH_2COOH$

     $CH_3COCHCOOCH_3$
            |
          CH_2COOCH_3

    $NaOOCCH_2CH_2COONa$

c. $CH_3COCH_2CH_2COCH_3$

     $CH_3COCHCOOCH_3$
            |
          CH_2COCH_3

    $CH_3COCH_2CH_2COONa$

d. $CH_3COCH_2COCH_2CH_3$

     $CH_3COCHCOOCH_3$
            |
          $COCH_2CH_3$

    $CH_3COONa + CH_3CH_2COONa$

e. $CH_3CO(CH_2)_4COCH_3$

     $CH_3COCHCOOCH_3$
          $(CH_2)_2$
            |
     $CH_3COCHCOOCH_3$

    $NaOOC(CH_2)_4COONa$

f. $CH_3COCH_2CHCH_2COOH$
              |
            COOH

     $CH_3COCHCOOCH_3$
            |
          $CHCOOCH_3$
            |
          CH_2COOCH_3

    $NaOOCCH_2CHCH_2COONa$
               |
             COONa

12-3   a. $EtMeC(COOEt)_2$

b. $CH_3COCH(COOCH_3)_2$

c. $(CH_3)_2CHCH(OH)C(CH_3)_2CHO$

d. $CH_3CH=C(CH_3)CH_2COOEt$

e. $CH_3CH_2COCH(CH_3)COOEt$

f. $CH_3CH(OH)CH_2CH_2OH$

g. $CH_3NHCH(CH_3)CH_2COOEt$

h. $H_2NCOCH_2CONH_2$

i. $(CH_3)_2C(OH)CH(CH_3)COOEt$

j. $C_6H_5CH=C(CH_3)CHO$

k. $C_6H_5CH=CHCOONa$

l. $EtOOCCH_2CH(COOEt)CH_2NO_2$

m.

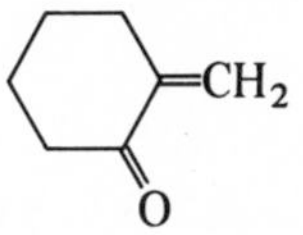

n.

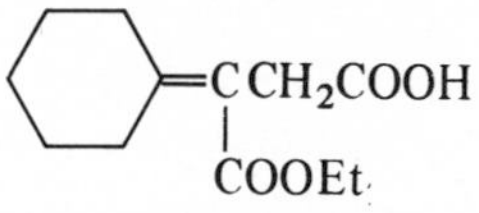

o. $CH_3CH_2CH_2COOEt$

p. $C_6H_5CH=C(COOEt)COCH_3$

q.

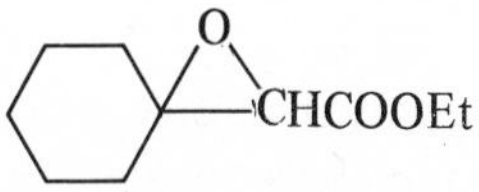

r.

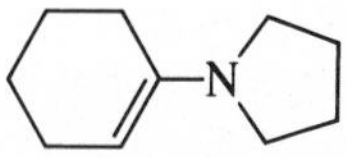

s.

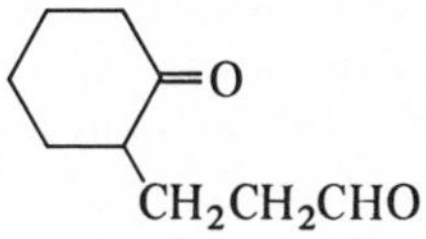

t. $C_6H_5COCH_2CH_2CH(CH_3)_2$

u. $(N\equiv C)_2CHCH(CH_3)CH_2COOEt$

v. $(CH_3OOC)_2CHCH_2CH_2CHO$

w. $CH_3COCHBrCOOEt \xrightarrow{Br_2} CH_3COCBr_2COOEt$

x.

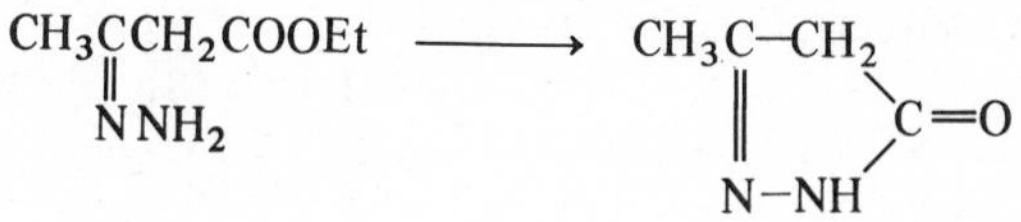

y.

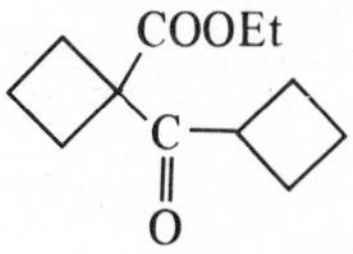

z.

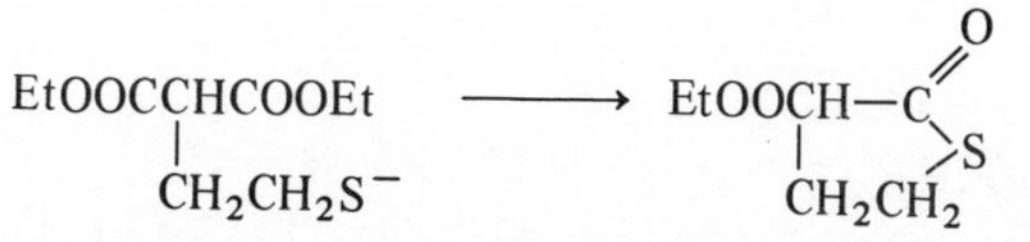

† Heating with alkali causes isomerization of double bond into the conjugated position.

375

$a'.$ $CH_3COCH-C(=O)$ (with O and CH_2CH_2 forming ring)

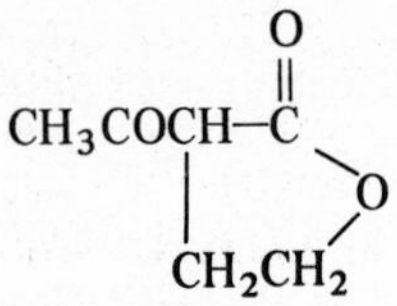

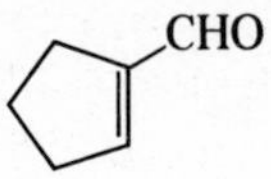

$b'.$ (cyclopentene)–CHO

$c'.$ (cyclohexane with $COOEt$ and $COOEt$)

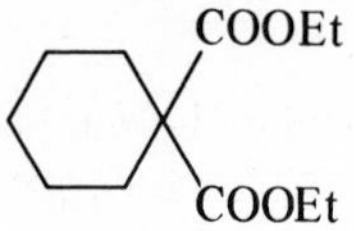

$d'.$ $CH_3COCH(CH_2CH_2CN)_2$

$e'.$ $Cl_3CCH_2CH_2CN$

$f'.$ (cyclopentanone with $CH_2CHCOOEt$, $C=O$, CH_3 group) $\xrightarrow{EtO^-}$ (bicyclic ketone with $COOEt$, O)

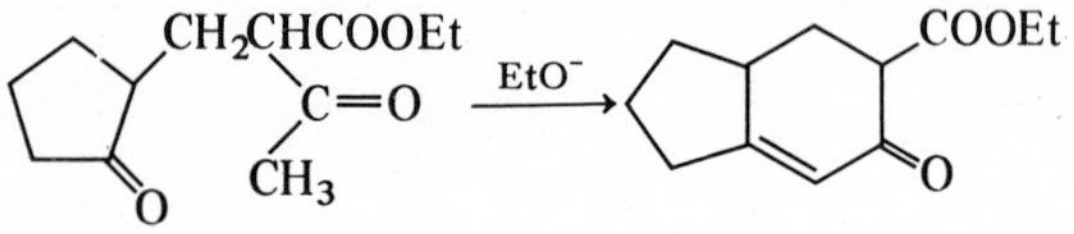

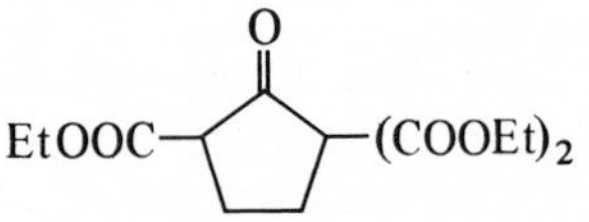

12-4

$a.$ $C_6H_5CH_2CONH_2 + NaOBr$

$b.$ $C_6H_5CH_2$, CH_3CH_2 $C=\ddot{N}$–OH

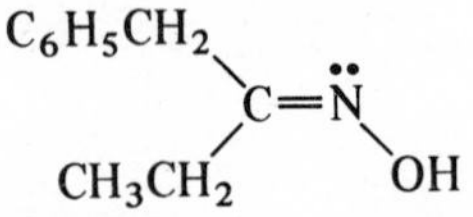

$c.$ $CH_3OOC(CH_2)_4CHO + (C_6H_5)_3P=CHCOOCH_3$

$d.$ $C_6H_5COCH(CH_3)CH_2\overset{+}{N}H_2CH_3$ $\overset{-}{C}l$

$e.$ $CH_3CHCHCH_2COCH_3$ (with C_6H_5 and NO_2 substituents)

$f.$ $EtOOC$–(cyclopentanone)–$(COOEt)_2$

$g.$ $EtOOCCH_2C(OH)CH_2COOEt$ (with $COOEt$) (Reformatsky)

$h.$ $(CH_3)_2C(OH)CH_2COCH_3$ (aldol)

$i.$ $C_6H_5C(CH_3)=C(COOEt)CH_2COOH$

$j.$ $C_6H_5CHO + N\equiv CCH_2COOEt$

$k.$ $C_6H_5CHO + $ concd $NaOH$

$l.$ $CH_3CH_2CHO + BrCH_2COOEt$

$m.$ $CH_3CH_2CH_2COOEt + (EtO)_2C=O$

12-5

$a.$ Acetoacetic ester will give a red coloration with ferric chloride.

$b.$ 2,4-Hexanedione gives a positive ferric chloride test.

$c.$ Only acetoacetic ester has an enol form which gives a ferric chloride test.

$d.$ 2,6-Octanedione gives a precipitate with bisulfite solution. Also, it gives an iodoform test.

$e.$ Ethyl diethylacetoacetate gives positive carbonyl tests, e.g., semicarbazone and phenylhydrazone.

$f.$ Methyl malonate gives a ferric hydroxamate test.

$g.$ Butyl malonate gives a ferric hydroxamate test.

$h.$ On boiling with strong NaOH solution, only the nitrile evolves ammonia.

12-6

$a.$ First extract the octanoic acid with sodium bicarbonate. Then extract with NaOH in which the butyl acetoacetate is acidic enough to form a soluble sodium salt. Extract the tributylamine with dilute acid.

$b.$ Extract the butyl hydrogen malonate with bicarbonate solution. Extract the nitrobutane with sodium hydroxide. Nitro compounds with an α hydrogen are acidic enough to form salts with sodium hydroxide. Then separate the 2-octanone as the bisulfite complex.

12-7

$a.$ CH_3COCH_2COOEt, $C_3H_7COCH(Et)COOEt$, $CH_3COCH(Et)COOEt$, $C_3H_7COCH_2COOEt$

$b.$ $HCOCH_2COOCH_3$, CH_3COCH_2COOEt

$c.$ $EtOOCCOCH(CH_3)COOEt$, $EtCOCH(CH_3)COOEt$

$d.$ $CH_3CH_2CH(CN)COOEt$

$e.$ $CH_3COCH(COOCH_3)_2$, $CH_3OOCCH_2COCH(COOCH_3)_2$, $CH_3COCH_2COOCH_3$, $CH_3OOCCH_2COCH_2COOCH_3$

$f.$ CH_2CH_2, CH_2CH with $C=O$ and $COOEt$

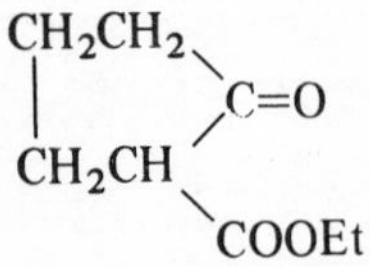

$g.$ $(CH_3)_2CHCOCH_2CN$

$h.$ $CH_3OOCC\equiv CCOCH_2COOCH_3$, $CH_3COCH_2COOCH_3$

12-8 Tertiary halides would tend to give elimination under the basic conditions while vinyl halides are very unreactive in S_N reactions.

12-9   Only 2-aminoheptane and 1-methylaminohexane are soluble in dilute acid. The former will evolve nitrogen when treated with nitrous acid. It will also react with benzenesulfonyl chloride to give an alkali-soluble product. Only ammonium hexanoate will give off the odor of ammonia with cold NaOH. Both valeronitrile and hexanamide will give off ammonia when boiled with NaOH; however, only the amide will evolve nitrogen with nitrous acid. Of the remaining two compounds, only hexyl acetoacetate will give a positive ferric chloride test.

12-10 $(CH_3)_2C(COOCH_3)_2$

12-11 $CH_3COCH_2COCH_3$

12-12   A, $CH_3COCH_2CH(COOH)_2$; B, $CH_3COCH_2CH_2COOH$; C, $HOOCCH_2CH_2COOH$

12-13   *a.*

$$CH_3\underset{\underset{\text{OH}}{|}}{C}=CH\underset{\underset{O}{||}}{C}CH_3 \longleftrightarrow CH_3\overset{+}{\underset{\underset{\text{OH}}{|}}{C}}-CH=\underset{\underset{^-O}{|}}{C}CH_3 \longleftrightarrow CH_3\underset{\underset{^+\text{OH}}{|}}{C}-CH=\underset{\underset{^-O}{|}}{C}CH_3 \longleftrightarrow CH_3\underset{\underset{\text{OH}}{|}}{C}=CH\overset{+}{\underset{\underset{^-O}{|}}{C}}CH_3$$

b.

$$CH_3O\underset{\underset{O}{||}}{C}\overset{-}{C}HC\equiv N \longleftrightarrow CH_3O\underset{\underset{O^-}{|}}{C}=CHC\equiv N \longleftrightarrow CH_3O\underset{\underset{O}{||}}{C}CH=C=\overset{-}{N}$$

Plus normal ester and C≡N resonance forms

12-14   Removal of a proton from nitromethane gives rise to the following anion:

$$^-CH_2-\overset{+}{\underset{\underset{^-O}{|}}{N}}=O \longleftrightarrow CH_2=\overset{+}{\underset{\underset{^-O}{|}}{N}}-O^-$$

Thus alkylation may occur either on carbon or oxygen.

12-15   *a.* Hexane, ethanol, acetoacetic ester, 3-acetyl-2,4-pentanedione, acetic acid.
          b. Cyclopentane, cyclopentadiene, 1-butanol, nitromethane, formic acid, CF_3COOH.

12-16   *a.* $CH_3\underset{\underset{\text{OH}}{|}}{C}=O$, $CH_3\overset{\overset{O}{||}}{C}-OH$, $CH_2=C(OH)_2$ *b.* $CH_3\overset{\overset{O}{||}}{C}-NH_2$, $CH_3\underset{\underset{\text{OH}}{|}}{C}=NH$, $CH_2=\underset{\underset{\text{OH}}{|}}{C}-NH_2$

          c. CH_3-NO_2, $CH_2=\underset{\underset{+}{\overset{\overset{O^-}{|}}{N}}}{}-OH$

          d. $CH_3COCH_2N=O$, $CH_3COCH=N-OH$, $CH_2=\underset{\underset{\text{OH}}{|}}{C}-CH_2N=O$, $CH_3-\underset{\underset{\text{OH}}{|}}{C}=CH-N=O$, $CH_2=\underset{\underset{\text{OH}}{|}}{C}-CH=N-OH$

12-17   *a.* Convert malonic ester to the diethyl derivative by means of EtONa and EtI. Condense this with urea using EtONa as a catalyst. Urea can be made from carbon dioxide and ammonia.
          b. Add $CH_3CH_2CH_2MgX$ to acetaldehyde and convert product to 2-bromopentane with HBr. Alkylate malonic ester with alkyl halide and condense with urea.

12-18   *a.* $CH_3COCH_2COCH_3$, CH_3COCH_2COOEt, $(CH_3)_2\overset{\overset{OH}{|}}{C}CH_2COCH_3$
          b. Isobutyl-$COCH_2COEt$, isobutyl-$\underset{\underset{CH_3}{|}}{C}(OH)CH_2CO$-isobutyl, $EtCO\underset{\underset{CH_3}{|}}{C}HCOOEt$

12-19 $EtCH(COOH)_2$ or $(CH_3)_2C(COOH)_2$

12-20   ▭–$(COOH)_2$ ▭–COOH

12-21   ⬠   or   ◻ (with CH_3)

12-22   *a.* Run a Claisen condensation and then hydrolyze this product with HCl.
          b. Run a Dieckmann condensation to get 2-carboethoxy-1-cyclopentanone. Alkylate on C-2 using EtONa followed by CH_3I. Acid hydrolysis causes simultaneous decarboxylation. Reduce ketone with Pt and H_2 to product.

c. Esterify oxalic acid. A Claisen condensation of diethyl succinate with ethyl oxalate followed by acid hydrolysis with the loss of one carboxyl group will give product.

d. Condense malononitrile with methyl formate and add HCN to the resulting aldehyde.

e. Run a Claisen condensation with ethyl butyrate. With the resulting product run a reductive alkylation using methylamine. Hydrolyze the ester to acid.

f. Hydrate propene to 2-propanol and oxidize to acetone. Condense acetone with ethyl oxalate (part c) using Claisen conditions.

g. Run an aldol condensation with CH_3CHO and oxidize with Tollens' reagent.

h. Esterify pimelic acid and run a Dieckmann reaction on the diester. Acid hydrolysis of this product cleaves the ester with concurrent decarboxylation to give cyclohexanone. Reduce this, dehydrate and oxidize with cold, dil $KMnO_4$.

i. Michael reaction of $CH_2=CHCHO$ and CH_3NO_2 followed by reductive alkylation with large excess of NH_3.

j. Michael reaction of $CH_2=CHCN$ with CH_3NO_2.

k. Esterify acetic acid and run Claisen condensation to obtain CH_3COCH_2COOEt. Treat this with EtONa and then CH_3COCl and hydrolyze with acid to $CH_3COCH_2COCH_3$. Add this to $CH_2=CHCN$ in a Michael reaction.

l. Condense acetone with HCHO and dehydrate.

m. Add HBr to propene and convert product to Grignard reagent. Treat this with CO_2 and esterify to $(CH_3)_2CHCOOEt$. Claisen condensation of this using $(C_6H_5)_3C^-K^+$ as a base followed by heating with H_2NNH_2 gives product.

n. Run Dieckmann reaction with ethyl adipate; alkylate 2 equiv of this with 1 equiv of $BrCH_2CH_2Br$ using EtONa. Acid hydrolysis and concurrent decarboxylation gives product.

o. Alkylate CH_3COCH_2COOEt with 2 equiv of CH_3COCl. Acid hydrolysis gives $(CH_3CO)_3CH$. Make Na salt of this with EtONa and alkylate with CH_3I.

p. A Claisen condensation between ethyl butyrate and EtOOCCOOEt followed by acid hydrolysis gives the product.

q. Treat acetone with $BrZnCH_2COOCH_3$ to get a β-hydroxy ester which upon $LiAlH_4$ reduction gives the product.

r. Esterify α-chloropropionic acid. Condense this ester with 2-pentanone using $(CH_3)_3COK$ (how would you prepare this?). The resulting glycidic ester upon saponification and heating loses CO_2 to give the product.

s. A Claisen condensation of CH_3CH_2COOEt followed by alkylation with allyl chloride using EtONa as a base gives

$$CH_3CH_2COC(CH_3)COOEt.$$ This, upon acid hydrolysis and loss of CO_2, gives the product
$$\overset{|}{C}H_2CH=CH_2$$

t. Prepare an enamine of butanal and react with $CH_2=CHCOOEt$ to obtain the product after acid hydrolysis.

u. Condense the given keto ester with OCHCOCHO.

v.

w. Treat 2-bromo-1-cyclohexanone with $(CH_3CH_2CH_2)_3B$ in the presence of $(CH_3)_3COK$ to get 2-propyl-1-cyclohexanone. This in a Wittig reaction with $(C_6H_5)_3P=CH_2$ gives the product.

378

x.

(see part *w*)

$\xrightarrow{\text{NaNH}_2}$ $\xrightarrow[\text{(Sec. 12-2}c)]{\text{CH}_3\text{COCH}_2\text{CH}_2\overset{+}{\text{NHR}}_2\ \overset{-}{\text{Cl}}}$ $\left[\ \text{structure}\ \right]$ $\longrightarrow$ product

12-23

a. $\text{C}_6\text{H}_5\text{CHO} + \text{CH}_3\text{CHO} \xrightarrow{\text{NaOEt}}$ product

b. $\text{C}_6\text{H}_5\text{CHO} + \text{CH}_2(\text{COOEt})_2 \xrightarrow{\text{R}_2\text{NH}} \xrightarrow{-\text{H}_2\text{O}} \text{C}_6\text{H}_5\text{CH}=\text{C}(\text{COOEt})_2 \xrightarrow[(-\text{CO}_2)]{\text{H}^+,\text{H}_2\text{O}}$ product

c. $\text{C}_6\text{H}_5\text{CHO} \xrightarrow{\text{BrZnCH}_2\text{COOEt}} \text{C}_6\text{H}_5\text{CH(OH)CH}_2\text{COOEt} \xrightarrow{\text{dehydrate}} \xrightarrow{\text{hydrolyze}}$ product

d. $\text{C}_6\text{H}_5\text{CHO} + (\text{CH}_3\text{CO})_2\text{O} \xrightarrow{\text{CH}_3\text{COONa},\ \Delta}$ product

e. $\text{C}_6\text{H}_5\text{CHO} + (\text{C}_6\text{H}_5)_3\text{P}=\text{CHCOOEt} \xrightarrow{\text{H}^+,\ \text{H}_2\text{O}}$ product

f. $\text{C}_6\text{H}_5\text{CHO} \xrightarrow{[\text{O}]} \text{C}_6\text{H}_5\text{COOH} \xrightarrow{\text{esterify}} \xrightarrow[\text{EtONa}]{\text{CH}_3\text{COOEt}} \text{C}_6\text{H}_5\text{COCH}_2\text{COOEt} \xrightarrow{\text{H}_2,\ \text{Pt}} \xrightarrow{\text{dehydrate}}$ product

12-24

a. $\text{CH}_3\text{CO}\overset{-}{\text{C}}\text{HCOOEt} \xrightarrow{\text{I—I}} \text{CH}_3\text{COCHICOOEt} + \text{I}^-$

$\text{CH}_3\text{CO}\overset{-}{\text{C}}\text{HCOOEt} + \text{CH}_3\text{COCHCOOEt} \longrightarrow \begin{array}{c}\text{CH}_3\text{COCHCOOEt}\\|\\\text{CH}_3\text{COCHCOOEt}\end{array} \xrightarrow[\Delta]{\text{H}_3\text{O}^+}$ product

b. $\text{CH}_2(\text{COOEt})_2 \xrightarrow{\text{NaOEt}} \xrightarrow{\text{EtI}} \text{EtCH(COOEt)}_2 \xrightarrow{\text{NaOEt}} \xrightarrow{\text{I}_2} \xrightarrow[\Delta]{\text{H}_3\text{O}^+}$ product

12-25

a. $\xrightarrow{\text{OH}^-}$ $\rightleftharpoons$ $\rightleftharpoons \text{NaOOC(CH}_2)_4\text{COOEt}$

b. $\overset{-}{\text{C}}\text{H(CN)}_2 + \text{H}_2\text{C}=\text{O} \rightleftharpoons (\text{NC})_2\text{CHCH}_2-\text{O}^- \longrightarrow (\text{NC})_2\text{CHCH}_2\text{OH} \xrightarrow{-\text{H}_2\text{O}}$

$[(\text{NC})_2\text{C}=\text{CH}_2] \xrightarrow{\overset{-}{\text{C}}\text{H(CN)}_2} (\text{NC})_2\overset{-}{\text{C}}-\text{CH}_2\text{CH(CN)}_2 \longrightarrow$ product

c.

$+ \text{CH}_3\text{CH}=\text{CHCOCH}_3 \xrightarrow{\text{Michael}} \xrightarrow{\text{H}^+}$

$\xrightarrow[\text{aldol}]{^-\text{OEt}} \xrightarrow{\text{H}^+}$ $\xrightarrow{-\text{H}_2\text{O}}$ product

d. $\text{CH}_2=\text{CHCH}_2\text{COOEt} + \text{EtO}^- \rightleftharpoons [\text{CH}_2=\text{CH}\overset{-}{\text{C}}\text{HCOOEt} \longleftrightarrow \overset{-}{\text{C}}\text{H}_2-\text{CH}=\text{CHCOOEt}]$
$\longrightarrow \text{CH}_3\text{CH}=\text{CHCOOEt} + \text{EtO}^-$

379

e. $CH_3CH=CHCH=CHCOOEt \longleftrightarrow CH_3\overset{+}{C}HCH=CHCH=\overset{O^-}{\overset{|}{C}}OEt \xrightarrow{\bar{C}H(COOEt)_2}$

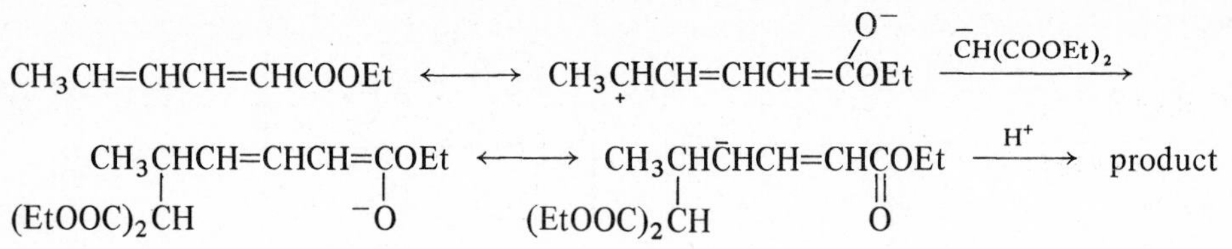

$\underset{(EtOOC)_2\overset{|}{C}H}{CH_3\overset{|}{C}HCH=CHCH=}\overset{O^-}{\overset{\|}{C}}OEt \longleftrightarrow \underset{(EtOOC)_2\overset{|}{C}H}{CH_3\overset{|}{C}H\bar{C}HCH=CHCOOEt} \xrightarrow{H^+} product$

f. $CH_3COCH_3 \underset{slow}{\overset{OH^-}{\rightleftharpoons}} CH_3CO\bar{C}H_2 \longleftrightarrow \underset{O^-}{CH_3\overset{|}{C}=CH_2} \xrightarrow{Br_2} \underset{O}{CH_3\overset{\|}{C}CH_2Br} + Br^-$

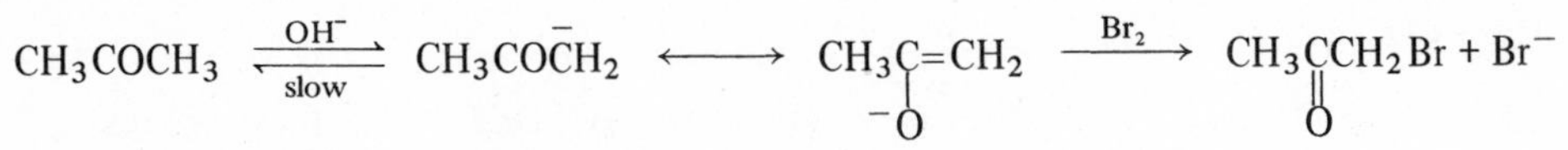

g. $CH_3COCH_3 \overset{H^+}{\rightleftharpoons} \underset{\overset{+}{O}H}{CH_3\overset{\|}{C}-CH_2-H} \underset{slow}{\xrightarrow{H_2O}} \underset{OH}{CH_3\overset{|}{C}=CH_2} + H_3O^+ + \xrightarrow{Br_2}$

$\underset{\overset{+}{O}H}{CH_3\overset{\|}{C}CH_2Br} + Br^- \xrightarrow{-H^+} CH_3COCH_2Br$

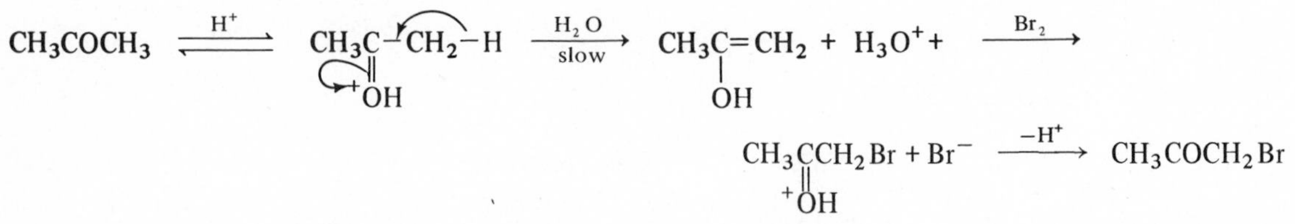

h. $ClCH_2CH_2CH_2CN \xrightarrow{OH^-} ClCH_2CH_2\bar{C}HCN \longrightarrow \underset{CH_2}{\overset{CH_2—CHCN}{|\quad\quad\;}} \xrightarrow[OH^-]{H_2O} \underset{CH_2}{\overset{CH_2—CHCOOH}{|\quad\quad\quad}}$

i. [cyclohexanone] $\xrightarrow{EtONa}$ [cyclohexanone anion] $\longleftrightarrow$ [cyclohexene-O^-] $\xrightarrow{O=C(OEt)_2}$

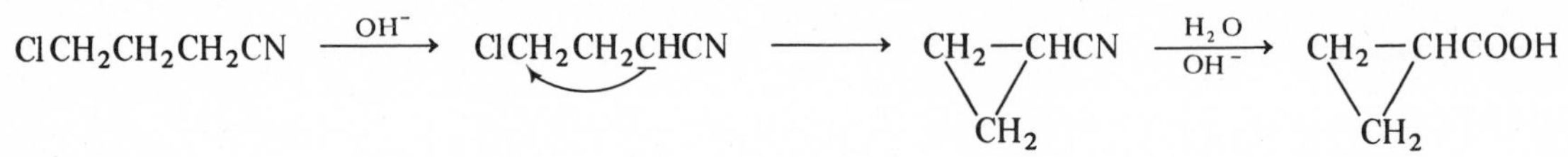

[cyclohexenyl-OCOEt] + [cyclohexanone-COOEt]

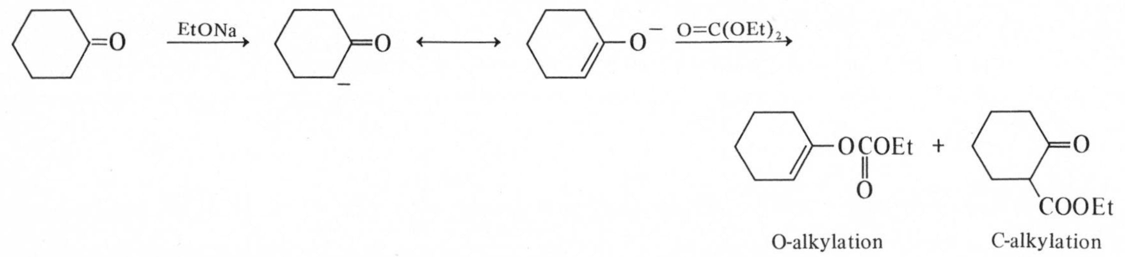

O-alkylation   C-alkylation

j. $BrCH_2CH=CHCOOEt \xrightarrow{Zn} Br\overset{+}{Z}n\bar{C}H_2CH=CHCOOEt \xrightarrow{(CH_3)_2CO}$

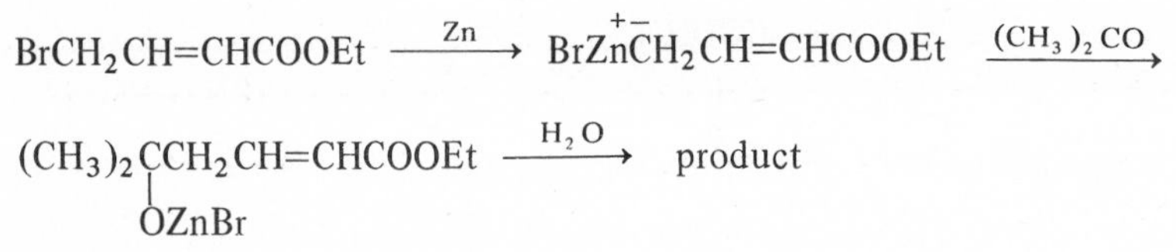

$\underset{OZnBr}{(CH_3)_2\overset{|}{C}CH_2CH=CHCOOEt} \xrightarrow{H_2O} product$

k. $HOCH_2C(CH_3)_2COOH \xrightarrow{H^+} [\overset{+}{C}H_2C(CH_3)_2COOH \longrightarrow CH_3CH_2\overset{+}{C}(CH_3)COOH] \xrightarrow{-H^+}$

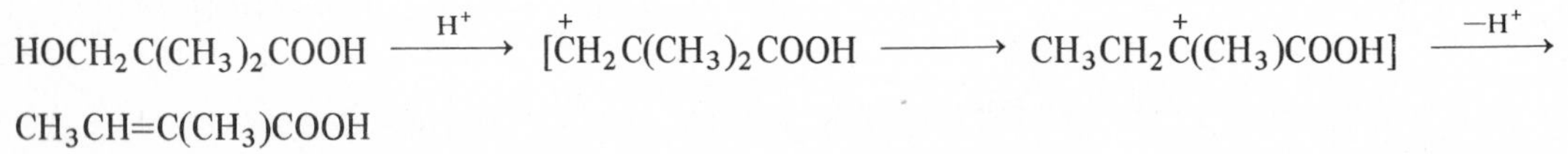

$CH_3CH=C(CH_3)COOH$

l. Tartaric acid $\xrightarrow[-H_2O]{H^+} [\underset{OH}{HOOCCH\overset{|}{C}HCOOH}]^+ \longrightarrow \underset{O}{HOOCCH_2\overset{\|}{C}COOH} \xrightarrow{-CO_2}$ pyruvic acid

m. $(CH_3)_2C=O + EtOOC\bar{C}HCH_2COOEt \xrightarrow[Stobbe]{} (CH_3)_2\overset{HO}{\overset{|}{C}}CH(COOEt)CH_2\overset{O}{\overset{\|}{C}}-OEt \longrightarrow$ product

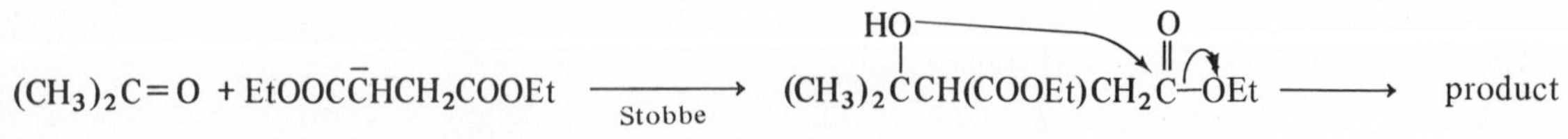

n. $CH_2=CBrCOOEt + \bar{C}H(COOEt)_2 \longrightarrow \underset{(EtOOC)_2\overset{|}{\underset{H}{C}}}{CH_2\overset{Br}{\overset{\curvearrowleft}{C}}HCOOEt} \xrightarrow{\bar{O}Et} \xrightarrow[-CO_2]{H_3O^+} product$

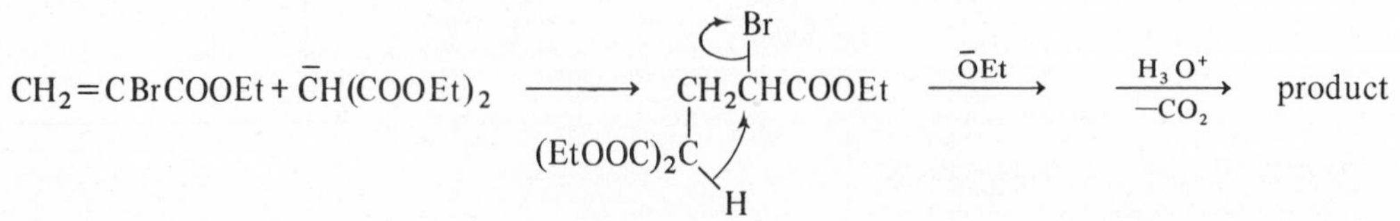

o. $CH_3CH=CHCHO \rightleftharpoons \underset{H-O}{CH_3\overset{|}{C}H-CH_2CHO} \underset{\substack{reverse \\ aldol}}{\rightleftharpoons} \underset{O}{CH_3\overset{\|}{C}H} + \bar{C}H_2CHO \longrightarrow 2CH_3CHO$

380

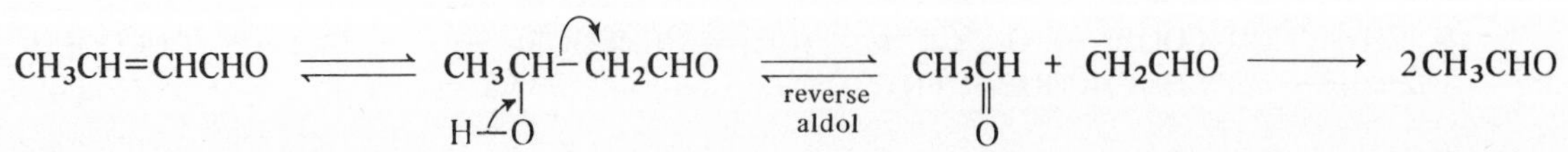

p. $CH_2=CHCOCH$ + (2-methylcyclohexanone) $\xrightarrow{-H_2O}$ (structure with CH₃, H, O, COOEt) $\xrightarrow[\text{Michael}]{^-OEt}$ $\xrightarrow{H^+}$ product

with $COOEt$ on the left reagent

q. $\xrightarrow[\longleftarrow]{^-OEt}$ (structure: O, OEt, CH₃, COOEt) $\longrightarrow$ (structure: COOEt, CH₂, CH₃, COOEt) $\underset{\text{proton transfer}}{\rightleftharpoons}$ (structure: COOEt, CH, CH₃, C—OEt, O, H) $\longrightarrow$ product

r. $C_6H_5CH=CHCOCH=CHC_6H_5$ + $^-CH(CN)COOMe$ $\xrightarrow{\text{Michael}}$

(cyclic structure: O, C, H₂C, CH, CH, CHC₆H₅, C₆H₅, C, H, NC, COOMe) $\xrightarrow[\text{Michael}]{^-OMe}$ $\xrightarrow{H^+}$ product

s. $\longrightarrow$ (morpholine-cyclopentene structure, O, N) + :CCl_2 $\longrightarrow$ (structure O, N, Cl, Cl) $\longrightarrow$ (structure O, N⁺, Cl) $\xrightarrow{H_2O}$ product

$\underset{CHCl_3}{(CH_3)_3COK}$

t. (2-methylcyclohexanone anion, CH₃, O) + $HC\equiv CCCH_3$ (with O) $\xrightarrow{\text{Michael}}$ (structure: CH₃, CH, CH, C=O, O, CH₂, H) $\xrightarrow[-H_2O]{\text{aldol} \atop Na}$ product

12-26 $CH_3COCH_2COCH_2CH_3$
12-27 $N\equiv CCH_2COOCH_2CH_3$
12-28   B, CH_2CH_2 ; C, $HOCH_2CH_2OH$

(structure with O, O, C, O)

12-29 $CH_3CCH_2CCH_3$ (with two O, $\delta = 3.63$, $\delta = 2.0$, 15%) $\rightleftharpoons$ $CH_3C=CHCCH_3$ (with HO, O, $\delta = 1.83$, $\delta = 5.57$, $\delta = 15.2$, 85%)

12-30   M, $CH_2(COOEt)_2$;   N, $C_6H_5CH_2Br$;   O, $C_6H_5CH_2CH(COOEt)_2$;   P, $C_6H_5CH_2CH_2COOH$
12-31 $CH_3CH_2CH(COOEt)_2$

13-1

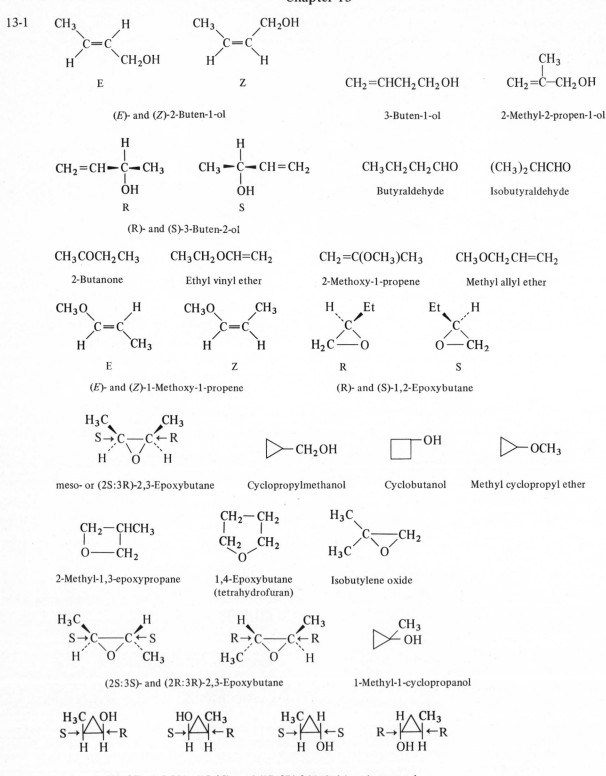

13-2   *a.* R and S forms both optically active.

    b. Only one optically inactive structure.

    c. Cis and trans forms both optically inactive.

    d. R and S forms both optically active.

    e. Two sets of *dl* pairs (four isomers) all optically active.

    f. Only one structure with no optical activity.

    g. Two isomers (syn and anti) both optically inactive.

    h. Three possible isomers all optically inactive: cis-cis (Z,Z), trans-trans (E,E), and cis-trans (Z,E). There is only one possible cis-trans isomer (see answer to Prob. 4-17e). 2,4-Heptadiene, on the other hand, can exist as four possible isomers all optically inactive.

    i. Four isomers all optically active: cis-R, cis-S, trans-R, trans-S.

    j. Three isomers possible: RR and SS forms both optically active and the optically inactive meso form (RS).

    k. Eight possible isomers all optically inactive: trans-trans-trans (E,E,E); trans-trans-cis (E,E,Z); trans-cis-trans (E,Z,E); cis-trans-trans (Z,E,E); trans-cis-cis (E,Z,Z); cis-trans-cis (Z,E,Z); cis-cis-trans (Z,Z,E); and cis-cis-cis (Z,Z,Z). 2,4,6-Octatriene on the other hand can exist only as six optically inactive isomers (the E,Z,Z and Z,Z,E forms above are the same in this case).

    l. Four optically active forms (two *dl* pairs).

    m. Four isomers possible: R,R(I) and S,S(II) forms both optically active; SsR(III) and SrR(IV) forms are optically inactive (meso). I and II constitute a *dl* pair in which C-2 has only one possible configuration since it has two like groups attached to it. In compounds III and IV there is a plane of symmetry through carbon-2 which is thus a pseudoasymmetric carbon in both compounds and its configuration is designated by small s and r, respectively, based on the sequence rule: R precedes S (see Sec. 13-2b-3).

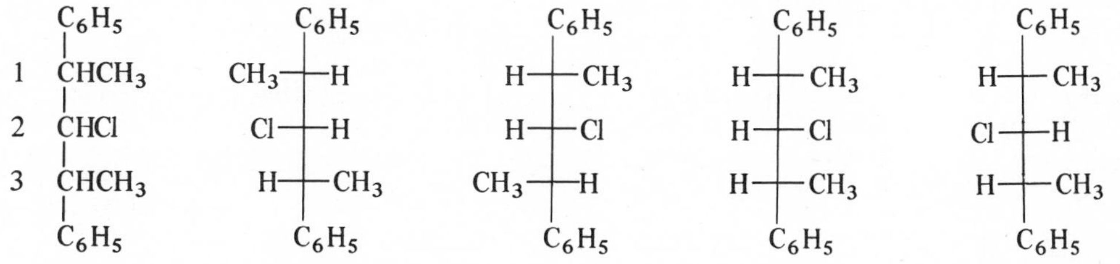

         I   II   III   IV

    n. Eight optically active forms (4 *dl* pairs).

    o-q. Two optically active forms in each case. All are examples of molecular asymmetry (see Sec. 13-4).

    r. Two isomers $(Z$ and $E)$ both optically inactive.

    s. Four optically active forms (two *dl* pairs):

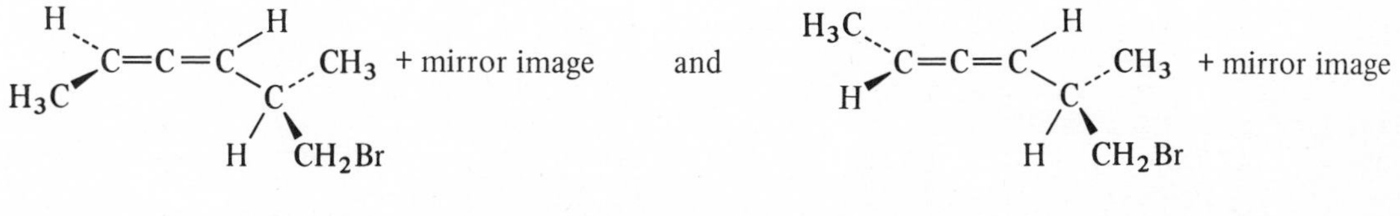

    t. See *o, p,* and *q* above.

    u. Three isomers possible: Two optically active (I, II) and a meso, optically inactive (III).

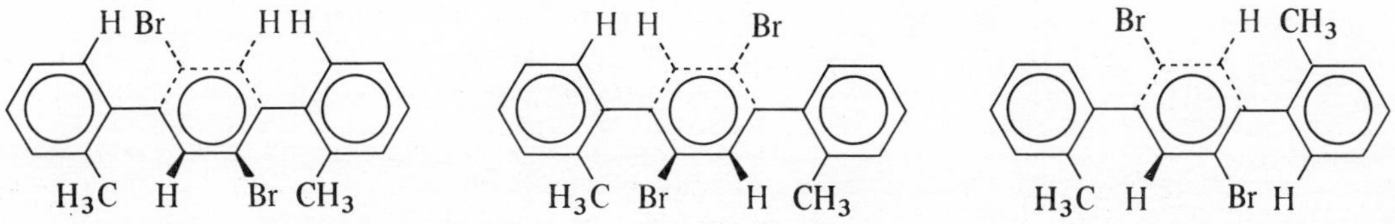

        I   II   III

v. Eight optically active isomers (four *dl* pairs):

A = —C≡C—C≡CH   and   B = —CH₂COOH

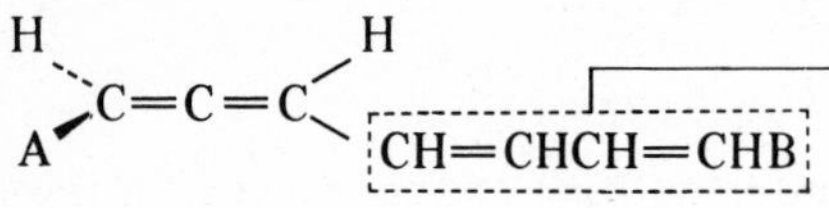

$$\begin{array}{c} H \\ \diagdown \\ A \end{array} C=C=C \begin{array}{c} H \\ \diagup \\ \diagdown \end{array} \boxed{CH=CHCH=CHB} \longrightarrow$$

cis-cis(Z, Z)	plus
cis-trans(Z, E)	mirror
trans-cis(E, Z)	images
trans-trans(E, Z)	

13-3 *a.*

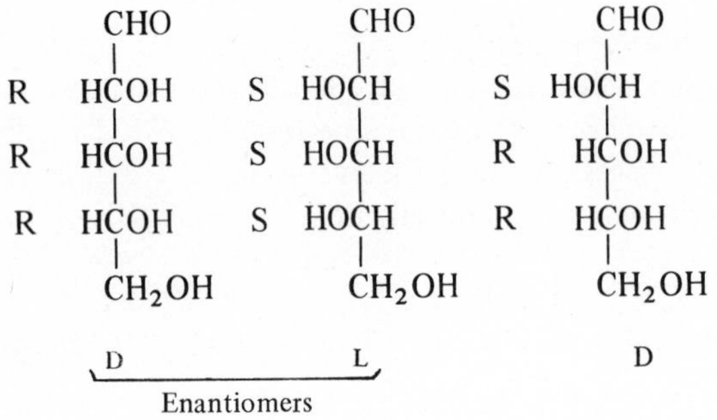

```
        CHO              CHO              CHO
  R   HCOH        S   HOCH        S   HOCH
  R   HCOH        S   HOCH        R   HCOH
  R   HCOH        S   HOCH        R   HCOH
       CH₂OH           CH₂OH           CH₂OH
       └── D ──────── L ──┘            D
           Enantiomers
```

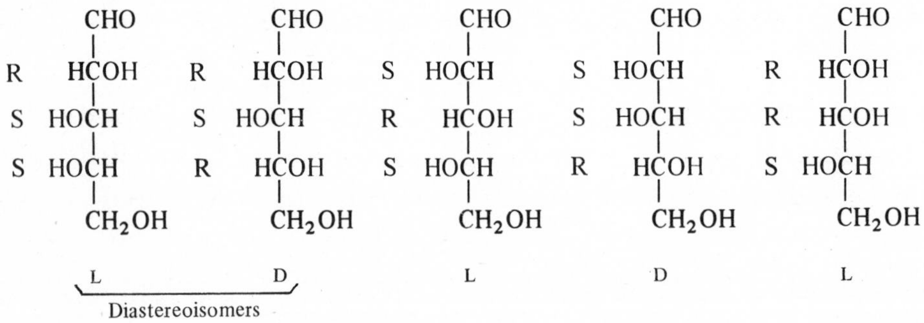

```
        CHO            CHO            CHO            CHO            CHO
  R   HCOH      R   HCOH      S   HOCH      S   HOCH      R   HCOH
  S   HOCH      S   HOCH      R   HCOH      S   HOCH      R   HCOH
  S   HOCH      R   HCOH      S   HOCH      R   HCOH      S   HOCH
       CH₂OH          CH₂OH          CH₂OH          CH₂OH          CH₂OH
       └── L ──────── D ──┘          L              D              L
            Diastereoisomers
```

b.

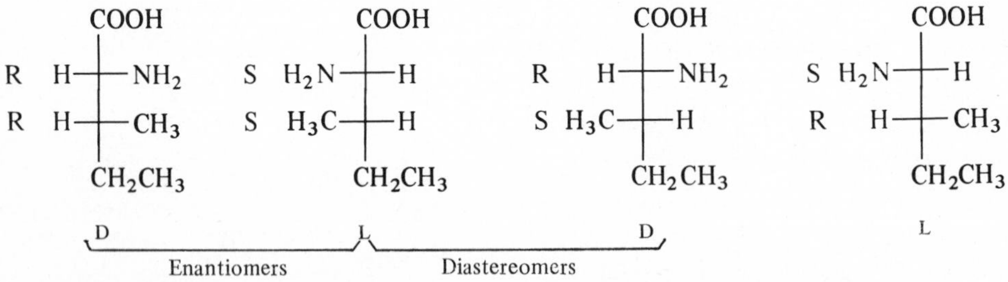

```
        COOH              COOH              COOH              COOH
  R  H──NH₂        S  H₂N──H        R  H──NH₂        S  H₂N──H
  R  H──CH₃        S  H₃C──H        S  H₃C──H        R  H──CH₃
      CH₂CH₃            CH₂CH₃            CH₂CH₃            CH₂CH₃
      └── D ──────── L ─┘└────── D ───────┘            L
          Enantiomers        Diastereomers
```

c.

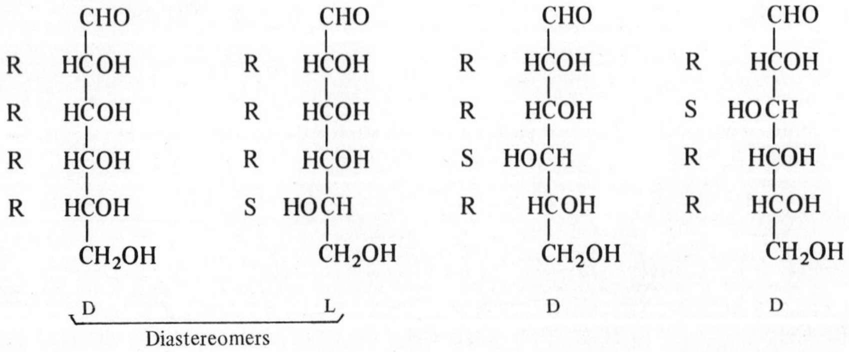

```
        CHO              CHO              CHO              CHO
  R   HCOH        R   HCOH        R   HCOH        R   HCOH
  R   HCOH        R   HCOH        R   HCOH        S   HOCH
  R   HCOH        R   HCOH        S   HOCH        R   HCOH
  R   HCOH        S   HOCH        R   HCOH        R   HCOH
       CH₂OH           CH₂OH           CH₂OH           CH₂OH
       └── D ──────── L ──┘            D              D
           Diastereomers
```

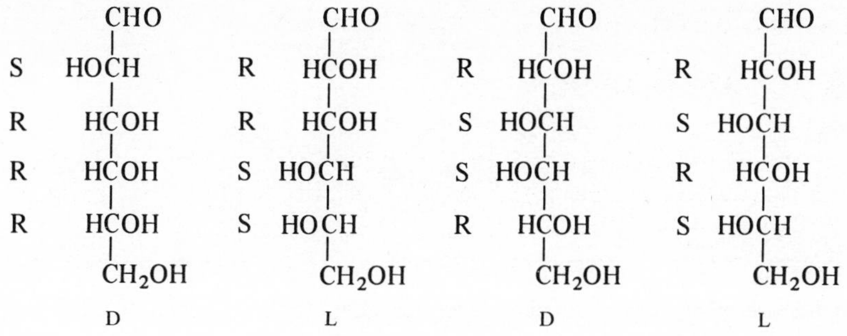

$$
\begin{array}{cccc}
\text{CHO} & \text{CHO} & \text{CHO} & \text{CHO} \\
\text{S } \; HOCH & \text{R } \; HCOH & \text{R } \; HCOH & \text{R } \; HCOH \\
\text{R } \; HCOH & \text{R } \; HCOH & \text{S } \; HOCH & \text{S } \; HOCH \\
\text{R } \; HCOH & \text{S } \; HOCH & \text{S } \; HOCH & \text{R } \; HCOH \\
\text{R } \; HCOH & \text{S } \; HOCH & \text{R } \; HCOH & \text{S } \; HOCH \\
\text{CH}_2\text{OH} & \text{CH}_2\text{OH} & \text{CH}_2\text{OH} & \text{CH}_2\text{OH} \\
\text{D} & \text{L} & \text{D} & \text{L}
\end{array}
$$

Eight other stereoisomers: Mirror images of each of the above (eight *dl* pairs total).

d.

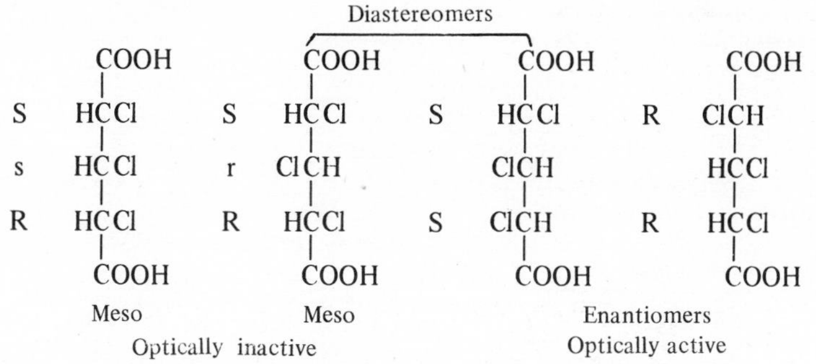

Diastereomers

$$
\begin{array}{cccc}
\text{COOH} & \text{COOH} & \text{COOH} & \text{COOH} \\
\text{S } \; HCCl & \text{S } \; HCCl & \text{S } \; HCCl & \text{R } \; ClCH \\
\text{s } \; HCCl & \text{r } \; ClCH & \quad ClCH & \quad HCCl \\
\text{R } \; HCCl & \text{R } \; HCCl & \text{S } \; ClCH & \text{R } \; HCCl \\
\text{COOH} & \text{COOH} & \text{COOH} & \text{COOH} \\
\text{Meso} & \text{Meso} & & \\
& & \text{Enantiomers} & \\
\text{Optically inactive} & & \text{Optically active} &
\end{array}
$$

13-4 *a.*

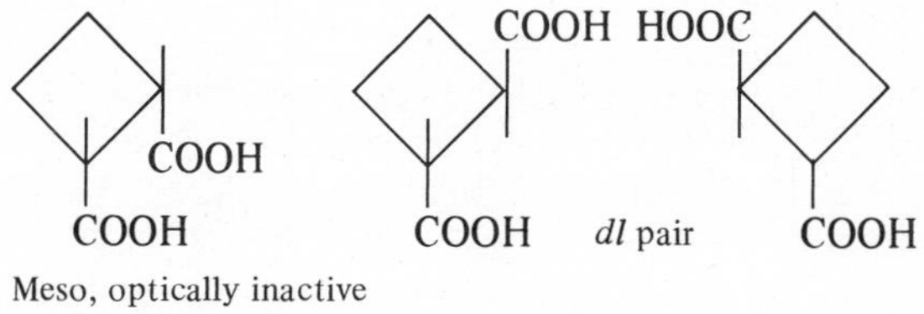

Meso, optically inactive   *dl* pair

b.

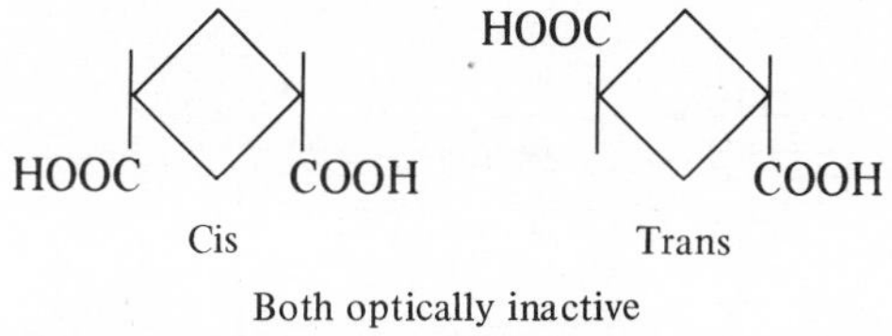

Cis   Trans

Both optically inactive

c.

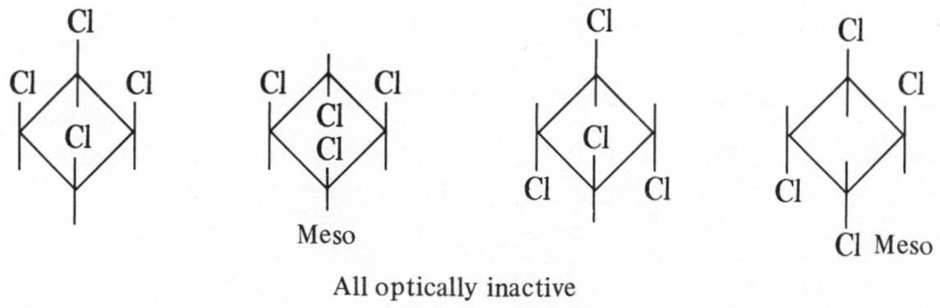

Meso

All optically inactive

Cl Meso

d.

Meso   Meso,   *dl* pair

Optically inactive

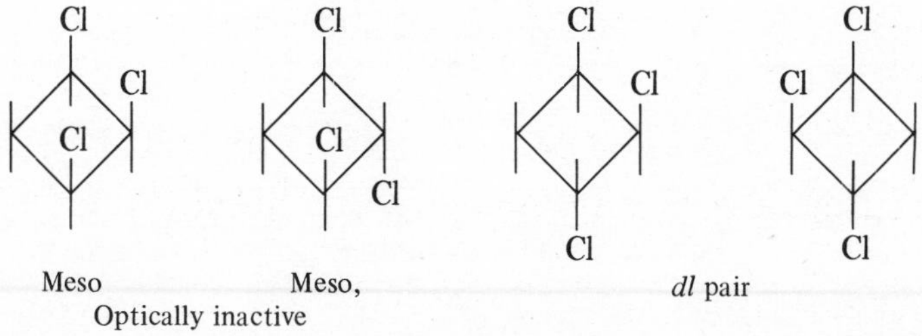

e. Let A = C_6H_5- and B = HOOC−

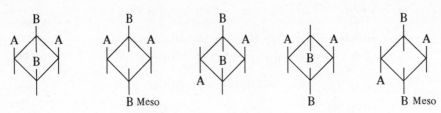

All optically inactive. Note point of symmetry in last structure.

f.

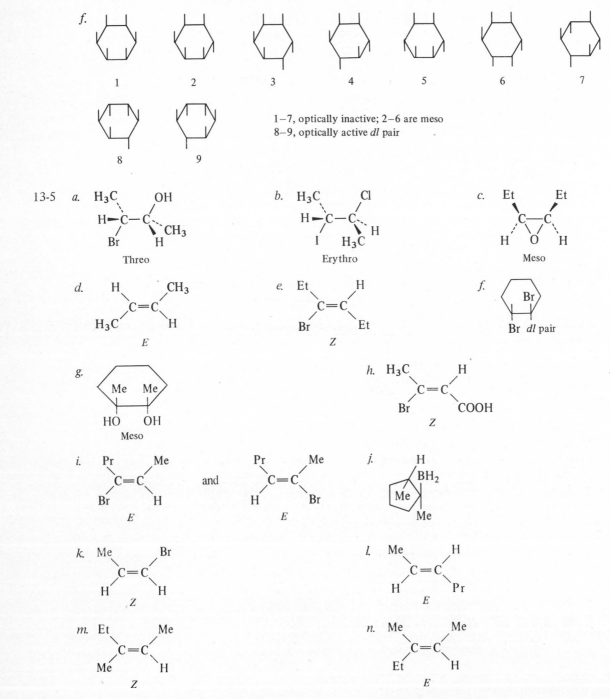

1 2 3 4 5 6 7

8 9

1−7, optically inactive; 2−6 are meso
8−9, optically active *dl* pair

13-5 *a.*

H₃C, OH
H─C─C⋯CH₃
Br   H
Threo

b.

H₃C, Cl
H─C─C⋯H
I   H₃C
Erythro

c.

Et   Et
C─C
H O H
Meso

d.

H, CH₃
C=C
H₃C   H
E

e.

Et   H
C=C
Br   Et
Z

f.

Br
Br *dl* pair

g.

Me   Me
HO   OH
Meso

h.

H₃C, H
C=C
Br   COOH
Z

i.

Pr   Me
C=C
Br   H
E

and

Pr   Me
C=C
H   Br
E

j.

H
BH₂
Me
Me

k.

Me   Br
C=C
H   H
Z

l.

Me   H
C=C
H   Pr
E

m.

Et   Me
C=C
Me   H
Z

n.

Me   Me
C=C
Et   H
E

13-6 *a.* $RCH_2COOH \xrightarrow{P, Br_2} RCHBrCOOH^*$

b. $RCOR' + R''MgX \longrightarrow \xrightarrow{H^+} R''R'RCOH^*$

$c.$ $CH_3CH=CH_2$ $\xrightarrow{HOCl}$ $CH_3\overset{*}{C}H–CH_2Cl$
$\quad\quad\quad\quad\quad\quad\quad\quad\quad\quad\quad\;\;\; |$
$\quad\quad\quad\quad\quad\quad\quad\quad\quad\quad\quad\;\;\; OH$

$d.$ CH_3⟨cyclohexanone⟩$=O$ $\xrightarrow[\Delta]{HNO_3}$

$\quad\quad\quad\quad\quad\quad\quad\quad HOOCCH_2\overset{*}{C}HCH_2CH_2COOH$
$\quad\quad\quad\quad\quad\quad\quad\quad\quad\quad\quad\quad |$
$\quad\quad\quad\quad\quad\quad\quad\quad\quad\quad\quad\quad CH_3$

$e.$ CH_3CHO $\xrightarrow{NaOH}$ $CH_3\overset{*}{C}H(OH)CH_2CHO$

$f.$ $\begin{matrix} R \\ \;\;\;\; \diagdown \\ \quad C=CHR'' \\ \;\;\;\; \diagup \\ R \end{matrix}$ $\xrightarrow{HB}$ $\xrightarrow{H_2CrO_4}$ $R–\overset{*}{C}H–CR''$
$\quad\quad\quad\quad\quad\quad\quad\quad\quad\quad\quad\quad\quad | \quad\;\; ||$
$\quad\quad\quad\quad\quad\quad\quad\quad\quad\quad\quad\quad\quad R' \;\;\; O$

$g.$ $RN–R'' + H_2O_2$ $\longrightarrow$ $\begin{matrix} \quad\quad O^- \\ \quad\quad | \\ R–*N^+–R'' \\ \quad\quad | \\ \quad\quad R' \end{matrix}$
$\;\;\;\;\; |$
$\;\;\;\;\; R'$

$h.$ $RCHO + BrCH_2COOEt$ $\xrightarrow{Zn}$ $\xrightarrow{H^+}$ $R\overset{*}{C}H(OH)CH_2COOEt$

$i.$ $(EtOOC)_2CH_2 + RCH=CHCOOEt$ $\xrightarrow{NaOEt}$ $\xrightarrow{H^+}$ $R\overset{*}{C}HCH\,COOEt$
$\quad\quad\quad\quad\quad\quad\quad\quad\quad\quad\quad\quad\quad\quad\quad\quad\quad\quad\quad\quad |$
$\quad\quad\quad\quad\quad\quad\quad\quad\quad\quad\quad\quad\quad\quad\quad\quad\quad\quad\quad\quad CH(COOEt)_2$

$j.$ $RCOR' + NH_3$ $\xrightarrow[Ni]{H_2}$ $R\overset{*}{C}H(NH_2)R'$

13-7　Yes. No.

13-8　$a.$ No inversion or racemization since no reaction takes place at an asymmetric carbon.

$b.$ Considerable racemization would occur on the carbon holding the Cl atom. Secondary halides tend to react by a mixed mechanism. Part S_N1, part S_N2.

$c.$ A racemic mixture would result.

$d.$ S_N1 mechanism; hence, a racemic mixture.

$e.$ Some racemization for same reason as in b.

$f.$ No inversion or racemization since no reaction occurs at an asymmetric center.

$g.$ Inversion.

$h.$ Racemization. Although each reaction of I^- in displacing an I^- would go with inversion, many such reactions could occur at each asymmetric carbon atom so that finally a 50:50 mixture of d and l isomers would be obtained.

$i.$ Retention of configuration. Product would be mostly $CH_3CH(OH)CH_2I$.

$j.$ S_N1 with racemization.

$k.$ Inversion.

$l.$ No inversion or racemization since no reaction occurs at an asymmetric center.

13-9　There is an equal chance that either carboxyl group will esterify. $meso$-Tartaric acid will give an equal mixture of two isomers (a dl pair) whereas D-tartaric acid will give only one isomer. D L-Tartaric acid will yield a D L mixture of products.

13-10　3-Methylhexane; 3-methyl-1-pentene; $CH_3CHDCH_2CH_3$

13-11　⟨cyclobutane⟩—COOH

13-12　$CH_3CH=C=CHCOOH$

13-13　$CH_3CH=C(CH_3)COOH$

13-14　$CH_3CH(OH)CH(OH)CH_3$

13-15　$HC\equiv CCH(Et)COOH$ or $EtCH=C=CHCOOH$

13-16　A, $HC\equiv CCH(Cl)CH_3$; B, $CH_3C\equiv CCH_2Cl$

13-17　3-Methyl-1-pentyn-3-ol

13-18   *a.*

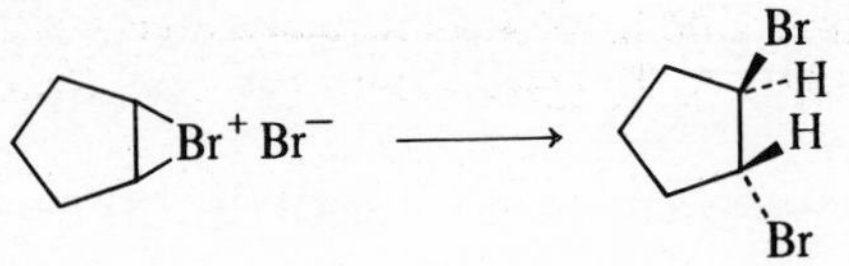

dl pair

b.

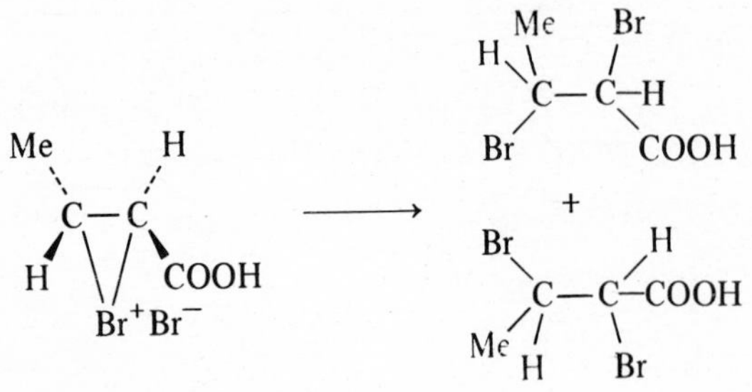

+

dl pair

c.

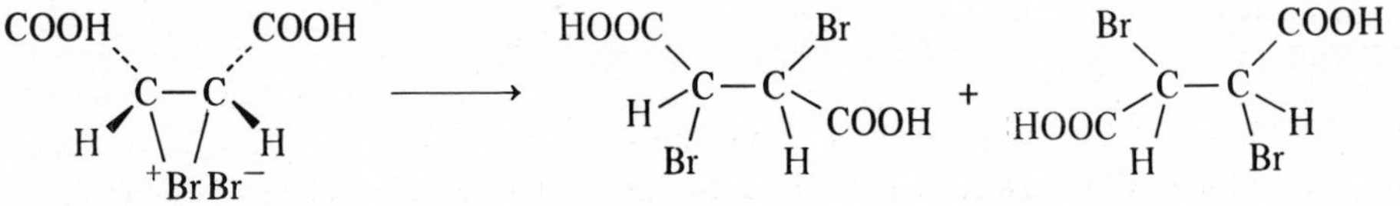

dl pair

d.

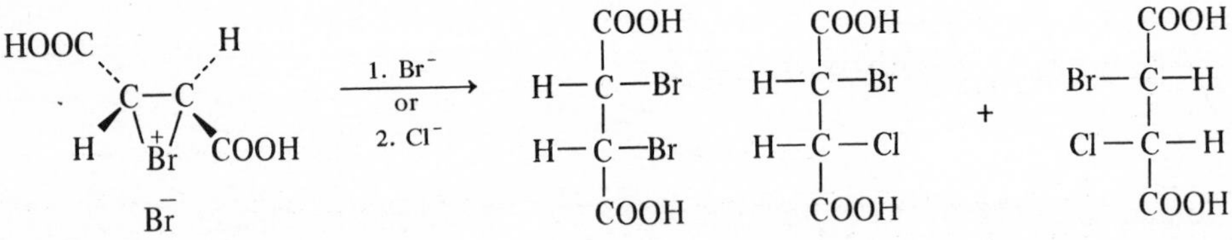

1. Br^-
or
2. Cl^-

+

Via 1: meso   Via 2: dl pair

If the Cl^- concentration is high, a high yield of the *dl* mixture is obtained. Some meso product is formed in any case.

13-19   *a.* Treat the mixture of the two amines with an optically pure acid (for example, D-tartaric acid). Recrystallize the resulting mixture of diastereoisomeric salts until they are pure. Then decompose the salts with a strong base, NaOH; extract the amines from their separate solutions with ether and purify by distillation.

b. Separated by fractional crystallization.

c. Treat the mixture of octanols with succinic or phthalic anhydride. This gives a half acid of the type dl-$ROOCCH_2CH_2COOH$. This then will form a salt easily with an optically pure base. The diastereoisomeric salts can be separated by crystallization, decomposed with strong acid and the two separate esters hydrolyzed to yield the two optically pure alcohols.

13-20   *a.* Trans addition occurs. After the addition of a proton a double bond still exists between the two carbons so that no rotation is possible and hence only one isomer is obtained.

b. Enolization causes racemization of C-3.

c. The bulky neopentyl group prevents back-side S_N2 attack by OH^-; however, front-side attack by Ag^+ pulls halide off for S_N1 reaction.

d. As illustrated in Sec. 6-4*b*, as the OR bond is broken a new bond forms between R and Cl before rearrangement can occur. With HCl, a carbonium ion intermediate rearranges before a bond with Cl can form.

e.

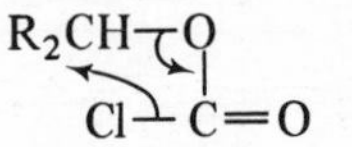

An S_N1 reaction occurs where Cl^- attacks the asymmetric carbon from the same side as $O-C=O$ leaves.

f. 2-Methylbutanoic acid is racemized via

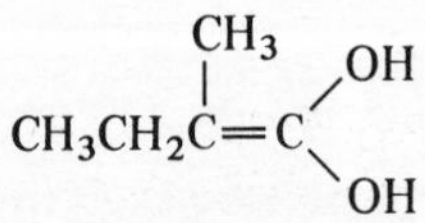

388

g. Formation of cyclic intermediate by neighboring-group participation of COOH results in inversion of configuration at C* (Ag$^+$ helps pull off Cl also). Back-side attack by H_2O on C* inverts the configuration again leading to overall retention.

13-21

Trans-1,2	Cis-1,2	Trans-1,3	Cis-1,3	Trans-1,4	Cis-1,4
a. e,e	*a,e* or *e,a*	*a,e* or *e,a*	*e,e*	*e,e*	*a,e* or *e,a*
b. a,a or *e,e*	*a,e* or *e,a*	*a,e* or *e,a*	*a,a* or *e,e*	*a,a* or *e,e*	*a,e* or *e,a*
$4 \times 0.9 - 1 \times 0.9 = 2.7$ kcal (trans)		$2 \times 0.9 - 2 \times 0.9 = 0$ kcal (trans)		$4 \times 0.9 - 0 = 3.6$ kcal (trans)	
$3 \times 0.9 - 3 \times 0.9 = 0$ kcal (cis)		$4 \times 0.9 + 1.8 - 0 = 5.4$ kcal (cis)		$2 \times 0.9 - 2 \times 0.9$ kcal $= 0$ kcal (cis)	
ΔE cis-trans $= 2.7 - 0.9 = 1.8$ kcal		ΔE cis-trans $= 2 \times 0.9 - 0 = 1.8$ kcal		ΔE trans-cis $= 2(0.9) - 0 = 1.8$ kcal	
c. dl pair	meso	*dl* pair	meso	(neither meso nor *dl* pair)	
d. a,a and *e,e*, dissym	*a,e* and *e,a*, dissym, mirror images	*a,e*, superimposable on *e,a*, dissym	*a,a* and *e,e*, nondissym	*a,a* and *e,e*, nondissym	*a,e* and *e,a*, nondissym

Reasons for optical inactivity: trans-1,2 because each conformer acts as a *dl* pair; cis-1,2 because *a,e–e,a* conformers represent a *dl* pair; trans-1,3 because the conformer (*a,e* and *e,a* are superimposable in this case) exists as a *dl* pair; cis-1,3, trans-1,4, and cis-1,4 because both conformers in each case are nondissymmetric (plane of symmetry present).

13-22 *a.* $\log K = \dfrac{\Delta G}{2.3\ RT} = \dfrac{(-1,800)}{-2.3(2)(298)}$ $\qquad K = 20.4$

b. Assuming ΔG values are additive, equatorial CH_3 is preferred over equatorial by $1.7 - 0.7 = 1.0$ kcal. Thus $\Delta G_{CH_3(e),OH(a)} = 1.0$ kcal and $K = 5.4$.

13-23 Only in the boat conformation can both *tert*-butyl groups not be forced into a very unfavorable axial position.

13-24

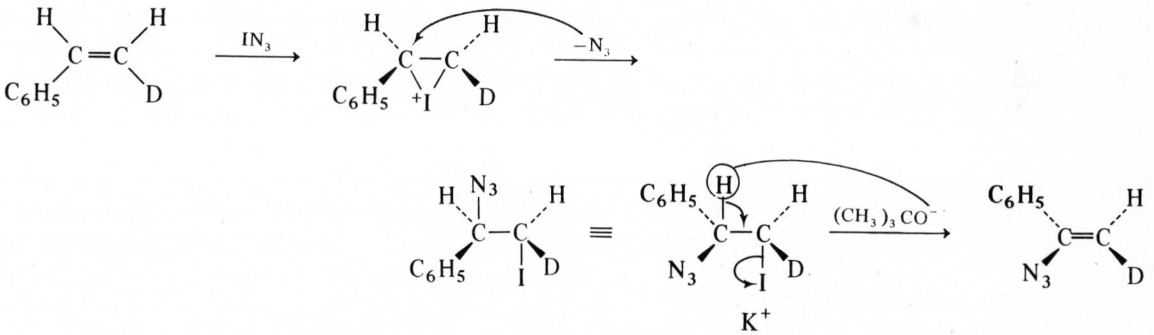

13-25 *a.*

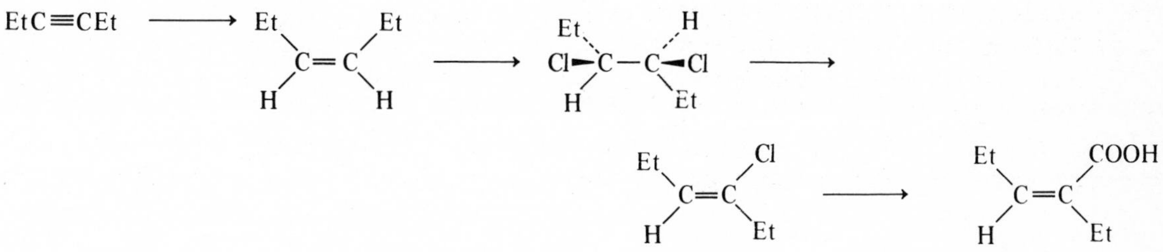

b. (R)CH$_3$CH$_2$CHOTs $\xrightarrow{\ominus}$ (S)CH$_3$CH$_2$CHICH$_3$ $\longrightarrow$ *trans*-2-butene + (R)-2-butanol
　　　　|
　　　CH$_3$

c. (R)-C$_6$H$_5$CHClCH$_3$ $\longrightarrow$ (S)-C$_6$H$_5$CHCH$_3$ $\longrightarrow$ (S)-C$_6$H$_5$CHCH$_3$ $\longrightarrow$ (R)-C$_6$H$_5$CHBrCH$_3$
　　　　　　　　　　　　　　　　|　　　　　　　　　　　　|
　　　　　　　　　　　　　OOCCH$_3$　　　　　　　　　OH

d.

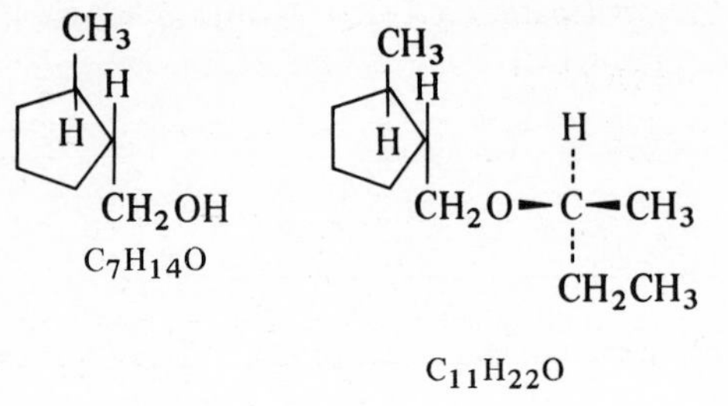

CH₃ ... structures with C₇H₁₄O and C₁₁H₂₂O

e.

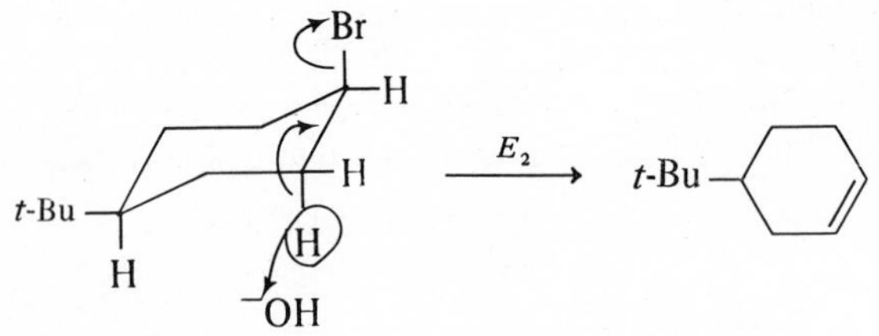

f. *E*2 elimination as in *e* is not possible with trans isomer since the *t*-butyl group's preference for equatorial position prevents trans-diaxial orientation of Br and H on an adjacent C which is necessary for *E*2.

13-26 *a.* Make Grignard reagent and treat with CO_2.

b. Convert to acid chloride with $SOCl_2$; then treat with NH_3 to get amide. Hofmann reaction with NaOCl gives product.

c. Add HOCl to get *trans*-1-chloro-2-hydroxy derivative. Treat this with PCl_3 to get *cis*-dichloro derivative.

d. Convert with PCl_3 to get (S)-2-chlorobutane. Gabriel's synthesis gives (R) amine.

e. Convert acetylene with methyl iodide to 2-butyne. Reduce this with Na and ND_3 to the *trans*-2-butene. Add Cl_2.

f. Convert with PCl_3 to (S)-2-chlorooctane and treat this with CN^- to get the (R)-nitrile. Reduce with Pd and H_2.

g. Convert (R)-2-butanol to (S)-2-chlorobutane with PCl_3 and treat this with $NaCH(COOEt)_2$ to get (R)-alkyl derivative. Make an Na salt of this with EtONa and alkylate with CH_3I. Acid hydrolysis with concurrent decarboxylation gives the product. Recrystallization separates the mixture of diastereoisomers.

h. Add HBr to 1-butene, make a Grignard reagent and treat with CO_2. Resolve the *dl* mixture of 2-methylbutanoic acids with an optically pure amine. Reduce the S isomer with $LiAlH_4$ to the alcohol and convert this to a Grignard reagent. React this with ethylene oxide and then oxidize the resulting alcohol to an aldehyde. Add HCN and hydrolyze. Separate the resulting mixture of diastereoisomers by crystallization.

i. Convert 1-butene to 2-butanol and oxidize this to methyl ethyl ketone. Reduce this with either Pt + D_2 or $LiAlD_4$. Separate this mixture of deuterobutanols by method shown in Prob. 13-19*c*, using phthalic anhydride and then *d*-α-phenylethylamine for the resolution. Convert 1-butene to 2-butene by addition of HBr and then dehydrohalogenation. Convert 2-butene to 2-butyne. Reduce this with (Pd + H_2), add Cl_2, and dehydrohalogenate to get (*E*)-2-chloro-2-butene. Make Li derivative and treat with CO_2. Esterify this acid with optically pure alcohol made above.

j. Make 2-chloro-2-butene as in *i*. Make Li derivative and add this to acetaldehyde. Convert the resulting alcohol to the chloro derivative with $SOCl_2$ to avoid rearrangement. Convert to the amine via Gabriel's method and resolve using *d*-mandelic acid.

13-27 *a.*

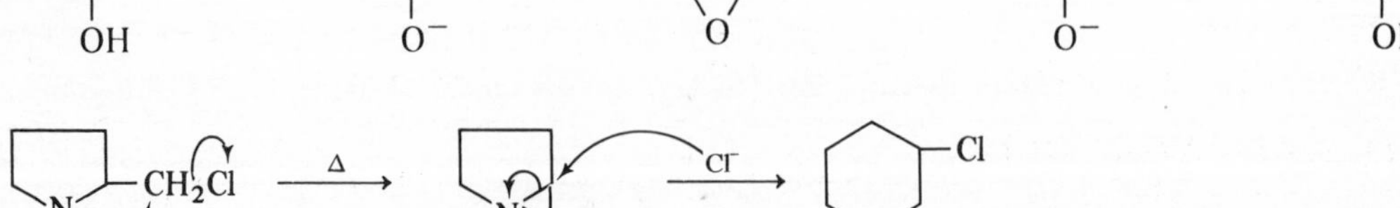

b.

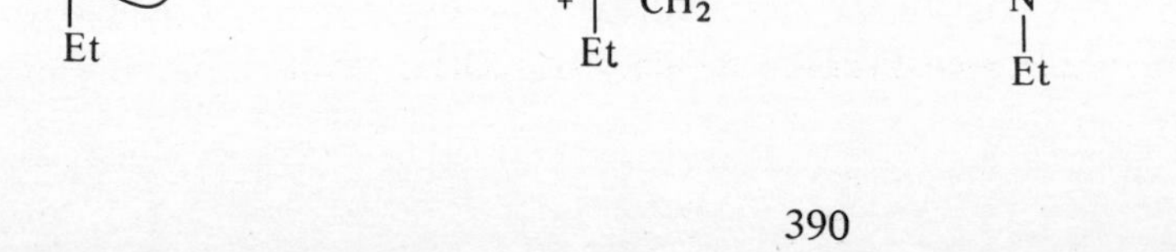

c. 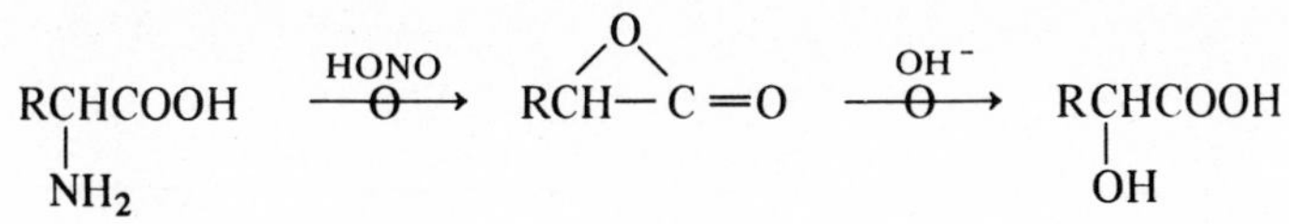 $RCHCOOH$ (with NH_2) $\xrightarrow{HONO}$ $RCH\overset{O}{-}C=O$ $\xrightarrow{OH^-}$ $RCHCOOH$ (with OH)

d.

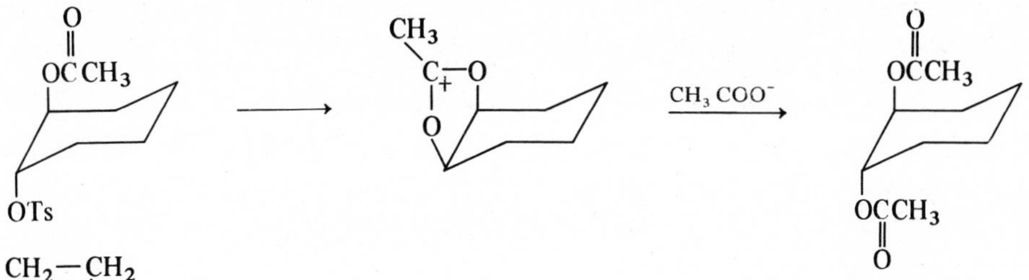

13-28 a. 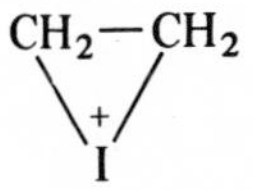 $CH_2—CH_2$ (bridged $\overset{+}{I}$)

b.

$\xrightarrow{CH_3OH}$ $CH_3\overset{I}{C}H\overset{OCH_3}{C}HCH_3$

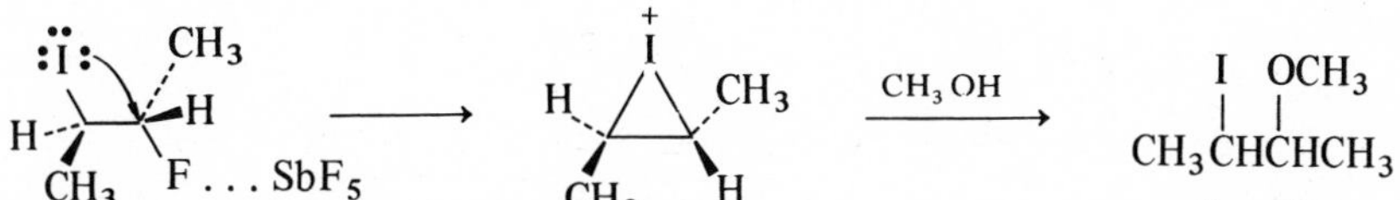

dl erythro

13-29 a. Erythro   b. Threo

13-30 a.

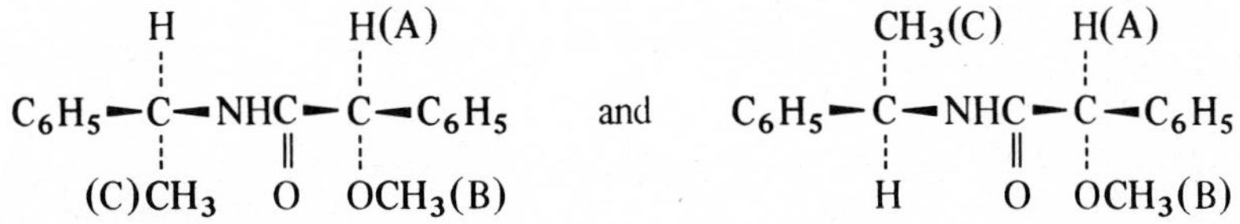

$$C_6H_5\overset{H}{-}C\overset{(C)CH_3}{-}NHC\overset{O}{\underset{\|}{-}}C\overset{H(A)}{\underset{OCH_3(B)}{-}}C_6H_5$$

and

$$C_6H_5\overset{CH_3(C)}{-}C\overset{H}{-}NHC\overset{O}{\underset{\|}{-}}C\overset{H(A)}{\underset{OCH_3(B)}{-}}C_6H_5$$

b. (A) at $\delta = 4.28$ and 4.21; (B) at $\delta = 2.67$ and 2.62; (C) at $\delta = 0.99$ and 0.86. Peak areas in each set (A, B, or C) are in a $90:10$ ratio since the two diastereomers would be produced in that ratio starting from 80 percent optically pure amine (for example, 80 parts d- + 20 parts dl-amine $\xrightarrow{\text{(R)-acid chloride}}$ 90 parts d-(R)-amide and 10 parts l-(R)-amide).

c. 60 percent optically pure amine. Racemic amine would give equal amounts of diastereomers and a 1:1 area ratio in the nmr signals at $\delta = 4.28$ and 4.21.

13-31 a. $-1.64°$ b. $-0.92°$ c. $+2.0°$

13-32

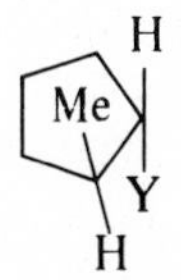

a. $Y = -OH$ b. $Y = -NH_2$

c. $Y = -CH$ (with $\overset{\|}{O}$ below, i.e. $-CHO$)   d. $Y = -CH_2CO_2Et$

13-33 a. 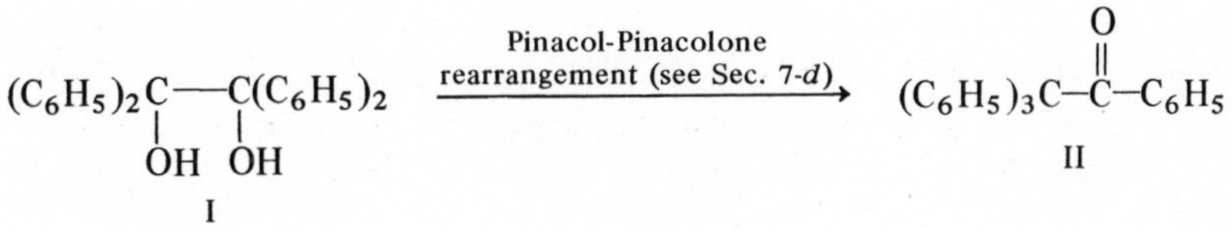 $(C_6H_5)_2C—C(C_6H_5)_2$ (with OH OH) $\underset{\text{rearrangement (see Sec. 7-d)}}{\xrightarrow{\text{Pinacol-Pinacolone}}}$ $(C_6H_5)_3C\overset{O}{\underset{\|}{-}}C—C_6H_5$

I   II

b.

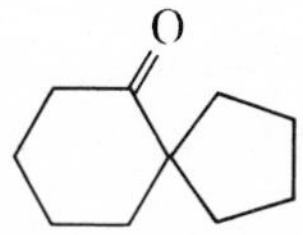

Via Pinacol-Pinacolone rearrangement

c. 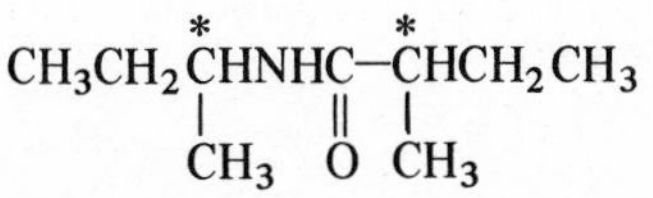 $CH_3CH_2\overset{*}{C}HNHC—\overset{*}{C}HCH_2CH_3$ (with CH_3 O CH_3)   Beckmann rearrangement (see Sec. 11-4f)

The asterisk indicates that both asymmetric carbons would have S configuration.

391

d. $(C_6H_5)_2\overset{|}{\underset{OH}{C}}-CH_2OH$ $\xrightarrow[\text{2. }H_2O]{\text{1. }H^+}$ $(C_6H_5)_2\overset{+}{C}-CH_2OH$ $\xrightarrow[\text{migration}]{\text{hydride}}$ $(C_6H_5)_2CH-\overset{+}{C}HOH$ $(C_6H_5)_2CH\overset{O}{\overset{\|}{C}}H$

e. $sec\text{-Butyl}-\overset{R}{\underset{\underset{O}{\|}}{C}}-\overset{+}{\ddot{N}}=N=\overset{-}{\ddot{\ddot{N}}}:$ $\xrightarrow{-N_2}$ $sec\text{-Butyl}-\overset{R}{\underset{\underset{O}{\|}}{C}}\overset{\curvearrowright}{\ddot{N}}$ $\longrightarrow$ $O=C=N-sec\text{-Butyl}$ R

f.

$Et\overset{Me}{\underset{C_6H_5}{+}}\overset{O}{\overset{\|}{C}}-\overset{\bar{\ \ }}{\ddot{C}}H-\overset{+}{\ddot{N}}\equiv N$ $\xrightarrow{-N_2}$ $Et\overset{Me}{\underset{C_6H_5}{+}}\overset{O}{\overset{\|}{C}}-\ddot{C}H$ $\xrightarrow[\text{configuration}]{\text{migration with}\atop\text{retention of}}$

     I   A carbene

$Et\overset{Me}{\underset{C_6H_5}{+}}\overset{\bar{\ \ }}{\ddot{C}}H-\overset{+}{C}=O \leftrightarrow Et\overset{Me}{\underset{C_6H_5}{+}}CH=C=O$ $\xrightarrow{H_2O}$ $Et\overset{Me}{\underset{C_6H_5}{+}}CH_2COOH$

                              A ketene   II

Note that II is an S isomer but the relative configuration at the asymmetric carbon in I and II has been retained.

g. $CH_3\overset{H}{\underset{CH_2CH_3}{+}}\overset{O}{\overset{\|}{O\text{C}}CH_3}$ See answer to Prob. 11-24*i* for mechanism. There is retention of configuration in the migrating (R)-*sec*-butyl group.

h. $\text{HOOC}\overset{NH_2}{\underset{H\ \ Me}{\underset{Me}{[}}}$ Via Hofmann mechanism (see Sec. 10-3*d*) $\xrightarrow{\Delta}$

(structure with $O=C-NH$ ring, Me) $\equiv$ (bicyclic structure with Me, Me, O, N-H)

                                                                II

13-34   *a.* Tropolone (enol form) $\rightleftharpoons$ (cycloheptadienone-dione structure) $\xrightarrow{\text{KOH}}$ (structure with OH, O^-, O) $\longrightarrow$

                              Benzilic acid rearrangement

(cyclohexadiene with COOH, O^-) $\rightleftharpoons$ (cyclohexadiene with COO^-, OH) $\xrightarrow{H_2O}$ (benzene with COOH)

b. $C_6H_5\overset{H}{\underset{Me}{+}}\overset{\curvearrowright NMe_2}{\underset{\underset{\underset{\underset{O}{\|}}{C_6H_5}}{CHC}}{}}$ $\xrightarrow[\text{rearrangement}]{\text{Stevens}}$ $C_6H_5\overset{H}{\underset{Me}{+}}\overset{NMe_2}{\underset{\underset{\underset{O}{\|}}{CC_6H_5}}{CH}}$

                                                 Retention of
                                                 configuration

c. Diallyl ether $\xrightarrow{C_6H_5Li}$ $CH_2=CHCH-\overset{..}{\underset{\underset{Li^+}{}}{O}}-CH_2CH=CH_2$ $\xrightarrow[\text{rearrangement}]{\text{Wittig}}$

                                        $^-O-\overset{}{\underset{\underset{CH_2CH=CH_2}{|}}{C}}HCH=CH_2$ $\xrightarrow{H^+}$ $CH_2=CHCH_2\overset{OH}{\underset{}{C}}HCH=CH_2$ (*dl*)

d. 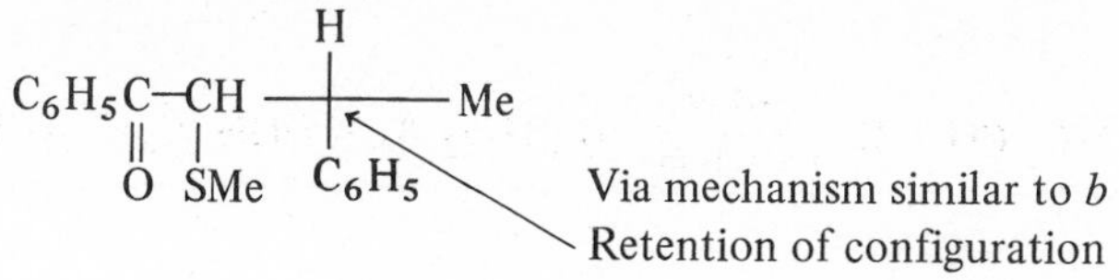 C_6H_5C-CH ... with structure, C_6H_5 attached, Me, H

$$\underset{\underset{O}{\parallel}\;\;\underset{SMe}{\mid}}{C_6H_5C-CH}\;-\!\!\!\!\underset{C_6H_5}{\overset{H}{\mid\!\!\!\!\mid}}\!\!-Me$$

Via mechanism similar to b
Retention of configuration

e. 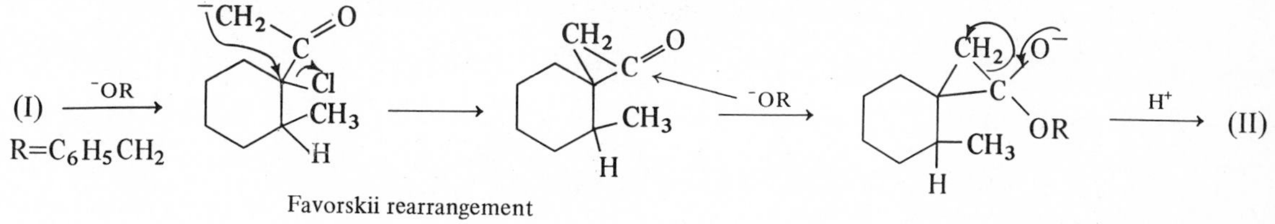 (I) $\xrightarrow{\;^-OR\;}$ [structure] $\longrightarrow$ [structure] $\xrightarrow{\;^-OR\;}$ [structure] $\xrightarrow{\;H^+\;}$ (II)

$R=C_6H_5CH_2$

Favorskii rearrangement

f. 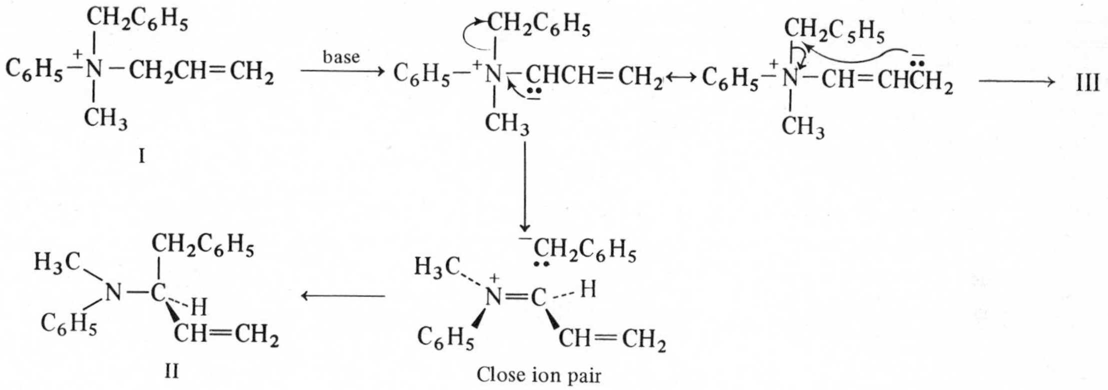 $C_6H_5-\overset{+}{\underset{\underset{CH_3}{\mid}}{\overset{\overset{CH_2C_6H_5}{\mid}}{N}}}-CH_2CH\!=\!CH_2$ $\xrightarrow{\;base\;}$ $C_6H_5-\overset{+}{\underset{\underset{CH_3}{\mid}}{\overset{\overset{CH_2C_6H_5}{\mid}}{N}}}-\overset{..}{C}HCH\!=\!CH_2 \longleftrightarrow C_6H_5-\overset{+}{\underset{\underset{CH_3}{\mid}}{\overset{\overset{CH_2C_5H_5}{\mid}}{N}}}-CH\!=\!CH\overset{..}{C}H_2 \longrightarrow$ III

I

[structure II] $\longleftarrow$ [Close ion pair structure]

II   Close ion pair

Note that I $\longrightarrow$ II is stereospecific and that asymmetry at N in I has in effect been transferred to C in II.

13-35 sec-Butyl$\xrightarrow{(1)}$Hg$\xrightarrow{(2)}sec$-Butyl

a. Inversion at (1) $\longrightarrow$ (S)-sec-butyl-HgBr + (RS)-sec-butyl-HgBr
   Inversion at (2) $\longrightarrow$ (R)-sec-butyl-HgBr + (RS)-sec-butyl-HgBr
   Net optical activity in products is zero since a racemic mixture is the net result.
b. Retention at (1) $\longrightarrow$ (R)-sec-butyl-HgBr + (RS)-sec-butyl-HgBr
   Retention at (2) $\longrightarrow$ (R)-sec-butyl-HgBr + (RS)-sec-butyl-HgBr
   Net optical activity in products is one-half the original optical activity in the (R)-sec-butylmercuric bromide used. This happens to be the experimentally observed stereochemical pathway (retention) in this reaction.
c. Racemization at (1) $\longrightarrow$ (RS)-sec-butyl-HgBr + (RS)-sec-butyl-HgBr
   Racemization at (2) $\longrightarrow$ (R)-sec-butyl-HgBr + (RS)-sec-butyl-HgBr
   Net optical activity in products is one-fourth the original optical activity in the (R)-sec-butylmercuric bromide used.

13-36   Four C_3 axes (along each C—H bond), three C_2 (passing through C and bisecting the H—C—H angle), and six σ (imagine CH_4 placed in a cube; a plane passing through opposite edges of the cube is a σ plane).

13-37   a. C_1 b. S_2 c. D_{6h} d. C_s
         e. C_2 f. D_{3h} g. D_{2d} h. D_{2h}
         i. C_{2v} j. C_{2h} k. D_2 l. D_{3d}
         m. S_4 n. $C_{\infty v}$ o. $D_{\infty v}$

14-1 *a.* 3-bromonitrobenzene (Br, NO$_2$)   *b.* o-(CH$_3$)(CH$_2$CH$_3$)benzene   *c.* H$_3$C—C$_6$H$_4$—CH$_3$ (para)

d. Same as *a*   *e.* H$_3$C, Br, CH$_3$, H$_3$C, Br substituted benzene   *f.* phenylcyclohexane

g. biphenyl   *h.* C$_6$H$_5$CH$_2$CN   *i.* C$_6$H$_5$CH$_2$CH(CH$_3$)$_2$

j. p-(CH$_2$CH=CH$_2$)(CH=CH$_2$)benzene   *k.* C$_6$H$_5$CH=CHCH=CHC$_6$H$_5$ *l.* o-dichlorobenzene (Cl, Cl)

m. O$_2$N—C$_6$H$_4$—CH(CH$_3$)$_2$ *n.* CH$_3$COCHCOOEt with C$_6$H$_5$ *o.* $C_6H_5CH_2\big(C=C\big)\,H,\ H,\ COOCH_2C_6H_5$

14-2 *a.* (CH$_3$)$_3$CCl, (CH$_3$)$_3$COH, (CH$_3$)$_2$C=CH$_2$, or isobutyl compounds which will rearrange.
b. C$_6$H$_5$CH$_2$MgBr   *c.* Excess KMnO$_4$ *d.* C$_6$H$_5$CH$_2$Cl
e. Cr$_2$O$_3$, 550°C   *f.* CH$_3$Cl   *g.* Cu
h. Zn–Hg, HCl   *i.* C$_6$H$_5$CHO, EtONa   *j.* CH$_2$=CHCOOH

14-3 *a.* Methyl benzoate   *b.* Benzamide   *c.* Benzoyl chloride
d. Benzoic anhydride   *e.* Phenyl benzoate   *f.* Benzyl benzoate

14-4 *a.* C$_6$H$_6$ $\xrightarrow{\text{Br}_2,\ \text{Fe}}$ C$_6$H$_5$Br $\xrightarrow{\text{Mg}}$ C$_6$H$_5$MgBr $\xrightarrow{\text{Et}_2\text{SO}_4}$ C$_6$H$_5$Et (or directly via Friedel-Crafts)

b. C$_6$H$_5$CH$_3$ $\xrightarrow{\text{Cl}_2,\ \text{uv light}}$ C$_6$H$_5$CH$_2$Cl $\xrightarrow{\text{OH}^-}$ C$_6$H$_5$CH$_2$OH

c. C$_6$H$_5$COOH $\xrightarrow{\text{SOCl}_2}$ $\xrightarrow{\text{NH}_3}$ C$_6$H$_5$CONH$_2$ $\xrightarrow{\text{P}_2\text{O}_5\ \text{(dehyd)}}$ C$_6$H$_5$CN

d. C$_6$H$_5$CH=CH$_2$ $\xrightarrow{\text{Br}_2}$ $\xrightarrow{\text{alc KOH, heat}}$ C$_6$H$_5$C≡CH

e. C$_6$H$_6$ $\xrightarrow{\text{I}_2,\ \text{HNO}_3}$ C$_6$H$_5$I $\xrightarrow{\text{Cu, }\Delta}$ C$_6$H$_5$–C$_6$H$_5$

f. C$_6$H$_5$MgBr (from *a*) $\xrightarrow{\text{CO}_2}$ C$_6$H$_5$COOH $\xrightarrow{\text{C}_6\text{H}_5\text{CH}_2\text{OH (from }a)}$ C$_6$H$_5$COOCH$_2$C$_6$H$_5$
g. Make the Grignard reagent and treat it with an acid (any proton source).

h. EtOH $\xrightarrow{\text{HBr}}$ $\xrightarrow{\text{Mg}}$ EtMgBr $\xrightarrow{\text{C}_6\text{H}_5\text{CH}_2\text{CN}}$ C$_6$H$_5$CH$_2$C=NMgBr (Et) $\xrightarrow{\text{H}_3\text{O}^+}$ C$_6$H$_5$CH$_2$C=O (Et)

i. C$_6$H$_5$CH$_3$ $\xrightarrow{\text{Cl}_2,\ \text{uv}}$ C$_6$H$_5$CH$_2$Cl $\xrightarrow{\text{Mg}}$ $\xrightarrow{\text{CH}_2=\text{CHCH}_2\text{Cl}}$ C$_6$H$_5$CH$_2$CH$_2$CH=CH$_2$

j. Reduce the ester to benzyl alcohol with LiAlH$_4$ and treat with HI.

k. C$_6$H$_5$COOH (from *f*) $\xrightarrow{\text{EtOH, H}^+}$ C$_6$H$_5$COOEt $\xrightarrow[\text{(from }a)]{\text{C}_6\text{H}_5\text{MgBr}}$ $\xrightarrow{\text{H}^+}$ (C$_6$H$_5$)$_3$C–OH

l. Oxidize with permanganate or dichromate.

m. C$_6$H$_5$COCH$_3$ $\xrightarrow{\text{Mg(Hg)}}$ $\xrightarrow{\text{HOH}}$ C$_6$H$_5$–C(OH)(CH$_3$)–C(OH)(CH$_3$)–C$_6$H$_5$ $\xrightarrow[\text{(dehyd)}]{\text{HBr, }\Delta}$ CH$_2$=C(C$_6$H$_5$)–C(C$_6$H$_5$)=CH$_2$

n. C_6H_6 $\xrightarrow{V_2O_5,\,O_2}$ maleic anhydride $\xrightarrow{Br_2}$ $\xrightarrow{OH^-}$ tartaric acid

o. $C_6H_5CH=CH_2$ $\xrightarrow{B_2H_6}$ $\xrightarrow{H_2O_2,\,OH^-}$ $C_6H_5CH_2CH_2OH$

p. C_6H_5CHO $\xrightarrow{(C_6H_5)_3P=CHCH_3}$ $C_6H_5CH=CHCH_3$

q. *p*-Xylene $\xrightarrow{XeF_2,\,HF}$ 2,5-dimethylfluorobenzene $\xrightarrow{excess\ KMnO_4}$ product

r. $C_6H_5CH_2MgCl$ (from *i*) $\xrightarrow{CO_2}$ $C_6H_5CH_2COOH$ $\xrightarrow{EtOH,\,H^+}$ $C_6H_5CH_2COOEt$ $\xrightarrow{^-OEt}$ product

14-5   *a.* A, C_6H_5MgBr; B, $C_6H_5CH(OH)CH_3$; C, $C_6H_5CH(MgBr)CH_3$; D, $C_6H_5CH(CH_3)CH_2CH_2OH$;
     E, $C_6H_5CH(CH_3)CH_2CH_2COOH$; F, $C_6H_5CH(CH_3)CH_2CH_2COCl$; G, $C_6H_5CH(CH_3)CH_2CH_2CONH_2$;
     H, $C_6H_5CH(CH_3)CH_2CH_2NH_2$; I, $C_6H_5CH(CH_3)CH_2CH_2NHCOCH_3$.

   b. A, $C_6H_5CH_2Cl$; B, $C_6H_5CH_2CN$; C, $C_6H_5CH_2COCH_2CH_3$; D, $C_6H_5CH_2\overset{\overset{\displaystyle OH}{|}}{C}(Et)CH_2COOEt$

   e. A, $C_6H_5COCH_2COOCH_3$; B, $C_6H_5COCH(CH_3)COOCH_3$;
     C, $\underset{C_6H_5COC(CH_3)COOCH_3}{\overset{C_6H_5COC(CH_3)COOCH_3}{|}}$ D, $C_6H_5COCH(CH_3)CH(CH_3)COC_6H_5$;
     E, $C_6H_5CH_2CH(CH_3)CH(CH_3)CH_2C_6H_5$.

14-6   *a.* Benzene will dissolve in fuming sulfuric acid.
   b. Cyclohexene will decolorize a solution of Br_2 in CCl_4.
   c. Benzyl chloride reacts rapidly with alc $AgNO_3$.
   d. Benzoic acid (a solid) is soluble in dilute bases.
   e. Except for the fact that the amide is a solid, it is difficult to distinguish between the pair by a chemical test. The amide will hydrolyze more rapidly in alkaline solution to yield NH_3.
   f. The methyl ketone will give a positive iodoform test.

14-7   *p*-Xylene will yield only one mononitro compound; *o*-xylene will yield two; *m*-xylene will yield three. The three trimethylbenzenes (1,3,5-, 1,2,3-, and 1,2,4-trimethylbenzene) could be distinguished since they would yield one, two, and three mononitro compounds, respectively. However, both *o*- and *m*-bromochlorobenzene would yield four different mononitro compounds and could not be distinguished on this basis.

14-8   *a.*

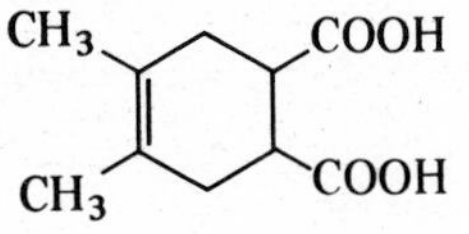

   b.

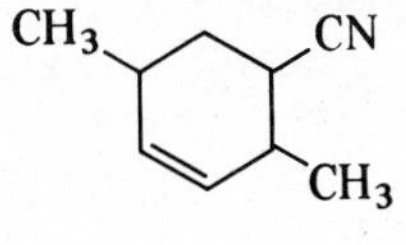

   c.

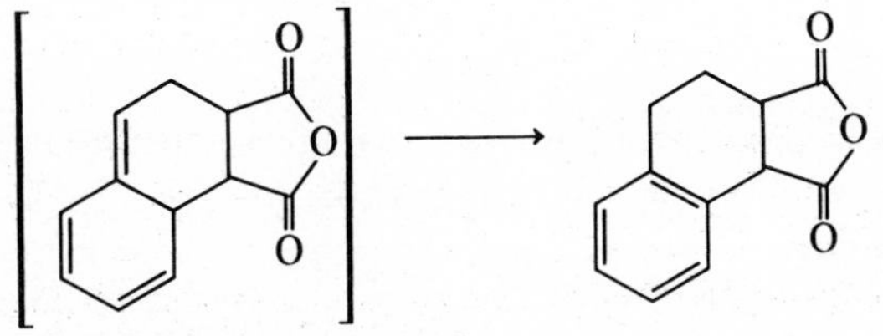

Isomerizes to aromatic system

   d.

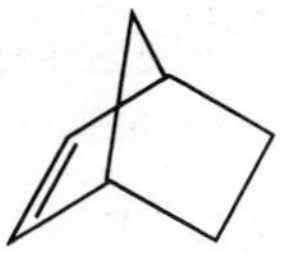

   e.

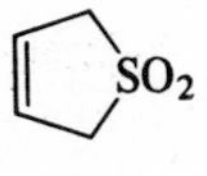

   f.

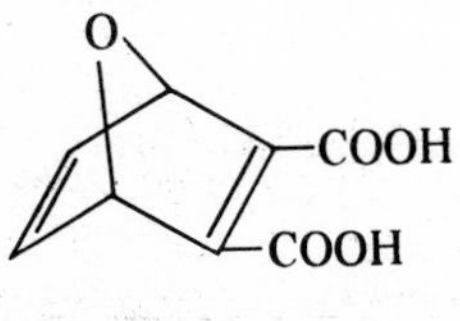

14-9 $CH_3COCOCH_3$, CH_3COCHO, $OHCCHO$

14-10   *a.* *p*-Xylene   *b.* Ethylbenzene, *m*-xylene   *c.* Mesitylene
     d. Propylbenzene, cumene, 1,2,4-trimethylbenzene   *e.* *o*- and *m*-Ethyltoluene

14-11   *m*-Ethyltoluene

14-12   Isopropyl phenylacetate

14-13   *a.* First extract the methanol with water. Then extract the acid with dilute alkali, and finally, extract the cyclohexanone with saturated sodium bisulfite.

b. Extract the trihydroxybenzene with water. Then extract the two amines with dilute acid and separate them by the Hinsberg method.

14-14   *a.* Benzene + LiOH   *b.* $C_6H_5C(CH_3)_2$ with OLi below   *c.* Benzene + CH_3OLi

d. $C_6H_5-N=C-C_6H_5$ with OLi below   *e.* $C_6H_5CCH_3$ with =O/NLi   *f.* $C_6H_5-S=O$ with OLi below

g. Benzene + LiSH   *h.* $C_6H_5CHC_6H_5$ with OLi below

14-15   *a.* *tert*-Butylbenzene   *b.* *tert*-Butylbenzene
         c. *tert*-Amylbenzene   *d.* 1-Phenyl-1-methylcyclohexane
         e. Phenylcyclopentane   *f.* Phenylcyclohexane

14-16   *a.* First oxidize benzene to maleic anhydride; then hydrogenate this to succinic anhydride. Benzene with succinic anhydride in the Friedel-Crafts reaction gives the desired product.

b. Add C_6H_5MgBr to cyclohexanone and dehydrate. Mild oxidation will yield a keto acid that can be reduced by the Clemmensen method to give the product.

c. Treat the diene with maleic acid and dehydrogenate by heating with Pt to get 1,2-dicarboxy-4,5-dimethyl-benzene. Oxidize the methyl groups with $KMnO_4$.

d. Make the acid chloride of maleic acid and treat it with dimethylcadmium. Condense the resulting 1,2-diacetylethene with 2,3-diphenyl-1,3-butadiene. Dehydrogenate.

e. Run a Reformatsky reaction between methyl phenyl ketone and methyl 2-bromopropanoate.

f. Convert benzyl chloride to $C_6H_5CH_2MgCl$ and add ethyl acetate. Hydrolyze.

g. Treat phenylacetaldehyde with 4-phenylbutylmagnesium bromide. Reduce the alcohol via the halide or alkene. Make the starting aldehyde from benzyl cyanide, and the Grignard from the Friedel-Crafts product of the reaction of benzene with succinic anhydride.

h. Oxidize cyclohexanol to cyclohexanone and treat with C_6H_5MgBr. Dehydrate resulting alcohol.

i. Prepare $C_6H_5CH_2CHO$ as in *g*, run aldol condensation and dehydrate.

j. Condense $C_6H_5COCH_3$ with C_6H_5CHO to get $C_6H_5COCH=CHC_6H_5$. A Michael addition reaction of this with acetone followed by an intramolecular aldol condensation (Robinson annelation) will give product.

14-17   *a.* Two isomers, carboxyl groups trans. A *dl* pair.

b. Two isomers; a *dl* pair. Carboxyl groups cis.

c. Acrolein and the phenylbutadiene could come together in two ways. One would place the CHO and phenyl groups 1,2 to each other; the other would place them 1,3. Four isomers; two *dl* pairs.

d. None optically active; both molecules are symmetric.

e. Same as *a*.

f. Two isomers; a *dl* pair.

14-18   *a.* A,   B,   and   C, Cyclic acetals

b. A,   B,   C,

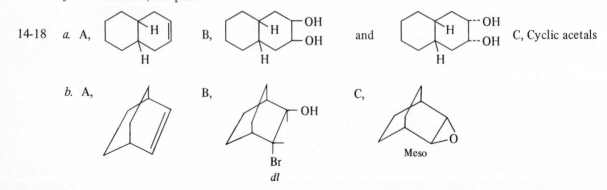

14-19   A, 3-cyclohexene-1-carboxylic acid; B, $HOOCCH_2CH(COOH)CH_2CH_2COOH$

14-20 $BuLi + O=C=O \longrightarrow BuC=O \xrightarrow{BuLi} \left[Bu-\overset{\overset{OLi}{|}}{\underset{\underset{OLi}{|}}{C}}-Bu \right] \xrightarrow{H^+} Bu-\overset{O}{\overset{||}{C}}-Bu + LiOH$

with OLi below the first $BuC=O$.

14-21 *a.* A, Cl—⟨C6H4⟩—MgBr; B, Cl—⟨C6H4⟩—$CH_2C_6H_5$; C, CH_3CH—⟨C6H4⟩—$CH_2C_6H_5$ (with OH below CH_3CH)

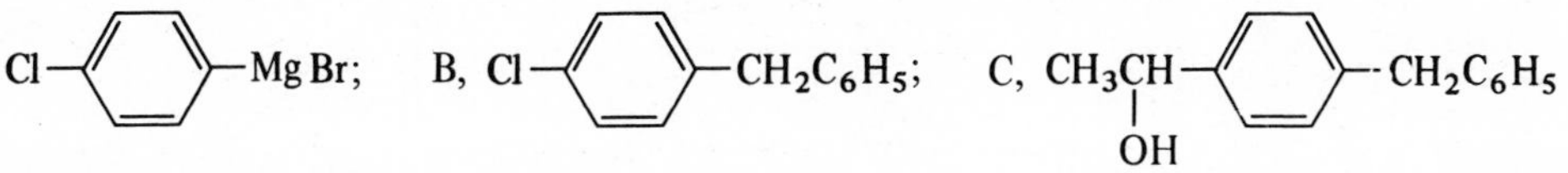

     b. A, $C_6H_5COCH_2CH_2CH_2CH_3$; B, $C_6H_5(CH_2)_4CH_3$

     c. A, Cl—⟨C6H4⟩—CH_2COOH; B, methyl ester of A; C, Cl—⟨C6H4⟩—$\overset{\overset{OH}{|}}{C}(C_6H_5)_2$

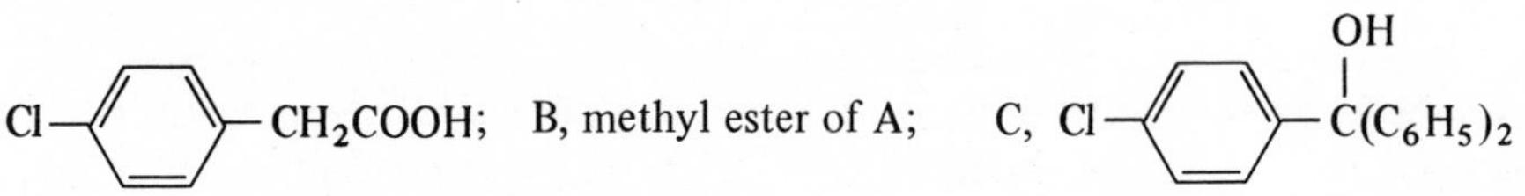

14-22 *a.* Bromination of cyclohexanone yields 2-bromo-1-cyclohexanone. Treat this with 9-BBN-C_6H_5 (prepared from 9-BBN and C_6H_5Li) using $(CH_3)_3$COK as a base to get product.
     b. React 9-BBN with butyllithium. Treat resulting product with $C_6H_5COCH_2Br$ ($C_6H_5COCH_3 + Br_2$) and $(CH_3)_3$COK to get final product.
     c. React *p*-methylphenyllithium with 9-BBN. Treat this product with $BrCH_2COOEt$ and $(CH_3)_3$COK followed by acid hydrolysis to get product.

14-23

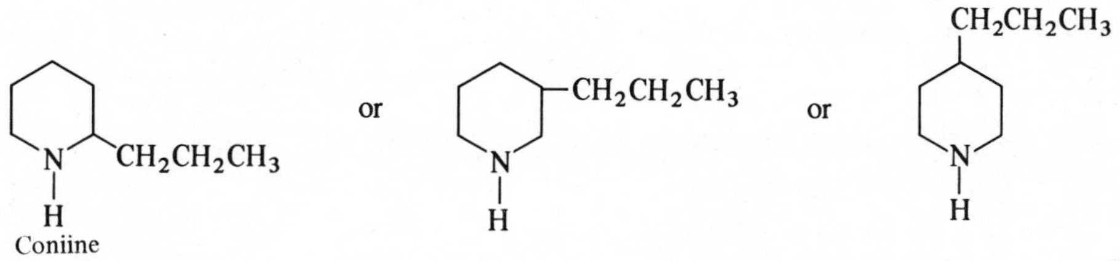

    Coniine   or   or

14-24 *a.* The Grignard reagent could have the normal structure, but the attack by formaldehyde could occur on an ortho position:

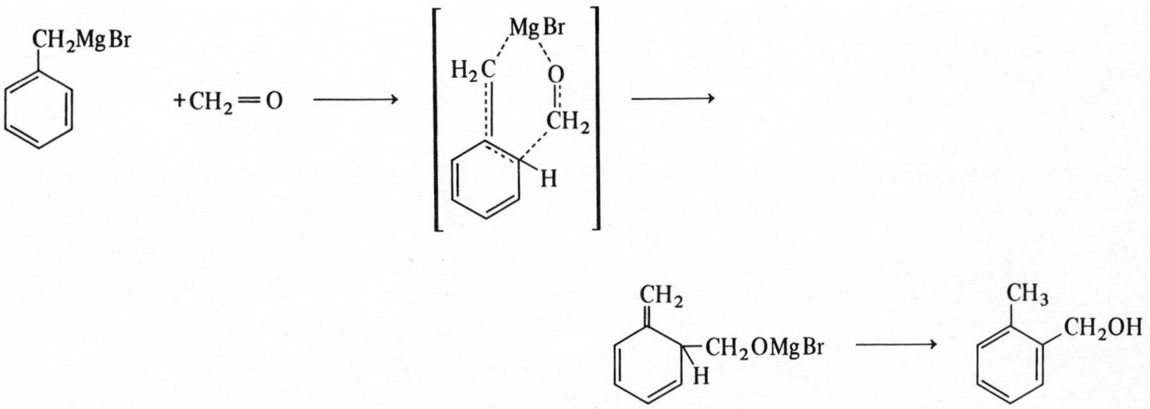

14-25 *a.* (1) $Cl_2 \rightleftharpoons 2Cl\cdot$, (2) $Cl\cdot + C_6H_5CH_3 \rightleftharpoons HCl + C_6H_5CH_2\cdot$,
     (3) $C_6H_5CH_2\cdot + Cl_2 \rightleftharpoons C_6H_5CH_2Cl + Cl\cdot$.
     b. $C_6H_5\overset{\cdot}{C}HCH_3$ is the more-stable free radical because of resonance stabilization by the benzene ring which is not possible with the $C_6H_5CH_2CH_2\cdot$ species. $2°$ hydrogens are more easily abstracted than $3°$ hydrogens.
     c. $C_6H_5CH(Cl)CH_3$ is formed predominantly since the reaction proceeds through the most stable carbonium ion which is $C_6H_5\overset{+}{C}HCH_3$.

14-26 *a.*

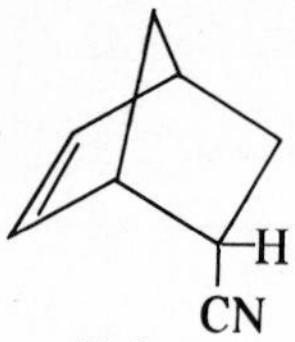

Endo 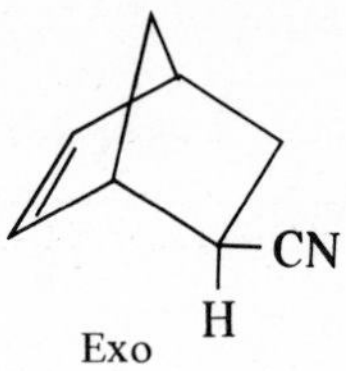   Exo

b.

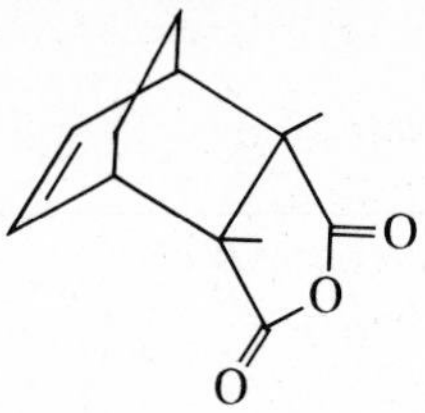

Endo 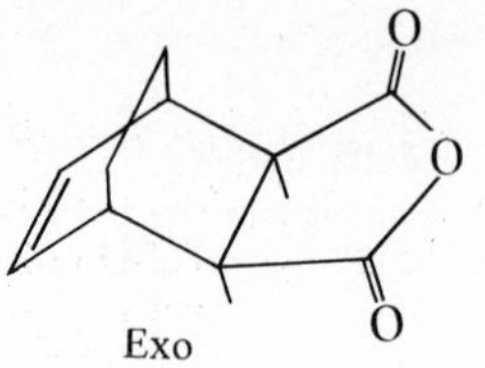   Exo

14-27   X, *p*-bromotoluene; Y, *m*-bromotoluene; Z, *o*-bromotoluene

14-28   A,

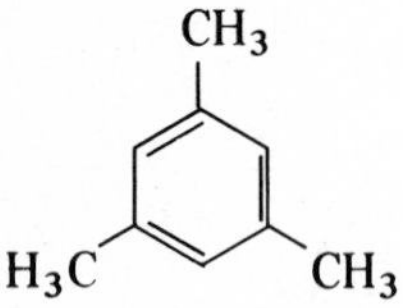

B,

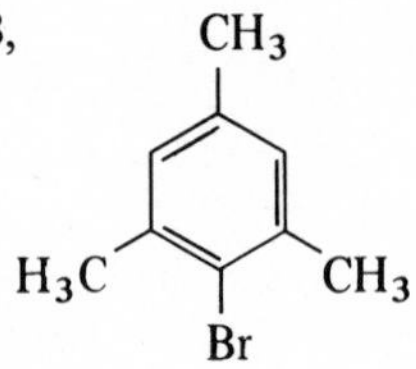

C,

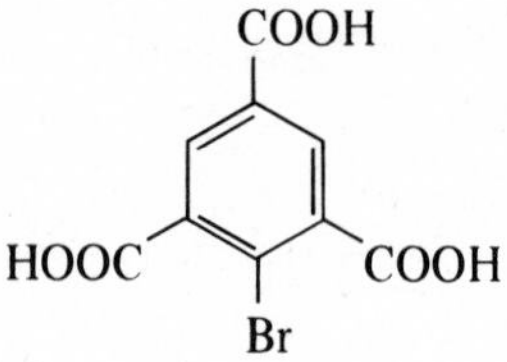

14-29   D, CH_3—⟨ ⟩—$C{\equiv}N$

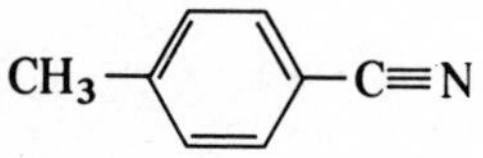

14-30   E, $C_6H_5C{\equiv}CH$; F, $C_6H_5COCH_3$

14-31   I, $C_6H_5CH_2CH_2OOCCH_3$; II, $C_6H_5CH_2OOCCH_3$; III, $C_6H_5CH_2COCH_2CH_3$; IV, $C_6H_5CH_2C(CH_3)_2OH$;
        V, $(C_6H_5)_2CHCOOH$; VI, $C_6H_5CH_2COCH_3$; VII, *p*-chlorophenyl methyl ketone; VIII, $C_6H_5CH(Br)CH_3$;
        IX, $C_6H_5CH_2CH_2CH_2Br$; X, 1-phenyl-1-methylcyclopropane.

14-32   M, $(C_6H_5)_2\underset{\overset{|}{OH}}{C}CH_2CH_3 \xrightarrow[\text{acid}]{\text{strong}} (C_6H_5)_2\underset{+}{C}CH_2CH_3$

Nmr spectrum is that observed for the carbonium ion.

14-33   A, HO—⟨ ⟩—$COOH$;   B, CH_3O—⟨ ⟩—CH_2OH;   C, CH_3O—⟨ ⟩—CH_2COOH

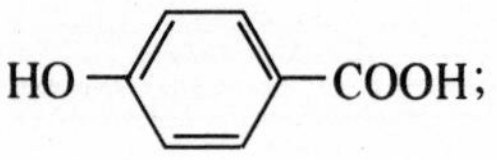

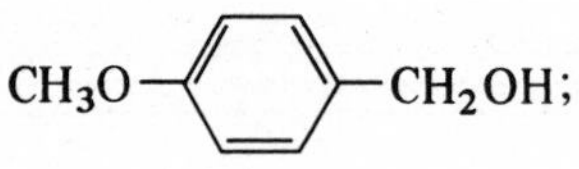

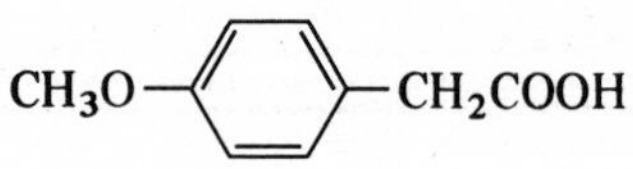

The two hydrogens on the OH and COOH groups are undergoing such rapid exchange that the nmr spectro-meter "sees" a time-averaged position for them and they appear as a singlet at $\delta = 9.75$. The doublet at $\delta = 8.2$ is the two hydrogens ortho to the OH group while the doublet at $\delta = 7.2$ is the two hydrogens ortho to the COOH group.

14-34

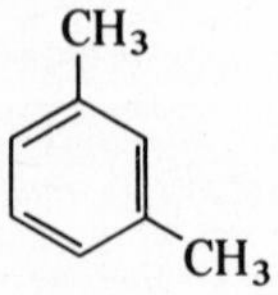

14-35

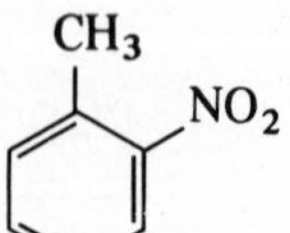

14-36

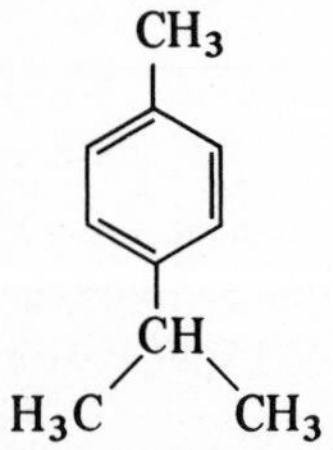

398

Chapter 15

15-1 *a, c, g, i*: ortho-para substitution.
      b, d, e, f, h: meta substitution.

15-2 *a.* 1,2-Dichloro-3-nitrobenzene   *b.* 4-Bromo-1,3-dinitrobenzene
            1,2-Dichloro-4-nitrobenzene
      c. 3-Nitro-4-methylbenzoic acid   *d.* 3,5-Dinitrobenzoic acid
      e. 4-Nitro-3-ethylbenzamide   *f.* 2,4-Dichloronitrobenzene
            2-Nitro-5-ethylbenzamide

15-3 *a.*

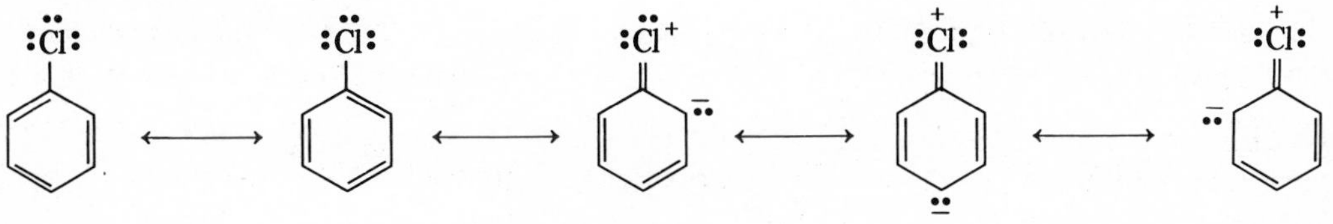

b.

Two kekule forms each

Two kekule forms each

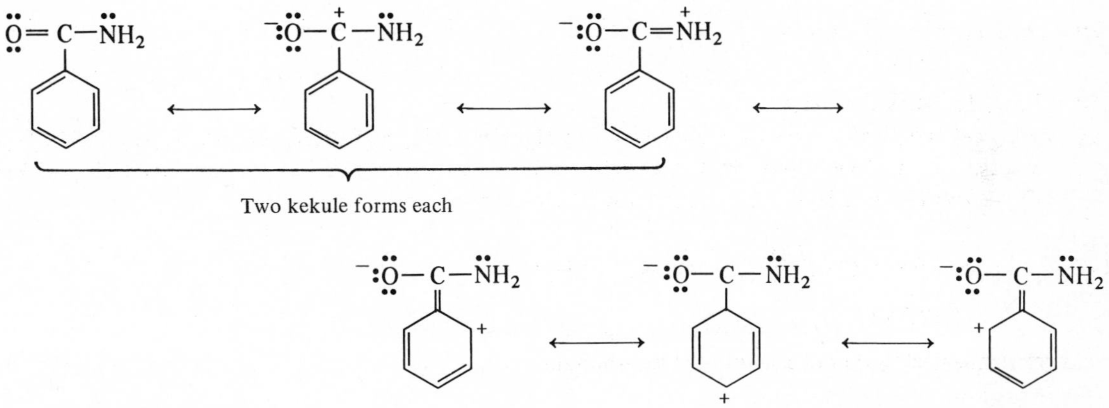

c.

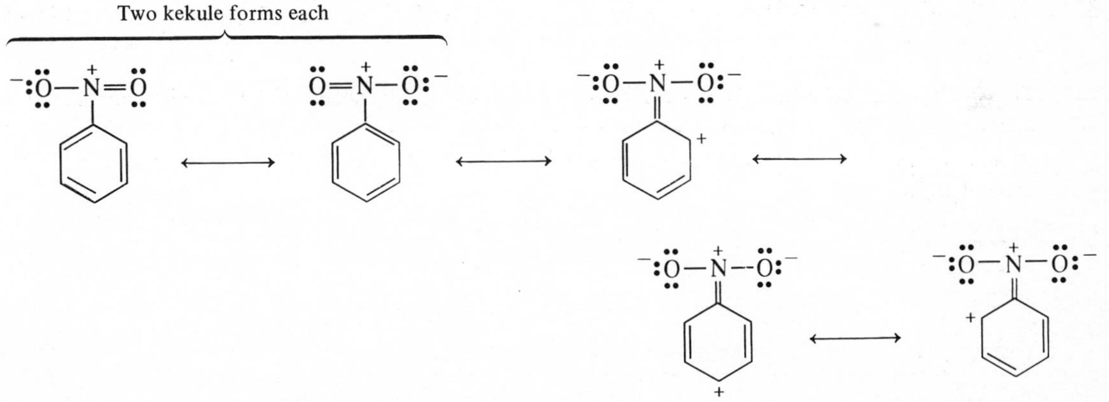

d.

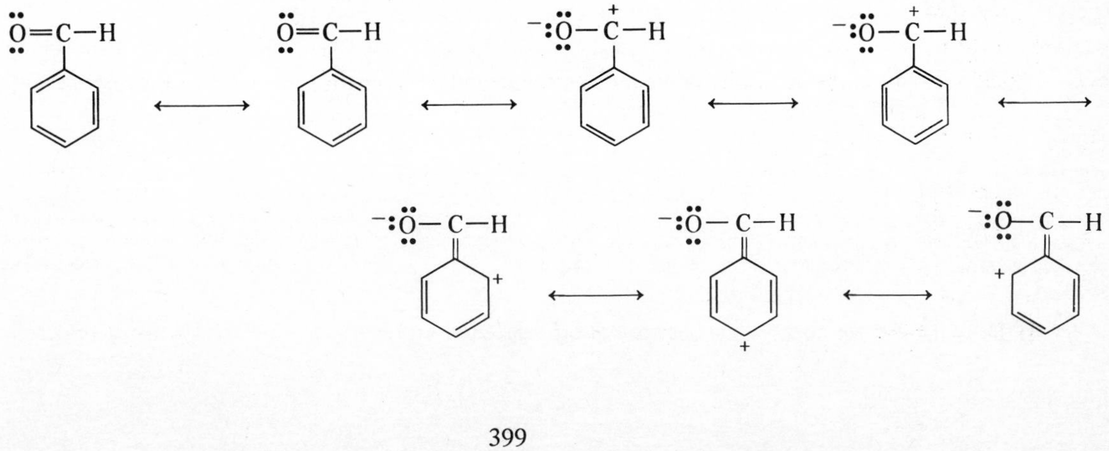

15-4

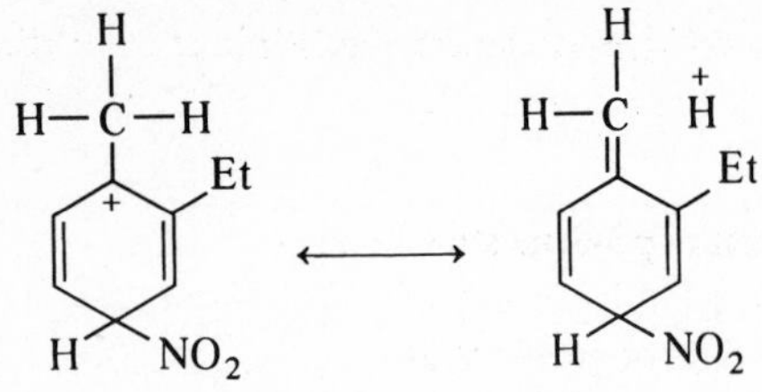

Two more hyperconjugated forms 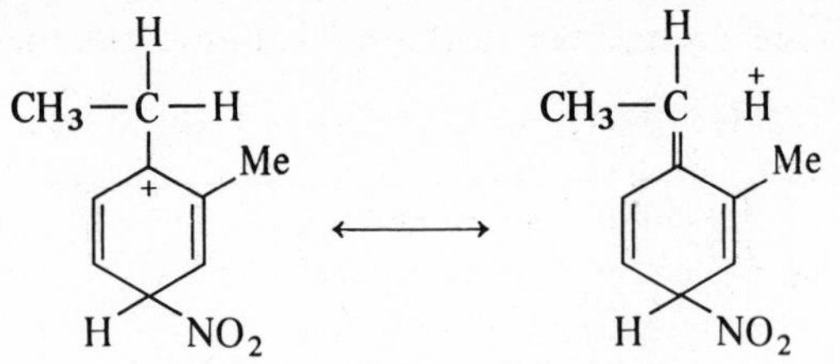   One more hyperconjugated form

     The first hybrid has one more contributing structure (hyperconjugation) than the second. By assuming that the transition state for the nitration resembles the intermediate, it follows that the first will have a lower activation energy.

15-5   *a.*

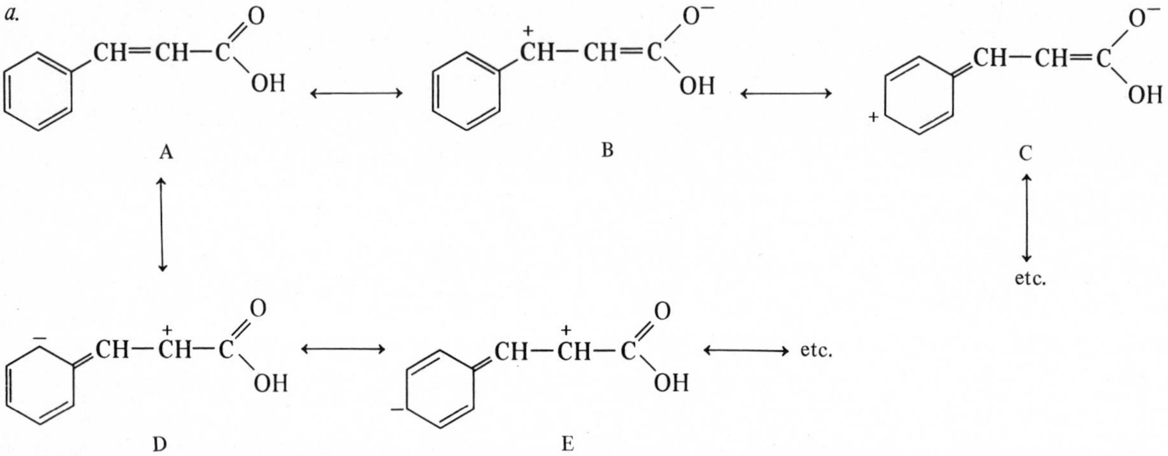

A   B   C

etc.

D   E

b.   Release of electrons to the ring of the type shown in D and E above also stabilizes the transition states in the case of ortho and para substitution.

15-6   *a.*   *b.*   Ortho-para-directing.

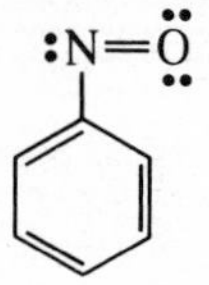

c.

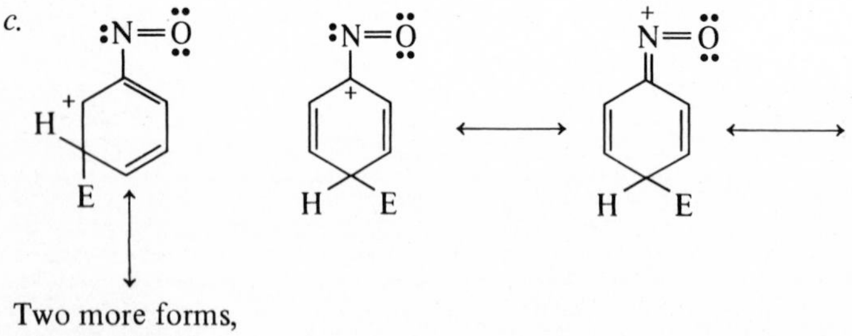

Two more forms, both with + in ring

Two more forms, both with + in ring

15-7   In acid solution aniline is present as 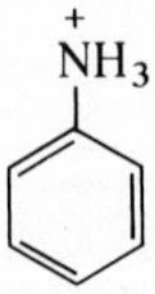 . Ortho and para substitution must go through an unfavorable

transition state: 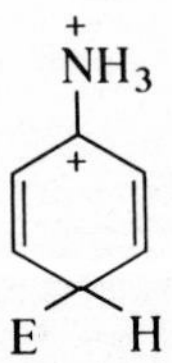 while meta substitution has a more favorable transition state:

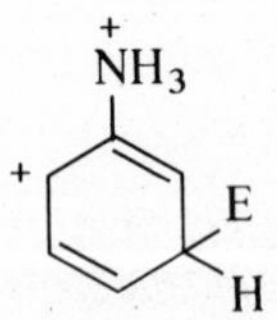

In the latter the two positive charges are not on adjacent atoms.

400

15-8 Product is mainly α. Stabilization of transition state by better charge delocalization is possible at α position:

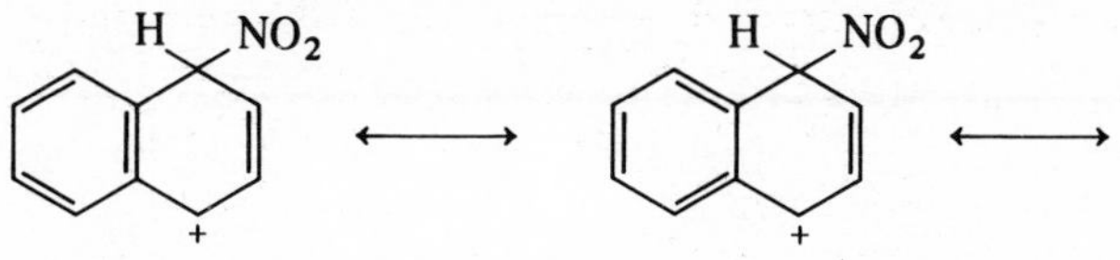

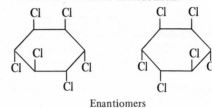

five other possible structures

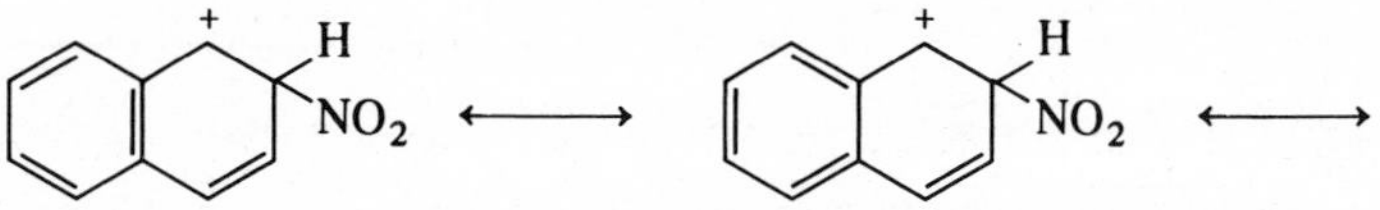

four other possible structures

15-9 *t*-Butyl groups ortho to each other produce a great amount of strain due to steric interactions.

15-10 *a.* Nine

b.

Enantiomers

c.

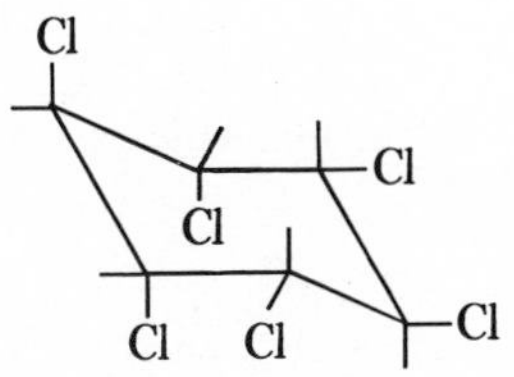

d.

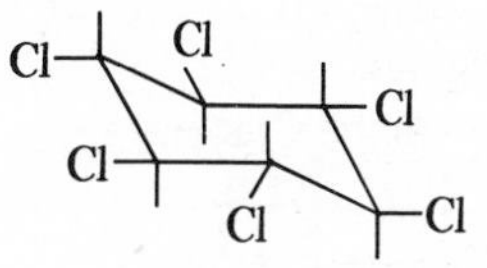

15-11 *a.* In A, isomerization of the propyl group occurs to give cumene. In B, the chlorine substitutes on the carbon attached to the ring.

b. In A, the strongly electron-attracting group (NO_2) prevents a Friedel-Crafts reaction. In B, oxidation of the side chain starts on the carbon attached to the ring and the side chain is broken at this point to give a carboxyl group attached to the ring. In C, Pd is better than Pt for hydrogenation of aromatic compounds because Pt may cause ring bonds to hydrogenate.

c. In A, nitro is meta-directing. In B, catalytic reduction would remove the halogen by hydrogenolysis. Iron plus dilute acid will accomplish reduction of nitro without bothering the chlorine. In C, aromatic chlorides do not form Grignard reagents except under special conditions. Also, the active hydrogen on NH_2 would destroy Grignard reagent.

15-12 A, cyclodecyne;   B, $HOOC(CH_2)_8COOH$

15-13 Cyclopentadiene

15-14 *a.* Approximately equal parts of meta- and para-aminotoluene (*m*- and *p*-toluidine).

b. Approximately equal parts of 1- and 2-substituted naphthalene.

c. Ortho-, meta-, and para-aminotoluene are formed with roughly twice as much meta as ortho or para.

d.

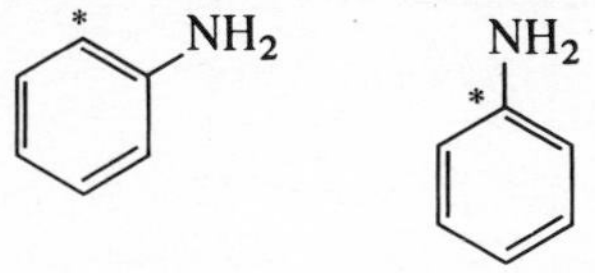

15-15 1,6-Cyclodecadiene.

15-16

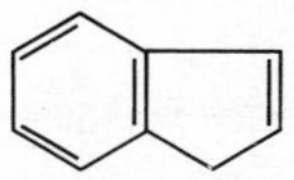

15-17 *a.* Plot log K for benzoic acids on the X axis and log K for phenylacetic acids on the Y axis and draw the best line through the points. The slope m is defined as $m = (Y_2 - Y_1)/(X_2 - X_1)$. This can be estimated by taking two points, $X_1 Y_1$ and $X_2 Y_2$, and substituting into this formula. Plotting log K_a for the phenylacetic acids against either log K_a for benzoic acids or σ gives a slope of approximately 0.55.

b.

1. Plot log K for phenylacetic acids against σ. Using σ_m from Table 15-2 and reading from the plot, log K_a for the meta-fluoro compound is -4.11.
2. From Table 15-3, ρ is 2.11. Hence log $K_{\text{3-nitrophenol}} = (2.11)(0.71) + \log 1 \times 10^{-10} = -8.5$.
3. Log $K_{\text{4-phenylbenzoic acid}} = (1.00)(-0.01) + \log 6.27 \times 10^{-5} = -4.2$.
4. Log $K_{\text{4-methoxyphenol}} = (2.11)(-0.27) + \log 1 \times 10^{-10} = -10.6$.

c.

1. ρ from Table 15-3 $= -1.70$; hence substituents with negative σ values increase the rate of hydrolysis. Hence 4–CH_3 hydrolyzes faster than 4–NO_2.
2. ρ from Table 15-3 $= 2.46$. 3–Br with $\sigma = 0.37$ saponifies faster than 3–CH_3 with $\sigma = -0.07$.

d. 1, 2, and 4.

15-18
$$
\begin{array}{c}
CH_3C{=}CH \\
| \qquad \diagdown \\
\qquad\qquad NH \\
| \qquad \diagup \\
CH_3C{=}CH
\end{array}
$$

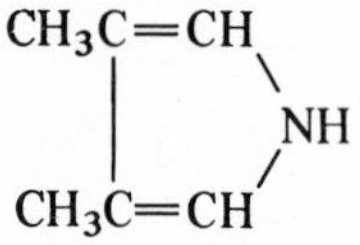

15-19 Cyclohexylacetylene

15-20 *a.* Essentially 0   *b.* 6.7   *c.* 36.7
d. 3.4 (using 1-butene) or 3.6   *e.* 36.0
(using 1,4-pentadiene)

15-21 Ortho- and para-iodophenol. Chlorine is more electronegative than iodine so that the molecule is polarized as I^+Cl^-.

15-22 Electrophilic aromatic substitution:

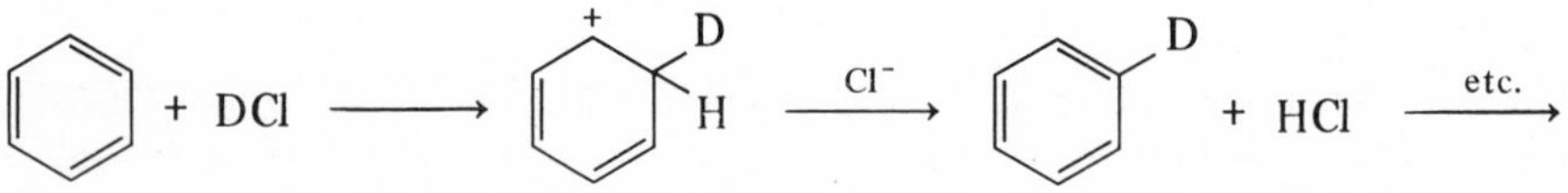

15-23 Anisole > toluene > benzene > chlorobenzene > nitrobenzene. Compare σ values for substituents.
15-24 *a.* o-p   *b.* m   *c.* m

15-25 *a.* Slope is: 2.2_σ. The leaving group in hydrolysis is $X{-}\langle\!\!\bigcirc\!\!\rangle{-}O^-$ Substituents which delocalize the negative charge stabilize the leaving group, making it a better leaving group.
b. Slope is: -11.4_{σ^+}. Strong electron-releasing groups help delocalize charge in the transition state:

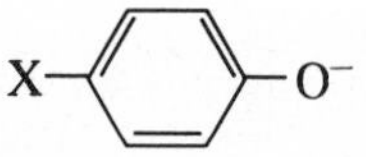

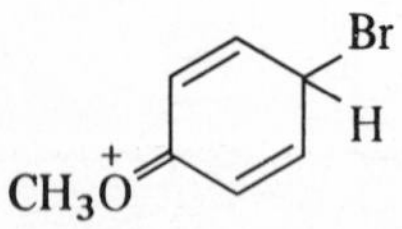

c. Slope is: 4.0_{σ^-}. Strong electron-delocalizing group stabilizes the transition state:

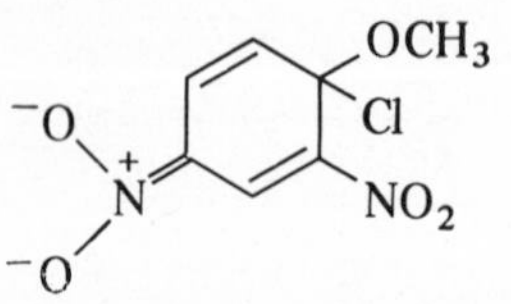

15-26 F has both strong inductive effect and a strong resonance effect (electron-releasing). From the meta position the effect is primarily inductive. From the para position the inductive effect is almost canceled by the resonance effect. CH_3O is like F except that its electron-releasing effect via resonance is stronger and its electron-attracting effect via induction weaker. The $\overset{+}{N}(CH_3)_3$ group cannot accept or release electrons via resonance. Its only effect is inductive. This effect falls off with distance so that $\sigma_m > \sigma_p$. CN has a strong electron attraction via resonance so that in the para position positive electron attraction occurs via resonance and inductive effects. From the meta position only the inductive effect is significant.

15-27

15-28 *a.* The nitrogen atom has sp^2 hybridization and is planar. In resonance interaction between the nitro group and the ring, proper overlap of the nitrogen and carbon *p* orbitals can take place only if the nitro group is coplanar with the benzene ring.

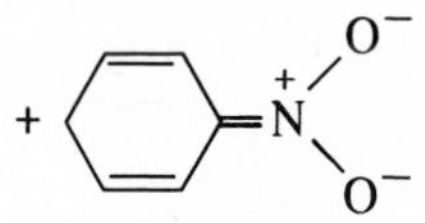

b. Ethyl ether has a greater dipole moment than ethyl vinyl ether because of the absence of resonance interaction between oxygen electrons and the π bond of the vinyl group.

$$Et-\overset{..}{\underset{..}{O}}-CH=CH_2 \longleftrightarrow Et-\overset{+}{\underset{..}{O}}=CH-\bar{C}H_2$$

c. The unbonded electrons of the nitrogen atom in pyrrole are involved in a resonance interaction with the π bonds of the carbon atoms. In order for pyrrole to form a salt, the resonance energy of the system must be overcome because these electrons are now localized in an N—H bond. Pyridine, with localized electrons on nitrogen, is the stronger base.

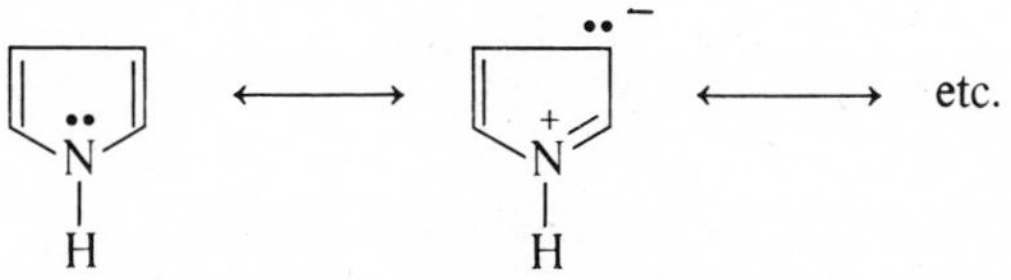

etc.

d. The acidity is related to the stability of the anion which is formed. Fluorene is the stronger acid because of extra electron delocalization illustrated by the canonical forms shown below.

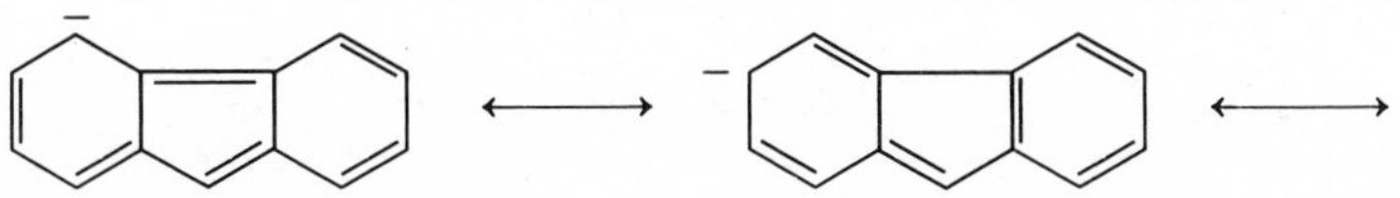

negative charge delocalized on every carbon atom

e. In order for electrons from the amino group nitrogen atom to be delocalized by the nitro group, the latter must be able to become coplanar with the benzene ring. This coplanarity is difficult in the 3,5-dimethyl derivative and consequently it is a stronger base.

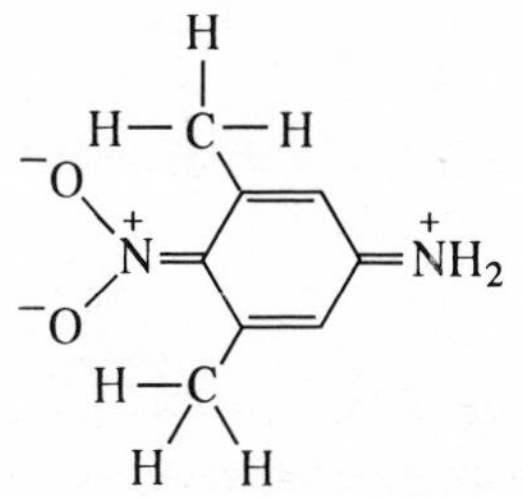

Steric repulsion between NO_2 and CH_3

f. This allows overlap of oxygen lone-pair electrons with those of the π system of the ring: 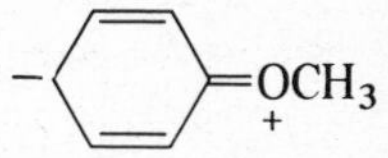 with resulting resonance stabilization.

g. Quinoline is a stronger base because the lone pair of electrons on N are in an orbital which is perpendicular to the π-electron system of the ring and hence are localized. In indole, the lone pair of electrons on N are in an orbital which can overlap with the π-electron system of the rings and are delocalized.

h.

The negative charge of indene anion is delocalized over a larger area than that of toluene. Hence it is more stable.

Chapter 16

16-1   *a.* 2-Propanethiol; isopropyl mercaptan   *b.* 2-(Ethylthio)propane; ethyl isopropyl sulfide
      c. Methylsulfonylbenzene; methyl phenyl sulfone   *d.* Hexanesulfonyl chloride; same
      e. Ethyl benzenesulfonate; same   *f.* Methylsulfinylmethane; dimethyl sulfoxide
      g. Propanethioic acid; thiopropionic acid   *h.* 4-Nitrobutanethial; δ-nitrothiobutyraldehyde
      i. *p*-Nitrobenzenesulfonic acid; same
      j. *p*-Nitrophenyldithiobenzene; *p*-nitrophenyl phenyl disulfide
      k. 1-Phenyl-2-propanethione; phenyl thioacetone   *l.* 2,4,6-Trinitrobenzenethiol; 2,4,6-trinitrothiophenol

16-2   *a.* *p*-Toluenesulfonic acid   *b.* *m*-Bromobenzenesulfonyl chloride
      c. *o*-Ethylbenzenethiol   *d.* *p*-Nitrotoluene
      e. 3,4-Dichloronitrobenzene   *f.* *o*-Nitrobenzyl chloride
      g. Dibenzyl disulfide   *h.* Methyl 3,4-dimethylbenzenesulfonate
      i. *p*-Ethylphenylhydroxylamine   *j.* 4-(3-nitrophenyl)butanethioic acid
      k. α,*o*-Dinitrotoluene   *l.* 3,3'-Dimethylazobenzene
      m. *p*-Nitrososulfinic acid   *n.* *N*-Phenyl-*p*-toluenesulfonamide
      o. Thiourea   *p.* Diallyl sulfone

16-3   *a.* $CH_3SSCCH_3 + Br^-$ *b.* $C_6H_5SEt + NaBr$ *c.* (R)-2-Butanethiol + Br^-
      d. $EtSSEt + H_2O + NaI$ *e.* $Et_3S^+Br^-$ *f.* p-$CH_3C_6H_4SO_2H$
      g. $C_6H_5SO_2Et$ *h.* $MeSC_6H_5 + CH_2{=}CH_2 + H_2O$ *i.* p-$HOC_6H_4NO_2$; S_N
      j. m-$C_6H_4(NO_2)_2$; S_E *k.* $EtSMgBr + C_2H_6$ *l.* m-$O_2NC_6H_4SO_2CF_3$; S_E
      m. $EtSO_2MgBr$ *n.* $Et_2SO + H_2O$ *o.* $(C_6H_5)_2SO_2 + H_2O$
      p. $C_6H_5SO_2NHCH_3$
      r. No reaction   *q.* (R)-α-(Ethylthio)ethylbenzene
      t. p-$O_2NC_6H_4SC_6H_5 + NaCl$; S_N *s.* $C_6H_5SH + NaCl + Me_2C{=}CH_2$
      v. $C_6H_5NH_2$; S_N overall via benzyne   *u.* o-$HOC_6H_4NO_2$; S_N

      w. O_2N—⟨ ⟩—$CH(COOEt)_2$; S_N (with NO_2 substituent)   *x.* ⟨ ⟩S + 2NaBr

                                                         Thiacyclopentane

y. structure (Me, Me on two carbons bonded to S; Pr, Pr substituents; H atoms)

z. $EtN\overset{+}{H_3}\,^-O_3SC_6H_5$ *a'.* $(NH_2)_2C{=}\overset{+}{N}HMe + MeSO_4^-$

b'. $(NH_2)_2C{=}\overset{+}{S}CH_2C_6H_5$ Cl^- *c'.* structure (Me, SNa; OH, Me substituents)

d'. structure (diaryl thioketone; NO_2 substituent)

e'. $C_6H_5SO_2Cl + MgBrCl$ *f'.* cyclohexene oxide CH_2 + $(CH_3)_2S$

16-4 *a.* Nitrate chlorobenzene and separate *o-p* isomers by crystallization.

b. Chlorinate nitrobenzene.

c. Make toluene by Friedel-Crafts; nitrate toluene and separate isomers. Oxidize the methyl group to COOH.

d. Friedel-Crafts alkylation followed by sulfonation.

e. Sulfonate benzene, convert to sulfonyl chloride, and add *sec*-butyl alcohol.

f. Sulfonate benzene, brominate, and convert to sulfonamide via PCl_5 and NH_3.

g. Oxidize toluene from *c* to benzoic acid and nitrate.

h. Reduce *p*-nitrotoluene from *c* to the hydroxylamine with Zn and NH_4Cl; then oxidize to the nitroso compound with dichromate.

i. Treat *p*-chloronitrobenzene from *a* with methoxide.

j. Friedel-Crafts alkylation of benzene to ethylbenzene; sulfonate this and then brominate, and finally desulfonate by heating with water.

k. Sulfonate toluene from *c* and fuse with NaOH.

l. Reduce $C_6H_5SO_2Cl$ from *e* and form salt of the thiol with NaOH.

m. Chlorinate benzene, then sulfonate with fuming H_2SO_4.

n. Brominate benzene, then sulfonate with fuming H_2SO_4; chlorinate with Cl_2 and Fe.

o. Friedel-Crafts alkylation to ethylbenzene followed by Friedel-Crafts acylation using $CH_3CH_2CH_2COCl$.

p. Reduce *m*-chloronitrobenzene from *b* with Zn, NaOH, CH_3OH.

q. Chlorinate *p*-nitrotoluene from *c* in the α position; then use the *p*-nitrobenzyl chloride for alkylating benzene in the Friedel-Crafts reaction.

r. Brominate *p*-nitrotoluene from *c*; then oxidize the methyl group to COOH.

s. Treat *p*-chloronitrobenzene from *a* with C_6H_5SNa from *l*.

t. Treat *p*-ethylbenzenesulfonic acid from *j* with PCl_5. Reduce the sulfonyl chloride to the sulfinic acid with Zn and H^+ and nitrate this product.

u. Treat $(C_6H_5)_2S$ with PrBr to get $(C_6H_5)_2\overset{+}{S}CH_2CH_2CH_3Br^-$; treat with $(CH_3)_3CLi$ to get the sulfur ylide $(C_6H_5)_2S{=}CHCH_2CH_3$. Treat ylide with benzaldehyde to get desired product. C_6H_5CHO may be prepared via Rosenmund reduction of C_6H_5COCl which may be obtained from benzene in a few easy steps.

16-5 *a.* α-Nitrotoluene is soluble in dil NaOH.

b. Aniline is soluble in dil HCl.

c. Benzamide yields nitrogen gas with HONO.

d. Nitrohexane is soluble in dil NaOH.

e. Nitrohexane is soluble in dil NaOH.

f. Nitrobenzene will give a positive ferrous hydroxide test.

g. The acid liberates CO_2 from $NaHCO_3$.

h. The thiol is soluble in dil NaOH.

i. The sulfonic acid liberates CO_2 from $NaHCO_3$.

j. The acid is soluble in cold water (the halide reacts very slowly).

k. The acid liberates CO_2 from $NaHCO_3$.

l. Hinsbert test. Diethylamine forms a product with benzenesulfonyl chloride that is insoluble in dil NaOH.

16-6 *a.* Dehydration, polymerization, etc., would accompany sulfonation.

b. A, add Fe catalyst; also, the *o- and p-* chloroalkylbenzenes are difficult to separate. B, the aromatic Cl is unreactive toward NaSH. C, HNO_3 oxidizes thiols.

c. Bromine is ortho-para-directing for electrophilic substitution in A. In B, the nitro group is not unreactive with Grignard reagents. A ketone forms in C.

d. Friedel-Crafts reactions fail with electron-poor rings as in nitrobenzene in A. In B, the nitro group would also be reduced. The chlorination in C will yield the α isomer.

16-7 *a.* Make butyl bromide by conventional methods. Treat this with sodium sulfide and oxidize with 1 equiv H_2O_2.

b. $EtOH \xrightarrow{\text{dichromate}} CH_3CHO \xrightarrow{\text{NaOH}} CH_3CH{=}CHCHO \xrightarrow{\text{LiAlH}_4} CH_3CH{=}CHCH_2OH$

$\xrightarrow{\text{HBr}} \xrightarrow{\text{NaSH}} \xrightarrow{\text{HNO}_3} EtSO_3H \xrightarrow{\text{NaOH}} \xrightarrow{\text{PCl}_5} EtSO_2Cl \longrightarrow$ product

c. Make *sec*-butyl bromide by conventional methods and treat with Na_2S_2.

d. Alkylate benzene with EtBr and $AlCl_3$; nitrate the product and isolate the para isomer; reduce with 1 equiv Zn and NaOH.

e. Use the Friedel-Crafts acylation reaction to make propylbenzene by way of $C_6H_5COCH_2CH_3$. Dinitrate the product.

f. Make cumene by means of a Friedel-Crafts reaction between benzene and isopropyl chloride. Nitration of cumene will yield principally the para isomer.

16-8 *a.* 3,4,5-Trimethylbenzenesulfonyl chloride

b. A, $C_6H_5SO_2NHCH(CH_3)CH_2CH_3$; B, $CH_3CH(NH_2)CH_2CH_3$; C, $CH_3CH(OH)CH_2CH_3$

c. α-Nitrotoluene *d.* *p*-Nitrotoluene *e.* *p*-$CH_3C_6H_4CH{=}NOH$

f. *p*-Dinitrobenzene *g.* $C_6H_5N{=}NC_6H_5$

h. A, *m*-$CH_3C_6H_4CH(NO_2)CH_3$; B, *m*-$C_6H_4(COOH)_2$

16-9 *a.* If a substitution takes place ortho or para to an electron-releasing group, the reaction is most likely electrophilic substitution. Consider the charge distribution in the transition state.

b. The electrophile is O_2N^+. The leaving group is H^+.

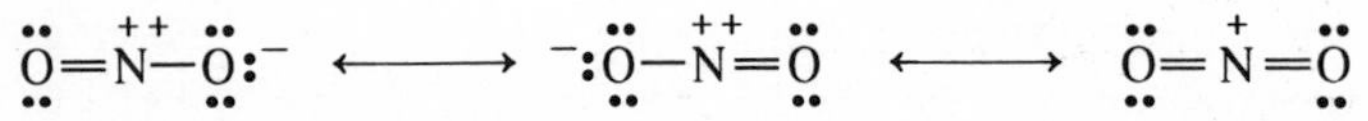

c. In order to get the nitronium ion, $^+NO_2$, consider protonation of $HO{-}NO_2$ by $HO{-}SO_2{-}OH$ followed by the removal of H_2O. However, water can also accept a proton from sulfuric acid with formation of the hydronium ion.

$$HO{-}NO_2 + HOSO_3H \xrightleftharpoons{\text{proton transfer}} H_2\overset{+}{O}{-}NO_2 + HSO_4^- \rightleftharpoons H_2O + NO_2^+ + HSO_4^-$$

$$H_2O + H_2SO_4 \xrightleftharpoons{\text{proton transfer}} H_3O^+ + HSO_4^-$$

The overall reaction would be: $HNO_3 + 2H_2SO_4 \rightleftharpoons NO_2^+ + 2HSO_4^- + H_3\overset{+}{O}$

d. The para atom will probably have a tetrahedral structure.

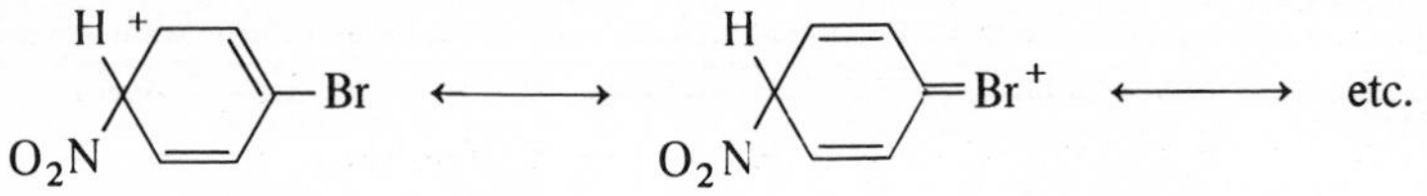

16-10 *a.*

and others

and others

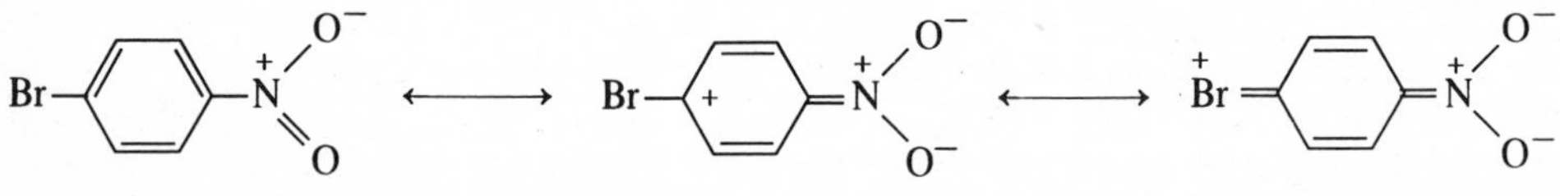

The para should be more easily displaced because of the low electron density at sites ortho and para to the nitro group.

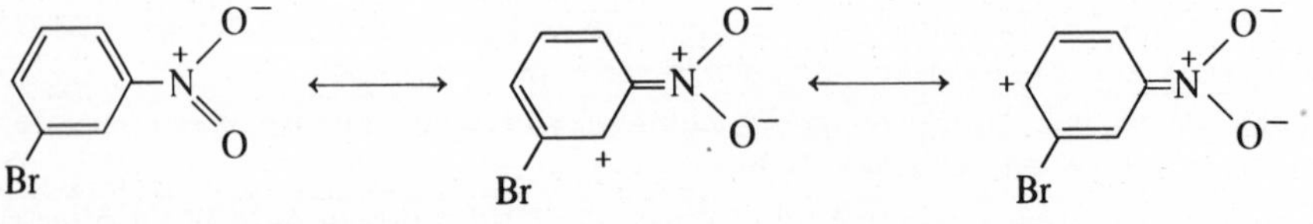

b. The oxygen atoms are coplanar with the ring. The carbon atom involved in substitution is sp^3.

c. (1) The second because two nitro groups are activating the halogen. (2) The second. Justify this by drawing resonance structures for the intermediate. (3) The first, because it is difficult for the nitro group in the second to be coplanar with the benzene ring. This lack of coplanarity will increase the energy of activation for formation of the intermediate.

d. Dinitrate chlorobenzene and treat with CH_3SNa.

16-11 When sulfinic acids ionize, two of the oxygen atoms become equivalent and asymmetry is lost.

16-12 *a.* Extract with water to remove the sulfonic acid, then with dil NaOH to remove the carboxylic acid.

b. Extract with dil HCl to remove the amine, then with dil $NaHCO_3$ to remove the acid. Finally wash with strong NaOH solution to remove the α-nitrotoluene.

16-13   a. Use an excess of concd H_2SO_4 for the forward direction and an excess of superheated steam for the reverse direction.

   b. For practical purposes this should be irreversible because any water that is formed is converted to H_2SO_4 by the reaction with SO_3.

16-14   a. Yes. Attack is by the electrophilic agent SO_3.

   b.

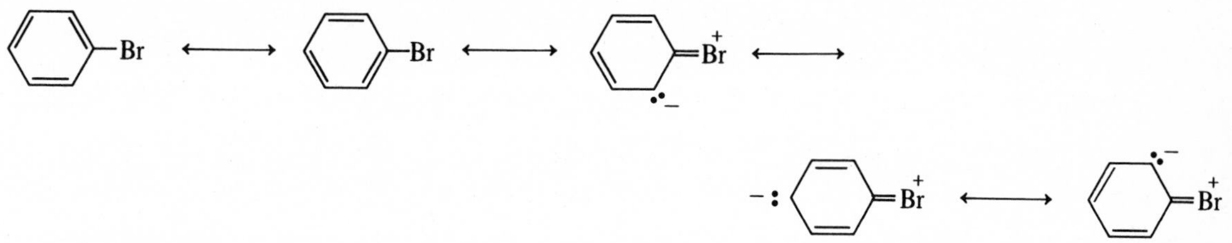

   c.

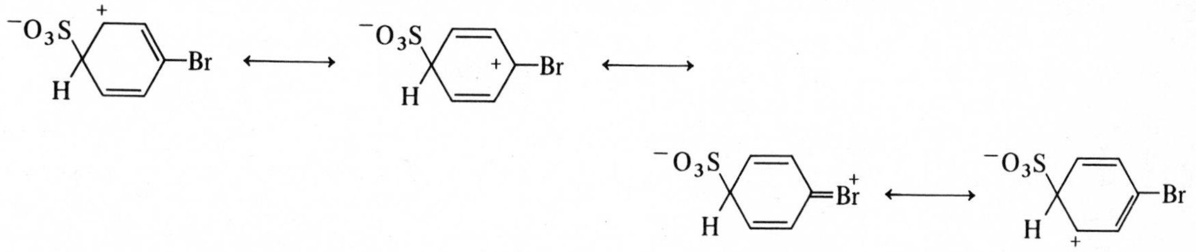

   d. 3,4-Dibromobenzenesulfonic acid.

   e. Take product from d and desulfonate by heating with steam.

16-15   a. A low sulfonation temperature will favor a high ortho-para ratio.

   b. Treat toluene with excess $ClSO_3H$. Reaction of resulting sulfonyl chloride with NH_3 followed by oxidation of CH_3 to COOH will give $o\text{-}HOOCC_6H_4SO_2NH_2$ which upon heating will give saccharin.

16-16   a. X,  Thiophene   Y,

   b. Thiophene has a fairly high resonance energy (29 kcal/mol according to the values given).

   c.

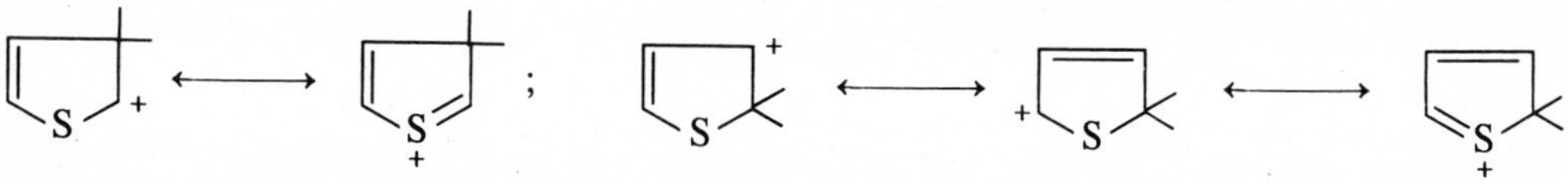

   d. The substitution would take place α to the sulfur atom because of a better charge distribution in the transition state (as shown for the intermediate in part c).

16-17   a.

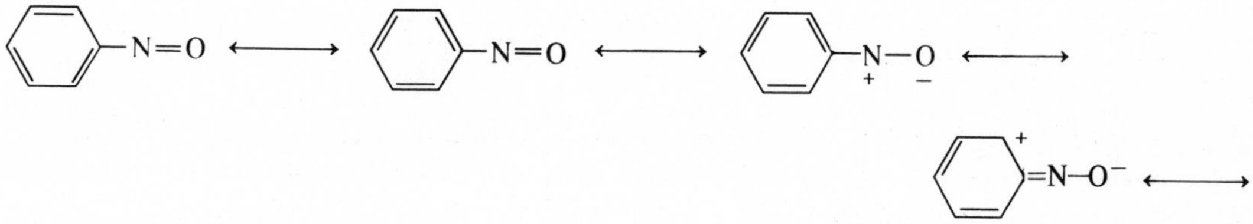

   b. Difference 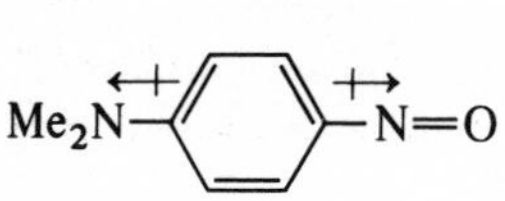   Opposing group moments.

   c. See answer to Prob. 15-6.

   d. Intermediate is resonance-stabilized, for example,

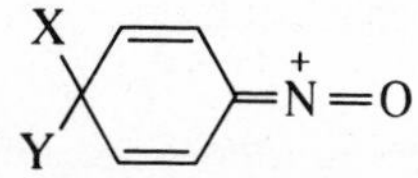

407

16-18

a.

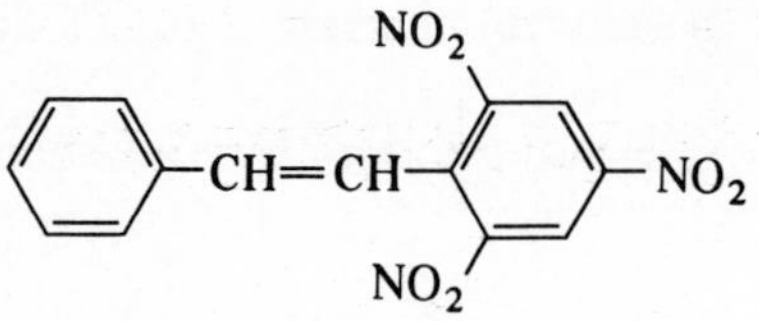

b. and c.

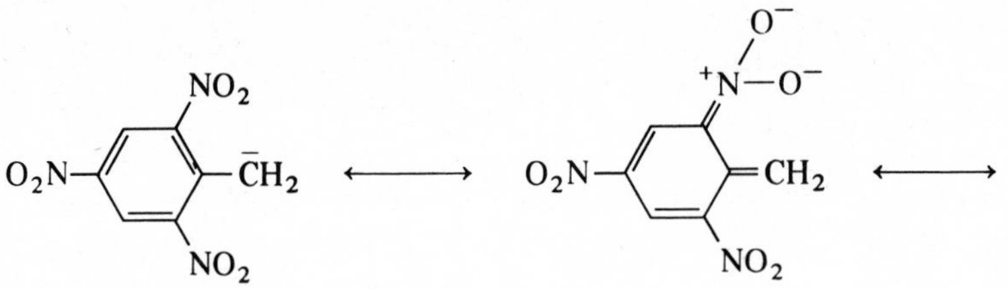

Anion stabilized by resonance.

16-19 a. Positive ρ. The S_N2 reaction between benzyl chloride and iodide ion in acetone.

b. Negative ρ. The S_N1 reaction between $(C_6H_5)_2CHCl$ and water.

c. Negative ρ. The basicity of sodium phenoxide in water.

d. Positive ρ. The acidity of $C_6H_5\overset{+}{N}H_3$ salts.

16-20 a. The *tert*-butyl group is more effective in electron release than the methyl group is. The observation is rationalized on the basis that the inductive effect in the former (three methyls vs. three hydrogens) is more important than hyperconjugation in the latter (however, the difference between values may be within experimental error).

b. The inductive and resonance effects of chlorine are opposite in sign. When chlorine is para, an electron-releasing resonance effect cancels out some of the electron-withdrawing inductive effect. When chlorine is in a meta position, the resonance effect has relatively little influence on positions which are ortho or para to chlorine.

16-21 Nitrate benzene, reduce, and acetylate to get $C_6H_5NHCOCH_3$ (acetanilide). Treat acetanilide with excess chlorosulfonic acid to get $p\text{-}ClO_2SC_6H_4NHCOCH_3$ which upon treatment with NH_3 and hydrolysis yields sulfanilamide. Marfanil may be prepared by oxidizing toluene to C_6H_5COOH and converting this to the amide. Reduction of benzamide with $LiAlH_4$ yields benzylamine which, if acetylated and treated as above with acetanilide, will yield Marfanil.

16-22 Hydrolysis of methyl acetate proceeds with attack of OH^- at the carbonyl carbon with cleavage occurring between oxygen and the acyl group. With methyl sulfonate, an S_N2 attack by OH^- occurs at the methyl carbon.

16-23 a. The order of arrangement of groups is: $-CH_2COOH$, $-CH_2CH_3$, $-CH_3$, $(:)$.

b. The dimethiodide would have one meso isomer and one *dl* pair.

16-24

a.

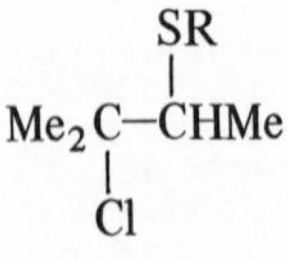

b.

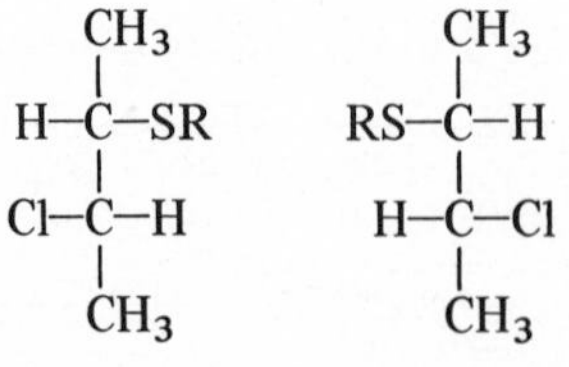

dl-threo

c.

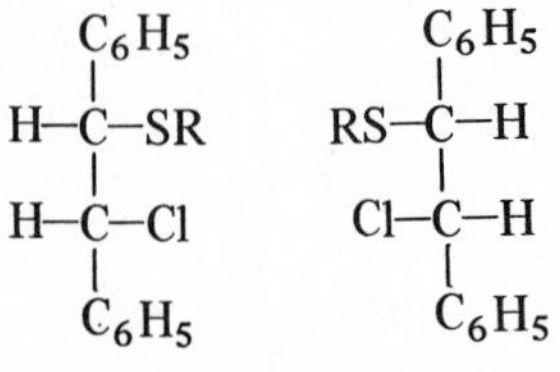

dl-erythro

16-25 a. According to the energy diagram, the rate-determining step takes place before the C–H or C–D bond is broken. Therefore, no isotope effect should be observed.

b. The isotope effect indicates that the C–H or C–D bond is being broken in the transition state. Consequently the energy diagram should be drawn with a higher barrier from intermediate to products than from intermediate to reactants.

16-26

a.

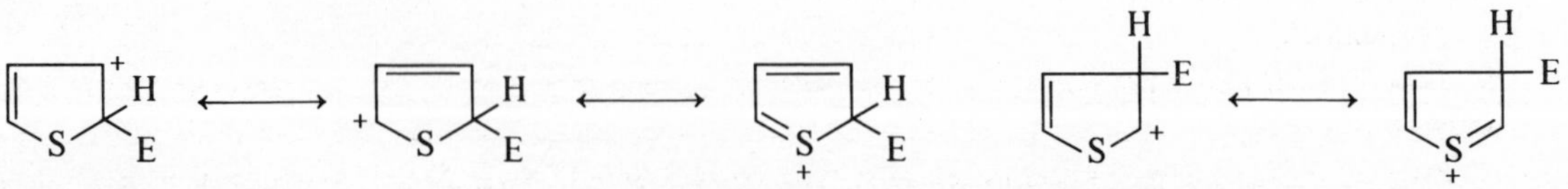

The first is better because of an additional canonical structure of equal significance.

b.

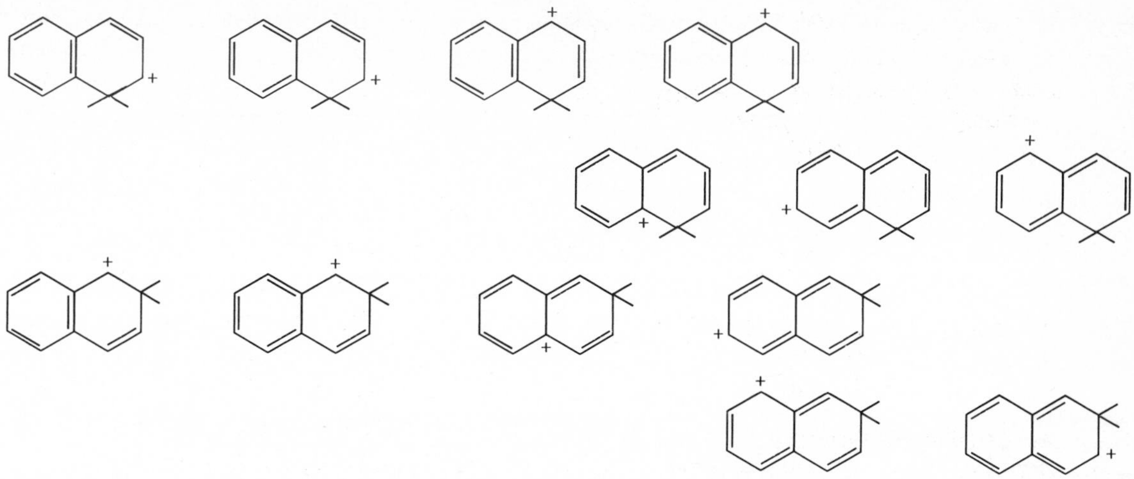

The top ion is more stable because of both more forms and more "aromatic sextets."

c.

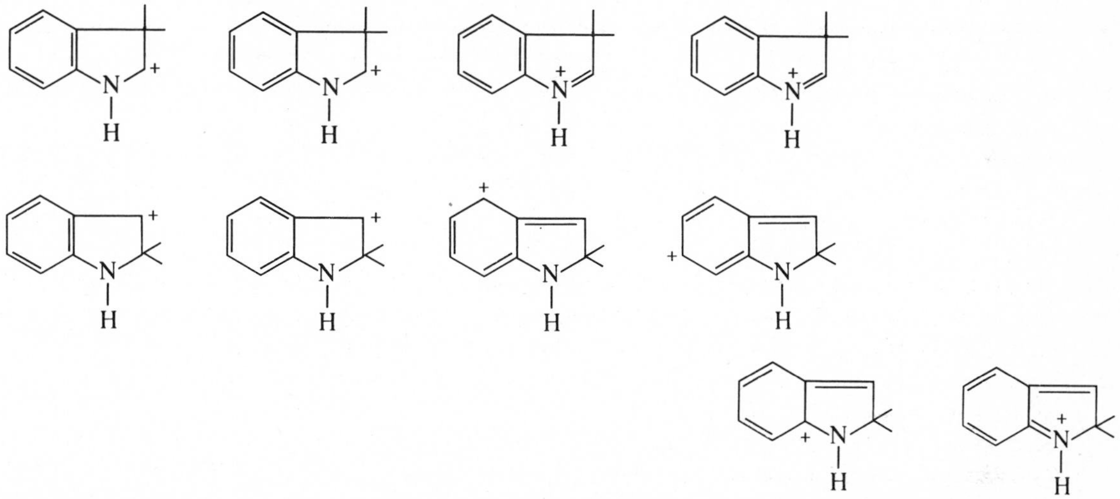

Our experimental evidence shows that substitution proceeds mainly via the top intermediate. Although the number of resonance forms which we can draw for this ion is less than that for the other, the discrepancy must be overbalanced by the fact that the structures are more significant (note that all of the upper forms have complete aromatic sextets in the six-membered ring).

d.

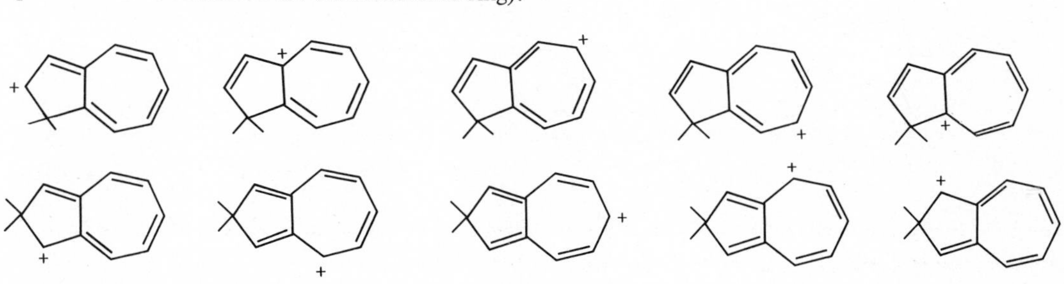

The top ion is more stable because the aromaticity of the seven-membered ring is kept more intact.

16-27   *a.* Two optically inactive isomers, one with cis oxygens and the other with trans. In the cis form the oxygen atoms are axial-equatorial and in the trans form they are equatorial-equatorial (or axial-axial in the less stable conformation).

b. Four optically active isomers; one *dl* pair with cis oxygen and methyl, another pair with trans oxygen and methyl.

c. Two optically inactive meso isomers, one with oxygen and methyls all cis and one with cis methyls and trans oxygen; two optically active isomers, a *dl* pair with the methyls trans to each other.

16-28   a. Direct displacement of TsO$^-$ by AcO$^-$ would lead to erythro (II), and carbonium ion formation by ionization of I would lead to a mixture of erythro and threo (II). Neither of these processes is operating, and therefore it is postulated that the benzene ring participates in the reaction (below) resulting in two inversions at the site.

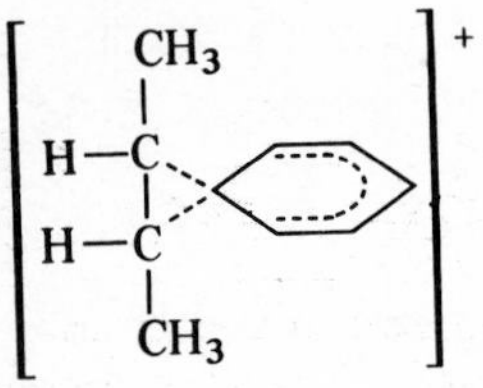

b. The intermediate shown above is symmetric starting from threo (I), but the intermediate resulting from erythro (I) is not symmetric and leads to optically active erthyro (II).

c. A p-OCH$_3$ group can better stabilize the intermediate carbonium ion intermediate.

16-29   a. Electrophilic attack by nitronium ion at the 1-position with the isopropyl carbonium ion the leaving group.

b. Electrophilic attack by nitronium ion with bromonium ion the leaving group.

c. Oxidation of the thiol to the disulfide by nitric acid.

d. An S_N2 reaction between CN$^-$ and butyl sulfate with C$_4$H$_9$SO$_4^-$ the leaving group.

e. Electrophilic aromatic substitution initiated by a proton transfer to the aromatic ring, giving rise to the same intermediate and transition state as that for the sulfonation reaction.

f.

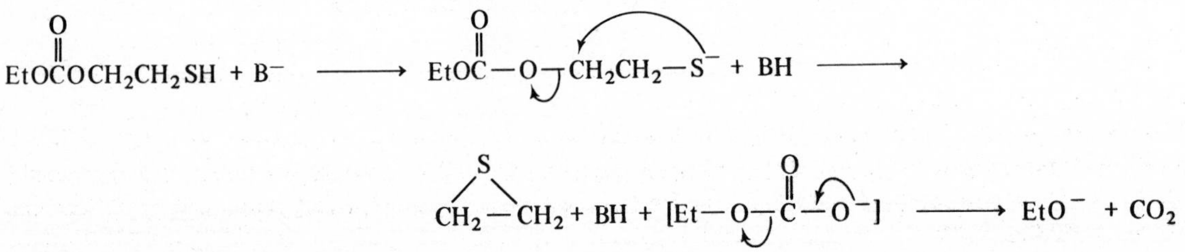

g. Racemization in the first suggests a carbonium ion mechanism. This is consistent with the structures because the first has much better possibilities for charge distribution. The second is an S_N2 reaction.

h. Nucleophilic aromatic substitution, with the sulfite ion the leaving group.

16-30

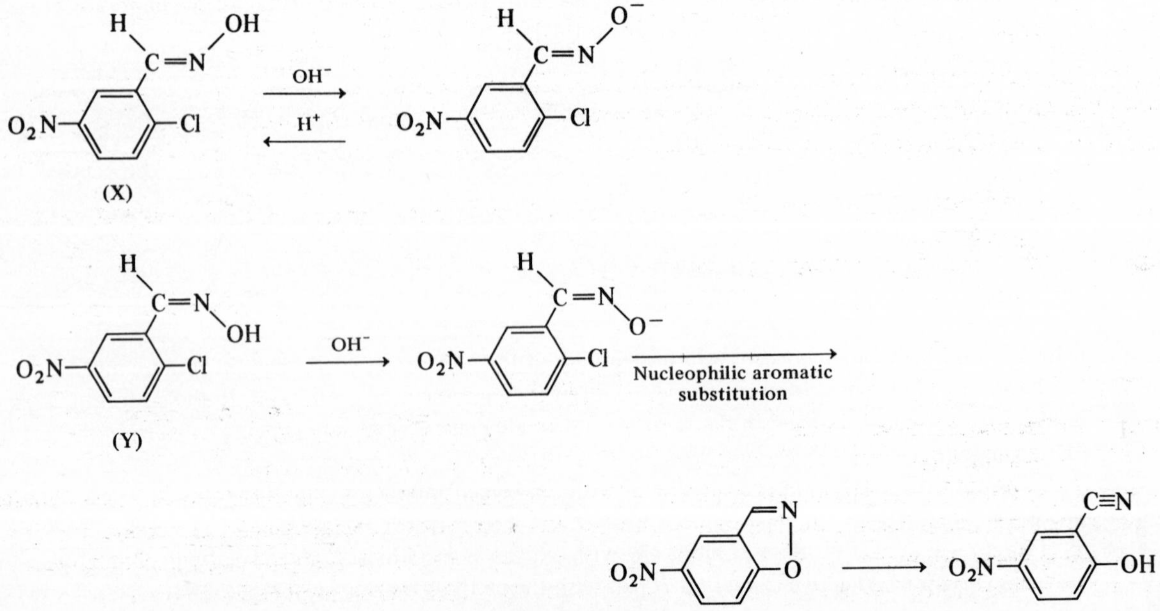

410

16-31 *a.* The nitrogen atom and all of the carbon atoms have sp^2 hybridization. Consequently the aromatic ring, hydrogen atoms, and the aluminum atom all lie in the same plane. The aluminum atom is tetrahedral.

b. In electrophilic aromatic substitution reactions there is an acid (Lewis or proton) present which will coordinate with or protonate the nitrogen atom of pyridine. The aromatic ring, having a positively charged nitrogen atom, is then unlikely to act as an electron donor in a substitution reaction. Two α-methyl groups, even though they are electron-donating, do not lead to a reactive ring. However, when two *tert*-butyl groups are in the α position, they may provide steric hindrance to Lewis acid coordination with the nitrogen atom, and the ring (now without a positive charge) will be more reactive as an electron donor.

c. 3-Bromopyridine is formed. In drawing electronic structures for the intermediates, one set (for 2 or 4 substitution) will have the positive charge on the nitrogen atom and will represent a higher energy than the other.

d. In considering the 2 position v. the 3 position, the intermediate formed from the nucleophilic attack of NH_2^- at the former will have a negative charge on the electronegative nitrogen atom in one canonical form, and hence will be favored over the other intermediate.

e. A proton is readily removed from the methyl group of 2-methylpyridine because of the stable anion which is formed (draw resonance structures). The removal of any other proton from either reactant would yield an anion of very high energy because the electron pair would be localized on the carbon atom. Pyridine reacts with phenyllithium to yield 2-phenylpyridine.

16-32 It must be emphasized that the dichlorination of nitrobenzene takes place stepwise. The problem then resolves itself into a prediction of the nature of electrophilic substitution of *m*-chloronitrobenzene. The chlorine will control the next step because it can delocalize the positive charge in the transition state.

16-33 *a.* 2,4,6-Trinitrotoluene *b.* $(CH_3)_2CHSOCH(CH_3)_2$ *c.* $(CH_3)_2CHSH$
       d. $HSCH_2CH_2CH_2SH$ *e.* $C_6H_5CH_2SCH_2COOH$ *f.* $p\text{-}CH_3NHC_6H_4NO_2$
       g. $CH_3CH_2C(SO_2CH_2CH_3)_2$
              $|$
              CH_3

16-34 A, $p\text{-}CH_3C_6H_4SO_2Cl$; B, $CH_3CH_2NH_2$; C, $p\text{-}CH_3C_6H_4SO_2NHEt$
16-35 A, C_6H_5Cl; B, $p\text{-}O_2NC_6H_4Cl$; C, 2,4-dinitrochlorobenzene; D, 2,4-dinitroanisole
16-36 X, $CH_2{=}CHSO_2CH_3$; Y, $CH_2{=}CHSCH_3$

16-37 A,   B, $C_6H_5{-}N{=}\overset{+}{N}{-}C_6H_5$
                                        $|$
                                        $O{-}$

Chapter 17

17-1 *a.* *N*-Methylaniline   *b.* *p*-Aminobenzylamine
      c. *m*-Toluidine   *d.* *p*-Aminoazobenzene
      e. *m*-Chlorobenzenediazonium chloride   *f.* 3,4-Dimethylbenzenediazonium fluoroborate
      g. *N,N*-Dimethyl-*p*-toluidine hydrobromide *h.* tris-*p*-Dimethylaminophenylamine
      i. *p*-Nitroacetanilide   *j.* Benzanilide (*N*-phenylbenzamide)
      k. *o*-Phenylenediamine   *l.* 2,2′,4-Trichlorohydrazobenzene

17-2   *a.* Forms *p*-nitroaniline and aniline hydrochloride (competition for a proton)

     b. Forms 3,3'-dimethylbenzidine   *c.* Forms phenylbenzylammonium chloride

     d. No reaction, diphenylamine is too weak a base   *e.* Forms the Schiff base, *p*-CH$_3$C$_6$H$_4$N=CHCH$_2$CH$_2$CH$_3$

     f. Forms sodium phthalimide   *g.* Forms *o*-tolunitrile, *o*-CH$_3$C$_6$H$_4$CN

     h. Forms *m*-fluorobromobenzene   *i.* Forms *p*-(*N*,*N*-dimethylamino)azobenzene

     j. Forms *p*-tolylhydrazine.

17-3   *a.* Make benzyl chloride from toluene (Cl$_2$ and light). Use this in a Gabriel synthesis.

     b. Nitrate toluene, oxidize the side chain, and reduce the nitro group.

     c. Nitrate toluene to get *o*-nitrotoluene. After oxidizing the CH$_3$ group, convert nitro into amino by reduction, then run a Sandmeyer reaction with CuCl.

     d. Nitrate benzene, reduce to aniline, acetylate to acetanilide, nitrate and separate the para isomer, hydrolyze the amide, and reduce the nitro group.

     e. Prepare *o*-nitroaniline as in *d*, diazotize, and treat with CuCN.

     f. Oxidize toluene to benzoic acid, convert to the acid chloride with SOCl$_2$, and add aniline (made in *d*).

     g. Successively nitrate and brominate benzene, then reduce the nitro group under mild conditions (SnCl$_2$ and HCl).

     h. Diazotize the product from *g* and treat with NaBF$_4$, Δ.

     i. Nitrate benzoic acid (part *f*), reduce, diazotize, treat with CuCN (Sandmeyer), and hydrolyze.

     j. Convert nitro in *p*-nitrotoluene (part *b*) to carboxyl as in *i*.

     k. Proceed as in *i* omitting hydrolysis step.

     l. Brominate *p*-nitroaniline (made in *d*), convert NH$_2$ to COOH as in *i*, reduce nitro, diazotize, and treat with H$_3$PO$_2$.

     m. Heat CS$_2$ with excess aniline (made in *d*).

     n. React benzyl chloride (part *a*) with aniline (part *d*) and nitrosate (can also make *N*-benzylaniline by reductive alkylation with C$_6$H$_5$NH$_2$ and C$_6$H$_5$CHO).

     o. Friedel-Crafts acylation of benzene with propionyl chloride, reduce ketone to propylbenzene with Zn and HCl, nitrate and isolate the ortho isomer. Reduce to the hydrazo compound with Zn and NaOH. Heat with acid to get the benzidine rearrangement.

     p.

     q. Prepare *m*-nitroaniline by Na$_2$S reduction of *m*-dinitrobenzene, diazotize, and couple with Cu.

     r. Proceed as in part *p* using *m*-bromonitrobenzene (made in *g*).

     s. Treat *p*-nitrobenzoic acid (made in *b*) with SOCl$_2$. React acid chloride with Et$_2$Cd to get ketone.

     t. Acetylate benzene by Friedel-Crafts, nitrate, and reduce (Wolff-Kishner followed by H$_2$, Ni).

     u. Heat *p*-toluidine hydrochloride with *p*-toluidine under pressure to make the bis amine. Treat with Li, then with *p*-iodotoluene in the presence of CuI.

     v. One possible sequence involves the following intermediates: *o*-dichlorobenzene, 3,4-dichloronitrobenzene, 2-chloro-4-nitroaniline, 2-chloro-4-nitro-6-bromoaniline, 3-chloro-5-bromonitrobenzene, product.

     w. *o*-Nitroaniline (made in *e*), *o*-nitrochlorobenzene, 2-nitro-4-bromochlorobenzene, product.

     x. Bromobenzene, *p*-bromoacetanilide, 2-chloro-4-bromoacetanilide, product.

     y. *p*-Methylacetanilide (made in *d*), 2-nitro-4-methylacetanilide, 3-nitro-4-fluorotoluene, product.

     z. Use the Gomberg reaction between *p*-nitrobenzenediazonium chloride (diazotize *p*-nitroanilide made in *d*) and toluene; oxidize to the acid; convert to the amine via the amide and NaOBr.

17-4   *a.* 3 > 1 > 4 > 2   *b.* 1 > 4 > 3 > 2   *c.* 6 > 2 > 3 > 1 > 5 > 4

17-5   *a.* A, C$_6$H$_5$$\overset{+}{\text{N}}$(CH$_3$)$_3Cl^-$ B, (CH$_3$)$_2$N—⟨○⟩—CH$_2$Cl   C, CH$_3$NHCH—⟨○⟩—CH$_3$
                                                                                                                                    Cl

           D, HOOC—⟨○⟩—COOH

b. E, C_6H_5NCOEt   F, C_6H_5NHMe
            |
            Me

c. G, H_2N—⟨O⟩—CHEt   H, diazonium salt of G   I, *sec*-butylbenzene
                   |
                   Me

     J. 4,4'-di-*sec*-butylbiphenyl   K, HOOC—⟨O⟩—⟨O⟩—COOH

d.

X Z on biphenyl with Y

Substituents	L	M	N	O	P	Q	R
X	NO_2	Br	NO_2	Br	NH_2	NO_2	Br
Y	—$CONH_2$	—$CONH_2$	NH_2	NH_2	NH_2	Br	Br
Z	Br	NO_2	Br	NO_2	Br	Br	NO_2

17-6   *a.* Make 3-methylpentanoic acid by the malonic ester synthesis, and *p*-nitroaniline as in 17-3*d*. Make the amide by way of the acid chloride.

b. Dinitration and α-chlorination of toluene to 2,4-dinitrobenzyl chloride; use this in an acetoacetic ester synthesis followed by cleavage with concd NaOH and finally reduction of the nitro groups.

c. Nitrate benzyl chloride to *p*-nitrobenzyl chloride and use this in an acetoacetic ester synthesis followed by hydrolysis and decarboxylation.

d. Make isobutyrylbenzene by Friedel-Crafts acylation, nitrate (meta) and reduce (Wolff-Kishner followed by H_2, Ni).

e. Convert the labeled benzoic acid to the amine by way of the amide and the Hofmann reaction. The amine can be diazotized and treated with KI to yield the product.

f. Nitrate the benzoic acid, then remove the carboxyl group either by way of the amine or by fusion with NaOH. The nitro group can be converted to fluoro by way of a diazonium reaction.

g. Acetylate the aniline (see part *e*), nitrate, and isolate the ortho isomer. Remove the amino group via diazotization and H_3PO_2, then convert nitro to hydroxyl.

h. Convert the carboxyl group to the amide, then to the amine with NaOBr. Acetylate, nitrate (para), hydrolyze, diazotize, H_3PO_2, reduce, diazotize, xanthate.

i. Start by way of aniline (as in part *h*), nitration (para), dibromination, etc.

j. Convert nitrobenzene to phenol via the diazonium salt; make Na salt of phenol and treat with propyl iodide to get propyl phenyl ether. Dinitration and selective reduction of the *o*-nitro group with Na_2S gives product.

k. Pinacol (from acetone plus Mg) is dehydrated to diene and reacted with maleic anhydride (Diels-Alder), and the product dehydrogenated to 4,5-dimethylphthalic anhydride. Convert this to the half acid—half amide followed by Hofmann reaction.

l. Use the following sequence: benzene, phenol, cyclohexanol, cyclohexene, cyclohexene oxide, *trans*-2-deuterio-1-cyclohexanol (with $LiAlD_4$), *trans*-tosylate, product.

m. Prepare cyclohexene as in (1); hydroboration (with B_2D_6) followed by H_2O_2,OH^- to *cis*-2-deuterio-1-cyclohexanol; then tosylation and S_N2 attack by NH_2^-.

n. Monobromination of *p*-methylacetanilide (17-3*d*) followed by conversion of amide group to carboxyl group via the diazonium salt—Sandmeyer reaction yields 2-bromo-4-methylbenzoic acid. Esterification of this followed by side-chain (CH_3) chlorination yields the benzyl chloride derivative which can be used in a malonic ester synthesis, followed by hydrolysis, decarboxylation, Hell-Volhard-Zelinsky reaction, and NH_3 treatment to give product.

17-7   a. *p*-Aminobenzoic acid is soluble in either dil HCl or NaOH.
       b. NH_3 gas will be liberated from the salt by cold dil NaOH.
       c. Aniline is soluble in dilute acids.
       d. The aniline derivative in the Hinsberg test will be soluble in dil NaOH.
       e. Same as *d*.
       f. Reaction of the diazonium salt of aniline with phenols gives highly colored coupling products.

17-8   a. The aromatic amine in A will be oxidized to a quinone. Nitration in B must be preceded by acetylation. The nitro group in C will react with $LiAlH_4$.
       b. The cyclic acetal in A will hydrolyze under acid conditions. In B, no displacement of aromatic halogen will occur except under very vigorous conditions.
       c. Amino group in A will react with both MeCl and $AlCl_3$. In B toluene is not active enough (no $-NR_2$ or $-OH$) for attack by the diazonium salt (a weak electrophile).

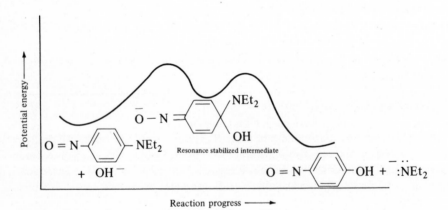

17-9   a.

       b.

       c.

17-10   a. A, nucleophilic substitution, S_N2; B, electrophilic aromatic substitution; C, nucleophilic substitution.
        b. Remove the salt with water, and separate the amines with the Hinsberg reagent. The compound you want, $C_6H_5NR_2$, will not react and can be extracted with dil HCl.
        c. Fig. P17-10c.

Figure A17-10c.

17-11   a. In aniline there is delocalization of the unshared nitrogen atom electrons by resonance with the ring.
        b. Consider resonance interactions between the groups.

414

c. An adjacent carbonyl group is more effective in delocalizing the electron pair on nitrogen than the phenyl group is.

d. Hydrogen bonding in the second compound decreases the concentration of hydroxide ion. The first compound is completely ionic.

e. In comparable situations, nitrogen is more basic than oxygen because of smaller nuclear charge.

f. Resonance delocalizations over two rings is more effective in reducing electron density at N.

17-12　*a.* The benzenediazonium salt forms a poor carbonium ion by loss of nitrogen. The positive charge is localized (no resonance stabilization).

　　　b. Refer to Sec. 5-2*c.*

　　　c. This would involve a nucleophilic displacement on carbon, with N_2 the leaving group, or an *E2* elimination reaction to form an olefin.

17-13　In answering parts *a* and *b* to this question, consider the phenoxide ion and the diazonium ion as the reacting species. A high pH favors the former but not the latter, and vice versa for a low pH.

c. 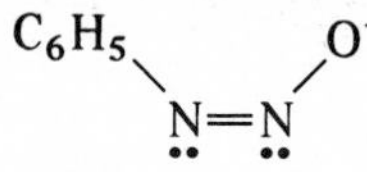　(cis) and 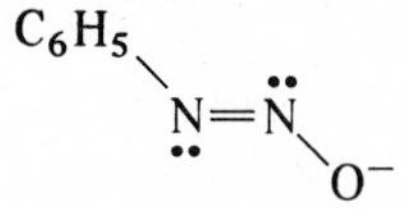　(trans)

17-14　*a.*

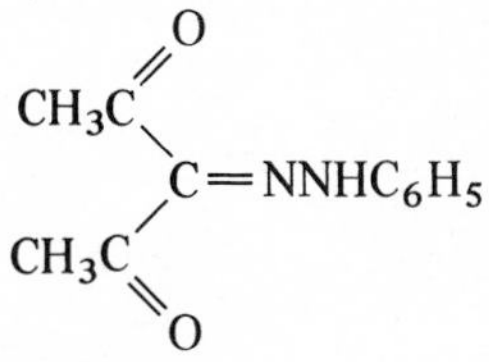

　　　c. Electrophilic

　　　b. A proton shift is involved (to nitrogen from carbon)

　　　d. Consider the most stable carbonium ion

17-15　*a.* α-Vinylpyridine

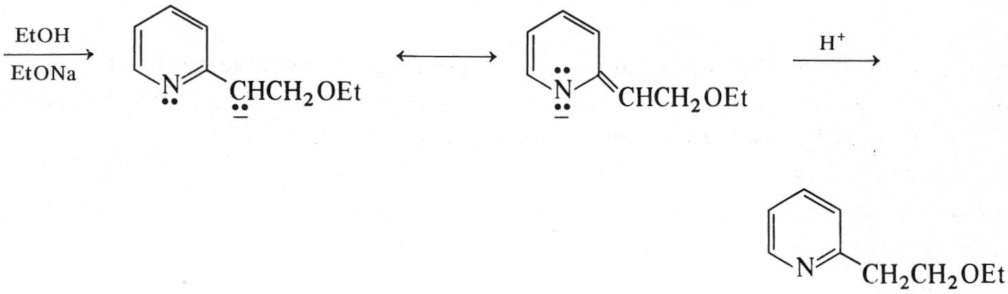

b. The nitrogen atom in the pyridine derivative is more electronegative than carbon and consequently is better able to delocalize the negative charge.

c. Positive.

17-16　*a.* 2,6-Diiodo-4-nitrophenylammonium chloride

　　　b. B, C_6H_5NHCSH　　　C, $C_6H_5N{=}C{=}S$　　　D, $C_6H_5NHCNHC_6H_5$
　　　　　　　　　 ‖ 　　　　　　　　　　　　　　　　　　　　　　　　 ‖
　　　　　　　　　 S 　　　　　　　　　　　　　　　　　　　　　　　　 S

　　　E, $C_6H_5N{=}C{=}NC_6H_5$　　　F, $C_6H_5NHCNHC_6H_5$
　　　　　　　　　　　　　　　　　　　　　　　　　 ‖
　　　　　　　　　　　　　　　　　　　　　　　　　 O

c. G,

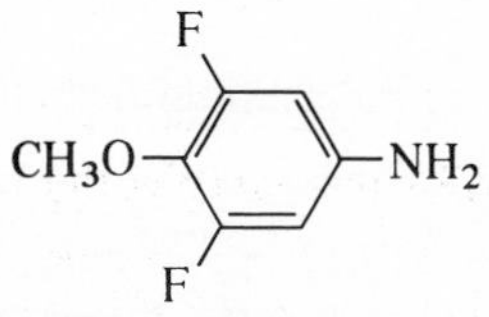

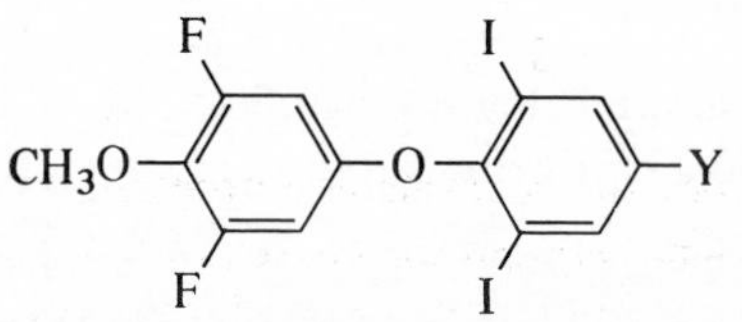

H, $Y = -CH$　(with =O)

I, $Y = -CH_2CHCOOH$
　　　　　　　　　|
　　　　　　　　 NH_2

When both fluoro groups in compound I are replaced by iodo groups, the resulting compound is thyroxine.

d. J, (structure)   K, (structure)

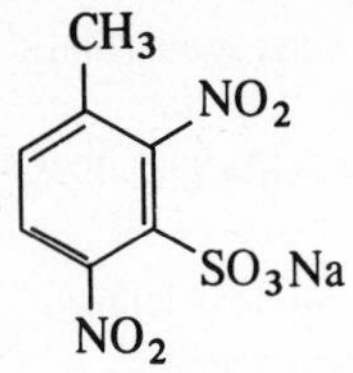

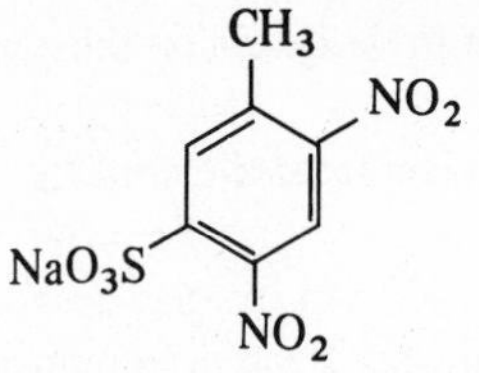

17-17 One alternative in rearrangements of this type involves fragmentation and recombination of the fragments. If this occurred, the indicated reaction would yield three products (provide structures for these). Only two products are obtained in practice, and consequently a concerted process (bonds being made and broken at the same time) would be a proper approach for mechanistic interpretation.

17-18 *a.* The rate-controlling step in bromine addition is the transfer of bromonium ion to the double bond. Since this involves generation of a positive charge at the reaction site, a nitro group would slow down the process because of an unfavorable coulombic interaction.

b. The pinacol rearrangement involves carbonium ion formation and possibly anchimeric (neighboring group) assistance. The former is better for both.

c. Consider distribution of negative charge in nucleophilic substitution.

d. Assume that both transition states are of the same energy. Then the ground state with the higher energy will lead to greater reactivity. The *tert*-butyl group is conformation-controlling; therefore the axial halogen is more reactive.

17-19 A, H_2N—(structure)—NH_2 B, (structure)—$NHNH_2$

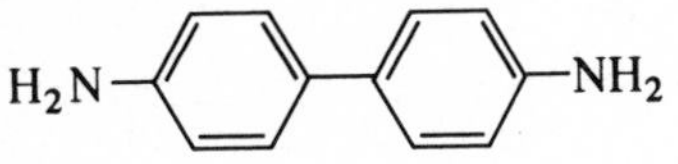

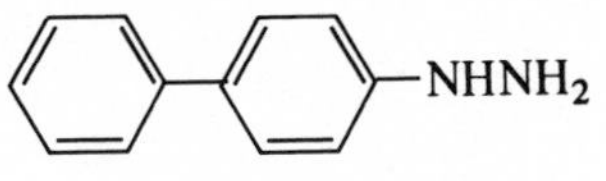

(Hydrazines are cleaved by vigorous reducing conditions)

C, (structure)—NH—(structure)—NH_2 D, (structure)—$NHNH$—(structure)

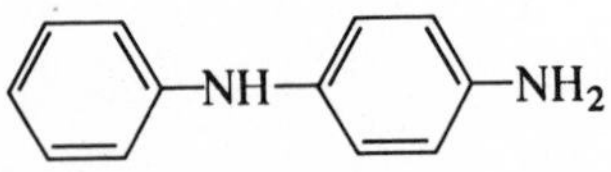

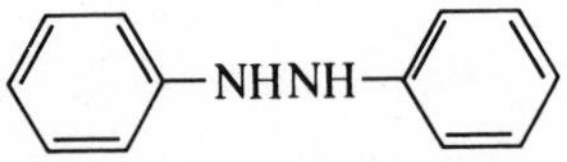

(Undergoes the benzidine rearrangement to yield salt M)

17-20

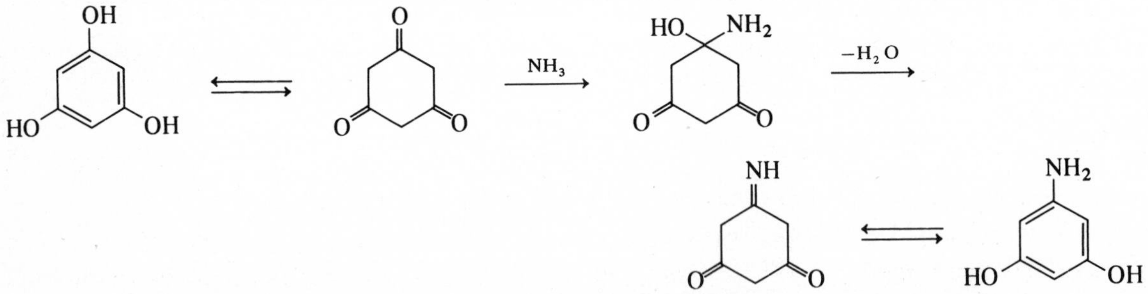

17-21 *a.* 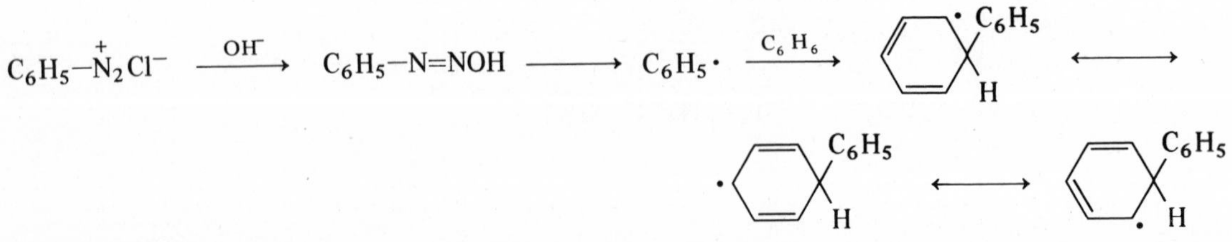 C_6H_5—$\overset{+}{N}_2Cl^-$ $\xrightarrow{OH^-}$ C_6H_5—$N{=}NOH$ $\longrightarrow$ $C_6H_5\cdot$ $\xrightarrow{C_6H_6}$ (structure) $\longleftrightarrow$ (structures)

b. If structures such as the one shown below are significant in stabilizing a free-radical intermediate, then the meta production will be slower than the ortho.

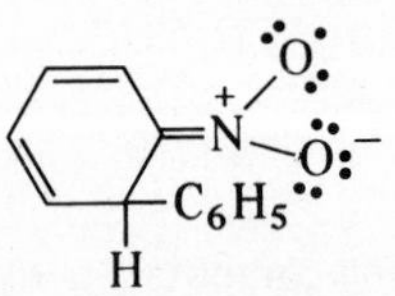

416

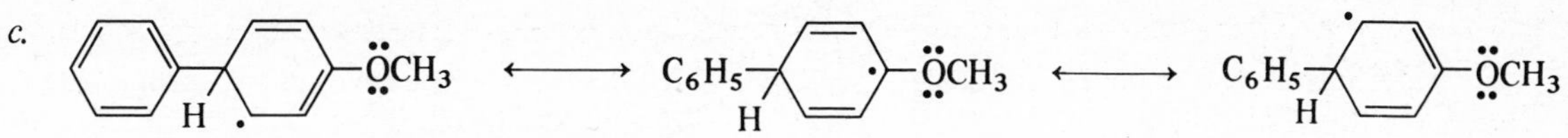

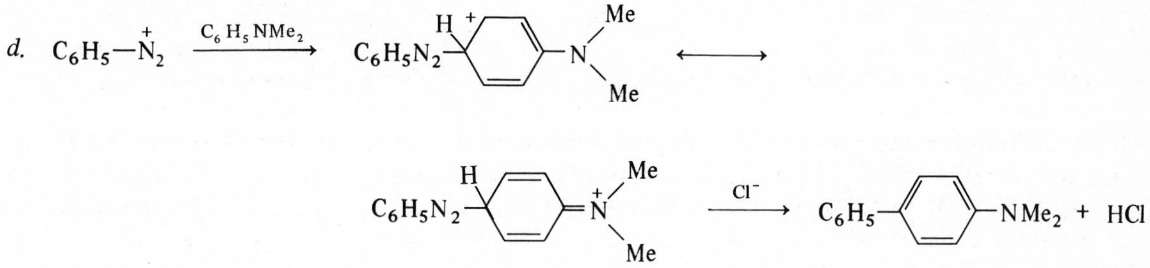

17-22 *a.* Fig. P17-22*a*.

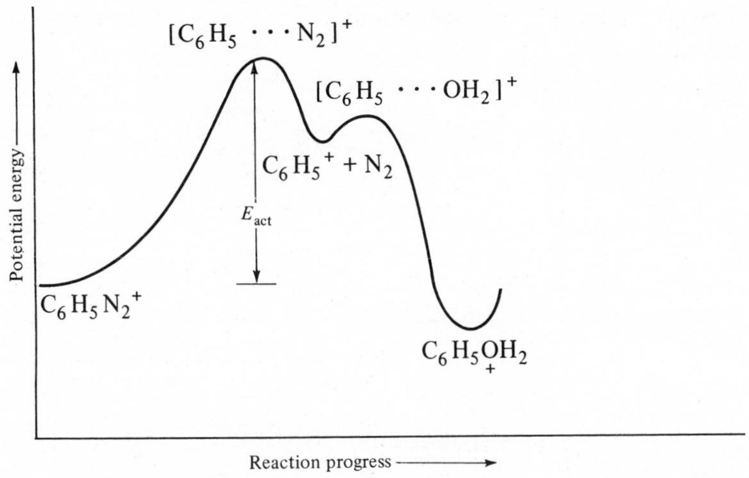

Figure A17-22a.

b. In the slow step, a positive charge is generated on the benzene ring. The formation of this carbonium ion is hindered because the para nitro group has already reduced the electron-density of the ring, particularly at the position where the nitrogen leaves.

c. The methoxyl group, through resonance, increases the double-bond character of the bond between the ring and the first nitrogen atom. To be sure you understand this, draw the significant contributing structures. In terms of the energy diagram, the net result is the lowering of the energy of the diazonium ion.

17-23 *a.* Methyl *p*-nitrobenzoate, methyl benzoate, methyl *p*-methylbenzoate.

b. Electron-releasing substituents should slow down the reaction. If k_o were known, the ratio could be calculated from the expression:

$$\frac{\log k' - \log k_o}{\log k'' - \log k_o} = \frac{\sigma'}{\sigma''}$$

17-24 A, *o*-$EtC_6H_4NH_2$; B, $C_6H_5CH(NH_2)CH_3$; C, C_6H_5NHEt; D, $C_6H_5CH_2CH_2NH_2$

17-25 X, diphenylamine; Y, *N*-nitrosodiphenylamine

17-26 I, *p*-$ClC_6H_4NH_2$; II, $C_6H_5N(CH_3)_2$; III, *p*-$EtOC_6H_4NH_2$; IV, *m*-$H_2NC_6H_4NH_2$; V, *m*-$O_2NC_6H_4NH_2$; VI, $C_6H_5N(CH_3)COCH_3$; VII, *p*-$CH_3C_6H_4NHCOCH_2COCH_3$

17-27 A, *p*-$CH_3C_6H_4NH_2$

17-28 Z, $C_6H_5NHCH_2CH_2OH$

18-1   *a.*   *p*-Methylanisole or *p*-methoxytoluene   *b.*   Phenyl propanoate or phenyl propionate

       c.   *p*-Hydroxyacetanilide   *d.*   2,4-Xylenol

       e.   *p*-Allylphenol   *f.*   Phenylhydroquinone

       g.   2,4,5-Trichlorophenoxyacetic acid   *h.*   2- or β-Naphthol

       i.   Sodium *o*-fluorophenoxide   *j.*   Methyl salicylate (oil of wintergreen)

       k.   2-Methoxybenzoic acid (*o*-anisic acid)   *l.*   Acetylsalicyclic acid (aspirin)

       m.   *p*-Isobutoxybenzaldehyde   *n.*   Potassium picrate

       o.   Tri-*p*-cresyl phosphate

18-2   *a.* $C_6H_5ONa + p\text{-}ClC_6H_4OH \longrightarrow C_6H_5OH + p\text{-}ClC_6H_4ONa$

       b. $C_6H_5ONa + Me_2SO_4 \longrightarrow C_6H_5OMe + MeSO_4Na$

       c. $m\text{-}MeC_6H_4OH + EtMgBr \longrightarrow m\text{-}MeC_6H_4OMgBr + CH_3CH_3$

       d. $C_6H_5OH + CH_2{=}C{=}O \longrightarrow C_6H_5OOCCH_3$ *e.* $p\text{-}HOC_6H_4OH + 2Zn \longrightarrow C_6H_6 + 2ZnO$

       f. 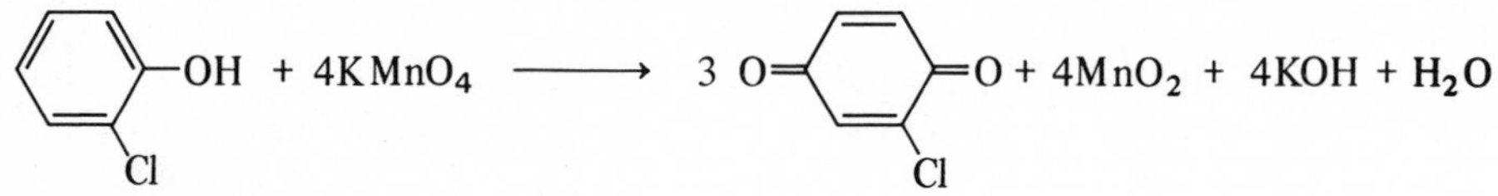 3 $\text{—OH} + 4KMnO_4 \longrightarrow 3$ $+ 4MnO_2 + 4KOH + H_2O$

       g. $C_6H_5NMe_2 + HONO \longrightarrow p\text{-}ONC_6H_4NMe_2 + H_2O$

       h. $C_6H_5OCH_3 + HI \longrightarrow C_6H_5OH + CH_3I$

       i. $p\text{-}CH_3C_6H_4OOCCH_3 \xrightarrow{\text{AlCl}_3}$

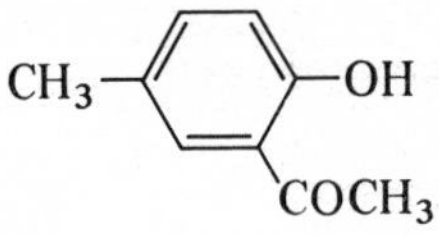

       j. $o\text{-}CH_3C_6H_4OH + 3KOH + CHCl_3 \longrightarrow$ 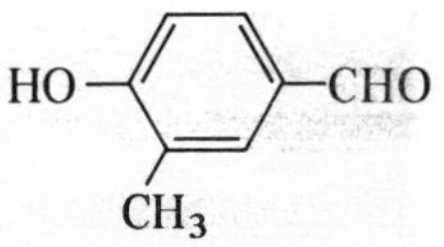 $+ 3KCl + 2H_2O$

       k. $p\text{-}CH_3C_6H_5OCH_3 + 2KMnO_4 \longrightarrow p\text{-}KOOCC_6H_4OCH_3 + 2MnO_2 + KOH + H_2O$

       l. $C_6H_5OTl + C_6H_5SO_2Cl \longrightarrow C_6H_5OSO_2C_6H_5 + TlCl$

       m. $o\text{-}CH_3C_6H_4OH + C_6H_5N_2Cl \longrightarrow C_6H_5N{=}N{-}$ $\text{—OH} + HCl$

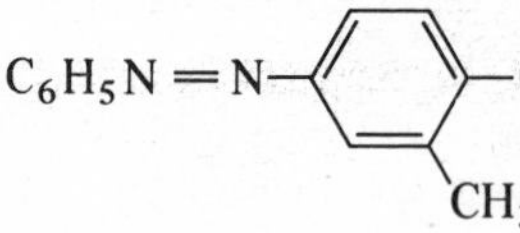

       n. $m\text{-}HOC_6H_4OH + NaOH + CO_2 \longrightarrow HO{-}$ $\text{—COONa} + H_2O$

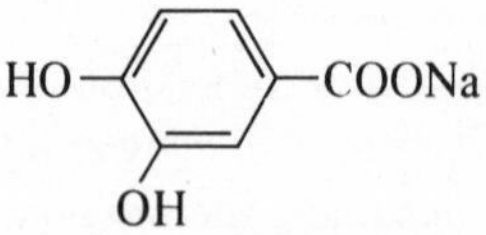

18-3   *a.*   Run a Fries rearrangement on the ester, then reduce the carbonyl to $-CH_2-$ with zinc amalgam and HCl.

       b.   Dinitrate chlorobenzene, then treat with KOH.

       c.   Use Gatterman reaction.

       d.   Houben-Hoesch synthesis followed by Clemmensen reduction.

       e.   Hydrogenate the ring with Pt, H_2; then oxidize the alcohol to the ketone.

       f.   Treat *p*-xylene with $Tl(OOCCF_3)_3$ followed by $Pb(OAc)_4$ and then dil NaOH.

       g.   Use Kolbe synthesis followed by acetylation with Ac_2O.

       h.   Convert aniline to phenol via the diazonium salt, then introduce the formyl group with the Reimer-Tiemann reaction.

       i.   Convert phenol to sodium phenoxide, make the allyl ether with allyl bromide, and heat (Claisen rearrangement).

       j.   Dichromate oxidation of 2,5-xylenol made in *f.*

18-4 *a.* C_6H_5COOH *b.* *o*-Nitrosation   *c.* *p*-Nitrosation
     d. $C_6H_5CH_2OH$ *e.* $p\text{-}HOC_6H_4SO_3H$ *f.* *N*-Nitrosation
     g. Mixture of alcohols and olefins   *h.* 1,2-Diphenyl-1-propanone
     i. *o*-Nitrosation

18-5 *a.* *p*-Cresol is soluble in dil NaOH.
     b. Dissolve in water and add CO_2. The phenol will come out of solution.
     c. The amine is soluble in dil HCl.   *d.* The acid is soluble in dil $NaHCO_3$.
     e. Test for halogen with the Beilstein test (green flame on hot copper wire).
     f. Salicylic acid is soluble in dil NaOH.
     g. *o*-Aminophenol is soluble in both dil HCl and dil NaOH.
     h. 2,4-Dinitrophenol is acidic enough to dissolve in dil $NaHCO_3$.
     i. Phenol gives a positive $FeCl_3$ test.

18-6 *a.* Extract phenol with aq NaOH.   *b.* The nitroheptane can be extracted with aq NaOH.
     c. The ortho compound will steam distill.   *d.* 3-Nitro-4-hydroxytoluene will steam distill.
     e. Extract salicylic acid with dil $NaHCO_3$.
     f. Dissolve in water and add CO_2. The phenol will come out of solution.

18-7 D, benzene ring with —OH and CH_2CHO substituents

     E, CH_2O

     F, benzene ring with —OCH_3 and CH_2CHO substituents

     H, benzene ring with —OCH_3 and CH_2CN substituents

     J, benzene ring with —OCH_3 and $(CH_2)_3CH_3$ substituents

     K, benzene ring with —OH and $(CH_2)_3CH_3$ substituents

     L, $C_6H_5(CH_2)_3CH_3$

18-8 *a.* A, HCl will not cause replacement of OH by Cl. B, Reimer-Tiemann requires a strong electron-releasing group on the ring, such as —O^-. C, both groups direct electrophilic substitution to the other position.
     b. A, under these conditions the aldehyde group will either be oxidized or will undergo a Cannizzaro reaction. C, Grignard reagent will react with phenolic hydrogen.
     c. A, the NH_2 group will also be acetylated. B, the NH_2 group would be acetylated in the rearrangement reaction plus reacting with the $AlCl_3$ initially.

18-9 A, CH_3O—benzene ring with —COOEt and NO_2 substituents

     B, CH_3O—benzene ring with —COOH and NO_2 substituents

     C, CH_3O—benzene ring with —COOH and OH substituents

     D, HO—benzene ring with —COOH and OH substituents

18-10 *a.* The key step is the reaction of phenylmagnesium bromide with cyclohexanone.
     b. The key step is a Claisen rearrangement of the allyl ether of guaiacol (*o*-methoxyphenol).
     c. Treat cyclobutanecarboxylic acid (from malonic ester and trimethylene bromide) with $SOCl_2$ to form the acyl halide. Add phenol to get the phenyl ester.
     d. The carbon skeleton is formed from a Fries rearrangement (para) of phenyl propionate. Reductive amination of the ketone gives the final product.
     e. Chlorinate *p*-nitrochlorobenzene and convert the —NO_2 to —OH via the diazonium salt. React the sodium salt of the resulting 3,4-dichlorophenol with chloroacetic acid and then convert the acid to the final phenyl ester product.

f. Friedel-Crafts acylation of benzene with succinic anhydride, followed by Clemmensen reduction, para hydroxylation (use the TTFA method, Sec. 18-2*e*), ethyl ether formation, reduction of –COOH to CH_2OH and dehydration.

g. Key steps are alkylation of *m*-methylanisole with *t*-butyl chloride followed by dinitration.

h. The key step is the Diels-Alder reaction of 1,2-diphenyl-1,3-butadiene with maleic anhydride. The diene could be prepared by a mixed aldol condensation of C_6H_5CHO and $C_6H_5CH_2CHO$ followed by a Wittig reaction using $CH_2=P(C_6H_5)_3$.

18-11 *a.*

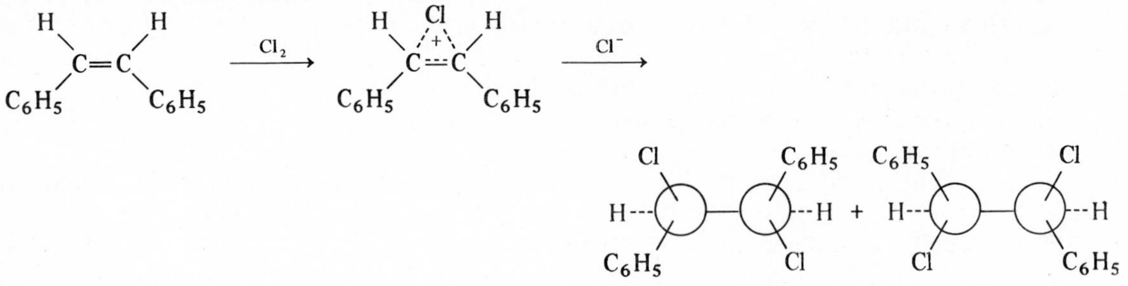

dl pair

b. One.   *c.* One meso product.

18-12 A, catechol   B, 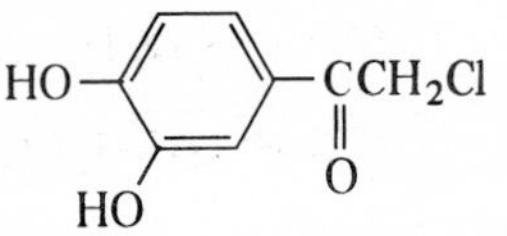   C,

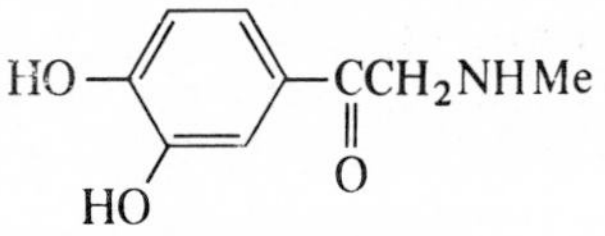

18-13

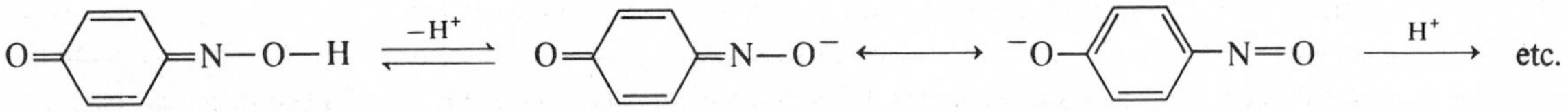

18-14 *a.*

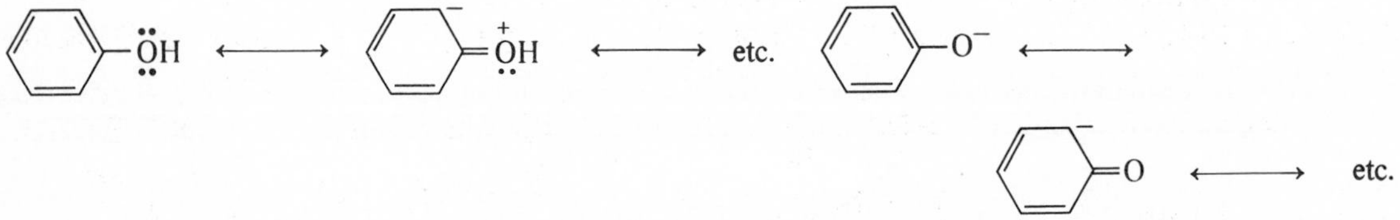

b. The phenoxide ion formation represents a small energy change because of charge distribution in the ring.

c. In nitrobenzene both inductive and resonance effects operate in the same direction (negative end toward nitro). In aniline the dominant resonance effect puts the negative end away from NH_2. In chlorobenzene the dominant inductive effect puts the negative end toward the Cl.

d.

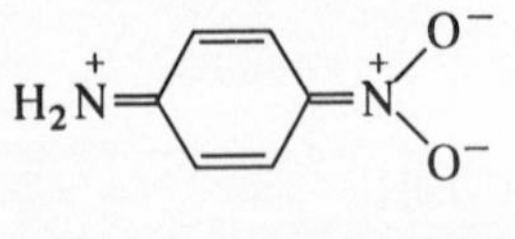

Highly contributing resonance structure with large charge separation.

18-15 *a.* *p*-Nitrophenol because NO_2 group can participate in distribution of negative charge.

b. *p*-Nitrophenol because *m*-NO_2 group can act only by an inductive effect.

c. *m*-Cresol because there is no resonance stabilization of $C_6H_5CH_2O^-$.

d. TNT because nitro group can distribute the negative charge in $ArCH_2{}^-$.

e. *p*-Nitrophenol because the inductive effect of the methyl groups strengthens the O–H bond.

f. *p*-Methoxybenzoic acid because of the greater resonance stabilization of the carboxylate ion compared to the phenoxide ion.

18-16 *a.* Under base catalysis the phenoxide ion which is involved is more reactive than the free phenol toward formaldehyde. Under acid catalysis, $H_2C=\overset{+}{O}H \longleftrightarrow H_2\overset{+}{C}-OH$, is formed which is more reactive than formaldehyde toward phenol.

b. Electrophilic.

420

c. Reactive electrophile is $o\text{-}HOC_6H_4CH_2{}^+$ ⟷ $HO\overset{+}{=}$ (cyclohexadienyl with CH_2) ⟷ etc.

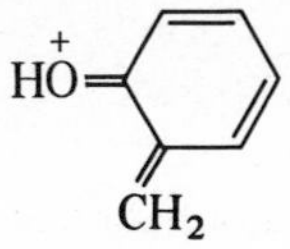

d.

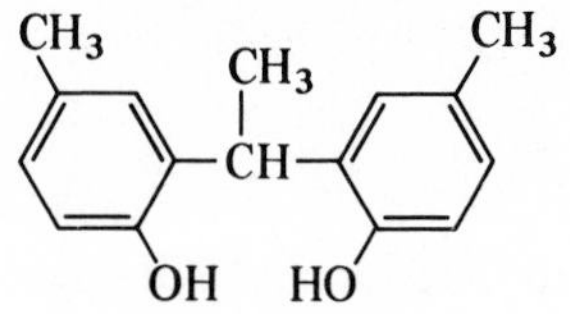

18-17 *a.* Chalcone is $C_6H_5COCH{=}CHC_6H_5$.

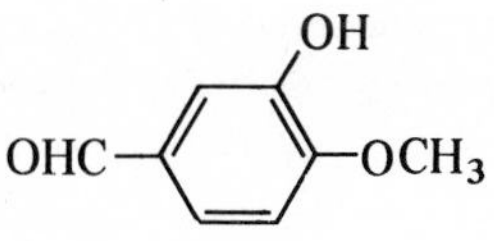

b. A, $HO-$ (ring with OH, OH, and $COCH_3$)   B, $OHC-$ (ring with OH, OCH_3)   V, $OHC-$ (ring with NO_2, OCH_3)

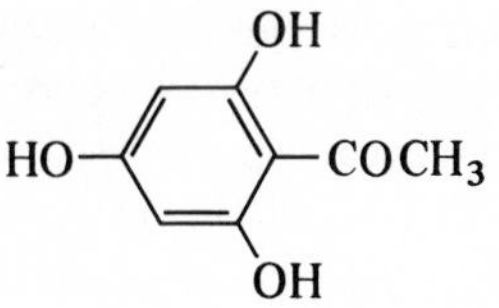

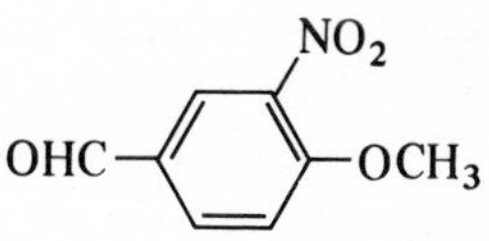

18-18 *a.*

ring with OH, $-\underset{\underset{C_6H_5}{|}}{CH}CH{=}CH_2$

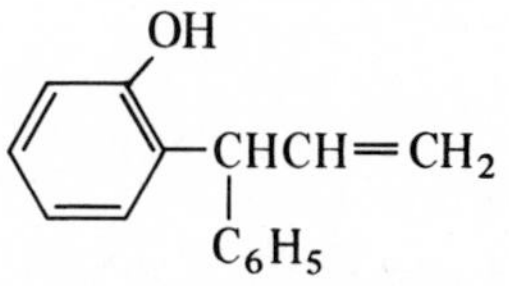

b. $HOCH{=}CHCH_2CH{=}CH_2$ Enol
$$\updownarrow$$
$O{=}CHCH_2CH_2CH{=}CH_2$

18-19 *a.* Run the reaction in the presence of some compound like toluene. If the $R-\overset{+}{C}{=}O$ species rearranges intermolecularly then it should react with the toluene present, forming some o- or $p\text{-}CH_3C_6H_4COR$.

b. Since the rearrangement involves attack by an electrophilic species $(R-\overset{+}{C}{=}O)$ the order of reactivity would be $p\text{-}CH_3C_6H_4OOCCH_3 > C_6H_5OOCCH_3 > p\text{-}O_2NC_6H_4OOCCH_3$.

18-20 *a.* $CHCl_3 + OH^- \rightleftharpoons :CCl_3{}^- + H_2O$ $\qquad$ $:CCl_3{}^- \longrightarrow :CCl_2$ (dichlorocarbene) $+ Cl^-$

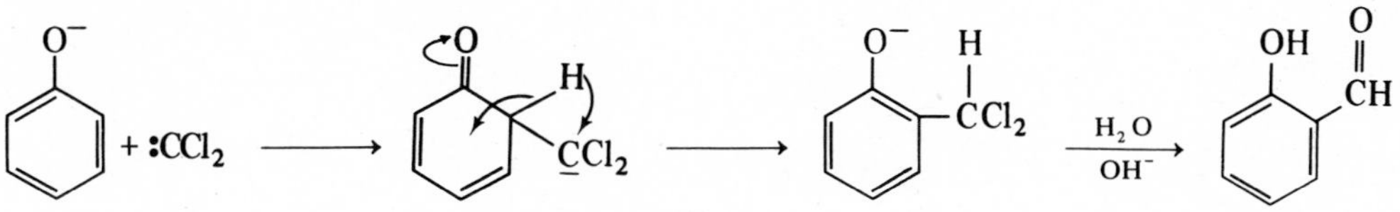

Hydrolysis

b. Products A and C are formed by the general mechanism shown in part *a*. Product B is formed by initial attack of $:CCl_2$ at the para position. Product D is formed by initial addition of $:CCl_2$ to a C=C followed by ring expansion.

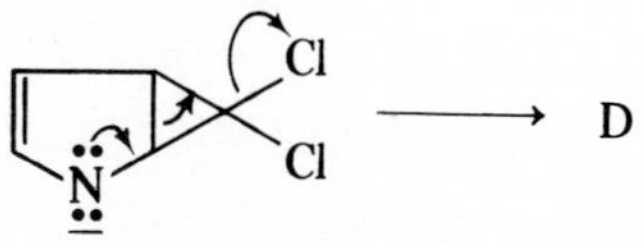

$\longrightarrow$ D

18-21 A is phenyl acetate formed by attack of CH_3CN at the phenolic oxygen resulting in O-acylation rather than ring C-acylation. B is methyl o-hydroxyphenyl ketone formed via Fries rearrangement of A.

18-22 *a.*

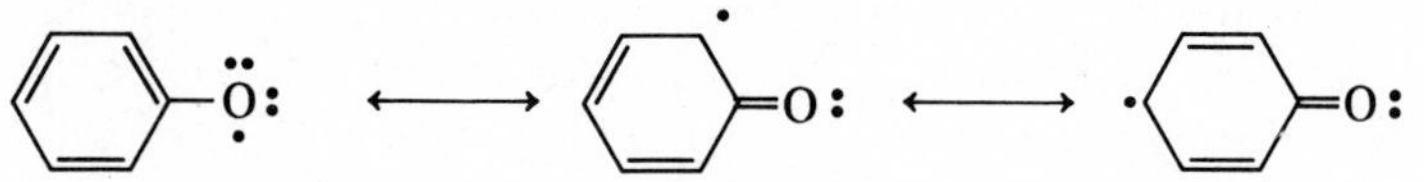

(A) ⟷ (B) ⟷ (C)

b. C + phenol $\xrightarrow{\text{para attack}}$ 4,4'-dihydroxybiphenyl

$\quad$ A + phenol $\xrightarrow{\text{ortho attack}}$ 2-hydroxydiphenyl ether

c.

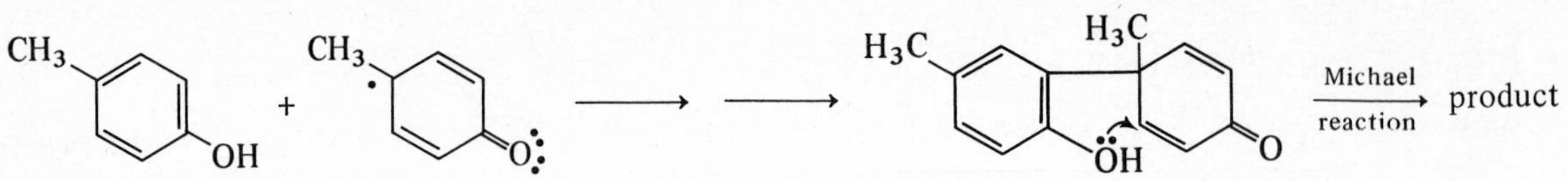

18-23   *a.* No solution for ortho.
      b. 13.2.
      c. 11.7.

18-24   *a.* Electron-withdrawing substituents have a positive σ value.
      b. The σ for *p*-amino is -0.66 and for *p*-acetamido is -0.02. If the value for *p*-OH is -0.36, it might be expected that the sign for *p*-acetoxy might change. This would be reasonable because the electron-withdrawing inductive effect would have increased greatly and at the same time the resonance contributions between oxygen and the ring would have been diminished.
      c. The presence of the second phenyl ring would lower the average resonance contribution of oxygen electrons to a ring. Hence the σ value for *p*-O$-$C$_6$H$_5$ should be less negative. The actual value for σ is -0.03.

18-25   *a.* In the transition state for this process the negative charge is being dispersed (eventually lost). This would indicate a negative ρ value.
      b. Because of attack by the negatively charged hydroxide ion, the transition state involves generation of negative charge, and hence the value of ρ would be positive.

18-26   A, *p*-cresol;   B, *o*-cresol.

18-27   C, *p*-dimethoxybenzene;   D, *p*-methoxyphenol;   E, *o*-methoxyphenol (guaicol);   F, catechol.

18-28   I, *p*-chlorophenol;   II, *o*-nitrophenol;   III, 2,6-dimethylphenol;   IV, *p*-bromoanisole;   V, 2-isopropyl-5-methylphenol. From the data given, other ring-substituted isomers for III and V would be possible.

18-29   G, *p*-CH$_3$OC$_6$H$_4$OOCN(CH$_3$)$_2$;   D, *p*-CH$_3$OC$_6$H$_4$OH; H, (CH$_3$)$_2$NH.

18-30   X, 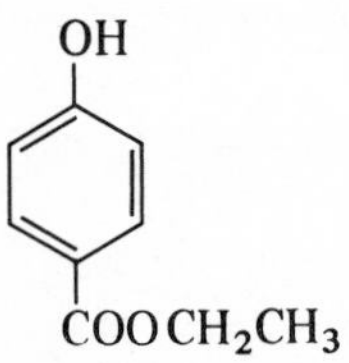   Y, 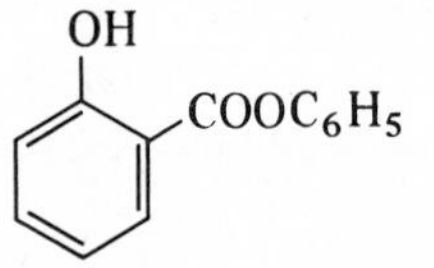   Z,

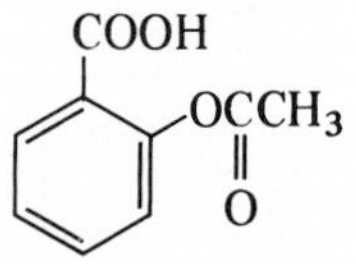

Chapter 19

19-1   *a.* 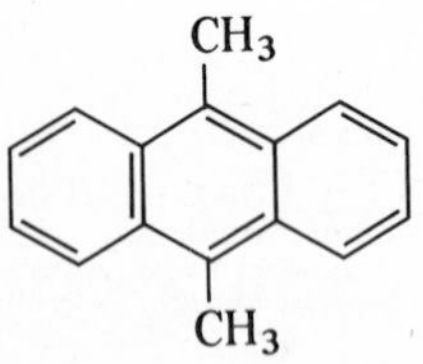   *b.* 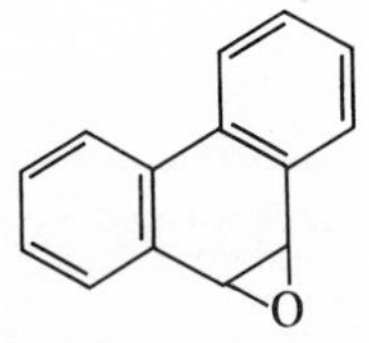   *c.*

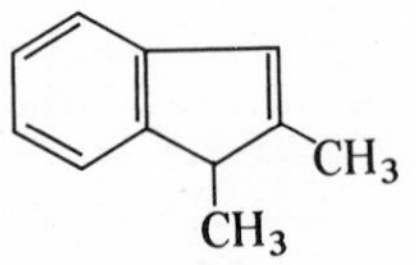

d. 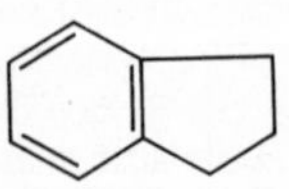   *e.* 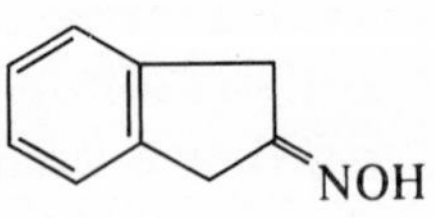   *f.*

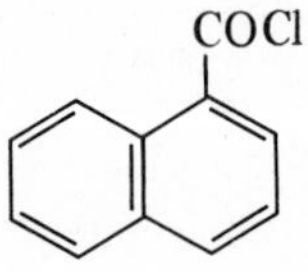

g.    *h.* 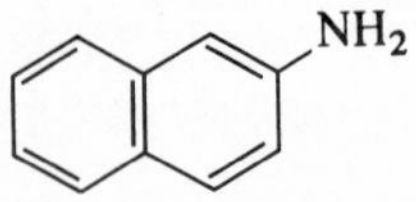   *i.*

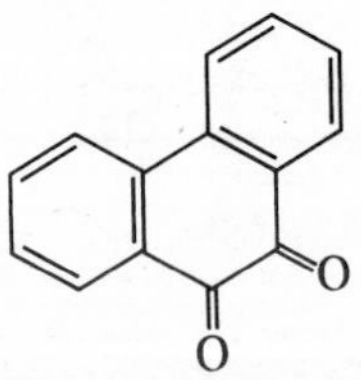

j.

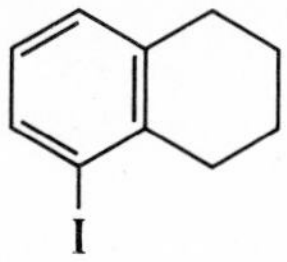

k.

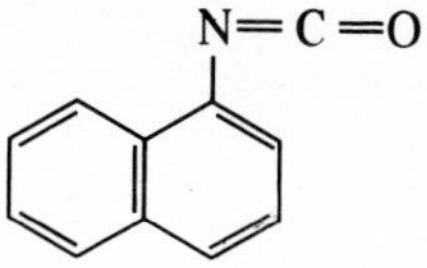

l. $C_6H_5CCH_2COOEt$
$\quad\quad\ \overset{\|}{O}$

m. O_2N——$CH_2{=}CHCOOH$

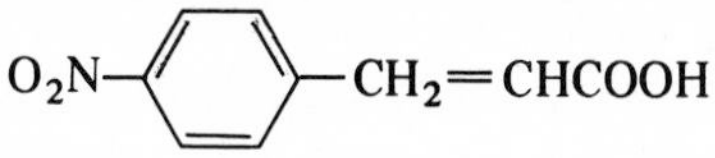

n.

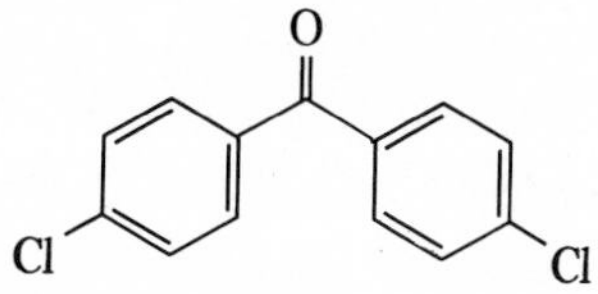

o. SO_3H

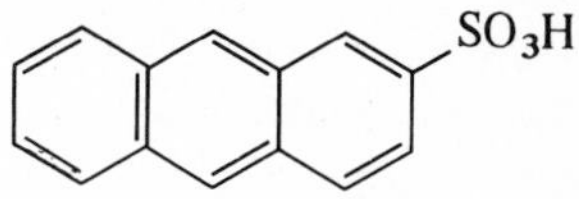

p. p-$OHCC_6H_4CHO$

q. $C_6H_5CHCl_2$

r. $C_6H_5C{=}NNHCNH_2$
$\quad\quad\ \ \underset{n\text{-}C_3H_7}{|}\ \ \overset{\ \ \|}{O}$

s.

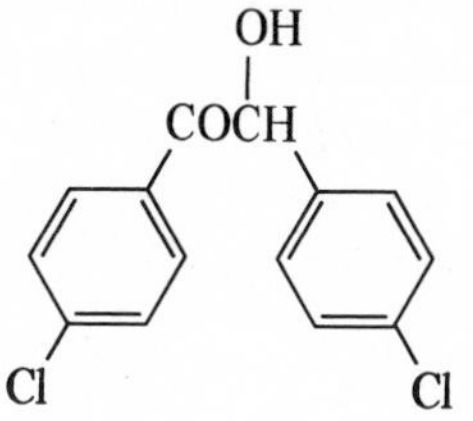

t.

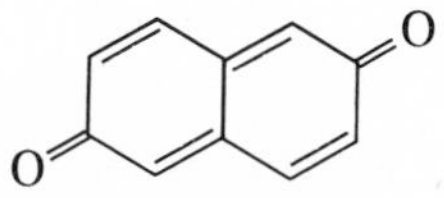

u.

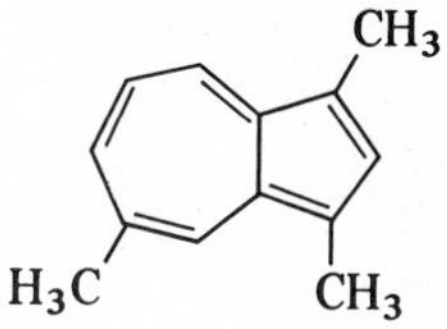

v.

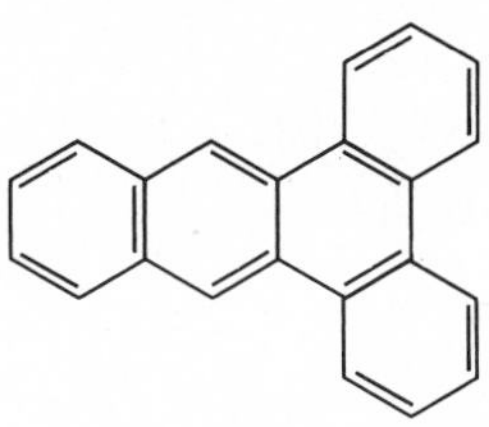

w.

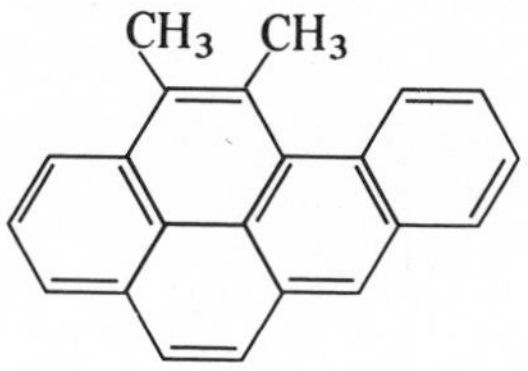

x. $C_6H_5CCH_2CH_2CONH_2$
$\quad\quad\ \overset{\|}{O}$

y. C_6H_5——C_6H_5
$\quad\quad\quad CH_3OOC \quad\ \ COOCH_3$

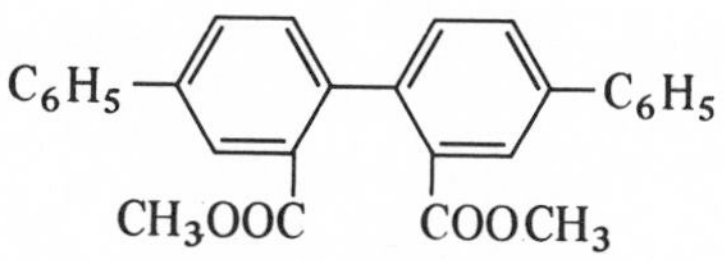

z. $CH_2{=}CHCH_2 \quad\quad CH{=}CH_2$

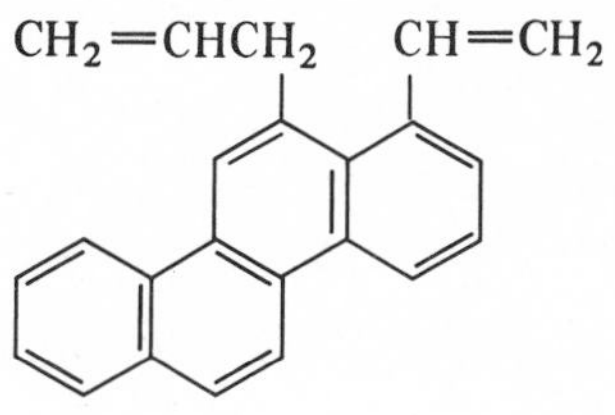

19-2   *a.* $C_6H_5COONH_4 + Ag$

b. $C_6H_5COONa + C_6H_5CH_2OH$

c. C_6H_5COCl

d. p-$CH_3C_6H_4CH{=}NNHC_6H_5$

e. $C_6H_5CH_2CH{=}CCHO$
$\quad\quad\quad\quad\quad\quad\ \underset{C_6H_5}{|}$

f. $C_6H_5CH_2CH\overset{O-CH_2}{\underset{O-CH_2}{\big\langle}}$

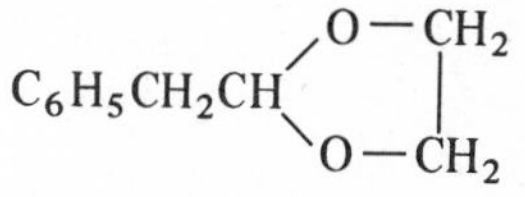

g. $[(C_6H_5)_2CHO]_4AlLi$

h. $(C_6H_5)_2C{=}NNHCONH_2$

i.   Phthalic acid

j.   2-Nitrophthalic acid

k.   1-Allyl-2-naphthol

l.   2-Aceto-1-naphthol

m. $O{=}$${=}O$

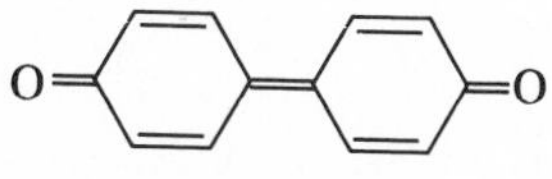

n.

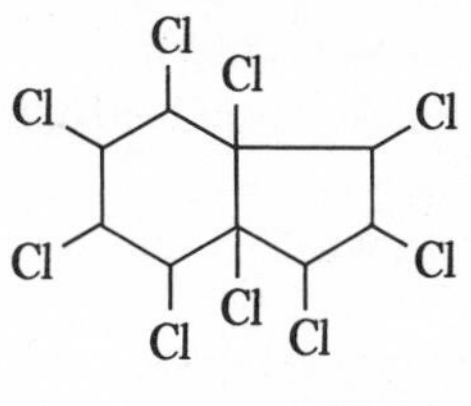

o.

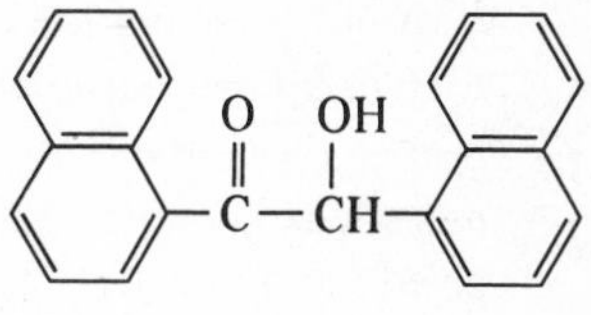

p. CN

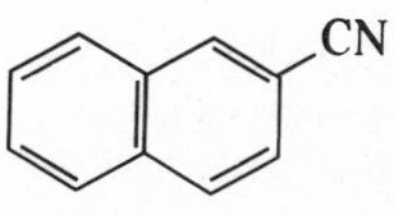

q.   4-Amino-1-naphthalenesulfonic acid

r. $\underset{\displaystyle \text{O OH}}{C_6H_5CH_2CH_2\overset{\parallel}{C}-\overset{\mid}{C}HCH_2CH_2C_6H_5}$

s. $(p\text{-}EtC_6H_4)_2\underset{\displaystyle \overset{\mid}{OH}}{CCOOH}$

t. 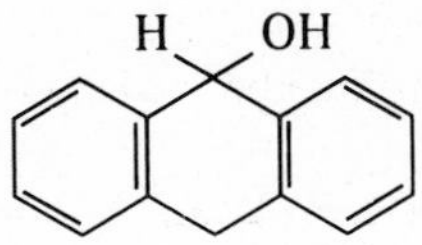 (after H^+)

u. $(C_6H_5NH)_2C{=}S$

v. No reaction

w. $C_6H_5CHBrCH_2CHO$

x. *o*- and *p*-Nitrophenyl benzoates

y.

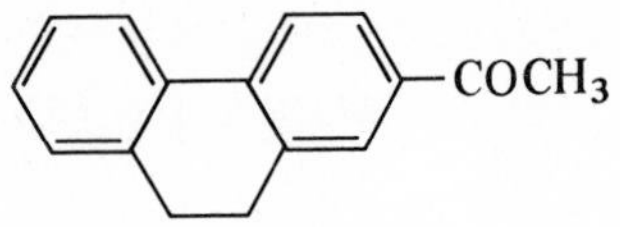

z. $C_6H_5CH{=}\underset{\displaystyle \overset{\mid}{C_6H_5}}{CCHO}$

a'.

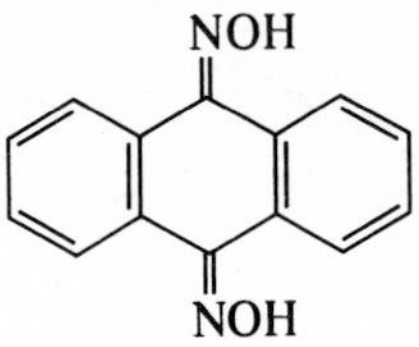

b'.

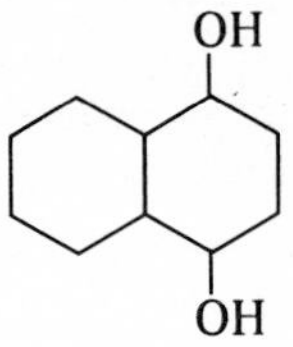

c'. HO— —OH with Cl

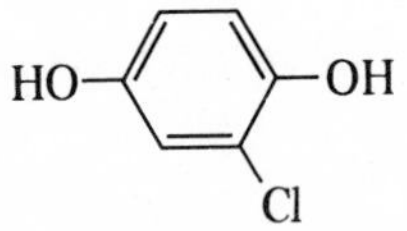

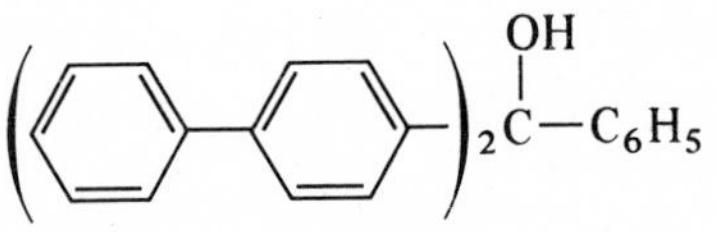

d'. $\left(\text{biphenyl} \right)_2 \overset{\displaystyle \text{OH}}{\underset{\mid}{C}}{-}C_6H_5$

e'. naphthalene with CN

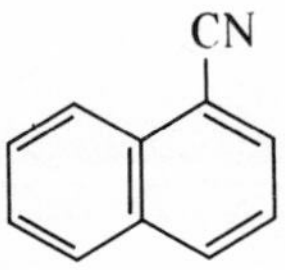

f'. $p\text{-}CH_3C_6H_4CH_2OH + EtOH$ (after H^+)

19-3
a. React EtMgBr with the nitrile and hydrolyze.
b. Friedel-Crafts acylation with butyryl chloride.
c. Pyrolysis of the Ca^{2+} salt of benzoic acid.
d. C_6H_5MgBr plus C_6H_5COOEt will yield the desired product.
e. Nitrate in the α position, then reduce the nitro group with Sn and HCl.
f. Sulfonate at 160°C to get β isomer; fuse with NaCN and hydrolyze the nitrile to the carboxylic acid.
g. Chloromethylate naphthalene in the α position with CH_2O, HCl, and $ZnCl_2$; treat with KCN, and hydrolyze the nitrile to the acid.
h. Treat naphthalene with butyryl chloride and $AlCl_3$ in nitrobenzene solvent; Wolff-Kishner reduction of β-butyrylnaphthalene yields the product.
i. Dinitrate naphthalene to the 1,8 isomer and reduce one nitro group using Na_2S in alcohol.
j. Aldol condensation of acetophenone yields product.
k. Pinacol rearrangement of benzopinacol made from benzophenone (Mg·Hg).
l. Use the Gattermann-Koch synthesis.
m. *p*-Benzoquinone in a Diels-Alder reaction with 2,3-dimethyl-1,3-butadiene followed by heating with Pt will give final product.
n. Esterify by the Fischer method.
o. Commercially, toluene is oxidized directly to C_6H_5CHO with MnO_2, with air and V_2O_5, or by dichlorination of the side chain and hydrolysis. The best laboratory route would involve oxidation of toluene to benzoic acid, conversion to the acid chloride, and reduction via the Rosenmund method.
p. Phenol + acetyl chloride gives phenyl acetate; then Fries rearrangement.
q. The final step is the benzylic acid rearrangement of 9,10-phenanthrenequinone.

r. *o*-Xylene $\xrightarrow{[O]}$ phthalic acid $\xrightarrow{LiAlH_4}$ diol $\xrightarrow{2HCl}$ $\xrightarrow{2NaCN}$ dinitrile $\xrightarrow{2\ EtOH,\ H_2SO_4}$ diester $\xrightarrow{Dieckmann}$ $\xrightarrow{H^+,\ H_2O,\ \Delta\ (-CO_2)}$ $\xrightarrow{MeMgI}$ $\xrightarrow{-H_2O}$ product

s. Stephen's (SnCl$_2$, HCl) or LiAlH$_4$ reduction of the nitrile gives the aldehyde.

t. Claisen-Schmidt reaction of *p*-tolualdehyde and acetone followed by H$_2$, Pt reduction.

u. Prepare tetralone as in Sec. 19-2*a*. React cyclopentylmagnesium bromide with 1-tetralone, dehydrate, and dehydrogenate.

v. Nitrate to 4-nitrobiphenyl, reduce to amine, diazotize, and treat with KI to get 4-iodobiphenyl; then perform the Ullmann reaction with Cu to couple to *p*-quaterphenyl.

w. Chalcone (prepared Prob. 19–22) reacted with malonic ester in a Michael condensation is the last step.

x. 1-Methylnaphthalene $\xrightarrow[AlCl_3]{}$ $\xrightarrow{HF}$ $\xrightarrow[HCl]{Zn(Hg)}$ $\xrightarrow[\Delta]{Pd}$ product

y. *p*-CH$_3$C$_6$H$_4$CHO $\xrightarrow{CN^-}$ *p*-CH$_3$C$_6$H$_4$C—CHC$_6$H$_4$CH$_3$-*p* $\xrightarrow{HNO_3}$ $\longrightarrow$ product
  $\quad\quad\quad\quad\quad\quad\quad\quad\quad\quad\quad\quad\quad\quad$ ‖ |
  $\quad\quad\quad\quad\quad\quad\quad\quad\quad\quad\quad\quad\quad\quad$ O OH

19-4 *a.* The naphthoic acid is soluble in NaHCO$_3$ solution. Phenols are not.

b. Catechol reduces Tollens' reagent.

c. Phenylacetaldehyde will reduce Fehling's solution.

d. Phenacyl chloride will give a Beilstein test.

e. The phenol is soluble in dil NaOH.

f. The quinone is colored, the hydroquinone is not.

g. 2-Nitrodecalin has hydrogen on α carbon and thus is soluble in dil NaOH.

h. Tetralin is soluble in fuming H$_2$SO$_4$. Decalin is not.

19-5 *a.* Extract with NaOH, leaving the two ketones. Separate methyl benzyl ketone from propiophenone with NaHSO$_3$. Acidify the basic solution from the first step and steam distill. The *p*-nitrophenol remains in the distilling flask. Separate the benzoic acid from phenol by dissolving the former in NaHCO$_3$.

b. Extract the organic layer with HCl removing the two amino-containing compounds. The *N*-ethyl-*p*-aminobenzoic acid can then be isolated because it is soluble in NaOH. The α-phenylsuccinic acid and the phenyl malonate which were not extracted by HCl can be separated because the former is soluble in NaOH.

19-6

19-7 *a.* Methane, propene, 1,3-pentadiene, toluene, diphenylmethane.

b. Phthalimide, propionic acid, benzoic acid, *o*-chlorobenzoic acid, 2,4-dinitrobenzoic acid.

c. β-Phenylethanol, *p*-cresol, *p*-methylbenzenethiol, phenylacetic acid, benzenesulfonic acid.

d. —SO$_2$R, —Cl, —CH$_3$, —NH$_2$, —NHCH$_3$.

e. Ethyl *o*-methylbenzoate (because of steric factors), ethyl *p*-methylbenzoate, ethyl benzoate, ethyl *p*-nitrobenzoate.

19-8 *a.*

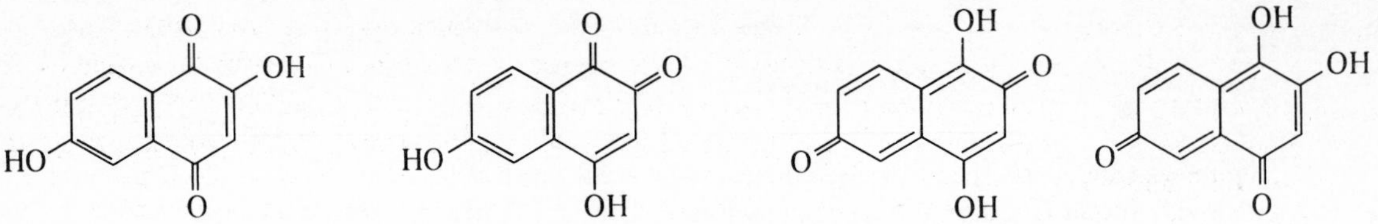

Note it is not possible to draw a conjugated double-bond system if the carbonyl groups are meta.

b.

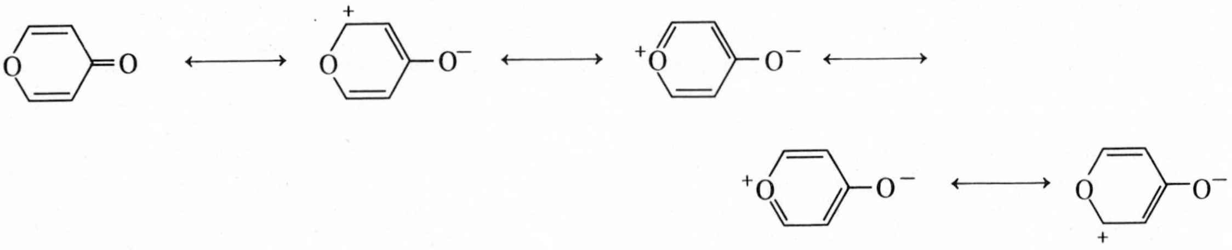

The extensive resonance stabilization and delocalization of charge in the above system makes γ-pyrone have a resonance energy more like benzene and very unreactive toward nucleophilic attack at its carbonyl group.

c.

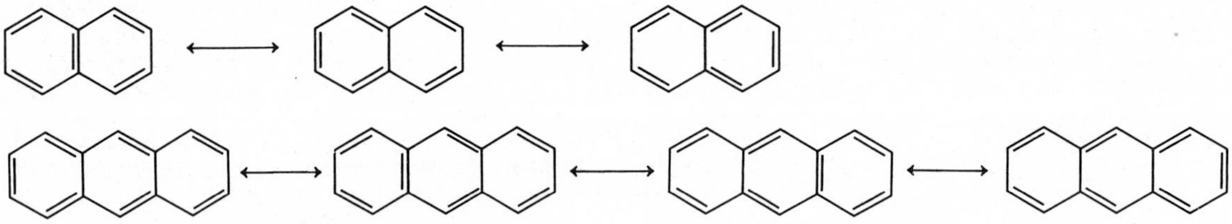

d.

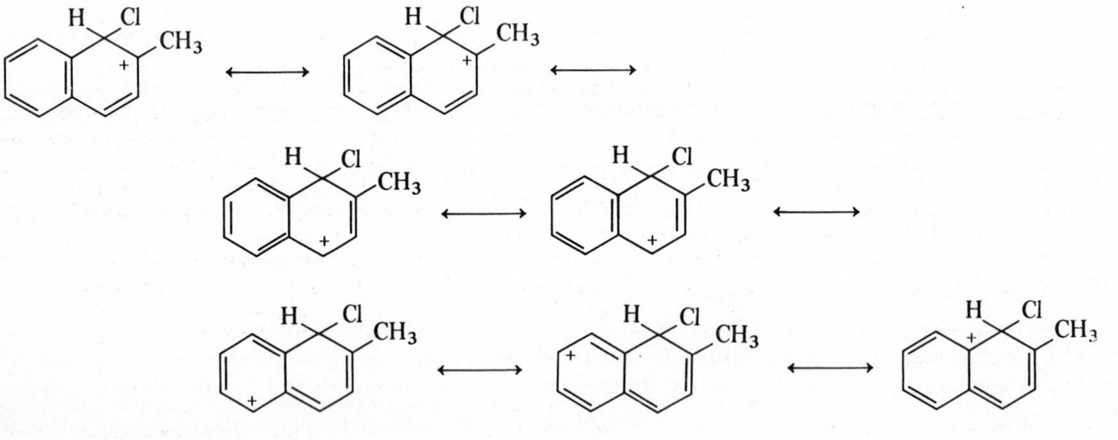

Note that hyperconjugation is possible.

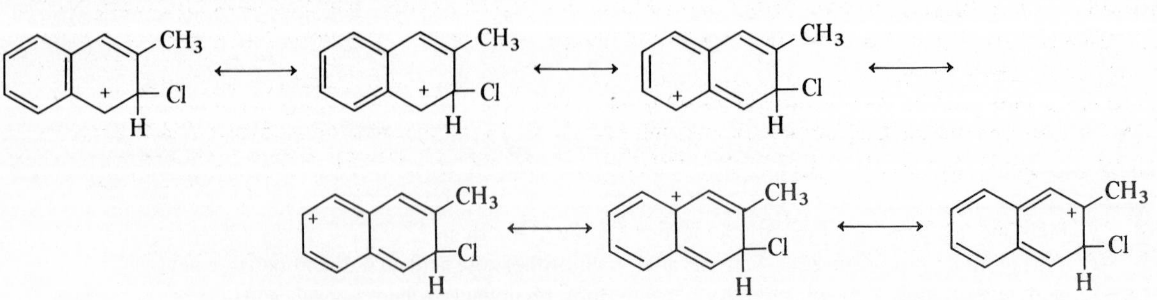

Attack at C-1 preferred because intermediate has more significant resonance structures (more structures with benzene ring intact plus more hyperconjugation structures).

19-9 The large isopropyl group blocks the 1 and 3 positions by steric hindrance. If nitrobenzene is used as a solvent, substitution will not occur at α positions 4, 5, or 8 because of steric hindrance of the hydrogen on the adjacent α position. The 6 position is activated by hyperconjugation of the one CH bond in isopropyl, but the 7 position is not (try drawing conjugated bonds with the electron pair from the CH of the isopropyl group at the 6 and 7 positions).

19-10   *a.* 5-Nitro-1-naphthoic acid
        b. 1-Acetylamino-2-nitronaphthalene and 1-acetylamino-4-nitronapththalene
        c. 6-Nitro-1,4-di-tert-butylnaphthalene
        d. 1-Nitro-2-naphthol

19-11   A, p-CH$_3$C$_6$H$_4$CH$_2$CHO;   B, o-CH$_3$CH$_2$COC$_6$H$_4$CHO;   C, p-CH$_3$COCH$_2$C$_6$H$_4$CHO

19-12   A, C$_6$H$_5$CH$_2$COCH$_2$C$_6$H$_5$;   B, C$_6$H$_5$CH$_2$NHCOCH$_2$C$_6$H$_5$

19-13   A, 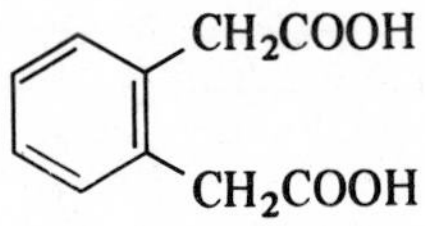
   B,

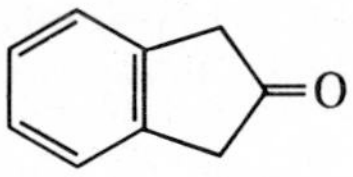

19-14

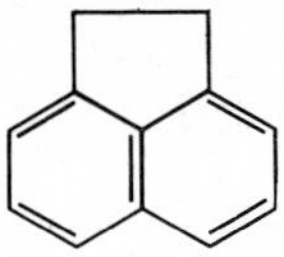

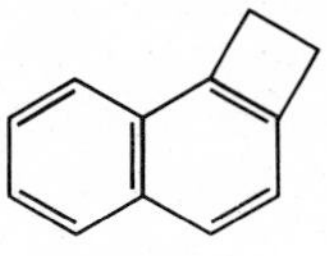

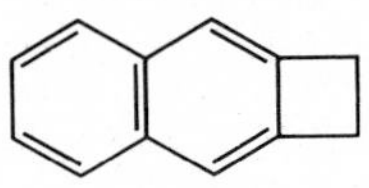

19-15   A, 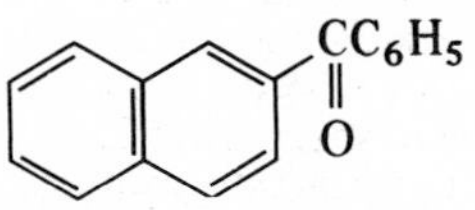
or α isomer
   B, 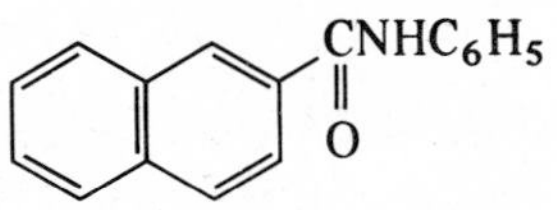
or α isomer

C, Aniline;   D, β-naphthoic acid, or α isomer;   E, phenol   F, 1,2,4-benzenetricarboxylic acid (or 1,2,3 isomer)

19-16   A, 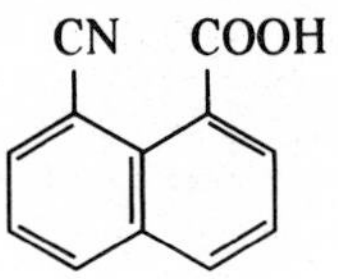
   B, 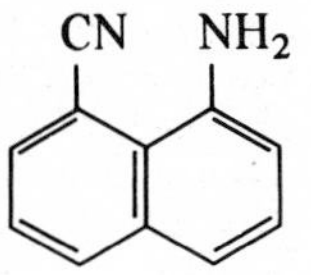
   C, 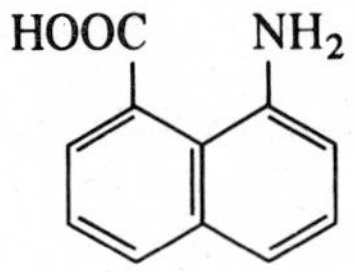
   D,

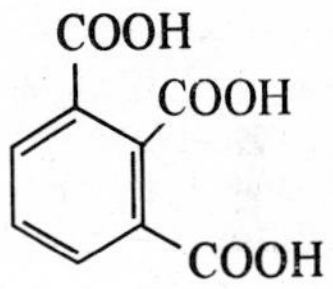

19-17   A, 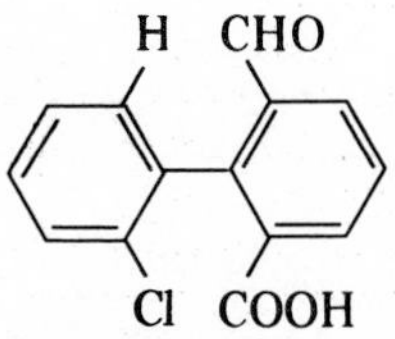
   B,

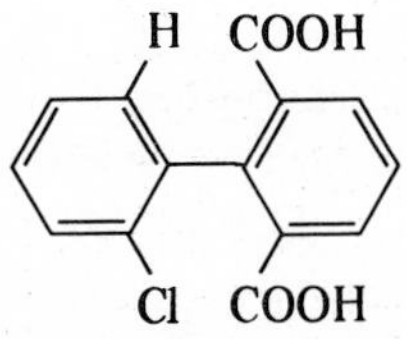

19-18   *a.* In A, protect the amino group by acetylation. In C, the Reimer-Tiemann fails for electron-poor rings.
        b. In A, Grignard reagents react with nitro groups. In B, the sulfuric acid will cause dehydration and polymerization.
        c. In B, acid conditions would hydrolyze the acetal.
        d. In A, Tollens' reagent will oxidize quinones. In B, bromine would add to the quinone ring.

19-19   *a.* A, o-CH$_3$C$_6$H$_4$N$_2^+$BF$_4^-$;   B, o-CH$_3$C$_6$H$_4$F;   C, o-FC$_6$H$_4$COOH

        b. D, p-HOOCC$_6$H$_4$SC—OEt;   E, p-HOOCC$_6$H$_4$SH
                               ‖
                               S

        c. F, C$_6$H$_5$OEt;   G, p-EtOC$_6$H$_4$CHO;   H, p-EtOC$_6$H$_4$CH=NC$_6$H$_5$;   I, p-EtOC$_6$H$_4$CH$_2$NHC$_6$H$_5$;
           J, p-EtOC$_6$H$_4$CH$_2$N(NO)C$_6$H$_5$;   K, p-EtOC$_6$H$_4$CH$_2$N(NH$_2$)C$_6$H$_5$
        d. L, p-BrC$_6$H$_4$CHO;   M, p-BrC$_6$H$_4$CH$_2$OH;   N, p-BrC$_6$H$_4$COOH

e. O, [structure: cyclopentanone with COOCH₃ substituent] *P,* [structure: δ-valerolactam, piperidinone N-H] *Q,* [structure: δ-valerolactone] *f.* R, [structure: bicyclic dione with two C₆H₅ groups]

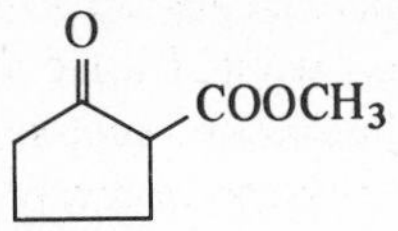

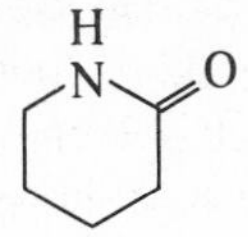

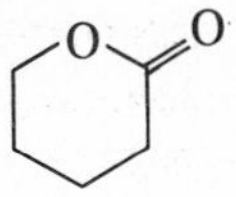

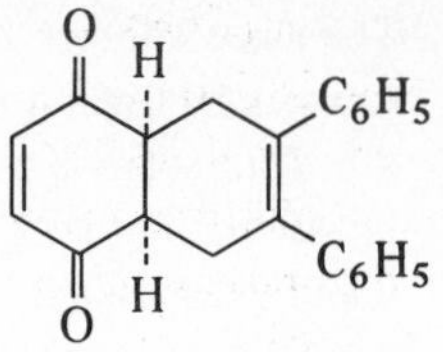

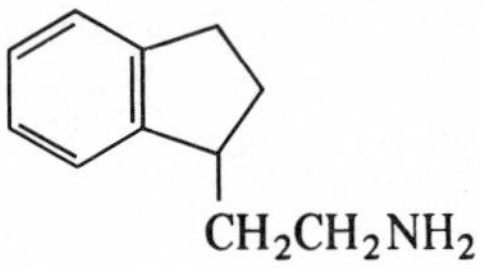

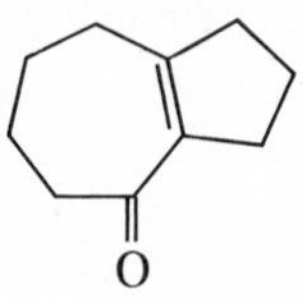

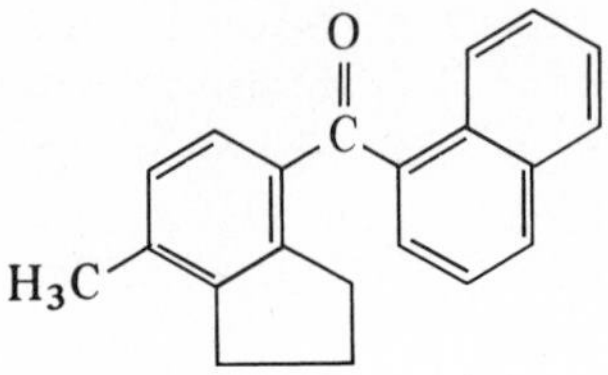

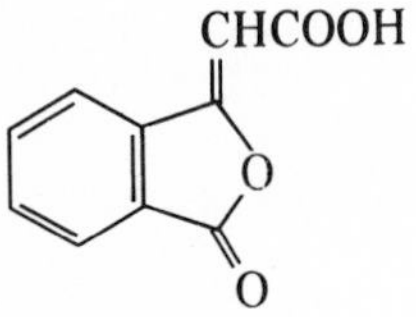

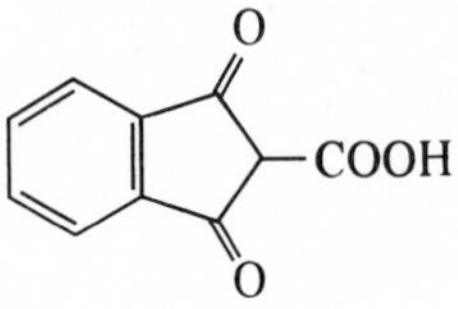

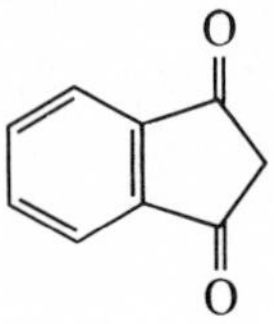

g. S, [structure: indane with CH₂CH₂NH₂ substituent]   *h.* T, [structure: azulenone-type bicyclic ketone]   *i.* U, [structure: methyl-indane linked to C=O and naphthalene]

j. V, [structure: phthalide-type with =CHCOOH]   W, [structure: indandione with COOH]   X, [structure: indandione]   Y, [structure: 2,2-dimethylindane]

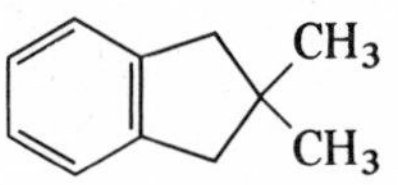

19-20   In 1,2-addition an aromatic ring cannot be the end product. There is about 36 kcal/mol resonance energy to be gained by 1,4 addition.

19-21   *a.* 2-Picoline is much more reactive.

   b. $C_6H_5CH=CH-C_5H_4N$.

   c. B is the quaternary salt; the aldehyde condenses with the C-methyl group.

   d. The base removes a proton from the methyl group (compare the aldol condensation).

   e. B is more reactive because the intermediate negative charge is better distributed when the nitrogen is quaternary.

19-22   *a.* No heavy oxygen appears in the alcohol; therefore the alkyl-oxygen bond is not broken during the ester hydrolysis. The appearance of heavy oxygen in unreacted ester indicates that the process involves an intermediate such as the one shown below, which has a sufficiently long lifetime to permit proton transfer and loss of H_2O or H_2O^{18}.

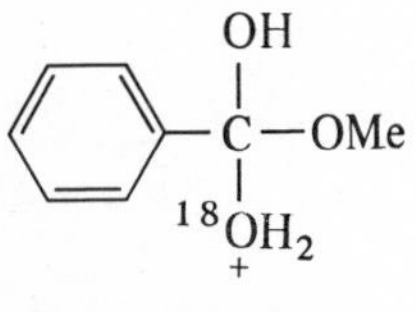

[structure: phenyl-C(OH)(OMe)($^{18}OH_2^+$)]

   b. The heavy oxygen appears in the alcohol, indicating rupture of an alkyl-oxygen bond during hydrolysis. If the reaction proceeded by way of a carbonium ion, the isotope and racemization observations would be expected.

   c. Under the conditions of the reaction, it would be expected that acid-catalyzed ester exchange would be taking place. No new product would result. The formation of dimethyl ether could occur by nucleophilic displacement of methanol on the methyl group of a protonated ester.

   d. Give the structure for protonated methyl benzoate (proton on the carbonyl oxygen atom) and show how resonance can delocalize the positive charge. When two ortho-methyl groups are present, the protonated carbomethoxy group cannot be coplanar with the ring. However, if it loses a molecule of water, there can again be ring involvement in delocalization of positive charge. Base a mechanism on these standards.

19-23   *a.* Two. Hydrogens on the two central carbon atoms may be arranged cis or trans with respect to each other. Neither form is optically active.

   b. Eight, four *dl* pairs. All eight are optically active. Four isomers have cis ring junctures and four isomers have trans ring junctures (see part *a*).

19-24   *a.* $\xrightarrow[1]{NaOCH_3}$ $\xrightarrow[2]{CH_3I}$ $\xrightarrow[3]{BrCH_2COOCH_3, Zn}$ $\xrightarrow[9]{SOCl_2}$ $\xrightarrow[10]{CH_2N_2}$ $\xrightarrow[11]{CH_3OH, Ag_2O}$ $\xrightarrow[12]{NaOCH_3}$ $\xrightarrow[13]{H_3O^+, \Delta}$

   b. One $COOCH_3$ group is attached to a 3° center and these groups slow its hydrolysis greatly.

428

c. Compound A has one asymmetric carbon atom, but racemization will take place during the first two synthetic steps. The number of isomers possible then would be four because there would be two asymmetric carbons. Compound C contains two asymmetric centers which give two pairs of enantiomorphs separable by crystallization. However, steps 4, 5, and 6 lead to D and even though stereoselective conditions were used, some racemization could occur. Thus resolution would best be made first at D since none of reactions 7 through 13 are occurring at asymmetric centers.

19-25 *a.*

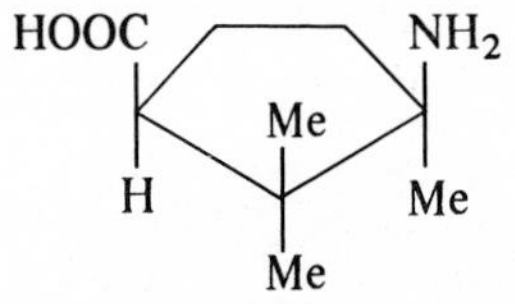

b.

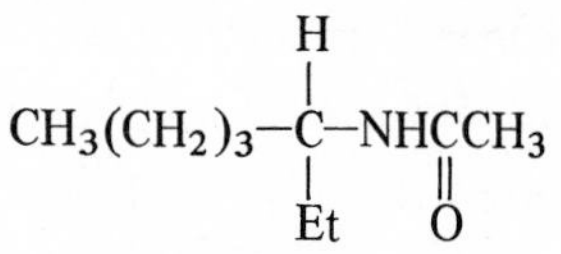

$$CH_3(CH_2)_3-\underset{\underset{Et}{|}}{\overset{\overset{H}{|}}{C}}-NH\underset{\overset{\|}{O}}{C}CH_3$$

c. Cyclopentene + cyclopentanol

d. Phenylacetone

e. Diphenylacetaldehyde

f. $CH_3CH_2N{=}C{=}O$

g.

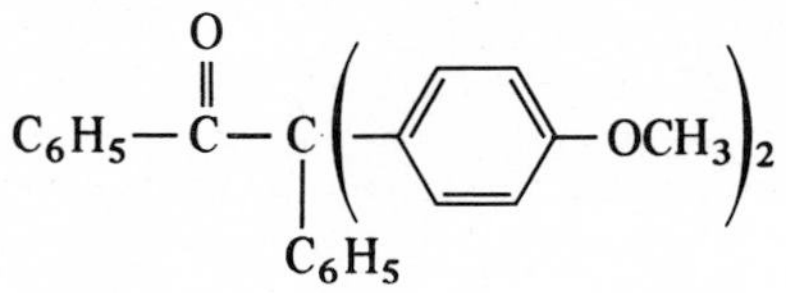

$$C_6H_5-\overset{\overset{O}{\|}}{C}-\underset{\underset{C_6H_5}{|}}{C}\left(\overset{}{\underset{}{}}-OCH_3\right)_2$$

h.

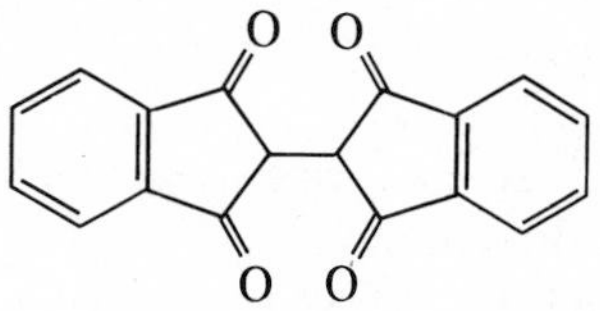

Via benzilic acid rearrangement and decarboxylation

19-26 Phenanthrene has a slightly higher resonance energy than anthracene. This could be anticipated on the basis of more-significant resonance structures which can be provided for the former (and possibly with the fact that in these structures there are more benzenoid rings).

19-27

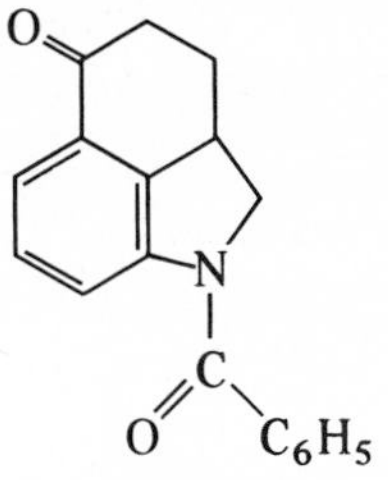

A

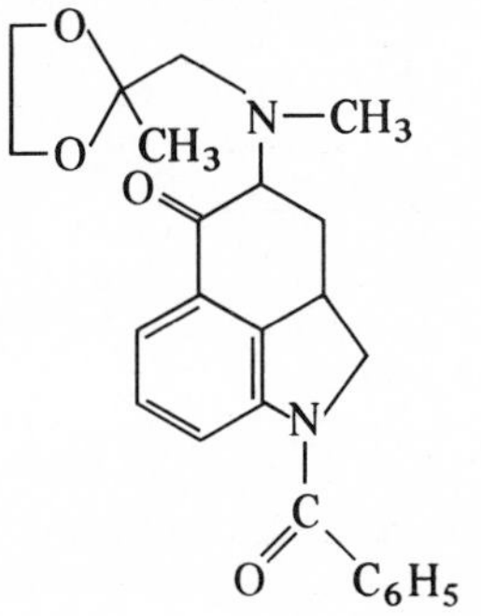

B

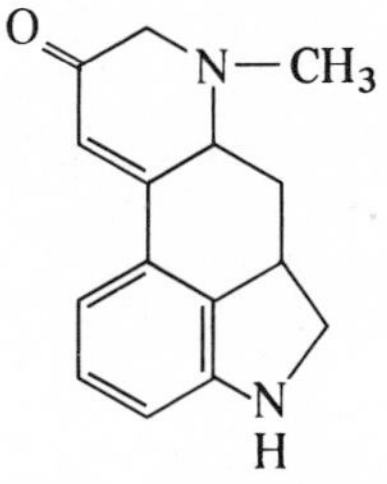

C

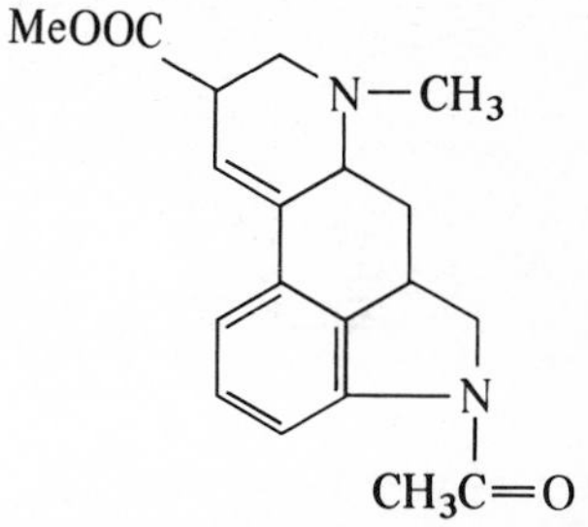

D

19-28 *a.* Mechanism is similar to that shown in Prob. 13-34 except initial attack is by CH_3O^- instead of OH^- and CH_3OH serves as a source of protons.

b.

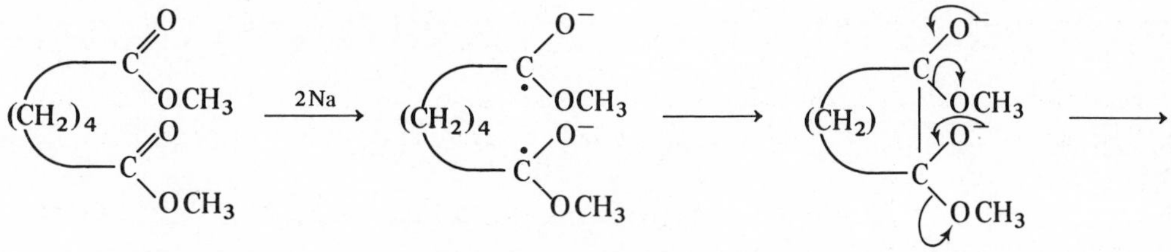

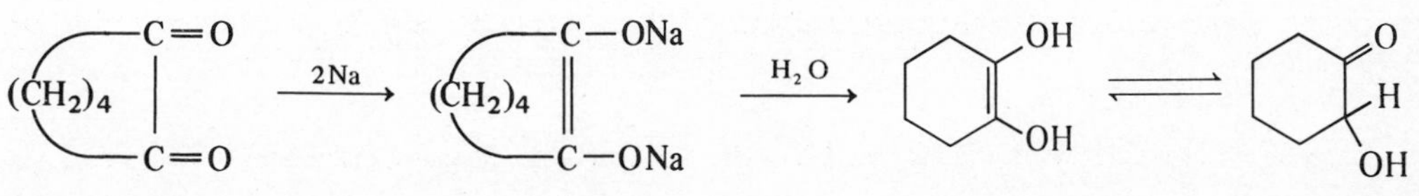

19-29 *a.* Cyanide ion is able to attack the carbonyl group and once attached, the CN group with its electron-withdrawing power increases the acidity of the aldehydic hydrogen and stabilizes the anion intermediate:

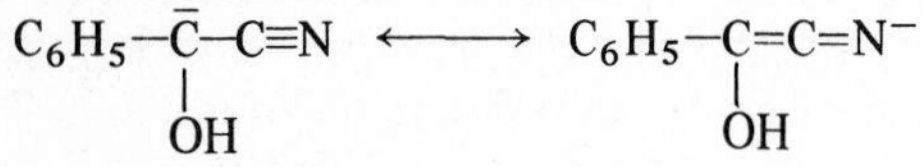

$$C_6H_5-\overset{\overline{}}{\underset{OH}{C}}-C{\equiv}N \longleftrightarrow C_6H_5-\underset{OH}{C}{=}C{=}N^-$$

b. Rate $= k\,[C_6H_5CHO]^2\,[CN^-]$

c. This reaction demonstrates the reversibility of the benzoin condensation.

Keto form   Enol form

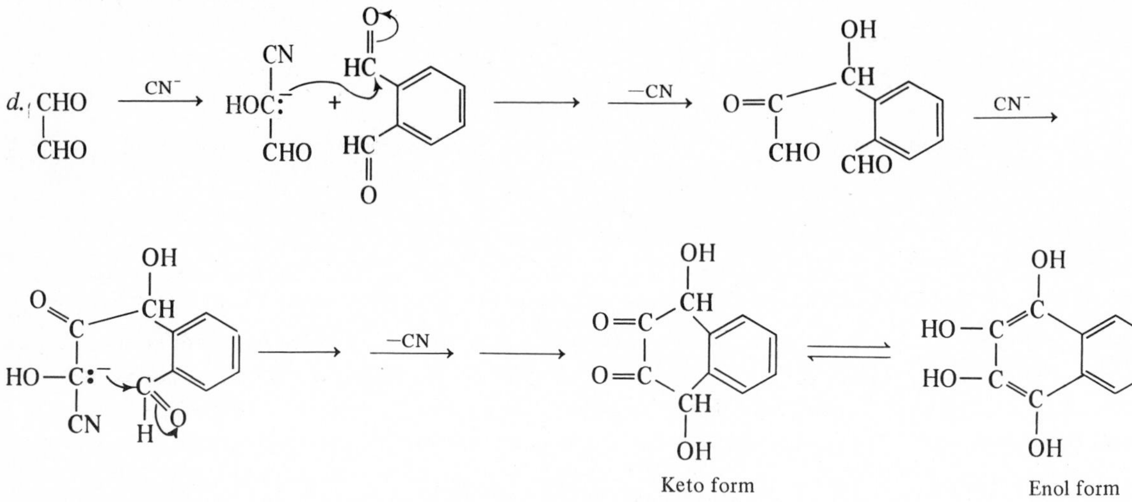

19-30 *a.* Use an excess of acetone or some other ketone or quinone (see part *b*) as a hydride ion acceptor.

b. Hydrogen as hydride ion is transferred from C-1 in the alcohol to *p*-benzoquinone, reducing it finally to hydroquinone.

c. The stereospecificity of this reaction results from the favored geometry in the transition state which place the bulky *t*-butyl and phenyl groups as far apart as possible (trans oriented):

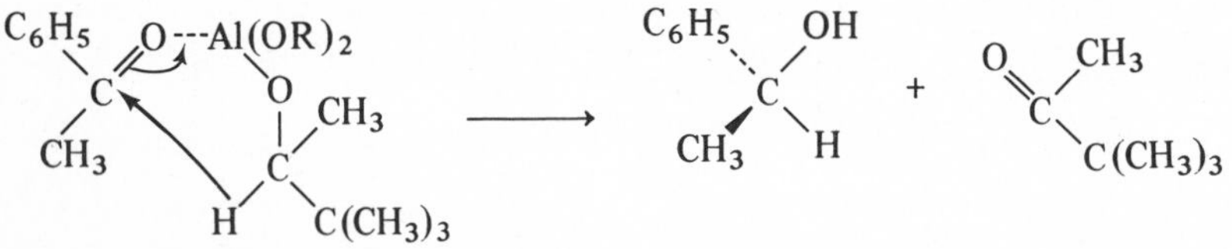

19-31 A, *p*-phenylenediacetate;   B, acetic acid;   C, hydroquinone;   D, *p*-benzoquinone

19-32 E, indan;   F, 2-indanone;   G, 1,3-indanedione

19-33 H, *p*-anisaldehyde;   I, *p*-hydroxybenzaldehyde;   J, *p,p'*-dimethoxybenzoin;   K, $(p\text{-}CH_3OC_6H_5)_2{-}\underset{OH}{\overset{}{C}}COOH$

19-34 L, $(CH_3)_2N{-}\underset{O}{\overset{}{C}}O{-}$⟨⟩$-OCH_3$ M, $HO{-}$⟨⟩$-OCH_3$

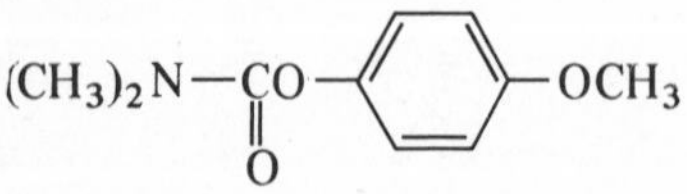

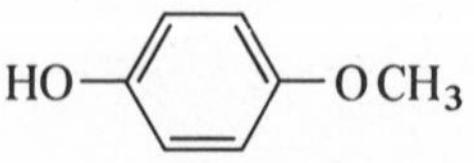

19-35 I, butyrophenone;   II, 9,10-dihydroanthracene;   III, benzyl benzoate;   IV, β-phenylethyl acetate;
      V, *p*-ethoxyacetanilide

19-36 X, 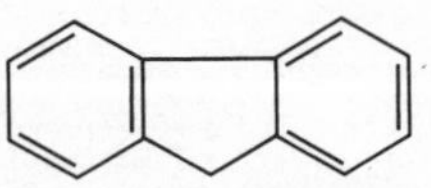   Y, 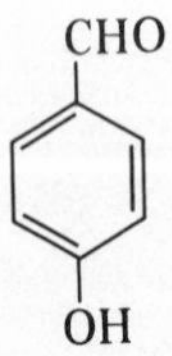   Z,

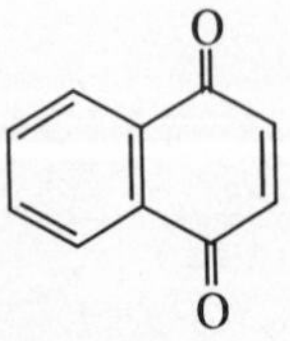

Chapter 20

20-1 *a.* H$_2$N(CH$_2$)$_4$CHCONHCHCONHCHCOOH
               | | |
               NH$_2$ CH$_3$ CH$_2$SH

b. HOOCCH$_2$CH$_2$CHCONHCHCONHCHCOOH
                        | | |
                      NH$_2$ CH$_2$ CH(OH)CH$_3$
                              CH(CH$_3$)$_2$

c. H$_2$NCH$_2$CONHCHCONHCHCONH$_2$
                 | |
                CH$_3$ CH$_3$

d.
                        HOOC
H$_2$NCHCONHCHCON⟨ring⟩
   | |
  CH$_2$OH CH$_2$
            |
        CH$_2$SCH$_3$

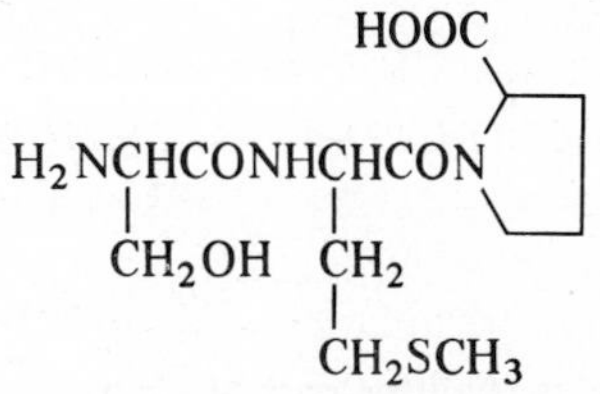

e.
                   CH$_2$—⟨C$_6$H$_4$⟩—OH
H$_2$NCHCONHCHCONHCHCOOH
    | |
CH$_3$CH$_2$CHCH$_3$ (CH$_2$)$_4$NH$_2$

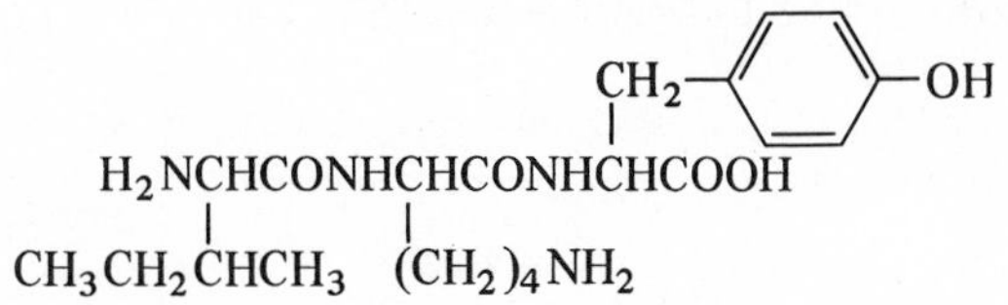

f. H$_2$N(CH$_2$)$_4$CHCONHCH$_2$CONHCH$_2$CONH$_2$
                 |
                NH$_2$

20-2 *a.* Glycine
b. Isoleucine, threonine, hydroxyproline
c. Cystine
d. All other amino acids not listed in parts *a* through *c*

20-3 *a.* 4 *b.* 4 *c.* 16 *d.* None *e.* 16 *f.* 16

20-4 Alanine will exist more nearly in the zwitterion form and will show only a small migratory aptitude. The second amino group of lysine will make its water solution basic, and the cation will migrate to the cathode. Aspartic acid will ionize to give an anion that will migrate toward the anode:

$$\text{HOOCCH}_2\text{CHCOO}^- + \text{H}_2\text{O} \rightleftharpoons {}^-\text{OOCCH}_2\text{CHCOO}^- + \text{H}_3\text{O}^+$$
          | |
        $^+$NH$_3$ $^+$NH$_3$

20-5 *a.* (CH$_3$)$_2$CHCHCOO$^-$ + HCl ⟶ (CH$_3$)$_2$CHCHCOOH
               | |
             $^+$NH$_3$ $^+$NH$_3$Cl$^-$

b. RCHCOO$^-$ + NaOH ⟶ RCHCOO$^-$Na$^+$ + H$_2$O
   | |
 $^+$NH$_3$ NH$_2$

c. $^-$ClH$_3$$\overset{+}{\text{N}}CH_2$COOH + SOCl$_2$ ⟶ $^-$ClH$_3$$\overset{+}{\text{N}}CH_2$COCl + HCl + SO$_2$

d. CH$_3$CHCOOH + NaNO$_2$ + HCl ⟶ CH$_3$CHCOOH + N$_2$ + NaCl + H$_2$O
       | |
     NH$_2$ OH

e. CH$_3$CHO + NH$_4$CN ⟶ CH$_3$CHCN + H$_2$O
                          |
                        NH$_2$

f. CH$_2$(COOEt)$_2$ + HONO ⟶ O=NCH(COOEt)$_2$ + H$_2$O

g.
          O   O
          ‖   ‖
          C   C
C$_6$H$_4$⟨ ⟩NK + ClCH$_2$COOEt ⟶ C$_6$H$_4$⟨ ⟩NCH$_2$COOEt + KCl
          C   C
          ‖   ‖
          O   O

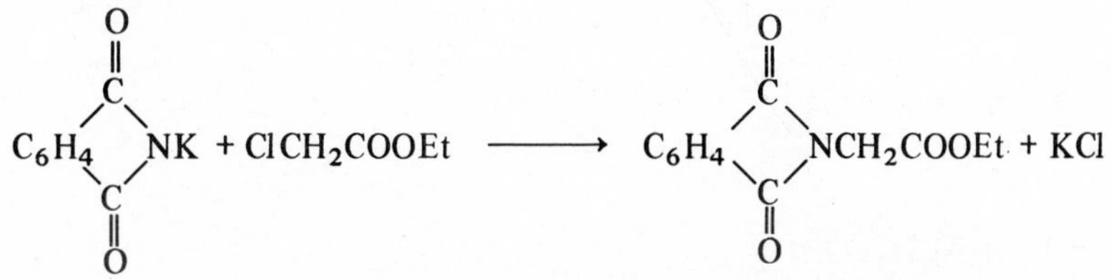

h. C$_6$H$_5$CH$_2$CHCOCl + CH$_3$NH$_2$ ⟶ C$_6$H$_5$CH$_2$CHCONHCH$_3$ + HCl
               | |
           NHCOCH$_3$ NHCOCH$_3$

i. $\underset{\overset{|}{NHCOCH_3}}{HC(COOEt)_2} + NaOEt \longrightarrow \underset{\overset{|}{NHCOCH_3}}{NaC(COOEt)_2} + EtOH$

j. $2\underset{\overset{|}{NH_2}}{CH_3CHCOOH} \xrightarrow{\Delta} \underset{\underset{O=C-NH}{\overset{NH-C=O}{}}}{CH_3CH\qquad CHCH_3}$

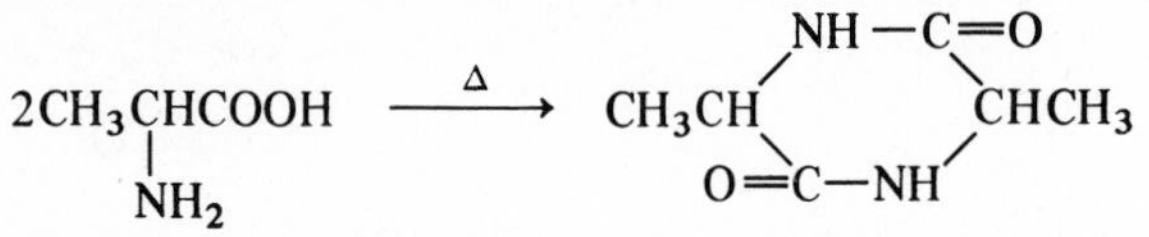

k. $H_2N(CH_2)_4COOH \longrightarrow \underset{\underset{NH}{|\_\_\_\_\_|}}{CH_2(CH_2)_3C=O}$

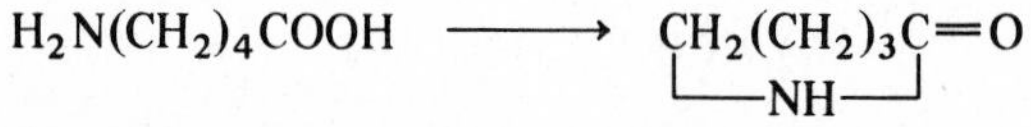

l. Leu·Met·Asp + HCl $\longrightarrow$ hydrochlorides of the three amino acids

20-6 The proton covers the lone pair of electrons of the amino group, eliminating its nucleophilic character for reaction with anhydrides, etc. Addition of an equivalent of base to free the amino group speeds up the reaction.

20-7 *a.* $\underset{\overset{|}{NH_2}}{CH_3CHCOOH} + 2CH_2{=}C{=}O \longrightarrow \underset{\overset{|}{NHCOCH_3}}{CH_3CHCOOCCH_3}$

b. $\underset{\overset{|}{NH_2}}{CH_3CHCOOH} + HCl + CH_3OH \longrightarrow \underset{\overset{|}{{}^+NH_3Cl^-}}{CH_3CHCOOCH_3} + H_2O$

c. $\underset{\overset{|}{NH_2}}{CH_3CHCOOH} + CH_3N{=}C{=}S \longrightarrow \underset{\overset{|}{NHCSNHCH_3}}{CH_3CHCOOH}$

d. $\underset{\overset{|}{NH_2}}{CH_3CHCOOH} + (CH_3O)_2SO_2 \longrightarrow \underset{\overset{|}{N(CH_3)_2}}{CH_3CHCOOH} + H_2SO_4$

e. $\underset{\overset{|}{NH_2}}{CH_3CHCOOH} + C_6H_5CH_2OCOCl \longrightarrow \underset{\overset{|}{NHCOOCH_2C_6H_5}}{CH_3CHCOOH} + HCl$

f. $\underset{\overset{|}{NH_2}}{CH_3CHCOOH} + CH_3N{=}C{=}O \longrightarrow \underset{\overset{|}{NHCONHCH_3}}{CH_3CHCOOH}$

g. $\underset{\overset{|}{NH_2}}{CH_3CHCOOH} + (CH_3)_3COCOCl \longrightarrow \underset{\overset{|}{(CH_3)_3COCONH}}{CH_3CHCOOH}$

h.

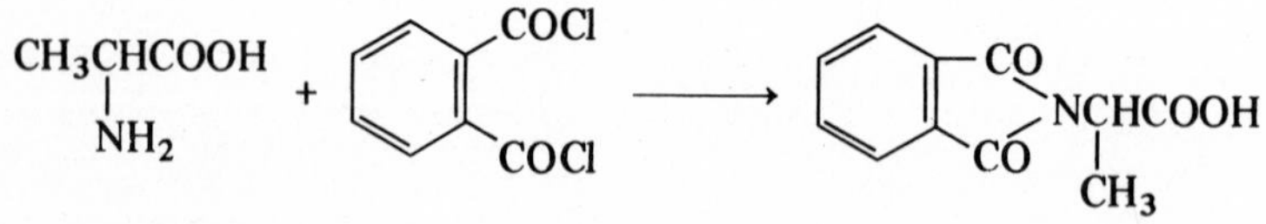

i.

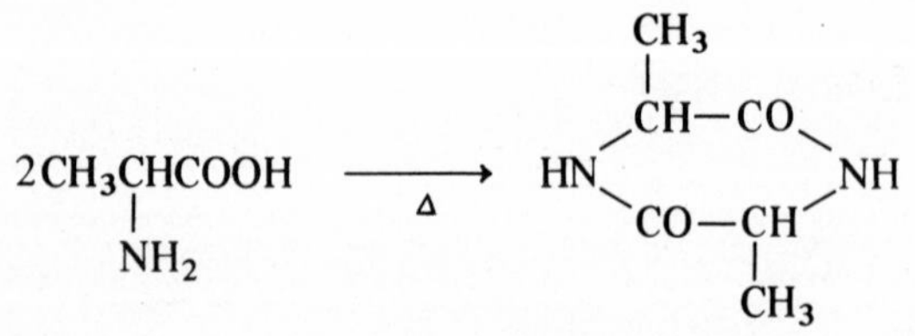

20-8 $H_3\overset{+}{N}CH_2COO^-, H_3\overset{+}{N}CH_2COOH, \underset{\overset{|}{OH}}{H_3\overset{+}{N}CH_2\overset{+}{C}OH}$

20-9 $\overset{+}{H_3N}CH_2COOH \xrightarrow{OH^-} \overset{+}{H_3N}CH_2COO^- \xrightarrow{OH^-} H_2NCH_2COO^- \xrightarrow[base]{strong} \overset{-}{HN}CH_2COO^-$

20-10

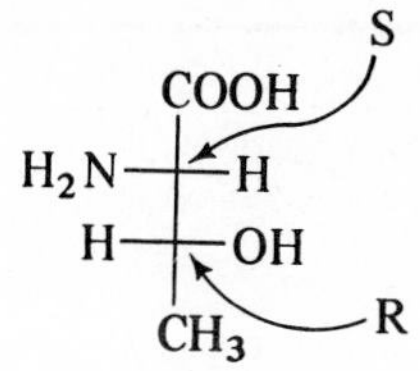

20-11

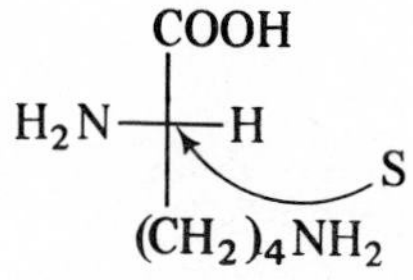

20-12 About 22 million.

20-13 Alkaline conditions enhance enol formation which involves double-bond formation at the asymmetric carbon atoms adjacent to the carbonyl group in the peptide linkages, thus racemizing those asymmetric centers:

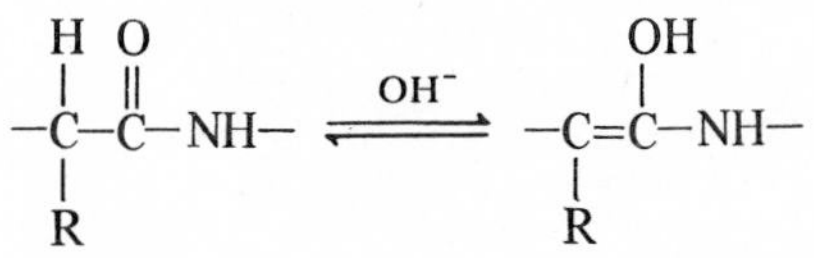

20-14 Nitrogen is less electron-attracting than oxygen; hence the carbon in the amide group is less electron-deficient and reacts slower with nucleophilic groups.

20-15 *a.* Nitrogen is evolved on treating aspartic acid with HONO.
b. Acetylphenylalanine is insoluble in dil HCl.
c. Threonine gives a positive iodoform test.
d. Glycine ethyl ester is alkaline toward litmus.

20-16 *a.* Treat valine with phthalyl chloride to cover the amino group. Then make acid chloride with $SOCl_2$. React this acid chloride with alanine in the presence of 1 equiv of NaOH. Treat this molecule with hydrazine to displace the phthalyl unit.
b. Cover the amino group in leucine with carbobenzoxy chloride and make the acid chloride of this unit. React this with glycine, isolate product, and make acid chloride. Treat valine with this acid chloride and hydrogenate to get the product.
c. Convert 1-propanol to α-chloropropionyl chloride via oxidation, HVZ reaction, and $SOCl_2$. React this chloroacyl halide with esterified glycine, treat the resulting product with NH_3, and hydrolyze the ester to get alanylglycine.
d. Prepare *t*-butyl alcohol and treat with phosgene to get $(CH_3)_3COCOCl$. Use this to protect the amino groups in glycine, alanine, leucine, and valine. Then proceed as outlined in Sec. 20-3*d*.

20-17

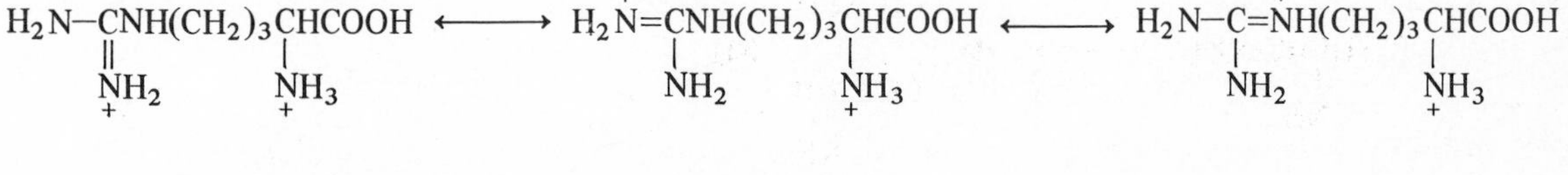

20-18 *a.* React ethyl iodoacetate with potassium phthalimide and hydrolyze with sodium hydroxide solution.
b. Treat malonic ester with nitrous acid and hydrogenate catalytically in acetic anhydride to get acetylaminomalonic ester. Make sodium salt of this with EtONa and alkylate with ethyl iodide. Acid hydrolysis also causes decarboxylation to give product.

c. Treat with nitrous acid.

d. Make glycine as in *a.* Convert benzoic acid to the acid chloride with $SOCl_2$ and allow this to react with glycine in the presence of 1 equiv of base.

e. Heat the anhydride with NH_3 to convert to the imide. Treat this with NaOCl.

f. Use the acetylaminomalonic ester method illustrated in *b.*

g. One could use the acetylaminomalonic ester method or one could use the Strecker method.

h. Dehydrogenate to the aldehyde and add NH_3 + HCN.

i. Use ethyl bromoacetate in the acetylaminomalonic ester method.

j. Make CH_3SH from methyl iodide and KSH. Nucleophilic addition of this gives $CH_3SCH_2CH_2CHO$. Add NH_3 + HCN and hydrolyze.

20-19

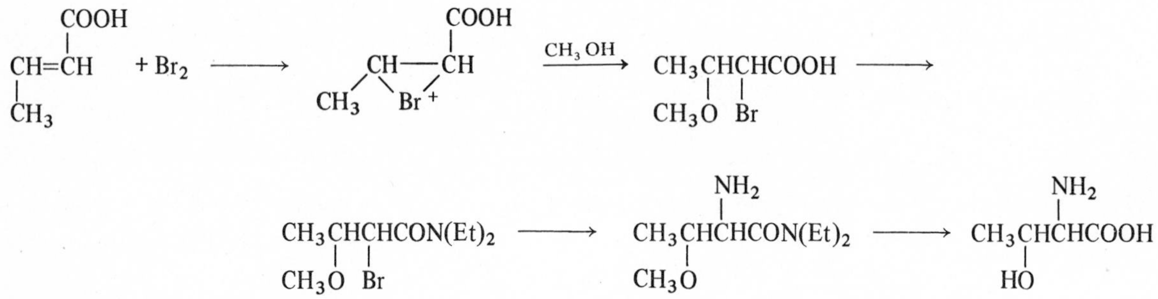

20-20

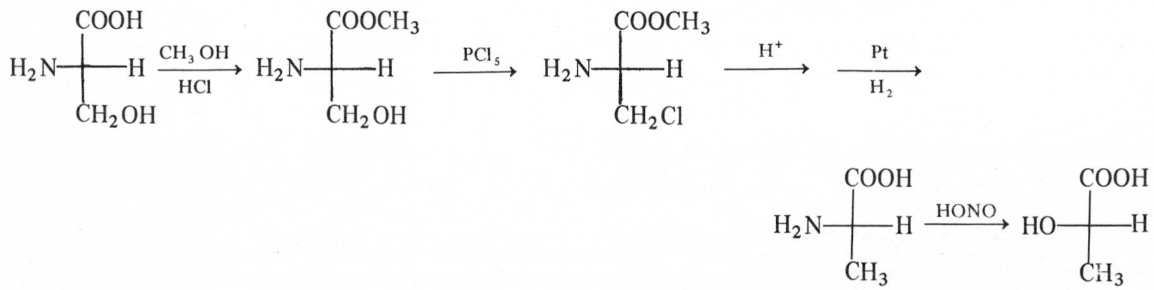

Assumption would be that none of the five reactions change the configuration at the asymmetric center.

20-21 Proline.

20-22 Alanine.

20-23 $CH_3CH(COOH)N(CH_3)_2$.

20-24 Gly·Gly·Ala, Gly·Ala·Gly.

20-25 Apparent MW $= 55.85(^{100}/_{0.34}) = 16,426$. Actual determination by sedimentation rates indicates a MW of 68,000. Thus, four atoms of iron are associated with each molecule.

20-26 For one atom Fe per molecule, the minimum MW would be 12,988; for one atom S per molecule, the minimum MW would be 2,167. $^{12,988}/_{2,167} \cong 6$ S atoms per molecule of cytochrome C.

20-27 Phe·Leu·Ser·Ala.

20-28 *a.* 3.   *b.* 32.

20-29

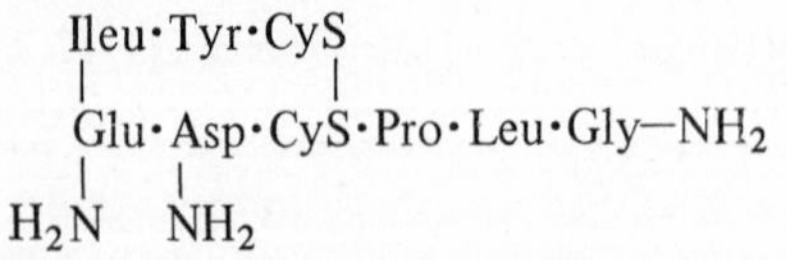

Ileu·Tyr·CyS
| |
Glu·Asp·CyS·Pro·Leu·Gly—NH$_2$ The NH$_2$ indicates a terminal amide group
| |
H$_2$N   NH$_2$

20-30 Ser·Tyr·Ser·Met·Glu·His·Phe·Arg·Try·Gly·Lys·Pro·Val·Gly·Lys·Lys·Arg·Arg·Pro·Val·Lys·Val·Tyr·Pro·
Asp·Gly·Ala·Glu·Asp·Glu·Leu·Ala·Glu·Ala·Phe·Pro·Leu·Glu·Phe.
|
NH$_2$

$$\begin{array}{ccc} NH_2 & NH_2 & NH_2 \\ | & | & | \end{array}$$

Ser·Leu·Tyr·Glu·Leu·Glu·Asp·Tyr·Cy Asp

Chain A

Chain B

Phe·Val·Asp·Glu·His·Leu·Cy·Gly·Ser·His·Leu·Val·Glu·Ala· Leu

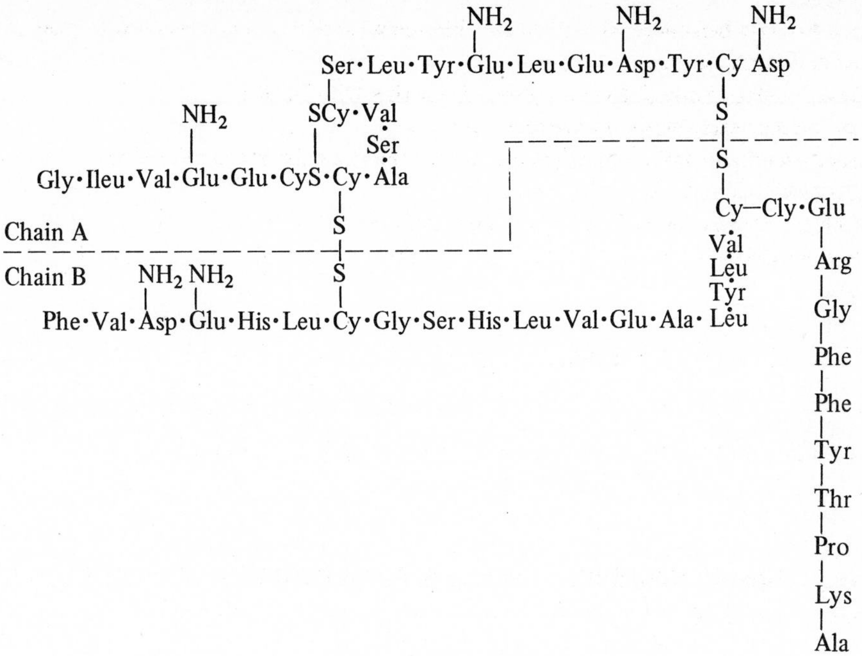

Chapter 21

21-1 *a.*

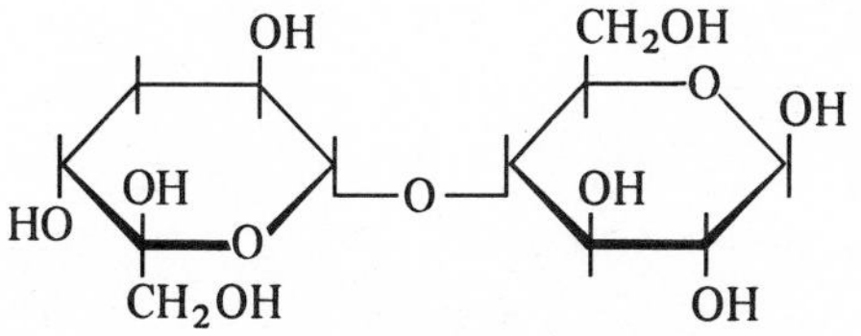

Lactose 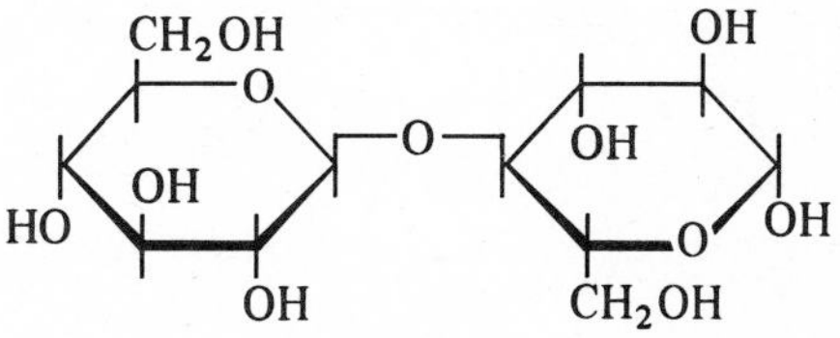   Cellobiose

b. Both will reduce Tollens' reagent.

21-2 *a.* Treat D-glucopyranose with methanol and HCl (anhydrous).

b. Treat D-galactopyranose with dimethyl sulfate and sodium hydroxide to get the 1,2,3,4,6-pentamethyl derivative.

c. The 3,4,5 carbon atoms of D-glucose have the same configuration as the 2,3,4 carbon atoms of D-arabinose. Consequently the Wohl degradation of D-glucose, which removes the aldehydic carbon atom, would be appropriate. See the sequence of Sec. 21-3*e.*

d. Treat the sugar with an acid solution of excess phenylhydrazine.

e. If the carbon chain of D-arabinose were lengthened by one carbon atom, a pair of epimers would result, one of which was D-glucose (what would the other be?). To carry out the conversion use the cyanohydrin synthesis illustrated in Sec. 21-3*d.*

f. Oxidize D-galactose with Br_2 in the presence of calcium carbonate. Acidification of the mixture will yield the lactone.

g. Oxidize D-galactose with nitric acid.

h.

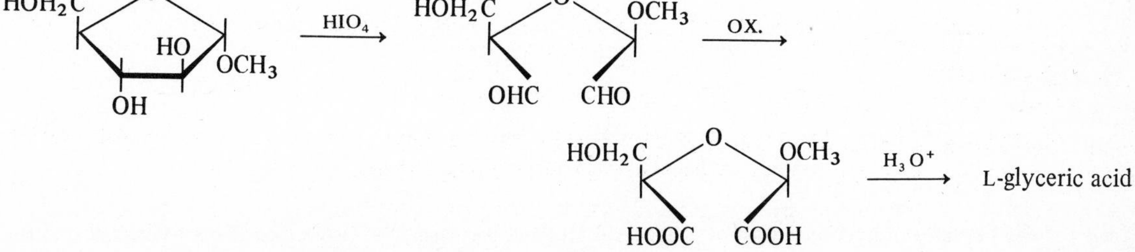

i. Treat α-D-glucofuranose with acetone and $ZnCl_2$. The derivative will be formed at the 1,2 and 5,6 positions.

j. Treat D-ribofuranose with 1-butanol and dry HCl to get butyl D-ribofuranoside. Treat this with diethyl sulfate and NaOH.

k. Use dimethyl sulfate and NaOH followed by hydrolysis of the acetal linkage with dil HCl.

l. Treat D-mannopyranose with propionic anhydride.

m. Treat D-xylofuranose with phenylhydrazine.

n. Use Wohl degradation.

o. Proceed as in part *e* isolating the other epimer.

p. Oxidize L-glucose with nitric acid.

q. Oxidize D-glucose with Br_2.

r. Treat α-D-ribopyranose with acetone and $ZnCl_2$.

21-3 Regardless of whether D- or L-galactose is oxidized with nitric acid, the same aldaric acid (optically inactive) is formed.

21-4 D-Glucofuranose, D-mannofuranose, D-mannopyranose, D-fructofuranose, D-fructopyranose.

21-5 32.

21-6 Since there is no methylation of the 5 position, it is probable that the glucose units are in the pyranose form. The 2,3,4,6-tetra-O-methylglucose must be on the end of a chain of molecules because only the glucoside bond is available for attachment to a chain. Most of the molecule consists of glucose units joined from 1- to 4-positions because of the presence of large amounts of 2,3,6-tri-O-methylglucose. The 2,3-di-O-methylglucose must be involved in branching of the main chain from the 6 position.

21-7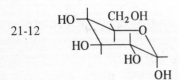

21-8 D-Fructofuranose, α or β.

21-9 D-Allopyranose, α or β.

21-10 A, D-xylose; B, D-arabinose;   C, D-ribose;   D, D-lyxose. D-Arabinose and D-lyxose give the same pentahydroxy compound in part *c*.

21-11 *a.* The pyranose will reduce Fehling's solution, the pyranoside will not.

b. The 2,3-diol will reduce periodic acid to iodic acid. The iodic acid will give a precipitate with silver nitrate.

c. The pentanal will give a Schiff test for aldehydes.

d. D-Glucose will reduce Fehling's solution.

e. Maltose will reduce Fehling's solution.

f. Propionaldehyde will give a positive Schiff test.

g. Lyxose will reduce Fehling's solution.

h. Ribopyranose will reduce Fehling's solution.

i. Glucaric acid would dissolve in aqueous sodium hydroxide.

j. Use a polarimeter.

21-12

If the hemiacetal hydroxyl group is placed in an axial position, the three remaining hydroxyl groups and the —CH_2OH group can occupy favorable equatorial positions.

21-13 α-D-Allopyranose.

21-14 *a.* *b.*

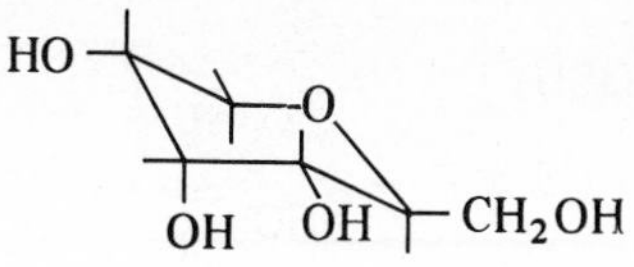

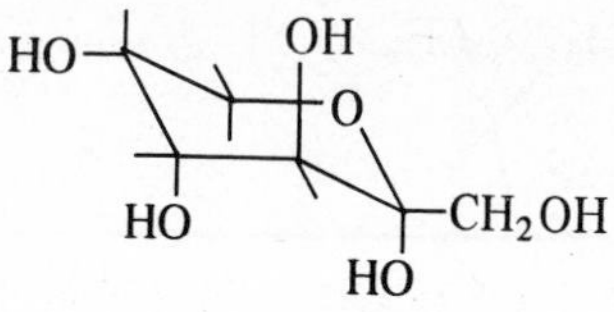

c. 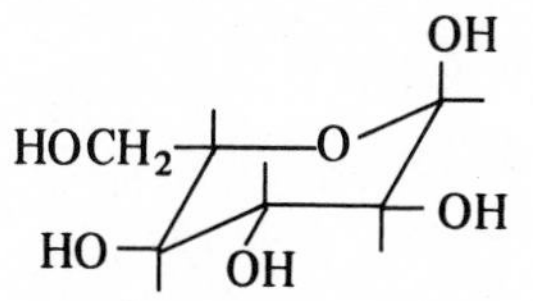 *d.*

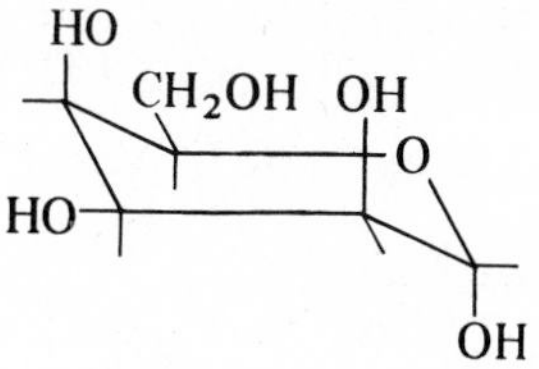

e. 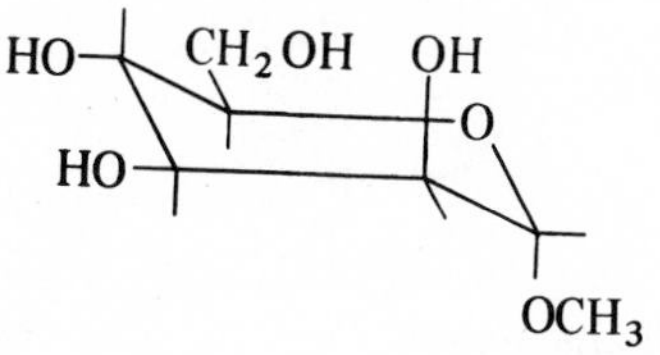 *f.*

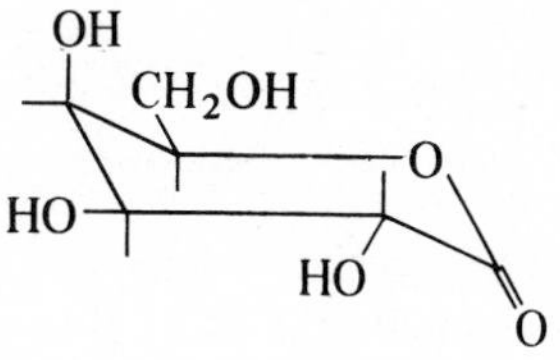

21-15 *a.* The reaction of cyclohexene oxide with water leads to a trans diol; the permanganate oxidation of cyclohexene to a cis diol. The cyclic ester with boric acid should be favored by the reaction with a cis diol because of the steric relationships of the hydroxyl groups (equatorial-equatorial).

b. D-Xylose and L-xylose.

c. In α-D-glucopyranose the 1- and 2-hydroxyl groups are cis. Since the number 1 carbon is in the form of a hemiacetal, it can isomerize to the β form, and an equilibrium is established between α and β. In the latter, the 1- and 2-hydroxyl groups are no longer cis; consequently the conductivity must decrease as the isomerization occurs.

d. The β form can isomerize to the α form which can react with boric acid, increasing the conductivity until an equilibrium is established between the α and β forms.

21-16 *a.*

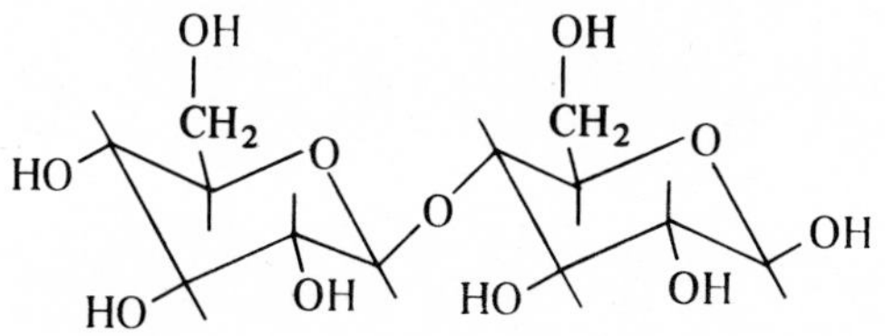

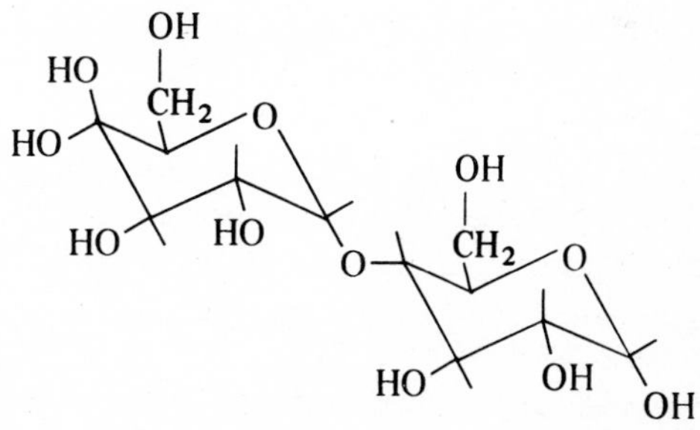

21-17

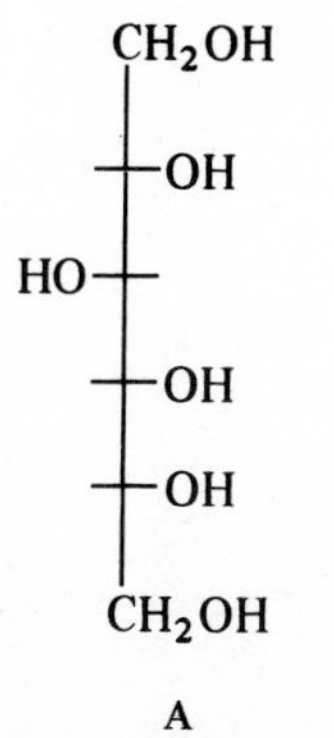

A 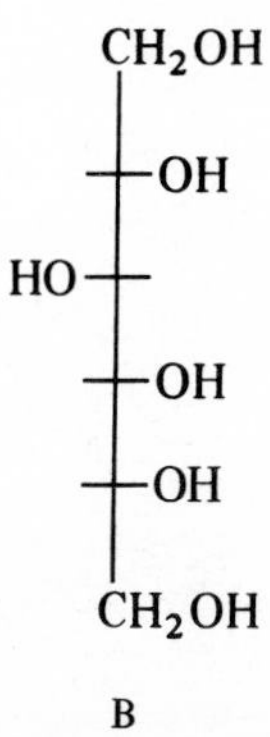   B 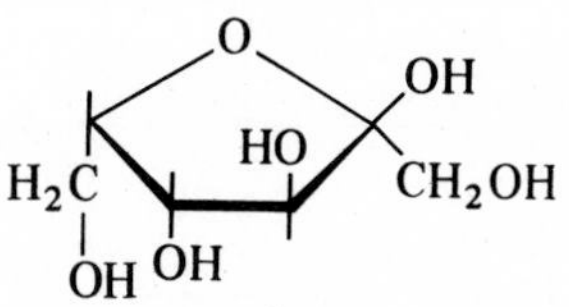   C 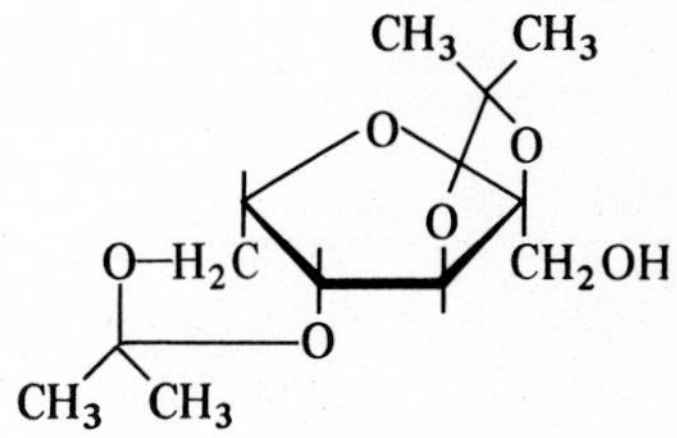   D

437

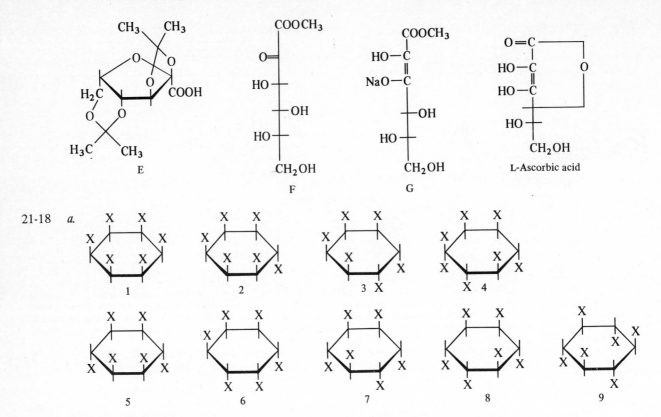

E

F

G

L-Ascorbic acid

21-18   *a.*

1   2   3   4

5   6   7   8   9

where X = OH. Compounds 7 and 8 are optically active and represent a *dl* pair. All the other compounds are meso (except 1 and 9) and optically inactive.

b.   All but compound 9.

c.   Compound 4.

index

In this index, a page number followed by a *p* (for example, 92*p*) indicates that a problem appears on that page that involves the use of the corresponding index entry. Otherwise, the page numbers refer to locations in the outline where that entry is discussed or mentioned. In order to quickly find problems and a discussion dealing with a particular topic, you should initially check the following comprehensive index entries: Isomerism; Nomenclature; Qualitative tests; Reaction mechanisms; Separation of organic compounds; Spectroscopy; Synthetic procedures. Also, you might check under the name of a specific functional group (e.g., Alkenes) and the subentries there (nomenclature of, preparation of, qualitative tests for, reactions of, and spectroscopy of, etc.). Finally, you might check under the name of a particular reaction (e.g., hydroboration, Wittig reaction).